MW01195458

The Theory of the Electromagnetic Field

PRENTICE-HALL PHYSICS SERIES

Consulting Editors
Francis M. Pipkin
George A. Snow

The Theory of the Electromagnetic Field

David M. Cook

Associate Professor of Physics
Lawrence University
Appleton, Wisconsin

PRENTICE-HALL, INC., Englewood Cliffs, New Jersey

Library of Congress Cataloging in Publication Data

Cook, David Marsden
The theory of the electromagnetic field.

(Prentice-Hall physics series)
Includes bibliographies.
1. Electromagnetic fields. I. Title.
QC665.E4C66 537 74–10842
ISBN 0–13–913293–7

Printed in the United States of America.

10 9 8 7 6 5 4 3 2 1

PRENTICE-HALL INTERNATIONAL, INC., *London*
PRENTICE-HALL OF AUSTRALIA, PTY. LTD., *Sydney*
PRENTICE-HALL OF CANADA, LTD., *Toronto*
PRENTICE-HALL OF INDIA PRIVATE LIMITED, *New Delhi*
PRENTICE-HALL OF JAPAN, INC., *Tokyo*

Contents

2

Charge and Current: The Specification of Arbitrary Distributions 50

3

The Electromagnetic Field: Its Definition and Its Effect on General Charge Distributions 74

4

The Electric Field Produced by Static Charges 91

8 Potential Theory 213

9 Properties of Matter I: Conduction 249

10 Properties of Matter II: Dielectric Polarization 260

11

Properties of Matter III: Magnetization 289

12

Time-Dependent Fields When Matter Is Present: Maxwell's Equations Revised 325

13

Plane Electromagnetic Waves in Linear Matter 347

14

Radiation from Prescribed Sources in Vacuum 399

15

Relativistic Formulation of Maxwell's Equations 426

Appendices

A

Linear Equations, Determinants, and Matrices 459

B

Binomial and Taylor Expansions 467

Preface

This book evolved concurrently with a two-term intermediate-level course that I have offered for upper-division undergraduates nearly every year since 1965, and its writing was motivated largely by a decision to treat the traditional topics of electricity and magnetism in a nontraditional order. The more significant departures from the traditional sequencing of topics include:

(a) Immediate parallel development of the phenomenologies both of charge and of current, using observed experimental properties to guide the selection of operational definitions for these fundamental entities.

(b) Early discussion of the differences among the common systems of units, emphasizing the way these differences are reflected in the definitions adopted for charge and current and simultaneously setting the stage for the later selection of the mksa system for most of the book.

(c) Introduction of the surface integral in the context of current calculations, where the flux of the current density represents something actually flowing across the surface and hence is more readily understood than the more abstract flux of, say, the electric field.

(d) Simultaneous introduction of both the electric and magnetic induction fields, thereby permitting a unified and compact treatment of particle trajectories and of forces and torques on general charge and current distributions.

(e) Postponement of the serious use of differential equations until after

Maxwell's equations have been fully developed, thereby making it possible for students to take electricity and magnetism concurrently with introductory differential equations.

(f) Full development of Maxwell's equations in vacuum before introducing the complications brought about by the presence of matter, regarding the modifications introduced by matter as an important special topic rather than as a part of the basic theory and (at the same time) permitting the student to become fully familiar with the fields in vacuum and with vector field theory in general before thrusting him into the subtle and sometimes confusing distinctions among the four fields used in the macroscopic theory of the fields in matter.

The flow chart following this preface displays the sequencing of topics in this book.

This flow chart can also be used as an aid to selecting that portion of the text to be covered if temporal constraints force an abbreviation of the course. The essential theory of the electromagnetic field in vacuum is covered in Chapters 1–6, which should be included in any course. (Sections 3-2, 4-6, and 5-6 can be omitted if desired.) The remaining chapters deal with a variety of applications, consequences, and extensions of the basic theory and, with the exception of Chapters 9–13, are largely independent of one another. The interrelationships among Chapters 7–15 are illustrated in the flow chart. If, for example, Section 12-8 is to be included, then the first four sections of Chapter 8 must also be included; if Chapter 13 or 14 is to be covered, then portions of Chapter 7 must be included; Section 7-4 can be omitted unless Section 13-6 is to be studied; and so on. The flow chart does not show that most of Chapter 8 (and particularly Section 8-7) can be studied any time after completion of Section 4-4.

Some of the features of this book are not apparent in the flow chart. In the first place, fewer than the average number of examples have been worked out explicitly in the text, partly to make the compactness of the basic theory the more apparent and partly to save some of these standard examples (which are almost invariably easier than most end-of-the-chapter problems) for assignment to students. When a standard example is set as a problem, the desired final result is either incorporated in the problem statement or included in the table of answers at the end of the book, and these problems are indexed along with the textual material. The text therefore does not suffer as a reference work because it fails to treat some of the standard examples explicitly. Secondly, the details of symmetry arguments, particularly in Sections 4-2 and 5-3, have been given more than average space, hopefully to prevent the student from developing either a gnawing mistrust of these powerful arguments or a willingness to use symmetry to support all sorts of outlandish claims. Thirdly, some of the problems draw specific attention to discussions in the periodical literature or in other texts, so that the student is encouraged to avoid complete reliance on any single source for his study of electricity and magnetism. Fourthly, throughout the text purely mathematical developments have been identified as such (and in some cases have been physically separated from the main text) so that the fundamental physics of electricity and magnetism stands out more prominently.

Finally, the growing availability of digital computers for instructional purposes and the growing sophistication of intermediate level students in the use of these computers have prompted the inclusion of several problems whose solution can best be carried out on a computer. Some of these problems simply use the computer to calculate a table of values of an unfamiliar function obtained as the analytic solution to the problem. Others describe an algorithm and lead the student through a numerical solution of a suggested problem, for example, the tracing of field lines for some distribution of point charges. In two cases (trajectory problems and Laplace's equation) a brief discussion of numerical approaches is incorporated in the text itself.

I conclude this Preface with an attempt to enumerate and acknowledge the several debts I owe to other individuals for their contributions—both direct and indirect—to this book. First, I wish to record my gratitude to several anonymous reviewers, to J. Bruce Brackenridge, to John R. Brandenberger, and to James S. Evans for numerous criticisms and suggestions that have led to improvements in the final version of this book; to John R. Merrill for many conversations about the uses of computers in physics instruction; and to dozens of students for their patience with endless drafts and for innumerable valuable suggestions. In addition I wish to acknowledge my debt to the numerous authors whose works on electricity and magnetism I have studied over the years. I am particularly aware of being influenced by the texts by E. M. Purcell; R. P. Feynman, R. B. Leighton, and M. Sands; and J. D. Jackson; and I have acknowledged these and other works that I can identify as having contributed in a general way to my own thinking by citing them more explicitly at appropriate points throughout this book. Further, I am grateful to Miss Jean St. Pierre for her expert and efficient typing of the manuscript. Last in order but first in importance, I owe to my wife and children a debt of a very different sort. They have patiently endured while the task of writing this book not only occupied all of my time but also went on and on and on. I recognize and am grateful for the love and understanding expressed by their patience.

DAVID M. COOK

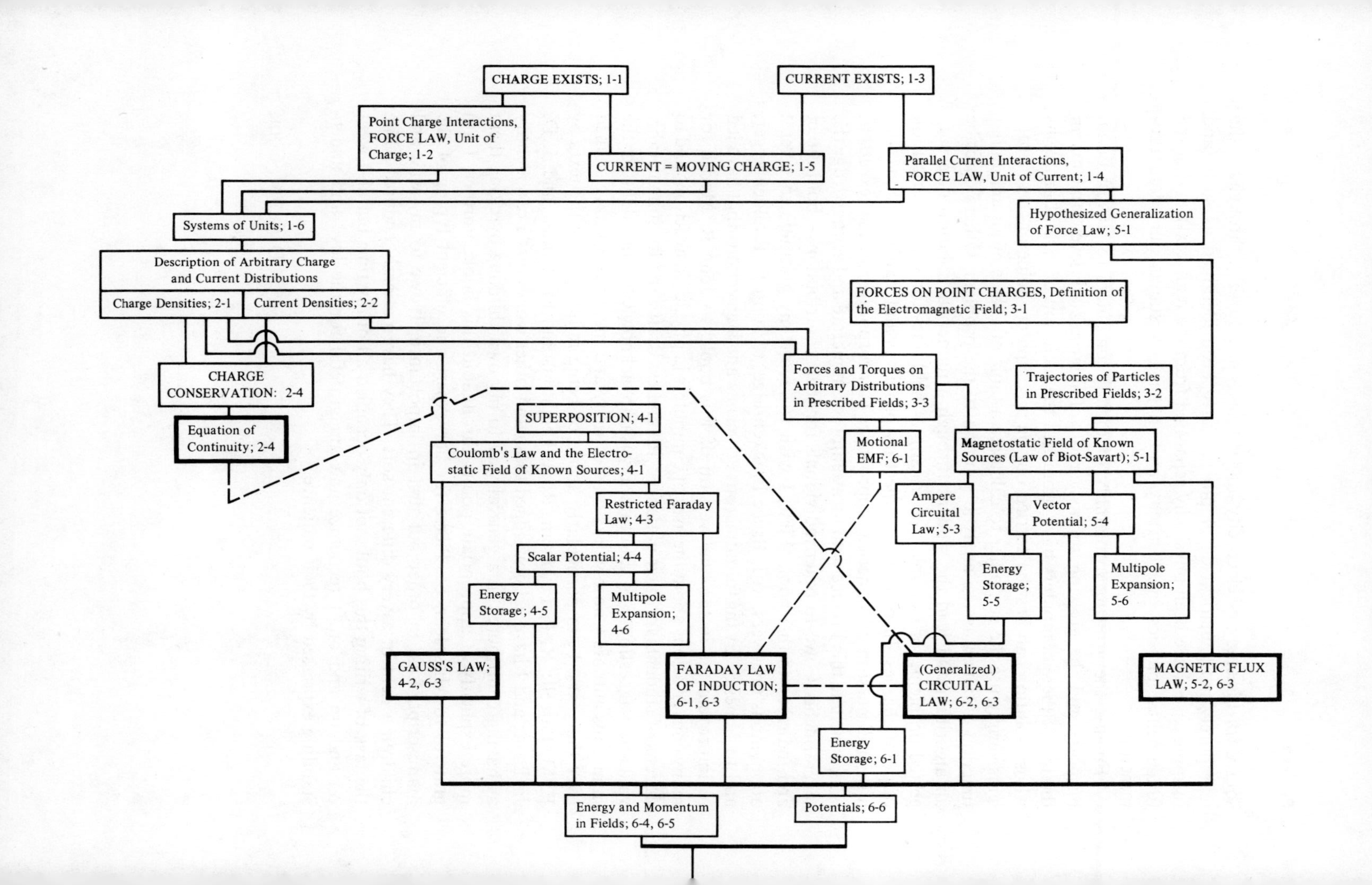
CHARGE EXISTS; 1-1
CURRENT EXISTS; 1-3
Point Charge Interactions, FORCE LAW, Unit of Charge; 1-2
CURRENT = MOVING CHARGE; 1-5
Parallel Current Interactions, FORCE LAW, Unit of Current; 1-4
Systems of Units; 1-6
Hypothesized Generalization of Force Law; 5-1
Description of Arbitrary Charge and Current Distributions
Charge Densities; 2-1
Current Densities; 2-2
FORCES ON POINT CHARGES, Definition of the Electromagnetic Field; 3-1
CHARGE CONSERVATION: 2-4
Forces and Torques on Arbitrary Distributions in Prescribed Fields; 3-3
Trajectories of Particles in Prescribed Fields; 3-2
SUPERPOSITION; 4-1
Equation of Continuity; 2-4
Motional EMF; 6-1
Magnetostatic Field of Known Sources (Law of Biot-Savart); 5-1
Coulomb's Law and the Electrostatic Field of Known Sources; 4-1
Restricted Faraday Law; 4-3
Ampere Circuital Law; 5-3
Vector Potential; 5-4
Scalar Potential; 4-4
Energy Storage; 4-5
Multipole Expansion; 4-6
Energy Storage; 5-5
Multipole Expansion; 5-6
GAUSS'S LAW; 4-2, 6-3
FARADAY LAW OF INDUCTION; 6-1, 6-3
(Generalized) CIRCUITAL LAW; 6-2, 6-3
MAGNETIC FLUX LAW; 5-2, 6-3
Energy Storage; 6-1
Energy and Momentum in Fields; 6-4, 6-5
Potentials; 6-6

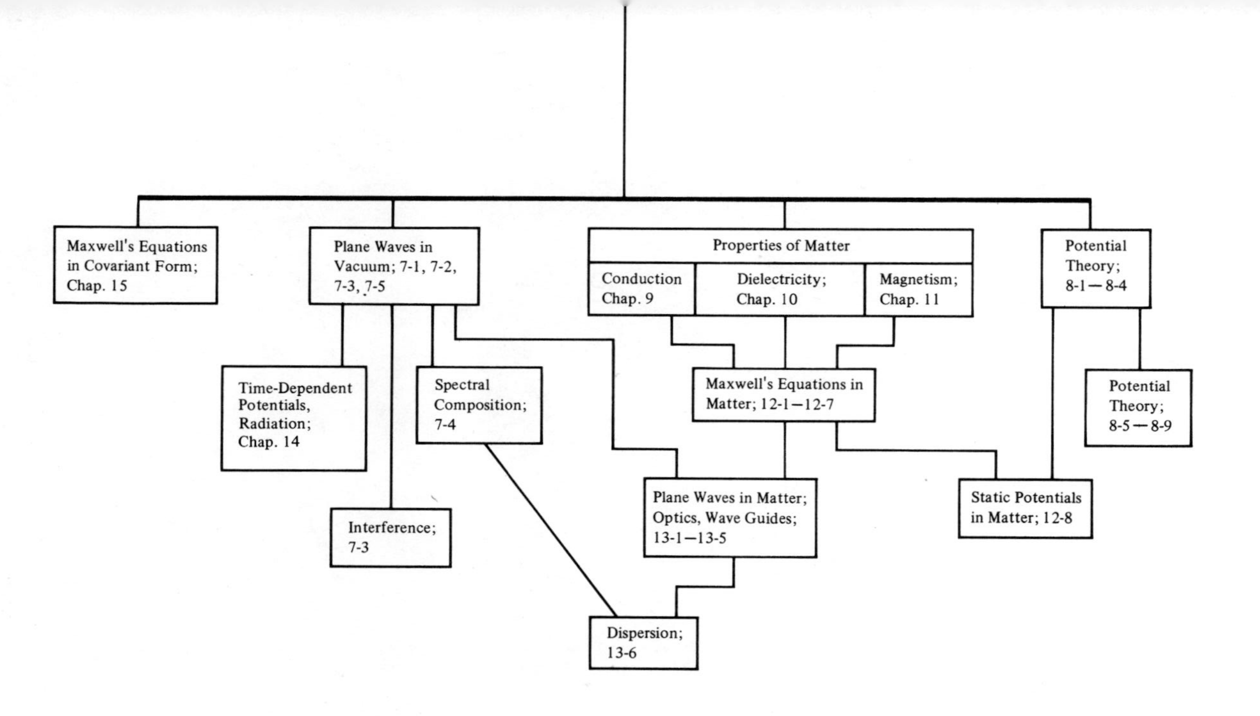
Maxwell's Equations in Covariant Form; Chap. 15
Plane Waves in Vacuum; 7-1, 7-2, 7-3, 7-5
Properties of Matter
Conduction Chap. 9
Dielectricity; Chap. 10
Magnetism; Chap. 11
Potential Theory; 8-1—8-4
Time-Dependent Potentials, Radiation; Chap. 14
Spectral Composition; 7-4
Maxwell's Equations in Matter; 12-1—12-7
Potential Theory; 8-5—8-9
Interference; 7-3
Plane Waves in Matter; Optics, Wave Guides; 13-1—13-5
Static Potentials in Matter; 12-8
Dispersion; 13-6

To my family:

Introduction

The major subdisciplines within classical physics are mechanics, electricity and magnetism, and thermodynamics. Although we cannot ignore completely the interactions among these subdisciplines, this book is concerned primarily with the classical theory of electricity and magnetism. Somewhat more accurately, we treat the classical theory of the electric and magnetic fields to be defined in Chapter 3. The physical phenomena that motivated the development of this theory were noticed and puzzled over in antiquity: Rubbed amber attracts bits of paper, and pieces of naturally occurring lodestone experience mutual forces of interaction. The development of the contemporary formal theory, however, did not begin until the late 1700's, and the experimental work of Coulomb (1785; Chapter 2), Ampere and Oersted (1800–1820; Chapter 2), and Faraday (1830–1860; Chapter 6) played a prominent role. Less than 100 years after Coulomb's work, the theory was brought to its present form by Maxwell (early 1860's; Chapter 6). Faraday and Maxwell discovered interrelationships between electric and magnetic phenomena, but a reexpression of Maxwell's theory in the terminology of Einstein's theory of special relativity (1905) revealed even deeper connections (Chapter 15) and we now view these initially distinct phenomena as a single phenomenon, which we call electromagnetism. Again more accurately, we now view the electric and magnetic fields as different aspects of a more inclusive electromagnetic field. Although the interrelationships supplied by the theory of relativity can be exploited to develop the classical theory of the electromagnetic field from a *very* small number of experimental observations, we nevertheless adopt a more traditional approach that requires a larger (but still small) number of experimental observations but does not require knowledge of transformations between frames of reference in relative motion. In Chapter 15, we shall show that the results of this development are consistent with the principles of relativity.

This book can be divided into two main sections. In the first section, we begin by enumerating the observed properties of charge and current and by using these properties to deduce suitable units for the quantitative measurement of charge and current (Chapter 1), continue by introducing the concepts of charge and current densities needed for describing the "state" of arbitrary charge distributions (Chapter 2) and by defining the electric and magnetic fields (Chapter 3), and conclude by examining the properties of these fields, first when they are time-independent (Chapter 4 and 5) and then when they are time-dependent (Chapter 6). Maxwell's equations, which are at once the goal of the first section of the book and the basis of the second section, are first stated in Section 6-3. From that point on, we shall be concerned with the consequences of Maxwell's equations and with their generalization to include the response of matter to externally applied fields. In the second main section of this book (Chapters 7–15), we examine potential theory, plane waves, properties of matter, radiation from accelerated particles, and the relativistic formulation of Maxwell's equations. The main course of the development in Chapters 1–6 and the interrelationships among the later chapters in the book are both shown in the flow chart following the preface. In this chart, items displayed entirely in capital letters are points of experimental contact, basic laws are placed in boxes with extra heavy borders, and each dotted line connects two boxes, the first of which provides a suggestion for (but not a deduction of) the second. The reader may find it useful to refer to this chart occasionally as he works his way through this book.

For many readers, this book probably represents a first (or perhaps a second) encounter with mathematical physics. It will therefore be necessary now and again to discuss mathematical techniques per se. To forestall the common (but wholly erroneous) conclusion that theoretical physics *is* mathematics, a special effort has been made to identify mathematical developments as such and to stress the role of experimental properties in guiding the development of the theory. In some cases (Chapter 0, Sections 2-3 and 2-5, and the Appendices) mathematical topics have been physically separated from the main text so that, once the reader has become facile with the mathematics, he can read the main text without being distracted from the physics by these mathematical digressions. The mathematical component of this text is important and the reader must strive for fluency in mathematical manipulation, but he cannot, on the other hand, allow the mathematics to overshadow the physics, which—after all—is what the mathematical model is constructed to represent.

Several additional comments and suggestions to guide the reader in his use of this book and to assist him in his study of the electromagnetic field follow:

(1) Throughout this book, equations are referred to by a two-part identification, the first part denoting the chapter or appendix in which the equation occurs. Equation (12-28) is the 28th equation in Chapter 12, Eq. (C-8) is the 8th equation in Appendix C, etc. Problems are referred to by a similar two-part identification prefixed with the letter P.
(2) Your attention is drawn to the identities in Appendix C, to the data in Appendix E, and to the final page of the index, on which you will find a quick directory to important formulas contained in the text proper.
(3) Even if you are already familiar with the preliminaries treated in Chapter 0, skim the chapter; you will find not only a review of background material but also the setting out of a notation whose meaning will subsequently be assumed without further definition.
(4) Fill in the omitted steps in each argument on your own initiative; only a few problems requesting the completion of these arguments are specifically included.
(5) Read *all* of the problems even if only a few are actually solved in detail; in total, these problems indicate something of the broad spectrum of circumstances to which the basic theory can be applied.
(6) Do not spend time laboriously evaluating standard integrals; use integral tables. Three of the many available tables are

H. B. DWIGHT, *Tables of Integrals and Other Mathematical Data* (The Macmillan Company, New York, 1961), Fourth Edition.

B. O. PIERCE and R. M. FOSTER, *A Short Table of Integrals* (Ginn and Company, Boston, 1956), Fourth Edition.

S. M. SELBY and R. C. WEAST, Editors, *CRC Standard Mathematical Tables* (Chemical Rubber Publishing Company, Cleveland, 1969), Seventeenth Edition.

(7) Refer regularly to other books on the same topics. Many may be found by looking in the subject area of the card catalog in your library or by browsing through the shelves in the proper area (537's and 538's in the Dewey decimal system; QC501–718 or so in the Library of Congress system). *Some* books at a level in the main more elementary than this book are

R. P. FEYNMAN, R. B. LEIGHTON, and M. SANDS, *The Feynman Lectures on Physics* (Addison-Wesley Publishing Company, Inc., Reading, Mass., 1964), Volume II.

D. HALLIDAY and R. RESNICK, *Physics* (John Wiley & Sons, Inc., New York, 1962), Second Edition.

A. F. KIP, *Fundamentals of Electricity and Magnetism* (McGraw-Hill Book Company, New York, 1969), Second Edition.

E. M. PURCELL, *Electricity and Magnetism* (McGraw-Hill Book Company, New York, 1965).

Books at a level comparable to that of this book include

D. R. Corson and P. Lorrain, *Introduction to Electromagnetic Fields and Waves* (W. H. Freeman and Company, San Francisco, 1962), First Edition.

J. R. Reitz and F. J. Milford, *Foundations of Electromagnetic Theory* (Addison-Wesley Publishing Company, Inc., Reading, Mass., 1967), Second Edition.

W. M. Schwarz, *Intermediate Electromagnetic Theory* (John Wiley & Sons, Inc., New York, 1964).

W. T. Scott, *The Physics of Electricity and Magnetism* (John Wiley & Sons, Inc., New York, 1962).

Finally, books at levels ranging from somewhat to very much more advanced than this book include

R. Becker and F. Sauter, *Electromagnetic Fields and Interactions* (Ginn/Blaisdell, Waltham, Mass., 1964), Volume I.

J. D. Jackson, *Classical Electrodynamics* (John Wiley & Sons, Inc., New York, 1962).

E. C. Jordan and K. G. Balmain, *Electromagnetic Waves and Radiating Systems* (Prentice-Hall, Inc., Englewood Cliffs, N.J., 1968), Second Edition.

L. D. Landau and E. M. Lifshitz, *The Classical Theory of Fields* (Addison-Wesley Publishing Company, Inc., Reading, Mass., 1962), Revised Second Edition.

P. Lorrain and D. R. Corson, *Electromagnetic Fields and Waves* (W. H. Freeman and Company, San Francisco, 1970), Second Edition.

J. B. Marion, *Classical Electromagnetic Radiation* (Academic Press, Inc., New York, 1965).

W. K. H. Panofsky and M. Phillips, *Classical Electricity and Magnetism* (Addison-Wesley Publishing Company, Inc., Reading, Mass., 1955).

W. R. Smythe, *Static and Dynamic Electricity* (McGraw-Hill Book Company, New York, 1968), Third Edition.

J. A. Stratton, *Electromagnetic Theory* (McGraw-Hill Book Company, New York, 1941).

The reader wanting assistance in the applications of computers to problems in electromagnetism is referred, for example, to

B. Carnahan, H. A. Luther, and J. O. Wilkes, *Applied Numerical Methods* (John Wiley & Sons, Inc., New York, 1969).

S. D. Conte, *Elementary Numerical Analysis* (McGraw-Hill Book Company, New York, 1965).

R. Ehrlich, *Physics and Computers* (Houghton Mifflin Company, Boston, 1973).

J. R. Merrill, *Computers in Physics* (Houghton Mifflin Company, Boston, 1975).

H. D. Peckham, *Computers, BASIC, and Physics* (Addison-Wesley Publishing Company, Inc., Reading, Mass., 1971).

and to the publications of Project COEXIST at Dartmouth College and of the Physics Computer Development Project at the Irvine Campus of the University of California; references to some of these publications are given at appropriate points in this text.

0

Mathematical and Physical Preliminaries

0-1 Scalars, Vectors, and Vector Algebra

Nearly every important physical quantity can be classified as a scalar, as a vector, or as a still more complicated entity called a tensor. For most of this book, we shall identify as a *scalar* any quantity (e.g., temperature, mass, density) that is completely determined by a magnitude alone, we shall identify as a *vector* any quantity (e.g., force, displacement, velocity) that is completely determined only if both a magnitude and a direction are specified, and we shall not identify tensors. We shall adopt these common definitions, however, only after recognizing that they fail to mention the following important property: True scalars, vectors, and tensors represent quantities (e.g., displacements) having a physical significance (e.g., direction and magnitude in space) that must remain unchanged even if a particular coordinate system used to represent these quantities by numbers is changed. That is, the physical content of a scalar, vector, or tensor must be invariant to such coordinate transformations as translations, rotations, and reflections. Consequently, the numbers used to express scalars, vectors, and tensors must exhibit very specific and determinable behavior under coordinate transformations. In this book, however, we shall explore this behavior only incompletely, particularly in P0-29 and in Chapter 15.

We shall adopt the usual notational conventions. Scalars will be denoted

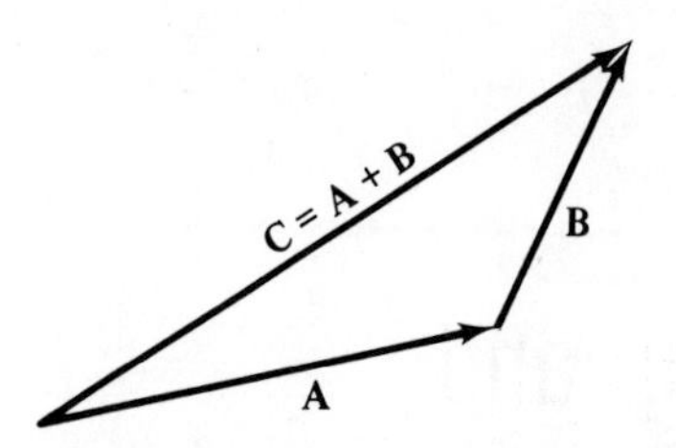

Fig. 0-1. Geometric addition of vectors.

by lightface letters, vectors by boldface letters, and the magnitude of a vector **A** by $|\mathbf{A}|$ or A. In figures the vector **A** will be represented by an arrow whose length is proportional to the magnitude of **A** and whose direction coincides with that of **A**.

The algebraic properties of scalars and vectors are assigned so that scalars and vectors combine mathematically in the same way that quantities represented by scalars and vectors combine physically. Without attempting any further motivation for our choice, we shall take *scalars* to be manipulated by the rules of ordinary algebra and we shall adopt the following properties for the algebraic manipulation of *vectors:*

(1) *Equality.* Two vectors **A** and **B** are defined to be equal, $\mathbf{A} = \mathbf{B}$, if and only if they have the same magnitude and direction. This definition does *not* require the two vectors to originate at the same point.
(2) *Addition.* A vector **C** is defined to be the sum of two other vectors **A** and **B**, $\mathbf{C} = \mathbf{A} + \mathbf{B}$, if and only if **A**, **B**, and **C** are related geometrically as the sides of a triangle (Fig. 0-1). If the vectors represent displacements, the sum **C** is evidently the single displacement representing the net effect of displacement **A** followed by displacement **B**.
(3) *Vector addition is commutative,* $\mathbf{A} + \mathbf{B} = \mathbf{B} + \mathbf{A}$. That is, the order in which the vectors are added does not affect the sum.
(4) *Vector addition is associative,* $\mathbf{A} + (\mathbf{B} + \mathbf{C}) = (\mathbf{A} + \mathbf{B}) + \mathbf{C}$. That is, the manner in which grouping is done to evaluate the sum of three (or more) vectors does not affect the final result.
(5) *Multiplication by a scalar.* A vector **B** is defined to be equal to the vector **A** times the scalar s, $\mathbf{B} = s\mathbf{A}$, if $|\mathbf{B}| = |s||\mathbf{A}|$ and the direction of **B** is that of **A** ($s > 0$) or opposite to that of **A** ($s < 0$). In particular if $s = -1$, **B** has the same magnitude as **A** but the opposite direction. Finally, if $s = 1/|\mathbf{A}|$, then $|\mathbf{B}| = 1$ and **B** is called a unit vector. It has the direction of **A** and is commonly denoted by $\hat{\mathbf{A}}$. Thus,

$$\hat{\mathbf{A}} = \frac{\mathbf{A}}{|\mathbf{A}|} = \frac{\mathbf{A}}{A}; \qquad \mathbf{A} = A\hat{\mathbf{A}} \tag{0-1}$$

(6) *Multiplication of a vector by a scalar is distributive,* both with respect to addition of scalars, $(s + t)\mathbf{A} = s\mathbf{A} + t\mathbf{A}$, and with respect to addition of vectors, $s(\mathbf{A} + \mathbf{B}) = s\mathbf{A} + s\mathbf{B}$.

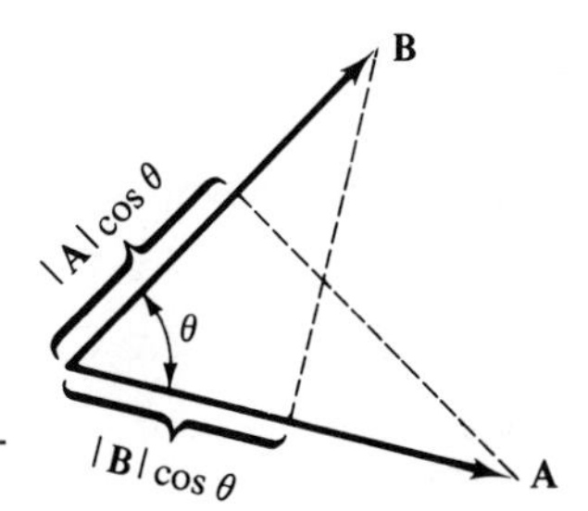

Fig. 0-2. Two geometric interpretations of the dot product.

(7) *Subtraction.* A vector **C** is defined to be the difference between two vectors **A** and **B**, $\mathbf{C} = \mathbf{A} - \mathbf{B}$, if $\mathbf{C} = \mathbf{A} + (-\mathbf{B})$.

(8) *Dot multiplication of vectors.* A scalar s is defined to be the dot product of two vectors **A** and **B**, $s = \mathbf{A}\cdot\mathbf{B}$, if $s = |\mathbf{A}||\mathbf{B}| \cos \theta$, where θ is the angle between **A** and **B** (Fig. 0-2). The dot product is commutative, $\mathbf{A}\cdot\mathbf{B} = \mathbf{B}\cdot\mathbf{A}$; is distributive with respect to vector addition, $\mathbf{A}\cdot(\mathbf{B} + \mathbf{C}) = \mathbf{A}\cdot\mathbf{B} + \mathbf{A}\cdot\mathbf{C}$; and is zero if **A** and **B** are nonzero and perpendicular. Geometrically $\mathbf{A}\cdot\mathbf{B}$ can be interpreted as $|\mathbf{A}|$ times the projection of **B** on **A** or as $|\mathbf{B}|$ times the projection of **A** on **B**. Finally, the dot product of a vector with itself is the square of the magnitude of the vector, $\mathbf{A}\cdot\mathbf{A} = |\mathbf{A}|^2 = A^2$.

(9) *Cross multiplication of vectors.* A vector **V** is defined to be the cross product of two vectors **A** and **B**, $\mathbf{V} = \mathbf{A} \times \mathbf{B}$, if $\mathbf{V} = |\mathbf{A}||\mathbf{B}| \sin \theta\hat{\mathbf{n}}$, where θ is the angle from **A** to **B** and $\hat{\mathbf{n}}$ is a unit vector in the direction determined by the thumb of the right hand when the fingers are extended along **A** and the palm faces **B** through the angle θ (Fig. 0-3). The cross product is *anti*commutative, $\mathbf{A} \times \mathbf{B} = -\mathbf{B} \times \mathbf{A}$; is distributive with respect to vector addition, $\mathbf{A} \times (\mathbf{B} + \mathbf{C}) = \mathbf{A} \times \mathbf{B} + \mathbf{A} \times \mathbf{C}$; and is zero if **A** and **B** are nonzero and parallel. Geometrically $|\mathbf{A} \times \mathbf{B}|$ is the area of the parallelogram bounded by **A** and **B**.

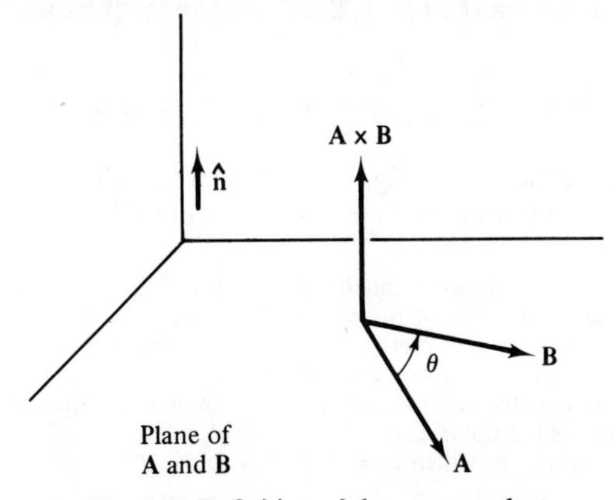

Fig. 0-3. Definition of the cross product.

(10) *Multiple vector products* can be evaluated by successive application of the definitions in (8) and (9). In particular, the triple scalar product, $\mathbf{A}\cdot(\mathbf{B}\times\mathbf{C})$, and the two (*unequal*) triple vector products, $(\mathbf{A}\times\mathbf{B})\times\mathbf{C}$ and $\mathbf{A}\times(\mathbf{B}\times\mathbf{C})$, occur frequently and the identities

$$\mathbf{A}\times(\mathbf{B}\times\mathbf{C}) = (\mathbf{A}\cdot\mathbf{C})\mathbf{B} - (\mathbf{A}\cdot\mathbf{B})\mathbf{C} \tag{0-2}$$

$$(\mathbf{A}\times\mathbf{B})\times\mathbf{C} = (\mathbf{A}\cdot\mathbf{C})\mathbf{B} - (\mathbf{B}\cdot\mathbf{C})\mathbf{A} \tag{0-3}$$

are extremely useful.[1]

PROBLEMS

P0-1. Prove geometrically that (a) $\mathbf{A}+\mathbf{B}=\mathbf{B}+\mathbf{A}$, (b) $(\mathbf{A}+\mathbf{B})+\mathbf{C} = \mathbf{A}+(\mathbf{B}+\mathbf{C})$, and (c) $\mathbf{A}\cdot(\mathbf{B}+\mathbf{C}) = \mathbf{A}\cdot\mathbf{B}+\mathbf{A}\cdot\mathbf{C}$.

P0-2. Interpret $\mathbf{A}\cdot(\mathbf{B}\times\mathbf{C})$ geometrically.

P0-3. Combine Eq. (0-2) with the anticommutativity of the cross product to derive Eq. (0-3). *Hint*: Rename the vectors.

P0-4. Let $\mathbf{A}$ be a known vector and $\mathbf{X}$ an unknown vector. (a) Describe in words the freedom remaining in $\mathbf{X}$ if $\mathbf{A}\cdot\mathbf{X}=s$ is fixed. (b) Describe in words the freedom remaining in $\mathbf{X}$ if $\mathbf{A}\times\mathbf{X}=\mathbf{c}$ is fixed. (c) Let both s and $\mathbf{c}$ be given. Show that $\mathbf{X}=(s\mathbf{A}+\mathbf{c}\times\mathbf{A})/A^2$.

Analytically, vectors are often conveniently represented by their components in some coordinate system. The *component* A_n of a vector $\mathbf{A}$ in the direction specified by a *unit* vector $\hat{\mathbf{n}}$ is defined by

$$A_n = \mathbf{A}\cdot\hat{\mathbf{n}} = A\cos\theta \tag{0-4}$$

where θ is the angle between $\mathbf{A}$ and $\hat{\mathbf{n}}$; equivalently, A_n is the projection of $\mathbf{A}$ in the direction $\hat{\mathbf{n}}$. If at some point in space we introduce a right-handed triad of mutually orthogonal unit vectors $\hat{\mathbf{e}}_1$, $\hat{\mathbf{e}}_2$, $\hat{\mathbf{e}}_3$,[2] then the vector $\mathbf{A}$ has the three components $A_i = \mathbf{A}\cdot\hat{\mathbf{e}}_i$, $i = 1, 2, 3$, and can itself be written as the sum of three vectors

$$\mathbf{A} = \sum_{i=1}^{3} A_i\hat{\mathbf{e}}_i = A_1\hat{\mathbf{e}}_1 + A_2\hat{\mathbf{e}}_2 + A_3\hat{\mathbf{e}}_3 \tag{0-5}$$

as illustrated geometrically in Fig. 0-4. All of the definitions and theorems summarized in the previous paragraph have corresponding expressions in

[1]The author finds the following mnemonic helpful: *Write down the product of the second vector with the dot product of the first and third vectors; then subtract the quantity obtained from this first term by exchanging the role of the two vectors occurring in parentheses in the original triple product.* This mnemonic sounds cumbersome, but it has the particular advantage of working equally well for both forms of the triple vector product.

[2]Right-handed triads are those in which $\hat{\mathbf{e}}_3 = \hat{\mathbf{e}}_1\times\hat{\mathbf{e}}_2$ and are almost universally preferred to left-handed triads, in which $\hat{\mathbf{e}}_3 = -\hat{\mathbf{e}}_1\times\hat{\mathbf{e}}_2$.

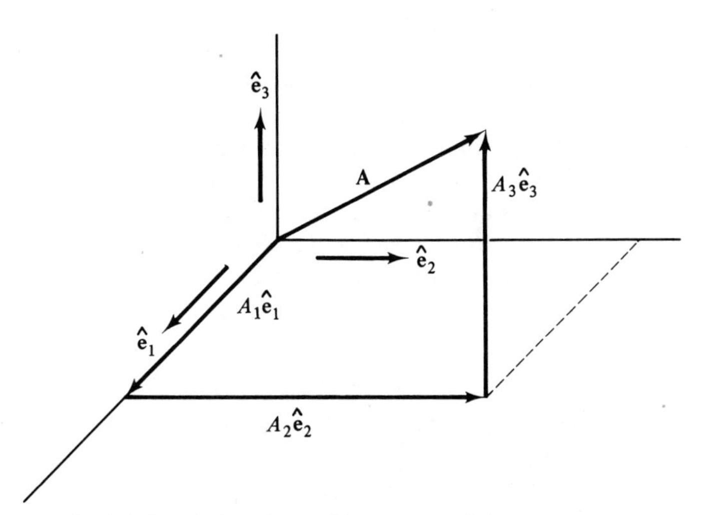

Fig. 0-4. Resolution of an arbitrary vector **A** into components.

terms of components. Equality of two vectors, for example, implies equality of the corresponding components,[3]

$$\mathbf{A} = \mathbf{B} \Longrightarrow A_1 = B_1, \qquad A_2 = B_2, \qquad A_3 = B_3 \tag{0-6}$$

A single vector equation is therefore equivalent to three separate scalar equations. Likewise,

$$\mathbf{C} = \mathbf{A} \pm \mathbf{B} \Longrightarrow C_1 = A_1 \pm B_1, \qquad C_2 = A_2 \pm B_2, \qquad C_3 = A_3 \pm B_3 \tag{0-7}$$

$$\mathbf{B} = s\mathbf{A} \Longrightarrow B_1 = sA_1, \qquad B_2 = sA_2, \qquad B_3 = sA_3 \tag{0-8}$$

and the commutative, associative, and distributive properties of vector algebra therefore reduce to a threefold application of the same properties of scalar algebra. Finally, in terms of components, one can show that

$$|\mathbf{A}| = \sqrt{(A_1)^2 + (A_2)^2 + (A_3)^2} \tag{0-9}$$

$$\mathbf{A} \cdot \mathbf{B} = A_1B_1 + A_2B_2 + A_3B_3 \tag{0-10}$$

$$\mathbf{A} \times \mathbf{B} = (A_2B_3 - A_3B_2)\hat{\mathbf{e}}_1 + (A_3B_1 - A_1B_3)\hat{\mathbf{e}}_2 + (A_1B_2 - A_2B_1)\hat{\mathbf{e}}_3 \tag{0-11}$$

$$= \begin{vmatrix} \hat{\mathbf{e}}_1 & \hat{\mathbf{e}}_2 & \hat{\mathbf{e}}_3 \\ A_1 & A_2 & A_3 \\ B_1 & B_2 & B_3 \end{vmatrix} \tag{0-12}$$

In Eq. (0-12), the vertical bars denote a determinant (Appendix A).

[3]The symbol $\Longrightarrow$ should be read "implies" or "implies that".

PROBLEMS

P0-5. A vector $\mathbf{A}$ is often usefully separated into a *longitudinal* part, $\mathbf{A}_{\|}$, parallel to a given unit vector $\hat{\mathbf{n}}$ and a *transverse* part, $\mathbf{A}_{\perp}$, perpendicular to $\hat{\mathbf{n}}$, where $\mathbf{A} = \mathbf{A}_{\|} + \mathbf{A}_{\perp}$. (a) Show that $\mathbf{A}_{\|} = (\mathbf{A}\cdot\hat{\mathbf{n}})\hat{\mathbf{n}}$ and $\mathbf{A}_{\perp} = \hat{\mathbf{n}} \times (\mathbf{A} \times \hat{\mathbf{n}})$. (b) Show that $\hat{\mathbf{n}} \times (\mathbf{A} \times \hat{\mathbf{n}}) = 0$ implies that $\mathbf{A} \times \hat{\mathbf{n}} = 0$.

P0-6. (a) Construct a table giving the nine possible dot products between two unit vectors chosen from $\hat{\mathbf{e}}_1$, $\hat{\mathbf{e}}_2$, and $\hat{\mathbf{e}}_3$. (b) Construct a similar table giving the nine possible cross products between two unit vectors chosen from $\hat{\mathbf{e}}_1$, $\hat{\mathbf{e}}_2$, and $\hat{\mathbf{e}}_3$. (c) Using these products and the algebraic properties of vectors, manipulate the expression

$$(A_1\hat{\mathbf{e}}_1 + A_2\hat{\mathbf{e}}_2 + A_3\hat{\mathbf{e}}_3) \otimes (B_1\hat{\mathbf{e}}_1 + B_2\hat{\mathbf{e}}_2 + B_3\hat{\mathbf{e}}_3)$$

where $\otimes$ represents either a dot or a cross, and derive Eqs. (0-10) and (0-11).

P0-7. (a) Show that

$$\mathbf{A}\cdot(\mathbf{B} \times \mathbf{C}) = \begin{vmatrix} A_1 & A_2 & A_3 \\ B_1 & B_2 & B_3 \\ C_1 & C_2 & C_3 \end{vmatrix}$$

and prove that the dot and the cross can be interchanged, i.e., that $\mathbf{A}\cdot(\mathbf{B} \times \mathbf{C}) = (\mathbf{A} \times \mathbf{B})\cdot\mathbf{C}$. (b) Prove that

$$(\mathbf{A} \times \mathbf{B})\cdot(\mathbf{C} \times \mathbf{D}) = \begin{vmatrix} \mathbf{A}\cdot\mathbf{C} & \mathbf{A}\cdot\mathbf{D} \\ \mathbf{B}\cdot\mathbf{C} & \mathbf{B}\cdot\mathbf{D} \end{vmatrix}$$

and then set $\mathbf{A} = \mathbf{C}$ and $\mathbf{B} = \mathbf{D}$ to show that $\sin^2\theta = 1 - \cos^2\theta$, where θ is the (arbitrary) angle between $\mathbf{A}$ and $\mathbf{B}$.

P0-8. Let $\mathbf{A} = 3\hat{\mathbf{e}}_1 + 2\hat{\mathbf{e}}_2 - 6\hat{\mathbf{e}}_3$ and $\mathbf{B} = \hat{\mathbf{e}}_1 + \hat{\mathbf{e}}_3$. Find $\hat{\mathbf{A}}$, $\mathbf{A}\cdot\mathbf{B}$, $\mathbf{A} \times \mathbf{B}$, and the angle between $\mathbf{A}$ and $\mathbf{B}$.

P0-9. Verify Eq. (0-2) by expressing all vectors in terms of their components and showing that both sides of the equation reduce to the same vector. *Hints*: (1) Choose the basic unit vectors wisely. For example, choose $\hat{\mathbf{e}}_3$ parallel to $\mathbf{C}$ and choose $\hat{\mathbf{e}}_2$ in the plane of $\mathbf{B}$ and $\mathbf{C}$. (2) Accept the theorem that, by virtue of their transformation properties, vector identities that are valid in one coordinate system are necessarily valid in any coordinate system obtained by arbitrary translation and/or rotation of the first system.

When specific components of a vector are needed, they will almost always be expressed in one of three commonly occurring coordinate systems. In *Cartesian coordinates*, a point in space is located by the three numbers (x, y, z) indicated geometrically in Fig. 0-5(a) and having the range $-\infty < x, y, z < \infty$; the mutually orthogonal unit vectors $\hat{\mathbf{i}}$, $\hat{\mathbf{j}}$, and $\hat{\mathbf{k}}$ at any point have, respectively, the direction of increasing x, y, and z at that point; and an

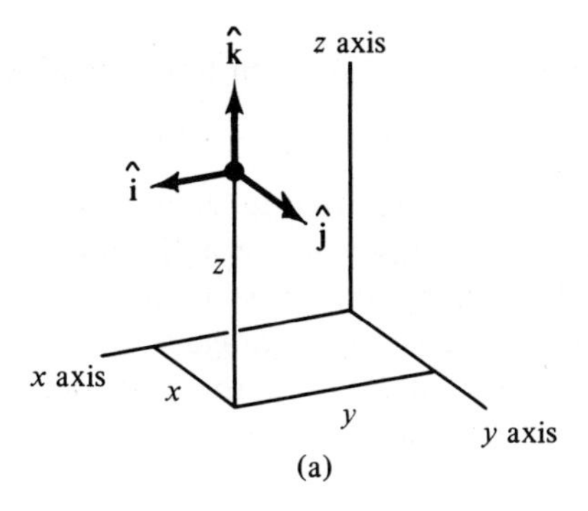

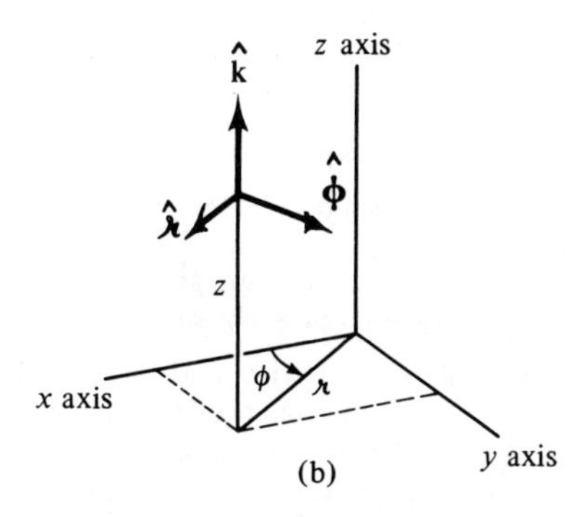

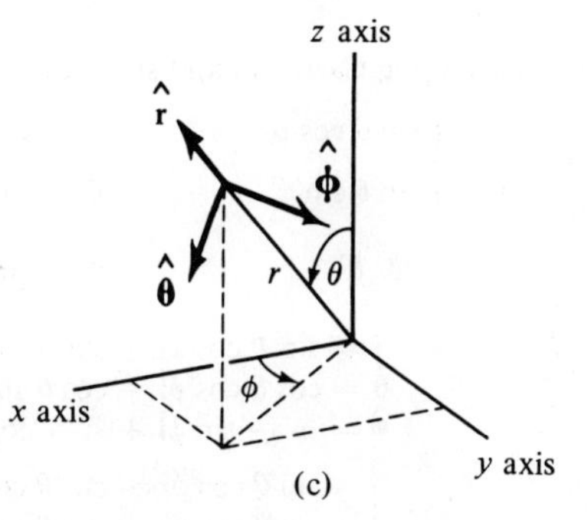

Fig. 0-5. Axes and unit vectors in (a) Cartesian coordinates, (b) cylindrical coordinates, and (c) spherical coordinates.

arbitrary vector **A** has the components $A_x = \mathbf{A}\cdot\hat{\mathbf{i}}$, $A_y = \mathbf{A}\cdot\hat{\mathbf{j}}$, and $\mathbf{A}_z = \mathbf{A}\cdot\hat{\mathbf{k}}$, in terms of which

$$\mathbf{A} = A_x\hat{\mathbf{i}} + A_y\hat{\mathbf{j}} + A_z\hat{\mathbf{k}} \quad \text{(Cartesian)} \tag{0-13}$$

In *cylindrical coordinates*, a point in space is located by the three numbers $(\imath, \phi, z)$ indicated geometrically in Fig. 0-5(b) and having the range $0 \leq \imath < \infty$, $0 \leq \phi < 2\pi$, $-\infty < z < \infty$; the mutually orthogonal unit vectors $\hat{\boldsymbol{\imath}}$, $\hat{\boldsymbol{\phi}}$, and $\hat{\mathbf{k}}$ at any point have, respectively, the direction of increasing $\imath$, ϕ, and z at that point; and an arbitrary vector **A** has the components $A_\imath = \mathbf{A}\cdot\hat{\boldsymbol{\imath}}$,

$A_\phi = \mathbf{A}\cdot\hat{\boldsymbol{\phi}}$, and $A_z = \mathbf{A}\cdot\hat{\mathbf{k}}$, in terms of which

$$\mathbf{A} = A_{\mathscr{r}}\hat{\boldsymbol{\mathscr{r}}} + A_\phi\hat{\boldsymbol{\phi}} + A_z\hat{\mathbf{k}} \quad \text{(cylindrical)} \tag{0-14}$$

$A_{\mathscr{r}}$, A_ϕ, and A_z are referred to, respectively, as the (cylindrical) radial, azimuthal, and axial components of **A**. Finally, in *spherical coordinates*, a point in space is located by the three numbers (r, θ, ϕ) indicated geometrically in Fig. 0-5(c) and having the range $0 \le r < \infty$, $0 \le \theta \le \pi$, $0 \le \phi < 2\pi$; the mutually orthogonal unit vectors $\hat{\mathbf{r}}$, $\hat{\boldsymbol{\theta}}$, and $\hat{\boldsymbol{\phi}}$ at any point have, respectively, the direction of increasing r, θ, and ϕ at that point; and an arbitrary vector

TABLE 0-1 Relationships Among Common Coordinates and Unit Vectors

(a) Relationships among Cartesian and cylindrical coordinates

$$x = \mathscr{r}\cos\phi \qquad \mathscr{r} = \sqrt{x^2 + y^2}$$

$$y = \mathscr{r}\sin\phi \qquad \phi = \tan^{-1}\left(\frac{y}{x}\right)$$

$$\hat{\boldsymbol{\mathscr{r}}} = \cos\phi\,\hat{\mathbf{i}} + \sin\phi\,\hat{\mathbf{j}} \qquad \hat{\mathbf{i}} = \cos\phi\,\hat{\boldsymbol{\mathscr{r}}} - \sin\phi\,\hat{\boldsymbol{\phi}}$$

$$\hat{\boldsymbol{\phi}} = -\sin\phi\,\hat{\mathbf{i}} + \cos\phi\,\hat{\mathbf{j}} \qquad \hat{\mathbf{j}} = \sin\phi\,\hat{\boldsymbol{\mathscr{r}}} + \cos\phi\,\hat{\boldsymbol{\phi}}$$

Expressions relating the Cartesian and cylindrical components of a vector **A** may be obtained by taking the dot products of **A** with the above relationships among unit vectors, e.g., $A_{\mathscr{r}} = \mathbf{A}\cdot\hat{\boldsymbol{\mathscr{r}}} = A_x\cos\phi + A_y\sin\phi$.

$$\frac{\partial\hat{\boldsymbol{\mathscr{r}}}}{\partial\phi} = \hat{\boldsymbol{\phi}} \qquad \frac{\partial\hat{\boldsymbol{\phi}}}{\partial\phi} = -\hat{\boldsymbol{\mathscr{r}}}$$

(b) Relationships among Cartesian and spherical coordinates

$$x = r\sin\theta\cos\phi \qquad r = \sqrt{x^2 + y^2 + z^2}$$

$$y = r\sin\theta\sin\phi \qquad \theta = \cos^{-1}\left[\frac{z}{\sqrt{x^2 + y^2 + z^2}}\right]$$

$$z = r\cos\theta \qquad \phi = \tan^{-1}\left(\frac{y}{x}\right)$$

$$\hat{\mathbf{r}} = \sin\theta\cos\phi\,\hat{\mathbf{i}} + \sin\theta\sin\phi\,\hat{\mathbf{j}} + \cos\theta\,\hat{\mathbf{k}}$$
$$\hat{\boldsymbol{\theta}} = \cos\theta\cos\phi\,\hat{\mathbf{i}} + \cos\theta\sin\phi\,\hat{\mathbf{j}} - \sin\theta\,\hat{\mathbf{k}}$$
$$\hat{\boldsymbol{\phi}} = -\sin\phi\,\hat{\mathbf{i}} + \cos\phi\,\hat{\mathbf{j}}$$

$$\hat{\mathbf{i}} = \sin\theta\cos\phi\,\hat{\mathbf{r}} + \cos\theta\cos\phi\,\hat{\boldsymbol{\theta}} - \sin\phi\,\hat{\boldsymbol{\phi}}$$
$$\hat{\mathbf{j}} = \sin\theta\sin\phi\,\hat{\mathbf{r}} + \cos\theta\sin\phi\,\hat{\boldsymbol{\theta}} + \cos\phi\,\hat{\boldsymbol{\phi}}$$
$$\hat{\mathbf{k}} = \cos\theta\,\hat{\mathbf{r}} - \sin\theta\,\hat{\boldsymbol{\theta}}$$

Expressions relating the Cartesian and spherical components of a vector **A** may be obtained by taking the dot products of **A** with the above relationships among unit vectors, e.g., $A_r = \mathbf{A}\cdot\hat{\mathbf{r}} = A_x\sin\theta\cos\phi + A_y\sin\theta\sin\phi + A_z\cos\theta$.

$$\frac{\partial\hat{\mathbf{r}}}{\partial\theta} = \hat{\boldsymbol{\theta}} \qquad \frac{\partial\hat{\mathbf{r}}}{\partial\phi} = \sin\theta\,\hat{\boldsymbol{\phi}}$$

$$\frac{\partial\hat{\boldsymbol{\theta}}}{\partial\theta} = -\hat{\mathbf{r}} \qquad \frac{\partial\hat{\boldsymbol{\theta}}}{\partial\phi} = \cos\theta\,\hat{\boldsymbol{\phi}}$$

$$\frac{\partial\hat{\boldsymbol{\phi}}}{\partial\theta} = 0 \qquad \frac{\partial\hat{\boldsymbol{\phi}}}{\partial\phi} = -\sin\theta\,\hat{\mathbf{r}} - \cos\theta\,\hat{\boldsymbol{\theta}}$$

A has the components $A_r = \mathbf{A}\cdot\hat{\mathbf{r}}$, $A_\theta = \mathbf{A}\cdot\hat{\boldsymbol{\theta}}$, and $\mathbf{A}_\phi = \mathbf{A}\cdot\hat{\boldsymbol{\phi}}$, in terms of which

$$\mathbf{A} = A_r\hat{\mathbf{r}} + A_\theta\hat{\boldsymbol{\theta}} + A_\phi\hat{\boldsymbol{\phi}} \quad \text{(spherical)} \tag{0-15}$$

A_r, A_θ, and A_ϕ are referred to, respectively, as the (spherical) radial, polar, and azimuthal components of **A**, and the z axis is often called the polar axis. Several relationships among these three sets of coordinates and unit vectors are summarized in Table 0-1. The following observations are pertinent to the use of vectors expressed in these coordinate systems: (1) Only in the Cartesian system are the unit vectors constant and independent of the point in space to which they apply. In the cylindrical and spherical systems, the unit vectors change direction as the point to which they apply moves about in space. (2) The rules—Eqs. (0-6)–(0-12)—for manipulating vectors expressed in terms of their components followed from the orthogonality and right-handedness of the basic unit vectors. Since the triads $(\hat{\mathbf{i}}, \hat{\mathbf{j}}, \hat{\mathbf{k}})$, $(\hat{\boldsymbol{\imath}}, \hat{\boldsymbol{\phi}}, \hat{\mathbf{k}})$, and $(\hat{\mathbf{r}}, \hat{\boldsymbol{\theta}}, \hat{\boldsymbol{\phi}})$ all exhibit these properties, Eqs. (0-6)–(0-12) apply in particular to the specific expressions in Eqs. (0-13)–(0-15). (3) The above discussion presupposed no specific choice for the coordinate origin or for the orientation of the x, y, and z axes; these features of the coordinate system will in general be selected for convenience in application to the problem at hand. Finally, (4) a number of different notations, particularly for the cylindrical coordinates, can be found in the literature, and suitable care must be exercised in comparing the writings of two or more authors when these coordinates are used.

PROBLEM

P0-10. Familiarize yourself with the entries in Table 0-1 and verify those in Table 0-1(a). *Note:* The expressions for $\partial\hat{\boldsymbol{\imath}}/\partial\phi$, $\partial\hat{\mathbf{r}}/\partial\theta$, and $\partial\hat{\mathbf{r}}/\partial\phi$ given in Table 0-1 can be conveniently used to obtain $\hat{\boldsymbol{\phi}}$ and $\hat{\boldsymbol{\theta}}$ by differentiating $\hat{\boldsymbol{\imath}}$ or $\hat{\mathbf{r}}$, which themselves can be written down in terms of $\hat{\mathbf{i}}$, $\hat{\mathbf{j}}$, and $\hat{\mathbf{k}}$ by inspection.

0-2 The Representation of Fields

Only a very few physical quantities are absolute constants. Most frequently, the value of a physical quantity (e.g., temperature, force) depends on where in space and when in time the quantity is determined. To denote the totality of values of a physical quantity at *all* points in some region of space-time, we introduce the word *field*, examples of which are the (scalar) temperature field in a room, the (vector) force field produced by a massive object, and the (scalar) probability field of a quantum mechanical particle.

For theoretical manipulations, fields are most usefully represented as (scalar or vector) functions of position and time. We introduce a coordinate system—probably one of those described in Section 0-1—and a clock, setting the origin of each for maximum convenience to the problem at hand. A point in space-time is then located by its three spatial coordinates, say (r, θ, ϕ), and its temporal coordinate t in the (arbitrarily selected) coordinate system, and the dependence of some field S on space-time is indicated by writing the space-time coordinates as arguments of a function, e.g., $S(r, \theta, \phi, t)$. This dependence may alternatively be indicated by recognizing that the single point P variously identified by (x, y, z), (ϱ, ϕ, z), or (r, θ, ϕ) could equally well be labeled by its *position vector* $\mathbf{r}$, which is the vector *from* the origin *to* the point P. The dependence of S on space-time might then be indicated by $S(\mathbf{r}, t)$. This more compact notation permits postponing introduction of $\mathbf{r}$ in one of the equivalent forms

$$\mathbf{r} = x\hat{\mathbf{i}} + y\hat{\mathbf{j}} + z\hat{\mathbf{k}} \quad \text{(Cartesian)} \tag{0-16}$$

$$= \varrho\hat{\boldsymbol{\varrho}} + z\hat{\mathbf{k}} \quad \text{(cylindrical)} \tag{0-17}$$

$$= r\hat{\mathbf{r}} \quad \text{(spherical)} \tag{0-18}$$

or in some other form until explicit selection of a coordinate system is required. It is, of course, not necessary for a field to depend on all four space-time coordinates, the time-independent or *static* field, denoted by $S(\mathbf{r})$, being particularly common.

Among the more important manipulations with scalar and vector fields are differentiation and integration. Differentiation of a *scalar* field is accomplished by the usual means. In Cartesian coordinates, the simple derivative of a *vector* field is that vector field whose components are the derivatives of the components of the field itself, e.g., if

$$\mathbf{Q}(\mathbf{r}, t) = Q_x(x, y, z, t)\hat{\mathbf{i}} + Q_y(x, y, z, t)\hat{\mathbf{j}} + Q_z(x, y, z, t)\hat{\mathbf{k}} \tag{0-19}$$

then

$$\frac{\partial \mathbf{Q}}{\partial y} = \frac{\partial Q_x}{\partial y}\hat{\mathbf{i}} + \frac{\partial Q_y}{\partial y}\hat{\mathbf{j}} + \frac{\partial Q_z}{\partial y}\hat{\mathbf{k}} \tag{0-20}$$

If the vector field is expressed in some other coordinate system, say,

$$\mathbf{Q}(\mathbf{r}, t) = Q_\varrho(\varrho, \phi, z, t)\hat{\boldsymbol{\varrho}} + Q_\phi(\varrho, \phi, z, t)\hat{\boldsymbol{\phi}} + Q_z(\varrho, \phi, z, t)\hat{\mathbf{k}} \tag{0-21}$$

then differentiation is more complicated, for the unit vectors may not be constant and each term must be differentiated as a product. Simple integration of a *scalar* field is accomplished by familiar procedures; in particular, we shall encounter the volume integral, variously denoted by

$$\int S(x, y, z)\, dx\, dy\, dz; \qquad \int S(\mathbf{r})\, dv; \qquad \int S(\mathbf{r})\, d^3\mathbf{r}$$

where the volume element has the expressions

$$dv = dx\, dy\, dz \quad \text{(Cartesian)} \tag{0-22}$$

$$= \varrho\, d\varrho\, d\phi\, dz \quad \text{(cylindrical)} \tag{0-23}$$

$$= r^2 \sin\theta\, dr\, d\theta\, d\phi \quad \text{(spherical)} \tag{0-24}$$

in the three common coordinate systems. In Cartesian coordinates, simple integration of a *vector* field produces another vector field whose components are the integrals of the components of the field being integrated. Caution must again be exercised in integrating vector fields expressed in other coordinate systems. More involved derivatives, particularly the gradient, the curl, and the divergence, and more involved integrals, particularly line integrals and surface integrals, will be introduced as our development proceeds.

PROBLEMS

P0-11. Let two points have position vectors $\mathbf{r}_1$ and $\mathbf{r}_2$, respectively. Obtain an expression for the vector pointing *from* point 1 *to* point 2.

P0-12. A scalar function of position is given in terms of the position vector $\mathbf{r}$ by $S(\mathbf{r}) = \mathbf{r}\cdot\mathbf{r}$. Write this function out more explicitly in Cartesian, cylindrical, and spherical coordinates.

P0-13. A scalar function S and a vector function $\mathbf{Q}$ are given in terms of two position vectors $\mathbf{r}$ and $\mathbf{r}'$ by

$$S(\mathbf{r}, \mathbf{r}') = \frac{1}{|\mathbf{r} - \mathbf{r}'|}, \qquad \mathbf{Q}(\mathbf{r}, \mathbf{r}') = \frac{\mathbf{r} - \mathbf{r}'}{|\mathbf{r} - \mathbf{r}'|^3}$$

Letting $\mathbf{r} = x\hat{\mathbf{i}} + y\hat{\mathbf{j}} + z\hat{\mathbf{k}}$ and $\mathbf{r}' = x'\hat{\mathbf{i}} + y'\hat{\mathbf{j}} + z'\hat{\mathbf{k}}$, write S and $\mathbf{Q}$ explicitly in terms of x, y, z, x', y', and z' and then show that

$$\mathbf{Q} = -\frac{\partial S}{\partial x}\hat{\mathbf{i}} - \frac{\partial S}{\partial y}\hat{\mathbf{j}} - \frac{\partial S}{\partial z}\hat{\mathbf{k}}$$

Although analytic representations of fields are essential to efficient theoretical manipulations, graphical representations often facilitate mental visualization of the fields. A graphical representation for a scalar field need convey only the magnitude of the field, and a diagram showing labeled *level contours*—lines or surfaces at all points of which the scalar field has the labeled value—suffices; the contour lines representing a particular (two-dimensional) scalar field are shown in Fig. 0-6. A graphical representation for a *vector* field, on the other hand, must convey both the magnitude and the direction of the field. One way to convey these features is to attach a vector to each of several representative points in space, drawing the vector at each point so that its direction coincides with the direction of the field and its length is proportional to the magnitude of the field; Fig. 0-7(a) shows a sample vector field represented in this way. Alternatively, a vector field can be mapped by drawing a set of *field lines*—lines that are everywhere tangent to the field vector. Let these field lines be spaced so that the number of lines crossing a unit area placed perpendicular to the field at some point is proportional to the magnitude of the field at that point. The resulting picture then conveys

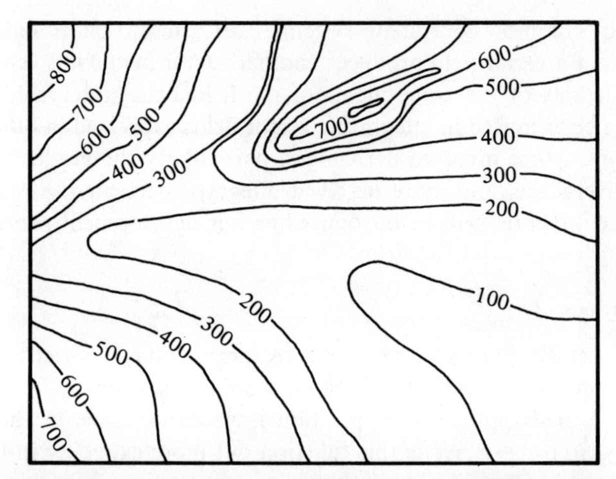

Fig. 0-6. Representation of a scalar field by contour lines. Physically, the field illustrated represents the altitude of points on the surface of a hypothetical earth and each contour line is labeled with the altitude (in feet) of points along that line. Wherever the contour lines are close together, the terrain is steep; wherever they are widely separated, the terrain slopes only gradually. In this figure, a mountain occurs near the upper right corner, a valley runs from center left to center top, and this same valley opens into a reasonably flat area in the lower right corner.

the direction of the field by the direction of the field lines and the magnitude of the field by the spacing of the field lines. Adherence to this convention on spacing may, of course, require that field lines start and stop at some points in space. The field shown in Fig. 0-7(a) may also be represented by the field lines shown in Fig. 0-7(b). We offer two warnings: (1) Graphical representations in *two* dimensions are not entirely satisfactory for conveying the details of fields in *three* dimensions, and (2) graphical representations show the fields only at representative points in space; the fields themselves have values at *all* points in space.

PROBLEMS

P0-14. Explain why the level contours representing a scalar field cannot intersect.

P0-15. Describe the (three-dimensional) level contours of a scalar field given as a function of the position vector $\mathbf{r}$ by $S(\mathbf{r}) = 1/|\mathbf{r} - \mathbf{r}_0|$, where $\mathbf{r}_0$ is a fixed vector.

P0-16. A scalar field is given as a function of the position vector $\mathbf{r}$ by $S(\mathbf{r}) = (\hat{\mathbf{k}} \cdot \mathbf{r})/r^3$. Obtain accurate drawings of half a dozen of the level contours of this field in the *y-z* plane. Consider both $S > 0$ and $S < 0$. Describe the three-dimensional contour diagram. *Hints*: (1) Use spherical coordinates.

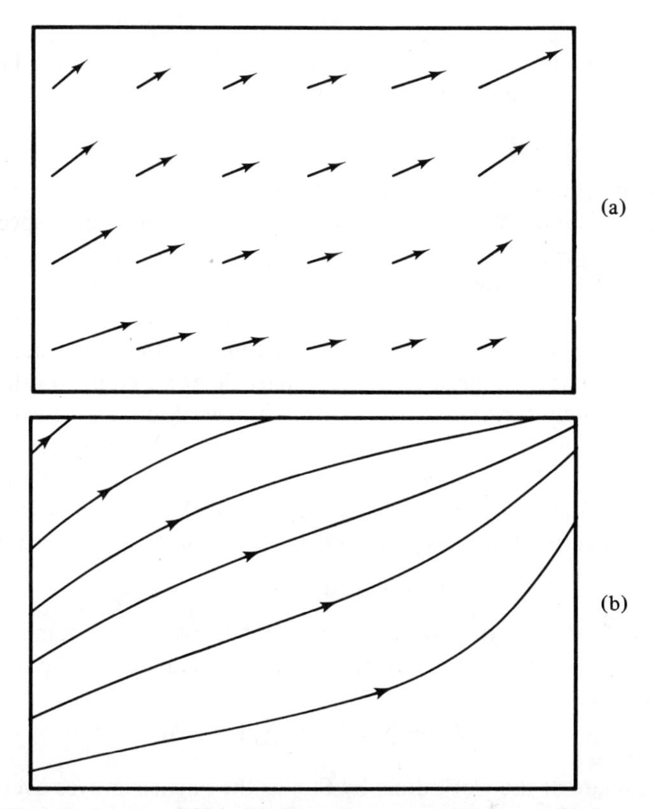

Fig. 0-7. A vector field in (a) the vector representation and (b) the field line representation. In (a) the vectors indicate both the magnitude and the direction of the field; in (b) the lines indicate the direction of the field and their separation indicates its magnitude.

(2) Plot the contour map on polar coordinate paper. (3) Use a computer or desk calculator.

P0-17. Show that the field lines representing the vector field $\mathbf{F}(\mathbf{r}) = \mathbf{r}/r^3$ are straight lines radiating away from the origin and then show that, in three dimensions, these field lines need be started only at the origin in order that $|\mathbf{F}|$ at every point P be proportional to the number of lines crossing a unit area oriented perpendicular to $\mathbf{F}$ at P.

0-3 Static Force Fields

In this section we shall summarize a number of concepts associated with static force fields. Throughout this section $\mathbf{F}(\mathbf{r})$ denotes a force—it may be the total force or only a portion of the total force—experienced by a particle located at point $\mathbf{r}$.

TORQUE

The torque $\mathbf{N}$ about the coordinate origin exerted by the force $\mathbf{F}(\mathbf{r})$ applied at the point $\mathbf{r}$ is defined by $\mathbf{N} = \mathbf{r} \times \mathbf{F}(\mathbf{r})$.

WORK AND THE LINE INTEGRAL

We ask first, how much work is done *on* a particle *by* the force field $\mathbf{F}(\mathbf{r})$ as the particle is moved from a point $\mathbf{r}_a$ to a point $\mathbf{r}_b$ along some specified path Γ?[4] Since in general the force will not have the same value at all points on the path, we seek an answer to this question by dividing the path into small *approximately straight* segments along any one of which the force may be regarded as *approximately* constant. If the center of the ith segment is located at $\mathbf{r}_i$, then the force experienced by the particle at all points on this segment is approximately $\mathbf{F}(\mathbf{r}_i)$. Now, *by definition* the work done by a *constant* force moved through some displacement along a *straight* line is the product of the displacement and the component of the force in the direction of the displacement. Thus, the work ΔW_i done by the force $\mathbf{F}(\mathbf{r})$ on the particle as the particle moves across the ith segment of its path is given approximately by

$$\Delta W_i \approx |\mathbf{F}(\mathbf{r}_i)||\Delta\mathbf{r}_i| \cos\theta_i = \mathbf{F}(\mathbf{r}_i)\cdot\Delta\mathbf{r}_i \tag{0-25}$$

where $\Delta\mathbf{r}_i$ is the net displacement represented by the ith segment and θ_i is the angle between $\mathbf{F}(\mathbf{r}_i)$ and $\Delta\mathbf{r}_i$ (Fig. 0-8). Summing contributions from all segments gives

$$W \approx \sum_i \Delta W_i = \sum_i \mathbf{F}(\mathbf{r}_i)\cdot\Delta\mathbf{r}_i \tag{0-26}$$

as an estimate of the work done by $\mathbf{F}(\mathbf{r})$ as the particle moves over the entire

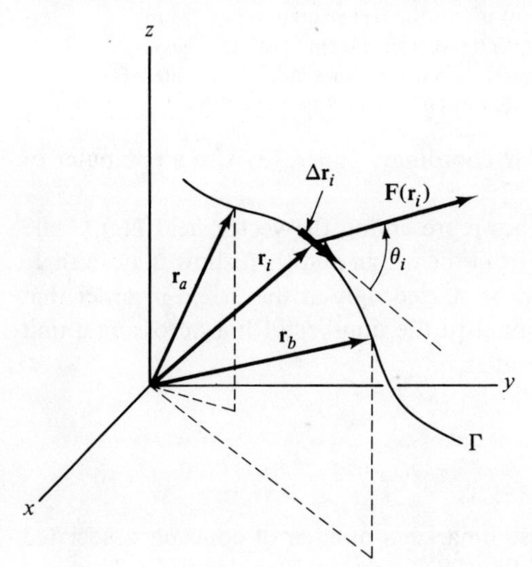

Fig. 0-8. The path of a particle moving in a force field.

[4]To bring about this motion, there will in general be forces acting on the particle in addition to those exerted by the specified force field.

path. We now refine the division of the path, making all segments smaller (and simultaneously increasing their number). As this refinement continues, the approximation becomes better, and in the limit, as $|\Delta \mathbf{r}_i| \longrightarrow 0$ for all i and the number of segments approaches infinity, the expression becomes exact, i.e.,

$$W = \lim_{|\Delta \mathbf{r}_i| \to 0} \sum_i \mathbf{F}(\mathbf{r}_i) \cdot \Delta \mathbf{r}_i = \int_\Gamma \mathbf{F}(\mathbf{r}) \cdot d\mathbf{r} \tag{0-27}$$

Equation (0-27) both defines the *line integral* and identifies the line integral of a force along some path with the work done *by* the force *on* a particle that moves along the path. In general, if a more explicit (analytic) evaluation of a line integral is needed, the integral must be written out in more detail, probably using one of the expressions

$$d\mathbf{r} = dx\, \hat{\mathbf{i}} + dy\, \hat{\mathbf{j}} + dz\, \hat{\mathbf{k}} \qquad \text{(Cartesian)} \tag{0-28}$$

$$= d\varkappa\, \hat{\boldsymbol{\varkappa}} + \varkappa\, d\phi\, \hat{\boldsymbol{\phi}} + dz\, \hat{\mathbf{k}} \qquad \text{(cylindrical)} \tag{0-29}$$

$$= dr\, \hat{\mathbf{r}} + r\, d\theta\, \hat{\boldsymbol{\theta}} + r \sin\theta\, d\phi\, \hat{\boldsymbol{\phi}} \qquad \text{(spherical)} \tag{0-30}$$

for the infinitesimal displacement $d\mathbf{r}$. In Cartesian coordinates, for example, we have that

$$W = \int_\Gamma [F_x(x, y, z)\, dx + F_y(x, y, z)\, dy + F_z(x, y, z)\, dz] \tag{0-31}$$

but *in general we can proceed no further until the path has been specified*, perhaps by stipulating relationships that determine both y and z along the path as functions of x. Once such relationships are available, we may reduce the line integral to an integral on a single variable and evaluate it by familiar techniques.

THE WORK-KINETIC ENERGY THEOREM

If the force $\mathbf{F}(\mathbf{r})$ in Eq. (0-27) is the resultant of *all* forces experienced by a particle of mass m at point $\mathbf{r}$, then by Newton's second law

$$\mathbf{F}(\mathbf{r}) = m \frac{d^2\mathbf{r}}{dt^2} = m \frac{d\mathbf{v}}{dt} \tag{0-32}$$

and we find that

$$W = \int_\Gamma m \frac{d\mathbf{v}}{dt} \cdot d\mathbf{r} = m \int \mathbf{v} \cdot d\mathbf{v} = \frac{1}{2} m v_b^2 - \frac{1}{2} m v_a^2 \tag{0-33}$$

where $\mathbf{v}_a$ ($\mathbf{v}_b$) is the velocity of the particle at the start (end) of the path. This result, called the *work-kinetic energy theorem*, equates the work done by the resultant of *all* forces acting on a particle to the *change* in kinetic energy of the particle.

CONSERVATIVE FORCE FIELDS AND POTENTIAL ENERGY

Although the line integral of the most general force field cannot be evaluated until after a path has been specified, many commonly occurring force fields permit an evaluation of $\int \mathbf{F}\cdot d\mathbf{r}$ if only the end points of the path are known. For these fields, the particular route by which the particle moves from start to finish does not affect the total work done on it by the force field. The line integral of these special fields is said to be *path-independent* and the fields themselves are said to be *conservative*. Symbolically, we can indicate the dependence of the line integral only on the initial and final positions $\mathbf{r}_a$ and $\mathbf{r}_b$ by writing

$$\int_{\Gamma} \mathbf{F}_c \cdot d\mathbf{r} = \int_{\mathbf{r}_a}^{\mathbf{r}_b} \mathbf{F}_c \cdot d\mathbf{r} = g(\mathbf{r}_b, \mathbf{r}_a) \tag{0-34}$$

where the subscript c has been added to $\mathbf{F}$ as a reminder that the expression applies only to conservative fields. Although the function $g(\mathbf{r}_b, \mathbf{r}_a)$ is in all cases a scalar, its specific form is determined by the particular force field involved. Equivalently (P0-19), a conservative force field satisfies

$$\oint_{\Gamma_c} \mathbf{F}_c \cdot d\mathbf{r} = 0 \tag{0-35}$$

where $\mathbf{\Gamma}_c$ is an *arbitrary* path and the circle on the integral sign signifies that $\mathbf{\Gamma}_c$ is a *closed* path starting from $\mathbf{r}_a$ and proceeding via some route back to $\mathbf{r}_a$. Thus, a conservative force field may be characterized either by the path independence of the corresponding line integral or by the vanishing of the line integral about an arbitrary closed path.

Conservative force fields (*and only conservative force fields*) admit the identification of an associated *scalar* potential energy function, the existence of which can be demonstrated by evaluating the work done on a particle by a conservative force field as the particle moves from $\mathbf{r}_0$ to $\mathbf{r}_a$ and on to $\mathbf{r}_b$. We can think of the journey as a single step, in which case the work done on the particle is $g(\mathbf{r}_b, \mathbf{r}_0)$, or as a sequence of two steps, in which case the work done on the particle is $g(\mathbf{r}_a, \mathbf{r}_0) + g(\mathbf{r}_b, \mathbf{r}_a)$. Since the total work must be independent of how we view the path, we conclude that these two evaluations must be equal or, with a minor algebraic rearrangement, that

$$g(\mathbf{r}_b, \mathbf{r}_a) = g(\mathbf{r}_b, \mathbf{r}_0) - g(\mathbf{r}_a, \mathbf{r}_0) \tag{0-36}$$

The line integral of a conservative force field between any two points is therefore given by the difference between a function evaluated at one point and the same function evaluated at the other point. To be sure, the function $g(\mathbf{r}, \mathbf{r}_0)$ depends on two points, but the point $\mathbf{r}_0$ is the same in both occurrences; it merely plays the role of a reference point. Relative to the point $\mathbf{r}_0$, $g(\mathbf{r}, \mathbf{r}_0)$ assigns a value to some (scalar) quantity at each point $\mathbf{r}$ in space. This scalar field is intimately associated with the vector field $\mathbf{F}_c(\mathbf{r})$, and it represents physically the work done by $\mathbf{F}_c(\mathbf{r})$ on a particle moved from $\mathbf{r}_0$ to $\mathbf{r}$. It is customary to suppress explicit indication of the reference point and (for reasons that will soon appear) to introduce a minus sign, defining the so-called *potential energy field* $U(\mathbf{r})$ by

$$U(\mathbf{r}) = -g(\mathbf{r}, \mathbf{r}_0) \tag{0-37}$$

in terms of which Eq. (0-36) yields the expression

$$\int_{\mathbf{r}_a}^{\mathbf{r}_b} \mathbf{F}_c \cdot d\mathbf{r} = -[U(\mathbf{r}_b) - U(\mathbf{r}_a)] \tag{0-38}$$

for the work done by a conservative force on a particle moved from $\mathbf{r}_a$ to $\mathbf{r}_b$. If $\mathbf{F}_c$ represents the resultant of *all* forces acting on the particle, Eq. (0-38) combines with Eq. (0-33) to give

$$\tfrac{1}{2}mv_a^2 + U(\mathbf{r}_a) = \tfrac{1}{2}mv_b^2 + U(\mathbf{r}_b) \tag{0-39}$$

which is recognizable as a statement of conservation of mechanical energy. [Had the minus sign not been inserted in Eq. (0-37), it would have appeared preceding each U in Eq. (0-39).]

The argument of the previous paragraph demonstrates only the *existence* of a (scalar) potential energy field associated with every (vector) conservative force field. An expression *determining* the scalar field from the vector field is obtained by setting $\mathbf{r}_b = \mathbf{r}$ and $\mathbf{r}_a = \mathbf{r}_0$ in Eq. (0-38) and then rearranging the equation; we find that

$$U(\mathbf{r}) = U(\mathbf{r}_0) - \int_{\mathbf{r}_0}^{\mathbf{r}} \mathbf{F}_c \cdot d\mathbf{r} \tag{0-40}$$

The value of $U(\mathbf{r}_0)$ remains arbitrary; only *differences* in potential energy are determined by the force field.

Equation (0-40) provides a means for calculating $U(\mathbf{r})$ when $\mathbf{F}_c(\mathbf{r})$ is known. A solution to the reverse problem—finding $\mathbf{F}_c(\mathbf{r})$ when $U(\mathbf{r})$ is known—can also be obtained. Since $U(\mathbf{r})$ is an integral of $\mathbf{F}_c(\mathbf{r})$, one might expect $\mathbf{F}_c(\mathbf{r})$ to be a derivative of $U(\mathbf{r})$. To see the nature of that derivative, we evaluate the difference in potential energy between two points $\mathbf{r}_a$ and $\mathbf{r}_b = \mathbf{r}_a + \Delta\mathbf{r}$, where $|\Delta\mathbf{r}|$ is small. Equation (0-38) then gives

$$U(\mathbf{r}_a + \Delta\mathbf{r}) - U(\mathbf{r}_a) = -\int_{\mathbf{r}_a}^{\mathbf{r}_a+\Delta\mathbf{r}} \mathbf{F}_c \cdot d\mathbf{r} \tag{0-41}$$

To obtain $F_{cx}(\mathbf{r})$, let $\Delta\mathbf{r} = \Delta x\,\hat{\mathbf{i}}$. Since the integral is path-independent, we can select a path from $\mathbf{r}_a$ to $\mathbf{r}_a + \Delta\mathbf{r}$ along which $d\mathbf{r} = dx'\,\hat{\mathbf{i}}$, and Eq. (0-41) then gives

$$\begin{aligned} U(x_a + \Delta x, y_a, z_a) &- U(x_a, y_a, z_a) \\ &= -\int_{x_a}^{x_a+\Delta x} F_{cx}(x', y_a, z_a)\,dx' \\ &\approx -F_{cx}(x_a, y_a, z_a)\,\Delta x \end{aligned} \tag{0-42}$$

Dividing Eq. (0-42) by Δx and then allowing $\Delta x \to 0$, we find that

$$F_{cx}(x_a, y_a, z_a) = -\left.\frac{\partial U}{\partial x}\right|_{x_a, y_a, z_a} \tag{0-43}$$

Since this expression applies at any point $\mathbf{r}_a$, the subscript a can be dropped, and we have finally that[5]

$$F_{cx}(\mathbf{r}) = -\frac{\partial U(\mathbf{r})}{\partial x} \tag{0-44}$$

Similar expressions for $F_{cy}(\mathbf{r})$ and $F_{cz}(\mathbf{r})$ are obtained by setting $\Delta\mathbf{r} = \Delta y\,\hat{\mathbf{j}}$ and $\Delta\mathbf{r} = \Delta z\,\hat{\mathbf{k}}$ in Eq. (0-41). In total we find that

$$\mathbf{F}_c = -\left(\hat{\mathbf{i}}\frac{\partial U}{\partial x} + \hat{\mathbf{j}}\frac{\partial U}{\partial y} + \hat{\mathbf{k}}\frac{\partial U}{\partial z}\right) \tag{0-45}$$

where all functions are understood to be evaluated at argument $\mathbf{r}$. Let us now introduce the *vector differential operator* known as the *gradient* operator, symbolized by $\boldsymbol{\nabla}$, and represented *in Cartesian coordinates* by

$$\boldsymbol{\nabla} = \hat{\mathbf{i}}\frac{\partial}{\partial x} + \hat{\mathbf{j}}\frac{\partial}{\partial y} + \hat{\mathbf{k}}\frac{\partial}{\partial z} \tag{0-46}$$

Acting on a scalar field $S(\mathbf{r})$, this operator produces a vector field known as the *gradient* of S and symbolized by $\boldsymbol{\nabla}S$; more explicitly,

$$\boldsymbol{\nabla}S = \hat{\mathbf{i}}\frac{\partial S}{\partial x} + \hat{\mathbf{j}}\frac{\partial S}{\partial y} + \hat{\mathbf{k}}\frac{\partial S}{\partial z} \tag{0-47}$$

A physical interpretation of this vector derivative will be explored briefly in Section 0-4; the derivative is introduced here because Eq. (0-45) can then be written in the particularly compact form

$$\mathbf{F}_c = -\boldsymbol{\nabla}U \tag{0-48}$$

The force exerted on a particle by a *conservative* force field is therefore given by the *negative* gradient of the associated potential energy field.

Several *equivalent* criteria for identifying a conservative force field can now be enumerated. We have already noted (1) that the line integral of a conservative force field is path-independent, (2) that this line integral about *any closed* path vanishes, and (3) that a conservative force field can be obtained as the (negative) gradient of an associated scalar (potential energy) field. To obtain a fourth criterion we use Eq. (0-48), finding that

$$\begin{aligned}\mathbf{F}_c\cdot d\mathbf{r} &= -\boldsymbol{\nabla}U\cdot d\mathbf{r} \\ &= -\left(\hat{\mathbf{i}}\frac{\partial U}{\partial x} + \hat{\mathbf{j}}\frac{\partial U}{\partial y} + \hat{\mathbf{k}}\frac{\partial U}{\partial z}\right)\cdot(dx\,\hat{\mathbf{i}} + dy\,\hat{\mathbf{j}} + dz\,\hat{\mathbf{k}}) \\ &= -\left(\frac{\partial U}{\partial x}dx + \frac{\partial U}{\partial y}dy + \frac{\partial U}{\partial z}dz\right) = -dU\end{aligned} \tag{0-49}$$

Thus, for a *conservative* force field, $\mathbf{F}_c\cdot d\mathbf{r}$ is an *exact differential.* Mathemat-

[5]Here (and in similar subsequent occurrences) the development would be more rigorous if we invoked the theorem of the mean and replaced Eq. (0-42) with

$$U(x_a + \Delta x, y_a, z_a) - U(x_a, y_a, z_a) = F_{cx}(x_a + \eta\,\Delta x, y_a, z_a)\,\Delta x$$

where $0 \leq \eta \leq 1$, before letting Δx approach 0.

ically, however, exactness of

$$\mathbf{F}_c \cdot d\mathbf{r} = F_{cx}\,dx + F_{cy}\,dy + F_{cz}\,dz \tag{0-50}$$

requires that[6]

$$\frac{\partial F_{cz}}{\partial y} - \frac{\partial F_{cy}}{\partial z} = 0, \qquad \frac{\partial F_{cx}}{\partial z} - \frac{\partial F_{cz}}{\partial x} = 0, \qquad \frac{\partial F_{cy}}{\partial x} - \frac{\partial F_{cx}}{\partial y} = 0 \tag{0-51}$$

Let us now define the *curl* of a vector field **Q** by

$$\nabla \times \mathbf{Q} = \left(\frac{\partial Q_z}{\partial y} - \frac{\partial Q_y}{\partial z}\right)\hat{\mathbf{i}} + \left(\frac{\partial Q_x}{\partial z} - \frac{\partial Q_z}{\partial x}\right)\hat{\mathbf{j}} + \left(\frac{\partial Q_y}{\partial x} - \frac{\partial Q_x}{\partial y}\right)\hat{\mathbf{k}} \tag{0-52}$$

where the notation is suggested by the result of a formal evaluation of $\nabla \times \mathbf{Q}$ when ∇ is replaced by the expression in Eq. (0-46). A physical interpretation of this vector derivative is explored briefly in Section 0-4; the derivative is introduced here because the condition in Eq. (0-51) for the exactness of $\mathbf{F}_c \cdot d\mathbf{r}$—which is the fourth condition for the "conservativeness" of $\mathbf{F}_c$—can then be written in the particularly compact form

$$\nabla \times \mathbf{F}_c = 0 \tag{0-53}$$

Some care must be used, however, in applying this criterion to fields that become infinite at some point in the region of interest.[7]

PROBLEMS

P0-18. A force field in two dimensions is given by the vector $\mathbf{F}(x, y) = 3x^2y\hat{\mathbf{i}} + xy\hat{\mathbf{j}}$. Evaluate the work done by this force field on a particle moved along the following paths in the x-y plane: (a) $(0, 0) \longrightarrow (0, 1) \longrightarrow (1, 1)$ along straight-line segments, (b) $(0, 0) \longrightarrow (1, 0) \longrightarrow (1, 1)$ along straight-line segments, (c) $(0, 0) \longrightarrow (1, 1)$ along the line $y = x$, and (d) $(0, 0) \longrightarrow (1, 1)$ along the parabola $y = x^2$. Points in the plane are denoted by (x, y).

P0-19. Show that the line integral of a vector field about an *arbitrary closed* path is zero if and only if the line integral between two arbitrary points is path-independent.

P0-20. Using an argument similar to that presented in the text, show that $F_{cy} = -\partial U/\partial y$.

P0-21. A force field in three dimensions is given by

$$\mathbf{F}(\mathbf{r}) = a\frac{\mathbf{r}}{r^3} = a\frac{x\hat{\mathbf{i}} + y\hat{\mathbf{j}} + z\hat{\mathbf{k}}}{(x^2 + y^2 + z^2)^{3/2}}$$

where a is a constant. By showing explicitly that $\mathbf{F} \cdot d\mathbf{r}$ is the exact differential of some function, show that **F** is a conservative force field and find the associated potential energy field.

[6] G. B. Thomas, Jr., *Calculus and Analytic Geometry* (Addison-Wesley Publishing Company, Inc., Reading, Mass., 1954), p. 530. (Or see *exact differentials* in the index of any calculus book.)

[7] See S. Y. Feng, *Am. J. Phys.* **37**, 616 (1969).

P0-22. Using the operator $\boldsymbol{\nabla}$ as given in Eq. (0-46), verify that formal evaluation of $\boldsymbol{\nabla} \times \mathbf{Q}$ leads to Eq. (0-52) and show that

$$\boldsymbol{\nabla} \times \mathbf{Q} = \begin{vmatrix} \hat{\mathbf{i}} & \hat{\mathbf{j}} & \hat{\mathbf{k}} \\ \dfrac{\partial}{\partial x} & \dfrac{\partial}{\partial y} & \dfrac{\partial}{\partial z} \\ Q_x & Q_y & Q_z \end{vmatrix}$$

Warning: One must use great care in evaluating $\boldsymbol{\nabla} \times \mathbf{Q}$ by this means in other coordinate systems; only the Cartesian unit vectors are constants. Compare P0-32.

P0-23. Familiarize yourself with vector identities (C-5)–(C-9) in Appendix C, and prove (C-5), (C-8), and (C-9).

P0-24. Show that, for a *conservative* force field, the field lines are perpendicular to the level contours—called here equipotential surfaces—of the associated potential energy field. *Hint*: Let $d\mathbf{r}$ in Eq. (0-49) represent the displacement between two infinitesimally separated points in the same equipotential surface. Physically, what must be the value of dU?

0-4 Coordinate-Free Definitions for the Gradient and the Curl

The gradient of a scalar field S and the curl and the divergence (Section 2-3) of a vector field $\mathbf{Q}$ play important roles in field theory and we shall subsequently need expressions for these quantities in non-Cartesian coordinates. We can, of course, obtain $\boldsymbol{\nabla} S$ and $\boldsymbol{\nabla} \times \mathbf{Q}$ in other coordinates by a tedious but direct transformation of Eqs. (0-47) and (0-52), but the same results can be obtained more easily if we invest a brief preliminary effort in developing coordinate-free definitions for the vector derivatives. Consider first the gradient. From Eq. (0-49) we infer that

$$dS = \boldsymbol{\nabla} S \cdot d\mathbf{r} \tag{0-54}$$

where dS is the (infinitesimal) change in S that results when the point at which S is evaluated undergoes the (infinitesimal) displacement $d\mathbf{r}$. Let $|d\mathbf{r}| = d\ell$ and let $\hat{\mathbf{n}}$ be a unit vector in the direction of $d\mathbf{r}$. Then $d\mathbf{r} = \hat{\mathbf{n}}\, d\ell$ and we find from Eq. (0-54) that

$$\boldsymbol{\nabla} S \cdot \hat{\mathbf{n}} = \frac{dS}{d\ell} \tag{0-55}$$

We now define the $\hat{\mathbf{n}}$-component of $\boldsymbol{\nabla} S$ to be $dS/d\ell$, called the *directional derivative* of S in the direction $\hat{\mathbf{n}}$. If we take $\hat{\mathbf{n}}$ successively to be $\hat{\mathbf{i}}$, $\hat{\mathbf{j}}$, and $\hat{\mathbf{k}}$ while simultaneously letting $d\ell$ be the corresponding elements of distance dx, dy, and dz, we recover the components of Eq. (0-47). Equation (0-55), however, is more general. For example, if θ is increased by $d\theta$ in Fig. 0-5(c), the point identified moves a distance $r\,d\theta$ in the θ direction. Taking $\hat{\mathbf{n}} = \hat{\boldsymbol{\theta}}$ and $d\ell = r\,d\theta$, we then quickly find from Eq. (0-55) that the $\hat{\boldsymbol{\theta}}$-component of the gradient in spherical coordinates is given by $\nabla S \cdot \hat{\boldsymbol{\theta}} = \partial S/r\,\partial\theta$. Similar additional manipulations with Eq. (0-55) lead easily to the expressions

$$\nabla S = \hat{\boldsymbol{\imath}}\frac{\partial S}{\partial \imath} + \hat{\boldsymbol{\phi}}\frac{1}{\imath}\frac{\partial S}{\partial \phi} + \hat{\mathbf{k}}\frac{\partial S}{\partial z} \tag{0-56}$$

in cylindrical coordinates and

$$\nabla S = \hat{\mathbf{r}}\frac{\partial S}{\partial r} + \hat{\boldsymbol{\theta}}\frac{1}{r}\frac{\partial S}{\partial \theta} + \hat{\boldsymbol{\phi}}\frac{1}{r\sin\theta}\frac{\partial S}{\partial \phi} \tag{0-57}$$

in spherical coordinates.

A physical interpretation of ∇S can also be inferred from Eq. (0-55): the $\hat{\mathbf{n}}$-component of ∇S measures the *rate* at which S changes *with respect to distance* in the direction $\hat{\mathbf{n}}$. The more rapidly S changes as an observer (or measuring instrument) moves away from some initial point in some direction $\hat{\mathbf{n}}$, the bigger will be the component of ∇S in that direction. In particular, since $\nabla S \cdot \hat{\mathbf{n}}$ has its maximum value when $\hat{\mathbf{n}}$ is parallel to ∇S (Why?), the gradient itself points in the direction of maximum increase of S, e.g., directly uphill in Fig. 0-6.

A coordinate-free definition for $\nabla \times \mathbf{Q}$ emerges from an evaluation of $\oint \mathbf{Q}\cdot d\boldsymbol{\ell}$ about a small closed path. (We shall now switch to the notation $d\boldsymbol{\ell}$ for an infinitesimal element of the path.) Consider, for example, $\oint \mathbf{Q}\cdot d\boldsymbol{\ell}$ about a small rectangle having sides Δx and Δy and lying in a plane parallel to and a distance z above the x-y plane (Fig. 0-9). Evaluating the integral in the counterclockwise direction as seen from a point above the path, we find that

$$\begin{aligned}\oint \mathbf{Q}\cdot d\boldsymbol{\ell} &= \int_x^{x+\Delta x} \mathbf{Q}(x', y, z)\cdot dx'\,\hat{\mathbf{i}} + \int_y^{y+\Delta y} \mathbf{Q}(x+\Delta x, y', z)\cdot dy'\,\hat{\mathbf{j}} \\ &\quad + \int_{x+\Delta x}^{x} \mathbf{Q}(x', y+\Delta y, z)\cdot dx'\,\hat{\mathbf{i}} + \int_{y+\Delta y}^{y} \mathbf{Q}(x, y', z)\cdot dy'\,\hat{\mathbf{j}} \\ &= \int_y^{y+\Delta y} [Q_y(x+\Delta x, y', z) - Q_y(x, y', z)]\,dy' \\ &\quad + \int_x^{x+\Delta x} [Q_x(x', y, z) - Q_x(x', y+\Delta y, z)]\,dx' \end{aligned} \tag{0-58}$$

where the integrals in the first form relate, respectively, to the segments labeled 1, 2, 3, and 4 in Fig. 0-9. Assuming that Δx and Δy are small, we now use the Taylor expansion (Appendix B) to approximate each integrand in Eq. (0-58), obtaining

$$\begin{aligned}\oint \mathbf{Q}\cdot d\boldsymbol{\ell} \approx \Delta x \int_y^{y+\Delta y} \left.\frac{\partial Q_y}{\partial x}\right|_{x, y', z} dy' \\ -\Delta y \int_x^{x+\Delta x} \left.\frac{\partial Q_x}{\partial y}\right|_{x', y, z} dx'\end{aligned} \tag{0-59}$$

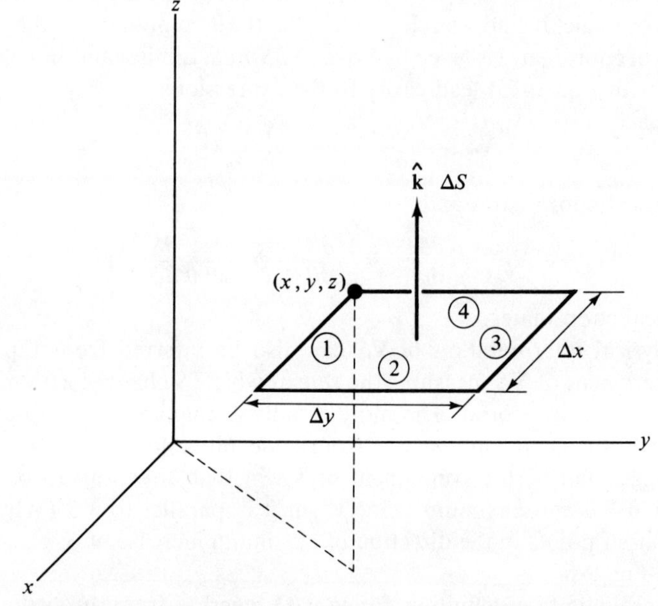

Fig. 0-9. A small rectangular path used to derive a coordinate free definition of $\nabla \times \mathbf{Q}$.

Finally, approximating each integral as we did in Eq. (0-42), we have

$$\oint \mathbf{Q}\cdot d\boldsymbol{\ell} \approx \left(\frac{\partial Q_y}{\partial x} - \frac{\partial Q_x}{\partial y}\right)\Delta x\,\Delta y = (\nabla \times \mathbf{Q})\cdot\hat{\mathbf{k}}\,\Delta S \qquad \text{(0-60)}$$

where the field $\mathbf{Q}$ is understood to be evaluated at the point (x, y, z) and $\Delta S = \Delta x\,\Delta y$ is the area of the surface bounded by the (small) path. More generally, if $\hat{\mathbf{n}}$ is a unit vector perpendicular to the plane of some small area ΔS and we represent the area by the vector $\hat{\mathbf{n}}\,\Delta S$, then we would find the value

$$\oint \mathbf{Q}\cdot d\boldsymbol{\ell} \approx [(\nabla \times \mathbf{Q})\cdot\hat{\mathbf{n}}]\,\Delta S \qquad \text{(0-61)}$$

for the line integral of $\mathbf{Q}$ about the path bounding ΔS. From this equation we then infer the coordinate-free definition

$$(\nabla \times \mathbf{Q})\cdot\hat{\mathbf{n}} = \lim_{\Delta S\to 0}\frac{1}{\Delta S}\oint \mathbf{Q}\cdot d\boldsymbol{\ell} \qquad \text{(0-62)}$$

for the $\hat{\mathbf{n}}$-component of $\nabla \times \mathbf{Q}$. We have above tacitly adopted a convention relating the direction of the vector $\hat{\mathbf{n}}$ to the direction in which the line integral is evaluated: *If the fingers of the right hand are positioned to point in the direction in which the line integral is evaluated and the palm faces the area bounded by the path, the thumb gives the direction of* $\hat{\mathbf{n}}$. Equation (0-62) is correct only if this *right-hand rule* is followed. Equation (0-62) reproduces the components of Eq. (0-52) if the path is taken successively to be the boundary of a small rectangle in a plane perpendicular to the x, y, and z axes, respectively, and it also yields the expressions

$$\nabla \times \mathbf{Q} = \left(\frac{1}{\imath}\frac{\partial Q_z}{\partial \phi} - \frac{\partial Q_\phi}{\partial z}\right)\hat{\boldsymbol{\imath}} + \left(\frac{\partial Q_\imath}{\partial z} - \frac{\partial Q_z}{\partial \imath}\right)\hat{\boldsymbol{\phi}} + \frac{1}{\imath}\left(\frac{\partial(\imath Q_\phi)}{\partial \imath} - \frac{\partial Q_\imath}{\partial \phi}\right)\hat{\mathbf{k}} \tag{0-63}$$

in cylindrical coordinates and

$$\nabla \times \mathbf{Q} = \frac{1}{r \sin\theta}\left(\frac{\partial(\sin\theta Q_\phi)}{\partial \theta} - \frac{\partial Q_\theta}{\partial \phi}\right)\hat{\mathbf{r}} + \left(\frac{1}{r\sin\theta}\frac{\partial Q_r}{\partial \phi} - \frac{1}{r}\frac{\partial(rQ_\phi)}{\partial r}\right)\hat{\boldsymbol{\theta}} + \frac{1}{r}\left(\frac{\partial(rQ_\theta)}{\partial r} - \frac{\partial Q_r}{\partial \theta}\right)\hat{\boldsymbol{\phi}} \tag{0-64}$$

in spherical coordinates with nearly equal ease.

A physical interpretation of $\nabla \times \mathbf{Q}$ can be inferred from Eq. (0-62): The $\hat{\mathbf{n}}$-component of $\nabla \times \mathbf{Q}$ measures a particular average, specifically $\oint \mathbf{Q}\cdot d\boldsymbol{\ell}/\Delta S$, of the tangential component of $\mathbf{Q}$ computed for a small path lying in a plane perpendicular to $\hat{\mathbf{n}}$. In effect, when $(\nabla \times \mathbf{Q})\cdot\hat{\mathbf{n}} \neq 0$, $\mathbf{Q}$ has a net tangential component about this path. Unfortunately, this statement is considerably less transparent than the corresponding statement for ∇S, and we present also a more specific example even though we risk oversimplification. Let $\mathbf{Q}$ represent the (two-dimensional) velocity field describing the motion of points on the surface of a phonograph record revolving about a vertical axis with constant angular velocity $\boldsymbol{\omega}$ [Fig. 0-10(a)]. The linear speed of a point a distance $\imath$ from the axis is $\omega\imath$ and the velocity field $\mathbf{Q}(\mathbf{r})$ expressed in cylindrical coordinates is given by $\mathbf{Q}(\mathbf{r}) = \omega\imath\hat{\boldsymbol{\phi}}$. Direct application of Eq. (0-63) now gives $\nabla \times \mathbf{Q} = 2\omega\hat{\mathbf{k}} = 2\boldsymbol{\omega}$. Thus, for this simple case, $\nabla \times \mathbf{Q}$ is *everywhere* twice the angular velocity. In particular $\nabla \times \mathbf{Q} = 0$ when the record is not rotating. Since the field lines of $\mathbf{Q}$ in this case are circles centered on the axis of rotation, these results suggest that $\nabla \times \mathbf{Q}$ is related to the tendency of field lines to circle (or curl!) about some point in the field, as in Fig. 0-10(b). To be correct, however, this statement must be further interpreted to mean what is said at the beginning of this paragraph. As the field $\mathbf{Q} = x^2\hat{\mathbf{j}}$ (for which $\nabla \times \mathbf{Q} = 2x\hat{\mathbf{k}}$) shows, $\nabla \times \mathbf{Q}$ can be nonzero even when the field lines are parallel and straight! One route to developing an intuitive feel for $\nabla \times \mathbf{Q}$ might be to sketch the field lines for several fields, e.g., $\mathbf{Q} = \omega\imath\hat{\boldsymbol{\phi}}$, $\mathbf{Q} = x^2\hat{\mathbf{j}}$, and $\mathbf{Q} = \mathbf{r}/r^3$, and then to allow the eye to trace around several small closed paths in these fields, mentally estimating the cumulative tangential component of $\mathbf{Q}$ for each path.

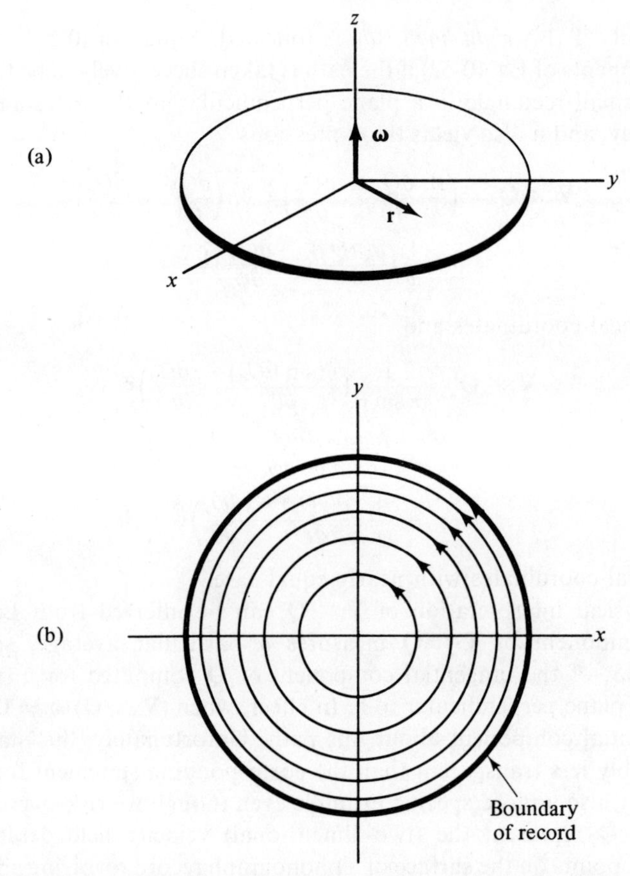

Fig. 0-10. A Rotating phonograph record. Part (a) shows the angular velocity of the record and part (b) shows the velocity field in the x-y plane.

PROBLEMS

P0-25. (a) Derive Eq. (0-56) by direct transformation of Eq. (0-47). *Hint*: Use Table 0-1. (b) Derive Eqs. (0-56) and (0-57) by the method based on Eq. (0-55).

P0-26. Use the definition of Eq. (0-62) to derive the $\hat{\mathbf{k}}$-component of Eq. (0-63).

P0-27. The force field $\mathbf{F}(\mathbf{r})$ is a *central force field* if it has the more explicit form $\mathbf{F}(\mathbf{r}) = f(r)\hat{\mathbf{r}}$ in spherical coordinates. This force is a function only of distance from some fixed point conveniently taken to be the origin and is everywhere directed along radial lines from that point. Show that every central force field is conservative and obtain an integral for the associated potential energy field. What is the potential energy field associated with the inverse square force field for which $f(r) = a/r^2$?

Supplementary Problems

P0-28. The *Kronecker delta* δ_{ij} and the three index symbol ϵ_{ijk} are defined by

$$\begin{aligned} \delta_{ij} &= 1, \quad i = j \\ &= 0, \quad i \neq j \end{aligned}$$

$$\begin{aligned} \epsilon_{ijk} &= +1, \quad (i, j, k) = (1, 2, 3), (3, 1, 2), (2, 3, 1) \\ &= -1, \quad \qquad \quad \ \, = (3, 2, 1), (1, 3, 2), (2, 1, 3) \\ &= 0, \quad \qquad \qquad = \text{anything else} \end{aligned}$$

With all indices assuming the values 1, 2, 3, show that

(a) $\hat{\mathbf{e}}_i \cdot \hat{\mathbf{e}}_j = \delta_{ij}$,
(b) $\mathbf{A} \cdot \mathbf{B} = \sum_i \sum_j \delta_{ij} A_i B_j$,
(c) interchange of any two indices changes the sign of ϵ_{ijk},
(d) $\hat{\mathbf{e}}_i \times \hat{\mathbf{e}}_j = \sum_k \epsilon_{ijk} \hat{\mathbf{e}}_k$,
(e) $(\mathbf{A} \times \mathbf{B})_i = \sum_j \sum_k \epsilon_{ijk} A_j B_k$,

and

(f) $\sum_k \epsilon_{ijk} \epsilon_{rsk} = \delta_{ir} \delta_{js} - \delta_{is} \delta_{jr}$.

Finally, (g) derive Eq. (0-2) by manipulating with these expressions. *Hint*: Begin by evaluating $[\mathbf{A} \times (\mathbf{B} \times \mathbf{C})]_k$.

P0-29. (a) A vector **A** *fixed in space* is described by its components (A_x, A_y, A_z) and $(A_{x'}, A_{y'}, A_{z'})$ in two Cartesian coordinate systems related as shown in Fig. P0-29. Show that the components must be related by

$$\begin{aligned} A_{x'} &= A_x \cos \phi + A_y \sin \phi \\ A_{y'} &= -A_x \sin \phi + A_y \cos \phi \\ A_{z'} &= A_z \end{aligned}$$

if the orientation and magnitude of **A** are to be unchanged by the rotation.

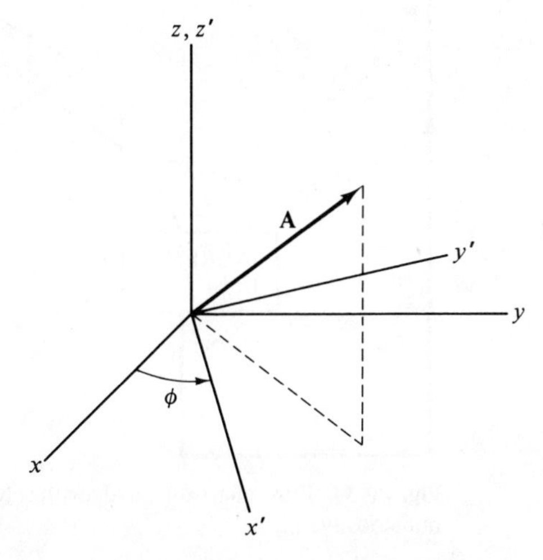

Figure P0-29

These are the rules for the transformation of the components of a vector under this rotation, and a three-component entity transforming in this way is said to be invariant to the transformation. (b) The prototype vector (think of displacement) is invariant to reflection of coordinates, i.e., to the transformation $x'' = -x$, $y'' = -y$, $z'' = -z$, which means that $A_{x''} = -A_x$, $A_{y''} = -A_y$, $A_{z''} = -A_z$. (Why?) Suppose **A** and **B** are invariant to reflection. How does the "vector" $\mathbf{C} = \mathbf{A} \times \mathbf{B}$ behave under reflection? Is **C** a vector?

P0-30. Equation (0-26) can be used to effect a numerical evaluation of $\int_\Gamma \mathbf{F} \cdot d\mathbf{r}$ on some path Γ. For simplicity consider a two-dimensional field,

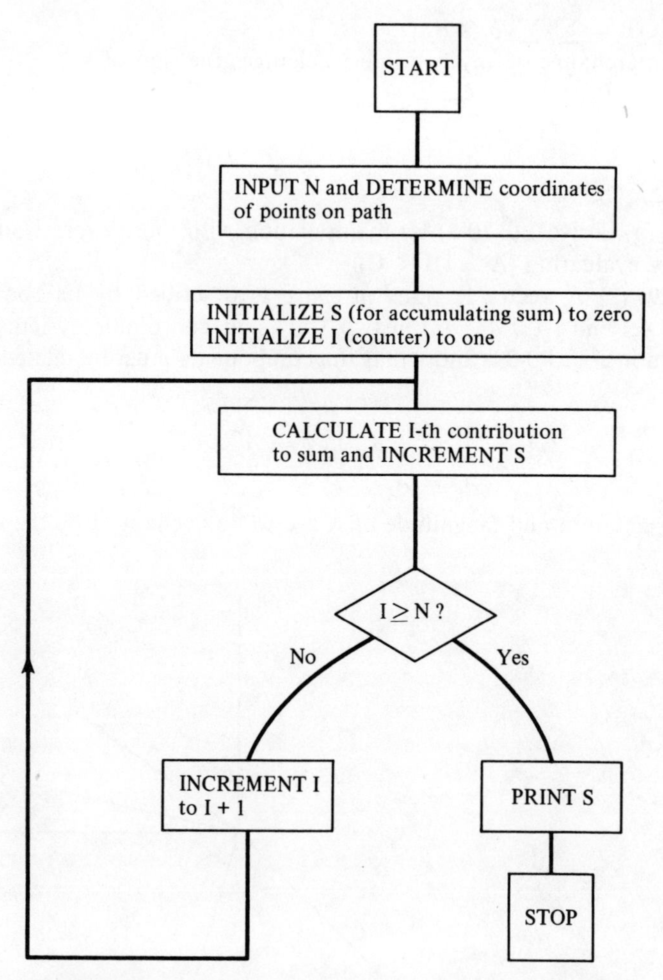

Fig. P0-30. Flow chart for an algorithm to evaluate a line integral numerically.

$\mathbf{F}(x, y) = F_x(x, y)\hat{\mathbf{i}} + F_y(x, y)\hat{\mathbf{j}}$, and let the path $\mathbf{\Gamma}$ lie in the x-y plane. Further, let the ith segment of $\mathbf{\Gamma}$ begin at (x_i, y_i) and end at (x_{i+1}, y_{i+1}), $1 \leq i \leq N$. Equation (0-26) then becomes

$$\sum_{\Gamma} \mathbf{F} \cdot d\mathbf{r} \approx \sum_{i=1}^{N} [F_x(\bar{x}_i, \bar{y}_i)\, \Delta x_i + F_y(\bar{x}_i, \bar{y}_i)\, \Delta y_i]$$

where $\bar{x}_i = \frac{1}{2}(x_{i+1} + x_i)$, $\bar{y}_i = \frac{1}{2}(y_{i+1} + y_i)$, $\Delta x_i = x_{i+1} - x_i$, $\Delta y_i = y_{i+1} - y_i$, and N is the number of segments into which $\mathbf{\Gamma}$ is divided. (a) Write a computer program to evaluate this integral for the force field in P0-18. A possible general strategy is depicted in Fig. P0-30. (b) Run your program for the paths in P0-18 and for several progressively larger values of N. *Optional*: (1) Try other force fields of your choosing and try several paths, including some closed paths. (2) Develop techniques and write programs for the numerical evaluation of other two- and three-dimensional line integrals, e.g., $\int S\, d\mathbf{r}$ and $\int \mathbf{Q} \times d\mathbf{r}$, where S is a scalar field and $\mathbf{Q}$ a vector field.

P0-31. Let Φ be a scalar function of position. Evaluate $\oint \Phi\, d\boldsymbol{\ell}$ about the path shown in Fig. 0-9 to show that

$$\oint \Phi\, d\boldsymbol{\ell} = (\hat{\mathbf{k}} \times \nabla\Phi)\, \Delta x\, \Delta y$$

and then infer that

$$\hat{\mathbf{n}} \times \nabla\Phi = \lim_{\Delta S \to 0} \frac{1}{\Delta S} \oint \Phi\, d\boldsymbol{\ell}$$

where $\hat{\mathbf{n}}\, \Delta S$ is the vector representing the area of the (plane) surface bounded by the (small) path about which the integral extends.

P0-32. Consider a general curvilinear coordinate system with coordinates (q_1, q_2, q_3) defined in terms of the Cartesian coordinates (x_1, x_2, x_3) by the functions $q_i = q_i(x_1, x_2, x_3)$, $i = 1, 2, 3$. We assume these equations can be inverted to give x_j as some function of the q's, $x_j = x_j(q_1, q_2, q_3)$, $j = 1, 2, 3$. Further, we introduce three quantities h_i, $i = 1, 2, 3$—which may be functions of the q's—such that $ds_i = h_i\, dq_i$, where ds_i is the physical distance that a point moves when its ith coordinate alone is changed from q_i to $q_i + dq_i$. Finally, we confine our attention to *orthogonal* systems, which in general means that the three families of surfaces defined by $q_i =$ constant, $i = 1, 2, 3$, are mutually orthogonal at every point in space and more specifically means (1) that the distance ds that a point moves when all three coordinates are simultaneously incremented is given by a sum of squares

$$ds^2 = h_1^2(dq_1)^2 + h_2^2(dq_2)^2 + h_3^2(dq_3)^2$$

with no cross terms (e.g., $dq_1\, dq_3$) and (2) that the unit vectors $\hat{\mathbf{e}}_i$ in the direction of increasing q_i are mutually orthogonal. Let q_1, q_2, q_3 be ordered so that $\hat{\mathbf{e}}_1, \hat{\mathbf{e}}_2, \hat{\mathbf{e}}_3$ form a right-handed set and let S and $\mathbf{Q}$ be a scalar and a vector field, respectively. Show that

(a) $dv =$ volume element $= h_1 h_2 h_3\, dq_1\, dq_2\, dq_3$.

(b) $\nabla S = \sum_i (1/h_i)(\partial S/\partial q_i)\hat{\mathbf{e}}_i$. *Hint:* Use Eq. (0-55).

(c)
$$\nabla \times \mathbf{Q} = \frac{1}{h_1 h_2 h_3} \begin{vmatrix} h_1 \hat{\mathbf{e}}_1 & h_2 \hat{\mathbf{e}}_2 & h_3 \hat{\mathbf{e}}_3 \\ \dfrac{\partial}{\partial q_1} & \dfrac{\partial}{\partial q_2} & \dfrac{\partial}{\partial q_3} \\ h_1 Q_1 & h_2 Q_2 & h_3 Q_3 \end{vmatrix}$$

Hint: Use Eq. (0-62) and note that the derivatives in row 2 are understood to operate only on the entries in row 3 and not on the entries in row 1. Finally, (d) determine the h's for cylindrical coordinates $(q_1, q_2, q_3) = (\imath, \phi, z)$ and for spherical coordinates $(q_1, q_2, q_3) = (r, \theta, \phi)$ and show that the expressions in parts (b) and (c) reduce to the expressions given in the text.

REFERENCES

G. ARFKEN, *Mathematical Methods for Physicists* (Academic Press, Inc., New York, 1966), Chapter 1.

G. E. HAY, *Vector and Tensor Analysis* (Dover Publications, Inc., New York, 1953).

J. B. MARION, *Classical Dynamics* (Academic Press, Inc., New York, 1970), Second Edition, Chapters 1 and 2.

H. M. SCHEY, *DIV, GRAD, CURL and All That* (W. W. Norton & Company, Inc., New York, 1973).

I. S. SOKOLNIKOFF and R. M. REDHEFFER, *Mathematics of Physics and Modern Engineering* (McGraw-Hill Book Company, New York, 1966), Second Edition, Chapters 4 and 6.

K. R. SYMON, *Mechanics* (Addison-Wesley Publishing Company, Inc., Reading, Mass., 1971), Third Edition, Chapter 3.

1

Charge and Current: From Qualitative Recognition to Quantitative Measurement

Every physical theory contains some (small) number of *fundamental* physical quantities, each of which is measured by following a carefully prescribed but *arbitrarily selected* procedure for comparing a given amount of the quantity to an *arbitrarily selected* standard amount. In contrast, *derived* quantities in a theory are defined in terms of the fundamental quantities without introducing any further arbitrary standards. In classical mechanics, length, time, and mass are almost universally chosen as the most convenient fundamental quantities. No such unanimity exists in electricity and magnetism. This theory is sometimes constructed without introducing any fundamental quantities beyond the three mechanical quantities, and it is sometimes constructed on a base obtained by supplementing the mechanical quantities with one or more fundamental and specifically electromagnetic quantities. The primary objectives of this chapter are (1) to provide a background into which most of the common approaches to electricity and magnetism can be placed and (2) to expose the approach adopted in the remainder of this book. To these ends, we imagine—at least initially—that we are discovering (electric) charge and (electric) current anew.[1] Emphasis is placed deliberately on the logical aspects of a development from the qualitative properties of

[1]Strictly, we should also include permanent magnetism in this list, but approaches based on this phenomenon are less common than those based on charge and current and, in any case, the essential ideas can be made clear from a consideration of charge and current alone.

each phenomenon to quantitative definitions both for charge and for current. The reader is cautioned against drawing too close a parallel between the actual historical development of these ideas and the development implied in this chapter. He is also alerted to realize that the experiments by which the properties quoted in this chapter are in fact most accurately confirmed are much more sophisticated than the experiments here implied.

1-1
The Phenomenon of Electric Charge

The first step in our study of electricity is the recognition that electric charge (hereafter simply charge) exists. Experimentally, rubbing a hard rubber rod with rabbit's fur or a glass rod with silk modifies the properties of the rod so that it is capable of deflecting the leaves of an electroscope (Fig. 1-1). The rubbed rod—and more generally any object capable of deflecting the leaves of an electroscope—is said to be *charged*, and we attribute this property to a physical entity called *charge* that we view to reside on the object. The electroscope itself is a primitive detector of charge.

Qualitative experiments on objects charged by rubbing suggest the existence of two kinds of charge. Let one charged object be brought up to an (initially uncharged) electroscope. A deflection of the leaves occurs. Now, leaving the first object near the electroscope, bring up a second charged object. *A priori* we expect that the deflection of the leaves may either decrease or increase. *Experimentally*, we find that nature admits both possibilities.[2] Thus, relative to a *particular* object, an array of second objects can be divided into two groups. Members of the first group increase the deflection and are

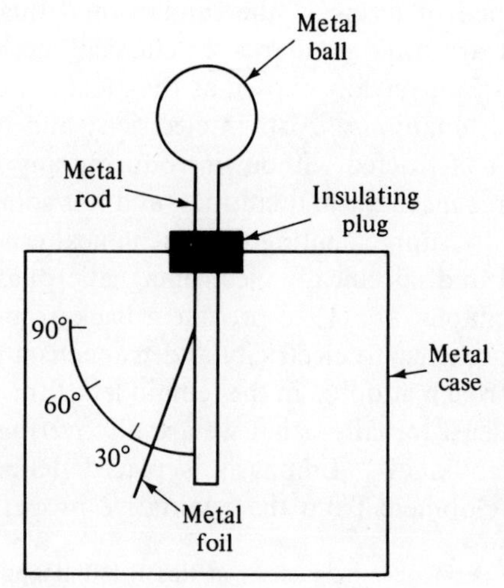

Fig. 1-1. An electroscope. This instrument consists of a vertical metal rod at the top of which is a metal ball and at the bottom of which is fastened a piece of metal foil. When a charged object is brought near the ball, the foil is deflected away from the rod. The numbers on the scale measure the deflection of the foil in degrees.

[2]In the analogous *gravitational* experiment, only one of the two a priori possibilities is realized in the physical world.

said to have the same kind of charge as the first object; members of the second group decrease the deflection and are said to have a kind of charge opposite to that of the first object. Furthermore, we find experimentally that the specific objects included in each group are the same, *regardless of which object is chosen as the first object.* We conclude that any charged object (hereafter simply any charge) can be placed unambiguously in one of two groups, that any two members of either group have *like* charges (each increases the deflection produced by the other), and that two charges drawn one from each group have *unlike* or *opposite* charges (each decreases the deflection produced by the other). To reflect algebraically the tendency of charges drawn from one group to cancel deflections produced by charges drawn from the other group, we designate the charge on members of one group as *positive* and that on members of the other group as *negative.* Convention takes the charge appearing on a hard rubber rod rubbed with rabbit's fur to be negative and that appearing on a glass rod rubbed with silk to be positive, but this firmly established choice is mere convention and might have been made the other way. In view of contemporary models of matter, which take macroscopic objects to be composed microscopically of very large amounts of charge of both signs, we must now slightly modify the above definition of a charged macroscopic object: A macroscopic object is *charged* when microscopically it has more charge of one sign than of the other. In reverse, an *uncharged* object is an object in which positive and negative charges exactly cancel, not an object containing no charge at all.

1-2 The Interaction of Point Charges

Although the recognition that charge exists is the first step, the specification of a quantitative definition for charge must precede the development of a quantitative science of charge. We could, of course, define charge as a fundamental quantity, say by picking a standard electroscope, supplying it with an arbitrarily marked scale to convert deflection into charge units, and specifying some standard way to present a charge to this instrument for measurement. The consequences of this approach, however, might be awkward expressions for physical laws involving charge (P1-10). We begin, instead, by exploring quantitatively the forces of interaction among charged objects, being careful not to make premature restrictive assumptions about the quantitative nature of charge. The experimental properties of these forces, which we intuitively expect to be conditioned by the amount of charge present, can then be used as a guide to the selection of a convenient definition for charge. We assume that the fundamental quantities—length, time, and (inertial) mass—and the derived quantity—force—are already measurable.

Turning then to quantitative aspects (but omitting descriptions of specific experiments), we discover first that the force of interaction between two arbitrary objects arbitrarily charged is not related in any simple way to the relative positions of the objects. When the objects are far apart, however, the interaction is more easily described. We therefore introduce the idealization of a point charge, realized physically when a charged object is viewed from a distance large compared to its own dimensions (so that it looks like a point), and at least initially we concentrate on the interactions between (idealized) point charges. If we further constrain these charges to be at rest relative to one another, we then discover the following *experimental* properties of the force of interaction between two point charges *in vacuum*:

(1) *The force between two point charges at rest is inversely proportional to the square of the separation of the charges.* Experimental tests of this statement are usually quoted by giving an upper limit on the deviation of n from 2 when the statement is written in the form force $\propto 1/(\text{separation})^n$. Let $n - 2 = \eta$. Contemporary experiments show that $|\eta| < (2.7 \pm 3.1) \times 10^{-16}$ over distances ranging from the macroscopic (tens of centimeters, cm) to the atomic (10^{-8} cm), that an inverse square law remains valid at nuclear distances (10^{-13} cm), but that this law may fail at the subnuclear level ($< 10^{-14}$ cm).[3]

(2) *The force between two point charges at rest is proportional to the magnitude of both charges.* Much as it might seem otherwise, experimental evidence supporting this observation can be obtained without presupposing the ability to measure charge. It is certainly intuitive to require that equal charges produce equal deflection when presented in some standard way to an electroscope. Thus, even though we cannot yet assign a number to any charge, we can nonetheless produce an array of equal charges. (Recall that we already are able to determine the sign of a charge.) Let us now require of our ultimate definition of charge the (intuitive) property that charge be additive, i.e., that several charges placed simultaneously in the same position be equivalent to a single object having a charge equal to the (algebraic) sum of the component charges. Then two equal charges placed close together and observed from a distance large compared both to their separation and to their dimensions will be equivalent to a point charge having twice the charge of either object separately. Thus, starting with a point charge having *any* magnitude, we can produce a point charge having twice (or three times or . . .) that magnitude, even though we do not know the magnitude of the first charge. We then discover that if one of two charges is replaced by a charge of twice (or three times

[3]E. R. Williams, J. E. Faller, and H. A. Hill, *Phys. Rev. Letters* **26**, 721 (1971); R. P. Feynman, R. B. Leighton, and M. Sands, *The Feynman Lectures on Physics* (Addison-Wesley Publishing Company, Inc., Reading, Mass., 1964), Volume II, Section 5-8. The first of these references contains a table showing how the upper limit on η—at least for macroscopic distances—has slowly been reduced since the measurements of Cavendish (1773; $|\eta| < 2 \times 10^{-2}$) and Coulomb (1785; $|\eta| < 4 \times 10^{-2}$).

or . . .) its magnitude, the force of interaction is doubled (or tripled or . . .).

(3) *The force between two point charges at rest acts along the line joining the two charges and is repulsive for charges of like sign and attractive for charges of opposite sign.* If one assumes space to be isotropic—the same in all directions—the force between two isolated static charges can have no other direction, for the direction of the line joining the point charges is the only distinguishable direction in space.

(4) *The force between two point charges at rest satisfies Newton's third law.*

In this chapter, we need primarily properties (1) and (2), which are expressed symbolically by the equation

$$F_{qq'} = k_1 \frac{qq'}{d^2} \tag{1-1}$$

where $F_{qq'}$ is the magnitude of the force between two charges[4] q and q' separated by a distance d and k_1 is a proportionality constant. The minor additions to Eq. (1-1) that incorporate properties (3) and (4) will be introduced in Chapter 4, the full statement being known as *Coulomb's law.*

Equation (1-1) has been experimentally inferred without having a unit of charge available; we needed only an unambiguous means to produce multiples of a given charge. The constant k_1, however, remains unspecified and provides the freedom to adapt Eq. (1-1) to whatever unit of charge is selected. We have assumed force and distance to be measurable. If a unit of charge *consistent with* but *independent of* Eq. (1-1) is also available, then all physical quantities in Eq. (1-1) are measurable, and k_1 must be found *experimentally*. If we choose to, however, we can use Eq. (1-1) to *define* the unit of charge by stipulating k_1 arbitrarily. If, for example, k_1 is given the (dimensionless) value unity, we find that

$$F_{qq'} = \frac{qq'}{d^2} \tag{1-2}$$

and the procedure for measuring an unknown charge Q involves the following steps:

(1) Make a second charge of equal magnitude Q using the procedure already outlined.

(2) Place this second charge a *measured* distance R from the first charge.

(3) *Measure* the force F_{QQ} between these two charges.

(4) Calculate the magnitude of Q from the expression $F_{QQ} = Q^2/R^2$, or $Q = \sqrt{F_{QQ}R^2}$.

(5) Determine the sign of the original charge by noting its interaction with a charge of known sign.

[4]We can, of course, give a symbol to a charge without knowing a specific number for the amount of charge represented by the symbol.

In this example, charge is a *derived* quantity having the (mechanical) dimensions of $\sqrt{\text{force}\cdot(\text{length})^2}$. The cgs-electrostatic system of units (cgs-esu), in which force is measured in dynes (dyn), length in centimeters, and time in seconds (sec), is based on exactly this procedure. The constant k_1 is set equal to 1 to define a unit of charge officially called the *statcoulomb* (statC) but more commonly referred to as the esu of charge. Other units of charge in common use will be discussed in Section 1-6; not all of these are defined by Coulomb's law. However the unit of charge is defined, we shall use the term *magnitude* for an amount of charge when its sign is ignored and the term *strength* for an amount of charge when its sign is included, e.g., a charge of strength -4 statC has a magnitude of 4 statC.

Once a procedure for measuring charge is selected, we can then produce an array of known charges which in turn can be used to calibrate the scale on an electroscope. The resulting instrument is, of course, considerably more convenient for laboratory measurements of charge than the primary procedure outlined in this section.

1-3
The Phenomenon of Current

Although it is well known that current and charge are related (and we shall soon incorporate this relationship), let us consider current for the moment as a separate phenomenon. The first step in a study of current is then the recognition of its existence. Experimentally, when a wire made of a suitable material (a conductor) is connected across the poles of the right sort of device (a battery or its equivalent), a small compass (or permanent magnet) placed near the wire manifests an interaction with the wire. Since an electroscope placed near the wire exhibits no effect, this new interaction is different from an electrostatic interaction and we shall attribute it to a *current* in the wire (even though we have at this point in the argument no convincing reason to think that the current flows *in* the wire rather than resides *on* the wire). We shall now examine the properties of this current, using the compass as a primitive detector.

To make this simple detector more useful, let one pole of a small compass needle be weighted slightly so that the needle assumes a vertical equilibrium position when supported from its center on a horizontal axis. (To avoid conceptually irrelevant complications, we assume that the magnetic effect of the earth on the compass is small compared to the effects we seek to examine.) Further introduce a scale with its zero position at the point of equilibrium. The resulting device is illustrated in Fig. 1-2 and, for want of a standard name, will be called a *currentscope*. This instrument is used by placing it behind a long vertical wire as shown in Fig. 1-3. When the poles of a battery are connected to the two ends of the wire, the compass needle will deflect to

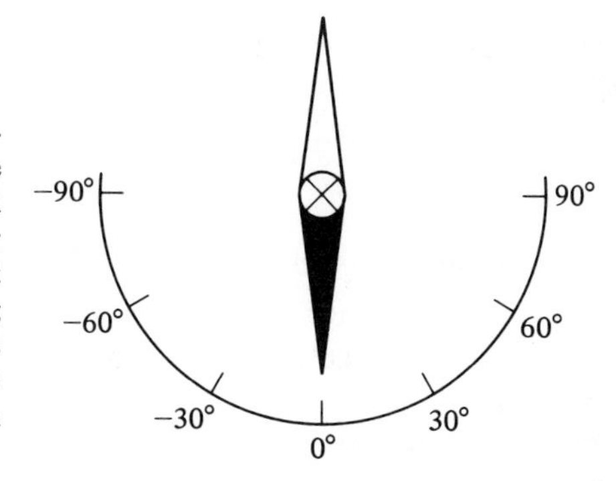

Fig. 1-2. Front view of a currentscope. The weighted pole of the compass needle is shown dark and the axis of rotation is perpendicular to the paper. When this instrument is placed near a current-carrying wire, the compass needle deflects left or right to an angle that can be determined from the numbers on the scale.

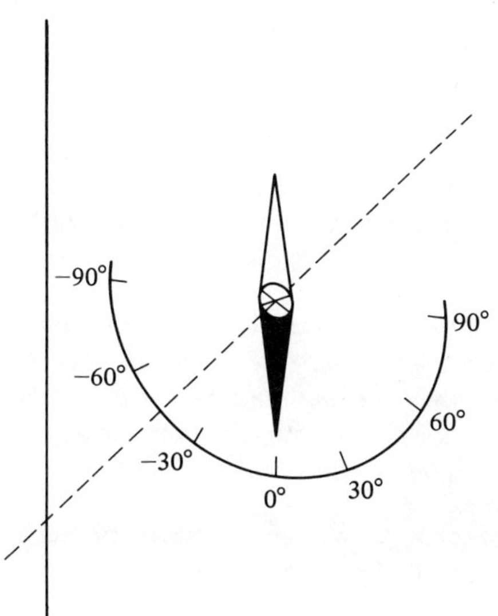

Fig. 1-3. A currentscope near a long, straight wire.

one side, manifesting the interaction we are attributing to a current in the wire. Experiment quickly reveals that the reading of the currentscope diminishes as the instrument is moved away from the wire. Thus, we adopt a standard distance from the wire to the instrument.

More detailed (but still qualitative) experiments with the currentscope result in the following observations:

(1) The reading of the currentscope is independent of where along the wire it is placed. The current is therefore the same at all points along the wire, and we can speak unambiguously of *the* current in the wire.
(2) If the connections to the battery are reversed, the currentscope deflects in the opposite direction, and we therefore adopt a convention taking one of the currents to be positive and the other to be negative. (The common convention will be introduced later in a different context; for now, we simply pick either convention.)

Together these two observations support the view that current flows along a wire.

1-4
The Interaction of Parallel Currents

Let us now develop a quantitative definition for current. The approach adopted parallels that used in determining a quantitative definition for charge *but is entirely independent of that development.* Just as two objects that an electroscope reveals to be charged experience an interaction with one another, two wires that a currentscope reveals to carry currents experience mutual forces. Experimentally, however, the force between two wires is not easily determined if the wires are arbitrarily oriented. Although we shall later consider the more general case, for the moment we shall restrict our attention to the interactions between long, thin, parallel wires and to points remote from either end of the wires. Repeating experiments first performed by Ampere, we then find *experimentally* that *the force* ***per unit length*** *between two stationary, parallel wires carrying (steady) currents*

(1) *is inversely proportional to the separation of the wires.*
(2) *is proportional to the magnitude of each current.*
(3) *is directed along the common perpendicular and is attractive for currents of like sign and repulsive for currents of opposite sign.*
(4) *satisfies Newton's third law.*

In this chapter, we need primarily properties (1) and (2), which are expressed symbolically by the equation

$$F_{II'} = k_2 \frac{II'}{s} \tag{1-3}$$

where $F_{II'}$ is the force *per unit length* between two long parallel wires carrying currents I and I', s is the (perpendicular) distance between the wires, and k_2 is a proportionality constant. The reformulation of Eq. (1-3) to include properties (3) and (4) and to include also interactions between currents in nonparallel wires is presented in Chapter 5.

As with the constant k_1 in Eq. (1-1), the constant k_2 permits Eq. (1-3) to be adapted to whatever unit of current is selected. Again we assume force and distance to be measurable. If a unit of current is also available independently of Eq. (1-3), then k_2 in Eq. (1-3) must be found *experimentally*. If we choose to, however, we can use Eq. (1-3) to *define* the unit of current by stipulating k_2 arbitrarily. If, for example, k_2 is given the (dimensionless) value 2, we find that

$$F_{II'} = 2\frac{II'}{s} \tag{1-4}$$

and the procedure for measuring an unknown current i involves the following steps:

(1) Construct a second wire carrying a current of the same strength.
(2) Place this second wire parallel to and a *measured* distance R from the first current.
(3) *Measure* the force *per unit length* F_{ii} experienced by either wire.
(4) Calculate the magnitude of the current from the expression $F_{ii} = 2i^2/R$, or $i = \sqrt{F_{ii}R/2}$.
(5) Determine the sign of i by noting its interaction with a current of known sign.

In this example, current is a *derived* quantity having the (mechanical) dimensions of $\sqrt{(\text{force})}$. (Remember that F_{ii} is a force *per unit length*.) The cgs-electromagnetic system of units (cgs-emu), in which force is measured in dynes, length in centimeters, and time in seconds, is based on exactly this procedure. The constant k_2 is set equal to 2 to define a unit of current officially called the *abampere* (abA) but more commonly referred to as the emu of current. Other units of current will be discussed in Section 1-6; not all of these are defined by Eq. (1-3). We shall apply the terms *magnitude* and *strength* to currents with meanings analogous to those given at the end of Section 1-2 for charges.

Once a procedure for measuring current is selected, we can then produce an array of known currents which in turn can be used to calibrate a currentscope.

1-5 Current as Charge in Motion

Consider now the following experiment. Suppose a wire runs from a suitable charged object properly placed near a calibrated electroscope past a calibrated currentscope to ground as shown in Fig. 1-4. When the switch is closed, not only do the leaves on the electroscope fall, implying a reducing charge on the object, but also the currentscope deflects as long as there is any charge remaining on the object. We therefore conclude that charge flows off of the object through the wire and that current is the same as charge in motion. By recording the readings of the electroscope and the currentscope as functions of time t, we further discover that the current $I(t)$ in the wire and the charge $q(t)$ measured by the electroscope are related by

$$I(t) = k_3 \frac{dq(t)}{dt} \tag{1-5}$$

where in the present context k_3 is an empirically determined proportionality constant whose value depends on the units in which $I(t)$ and $q(t)$ (and indeed

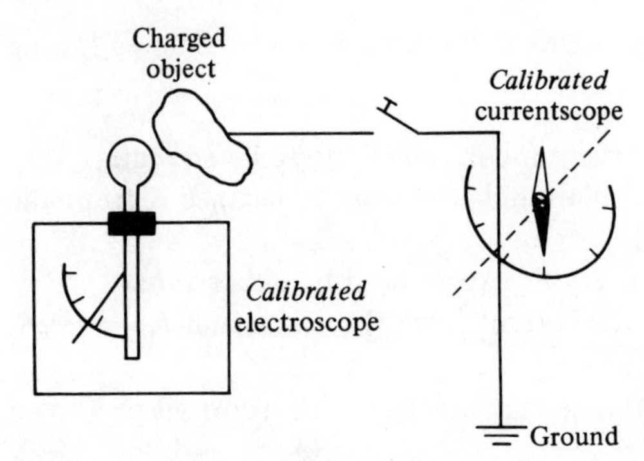

Fig. 1-4. Apparatus demonstrating that current is the same as charge in motion.

t) are measured. Although the connection expressed in Eq. (1-5) leads us to view moving charge wherever it occurs—possibly in wires but equally well, for example, within the body of an ionized gas or in the beam of a particle accelerator—as a current, we shall postpone exploring this broader interpretation until Chapter 2.

1-6 Units in Electricity and Magnetism

We now have three different expressions involving charge and current: (1) Coulomb's law for the force between two point charges,

$$F_{qq'} = k_1 \frac{qq'}{d^2} \tag{1-6}$$

(2) the analogous law for the force *per unit length* between two parallel wires,

$$F_{II'} = k_2 \frac{II'}{s} \tag{1-7}$$

and (3) the relationship

$$I = k_3 \frac{dq}{dt} \tag{1-8}$$

between charge and current. Unless we are willing to define t as well as I and q by these equations, there are at most[5] two quantities (I and q) to be defined. Therefore, at most two of the three constants k_1, k_2, and k_3 can be assigned arbitrary values. The third constant is fixed by the other two even though it takes an experiment to determine its numerical value. Several different systems of units, all in reasonably common use, will now be described.

CGS-ESU

The cgs [*c*entimeter-*g*ram(g)-*s*econd]-*e*lectro*s*tatic system of *u*nits adopts the cgs units of length, mass, and time and defines a unit of charge (the

[5]Remember that it is logically possible for I and q to be defined independently of these equations.

statcoulomb, statC) by setting $k_1^{\text{esu}} = 1$ and a unit of current (the *statampere*, statA) by setting $k_3^{\text{esu}} = 1$. Both of these units are mechanical, the statcoulomb having the dimensions of $\sqrt{\text{dyne}\cdot\text{cm}^2}$ and the statampere having the dimensions of statC/sec $= \sqrt{\text{dyne}\cdot\text{cm}^2/\text{sec}^2}$. Although it is at times convenient to regard k_1^{esu} to have the dimensions (dyne·cm²/statC²), this combination is actually dimensionless. In cgs-esu, the remaining constant k_2^{esu} is determined *empirically*. For subsequent elegance, we introduce another constant c by

$$k_2^{\text{esu}} = \frac{2}{c^2} \tag{1-9}$$

and write Eq. (1-7) in the form

$$F_{II'} = \frac{2}{c^2}\frac{II'}{s} \tag{1-10}$$

The dimensions of c are demonstrably those of velocity (P1-2). Measurement of c, in principle by measuring $F_{II'}$ when known currents are separated by a measured amount s but in practice by any of several more refined techniques, leads to the value

$$c = (2.997925 \pm .000003) \times 10^{10} \text{ cm/sec} \tag{1-11}$$

for which the approximate value 3×10^{10} cm/sec is often satisfactory. Although *unanticipated* and *accidental* at this point, this value is numerically equal to the speed of light, which explains why the 2 was inserted in Eq. (1-9). This appearance of the speed of light in the context of electricity and magnetism also suggests strongly that light and electromagnetism are closely connected, but the intimacy of that connection will not emerge more fully until Chapter 7.

CGS-EMU

The *cgs-e*lectro*m*agnetic system of *u*nits adopts the cgs units of length, mass, and time and defines a unit of current (the *abampere*, abA) by setting $k_2^{\text{emu}} = 2$ and a unit of charge (the *abcoulomb*, abC) by setting $k_3^{\text{emu}} = 1$. Both of these units are mechanical, the abampere having the dimensions of $\sqrt{\text{dyne}}$ and the abcoulomb having the dimensions of abA·sec $= \sqrt{\text{dyne}\cdot\text{sec}^2}$. Although it is at times convenient to regard k_2^{emu} to have dimensions (dyne/abA²), this combination is actually dimensionless. In cgs-emu, the remaining constant k_1^{emu} is determined *empirically*. For subsequent elegance, We set[6]

$$k_1^{\text{emu}} = c^2 \tag{1-12}$$

and write Eq. (1-6) in the form

$$F_{qq'} = c^2\frac{qq'}{d^2} \tag{1-13}$$

The dimensions of this c are also demonstrably those of velocity (P1-2). When

[6]At the moment this c is different from the c in Eq. (1-9).

measured by measuring the force between two known point charges at known separation (or by some equivalent more refined technique), the c here turns out to have the value given in Eq. (1-11), and our use of the same symbol is justified.

RATIONALIZED MKS UNITS

The rationalized mks [*m*eter(m)-*k*ilogram(kg)-*s*econd] system of units adopts the mks units of length, mass, and time and defines a unit of current (the *ampere*, A) by setting $k_2^{\text{mks}} = 2 \times 10^{-7}$ and a unit of charge (the *coulomb*, C) by setting $k_3^{\text{mks}} = 1$. Both of these units are mechanical, the ampere having the dimensions of $\sqrt{\text{N}}$, where N abbreviates newton, and the coulomb having the dimensions of $\text{A}\cdot\text{sec} = \sqrt{\text{N}\cdot\text{sec}^2}$. Although it is at times convenient to think of k_2^{mks} as having the dimensions N/A^2, in our approach this combination is actually dimensionless. The constant k_2^{mks} is often given the expression

$$k_2^{\text{mks}} = \frac{\mu_0}{2\pi} = 2\frac{\mu_0}{4\pi} \tag{1-14}$$

in which μ_0 is *defined* to have the value $4\pi \times 10^{-7}\ \text{N/A}^2$ and is called the *permeability* of free space. In mks units, the force *per unit length* between two parallel wires then has the expression

$$F_{II'} = \frac{\mu_0}{4\pi}\frac{2II'}{s} \tag{1-15}$$

The remaining constant k_1^{mks} is determined *empirically* and has the value

$$k_1^{\text{mks}} = (8.98755 \pm .00002) \times 10^9\ \text{N}\cdot\text{m}^2/\text{C}^2 \tag{1-16}$$

The value $9 \times 10^9\ \text{N}\cdot\text{m}^2/\text{C}^2$ is commonly employed unless extreme accuracy is needed. It is also common to write k_1^{mks} in terms of another constant ϵ_0 as follows:

$$k_1^{\text{mks}} = \frac{1}{4\pi\epsilon_0} \tag{1-17}$$

in which $\epsilon_0 = (8.85418 \pm .00002) \times 10^{-12}\ \text{C}^2/\text{N}\cdot\text{m}^2$ is called the *permittivity* of free space. Coulomb's law then assumes the form

$$F_{qq'} = \frac{1}{4\pi\epsilon_0}\frac{qq'}{d^2} \tag{1-18}$$

The word *rationalized* in the name of this system of units refers to the deliberate placement of an explicit factor of 4π in Coulomb's law. By this means, the 4π is suppressed in several subsequent relationships that are more commonly used than Coulomb's law.

OTHER SYSTEMS OF UNITS

Three other systems of units deserve brief mention. In the *Gaussian* system, the cgs units of length, mass, and time are supplemented with the esu of

TABLE 1-1 The Form of Eqs. (1-6)–(1-8) in Several Common Systems of Units

In this table

$$c = (2.997925 \pm .000003) \times 10^{10} \text{ cm/sec}$$

and must be expressed numerically in these units wherever it occurs in the table,

$$\mu_0 = 4\pi \times 10^{-7} \text{ N/A}^2$$
$$\epsilon_0 = (8.85418 \pm .00002) \times 10^{-12} \text{ C}^2/\text{N}\cdot\text{m}^2$$
$$\frac{1}{4\pi\epsilon_0} = (8.98755 \pm .00002) \times 10^9 \text{ N}\cdot\text{m}^2/\text{C}^2$$

System	*Coulomb's Law*	*Parallel Current Interaction*	*Current-Charge Connection*
cgs-esu	$\frac{qq'}{d^2}$	$\frac{1}{c^2}\frac{2II'}{s}$	$I = \frac{dq}{dt}$
cgs-emu	$c^2\frac{qq'}{d^2}$	$\frac{2II'}{s}$	$I = \frac{dq}{dt}$
Rationalized mks	$\frac{1}{4\pi\epsilon_0}\frac{qq'}{d^2}$	$\frac{\mu_0}{4\pi}\frac{2II'}{s}$	$I = \frac{dq}{dt}$
Gaussian	$\frac{qq'}{d^2}$	$\frac{1}{c^2}\frac{2II'}{s}$	$I = \frac{dq}{dt}$
Heaviside-Lorentz	$\frac{1}{4\pi}\frac{qq'}{d^2}$	$\frac{1}{4\pi c^2}\frac{2II'}{s}$	$I = \frac{dq}{dt}$

charge ($k_1^{\text{Gaussian}} = 1$) and the esu of current ($k_3^{\text{Gaussian}} = 1$). Although Gaussian units therefore are at this stage identical with cgs-esu, differences will subsequently appear in the way in which other quantities, particularly the magnetic field, are defined in the two systems. The rationalized counterpart of the Gaussian system is called the *Heaviside-Lorentz* system; it uses cgs units of length, mass, and time and the values $k_1^{\text{HL}} = 1/4\pi$ and $k_3^{\text{HL}} = 1$ to define units of charge and current; k_2^{HL} is then determined *experimentally* to have the value $1/2\pi c^2$. Finally, in the *natural* system, adopted principally by nuclear physicists, the velocity of light and a few other fundamental constants are given the value unity.

SUMMARY

The forms assumed by Eqs. (1-6)–(1-8) in the several systems of units described above are summarized in Table 1-1. Although we shall continue to indicate the form of subsequent definitions in each of these sets of units, this text will use the rationalized mks system of units.

CONVERSION OF UNITS

The need to convert from one system of units to another accompanies the common use of more than one system. We shall illustrate the technique for obtaining conversion factors by working out the factors relating cgs-esu and cgs-emu. Consider two point charges separated by a distance d. In cgs-esu, the force between these two charges is given by

$$F_{qq'}\,(\text{dyne}) = \frac{q_{\text{esu}}q'_{\text{esu}}}{d^2_{\text{cm}}} \tag{1-19}$$

and in cgs-emu this *same* force is given by

$$F_{qq'}\,(\text{dyne}) = c^2_{\text{cm/sec}}\frac{q_{\text{emu}}q'_{\text{emu}}}{d^2_{\text{cm}}} \tag{1-20}$$

where units have been indicated explicitly and q_{esu} and q_{emu}, for example, express the magnitude of the *same* physical charge in the two sets of units. Since the forces and the distances in Eqs. (1-19) and (1-20) are numerically the same, we infer the *numerical* equality

$$q_{\text{esu}}q'_{\text{esu}} = c^2_{\text{cm/sec}}q_{\text{emu}}q'_{\text{emu}} \tag{1-21}$$

or, since q_{esu} is proportional to q_{emu}, we find that

$$q_{\text{esu}} = c_{\text{cm/sec}}q_{\text{emu}} \tag{1-22}$$

Similarly, starting with two currents and the equations

$$F_{II'}\,(\text{dyne/cm}) = \frac{2I_{\text{esu}}I'_{\text{esu}}}{c^2_{\text{cm/sec}}s_{\text{cm}}} = \frac{2I_{\text{emu}}I'_{\text{emu}}}{s_{\text{cm}}} \tag{1-23}$$

we find that

$$I_{\text{esu}} = c_{\text{cm/sec}}I_{\text{emu}} \tag{1-24}$$

Equivalently, we obtain I_{esu} by the chain of argument expressed in the equation

$$I_{\text{esu}} = \frac{dq_{\text{esu}}}{dt_{\text{sec}}} = c_{\text{cm/sec}}\frac{dq_{\text{emu}}}{dt_{\text{sec}}} = c_{\text{cm/sec}}I_{\text{emu}} \tag{1-25}$$

Equations such as Eqs. (1-22) and (1-24) frequently generate confusion. These equations are *numerical* equalities. Equation (1-22), for example, states that the numerical value of some particular charge in esu is equal to $c_{\text{cm/sec}}$ ($\approx 3 \times 10^{10}$ in numerical value) times the numerical value of the *same* charge in emu. Thus, a charge of 1 abC is physically the same charge as a charge of about 3×10^{10} statC. The temptation to read this and similar equations the other way—i.e., 1 esu $= 3 \times 10^{10}$ emu—must be resisted; it is incorrect.

TABLE 1-2 Conversion Factors

In this table the number 3 arises from the speed of light and, in accurate work, should be replaced by 2.997925.

$$\begin{Bmatrix}\text{1 statcoulomb}\\ \text{1 statampere}\end{Bmatrix} = \frac{1}{3}\times 10^{-10}\begin{Bmatrix}\text{abcoulomb}\\ \text{abampere}\end{Bmatrix} = \frac{1}{3}\times 10^{-9}\begin{Bmatrix}\text{coulomb}\\ \text{ampere}\end{Bmatrix}$$

$$\begin{Bmatrix}\text{1 abcoulomb}\\ \text{1 abampere}\end{Bmatrix} = 10\begin{Bmatrix}\text{coulomb}\\ \text{ampere}\end{Bmatrix} = 3\times 10^{10}\begin{Bmatrix}\text{statcoulomb}\\ \text{statampere}\end{Bmatrix}$$

$$\begin{Bmatrix}\text{1 coulomb}\\ \text{1 ampere}\end{Bmatrix} = 3\times 10^{9}\begin{Bmatrix}\text{statcoulomb}\\ \text{statampere}\end{Bmatrix} = 0.1\begin{Bmatrix}\text{abcoulomb}\\ \text{abampere}\end{Bmatrix}$$

Conversion factors relating esu, emu, and mks units of charge and current are summarized in Table 1-2.

PROBLEMS

P1-1. Write a paragraph outlining a procedure for confirming property (2) of Section 1-4 before a quantitative measure of current is available.

P1-2. Show that the constant c in Eq. (1-10) and the constant c in Eq. (1-13) both have the dimensions of velocity.

P1-3. At one time, the ampere was defined by an experiment in electrolysis that involved measuring the mass of silver deposited per unit time in a standard silver voltameter. If we had adopted this definition, we would properly view the ampere as a fundamental quantity. Explain how this change affects the status of Eq. (1-15) and of the constant k_2^{mks}. The value assigned to k_2^{mks} in the text was, of course, chosen to make the present definition of the ampere agree with the earlier definition, so the numerical value of k_2^{mks} is not changed by the change in definition.

P1-4 An alternative way to state the definition of the ampere is as follows: *The ampere is that constant current which, when present in two parallel wires separated by* 1 m, *gives rise to a force of* 2×10^{-7} N/m *between the wires.* Formulate similar definitions for the statcoulomb and the abampere, which are defined by Eqs. (1-6) and (1-7), respectively, and for the coulomb, the statampere, and the abcoulomb, which are defined by Eq. (1-8).

P1-5. Show that the ampere is *exactly* one-tenth the abampere.

P1-6. (a) Derive the conversion factor(s) between esu and mks units of charge and current. (b) Derive the conversion factor(s) between emu and mks units of charge and current. (c) The charge on the proton is 1.60210×10^{-19} C. Determine this charge in statcoulombs and in abcoulombs.

P1-7. In what are called *modified* Gaussian units, charges are measured in statcoulombs ($k_1 = 1$) and currents are measured in abamperes ($k_2 = 2$). Determine the value of k_3 in modified Gaussian units and write out the corresponding entries in Table 1-1.

P1-8. Let an unused set of units—call them Cookian units—be defined as follows:

length	1 cookmeter = b cm
time	1 cooksecond = f sec
force	1 cookdyne = a dyne
charge	Set $k_1 = K_1$ in Eq. (1-6), defining the cookcoulomb
current	Set $k_2 = 2K_2$ in Eq. (1-7), defining the cookampere

where a, b, f, K_1, and K_2 are arbitrary but fixed. (Note, for example, that Cookian units reduce to cgs-esu if $a = b = f = K_1 = 1$ and $K_2 = 1/c^2$, with c in cm/sec.) With everything (*including* the speed of light) expressed in

Cookian units, show that

$$\frac{dq}{dt} = \sqrt{\frac{K_2 c^2}{K_1}} I$$

and infer (a) that any system of units in which $I = dq/dt$ must have $K_1/K_2 = c^2$ and (b) that, in mks units, $\mu_0 \epsilon_0 = 1/c^2$.

P1-9. What (numerically) is the ratio of the electrostatic force of repulsion to the gravitational force of attraction between two protons?

P1-10. (a) Given two particles with mass m and charge q suspended from a single point by strings of length ℓ, show that the angle θ between each string and the vertical satisfies $\tan \theta \sin^2 \theta = (q/q_0)^2$, where (in mks units) $q_0^2 = 16\pi\epsilon_0 mg\ell^2$ and g is the acceleration of gravity. (b) Obtain a graph of q/q_0 versus θ. *Hint*: Use a computer or desk calculator. (c) What deflection is produced by a charge of strength q_0? *Hint*: Estimate the angle from your graph or, using a short computer program that inputs an angle and then calculates and prints out the value of q/q_0, guess the solution, try it, and then refine your guess until you have found the solution to, say, four significant figures. (d) What (numerically) is q_0 for a reasonable apparatus, say $m = .5$ g and $\ell = 20$ cm? Compare your answer with the charge on the proton. *Optional*: Suppose a charge has strength q in coulombs and strength Q in foolcoulombs, the latter value being measured by making a second, equal charge, placing each charge on a separate small sphere of some *standard* mass m, suspending the spheres from strings of some *standard* length ℓ, and calculating Q from the *measured* angle θ (in radians) by the defining equation $Q = 2\theta/\pi$. Show that no charge can have a magnitude greater than 1 foolcoulomb, and find the (very awkward) form assumed by Coulomb's law if charges are measured in foolcoulombs and everything else is measured in mks units.

P1-11. Two uncharged aluminum spheres, each having a mass of 1 g, are suspended so their centers are 10 cm apart. By observing the parallelism of the supporting strings, an experimenter deduces that any forces of interaction between these spheres are weaker than about 10^{-5} N—actually a fairly coarse measurement. What maximum fraction of the total number of electrons in one sphere could be transferred to the other without upsetting this result? What does your answer imply about the precision with which apparently neutral matter in fact contains equal amounts of positive and negative charges? *Note*: The atomic number of aluminum is 13; its atomic weight is 27.

P1-12. A wire is free to slide without friction on rails that make an angle θ with the horizontal. It carries a current I opposite in direction but equal in magnitude to the current carried by a parallel wire at the bottom of the rails. If the sliding wire has length ℓ and mass m, find the separation s of the two wires at which the sliding wire will be in equilibrium. (See Fig. P1-12.)

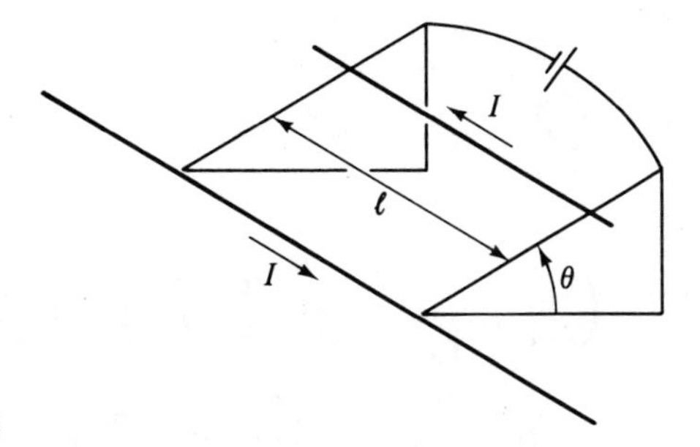

Figure P1-12

REFERENCES

J. C. MAXWELL, *A Treatise on Electricity and Magnetism* (Dover Publications, Inc., New York, 1954), Articles 1–6, 27–43, 371–375, 475–479, and 502–527. (This treatise is a republication in two volumes of a work whose third edition was originally published in 1891.)

M. NICOLA, "On the Definition of Electric Charge," *Am. J. Phys.* **40**, 189 (1972).

M. A. ROTHMAN, *Discovering the Natural Laws* (Doubleday & Company, Inc., Garden City, N.Y., 1972), Chapter 8.

2

Charge and Current: The Specification of Arbitrary Distributions

Much of our subsequent development is concerned with the properties and the behavior of various spatial distributions of (possibly moving) charge. We therefore need a means to specify the "state" of a general charge distribution. For the simplest distributions, which contain a small number of discrete and widely separated particles, the state is conveniently and satisfactorily specified by giving the charge, position, and velocity of each particle. Many of the distributions with which we must deal, however, "fill" some region of space by containing a very large number—say 10^{23}—of very small, densely packed, erratically moving, charged particles. Merely printing the charges, positions, and velocities of this many particles would keep a high-speed line printer on a computer busy for about 10^{14} years! Clearly, for describing space-filling distributions, we must replace charges, positions, and velocities of individual particles with more appropriate new concepts. The objectives of this chapter are (1) to define these concepts, called the *charge density* and the *current density*, and (2) to explore the mathematical apparatus that is useful for manipulating these concepts analytically.

We shall introduce the *macroscopic* description in terms of charge and current densities by relating it to the underlying *microscopic* description in terms of individual particles. The success of this transition from a microscopic to a macroscopic model is a consequence of the large differences that exist between the microscopic and the macroscopic scales of time and distance. On the microscopic scale, for example, macroscopic observations are extremely sluggish; at best they are able to resolve time intervals (say $\approx 10^{-4}$ sec) that are something like 10^{11} times as long as the period ($\approx 10^{-15}$ sec) of the thermal oscillations of the microscopic particles. Again, on the microscopic scale, macroscopic observations are extremely coarse; at best they are able to resolve spatial separations (say $\approx 10^{-2}$ cm) that are something like 10^6 times as large as the dimensions ($\approx 10^{-8}$ cm) of the space occupied by a single particle. Thus, what a *macroscopic* measurement detects as an *instant* of time is still long enough to include very many—perhaps 10^9—*microscopic* fluctuations; what a *macroscopic* measurement detects as a *point* in space still occupies sufficient volume to contain very many—perhaps 10^{15}—*microscopic* particles. Instead of responding directly to the microscopic details, macroscopic measurements therefore average these microscopic details over macroscopically small temporal and spatial intervals that are still large enough microscopically to include many fluctuations and many particles. The theoretical concepts appropriate to the description of macroscopic charge distributions are therefore averages of the microscopic concepts, and we shall make these averages more explicit in the next two sections.

The transition from microscopic to macroscopic descriptive concepts carries with it a change in our mental image of these charge distributions. In effect, the microscopic model of erratically moving, small particles is replaced by a more sedate macroscopic model in which we ignore not only the erratic motion of each particle but also the elemental discreteness of matter and of charge. Insofar as we think of them at all, particles are effectively at rest or perhaps in motion with some smoothly varying *drift velocity*. Further, matter and charge are spread continuously throughout the region occupied by the distribution. The result is a model of a macroscopic charge distribution that, except for the charge it carries with it, corresponds exactly with the classical model of a fluid. We therefore precede the definition of charge and current densities by introducing two fields that have been found useful for the macroscopic description of fluids, whether charged or not. Classify each microscopic particle in the fluid according to its mass and charge, particles of type a having mass m_a and charge q_a. The first field, called the *particle density field* $n^{(a)}(\mathbf{r}, t)$, measures at time t the number of particles of type a per unit volume in a small volume ΔV centered on the point $\mathbf{r}$; it is defined formally by

$$n^{(a)}(\mathbf{r}, t) = \lim_{\substack{\Delta V \to 0 \\ \text{about } \mathbf{r}}} \frac{\Delta N^{(a)}(t)}{\Delta V} \tag{2-1}$$

where $\Delta N^{(a)}(t)$ is the number of particles of type a in the volume ΔV at time t,[1] and (both here and hereafter) the limit $\Delta V \longrightarrow 0$ carries the proviso that

[1]Strictly, $\Delta N^{(a)}(t)$ is the (time) average of the number of particles of type a in the volume ΔV, the average being evaluated over a macroscopically small but microscopically large time interval centered at time t. Subsequent statements similar to the one to which this footnote applies must be similarly interpreted.

ΔV becomes macroscopically small but remains microscopically large. The second field, called the *velocity field* $\mathbf{v}^{(a)}(\mathbf{r}, t)$, gives the typical drift velocity of a particle of type a that happens at time t to be in the volume ΔV; it is defined formally by

$$\mathbf{v}^{(a)}(\mathbf{r}, t) = \lim_{\substack{\Delta V \to 0 \\ \text{about } \mathbf{r}}} \frac{1}{\Delta N^{(a)}(t)} \sum_i \mathbf{v}_i^{(a)}(t) \tag{2-2}$$

where $\mathbf{v}_i^{(a)}(t)$ is the *drift* velocity of the ith particle of type a in the volume ΔV at time t. Because these definitions involve macroscopically small but microscopically large temporal and spatial intervals, both $n^{(a)}(\mathbf{r}, t)$ and $\mathbf{v}^{(a)}(\mathbf{r}, t)$ are smoothly and slowly varying functions of $\mathbf{r}$ and t and, in particular, may even be constant in time and/or in space. Although the state of a charged fluid is more directly specified by the charge and current densities to be introduced in the next sections, we shall occasionally use $n^{(a)}$ and $\mathbf{v}^{(a)}$ to link charge and current densities to the underlying particulate model.

PROBLEMS

P2-1. A high-speed line printer may print on the order of 1200 lines per minute. Suppose each line can contain the charge, position, and velocity for two particles. (Remember that position and velocity are *vectors.*) (a) How long would it take to print out the state of a system containing 10^{23} particles? (b) How thick would the resulting pile of paper be?

P2-2. Estimate the period associated with the erratic oscillations of a microscopic particle. *Hint*: A typical thermal speed, say at room temperature, can be estimated from kinetic theory. How long does it take an electron (proton) to travel a typical microscopic distance, say 10^{-8} cm, at this speed?

2-1 Charge Density

Consistent with the macroscopic model developed in the above introductory paragraphs, we ignore the elemental discreteness of electric charges. In addition to being concentrated on small objects (point charges), charge may then be distributed more or less smoothly throughout volumes, over surfaces, and along lines. To facilitate a macroscopic description of each type of distribution, we therefore introduce three macroscopic charge densities: (1) a *volume* charge density, having mks units of coulombs per cubic meter (C/m^3) and defined by

$$\rho(\mathbf{r}, t) = \lim_{\substack{\Delta V \to 0 \\ \text{about } \mathbf{r}}} \frac{\Delta q(t)}{\Delta V} \tag{2-3}$$

(2) a *surface* charge density, having mks units of coulombs per square meter

(C/m²) and defined by

$$\sigma(\mathbf{r}, t) = \lim_{\substack{\Delta S \to 0 \\ \text{about } \mathbf{r}}} \frac{\Delta q(t)}{\Delta S} \tag{2-4}$$

and (3) a *linear* charge density, having mks units of coulombs per meter (C/m) and defined by

$$\lambda(\mathbf{r}, t) = \lim_{\substack{\Delta \ell \to 0 \\ \text{about } \mathbf{r}}} \frac{\Delta q(t)}{\Delta \ell} \tag{2-5}$$

where $\Delta q(t)$ is successively the charge in the volume ΔV, on the surface of area ΔS, and on the line of length $\Delta \ell$ at time t. In each case, the point $\mathbf{r}$ must lie in the volume, surface, or line element involved, and the element must become small compared to macroscopic dimensions while remaining large compared to microscopic dimensions. When these charge densities can be meaningfully defined (i.e., for space-filling macroscopic distributions), all three are smooth functions of $\mathbf{r}$ and t. A distribution is said to be *static* if the corresponding charge densities do not depend on time and *uniform* if the charge densities do not depend on the spatial coordinates.

Among other things, knowledge of the appropriate charge densities permits a calculation of how much charge $Q(t)$ is present at time t in any portion of the distribution. We simply break the portion of the distribution of interest into appropriate small elements, distinguished by an index i, and add up the contributions Δq_i from each element, passing to the limit as the extent of all elements becomes small and, correspondingly, the total number of elements becomes large. If the distribution is described by a volume charge density, for example, we find that

$$\begin{aligned} Q(t) &= \lim_{\Delta V_i \to 0} \sum_i \Delta q_i(t) \\ &= \lim_{\Delta V_i \to 0} \sum_i \rho(\mathbf{r}_i, t)\, \Delta V_i \\ &= \int \rho(\mathbf{r}, t)\, dv \end{aligned} \tag{2-6}$$

where ΔV_i is the volume of the ith element, $\mathbf{r}_i$ is a point in the ith element, dv is an infinitesimal volume element in a convenient coordinate system [see Eqs. (0-22)–(0-24)], and the volume integral extends over the portion of the charge distribution of interest (which may, of course, be the entire distribution). By a similar limiting process, we find for surface and line distributions that

$$Q(t) = \int \sigma(\mathbf{r}, t)\, dS, \qquad Q(t) = \int \lambda(\mathbf{r}, t)\, d\ell \tag{2-7}$$

where dS and $d\ell$ are the area and length of infinitesimal surface and line elements, respectively, and the integrals again extend over a portion (or perhaps over all) of the distribution. Here and subsequently, we shall combine integrals like those in Eqs. (2-6) and (2-7) by introducing an infinitesimal element of charge dq to stand for $\rho\, dv$, $\sigma\, dS$, or $\lambda\, d\ell$ as appropriate. Then, the integrals all have the form

$$Q = \int dq \tag{2-8}$$

where explicit indication of spatial and temporal dependences has been suppressed. Although we shall make no use of it, it is in fact possible to introduce a mathematical function, known as the *Dirac delta function*, in terms of which all types of charge distribution, including point charges, can be formally described with *volume* charge densities only. With this function, Eq. (2-6) alone is sufficient to cover all cases.

Finally, we relate the volume charge density to the particle density field defined in Eq. (2-1). Suppose the distribution of interest contains particles of several different types and let particles of type a with charge q_a be distributed in accordance with the particle density field $n^{(a)}(\mathbf{r}, t)$. Then the total charge $\Delta q(t)$ at time t in a volume ΔV centered at the point $\mathbf{r}$ is

$$\Delta q(t) = \sum_a q_a n^{(a)}(\mathbf{r}, t)\, \Delta V \tag{2-9}$$

and Eq. (2-3) gives

$$\rho(\mathbf{r}, t) = \sum_a q_a n^{(a)}(\mathbf{r}, t) \tag{2-10}$$

The ath term in this sum, of course, expresses the contribution of particles of type a to the total charge density.

PROBLEMS

P2-3. Let a total charge Q be uniformly distributed in succession throughout the volume of a sphere of radius R, on the surface of the same sphere, and along the perimeter of a circle of radius R. What is the charge density resulting in each case?

P2-4. According to quantum mechanics, the electron in the ground state of the hydrogen atom is characterized by a charge density

$$\rho(\mathbf{r}) = -\frac{q}{\pi a_0^3} e^{-2r/a_0}$$

where a_0 is the Bohr radius. Show that the total charge in this distribution is $-q$. *Optional:* Determine the fraction of this charge that lies within a sphere of radius R and obtain a careful graph of this fraction as a function of R/a_0. *Hint:* Use a computer to evaluate the function.

2-2
Current Density

To define a current density for specifying the *motion* of a charge distribution, we must first extend Eq. (1-8) so that the meaning of current is unambiguous even when the flow of charge does not follow the path defined

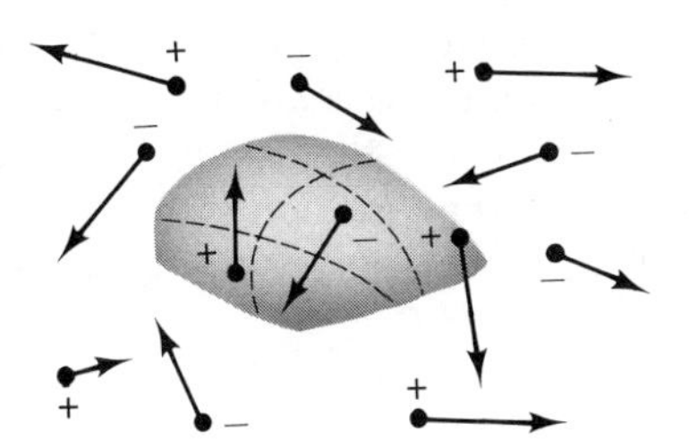

Fig. 2-1. View of a region of space filled with moving charged particles. An imaginary surface is shown in the region.

by a wire. Imagine that a region of space is occupied by a large number of charged particles, each particle moving about within this region in some smoothly varying way. (We ignore again the erratic microscopic motion.) Let some of these particles be positively charged and others negatively charged. Now, place an imaginary surface somewhere in the region of space occupied by these charges (Fig. 2-1) and let this surface be fixed in position. Charges of both signs move in both directions across this surface. In any given small (but macroscopic) interval of time Δt there may be a net transfer of charge through the surface from one side to the other. Let the net charge transported across the surface be evaluated as follows: Choose (arbitrarily) one direction through the surface as the positive direction. Then let positive charge moving in the positive direction and negative charge moving in the negative direction make positive contributions to the charge transported, and let positive charge moving in the negative direction and negative charge moving in the positive direction make negative contributions. Finally, identify the net charge transported ΔQ as the algebraic sum of these several contributions. Thus, a positive charge transport results when a net positive charge is transported in the positive direction across the surface; a negative charge transport results when a net positive charge is transported across the surface in the negative direction; etc. If the net charge ΔQ is transported across the surface in an elapsed time Δt, then [consistent with Eq. (1-8)] the *average* current $\bar{I}$ flowing across the surface in this time interval is given by

$$\bar{I} = \frac{\Delta Q}{\Delta t} \tag{2-11}$$

where we have set $k_3 = 1$, thereby fixing at least a part of the system of units. The *instantaneous* current $I(t)$ flowing across the surface at a given time t is then defined by allowing Δt to shrink to zero about the time instant t, i.e., by

$$I(t) = \lim_{\substack{\Delta t \to 0 \\ \text{about } t}} \frac{\Delta Q}{\Delta t} = \frac{dQ}{dt} \tag{2-12}$$

where the expression as a derivative is appropriate only if Q is interpreted as the total charge that has arrived at time t on the positive side of the surface by a route that passes through the surface. As always, the limit means more specifically that Δt must be made macroscopically small while simultaneously remaining microscopically large. With the sign conventions established above for measuring ΔQ, the current across the surface will be positive if positive charge is transported in the positive direction across the surface and so on. Convention takes the direction of current flow across the surface to be the direction of the equivalent positive charge transport.

PROBLEMS

P2-5. Suppose a current in a wire is carried solely by electrons. How many electrons are transported in 1 sec past a point in this wire if the current is 0.1 A? What conclusion do you draw about the legitimacy of regarding a (macroscopic) current as a flowing charged fluid?

P2-6. (a) A total charge Q is distributed uniformly on the perimeter of a circular ring of radius R. The ring is then set into rotation about its axis with angular speed ω (radian/sec). Determine the current represented by the rotating ring. (b) Suppose now the charge Q is concentrated on a particle that moves in a circle of radius R with angular speed ω. Under what conditions will the result of part (a) also give the current represented by the circulating particle?

Although the current across a surface is very often the answer to practical questions, a more convenient description of moving charge is formulated about a quantity known as the *current density*. We shall consider only the current density appropriate to distributions occupying some three-dimensional volume; a current density suited to the description of surface distributions of charge is explored in P2-25.

To define the *volume* current density, we begin by examining the state of affairs at time t within a small volume ΔV surrounding the point $\mathbf{r}$ in the arbitrary charge distribution of Fig. 2-1. A view of this volume is shown in Fig. 2-2. If there is any net transport of charge past the point $\mathbf{r}$, that transport, of course, occurs in some well-defined direction and we introduce a unit vector $\hat{\mathbf{n}}(\mathbf{r}, t)$, which may in general depend on $\mathbf{r}$ and t, having the direction of this transport. Further, let us introduce a small plane surface of area ΔS oriented with its plane perpendicular to $\hat{\mathbf{n}}$ and positioned so that some point on ΔS is at $\mathbf{r}$. Finally, let $\Delta I(t)$, determined as described in an earlier paragraph, be the current crossing ΔS at time t. The current density $\mathbf{J}(\mathbf{r}, t)$ at point $\mathbf{r}$ and time t in the illustrated distribution is then defined by

$$\mathbf{J}(\mathbf{r}, t) = \lim_{\Delta S \to 0} \frac{\Delta I(t)}{\Delta S} \hat{\mathbf{n}}(\mathbf{r}, t) \tag{2-13}$$

and is a *vector* quantity whose direction coincides with the direction of net (positive) charge transport and whose magnitude is the rate at which charge is transported (i.e., the current) across a surface of *unit* area oriented perpen-

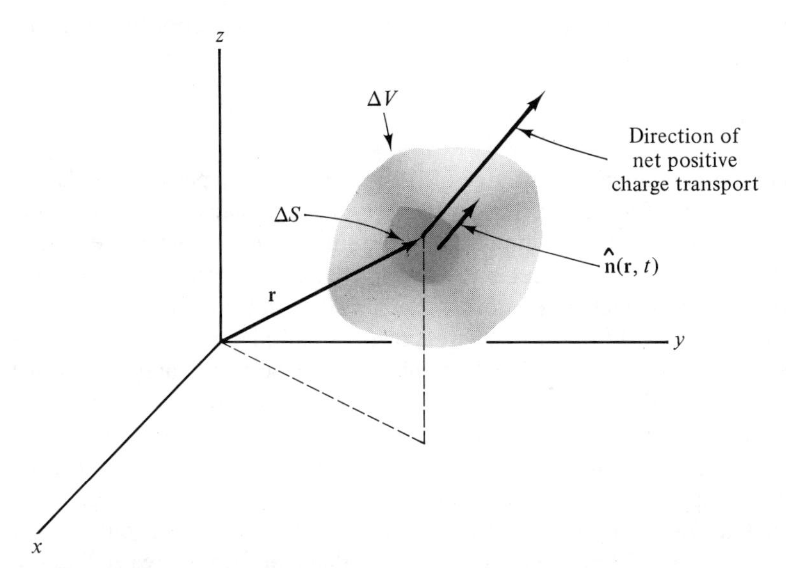

Fig. 2-2. View of a small volume about the point **r** in Fig. 2-1. Individual charges are not shown.

dicular to the direction of charge transport. The mks units of **J** are therefore amperes per square meter (A/m²). This current density characterizes the macroscopic motion of the charge distribution and knowledge of both $\mathbf{J}(\mathbf{r}, t)$ and of the (volume) charge density $\rho(\mathbf{r}, t)$ at some time t and at all points **r** in an arbitrary (volume) charge distribution fully determines the state of that charge distribution at time t.

PROBLEMS

P2-7. (a) Show that the current density in a wire carrying a total current I uniformly distributed over a cross-sectional area S is $\mathbf{J} = (I/S)\hat{\mathbf{t}}$, where $\hat{\mathbf{t}}$ is a unit vector tangent to the wire and in the direction of the current. (b) The product $I\,d\boldsymbol{\ell}$, where $d\boldsymbol{\ell} = d\ell\hat{\mathbf{t}}$ is a vector representing an infinitesimal element of the wire having length $d\ell$, will occur in our study of magnetism. Show that $I\,d\boldsymbol{\ell} = \mathbf{J}\,dv$, where dv is the volume of the element represented by $d\boldsymbol{\ell}$.

P2-8. A point radioactive source located at the origin emits N particles per second, each having charge q. If these particles are emitted uniformly in all directions, determine the current density at the point **r**. *Hint:* Express the answer in spherical coordinates.

P2-9. A total charge Q is distributed uniformly throughout the volume of a sphere of radius a. The sphere is then set into rotation with (constant) angular velocity ω (radian/sec) about a diameter, which it is convenient to take as the axis of a cylindrical coordinate system $(\imath, \phi, z)$. Express the current density in the region $|\mathbf{r}| \leq a$ in terms of cylindrical coordinates and unit vectors.

Among other things, knowledge of the current density permits a calculation of the current flowing across an arbitrary surface placed somewhere in the charge distribution. We simply break the surface into appropriate small elements, calculate the current across each element, sum the individual contributions, and then let the size of the elements become indefinitely small. First, however, we must find the current across a small plane surface, e.g., the surface of area ΔS in Fig. 2-3, whose plane is *not* perpendicular to $\mathbf{J}$. Let this surface be small enough to fit within a small volume throughout which at any instant of time t the current density can be regarded to have the *constant* value $\mathbf{J}(\mathbf{r}, t)$, where $\mathbf{r}$ locates some *fixed* point on the surface. Then, because $\mathbf{J}(\mathbf{r}, t)$ gives the direction of the net transport of charge, no (net) charge is transported across any surface whose plane is parallel to $\mathbf{J}$, and the current across ΔS is the same as the current across the auxiliary surface of area $\Delta S'$ obtained by projecting ΔS onto a plane perpendicular to $\mathbf{J}$ and passing through $\mathbf{r}$. But the current across the auxiliary surface is just $|\mathbf{J}|\,\Delta S'$. Thus, since $\Delta S' = \Delta S \cos\theta$, where θ is the angle between $\mathbf{J}$ and the normal to the surface ΔS, we find that the current ΔI across ΔS is given by

$$\Delta I = |\mathbf{J}|\,\Delta S \cos\theta = \mathbf{J}\cdot\Delta\mathbf{S} \tag{2-14}$$

where the vector $\Delta\mathbf{S}$ represents the surface of area ΔS and by convention is assigned a direction perpendicular to the plane of the surface and a magnitude equal to the area of the surface represented. Whichever of the two possible directions is assigned to $\Delta\mathbf{S}$, Eq. (2-14) will give a positive value for ΔI if a

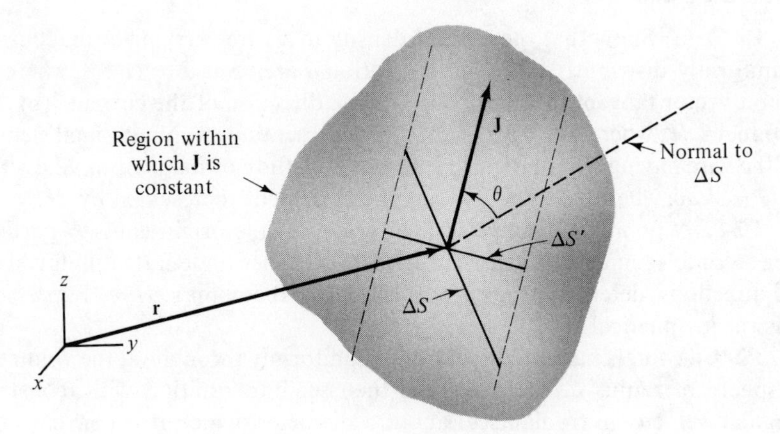

Fig. 2-3. Geometry for determining the current across a surface whose plane is not perpendicular to the current density.

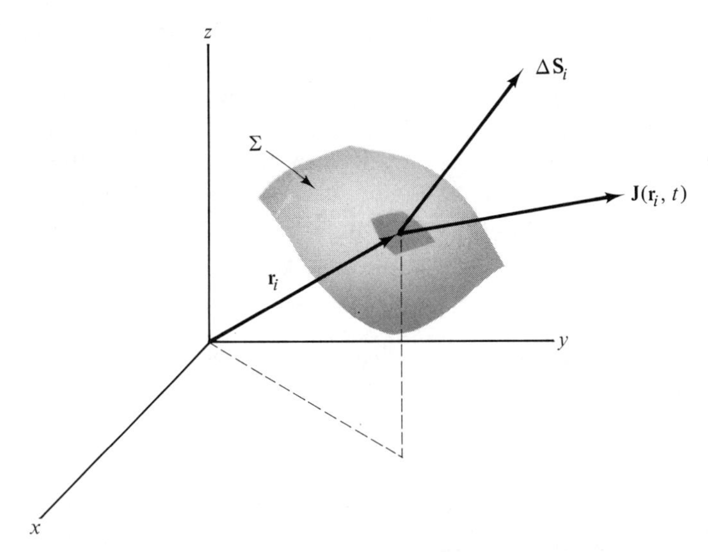

Fig. 2-4. An arbitrary surface in a region of moving charge.

net positive charge is transported across the surface in the direction of $\mathbf{\Delta S}$ and a negative value if a net positive charge is transported across the surface in a direction opposite to $\mathbf{\Delta S}$. In principle, either direction may be assigned to $\mathbf{\Delta S}$; in practice—at least in some special cases—particular choices have become conventional, and we shall introduce these conventions as we encounter them.

We are now ready to determine the current flowing at time t across an arbitrary surface placed in the charge distribution described by the current density $\mathbf{J}(\mathbf{r}, t)$. Let the surface be divided into small elements, each of which can be regarded as plane and over each of which the current density is approximately constant, and let the ith element of the surface be represented by a vector $\mathbf{\Delta S}_i$ defined as in the previous paragraph with the added proviso that all elements $\mathbf{\Delta S}_i$ be assigned vectors in the "same" direction.[2] One element of this surface is shown in Fig. 2-4, which also shows the current density $\mathbf{J}(\mathbf{r}_i, t)$ at the ith element of the surface at time t. In accordance with Eq. (2-14), the current across this element is given by $\mathbf{J}(\mathbf{r}_i, t)\cdot\mathbf{\Delta S}_i$, and the total current $I(t)$ across the arbitrary surface, call it Σ, is given by

$$I(t) = \lim_{|\Delta S_i|\to 0} \sum_i \mathbf{J}(\mathbf{r}_i, t)\cdot\mathbf{\Delta S}_i$$

$$= \int_\Sigma \mathbf{J}(\mathbf{r}, t)\cdot d\mathbf{S} \qquad (2\text{-}15)$$

[2]The meaning of this proviso is difficult to state for surfaces that may be curved. In essence the requirement might be stated by imagining some small creature walking around on one side of the surface. This creature must *always* look either up away from the surface or down through the surface to find the head of the nearest vector $\mathbf{\Delta S}_i$. If the creature must look sometimes up and sometimes down, the several vectors $\mathbf{\Delta S}_i$ do not point in the "same" direction. Fortunately, difficulties rarely arise on this question.

which, in addition to expressing the total current across the surface, defines formally what is called a *surface integral.* In particular, if the surface over which the integral extends is a plane surface lying in the x-y plane, $d\mathbf{S} = \pm dx\, dy\, \hat{\mathbf{k}}$ and the surface integral becomes an ordinary two-dimensional integral on the variables x, y. If I as given by Eq. (2-15) is positive, the conventional current is flowing across the surface in the direction of the vectors $d\mathbf{S}$; if $I < 0$, the conventional current flow is opposite to the vectors $d\mathbf{S}$.

PROBLEMS

P2-10. Evaluate $\int \mathbf{J}\cdot d\mathbf{S}$ over a *plane* surface of area S if $\mathbf{J}$ is constant.

P2-11. Determine the current crossing the plane surface bounded by a square of side $2a$ positioned with its plane parallel to the x-z plane and its center at the point $(x, y, z) = (0, b, 0)$ if

(a) $\mathbf{J}(\mathbf{r}) = \alpha\mathbf{r}$

(b) $\mathbf{J}(\mathbf{r}) = \beta(xy^2z\hat{\mathbf{i}} + x^2yz\hat{\mathbf{j}} + xyz^2\hat{\mathbf{k}})$

where a, b, α, and β are constants.

We shall conclude this section by obtaining an expression for the current density in terms of the particle density fields and velocity fields introduced in Eqs. (2-1) and (2-2). To this end, we examine first the contribution of particles of type a to the current across a small surface of area $\Delta\mathbf{S}$ placed in an arbitrary (macroscopic) charge distribution. Let $|\Delta\mathbf{S}|$ be small enough so that the surface itself can be entirely contained within a volume ΔV throughout which, at any fixed time t, $n^{(a)}(\mathbf{r}, t)$ and $\mathbf{v}^{(a)}(\mathbf{r}, t)$ can be regarded as constant. Then, at time t, the portion of the charge distribution composed of particles of type a has the appearance shown in Fig. 2-5. Now, let Δt be a time interval sufficiently small that (1) the density and velocity fields do not change appreciably in the interval from t to $t + \Delta t$ and (2) the volume $\Delta V'$ outlined with the broken line, whose slant height is $|\mathbf{v}^{(a)}(\mathbf{r}, t)|\,\Delta t$, lies wholly within ΔV. Under these conditions, the total charge $\Delta Q^{(a)}$ transported by particles of type a across $\Delta\mathbf{S}$ in time Δt is given by q_a times the number of particles in the volume $\Delta V'$, or by

$$\begin{aligned}\Delta Q^{(a)} &= q_a n^{(a)}(\mathbf{r}, t)\,\Delta V' \\ &= q_a n^{(a)}(\mathbf{r}, t)\,|\mathbf{v}^{(a)}(\mathbf{r}, t)|\,|\Delta\mathbf{S}|\cos\theta^{(a)}\,\Delta t \\ &= q_a n^{(a)}(\mathbf{r}, t)\mathbf{v}^{(a)}(\mathbf{r}, t)\cdot\Delta\mathbf{S}\,\Delta t \qquad (2\text{-}16)\end{aligned}$$

where $\theta^{(a)}$ is the angle between $\Delta\mathbf{S}$ and $\mathbf{v}^{(a)}(\mathbf{r}, t)$, and $\Delta V'$ is evaluated as the area of the base, $|\Delta\mathbf{S}|$, times the altitude, $d = |\mathbf{v}^{(a)}(\mathbf{r}, t)|\,\Delta t\cos\theta^{(a)}$. The total charge transported across $\Delta\mathbf{S}$ by charges of all types is now obtained by summing Eq. (2-16) over a, and the total current $\Delta I(t)$ by dividing that result by

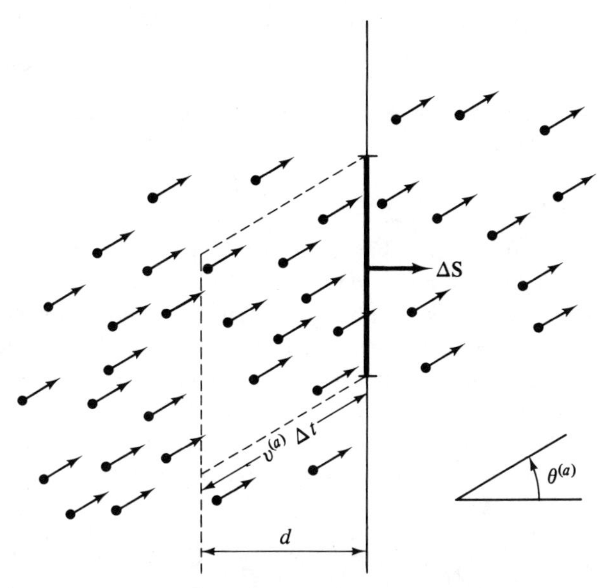

Fig. 2-5. A portion of a system of charged particles. Only particles of type a are shown, and the surface $\mathbf{\Delta S}$ is centered at the point $\mathbf{r}$. All of the particles shown have charge q_a and are moving with velocity $\mathbf{v}^{(a)}(\mathbf{r}, t)$.

Δt; we find that

$$\Delta I(t) = \left(\sum_a q_a n^{(a)}(\mathbf{r}, t)\mathbf{v}^{(a)}(\mathbf{r}, t)\right)\cdot\mathbf{\Delta S} \tag{2-17}$$

from which, by comparison with Eq. (2-14), we find that

$$\mathbf{J}(\mathbf{r}, t) = \sum_a q_a n^{(a)}(\mathbf{r}, t)\mathbf{v}^{(a)}(\mathbf{r}, t) \tag{2-18}$$

The ath term in this sum expresses the contribution of particles of type a to the total current density.

2-3 Mathematical Digression I: Stokes' Theorem and the Divergence Theorem

The surface integral defined in Eq. (2-15) shares with the line integral a very prominent role in the classical theory of fields. Although it has arisen in a context for which its value has a very direct physical connection with something—namely charge—that actually moves across a surface, the surface integral $\int_\Sigma \mathbf{Q}\cdot d\mathbf{S}$ of a (continuous) vector field $\mathbf{Q}(\mathbf{r})$ is itself a well-defined mathematical entity quite apart from the existence of any direct physical interpretation. Even though it rarely represents something that is physically moving across Σ, $\int_{\Sigma} \mathbf{Q} \cdot d\mathbf{S}$ is called the *flux* of $\mathbf{Q}$ across Σ. This integral has a number of mathematical properties that we digress to develop because of their subsequent utility.

STOKES' THEOREM

Let us first derive an identity that relates the line integral of a (suitably continuous) vector field $\mathbf{Q}$ about a *closed* path to a particular surface integral. Consider, for example, $\oint \mathbf{Q} \cdot d\boldsymbol{\ell}$ about the path $\boldsymbol{\Gamma}$ shown by the heavy line in Fig. 2-6. As illustrated, any (open) surface Σ bounded by this path can be divided into small elements. Further, the *sum* of the line integrals about each element is approximately the line integral about the original perimeter, because each *internal* line occurs twice in that sum, traversed once in each direction. The approximation improves as the size of all elements is reduced and we conclude that

$$\oint_{\Gamma} \mathbf{Q} \cdot d\boldsymbol{\ell} = \lim_{|\Delta \mathbf{S}_i| \to 0} \sum_i \oint_i \mathbf{Q} \cdot d\boldsymbol{\ell} \tag{2-19}$$

where the integral on the right is taken around the perimeter of the ith segment. Once the elements are small enough, however, we can use Eq. (0-61) to evaluate each integral under the sum in Eq. (2-19); we find that

$$\oint_{\Gamma} \mathbf{Q} \cdot d\boldsymbol{\ell} = \lim_{|\Delta \mathbf{S}_i| \to 0} \sum_i (\boldsymbol{\nabla} \times \mathbf{Q})^i \cdot \Delta \mathbf{S}_i \tag{2-20}$$

where $(\boldsymbol{\nabla} \times \mathbf{Q})^i$ is the value of $\boldsymbol{\nabla} \times \mathbf{Q}$ at some point on the ith surface element and $\Delta \mathbf{S}_i$ is the *vector* representing that surface element. The direction

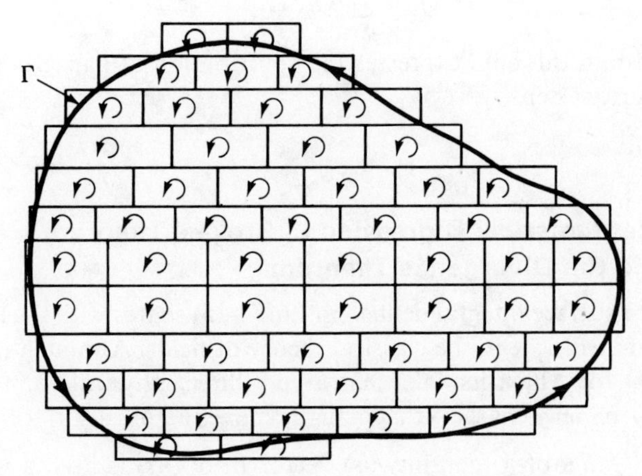

Fig. 2-6. Division of a large area into small segments. The line integral about the perimeter of the area is approximately the sum of the line integrals about each separate segment.

of $\Delta \mathbf{S}_i$ must, of course, be chosen to be consistent with the condition imposed on Eq. (0-61); that is, $\Delta \mathbf{S}_i$ must be directed *as the thumb of the right hand when the fingers point in the direction of* $d\boldsymbol{\ell}$ *and the palm faces the area* $\Delta \mathbf{S}_i$. The right-hand side of Eq. (2-20) now exactly defines the surface integral over the surface Σ and we have what is called *Stokes' theorem*,

$$\oint_\Gamma \mathbf{Q}\cdot d\boldsymbol{\ell} = \int_\Sigma (\nabla \times \mathbf{Q})\cdot d\mathbf{S} \tag{2-21}$$

Thus, the line integral of a (suitably continuous) vector field about a *closed* path is equal to the integral of the curl of the field over *any open* surface bounded by the path, *provided the direction assigned to the surface vectors is related to the direction of traverse of the path by the right-hand rule described above.* Although we have derived Eq. (2-21) assuming that $\mathbf{Q}$ depends only on $\mathbf{r}$, the theorem also applies if $\mathbf{Q}$ depends on other variables (e.g., t) provided these other variables can be treated as constants within the context of the theorem itself.

THE DIVERGENCE THEOREM

To obtain a starting point for deriving another identity involving a surface integral, let us evaluate the surface integral of a (suitably continuous) vector field $\mathbf{Q}$ over the *closed* surface that bounds a small rectangular parallelopiped of sides Δx, Δy, Δz. Let the lower back corner of the surface be at the point (x, y, z). Figure 2-7 illustrates the geometry. The surface integral consists of six contributions, one from each of the six faces. Denoting an integral over a closed surface by a circle superimposed on the integral sign, taking $d\mathbf{S}$ every-

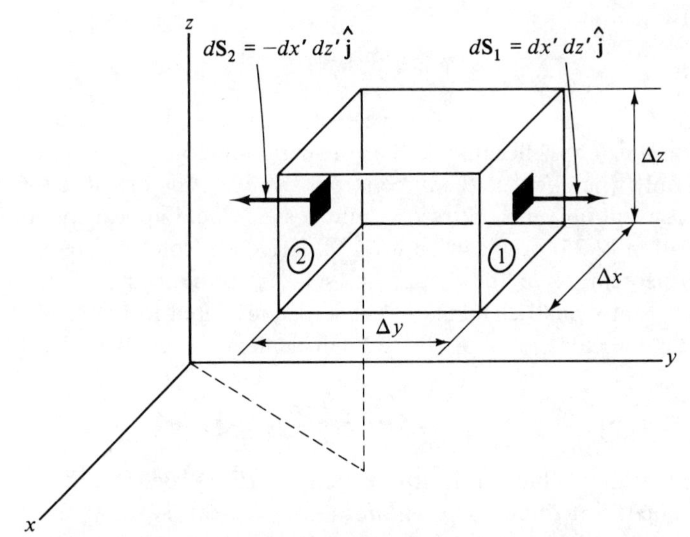

Fig. 2-7. A small rectangular parallelopiped whose surface is used in deriving the divergence theorem.

where in the direction of the *outward* normal, and writing explicitly only the contributions from the two faces parallel to the x-z plane, we find that

$$\begin{aligned}\oint \mathbf{Q}\cdot d\mathbf{S} &= \int_{①} \mathbf{Q}(x', y+\Delta y, z')\cdot dx'\,dz'\hat{\mathbf{j}} \\ &\quad + \int_{②} \mathbf{Q}(x', y, z')\cdot(-dx'\,dz'\hat{\mathbf{j}}) + \cdots \\ &= \int_x^{x+\Delta x} dx' \int_z^{z+\Delta z} dz'[Q_y(x', y+\Delta y, z') - Q_y(x', y, z')] + \cdots \\ &\approx \Delta y \int_x^{x+\Delta x} dx' \int_z^{z+\Delta z} dz' \frac{\partial Q_y(x', y, z')}{\partial y} + \cdots \\ &\approx \Delta x\,\Delta y\,\Delta z \frac{\partial Q_y}{\partial y} + \cdots \end{aligned} \tag{2-22}$$

where evaluation of Q_y at argument (x, y, z) has been suppressed in the final form. The remaining terms in this approximation combine to yield the total result

$$\oint \mathbf{Q}\cdot d\mathbf{S} \approx \left(\frac{\partial Q_x}{\partial x} + \frac{\partial Q_y}{\partial y} + \frac{\partial Q_z}{\partial z}\right) \Delta x\,\Delta y\,\Delta z \tag{2-23}$$

for the integral over the rectangular parallelopiped. Let us now define the *divergence* of the vector field $\mathbf{Q}$ by

$$\nabla\cdot\mathbf{Q} = \frac{\partial Q_x}{\partial x} + \frac{\partial Q_y}{\partial y} + \frac{\partial Q_z}{\partial z} \tag{2-24}$$

where the notation is suggested by a formal evaluation of $\nabla\cdot\mathbf{Q}$ when ∇ is replaced by the expression in Eq. (0-46). We then have from Eq. (2-23) the relationship

$$\oint \mathbf{Q}\cdot d\mathbf{S} \approx \nabla\cdot\mathbf{Q}\,\Delta V \tag{2-25}$$

where $\Delta V = \Delta x\,\Delta y\,\Delta z$ is the volume enclosed by the small surface over which the integral extends. Because of the convention adopted above, Eq. (2-25) is correct only when $d\mathbf{S}$ has the direction of the *outward* normal to the surface. [If the convention were reversed, a minus sign would appear in Eq. (2-25).]

Equation (2-25) now can be used to derive a second theorem similar to Stokes' theorem. Note first that any closed surface bounds a volume that can be divided into small elements of the sort considered in the previous paragraph. Let the index i number these small elements. Then, from Eq. (2-25) we find that

$$\sum_i \oint_{i\text{th element}} \mathbf{Q}\cdot d\mathbf{S} \approx \sum_i (\nabla\cdot\mathbf{Q})^i\,\Delta V_i \tag{2-26}$$

where ΔV_i is the volume of the ith element and $(\nabla\cdot\mathbf{Q})^i$ denotes the divergence of $\mathbf{Q}$ evaluated at some point within the ith element. Now, except for those surfaces that coincide with the surface of the finite volume, every surface element on the left in Eq. (2-26) occurs twice; these "inner" surface elements bound two different volume elements. Since the outward normal to each such surface element is oppositely directed for the two occurrences, these inner surface elements make no net contribution to the left-hand side of Eq. (2-26) and the sum reduces to an integral over the surface bounding the finite volume. In the limit as all $\Delta V_i \longrightarrow 0$, the right-hand side of Eq. (2-26) becomes a volume integral and the equation itself becomes exact; we have derived what is called the *divergence theorem*,

$$\oint \mathbf{Q}\cdot d\mathbf{S} = \int \nabla\cdot\mathbf{Q}\, dv \tag{2-27}$$

In words, the flux of a (suitably continuous) vector field $\mathbf{Q}$ *out of* a *closed* surface is equal to the integral of the divergence of the field over the volume bounded by the surface, *provided the surface vectors are all assigned the direction of the outward normal to the surface.* As with Stokes' theorem, the divergence theorem can also be applied to fields that depend on more variables than the three spatial coordinates by regarding the other variables to be constants insofar as the theorem itself is concerned.

Equation (2-25) also provides a coordinate-free definition of the divergence of a vector field, namely,

$$\nabla\cdot\mathbf{Q} = \lim_{\Delta V\to 0} \frac{1}{\Delta V} \oint \mathbf{Q}\cdot d\mathbf{S} \tag{2-28}$$

Equation (2-28) certainly reduces to Eq. (2-24) in Cartesian coordinates; it takes very little labor to verify that Eq. (2-28) leads to

$$\nabla\cdot\mathbf{Q} = \frac{1}{\imath}\frac{\partial(\imath Q_\imath)}{\partial \imath} + \frac{1}{\imath}\frac{\partial Q_\phi}{\partial \phi} + \frac{\partial Q_z}{\partial z} \tag{2-29}$$

in cylindrical coordinates and to

$$\nabla\cdot\mathbf{Q} = \frac{1}{r^2}\frac{\partial(r^2 Q_r)}{\partial r} + \frac{1}{r\sin\theta}\frac{\partial(\sin\theta\, Q_\theta)}{\partial\theta} + \frac{1}{r\sin\theta}\frac{\partial Q_\phi}{\partial\phi} \tag{2-30}$$

in spherical coordinates.

A geometric interpretation for $\nabla\cdot\mathbf{Q}$ can be inferred if we recognize that $\oint \mathbf{Q}\cdot d\mathbf{S}$ in effect counts the (net) number of field lines of $\mathbf{Q}$ that cross the closed surface. (Remember that, by convention, the number of field lines crossing a unit surface perpendicular to the field is proportional to the magnitude of the field.) In this reckoning, lines passing from inside to outside are regarded as positive; those going from outside to inside are negative. Thus, when $\oint \mathbf{Q}\cdot d\mathbf{S}$ over some surface differs from zero, field lines in balance either emerge from or terminate within the volume bounded by the surface. Arguing from Eq. (2-28), we conclude that a point at which $\nabla\cdot\mathbf{Q} \neq 0$ is a point at which at least some of the field lines of $\mathbf{Q}$ either start or stop. In reverse, if $\nabla\cdot\mathbf{Q} = 0$ at some point, field lines neither begin nor terminate at that point, and, even more generally, if $\nabla\cdot\mathbf{Q} = 0$ everywhere, the field lines of $\mathbf{Q}$ have neither beginning nor ending anywhere and can only close on themselves. The velocity field of points on a rotating phonograph record [Fig. 0-10(b)], for which $\mathbf{Q} = \omega r\hat{\boldsymbol{\phi}}$ and $\nabla \cdot \mathbf{Q} = 0$, is a simple example of this latter type.

A USEFUL INTEGRAL (SOLID ANGLE)

To illustrate a typical use of the divergence theorem, we now evaluate an integral that will be important to our development of Gauss's law. Let $\mathbf{r}_0$ be some point in space and Σ be a closed surface (Fig. 2-8). Consider the integral

$$G = \oint_{\Sigma} \frac{\mathbf{r} - \mathbf{r}_0}{|\mathbf{r} - \mathbf{r}_0|^3} \cdot d\mathbf{S} \tag{2-31}$$

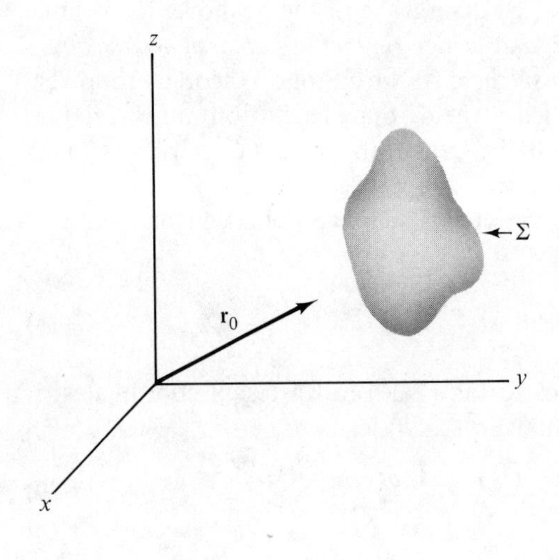

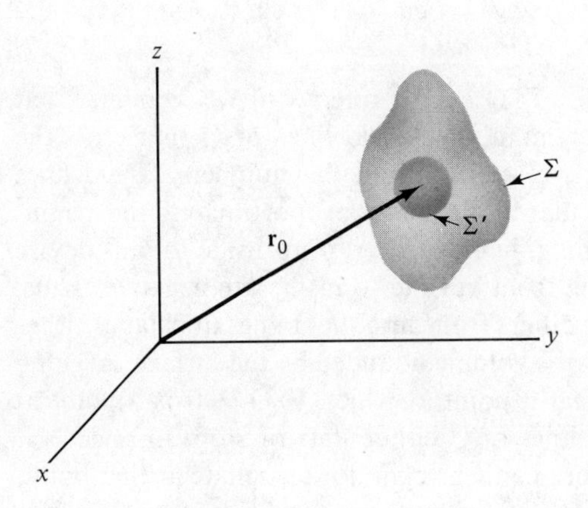

Fig. 2-8. A point $\mathbf{r}_0$ lying (a) outside of and (b) inside of a closed surface Σ.

where the integration variables are the components of $\mathbf{r}$ suitably constrained so that $\mathbf{r}$ always lies on Σ. Using the divergence theorem we can rewrite Eq. (2-31) as the volume integral

$$G = \int_V \nabla \cdot \frac{\mathbf{r} - \mathbf{r}_0}{|\mathbf{r} - \mathbf{r}_0|^3}\, dv \tag{2-32}$$

where V is the volume bounded by Σ and ∇ involves derivatives with respect to the components of $\mathbf{r}$. Direct evaluation shows that the integrand in Eq. (2-32) is zero except when $\mathbf{r} = \mathbf{r}_0$. Thus, if the point $\mathbf{r}_0$ lies *outside* the volume V, Eq. (2-32) gives $G = 0$. When $\mathbf{r}_0$ lies *inside* V, however, the evaluation of G is a bit more involved. We begin by dividing V into two regions by inserting a small sphere centered on $\mathbf{r}_0$ and denoted by Σ' [Fig. 2-8(b)]. Now, let V'' be the volume between Σ and Σ'. The point $\mathbf{r}_0$ lies outside V'' and, consequently, we have that

$$\int_{V''} \nabla \cdot \frac{\mathbf{r} - \mathbf{r}_0}{|\mathbf{r} - \mathbf{r}_0|^3}\, dv = 0 \tag{2-33}$$

We now use the divergence theorem to express Eq. (2-33) as a surface integral, the surface involved having two parts; we find that

$$\int_\Sigma \frac{\mathbf{r} - \mathbf{r}_0}{|\mathbf{r} - \mathbf{r}_0|^3} \cdot d\mathbf{S} + \int_{\Sigma'} \frac{\mathbf{r} - \mathbf{r}_0}{|\mathbf{r} - \mathbf{r}_0|^3} \cdot d\mathbf{S} = 0 \tag{2-34}$$

In both integrals $d\mathbf{S}$ stands for an *outward* normal, *where the direction outward is reckoned from an observation point within V''*. Thus, $d\mathbf{S}$ on Σ' in fact points *inward* toward $\mathbf{r}_0$. Since the first integral in Eq. (2-34) is G, we have that

$$G = -\int_{\Sigma'} \frac{\mathbf{r} - \mathbf{r}_0}{|\mathbf{r} - \mathbf{r}_0|^3} \cdot d\mathbf{S} \tag{2-35}$$

The integral to which we have reduced G, however, can now be quickly evaluated. Let R be the radius of the sphere Σ'. Then $|\mathbf{r} - \mathbf{r}_0|^3 = R^3$ at all points on the sphere and $(\mathbf{r} - \mathbf{r}_0)\cdot d\mathbf{S} = -R\,|d\mathbf{S}|$. (Remember that $d\mathbf{S}$ points *toward* $\mathbf{r}_0$ on Σ'.) Thus Eq. (2-35) reduces to

$$G = \frac{1}{R^2} \oint_{\Sigma'} |d\mathbf{S}| = 4\pi \tag{2-36}$$

since $\int |d\mathbf{S}|$ is merely the surface area of the sphere ($4\pi R^2$). Combining our two results, we have finally that

$$G = \oint_\Sigma \frac{\mathbf{r} - \mathbf{r}_0}{|\mathbf{r} - \mathbf{r}_0|^3} \cdot d\mathbf{S} = \begin{cases} 4\pi, & \mathbf{r}_0 \text{ inside } \Sigma \\ 0, & \mathbf{r}_0 \text{ outside } \Sigma \end{cases} \tag{2-37}$$

The integral in Eq. (2-31) is directly related to the geometric concept of *solid angle*, which is the three-dimensional analog of the familiar angle in the plane and measures the three-dimensional "opening" at the vertex of a general cone. The contribution $d\Omega$ of a small element $d\mathbf{S}$ of a broad surface Σ to the

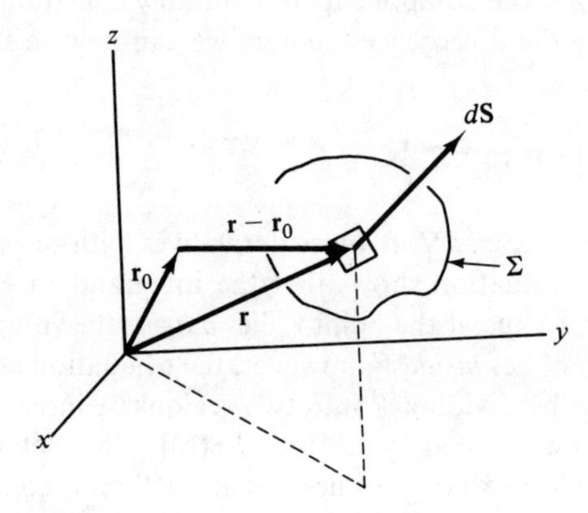

Fig. 2-9. Geometry for defining solid angle.

solid angle subtended by Σ from an observation point $\mathbf{r}_0$ is defined to be the quotient of the area of $d\mathbf{S}$ projected onto a plane normal to the line of sight and the square of the distance of the area from the observation point; that is,

$$d\Omega = \left(\frac{\mathbf{r} - \mathbf{r}_0}{|\mathbf{r} - \mathbf{r}_0|} \cdot d\mathbf{S}\right) \div |\mathbf{r} - \mathbf{r}_0|^2 \tag{2-38}$$

where the element $d\mathbf{S}$ is located at $\mathbf{r}$ and $\mathbf{r} - \mathbf{r}_0$ is a vector from the observation point to $d\mathbf{S}$ (Fig. 2-9). The solid angle subtended by the entire surface Σ is then given by

$$\Omega = \int_\Sigma d\Omega = \int_\Sigma \frac{\mathbf{r} - \mathbf{r}_0}{|\mathbf{r} - \mathbf{r}_0|^3} \cdot d\mathbf{S} \tag{2-39}$$

which in particular coincides with Eq. (2-31) if Σ is a closed surface. Interpreted geometrically, Eq. (2-37) states that the solid angle subtended by a *closed* surface Σ from an observation point inside (outside) Σ is 4π (0). (Visualization of this general statement may be aided by composing the analogous statement involving ordinary angles and closed curves in a plane. See P2-21.) Although solid angle is a dimensionless concept, it is common to quote solid angles with "units" of *steradians.*

PROBLEMS

P2-12. Familiarize yourself with identities (C-10)–(C-14) in Appendix C and prove (C-12). *Hint:* The notation of P0-28 is convenient, but not essential.

P2-13. Familiarize yourself with identities (C-20)–(C-25) in Appendix C and prove (C-20). *Hint:* See P0-31.

P2-14. Given that a force field $\mathbf{F}$ is conservative if $\oint \mathbf{F} \cdot d\boldsymbol{\ell}$ is zero for an *arbitrary* path, use Stokes' theorem to show that a force field is conservative if $\nabla \times \mathbf{F} = 0$.

P2-15. Derive another of the terms in Eq. (2-22) by applying the method illustrated in the text to the two faces in Fig. 2-7 that are parallel to the x-y plane.

P2-16. Show that formal evaluation of the dot product of the operator ∇ with a vector $\mathbf{Q}$ in Cartesian coordinates yields Eq. (2-24). *Warning: One must use great care in evaluating $\nabla \cdot \mathbf{Q}$ in other coordinate systems by this means. Only the Cartesian unit vectors are constants.*

P2-17. Using the coordinate-free definition in Eq. (2-28), derive the expression for $\nabla \cdot \mathbf{Q}$ in cylindrical coordinates.

P2-18. Use the divergence theorem to show that $\oint_\Sigma \mathbf{r} \cdot d\mathbf{S} = 3V$, where the integration variables are the components of the position vector $\mathbf{r}$ and V is the volume enclosed by the surface Σ.

2-4 The Equation of Continuity

During the conduct of electrostatic experiments, it is observed empirically that, whenever a neutral object is charged by rubbing, equal amounts of positive and negative charge appear. The net charge in the universe appears to be conserved, that is,

$$\frac{d}{dt} \int_{\text{entire universe}} \rho(\mathbf{r}, t)\, dv = 0 \tag{2-40}$$

where $\rho(\mathbf{r}, t)$ is the charge density at point $\mathbf{r}$ at time t. This expression of charge conservation, however, is less useful than one that applies to a more local region of space. An alternative expression is obtained by inquiring about the total charge in a closed (but not necessarily isolated) region of space. The charge $Q(t)$ at time t in a volume V enclosed by a surface Σ is given by

$$Q(t) = \int_V \rho(\mathbf{r}, t)\, dv \tag{2-41}$$

If we assume that there are no sources or sinks of charge within V, then the only way that the charge within V can change is by transport of charge across the bounding surface Σ. With $d\mathbf{S}$ being the conventional *outward* normal, the current flowing *into* V across Σ is given by

$$I_{\text{in}}(t) = -\oint_\Sigma \mathbf{J}(\mathbf{r}, t) \cdot d\mathbf{S} \tag{2-42}$$

where the minus sign appears because of the convention on the direction of $d\mathbf{S}$. [Compare Eq. (2-15).] Because $I_{\text{in}}(t)$ is also the rate at which the charge inside V is changing, we have that $I_{\text{in}} = dQ/dt$, or

$$-\oint_\Sigma \mathbf{J}(\mathbf{r}, t) \cdot d\mathbf{S} = \frac{d}{dt} \int_V \rho(\mathbf{r}, t)\, dv \tag{2-43}$$

Since $\int_V \rho(\mathbf{r}, t)\, dv$ is dependent only on t and not on $\mathbf{r}$, we can write $\partial/\partial t$ instead of d/dt in Eq. (2-43). Then, since the volume V is fixed in space, we can take the partial time derivative under the integral sign. With some further rearrangement of the resulting terms and with suppression of the explicit arguments $\mathbf{r}$ and t, we then find the *equation of continuity*,

$$\oint_\Sigma \mathbf{J} \cdot d\mathbf{S} + \int_V \frac{\partial \rho}{\partial t}\, dv = 0 \tag{2-44}$$

This equation expresses charge conservation as applied to nonisolated regions of space. In effect, it states that any current flowing into some closed volume must increase the charge in that volume by precisely the amount transported into the volume.

An equivalent and extremely useful alternative form of the equation of continuity may be obtained by applying Eq. (2-44) specifically to a *small* volume surrounding the point $\mathbf{r}$. We can then rewrite the first integral in Eq. (2-44) by using Eq. (2-25) and we can express the second integral more simply by assuming $\partial\rho/\partial t$ to be (approximately) constant throughout V—now more appropriately denoted by ΔV; Eq. (2-44) then becomes

$$\nabla \cdot \mathbf{J}\, \Delta V + \frac{\partial \rho}{\partial t}\, \Delta V \approx 0 \tag{2-45}$$

We now divide by ΔV and then allow ΔV to become arbitrarily small; the result is the *differential form* of the equation of continuity,

$$\nabla \cdot \mathbf{J} + \frac{\partial \rho}{\partial t} = 0 \tag{2-46}$$

In particular, if the currents are *steady* so charge does not accumulate anywhere (ρ independent of t), then

$$\nabla \cdot \mathbf{J} = 0 \quad \text{(steady currents)} \tag{2-47}$$

We shall need these results occasionally in the next three chapters but we postpone detailed exploration of the differential forms of the basic equations until Chapter 6 and later chapters.

PROBLEM

P2-19. The current density in a region of space is given by

$$\mathbf{J}(\mathbf{r}) = \alpha[\mathbf{r} \cdot \mathbf{r} - (\mathbf{r} \cdot \hat{\mathbf{k}})^2] e^{-\beta \mathbf{r} \cdot \hat{\mathbf{k}}} \hat{\mathbf{k}}$$

where α and β are constants. Imagine a volume bounded by a cylindrical surface having its axis coincident with the z axis and its lower face in the x-y plane. Let the cylinder have radius a and altitude b. If at time zero there is no net charge within this volume, determine how much charge it contains at a later time t.

2-5 Mathematical Digression II: Several Operators Involving ∇

The differential operator ∇ occurs in a variety of important contexts beyond the three first-order derivatives—∇S, $\nabla \times \mathbf{Q}$, and $\nabla \cdot \mathbf{Q}$—that have already been introduced. One additional first-order derivative is the operator $\mathbf{Q}\cdot\nabla$, where $\mathbf{Q}$ may be a constant vector or a spatially and temporally dependent vector field. This operator is a scalar operator, and it may act either on a scalar field S, giving $\mathbf{Q}\cdot\nabla S$ where ∇S is the ordinary gradient, or on another vector field $\mathbf{R}$, giving $(\mathbf{Q}\cdot\nabla)\mathbf{R}$, which is a vector quantity whose *Cartesian* components are $\mathbf{Q}\cdot\nabla R_x$, $\mathbf{Q}\cdot\nabla R_y$, and $\mathbf{Q}\cdot\nabla R_z$ but whose components in other coordinate systems are much less simple because of the spatial dependence of the unit vectors in non-Cartesian coordinates. Acting on a scalar, this operator usually assumes one of the forms

$$\mathbf{Q}\cdot\nabla = Q_x\frac{\partial}{\partial x} + Q_y\frac{\partial}{\partial y} + Q_z\frac{\partial}{\partial z} \tag{2-48}$$

$$= Q_\varrho\frac{\partial}{\partial \varrho} + \frac{Q_\phi}{\varrho}\frac{\partial}{\partial \phi} + Q_z\frac{\partial}{\partial z} \tag{2-49}$$

$$= Q_r\frac{\partial}{\partial r} + \frac{Q_\theta}{r}\frac{\partial}{\partial \theta} + \frac{Q_\phi}{r\sin\theta}\frac{\partial}{\partial \theta} \tag{2-50}$$

Note the property $(\mathbf{Q}\cdot\nabla)\mathbf{r} = \mathbf{Q}$ for any vector field $\mathbf{Q}$.

The three quantities $\nabla\cdot\mathbf{Q}$, ∇S, and $\nabla\times\mathbf{Q}$ can be the operand of the operator ∇ in at least five different ways. The first quantity is a scalar and has a gradient; the second and third quantities are vectors and have both a divergence and a curl. We shall comment on each of these second-order derivatives:

(1) $\nabla(\nabla\cdot\mathbf{Q})$ occurs in some vector identities but (so far as the author knows) has no special significance.

(2) $\nabla\cdot\nabla S$, commonly symbolized by $\nabla^2 S$, is called the *Laplacian* of S and plays a very important role in potential theory. Expressions for the operator ∇^2 in the three common coordinate systems can be obtained by direct substitution of the components of the gradient [Eqs. (0-47), (0-56), and (0-57)] into the expressions for the divergence of a vector [Eqs. (2-24), (2-29), and (2-30)]; we find

$$\nabla^2 = \frac{\partial^2}{\partial x^2} + \frac{\partial^2}{\partial y^2} + \frac{\partial^2}{\partial z^2} \tag{2-51}$$

$$= \frac{1}{\varrho}\frac{\partial}{\partial \varrho}\left(\varrho\frac{\partial}{\partial \varrho}\right) + \frac{1}{\varrho^2}\frac{\partial^2}{\partial \phi^2} + \frac{\partial^2}{\partial z^2} \tag{2-52}$$

$$= \frac{1}{r^2}\frac{\partial}{\partial r}\left(r^2\frac{\partial}{\partial r}\right) + \frac{1}{r^2\sin\theta}\frac{\partial}{\partial \theta}\left(\sin\theta\frac{\partial}{\partial \theta}\right) + \frac{1}{r^2\sin^2\theta}\frac{\partial^2}{\partial \phi^2} \tag{2-53}$$

Occasionally, the Laplacian $\nabla^2\mathbf{Q}$ of a vector arises; in Cartesian coordinates the components of this vector are $\nabla^2 Q_x$, $\nabla^2 Q_y$, and $\nabla^2 Q_z$, but expressions for its components in other coordinate systems are much less simple because of the spatial dependence of the unit vectors in non-Cartesian coordinates.

(3) $\nabla \times (\nabla S)$ is *always* zero if S is well behaved. This property is directly related to two properties of a conservative force field $\mathbf{F}_c$: $\nabla \times \mathbf{F}_c = 0$ and $\mathbf{F}_c = -\nabla U$. (See Section 0-3.) Either property, in fact, implies the other and our earlier discussion in effect proved the theorem: The curl of a vector field $\mathbf{Q}$ is zero if and only if that field is the gradient of an associated scalar field S. That is, $\nabla \times \mathbf{Q} = 0 \Leftrightarrow \mathbf{Q} = \nabla S$. This theorem, of course, merely states the existence of S; it does not provide a means to find S if $\mathbf{Q}$ is given.

(4) $\nabla \cdot (\nabla \times \mathbf{Q})$ is also *always* zero if $\mathbf{Q}$ is well behaved. This property suggests, but does not prove all aspects of, the theorem: The divergence of a vector field $\mathbf{R}$ is zero if and only if that field is the curl of another vector field $\mathbf{Q}$. That is, $\nabla \cdot \mathbf{R} = 0 \Leftrightarrow \mathbf{R} = \nabla \times \mathbf{Q}$. As with the analogous theorem in (3), this theorem merely states the existence of $\mathbf{Q}$; it does not provide a means to find $\mathbf{Q}$ if $\mathbf{R}$ is given.

(5) $\nabla \times (\nabla \times \mathbf{Q})$ occurs frequently and is usually simplified to the equivalent form $\nabla(\nabla \cdot \mathbf{Q}) - \nabla^2\mathbf{Q}$, interpretation of the term $\nabla^2\mathbf{Q}$ requiring special care in non-Cartesian coordinates.

PROBLEM

P2-20. Familiarize yourself with identities (C-15)–(C-19), (C-26), and (C-27) in Appendix C, and prove (C-18) and (C-19).

Supplementary Problems

P2-21. Let Γ be some path lying wholly in the x-y plane and let $\mathbf{r}_0$ be a point also in the x-y plane but not on Γ. Show that the (ordinary) angle θ subtended by Γ from the point $\mathbf{r}_0$ is given by

$$\theta = \hat{\mathbf{k}} \cdot \int_\Gamma \frac{\mathbf{r} - \mathbf{r}_0}{|\mathbf{r} - \mathbf{r}_0|^2} \times d\boldsymbol{\ell}$$

where the integration extends over points $\mathbf{r}$ on Γ, and show in particular that, if Γ is a *closed* path, θ is either 2π or 0 depending on whether $\mathbf{r}_0$ lies inside or outside Γ.

P2-22. Referring to the general orthogonal coordinate system described in P0-32, use the definition in Eq. (2-28) to show that

$$\nabla \cdot \mathbf{Q} = \frac{1}{h_1 h_2 h_3} \sum_i \frac{\partial}{\partial q_i}\left(\frac{h_1 h_2 h_3}{h_i} Q_i\right)$$

and combine this result with the expression in P0-32 for ∇S to show that

$$\nabla^2 = \frac{1}{h_1 h_2 h_3} \sum_i \frac{\partial}{\partial q_i}\left(\frac{h_1 h_2 h_3}{h_i^2} \frac{\partial}{\partial q_i}\right)$$

Finally, show that these results reduce to those given in the text for $\nabla \cdot \mathbf{Q}$ and ∇^2 in cylindrical and in spherical coordinates.

P2-23. Determine the form of the equation of continuity first in Cookian units (P1-8) and then in modified Gaussian units (P1-7).

P2-24. Let $\psi(\mathbf{r}, t)$ be the wave function describing a quantum mechanical particle of mass m. The probability for finding this particle in a volume V is given by $\int_V \psi^* \psi \, dv$. Given that ψ satisfies the Schrödinger equation

$$i\hbar \frac{\partial \psi}{\partial t} = -\frac{\hbar^2}{2m} \nabla^2 \psi + U\psi$$

where $\hbar$ is Planck's constant divided by 2π and U is the potential energy of the particle, show that

$$\frac{\partial}{\partial t} \int_V \psi^* \psi \, dv = -\oint_\Sigma \mathbf{J} \cdot d\mathbf{S}$$

where Σ bounds V, and find an explicit expression for $\mathbf{J}$. Evidently, quantum mechanical probabilities satisfy a continuity equation if the *probability current* $\mathbf{J}$ is properly identified. *Hint:* Use Green's theorem, Eq. (C-27).

P2-25. Currents flowing in a surface can be described by a current density of a slightly different sort than the one introduced in the text. Suppose that a current is flowing in the plane of the paper as indicated in Fig. P2-25. Then

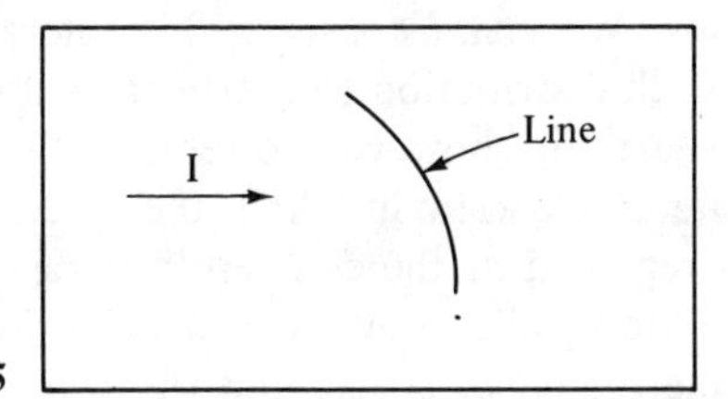

Figure P2-25

charge is transported across any line in this surface. The appropriate current density at any point in the surface is defined to have a direction determined by the direction in which positive charge is being transported at that point and a magnitude given by the rate at which charge is transported across a line of *unit* length lying in the surface and oriented perpendicular to the direction of the current density. (a) What are the dimensions of this current density? (b) Show that the current flowing across a line in the surface is given by

$$I = \int \mathbf{j} \cdot d\boldsymbol{\ell} \times \hat{\mathbf{n}}$$

where $\mathbf{j}$ is the surface current density, $d\boldsymbol{\ell}$ is an element of the line, $\hat{\mathbf{n}}$ is a unit vector normal to the surface, and the positive direction of current flow across the line is determined by the direction of $d\boldsymbol{\ell} \times \hat{\mathbf{n}}$. (c) Derive a continuity equation applicable to currents in a surface.

3

The Electromagnetic Field: Its Definition and Its Effect on General Charge Distributions

We shall next examine the interaction between two general charge distributions. We *could* consider this interaction as an indivisible whole, regarding each distribution to exert forces directly on the other. It is significantly more fruitful, however, to replace this concept of action at a distance with an alternative view in which the force exerted by one distribution on another is regarded as the end result of *two* successive effects. In brief, the first distribution, called the *source* distribution, produces an *electromagnetic field* in some region of space, and the second distribution then experiences forces by virtue of its interaction with this field rather than by virtue of a direct interaction with the first distribution. In effect, the field communicates forces of interaction from one distribution to another. To make this conceptual division of the interaction physically useful, however, we need detailed quantitative answers to the following questions:

(1) How is the electromagnetic field defined?
(2) What forces and torques are experienced by an arbitrary distribution placed in a *given* electromagnetic field?
(3) What electromagnetic field is established by a *given* source distribution?

Questions (1) and (2) are considered in this chapter; question (3) is treated in Chapters 4, 5, and 6.

3-1
Forces on Point Charges: A Definition of the Electromagnetic Field

The electromagnetic field at point **r** and time t is defined most directly by relating it to the force on a (point) *test charge* that happens at time t to be at point **r**. Since we are concerned only with the part of the total force that depends specifically on the charge of the test particle, such forces as the gravitational force, if present at all, must be subtracted from the total force in order to isolate the electromagnetic force. Even the electromagnetic force turns out experimentally to have two parts. The *electric force* is independent of the velocity of the particle and, in particular, is experienced by test charges at rest; the *magnetic force* depends on the velocity of the particle and is experienced only by test charges in motion. Experimentally, only the position and velocity of the test charge are important; accelerated particles experience no forces beyond those on particles moving with constant velocity.[1] Thus, the division of the electromagnetic force into velocity-independent and velocity-dependent parts is exhaustive and we write

$$\mathbf{F}_q(\mathbf{v}, \mathbf{r}, t) = \mathbf{F}_q^{\text{elec}}(\mathbf{r}, t) + \mathbf{F}_q^{\text{mag}}(\mathbf{v}, \mathbf{r}, t) \tag{3-1}$$

where **v** and **r** are the velocity and position at time t of a test particle with charge q, $\mathbf{F}_q$ is the electromagnetic force, $\mathbf{F}_q^{\text{elec}}$ is the electric part, and $\mathbf{F}_q^{\text{mag}}$ is the magnetic part. The objective in the remainder of this section is to introduce position- and time-dependent (but *not* velocity-dependent) fields from which each part of the electromagnetic force can be determined.

The *electric field* at the point **r**, t in space-time is measured (and hence defined) by the following operations: Place a test charge of strength q *at rest* at the point **r** at time t, and measure the force $\mathbf{F}_q^{\text{elec}}(\mathbf{r}, t)$ on this charge. The electric field $\mathbf{E}(\mathbf{r}, t)$ is then defined by the ratio $\mathbf{F}_q^{\text{elec}}(\mathbf{r}, t)/q$. Equivalently,[2] we set

$$\mathbf{F}_q^{\text{elec}}(\mathbf{r}, t) = q\mathbf{E}(\mathbf{r}, t) \tag{3-2}$$

Strictly, of course, Eq. (3-2) defines the electric field at **r**, t *when the test charge is present.* Since in general the test charge gives rise to new forces on whatever source distribution is establishing the field, its presence *may* cause a redistribution of the source charges so that the field with which the test charge interacts *may* differ from the field present at **r** before the test charge was introduced. To minimize the modifying influence of the test charge on the field,

[1] We ignore the very small forces resulting from the reaction of an accelerated particle to its own electromagnetic radiation. (See Section 14-6.)

[2] In the spirit of Chapter 1 we might relate **E** to $\mathbf{F}_q^{\text{elec}}$ by setting $\mathbf{F}_q^{\text{elec}} = k_4 q\mathbf{E}$, where k_4 assumes an arbitrary value if this expression *defines* **E** and an empirically determined value if **E** is defined independently of this expression. Further, k_4 may or may not have dimensions. All systems of units of which the author is aware *define* **E** by setting k_4 equal to the dimensionless constant unity, i.e., by Eq. (3-2).

we make the test charge small. Indeed, all uncertainty in the definition of $\mathbf{E}(\mathbf{r}, t)$ is removed if we allow q to become arbitrarily small, defining $\mathbf{E}(\mathbf{r}, t)$ by

$$\mathbf{E}(\mathbf{r}, t) = \lim_{q \to 0} \frac{\mathbf{F}_q^{\text{elec}}(\mathbf{r}, t)}{q} \tag{3-3}$$

If Eq. (3-3) is adopted, then Eq. (3-2) still gives the force on a small test charge placed at $\mathbf{r}$, t provided the influence of the test charge on the source can be neglected. By either equation, the electric field is a vector quantity whose direction at the point $\mathbf{r}$ is the same as the direction of the force on a positive charge placed at that point. The dimensions of $\mathbf{E}$ are those of force per charge and, in mks units, $\mathbf{E}$ is expressed in newtons per coulomb (N/C).

The magnetic force $\mathbf{F}_q^{\text{mag}}(\mathbf{v}, \mathbf{r}, t)$, which is experienced only by a *moving* test charge, is attributed to the interaction of the charge with a second field $\mathbf{B}(\mathbf{r}, t)$, called the *magnetic induction field* (or more briefly the *magnetic induction*). To guide our definition of $\mathbf{B}(\mathbf{r}, t)$, we explore the properties of $\mathbf{F}_q^{\text{mag}}(\mathbf{v}, \mathbf{r}, t)$ by projecting test charges in different directions and with different velocities through the point $\mathbf{r}$ at time t, measuring the magnetic force on the particle in each experiment.[3] If all of the test charges are small enough so as not to disturb whatever sources establish $\mathbf{B}(\mathbf{r}, t)$, we find experimentally that the magnitude of the magnetic force is proportional both to the speed of and to the charge on the test particle and that the direction of the magnetic force is always perpendicular to the velocity of the particle. All of these experimental properties follow if we assign to the point $\mathbf{r}$, t a magnetic induction field $\mathbf{B}(\mathbf{r}, t)$ that determines the magnetic force by

$$\mathbf{F}_q^{\text{mag}}(\mathbf{v}, \mathbf{r}, t) = q\mathbf{v} \times \mathbf{B}(\mathbf{r}, t) \tag{3-4}$$

Strictly, this expression defines $\mathbf{B}(\mathbf{r}, t)$ *in mks units*, and the mks unit of $\mathbf{B}$ therefore has the abbreviation N·sec/C·m = N/A·m—a combination that has long been called the weber/m^2 (Wb/m^2) but is now officially called the tesla (T).[4]

Although it is readily verified that the force determined by Eq. (3-4) exhibits the properties required by experiment, it is not as easily seen that Eq. (3-4) indeed leads to a means to measure the magnetic induction at some point $\mathbf{r}$, t in space-time. We shall describe such a procedure in this

[3]The experimental difficulty of making several different measurements all at a single time does not preclude our using such measurements to define $\mathbf{B}$ in principle. Actual measurement of both $\mathbf{E}$ and $\mathbf{B}$ is almost always accomplished by means other than those envisioned in the definitions.

[4]Again in the spirit of Chapter 1, we might relate $\mathbf{B}$ to $\mathbf{F}_q^{\text{mag}}$ by $k_5 q\mathbf{v} \times \mathbf{B}$, where k_5 assumes an arbitrary value if this expression *defines* $\mathbf{B}$ and an empirically determined value if $\mathbf{B}$ is defined independently of this expression. Further, k_5 may or may not have dimensions. Two different choices for k_5 are in common use. The Gaussian and Heaviside-Lorentz systems set k_5 equal to the *dimensional* constant $1/c$, where c is the speed of light; cgs-esu, cgs-emu, and rationalized mks units all assign to k_5 the *dimensionless* value unity, as in Eq. (3-4). The Gaussian unit of $\mathbf{B}$ is called the *gauss* (G); 1 T is exactly 10^4 G.

paragraph.[5] Unfortunately, Eq. (3-4) cannot be solved explicitly for **B**; the best we can do is compute the cross product of Eq. (3-4) with **v** and exploit the identity in Eq. (C-1) for expanding the triple vector product to find that

$$\mathbf{B} = -\frac{\mathbf{v} \times \mathbf{F}_q^{\text{mag}}}{qv^2} + \frac{(\mathbf{v} \cdot \mathbf{B})\mathbf{v}}{v^2} \tag{3-5}$$

where all arguments have been suppressed. Now, project a particular test charge through the point in question with two *different* velocities, $\mathbf{v}_1$ and $\mathbf{v}_2$, and measure the (magnetic) force $\mathbf{F}_{q1}^{\text{mag}}$ and $\mathbf{F}_{q2}^{\text{mag}}$ experienced in each case. We can then use Eq. (3-5), which gives

$$\mathbf{B} = -\frac{\mathbf{v}_1 \times \mathbf{F}_{q1}^{\text{mag}}}{qv_1^2} + \frac{(\mathbf{v}_1 \cdot \mathbf{B})\mathbf{v}_1}{v_1^2} \tag{3-6}$$

and also

$$\mathbf{B} = -\frac{\mathbf{v}_2 \times \mathbf{F}_{q2}^{\text{mag}}}{qv_2^2} + \frac{(\mathbf{v}_2 \cdot \mathbf{B})\mathbf{v}_2}{v_2^2} \tag{3-7}$$

to determine the magnetic induction field in the following way. From Eq. (3-7), one finds that

$$\mathbf{v}_1 \cdot \mathbf{B} = -\frac{\mathbf{v}_1 \cdot (\mathbf{v}_2 \times \mathbf{F}_{q2}^{\text{mag}})}{qv_2^2} + \frac{(\mathbf{v}_2 \cdot \mathbf{B})}{v_2^2}(\mathbf{v}_1 \cdot \mathbf{v}_2) \tag{3-8}$$

Hence, if we select $\mathbf{v}_2$ perpendicular to $\mathbf{v}_1$ (so that $\mathbf{v}_1 \cdot \mathbf{v}_2 = 0$), we find that $\mathbf{v}_1 \cdot \mathbf{B}$ is determined by the *measurable* first term in Eq. (3-8). Substituting that term into Eq. (3-6), we find an expression for **B** that involves only measured quantities, viz.,

$$\mathbf{B} = -\frac{\mathbf{v}_1 \times \mathbf{F}_{q1}^{\text{mag}}}{qv_1^2} - \frac{\mathbf{v}_1 \cdot (\mathbf{v}_2 \times \mathbf{F}_{q2}^{\text{mag}})}{qv_1^2 v_2^2}\mathbf{v}_1, \quad \mathbf{v}_1 \perp \mathbf{v}_2 \tag{3-9}$$

Although this expression, perhaps with the limit $q \longrightarrow 0$, might have been taken instead of Eq. (3-4) as a definition of **B**, it is unlikely that someone might arrive at Eq. (3-9) without first recognizing something like Eq. (3-4).

Combining Eqs. (3-2) and (3-4) with Eq. (3-1), we find finally that the electromagnetic force $\mathbf{F}_q$ on a charged particle in the *electromagnetic field* **E**, **B** is given (in mks units) by

$$\mathbf{F}_q(\mathbf{r}, t) = q\mathbf{E}(\mathbf{r}, t) + q\mathbf{v} \times \mathbf{B}(\mathbf{r}, t) \tag{3-10}$$

an expression often called the *Lorentz force*. Relative to the coordinate origin, this particle also experiences a torque $\mathbf{N}_q(\mathbf{r}, t)$ given by

$$\begin{aligned} \mathbf{N}_q(\mathbf{r}, t) &= \mathbf{r} \times \mathbf{F}_q(\mathbf{r}, t) \\ &= \mathbf{r} \times [q\mathbf{E}(\mathbf{r}, t) + q\mathbf{v} \times \mathbf{B}(\mathbf{r}, t)] \end{aligned} \tag{3-11}$$

[5]The procedure here described is patterned after the procedure described in *Foundations of Electromagnetic Theory* by J. R. Reitz and F. J. Milford (Addison-Wesley Publishing Company, Inc., Reading, Mass., 1967), Chapter 8, and is used here by permission of Addison-Wesley Publishing Company, Inc.

We conclude this section by pointing out that the electromagnetic field here defined is the field as observed in a specific frame of reference. We have used particles at rest and particles in motion with specific velocities in defining the fields and have therefore tacitly selected a frame of reference. The procedures we have described for measuring **E** and **B**, can, of course, be carried out in *any* frame of reference, so the fields are meaningful to any observer. The relationship between the fields as measured by two different observers is one aspect of special relativity and will be examined in Chapter 15.

PROBLEM

P3-1. Show that $1\ \text{T} = 10^4\ \text{G}$.

3-2 Trajectories of Particles in Prescribed Fields

Finding the trajectory of a charged particle in a prescribed electromagnetic field is important in designing particle accelerators and other apparatus for guiding and focusing beams of charged particles, in studying the behavior of ionized gases, and in investigating many other physical phenomena that involve charged particles. The starting point for calculating such a trajectory is Newton's second law, Eq. (0-32), in which the force on the particle is given by Eq. (3-10). The *equation of motion* of the particle in a given field **E**, **B** thus is

$$m\frac{d^2\mathbf{r}}{dt^2} = q\mathbf{E}(\mathbf{r}, t) + q\frac{d\mathbf{r}}{dt} \times \mathbf{B}(\mathbf{r}, t) \tag{3-12}$$

where m and q are the mass and charge of the particle and $\mathbf{r} = \mathbf{r}(t)$ is now the position of the particle as a function of time. Initial values of $\mathbf{r}$ and $d\mathbf{r}/dt$ must, of course, be given before the solution to Eq. (3-12) is unique.

PROBLEMS

P3-2. Show that a particle of mass m and charge q moving in a constant electric field, $\mathbf{E} = \mathbf{E}_0 =$ constant and $\mathbf{B} = 0$, follows a trajectory given by

$$\mathbf{r}(t) = \frac{qt^2}{2m}\mathbf{E}_0 + \mathbf{v}_0 t + \mathbf{r}_0$$

where $\mathbf{r}_0$ and $\mathbf{v}_0$ are the initial position and velocity. Describe the trajectory in words and note the similarity between this result and the trajectory of a particle moving in a uniform gravitational field.

P3-3. A constant electric field $\mathbf{E} = E\hat{\mathbf{j}}$, with $E > 0$, exists in the region of space $0 < x < a$. A particle of charge q and mass m is projected into this region from the origin with an initial speed v along the x axis. Show that the particle follows a trajectory that lies in the x-y plane and determine the y coordinate of the point at which the particle strikes the plane $x = b$, $b > a$. The situation described in this problem is a crude model of the beam in a cathode ray oscilloscope or television tube.

P3-4. Without solving Eq. (3-12), show that a particle projected into a constant magnetic induction, $\mathbf{B} = \mathbf{B}_0 =$ constant and $\mathbf{E} = 0$, with an initial velocity $\mathbf{v}_0 \perp \mathbf{B}_0$ moves with constant speed in a circle of radius $a = mv_0/qB_0$, where q and m are the charge and mass of the particle. Show also that the *angular* velocity of the particle is independent of its speed—a property crucially important to the functioning of the cyclotron. Finally, describe the trajectory if the initial velocity is *not* perpendicular to the field.

P3-5. Show that a particle of mass m and charge q moving in a constant magnetic induction, $\mathbf{B} = \mathbf{B}_0 =$ constant and $\mathbf{E} = 0$, follows a trajectory given by

$$\mathbf{r}(t) = \mathbf{r}_0 + (\mathbf{v}_0 \cdot \hat{\mathbf{b}})\hat{\mathbf{b}}t + \frac{\sin \omega t}{\omega}[\hat{\mathbf{b}} \times (\mathbf{v}_0 \times \hat{\mathbf{b}})] + \frac{1 - \cos \omega t}{\omega}\mathbf{v}_0 \times \hat{\mathbf{b}}$$

where $\mathbf{r}_0$ and $\mathbf{v}_0$ are the initial position and velocity, $\hat{\mathbf{b}}$ is a unit vector parallel to $\mathbf{B}_0$, and $\omega = qB_0/m$. Describe the trajectory in words.

P3-6. A region of space contains crossed **E** and **B** fields, i.e., a constant electric field **E** and a constant magnetic induction **B** with **E** perpendicular to **B**. A particle having charge q and mass m is projected with an initial velocity **v** in the direction of $\mathbf{E} \times \mathbf{B}$. Find the speed v_0 for which the particle moves through the region undeflected and describe qualitatively what happens if $v \neq v_0$. This simple arrangement is the basis of one type of velocity selector, used to separate particles of a specific velocity from a beam containing a wide spread of velocities.

Unfortunately, simple analytic solutions to Eq. (3-12) rarely exist. More often than not, numerical solutions must be sought, and many numerical procedures of varying sophistication and accuracy have been developed. Essentially, these procedures prescribe means to use Eq. (3-12) and the *given* initial conditions at $t = 0$ to estimate the position and velocity at a slightly later time $t = \Delta t$. The same method can then be applied to generate position and velocity at $t = 2\,\Delta t$ from those at $t = \Delta t$, and then at $t = 3\,\Delta t$ from those at $t = 2\,\Delta t$, and so on until (estimated) solutions have been obtained over some prespecified time interval. Underlying many of the common methods is a pair of equations obtained by formal integration of the definitions of velocity and of acceleration. For example, integrating $\dot{\mathbf{r}} = \mathbf{v}$ (dots symbolize time derivatives) over the interval $t \leq t' \leq t + \Delta t$, we find that

$$\mathbf{r}(t + \Delta t) = \mathbf{r}(t) + \int_t^{t+\Delta t} \mathbf{v}(t')\,dt' = \mathbf{r}(t) + \Delta t\,\langle \mathbf{v} \rangle_t^{t+\Delta t} \qquad (3\text{-}13)$$

where $\langle \mathbf{v} \rangle_t^{t+\Delta t}$ is the average value of $\mathbf{v}$ over the interval of integration. Similarly, we obtain

$$\mathbf{v}(t + \Delta t) = \mathbf{v}(t) + \Delta t \langle \mathbf{a} \rangle_t^{t+\Delta t} \tag{3-14}$$

from the definition $\dot{\mathbf{v}} = \mathbf{a}$, where $\mathbf{a}$ is the acceleration of the particle. Thus, if we know the average velocity over each time interval, we can use Eq. (3-13) to step progressively from $\mathbf{r}(t)$ to $\mathbf{r}(t + \Delta t)$ to $\mathbf{r}(t + 2\Delta t)$ to . . . ; similarly, Eq. (3-14) can be used to obtain $\mathbf{v}$ at a sequence of times. The *exact* average velocities and accelerations, however, are in general not available. We are forced to approximate, and the differences among various numerical methods

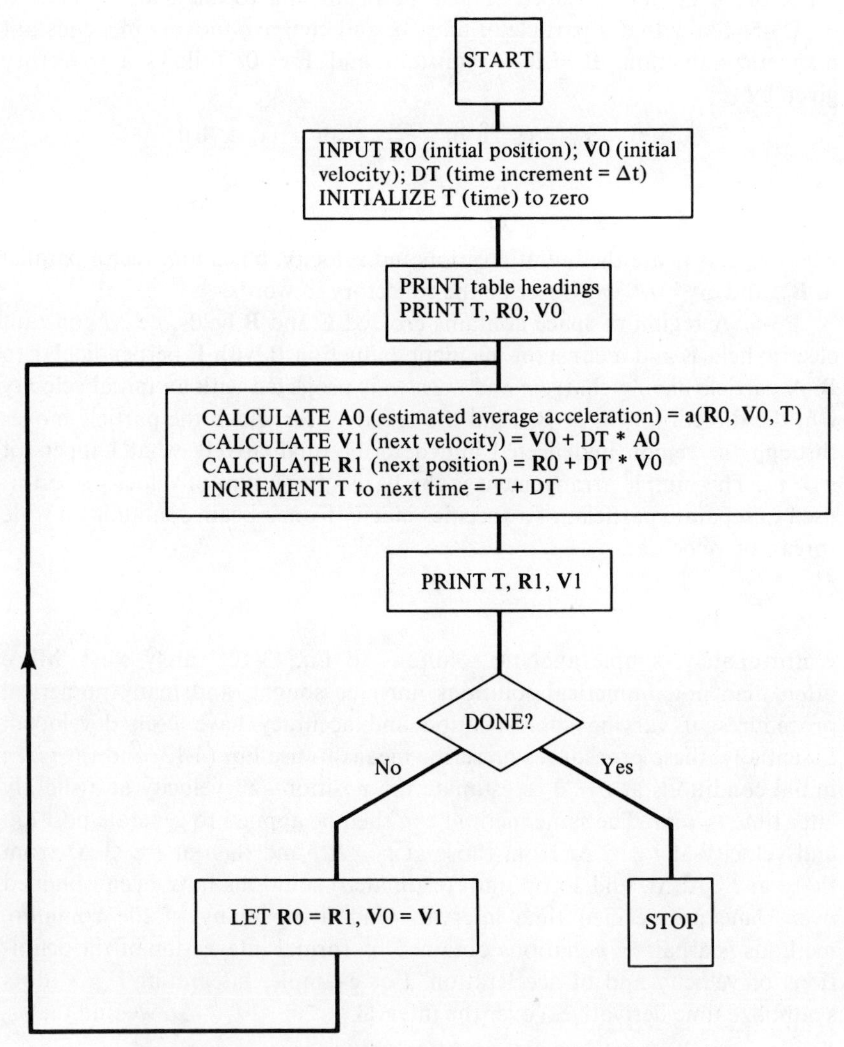

Fig. 3-1. Euler's method for solving ordinary differential equations numerically.

of solution frequently lie in the manner adopted to approximate the needed averages. The simplest approach, known as *Euler's method,* approximates the averages by their values at the lower end of the interval, i.e.,

$$\langle \mathbf{a} \rangle_t^{t+\Delta t} \approx \mathbf{a}(t), \qquad \langle \mathbf{v} \rangle_t^{t+\Delta t} \approx \mathbf{v}(t) \tag{3-15}$$

where, since $\mathbf{r}(t)$ and $\mathbf{v}(t)$ are known when these averages are needed, $\mathbf{a}(t) =$

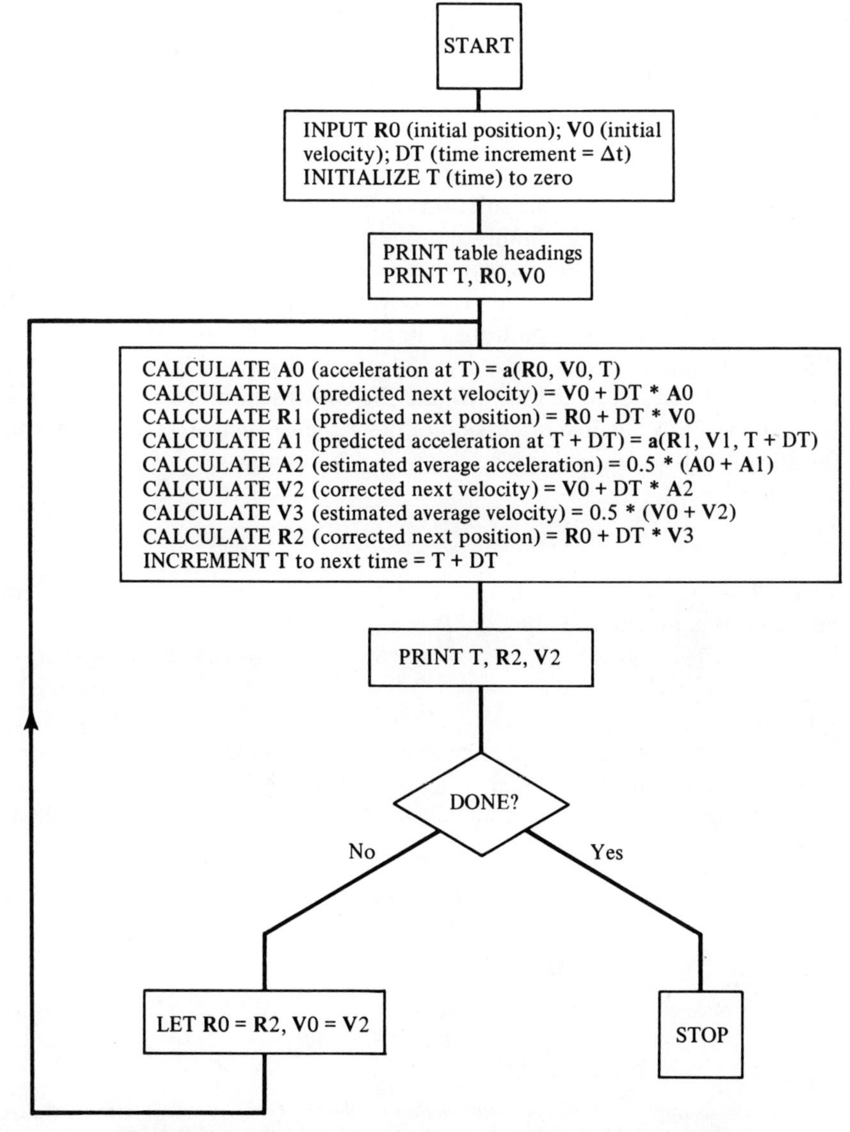

Fig. 3-2. A predictor–corrector scheme for solving ordinary differential equations numerically.

$a(\mathbf{r}(t), \mathbf{v}(t), t)$ is calculable directly from the equation of motion—Eq. (3-12) in the present context. Figure 3-1 shows a flow diagram that could be used as a guide to solving Eq. (3-12) either manually or automatically by Euler's method. (For compactness, vector quantities are referred to as vectors; an actual computation would involve separate evaluation of the three components.) An improved strategy—one of the so-called *predictor-corrector* schemes—involves regarding the Euler solution at each step as a prediction of the solution and then improving (or correcting) that prediction by using it to obtain a better estimate of the average acceleration and velocity over each time interval. The approach is presented in Fig. 3-2. These two methods and the many other methods described in books on numerical analysis all give solutions of improving accuracy as Δt is made smaller, provided only that Δt is not made so small that intrinsic roundoff errors in the computer dominate the accuracy of the solution.

Yet another means to obtain solutions to Eq. (3-12) involves constructing an electronic circuit in such a way that voltages within the circuit vary with time exactly as the position and velocity vary with time. The solution is then obtained by measuring or plotting voltages as functions of time. Circuits of this sort are said to *simulate* the physical system described by Eq. (3-12) and there exist devices designed to facilitate the construction of circuits to simulate Eq. (3-12). A detailed discussion of these *analog computers* would take us far afield, and we leave this discussion to other authors.[6]

PROBLEMS

P3-7. A particle having charge q and mass m is projected from the origin in a uniform magnetic induction $\mathbf{B} = B_0\hat{\mathbf{k}}$ with an initial velocity $\mathbf{v}_0 = \hat{\mathbf{i}}$. For simplicity, suppose $qB_0/m = 1$. (a) Show that the equation of motion reduces to $\ddot{x} = \dot{y}$, $\ddot{y} = -\dot{x}$, $\ddot{z} = 0$. (b) Show that the particle moves in the x-y plane, i.e., that $z(t) = 0$. (c) Use Euler's method with $\Delta t = 0.1$ to obtain the first five points on the trajectory of the particle in the x-y plane. *Optional:* Write and test a short computer program to solve this problem.

P3-8. (a) For Eq. (3-12) show that the three components of the acceleration are

$$a_x = \frac{q}{m}(E_x + \dot{y}B_z - \dot{z}B_y)$$

$$a_y = \frac{q}{m}(E_y + \dot{z}B_x - \dot{x}B_z)$$

$$a_z = \frac{q}{m}(E_z + \dot{x}B_y - \dot{y}B_x)$$

[6]See, for example, J. J. Blum, *Introduction to Analog Computation* (Harcourt Brace Jovanovich, Inc., New York, 1969); G. A. Korn and T. M. Korn, *Electronic Analog and*

(b) Write a computer program to implement Euler's method (Fig. 3-1) when the acceleration is given by these equations. (c) Using this program, explore the trajectories of particles moving in several different fields with several different initial conditions. Try $\mathbf{E} = \mathbf{r}/r^3$, $\mathbf{B} = 0$; $\mathbf{E} = \mathbf{r}/r^{3.1}$, $\mathbf{B} = 0$; $\mathbf{E} = 0$, $\mathbf{B} = B_0\hat{\mathbf{k}}$; $\mathbf{E} = E_0\hat{\mathbf{j}}$, $\mathbf{B} = B_0\hat{\mathbf{k}}$; etc. *Optional*: (1) Modify your program so that only every Nth point is printed out, thus permitting smaller time steps without excessive output. (2) Modify your program so that it plots the trajectory rather than prints a table of coordinates.

P3-9. Do P3-8 but use the predictor-corrector scheme of Fig. 3-2. *Optional*: Look up the fourth-order Runge-Kutta method in an available book on numerical analysis, write a program using this method, and test the program.

3-3 Forces and Torques on General Distributions in Prescribed Fields

In this section we shall consider the forces and torques arising on a general charge distribution when it is placed in a *known* electromagnetic field. For a distribution consisting of point charges, the ith of which has charge q_i and at time t is at $\mathbf{r}_i$ with velocity $\mathbf{v}_i$, the net (external) electromagnetic force $\mathbf{F}(t)$ and the net (external) torque $\mathbf{N}(t)$ are obtained by summing individual contributions, i.e.,

$$\mathbf{F}(t) = \sum_i q_i[\mathbf{E}(\mathbf{r}_i, t) + \mathbf{v}_i \times \mathbf{B}(\mathbf{r}_i, t)] \tag{3-16}$$

$$\mathbf{N}(t) = \sum_i \mathbf{r}_i \times q_i[\mathbf{E}(\mathbf{r}_i, t) + \mathbf{v}_i \times \mathbf{B}(\mathbf{r}_i, t)] \tag{3-17}$$

When the charge distribution is more easily described by particle densities $n^{(a)}(\mathbf{r}, t)$ and velocity fields $\mathbf{v}^{(a)}(\mathbf{r}, t)$ for each of several types of particle, the sum over all particles in Eq. (3-16), for example, is more appropriately expressed as

$$\mathbf{F}(t) = \sum_j \Delta\mathbf{F}_j(t) \tag{3-18}$$

where $\Delta\mathbf{F}_j$ is the force experienced by an element of the distribution occupying a volume ΔV_j centered at the point $\mathbf{r}_j$. Within ΔV_j at time t, however, there are $n^{(a)}(\mathbf{r}_j, t)\,\Delta V_j$ particles of type a, each of which experiences the force $q_a[\mathbf{E}(\mathbf{r}_j, t) + \mathbf{v}^{(a)}(\mathbf{r}_j, t) \times \mathbf{B}(\mathbf{r}_j, t)]$. Thus,

$$\begin{aligned}\Delta\mathbf{F}_j(t) &= \sum_a [n^{(a)}(\mathbf{r}_j, t)\,\Delta V_j]\{q_a[\mathbf{E}(\mathbf{r}_j, t) + \mathbf{v}^{(a)}(\mathbf{r}_j, t) \times \mathbf{B}(\mathbf{r}_j, t)]\} \\ &= [\rho(\mathbf{r}_j, t)\mathbf{E}(\mathbf{r}_j, t) + \mathbf{J}(\mathbf{r}_j, t) \times \mathbf{B}(\mathbf{r}_j, t)]\,\Delta V_j \end{aligned} \tag{3-19}$$

where we have recognized the charge density [Eq. (2-10)] and the current

---

Hybrid Computers (McGraw-Hill Book Company, New York, 1964); and D. I. Rummer, *Introduction to Analog Computer Programming* (Holt, Rinehart and Winston, New York, 1969).

density [Eq. (2-18)] in the sum over types of particle. Finally, summing Eq. (3-19) over all volume elements and allowing $|\Delta V_j| \rightarrow 0$ for all j, we find that

$$\mathbf{F}(t) = \int [\rho(\mathbf{r}, t)\mathbf{E}(\mathbf{r}, t) + \mathbf{J}(\mathbf{r}, t) \times \mathbf{B}(\mathbf{r}, t)]\, dv \tag{3-20}$$

where the integral extends over the region occupied by the distribution at time t. The expression

$$\mathbf{F} = \int (\rho\mathbf{E} + \mathbf{J} \times \mathbf{B})\, dv \tag{3-21}$$

obtained by suppressing all arguments is easier to remember and also supports more obviously the interpretation of $\rho\mathbf{E}$ as an *electric* force density, $\mathbf{J} \times \mathbf{B}$ as a *magnetic* force density, and the combination $\rho\mathbf{E} + \mathbf{J} \times \mathbf{B}$ as an *electromagnetic* force density. Similar arguments lead from Eq. (3-17) to the expression

$$\mathbf{N} = \int \mathbf{r} \times (\rho\mathbf{E} + \mathbf{J} \times \mathbf{B})\, dv \tag{3-22}$$

for the torque at time t on a general distribution placed in the fields $\mathbf{E}$, $\mathbf{B}$.

Although Eqs. (3-21) and (3-22) answer the primary question of this section, some alternative forms are frequently useful. If, for example, the charge distribution is more appropriately described by a surface or line charge density, it will be convenient to recognize in $\rho\, dv$ a charge element dq and to write the electric force in the form

$$\mathbf{F}^{\text{elec}} = \int \mathbf{E}\, dq \tag{3-23}$$

and the electric torque in the form

$$\mathbf{N}^{\text{elec}} = \int \mathbf{r} \times \mathbf{E}\, dq \tag{3-24}$$

Here dq can be interpreted as $\rho\, dv$, $\sigma\, dS$, or $\lambda\, d\ell$ as appropriate. Indeed, it is often convenient to understand these integrals in a very general way, allowing them to include sums over point charges in the distribution as well as integrals over distributed charge.

Other useful alternative forms are expressions for the magnetic parts of Eqs. (3-21) and (3-22) that apply more specifically to currents in wires. In that special case, the entity $I\, d\boldsymbol{\ell}$ replaces $\mathbf{J}\, dv$ (P2-7) and line integrals replace volume integrals. Thus, the magnetic force on a wire carrying current I is given by

$$\mathbf{F}^{\text{mag}} = I \int d\boldsymbol{\ell} \times \mathbf{B} \tag{3-25}$$

and the magnetic torque by

$$\mathbf{N}^{\text{mag}} = I \int \mathbf{r} \times (d\boldsymbol{\ell} \times \mathbf{B}) \tag{3-26}$$

Both integrals extend over the portion of the wire on which the force or torque is desired and may—indeed most frequently will—be extended over a closed circuit. In these integrals, the direction of $d\boldsymbol{\ell}$ coincides with the direction of $\mathbf{J}$ if I is to be a positive number and the wire must be supposed to have a small cross section so that the current element can be adequately treated as having zero cross section.

Equations (3-23) and (3-24) have particularly simple evaluations if $\mathbf{E}$ is *constant*. Under those conditions $\mathbf{E}$ can be taken outside the integral and we find for a general charge distribution in a *constant* $\mathbf{E}$-field that

$$\mathbf{F}^{\text{elec}} = \mathbf{E}\int dq = Q\mathbf{E} \tag{3-27}$$

where Q is the net charge in the distribution, and that

$$\mathbf{N}^{\text{elec}} = \left(\int \mathbf{r}\, dq\right) \times \mathbf{E} = \mathbf{p} \times \mathbf{E} \tag{3-28}$$

where $\mathbf{p}$, defined by

$$\mathbf{p} = \int \mathbf{r}\, dq \tag{3-29}$$

is called the (electric) *dipole moment* of the general distribution.[7] Both Q and $\mathbf{p}$ are characteristics of the charge distribution; *they do not depend in any way on the externally applied field.*

The general expressions for the magnetic force and torque on a complete circuit (closed line integrals) in a *constant* magnetic induction can also be fully evaluated. If $\mathbf{B}$ is constant, the magnetic force on a complete circuit is given by

$$\mathbf{F}^{\text{mag}} = \oint I\, d\boldsymbol{\ell} \times \mathbf{B} = I\left(\oint d\boldsymbol{\ell}\right) \times \mathbf{B} = 0 \tag{3-30}$$

regardless of the shape of the circuit. ($\oint d\boldsymbol{\ell}$ is the vector from the starting point to the end point of the path, which is the zero vector if the path is closed.) The magnetic torque on a circuit in a uniform magnetic induction is more difficult to evaluate, since it is more difficult to remove $\mathbf{B}$ from under the integral sign in Eq. (3-26). Recognizing that $d\boldsymbol{\ell}$ in Eq. (3-26) is identical with an increment in $\mathbf{r}$, we can replace $d\boldsymbol{\ell}$ by $d\mathbf{r}$. Then

$$\begin{aligned}\mathbf{N}^{\text{mag}} &= I\oint \mathbf{r} \times (d\mathbf{r} \times \mathbf{B}) \\ &= I\oint (\mathbf{r}\cdot\mathbf{B})\, d\mathbf{r} - I\left(\oint \mathbf{r}\cdot d\mathbf{r}\right)\mathbf{B} \\ &= I\oint (\mathbf{r}\cdot\mathbf{B})\, d\mathbf{r} \end{aligned}\tag{3-31}$$

[7]Equation (3-29) defines the electric dipole moment of a charge distribution in all systems of units with which the author is familiar

the final form following because $\mathbf{r}\cdot d\mathbf{r} = x\,dx + y\,dy + z\,dz = d(x^2 + y^2 + z^2)/2$ is an exact differential and $\oint \mathbf{r}\cdot d\mathbf{r}$, which extends over a *closed* path, is therefore zero. We now remove **B** from under the integral sign in Eq. (3-31). First, Eq. (C-20) with $d\boldsymbol{\ell}$ replaced by $d\mathbf{r}$ converts Eq. (3-31) to

$$\mathbf{N}^{\text{mag}} = I\int d\mathbf{S} \times \nabla(\mathbf{r}\cdot\mathbf{B}) \tag{3-32}$$

where the integral now extends over the surface bounded by the circuit. Remembering that **B** is constant and applying Eq. (C-15), we next find that

$$\mathbf{N}^{\text{mag}} = I\int d\mathbf{S} \times [(\mathbf{B}\cdot\nabla)\mathbf{r} + \mathbf{B} \times (\nabla \times \mathbf{r})] \tag{3-33}$$

Finally, recognizing that $(\mathbf{B}\cdot\nabla)\mathbf{r} = \mathbf{B}$ and $\nabla \times \mathbf{r} = 0$, we find that the torque on a current loop in a constant **B**-field is given by

$$\mathbf{N}^{\text{mag}} = \mathbf{m} \times \mathbf{B} \tag{3-34}$$

where

$$\mathbf{m} = I\int d\mathbf{S} \tag{3-35}$$

is called the (magnetic) *dipole moment* of the loop. If in particular the loop is plane and bounds a plane surface with unit normal $\hat{\mathbf{n}}$ and area S, $d\mathbf{S} = |d\mathbf{S}|\hat{\mathbf{n}}$ and $\mathbf{m} = IS\hat{\mathbf{n}}$. Thus, in this simple case **m** is directed normal to the plane of the loop and has a magnitude given by the product of the current in the loop and the area of the loop. (The usual right-hand rule relates the direction of I to that of $\hat{\mathbf{n}}$.) As with Q and **p** in the previous paragraph, **m** is a characteristic of the current distribution; *it does not depend in any way on the externally applied field.* It is shown in P3-19 that Eq. (3-35) can alternatively be written as a line integral[8]

$$\mathbf{m} = \tfrac{1}{2}I\oint \mathbf{r} \times d\mathbf{r} \tag{3-36}$$

The power input to a general distribution by an external electromagnetic field is the final quantity of interest in this section. For a collection of discrete

[8] In all systems of units with which the author is familiar, the magnetic dipole moment of a current distribution is defined so that Eq. (3-34) gives the torque experienced by the dipole in a constant magnetic induction. In cgs-esu, cgs-emu, and mks units, Eqs. (3-35) and (3-36) can be used to calculate the magnetic moment. In Gaussian and Heaviside-Lorentz units, however, the factor of c incorporated explicitly in the definition of **B** (see footnote 4) will also appear dividing **J** in Eq. (3-21). Thus, Eq. (3-26) becomes

$$\mathbf{N}^{\text{mag}} = \frac{I}{c}\int \mathbf{r} \times (d\boldsymbol{\ell} \times \mathbf{B})$$

and, in Gaussian and Heaviside-Lorentz units, the magnetic dipole moment must be defined by

$$\mathbf{m} = \frac{I}{c}\int d\mathbf{S} = \frac{I}{2c}\oint \mathbf{r} \times d\mathbf{r}$$

if Eq. (3-34) is to be preserved.

particles distinguished by an index i, the power input $P(t)$ at time t is given by

$$P(t) = \sum_i \mathbf{v}_i \cdot \mathbf{F}_i = \sum_i q_i \mathbf{v}_i \cdot \mathbf{E}(\mathbf{r}_i, t) \tag{3-37}$$

where $\mathbf{F}_i$ is the electromagnetic force on the ith particle and the magnetic force ultimately contributes nothing to the power because it is always perpendicular to $\mathbf{v}$ ($\mathbf{v}\cdot\mathbf{v} \times \mathbf{B} = 0$). When the charge distribution is described by particle density and velocity fields $n^{(a)}(\mathbf{r}, t)$ and $\mathbf{v}^{(a)}(\mathbf{r}, t)$, we follow an argument similar to that leading from Eq. (3-16) to Eq. (3-20). In the notation of that argument, the power input $\Delta P_j(t)$ to a volume element ΔV_j centered at $\mathbf{r}_j$ then is given by

$$\begin{aligned}\Delta P_j(t) &= \sum_a q_a n^{(a)}(\mathbf{r}_j, t)\mathbf{v}^{(a)}(\mathbf{r}_j, t)\cdot\mathbf{E}(\mathbf{r}_j, t)\,\Delta V_j \\ &= \mathbf{J}(\mathbf{r}_j, t)\cdot\mathbf{E}(\mathbf{r}_j, t)\,\Delta V_j \end{aligned} \tag{3-38}$$

Summing over all volume elements and allowing $|\Delta V_j| \longrightarrow 0$ for all j, we find the expression

$$P(t) = \int \mathbf{J}(\mathbf{r}, t)\cdot\mathbf{E}(\mathbf{r}, t)\,dv \tag{3-39}$$

for the power input to the entire distribution.

PROBLEMS

P3-10. (a) Interpreting the integral in Eq. (3-29) to include a sum over any point charges present, show that the dipole moment of a distribution consisting of two (rigidly connected) charges of strength q and $-q$ located at $\mathbf{r}_+$ and $\mathbf{r}_-$ is given by $\mathbf{p} = q\mathbf{a}$, where $\mathbf{a}$ is the vector *from* charge $-q$ *to* charge q. Qualitatively, what characteristic of a charge distribution is measured by the dipole moment? (b) Let this charge distribution be placed in a constant electric field $\mathbf{E}$ and oriented with $\mathbf{p}$ at an angle θ to $\mathbf{E}$. Draw a diagram showing the external force experienced by each charge and, arguing from this diagram, find the net force and the net torque on the distribution. Verify that your results agree with Eqs. (3-27) and (3-28). (c) Suppose this distribution, initially oriented as in part (b), is released from rest. Describe its subsequent motion qualitatively. Are there any positions of static equilibrium? If so, are these positions stable or unstable? (d) Let $\mathbf{r}_\pm = (x_0 \pm \frac{1}{2}a)\hat{\mathbf{i}} + y_0\hat{\mathbf{j}} + z_0\hat{\mathbf{k}}$ and let the field now be the (nonuniform) field $\mathbf{E}(\mathbf{r}) = E(x)\hat{\mathbf{i}}$. Assuming a to be sufficiently small, show that this distribution experiences the force $\mathbf{F} = p(dE/dx)|_0\hat{\mathbf{i}}$.

P3-11. The response of an arbitrary charge distribution to a constant external electric field is observed experimentally. Discuss what can be learned about the distribution from these observations.

P3-12. The electron cloud of one of the stationary states of the hydrogen atom in a constant external electric field along the z axis is described quantum mechanically by the charge density

$$\rho(r) = -\frac{q}{16\pi a_0^3}\left(1 - \frac{r}{a_0}\sin^2\frac{\theta}{2}\right)^2 e^{-r/a_0}$$

where q is the proton charge, a_0 is the Bohr radius, and spherical coordinates are employed. Calculate the electric dipole moment of this atom about the nucleus. *Hints*: (1) $\mathbf{r} = r(\sin\theta\cos\phi\hat{\mathbf{i}} + \sin\theta\sin\phi\hat{\mathbf{j}} + \cos\theta\hat{\mathbf{k}})$ in spherical coordinates. (2) Don't be afraid of integral tables.

P3-13. A rectangular circuit of length a and width b is placed in a constant magnetic induction $\mathbf{B}$ such that the normal to its plane makes an angle θ with the direction of $\mathbf{B}$. The circuit carries a current I that goes into the paper at the point marked $\times$ in Fig. P3-13 and emerges at the point marked $\cdot$. (a) Draw a diagram showing the external force experienced by each side of the

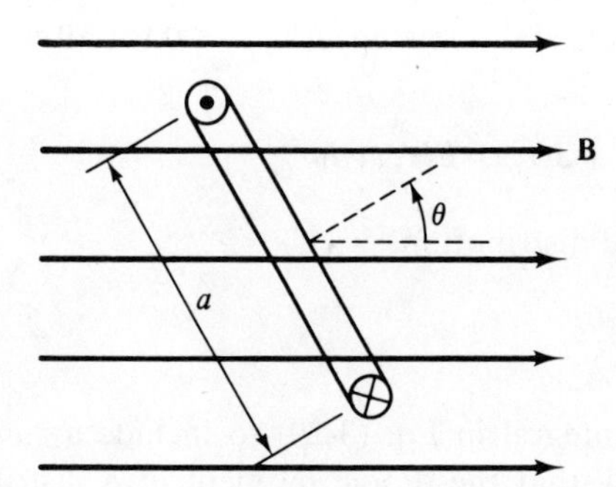

Figure P3-13

rectangle and, arguing from this diagram, find the net force and the net torque on the rectangle. Express the torque in terms of the magnetic moment of the current loop and show that your result can be written in the form of Eq. (3-34). (b) Let the rectangle now be released from rest. Describe its subsequent motion qualitatively. Are there any positions of static equilibrium? If so, are these positions stable or unstable?

P3-14. Calculate the magnetic dipole moment of a sphere of radius R carrying charge Q distributed uniformly over its surface and spinning about a diameter with angular velocity $\boldsymbol{\omega}$. *Hint*: Break the sphere into small current loops and sum the contribution of each.

Supplementary Problems

P3-15. A particle having charge q and mass m is projected from the origin with an initial velocity $\mathbf{v}_0 = v_{x0}\hat{\mathbf{i}} + v_{z0}\hat{\mathbf{k}}$ in a constant magnetic induction $\mathbf{B} = B_0\hat{\mathbf{k}}$. Find the trajectory of the particle and in particular determine the coordinates x_d and y_d of the point P at which the particle strikes a screen in the plane $z = d$. Describe how those coordinates depend on B_0. The analysis of this problem is essentially the analysis of the Busch method for measuring the charge-to-mass ratio of a particle. *Hint*: Express the coordinates of the point P in terms of the angle $\psi = qdB_0/2mv_{z0}$ and then interpret the angle geometrically.

P3-16. Solve the equation of motion for a particle of charge q and mass m moving in the combination of a constant magnetic induction $\mathbf{B} = B_0\hat{\mathbf{k}}$ and a constant electric field $\mathbf{E} = E_0\hat{\mathbf{j}}$. Let the particle start at the origin but have an arbitrary initial velocity in the x-y plane. Describe the trajectory for several different initial velocities and in particular examine the dependence of the points at which the particle crosses the x axis on the initial velocity. *Hint*: Show first that, when $\mathbf{r}$ satisfies Eq. (3-12) with these fields, $\mathbf{r}' = \mathbf{r} - [(\mathbf{E} \times \mathbf{B})/B^2]t$ satisfies $m\ddot{\mathbf{r}}' = q\dot{\mathbf{r}}' \times \mathbf{B}$, which is the equation solved in P3-5.

P3-17. Let the charge distribution to which Eq. (3-23) applies be confined to a small enough region of space about the point $\mathbf{r}_0$ that the terms

$$\mathbf{E}(\mathbf{r}, t) = \mathbf{E}_0 + (x - x_0)\frac{\partial \mathbf{E}}{\partial x}\bigg|_0 + (y - y_0)\frac{\partial \mathbf{E}}{\partial y}\bigg|_0 + (z - z_0)\frac{\partial \mathbf{E}}{\partial z}\bigg|_0$$

of the Taylor expansion provide an adequate approximation to $\mathbf{E}$ over the region occupied by the charge. Show that

$$\mathbf{F}^{\text{elec}} = Q\mathbf{E}_0 + (\mathbf{p}_0 \cdot \boldsymbol{\nabla})\mathbf{E}|_0$$

where $\mathbf{p}_0$ is the dipole moment of the distribution about $\mathbf{r}_0$.

P3-18. Show that the dipole moment of an arbitrary charge distribution is invariant to translation of the coordinate system if and only if the distribution has zero net charge.

P3-19. Use Eq. (C-24) to show that $\oint \mathbf{r} \times d\mathbf{r} = 2\int d\mathbf{S}$, thus verifying the equivalence of Eqs. (3-35) and (3-36).

P3-20. A magnetic dipole is located at the origin of a coordinate system with its dipole moment $\mathbf{m}$ oriented perpendicular to a constant magnetic induction field $\mathbf{B}$. Show that the work that must be done on the dipole to move it to an arbitrary point $\mathbf{r}$ in this field and rotate it so that $\mathbf{m}$ makes an angle θ different from 90° with $\mathbf{B}$ is given by $-\mathbf{m}\cdot\mathbf{B}$.

P3-21. (a) Show that the magnetic dipole moment $\mathbf{m}$ of a charge q moving in a circle and having mass μ is related to the angular momentum $\mathbf{L}$ of the charge by $\mathbf{m} = (q/2\mu)\mathbf{L}$. (b) According to the rotational analog of Newton's second law, the time rate of change of the angular momentum of an object is equal to the external torque acting on the object. Show that, when the charge in part (a) is placed in a constant magnetic induction field $\mathbf{B}$, its magnetic moment satisfies the equation of motion

$$\frac{d\mathbf{m}}{dt} = \frac{q}{2\mu}\mathbf{m} \times \mathbf{B}$$

and then, taking $\mathbf{B} = B\hat{\mathbf{k}}$ and $\mathbf{m}(0)$ to have magnitude m_0 and to lie in the x-z plane at an angle θ to the z axis, find $\mathbf{m}(t)$ and describe its behavior geometrically as a function of time.

P3-22. The force on a current loop in an arbitrary magnetic induction $\mathbf{B}(\mathbf{r})$ is given by Eq. (3-25). Suppose the loop is centered at $\mathbf{r}_0$ and is small enough so that the approximation

$$\mathbf{B}(\mathbf{r}) = \mathbf{B}_0 + (x - x_0)\frac{\partial \mathbf{B}_0}{\partial x_0} + (y - y_0)\frac{\partial \mathbf{B}_0}{\partial y_0} + (z - z_0)\frac{\partial \mathbf{B}_0}{\partial z_0}$$

where $\mathbf{B}_0 = \mathbf{B}(\mathbf{r}_0)$, can be accurately made at all points on the loop. Show that the force $\mathbf{F}$ on the loop is given by $\mathbf{F} = \boldsymbol{\nabla}_0(\mathbf{m}\cdot\mathbf{B}_0)$, where $\boldsymbol{\nabla}_0$ involves derivatives with respect to x_0, y_0, and z_0 and $\mathbf{m}$ is the magnetic dipole moment of the loop. This force on a dipole in an *inhomogeneous* magnetic induction is exploited in the Stern-Gerlach experiment to separate particles in a beam according to their magnetic moments. *Hints*: (1) Show first that

$$\mathbf{F} = \oint I\,d\boldsymbol{\ell} \times \left(x\frac{\partial \mathbf{B}_0}{\partial x_0} + y\frac{\partial \mathbf{B}_0}{\partial y_0} + z\frac{\partial \mathbf{B}_0}{\partial z_0}\right)$$

and then use Eq. (C-24) to show, for example, that

$$\oint I\,d\boldsymbol{\ell} \times \left(x\frac{\partial \mathbf{B}_0}{\partial x_0}\right) = \frac{\partial}{\partial x_0}\oint I\,d\boldsymbol{\ell} \times (x\mathbf{B}_0) = -\mathbf{m}\frac{\partial B_{0x}}{\partial x_0} + \hat{\mathbf{i}}\frac{\partial}{\partial x_0}(\mathbf{m}\cdot\mathbf{B}_0)$$

[See Eq. (3-35).] (2) Anticipating a result to be demonstrated in Chapter 5, note that every magnetic induction field $\mathbf{B}_0(\mathbf{r}_0)$ satisfies $\boldsymbol{\nabla}_0\cdot\mathbf{B}_0 = 0$.

4

The Electric Field Produced by Static Charges

In this and the next two chapters, we shall be concerned principally with the question, How is the electromagnetic field determined from its sources? We shall, however, also examine a number of consequences of the principal relationships between the fields and their sources. The three chapters treat static electric fields, static magnetic induction fields, and the time-dependent electromagnetic field in sequence. Throughout these chapters, we shall focus on determining the fields produced by given sources and on determining what sources must be set up to establish a required field. The forces that these fields may subsequently exert on other distributions of charge play no intrinsic role in our present considerations.

4-1 Coulomb's Law and the Electrostatic Field of Given Sources

The starting point for relating the static electric field to its sources is Coulomb's law for the force of interaction between two point charges, which we shall now write in a general vector form. Although Coulomb's law has the simplest form in a coordinate system whose origin coincides with one of the two point charges, we are here interested in obtaining a more general expression so that the choice of a coordinate origin can be dictated by its convenience to later problems. We therefore select now a coordinate system

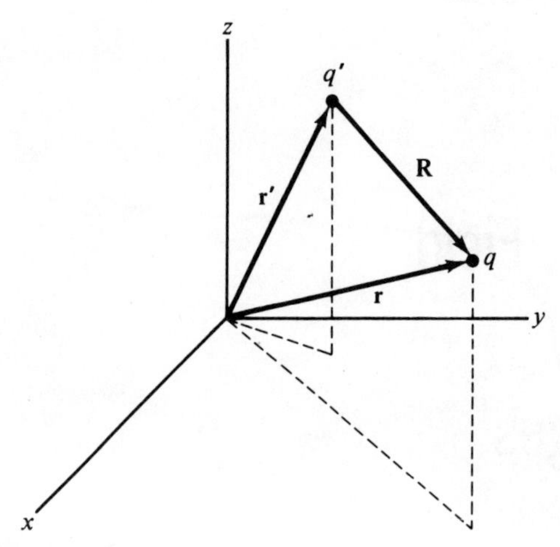

Fig. 4-1. Two point charges in an arbitrary coordinate system.

that may seem unnecessarily cumbersome for the shorter-range objectives of this paragraph. Let point charges q and q' be located at $\mathbf{r}$ and $\mathbf{r}'$, as shown in Fig. 4-1. Further, let $\mathbf{R}$ be the vector *from* $\mathbf{r}'$ *to* $\mathbf{r}$. The force $\mathbf{F}_q(\mathbf{r})$ on q due to the presence of q' (as given by the experimental properties outlined in Section 1-2) may then be expressed *in mks units* (which we now explicitly select) by

$$\mathbf{F}_q(\mathbf{r}) = \frac{1}{4\pi\epsilon_0}\frac{qq'}{R^2}\hat{\mathbf{R}} = \frac{qq'}{4\pi\epsilon_0}\frac{\mathbf{r}-\mathbf{r}'}{|\mathbf{r}-\mathbf{r}'|^3} \tag{4-1}$$

the second form following because $\mathbf{R} = \mathbf{r} - \mathbf{r}'$, $R = |\mathbf{r} - \mathbf{r}'|$, and $\hat{\mathbf{R}} = \mathbf{R}/R$. All of the observed features are incorporated in this law, including in particular the repulsive and attractive nature of the force. If q and q' have the same sign, the coefficient of $\hat{\mathbf{R}}$ is overall positive and the force on q has the direction of $\hat{\mathbf{R}}$, i.e., is repulsive. If q and q' have opposite signs, the coefficient of $\hat{\mathbf{R}}$ is negative and the force is attractive.

Before we can use Coulomb's law to relate a general electrostatic field to its sources, we need one additional *experimental* observation: The force between two given point charges is unaffected by the presence of still other charges in their vicinity. Thus, the force exerted on a charge q at $\mathbf{r}$ by an assembly of point charges q_i residing at points $\mathbf{r}_i$ is given (in mks units) by

$$\mathbf{F}_q(\mathbf{r}) = \frac{q}{4\pi\epsilon_0}\sum_i q_i \frac{\mathbf{r}-\mathbf{r}_i}{|\mathbf{r}-\mathbf{r}_i|^3} \tag{4-2}$$

which expresses the *principle of superposition*. Its content is more far-reaching than the limited superposition built into the definition of charge—earlier we took two point charges *at the same place* to superpose. In essence, there are two independent experimental observations underlying electrostatics: Coulomb's law *and* the principle of superposition.

The charge densities introduced in Section 2-1 now facilitate a reexpression of Eq. (4-2) that is more appropriate for finding the force exerted on a point charge q by a space-filling distribution of charge. The volume, surface,

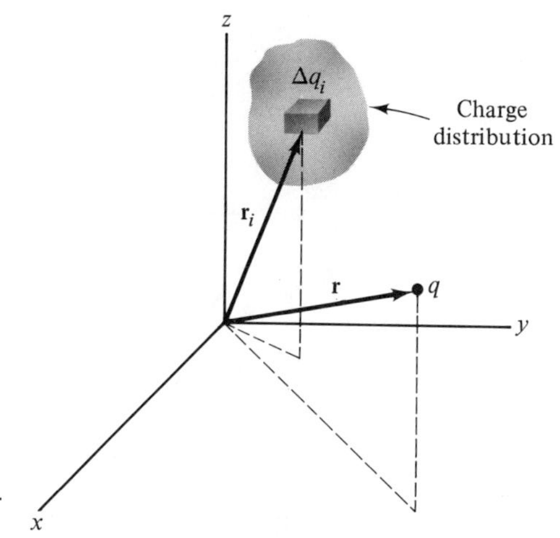

Fig. 4-2. An arbitrary charge distribution.

or line making up the source distribution can be divided into elements that are small enough to be regarded as point charges. If the ith element carries charge Δq_i, then the total force $\mathbf{F}_q(\mathbf{r})$ on q at $\mathbf{r}$ is simply a sum of contributions from each element, i.e.,

$$\mathbf{F}_q(\mathbf{r}) = \frac{q}{4\pi\epsilon_0}\sum_i \Delta q_i \frac{\mathbf{r}-\mathbf{r}_i}{|\mathbf{r}-\mathbf{r}_i|^3} \tag{4-3}$$

(See Fig. 4-2.) If all of the elements now become arbitrarily small, Eq. (4-3) is replaced by the integral

$$\mathbf{F}_q(\mathbf{r}) = \frac{q}{4\pi\epsilon_0}\int_{\substack{\text{charge}\\\text{distribution}}} \frac{\mathbf{r}-\mathbf{r}'}{|\mathbf{r}-\mathbf{r}'|^3}\,dq' \tag{4-4}$$

in which the limits are chosen so that the integration extends over the source charge distribution and $dq' = \rho(\mathbf{r}')\,dv'$, $\sigma(\mathbf{r}')\,dS'$, or $\lambda(\mathbf{r}')\,d\ell'$ as appropriate to the distribution. If, in fact, the distribution contains several portions, the integral in Eq. (4-4) includes a sum of integrals, each related to a single portion of the distribution, and the charge element may be differently expressed for different portions of the distribution. Further, interpreting the integral in Eq. (4-4) to include a sum over any point charges present, we can view Eq. (4-2) as a special case of Eq. (4-4).

The final step in obtaining an expression relating the field to its sources is to apply the definition in Eq. (3-3) to Eqs. (4-1), (4-2), and (4-4). We find from Eq. (4-1), for example, that the electric field $\mathbf{E}(\mathbf{r})$ established at a point $\mathbf{r}$ by a point charge q' located at $\mathbf{r}'$ is given by

$$\mathbf{E}(\mathbf{r}) = \frac{q'}{4\pi\epsilon_0}\frac{\mathbf{r}-\mathbf{r}'}{|\mathbf{r}-\mathbf{r}'|^3} \tag{4-5}$$

Similarly, Eq. (4-2) leads to

$$\mathbf{E}(\mathbf{r}) = \frac{1}{4\pi\epsilon_0}\sum_i q_i \frac{\mathbf{r}-\mathbf{r}_i}{|\mathbf{r}-\mathbf{r}_i|^3} \tag{4-6}$$

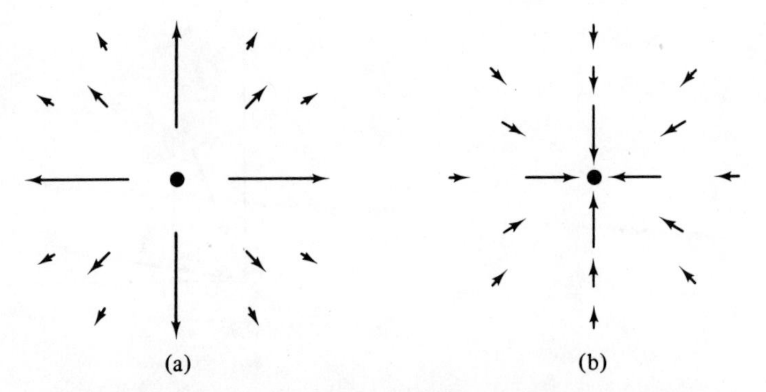

Fig. 4-3. The vector representation of the field of (a) a positive and (b) a negative point charge. The field vectors are directed everywhere away from a positive charge and toward a negative charge. The magnitude of the vectors is inversely proportional to the square of their distances from the charge.

for the field established at $\mathbf{r}$ by an array of point charges, and Eq. (4-4) leads to

$$\mathbf{E}(\mathbf{r}) = \frac{1}{4\pi\epsilon_0} \int_{\substack{\text{charge} \\ \text{distribution}}} \frac{\mathbf{r} - \mathbf{r}'}{|\mathbf{r} - \mathbf{r}'|^3} \, dq' \tag{4-7}$$

for the field established at $\mathbf{r}$ by a more general distribution of charge. Equations (4-5)–(4-7) not only provide the means explicitly to determine the field established by a *known* charge distribution, but they also express the relationship between the field and its sources even if the charge distribution itself is not known. In conventional terminology, the point $\mathbf{r}$, at which the field is evaluated, is called the *field* point and the point $\mathbf{r}'$, at which an element of the source is located, is called the *source* point. In more advanced texts, the convergence of the integral in Eq. (4-7) is explored in some detail, particularly when the point $\mathbf{r}$ lies *within* the source distribution.

We shall now present several examples of specific electric fields, beginning with the field of a single point charge, sometimes called a *monopole*. Let the charge have strength Q and be located at the origin. Then Eq. (4-5) gives

$$\mathbf{E}(\mathbf{r}) = \frac{Q}{4\pi\epsilon_0 r^2}\hat{\mathbf{r}} \tag{4-8}$$

for the resulting field. This field is at every point directed away from the source charge when $Q > 0$ and toward the source when $Q < 0$, and its magnitude varies as $1/r^2$. The field given by Eq. (4-8) for $Q > 0$ and for $Q < 0$ is shown in Figs. 4-3 and 4-4. Figure 4-4, with field lines emerging from positive charges and terminating on negative charges, illustrates graphically the statement that charge is the source of the electric field, but this observation can be taken seriously only after it has been proved that conventions relating the density of field lines to the strength of the field can be adhered to without

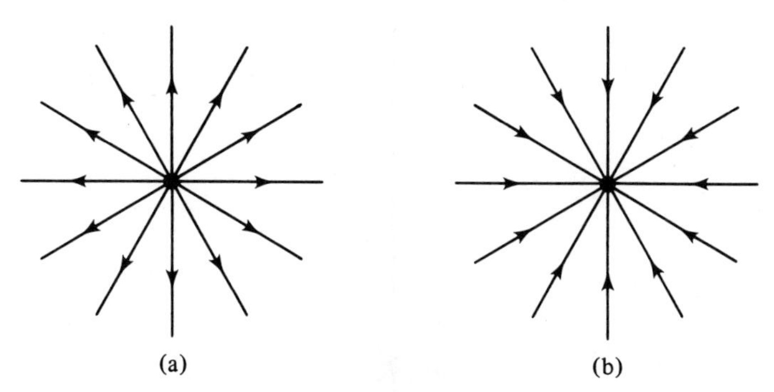

Fig. 4-4. The field line representation of the field of (a) a positive and (b) a negative point charge. The field lines are directed radially outward from a positive charge and radially inward toward a negative charge, and the magnitude of the field is conveyed by the spacing of the field lines.

starting or stopping field lines at points where there are no charges. (See P0-17.)

Another important charge distribution called the *electric dipole* is composed of two charges of the same magnitude but of opposite sign. The field lines of this dipole are shown graphically in Fig. 4-5. Near the positive charge, the field is directed radially away from the charge; near the negative charge, the field is directed radially toward the charge; at points on the perpendicular bisector of the line joining the charges, the field is parallel to the line joining the charges. (Why?) Analytically the field of this dipole is given by Eq. (4-6) or, in terms of the symbols introduced in Fig. 4-6, by

$$\mathbf{E}(\mathbf{r}) = \frac{q}{4\pi\epsilon_0}\left(\frac{\mathbf{r}-\frac{1}{2}\mathbf{a}}{|\mathbf{r}-\frac{1}{2}\mathbf{a}|^3} - \frac{\mathbf{r}+\frac{1}{2}\mathbf{a}}{|\mathbf{r}+\frac{1}{2}\mathbf{a}|^3}\right) \tag{4-9}$$

In this result, the charges q and $-q$ are taken to be at the points $(0, 0, \frac{1}{2}a)$ and $(0, 0, -\frac{1}{2}a)$ and the vector $\mathbf{a}$ is directed *from* the negative charge *to* the positive charge. Now, in most cases when the dipole is a reasonable approximation to a more complicated physically important distribution (e.g., an asymmetric molecule), the distance from the field point to the dipole is large compared to the separation of the charges, i.e., $r \gg a$. Using the binomial theorem (Appendix B) and keeping only the first nonzero term in an expansion in powers of a/r, we find that the field given by Eq. (4-9) can be approximated when $r \gg a$ by the simpler expression

$$\mathbf{E}(\mathbf{r}) = \frac{1}{4\pi\epsilon_0 r^3}[3(\mathbf{p}\cdot\hat{\mathbf{r}})\hat{\mathbf{r}} - \mathbf{p}] \tag{4-10}$$

$$= \frac{p}{4\pi\epsilon_0 r^3}[2\cos\theta\hat{\mathbf{r}} + \sin\theta\hat{\boldsymbol{\theta}}] \tag{4-11}$$

where

$$\mathbf{p} = q\mathbf{a} \tag{4-12}$$

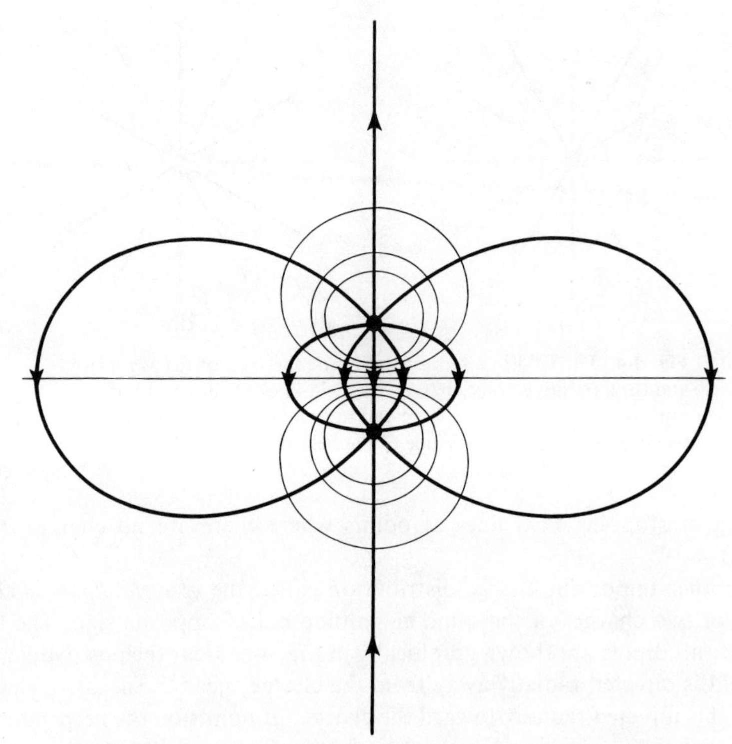

Fig. 4-5. Field lines of a dipole. (The light lines indicate equipotential surfaces, which will be discussed in Section 4-4; those equipotential surfaces lying very close to the point charges are not shown.)

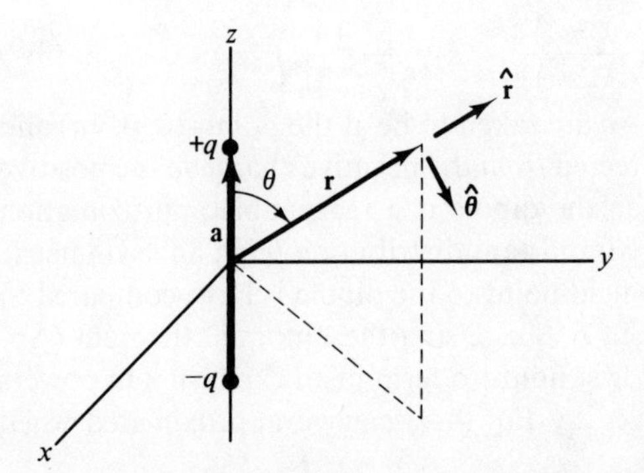

Fig. 4-6. Coordinate system for expressing the field of a dipole.

is the dipole moment of the source distribution (P4-2). In the present context, however, $\mathbf{p}$ pertains to the *source* distribution and determines the resulting field; in Section 3-3, $\mathbf{p}$ pertained to the *test* distribution and determined (among other things) the torque exerted on the distribution by a uniform electric field. It is a convenient coincidence that this single characteristic of a distribution plays these two very different roles. Although Eqs. (4-10) and

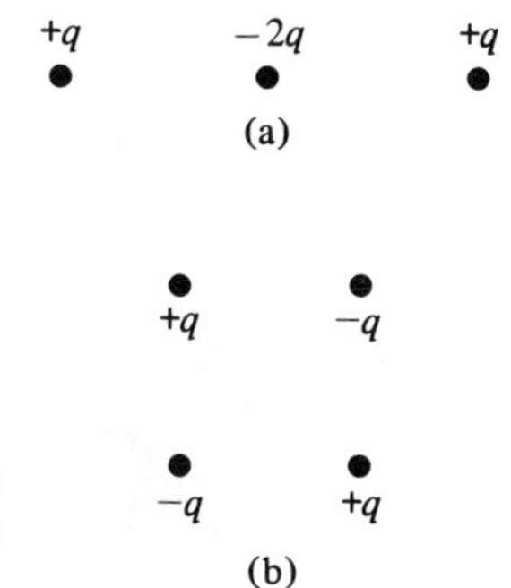

Fig. 4-7. Two quadrupoles: (a) the linear quadrupole and (b) a more general quadrupole.

(4-11) strictly apply only when $r \gg a$, it is sometimes useful to visualize an idealization in which a is so small that $r \gg a$ for every point in space. Then Eqs. (4-10) and (4-11) apply for all r, and the corresponding ideal charge distribution—called a *point dipole*—is characterized only by a dipole moment; one does not ask about the individual charges composing a *point* dipole. The field of a dipole varies inversely as the *cube* of r and is independent of the azimuthal angle ϕ, the latter property reflecting the invariance of the charge distribution to arbitrary rotation about the axis of the dipole.

One way to visualize the formation of a dipole is to think first of a monopole. Now place near this monopole a second monopole of opposite sign. The result is a dipole. Using this same sort of visualization, we can construct higher-order "-poles." Location of one dipole adjacent to a second of opposite dipole moment, for example, gives rise to a quadrupole, of which there are several sorts, since the direction of the displacement of the two dipoles has a more profound effect on the overall distribution than does the direction of the displacement of the two charges composing a dipole. Two different quadrupoles are illustrated in Fig. 4-7. Octupoles are made by placing two quadrupoles side by side, in one of which the charges have opposite signs to the corresponding charges in the other. And so it goes to higher moments. That a charge distribution of a particular character can be produced by com-

---

Fig. 4-8. (*Next 2 pages*) Field lines for several point charge distributions: (a) charge of $+2$ units on the left and -1 unit on the right; (b) charge of $+2$ units on the left and $+1$ unit on the right; (c) three equal charges at the corners of an *isosceles* triangle with vertices at $(x, y) = (1, 0)$, $(0, 1)$, and $(-1, 0)$; and (d) charge of $+2$ units on the left and two charges, each of -1 unit, in the upper and lower right corners. [The light lines indicate equipotential surfaces, which will be discussed in Section 4-4. Note particularly that the equipotential surfaces are perpendicular to the field lines and that in part (a) the marked equipotential is a true sphere in three dimensions. Because they are very close together in the immediate vicinity of point charges, the equipotential surfaces in those regions are not shown.]

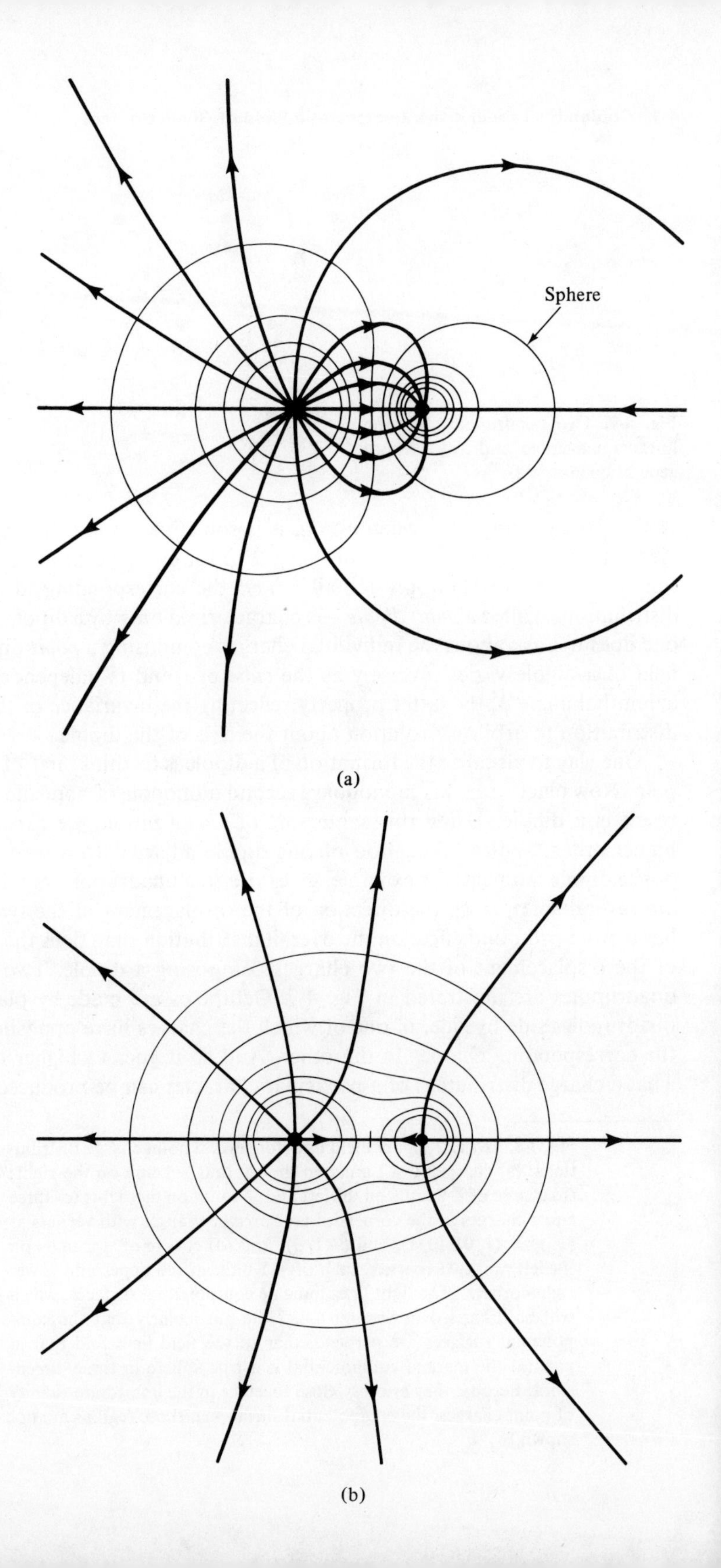

(a)

(b)

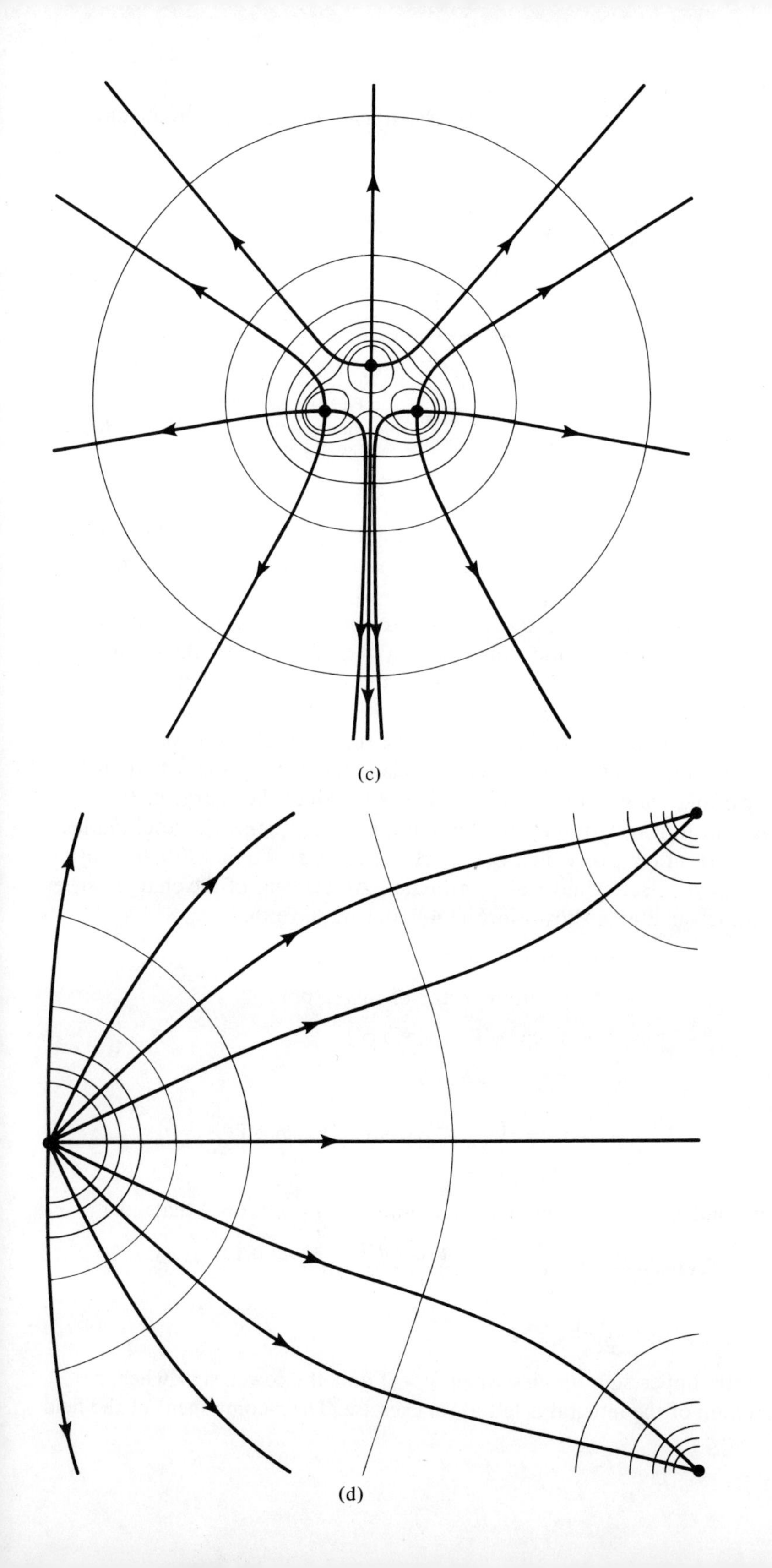
(c)
(d)

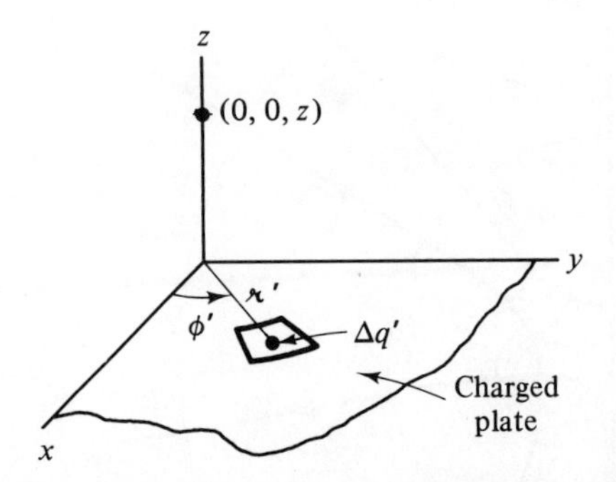

Fig. 4-9. Coordinate system for determining the field of a charged infinite plate.

bining two slightly displaced distributions of the next lower order suggests that the field of the higher-order "-pole" might be obtained by differentiating the field of the lower-order "-pole", and such a relationship indeed exists (P4-47).

The various multipole distributions described in the previous paragraphs require very specific relationships among the strengths of the component charges. The field lines for some representative distributions that do not conform exactly to any multipole are shown in Fig. 4-8.

To illustrate the use of Eq. (4-7), let the source distribution be an infinite two-dimensional plane sheet of charge characterized by a constant surface density σ and lying in the x-y plane (Fig. 4-9). Since the charge distribution is invariant to translation by any amount in the x-y plane, the resulting field cannot depend on x or y, i.e., $\mathbf{E}(x, y, z) = \mathbf{E}(0, 0, z)$. To find $\mathbf{E}(0, 0, z)$ using Eq. (4-7), we select cylindrical coordinates. An element of the charged sheet is then the small area shown in Fig. 4-9 and we have that

$$\begin{aligned}\mathbf{r} &= \text{position vector of } \textit{field} \text{ point} = z\hat{\mathbf{k}} \\ \mathbf{r}' &= \text{position vector of } \textit{source} \text{ point} \\ &= \imath' \cos\phi'\hat{\mathbf{i}} + \imath' \sin\phi'\hat{\mathbf{j}} \\ dq' &= \sigma \imath'\, d\imath'\, d\phi'\end{aligned}$$

Hence,

$$\begin{aligned}\mathbf{r} - \mathbf{r}' &= z\hat{\mathbf{k}} - \imath' \cos\phi'\hat{\mathbf{i}} - \imath' \sin\phi'\hat{\mathbf{j}} \\ |\mathbf{r} - \mathbf{r}'| &= [(\imath')^2 + z^2]^{1/2}\end{aligned}$$

and we find on substitution of these quantities into Eq. (4-7) that

$$\begin{aligned}\mathbf{E}(\mathbf{r}) &= \frac{\sigma}{4\pi\epsilon_0}\int_0^\infty \int_0^{2\pi} \frac{z\hat{\mathbf{k}} - \imath' \cos\phi'\hat{\mathbf{i}} - \imath' \sin\phi'\hat{\mathbf{j}}}{[(\imath')^2 + z^2]^{3/2}}\, \imath'\, d\phi'\, d\imath' \\ &= \pm\frac{\sigma}{2\epsilon_0}\hat{\mathbf{k}} \qquad (4\text{-}13)\end{aligned}$$

where the upper sign applies when $z > 0$ and the lower sign when $z < 0$. Evaluation of the integral is left as an exercise. The z-component of the field

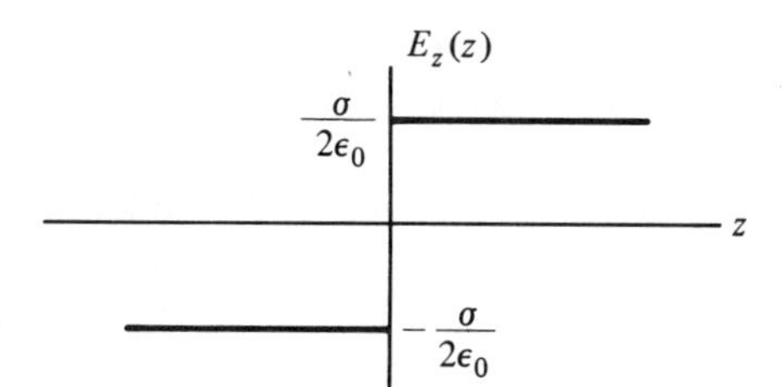

Fig. 4-10. The field of a uniformly charged infinite plate.

is shown in Fig. 4-10; it exhibits a discontinuity at $z = 0$ (at the sheet) but has everywhere the same magnitude. Thus, this (infinite) sheet establishes a constant field directed away from the sheet when $\sigma > 0$ and toward the sheet when $\sigma < 0$. Note that Eq. (4-13) was obtained by following a very systematic procedure that involved identifying the pertinent variables, substituting into the general expression of Eq. (4-7), and evaluating an integral. The reader is urged to adopt this approach whenever his problem involves setting up Eq. (4-7) for a charge distribution.

PROBLEMS

P4-1. Three point charges, each of strength 10^{-9} C, are placed one at each of three corners of a square 10 cm on a side. Calculate the electric field (magnitude *and* direction) at the fourth corner of the square. *Optional:* Write a short computer program that (1) accepts the coordinates and strength of, say, three point charges as input, (2) accepts the coordinates of a selected field point as input, (3) calculates and prints out the components of the electric field at the specified field point, and (4) returns to step (2). Test your program using the charge distribution in this problem and several different field points.

P4-2. Apply the binomial theorem (Appendix B) to expand Eq. (4-9) in powers of a/r to derive Eq. (4-10) and show that Eq. (4-11) follows from Eq. (4-10). *Hint:* See Table 0-1.

P4-3. Let a point charge Q be located at the origin and a dipole with dipole moment $\mathbf{p} = p(\cos\alpha\hat{\mathbf{r}} + \sin\alpha\hat{\boldsymbol{\theta}})$ be located at the point (r, θ, ϕ) in spherical coordinates. Determine the force experienced by this dipole and draw sketches of the three spherical components of the force as functions of α. Explain physically the origin of the $\hat{\boldsymbol{\theta}}$-component.

P4-4. Sketch the field lines of the linear quadrupole shown in Fig. 4-7(a).

P4-5. Draw a linear octupole consisting of four equally spaced charges, showing particularly the relative strengths of each charge. Note the appearance of the binomial coefficients.

P4-6. By direct integration, show that the electric field at a point a distance $\imath$ from an infinitely long line charge carrying a constant (linear) charge density λ is given by $\mathbf{E}(\mathbf{r}) = (\lambda/2\pi\epsilon_0\imath)\hat{\boldsymbol{\imath}}$, where the line defines the z axis and cylindrical coordinates are used. Don't be afraid of integral tables.

P4-7. A total charge Q is distributed on a plane circular ring of radius a with a linear charge density $\lambda(\phi) = \lambda_0(1 + \sin \phi)$, with ϕ defined as in Fig. P4-7. By direct integration, determine the electric field at the point $(0, 0, b)$ on the z axis. Express your answer in terms of Q and examine the field in the limit $b \gg a$.

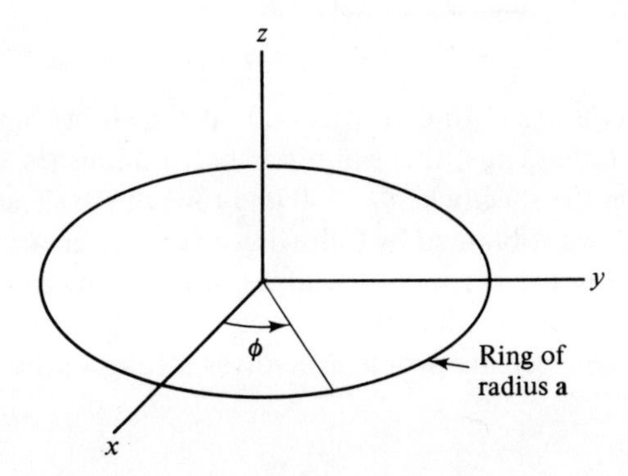

Figure P4-7

P4-8. A total charge Q is uniformly distributed over the surface of the region in the x-y plane bounded by the circle $x^2 + y^2 = a^2$. (a) Obtain an integral giving the electric field at the point $(0, y, z)$ in the y-z plane. (b) Evaluate the integral when the field point is on the z axis, i.e., for $y = 0$. (c) Draw a careful graph of $E_z(0, 0, z)$ as a function of z/a over the range $-\infty < z/a < \infty$. Use a desk calculator or computer. (d) Examine the result of part (b) in the limit $z \gg a$.

P4-9. A sphere of radius a carries a total charge Q uniformly distributed throughout its volume. By direct integration, determine the electric field at a point on a diameter (or extension thereof) at a distance b from the center of the sphere. Consider both $b < a$ and $b > a$. Express the results in terms of Q and show that the field at the specified point is the same as the field produced by a point charge located at the center with strength equal to the net charge lying inside a sphere of radius b. Sketch a graph of $|\mathbf{E}|$ versus b.

P4-10. Use Eqs. (4-5)–(4-7) to examine the field produced by a charge distribution of your own invention.

4-2 Gauss's Law

The basic information about electrostatics contained in Coulomb's law and in the principle of superposition can be reexpressed in a variety of ways. In this section we shall develop the first of two laws that are direct consequences of the information already presented. First, however, we must define the concept of the flux of the electric field across a specified surface Σ

placed in the field. This *electric flux*, denoted by Φ_e, is defined by the surface integral

$$\Phi_e = \int_\Sigma \mathbf{E}(\mathbf{r})\cdot d\mathbf{S} \tag{4-14}$$

and in effect counts the net number of lines of $\mathbf{E}$ piercing Σ. (See Section 2-3.) In particular, if $\mathbf{E}$ is constant and Σ is a plane surface of area $\mathbf{S}$, $\Phi_e = \mathbf{E}\cdot\mathbf{S}$ and is the product of the surface area and the component of $\mathbf{E}$ perpendicular to (the plane of) the surface. The electric flux may be either positive or negative and, in mks units, is expressed in $\text{N}\cdot\text{m}^2/\text{C}$. If Σ happens to be a *closed* surface (Fig. 4-11), we take $d\mathbf{S}$ to represent an *outward* normal and the flux Φ_e, now defined by $\oint_\Sigma \mathbf{E}\cdot d\mathbf{S}$, represents the net outward flux of $\mathbf{E}$ over the surface. Regions in which the field lines enter the surface make negative contributions to the flux while regions in which the field lines leave the surface make positive contributions.

Now, the flux of a static electric field across a *closed* surface has a simple evaluation. Consider first the flux of the electric field of a *point* charge q_0 located at $\mathbf{r}_0$ across an arbitrary closed surface Σ. The field is given by Eq. (4-5) with $\mathbf{r}' = \mathbf{r}_0$ and $q' = q_0$, and the required flux is then given by

$$\Phi_e = \frac{q_0}{4\pi\epsilon_0}\oint_\Sigma \frac{\mathbf{r}-\mathbf{r}_0}{|\mathbf{r}-\mathbf{r}_0|^3}\cdot d\mathbf{S} \tag{4-15}$$

We have, however, already evaluated the integral in Eq. (4-15). Substituting

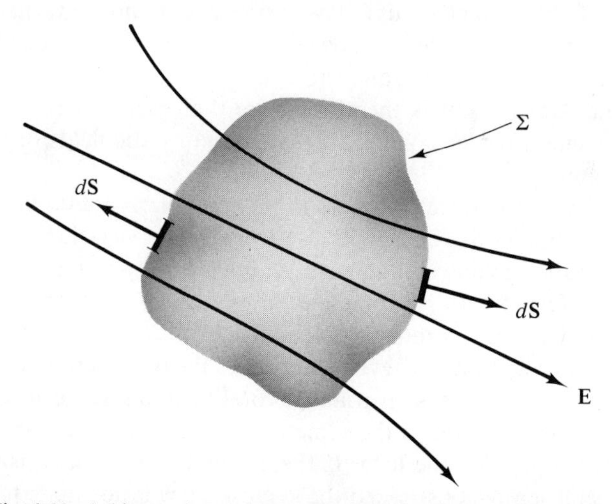

Fig. 4-11. Arbitrary closed surface in an electric field. The contribution to the net electric flux is negative along that portion of the surface where the field enters the enclosed volume and positive along that portion where the field emerges from the volume.

from Eq. (2-37) we find the very simple value

$$\begin{aligned} \Phi_e &= \frac{q_0}{\epsilon_0} \quad \text{if } q_0 \text{ is inside } \Sigma \\ &= 0 \quad \text{if } q_0 \text{ is outside } \Sigma \end{aligned} \tag{4-16}$$

regardless of the shape of Σ! [Note the disappearance of the factor 4π in Eq. (4-16)—a consequence of our earlier selection of rationalized units.] If the field is produced by a distribution of charge, the principle of superposition enables us to write immediately that

$$\Phi_e = \oint_\Sigma \mathbf{E} \cdot d\mathbf{S} = \frac{(\text{net charge } \textit{within } \Sigma)}{\epsilon_0} \tag{4-17}$$

When the charge distribution is described by a volume charge density, for example, Eq. (4-17) assumes the more specific form

$$\oint_\Sigma \mathbf{E} \cdot d\mathbf{S} = \frac{1}{\epsilon_0} \int_V \rho \, dv \tag{4-18}$$

where the volume integral extends over the volume V bounded by the closed surface Σ. Note particularly that Σ is arbitrary; Eq. (4-18) applies to any (closed) surface whatever, even if that surface is only an imaginary surface introduced for the purpose of exploiting this relationship. Equations (4-16)–(4-18) are forms of what is called *Gauss's law*, one of the basic laws of electricity. When the fields are *static*, Gauss's law is a direct consequence of Coulomb's law and merely expresses some of the information in Coulomb's law in a new form. In fact, Gauss's law carries over to nonstatic fields without change of form. That extension, however, can be made only on the basis of experimental observations beyond those supporting Coulomb's law, and the "generalized" Gauss's law is therefore more than merely a restatement of a portion of Coulomb's law. We shall examine nonstatic fields in more detail in later chapters.

Gauss's law in the form of Eq. (4-18) is particularly useful in at least two contexts. We shall consider first its use to calculate electric fields when the source distribution manifests sufficient symmetry that some properties of the field can be inferred without complete knowledge of the field. The essential premise of a symmetry argument is that the field established by some source distribution must exhibit whatever symmetries the distribution itself exhibits. If, for example, the source is invariant to rotation about some axis, the resulting field must be invariant to the same transformation. Thus, any symmetry of the source constrains the field. If these symmetries are extensive enough, the field is sufficiently constrained that Gauss's law is adequate to determine those properties of the field not completely fixed by the symmetries. We shall illustrate the approach by applying symmetry and Gauss's law to determine the field established by an infinitely long line charge with constant linear charge density λ. For simplicity let this line define the z axis of a cylindrical coordinate system. The most general field then has the form

$$\mathbf{E}(\mathbf{r}) = E_{\imath}(\imath, \phi, z)\hat{\boldsymbol{\imath}} + E_{\phi}(\imath, \phi, z)\hat{\boldsymbol{\phi}} + E_{z}(\imath, \phi, z)\hat{\mathbf{k}} \tag{4-19}$$

The source distribution, however, is invariant to rotation about the z axis, and the field exhibits this invariance only if $E_{\imath}$, E_{ϕ}, and E_z are independent of ϕ. In addition, the source distribution is invariant to translation along the z axis, and the field exhibits this invariance only if $E_{\imath}$, E_{ϕ}, and E_z are independent of z. The most general field consistent with these two symmetries then is

$$\mathbf{E}(\mathbf{r}) = E_{\imath}(\imath)\hat{\boldsymbol{\imath}} + E_{\phi}(\imath)\hat{\boldsymbol{\phi}} + E_{z}(\imath)\hat{\mathbf{k}} \tag{4-20}$$

But the source distribution is also invariant to reflection in the x-y plane, under which transformation $z \longrightarrow -z$ and $\hat{\mathbf{k}} \longrightarrow -\hat{\mathbf{k}}$; Eq. (4-20) is invariant to this transformation only if $E_z = 0$. Finally, the source distribution is invariant to reflection in the x-z plane, under which transformation $\phi \longrightarrow -\phi$, $\hat{\mathbf{j}} \longrightarrow -\hat{\mathbf{j}}$, and

$$\hat{\boldsymbol{\imath}} = \cos\phi\hat{\mathbf{i}} + \sin\phi\hat{\mathbf{j}} \longrightarrow \cos\phi\hat{\mathbf{i}} + \sin\phi\hat{\mathbf{j}} = \hat{\boldsymbol{\imath}}$$
$$\hat{\boldsymbol{\phi}} = -\sin\phi\hat{\mathbf{i}} + \cos\phi\hat{\mathbf{j}} \longrightarrow \sin\phi\hat{\mathbf{i}} - \cos\phi\hat{\mathbf{j}} = -\hat{\boldsymbol{\phi}}$$

but nothing else is changed; Eq. (4-20) is consistent with this invariance only if $E_{\phi} = 0$. Thus, the symmetries of the selected source distribution constrain the field to be no more complicated than[1]

$$\mathbf{E} = E_{\imath}(\imath)\hat{\boldsymbol{\imath}} \tag{4-21}$$

We now use Gauss's law to determine $E_{\imath}(\imath)$. Although the law applies to any closed surface, the most useful surfaces to choose are those that take advantage of at least some of the symmetries of the source. The appropriate surface for this problem might be described as a cylindrical fruit juice can of radius $\imath$ and length L with its axis coincident with the line of charge (Fig. 4-12). The flux of the electric field out of this surface (sometimes called a Gaussian surface or, especially when small, a Gaussian pillbox) possesses potentially three contributions:

$$\Phi_e = \oint \mathbf{E}\cdot d\mathbf{S} = \int_{\text{top}} \mathbf{E}\cdot d\mathbf{S} + \int_{\text{bottom}} \mathbf{E}\cdot d\mathbf{S} + \int_{\text{cylinder}} \mathbf{E}\cdot d\mathbf{S} \tag{4-22}$$

With $\mathbf{E}$ given by Eq. (4-21), however, $\mathbf{E}\cdot d\mathbf{S} = 0$ on the top and bottom surfaces ($d\mathbf{S} \,||\, \hat{\mathbf{k}}$). Only the integral over the cylindrical surface makes a contribution. On that surface, $d\mathbf{S} = \imath\, d\phi\, dz\, \hat{\boldsymbol{\imath}}$ and

$$\Phi_e = \int_{z_0}^{z_0+L} dz \int_0^{2\pi} d\phi\, E_{\imath}(\imath)\hat{\boldsymbol{\imath}}\cdot\imath\hat{\boldsymbol{\imath}} = 2\pi L \imath E_{\imath}(\imath) \tag{4-23}$$

since $\imath$ is constant (albeit arbitrary) insofar as the integral is concerned. Now, the charge enclosed within the selected surface is λL. Since Gauss's law re-

[1] As can be seen, the statement "because of symmetry, the field must be radially directed with a radial component dependent only on $\imath$", while true, nonetheless hides the rather extensive symmetry argument that supports it. The reader is urged to quote symmetry in support of various conclusions only after supplying the full argument.

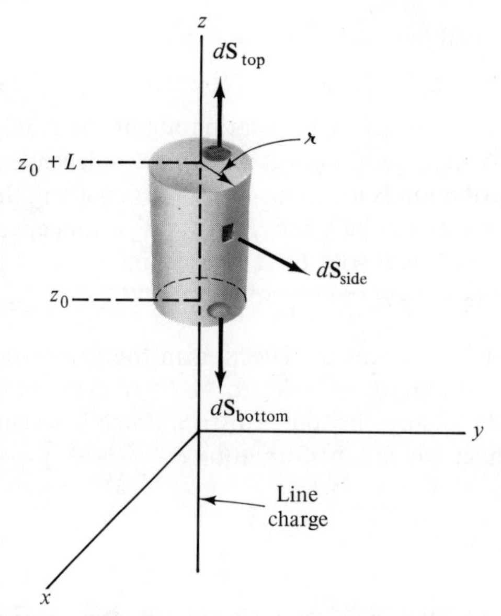

Fig. 4-12. A Gaussian pillbox for determining the field of a line charge.

quires that $\Phi_e = \lambda L/\epsilon_0$, Eq. (4-23) gives

$$2\pi L\imath E_{\imath}(\imath) = \frac{\lambda L}{\epsilon_0} \tag{4-24}$$

Thus $E_{\imath}(\imath) = \lambda/2\pi\epsilon_0\imath$ and from Eq. (4-21) we find that

$$\mathbf{E} = \frac{\lambda}{2\pi\epsilon_0\imath}\hat{\imath} \tag{4-25}$$

in agreement with the result obtained by direct integration in P4-6.

Gauss's law can also be fruitfully applied to some situations involving conductors, which are by definition pieces of matter whose intrinsic composition includes microscopic charges that are free to move in response to any forces applied to them. Thus, if a conductor is placed in a region of space initially containing a static electric field, these *free charges* will move in response to the field, thereby being redistributed within the conductor and in turn adding their own contribution to the field within (and without) the conductor. This adjustment of the charge distribution within the conductor will continue as long as there remains any field within the conductor to apply forces to the free charges. *Once the fields are again static, the field within the conductor must be zero;* any other value contradicts the requirement that the fields be static. Thus, when the fields outside a conductor are static, the region occupied by the conductor itself is free of fields. Further, the electric field at the surface of such a conductor must be normal to the surface, for any tangential component would cause a motion of free charges along the surface, again contradicting the requirement that the fields be static. These properties of the

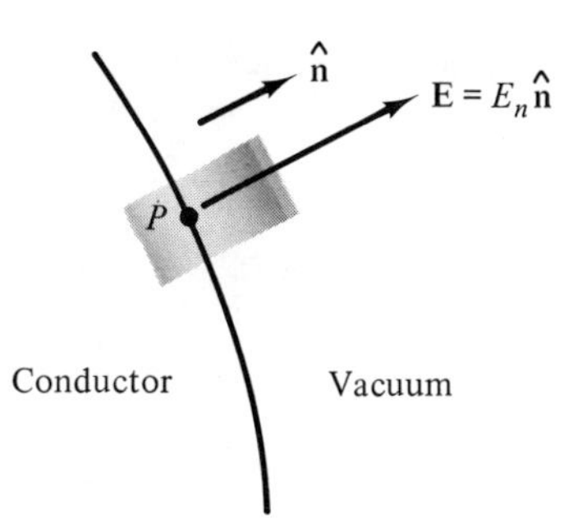

Fig. 4-13. A Gaussian pillbox for determining the field at the surface of a conductor.

fields in and around conductors facilitate the use of Gauss's law, and we shall illustrate by determining the field just outside a conducting surface at a point P where the (surface) charge density is σ. A suitable Gaussian pillbox is shown in Fig. 4-13; its sides consist of a cylindrical surface with its axis normal to the conducting surface at P and two plane surfaces, one on each side of the conducting surface. Further, the pillbox is chosen small enough to enclose an element of the conducting surface that is approximately plane and over which σ and $\mathbf{E}$ are both approximately constant. Let the cross-sectional area of the pillbox be ΔS and let $\hat{\mathbf{n}}$ be a unit vector in the direction of the *outward* normal to the conducting surface at P. Then the charge within the pillbox is $\sigma\,\Delta S$, and the flux out of the pillbox is $E_n\,\Delta S$, the latter following because only the plane surface of the pillbox outside the conductor contributes to the flux. (The field is zero inside the conductor and does not pierce the cylindrical walls of the pillbox outside the conductor.) Applying Gauss's law as expressed in Eq. (4-17), we find that $E_n\,\Delta S = \sigma\,\Delta S/\epsilon_0$ or $E_n = \sigma/\epsilon_0$ or finally that the field $\mathbf{E}$ just outside a conducting surface at a point P where the charge density is σ is given by

$$\mathbf{E} = \frac{\sigma}{\epsilon_0}\hat{\mathbf{n}} \tag{4-26}$$

where $\hat{\mathbf{n}}$ is a unit outward normal at the point P.

For many purposes, a differential form of Gauss's law is more convenient than the integral form, Eq. (4-18). Let Σ and V become arbitrarily small. In that limit, we can use the identity in Eq. (2-25) to write

$$\oint_{\Delta\Sigma} \mathbf{E}\cdot d\mathbf{S} \approx \nabla\cdot\mathbf{E}\,\Delta V \tag{4-27}$$

(We now denote the small surface and volume elements by $\Delta\Sigma$ and ΔV.) Further, if ΔV is small enough, ρ can be regarded as a constant throughout ΔV, and

$$\frac{1}{\epsilon_0}\int_{\Delta V} \rho\, dv \approx \frac{1}{\epsilon_0}\rho\,\Delta V \tag{4-28}$$

Finally, substituting Eqs. (4-27) and (4-28) into Eq. (4-18), dividing by ΔV, and passing to the limit $\Delta V \longrightarrow 0$ [in which limit the approximations in Eqs.

(4-27) and (4-28) become exact], we find the differential relationship

$$\nabla \cdot \mathbf{E} = \frac{\rho}{\epsilon_0} \tag{4-29}$$

which is equivalent to the integral expression in Eq. (4-18). Although we shall occasionally need this result in the rest of this chapter, we shall postpone detailed consideration of the differential form of the basic laws until Chapter 6.

PROBLEMS

P4-11. An infinitely long straight rod having a circular cross section of radius b is charged to a uniform volume charge density ρ. Using symmetry and Gauss's law, determine the electric field at a distance $\imath$ from the axis of the rod. Consider both $\imath < b$ and $\imath > b$. Sketch a graph of $E_\imath$ versus $\imath$.

P4-12. Use symmetry and Gauss's law to obtain the field established by an infinite plane sheet of charge characterized by a constant (surface) charge density σ. Compare your result with Eq. (4-13).

P4-13. Use Gauss's law to show that a region containing an electrostatic field but devoid of charge includes *no* point at which a test charge would be in stable equilibrium. *Hint:* What must the field lines look like near a point of stable equilibrium?

P4-14. Let all of space be uniformly charged to a volume charge density ρ. This distribution is invariant to arbitrary translations and to arbitrary rotations about an arbitrary axis. Symmetry therefore requires the field everywhere to be zero. Hence, any arbitrary closed surface lying in this distribution has no electric flux across it. But that surface certainly contains some charge, and we have an apparent contradiction of Gauss's law. Where is the error? Why is there no contradiction in applying symmetry arguments and Gauss's law to infinite sheets and infinite line charges?

P4-15. An uncharged conducting object has a hollow cavity in its interior. If a point charge q is placed in the cavity, prove that a charge $-q$ is induced on the surface of the cavity and a charge q is induced on the outer surface of the conductor.

P4-16. One of two plane parallel conducting plates of nonzero thickness carries a charge Q and the other carries a charge $-Q$. Each charge assumes a static distribution on the *surfaces* of its plate. The situation is shown in Fig. P4-16. (a) Neglect fringing at the ends (i.e., treat the plates as infinite) and use symmetry to argue that σ_1, σ_2, σ_3, and σ_4 must be constant, that $\sigma_2 = -\sigma_3$, that $\sigma_1 = -\sigma_4$, and that the electric field everywhere is parallel to $\hat{\mathbf{n}}$. (b) Use Gauss's law to show that the field in the region between the plates is uniform and given by $\mathbf{E} = (\sigma_2/\epsilon_0)\hat{\mathbf{n}}$. (c) Using the result in Eq. (4-13), sketch graphs of the $\hat{\mathbf{n}}$-component of the electric field established by *each* sheet of charge in Fig. P4-16 and then sketch a graph of the *net* field established by all

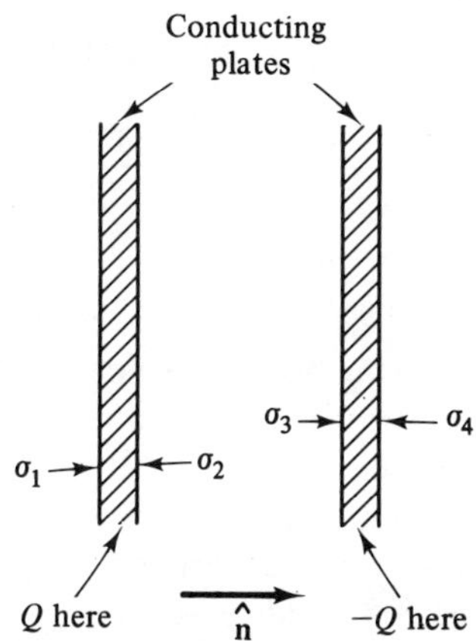

Figure P4-16

four sheets. Show that the field in the interior of each plate will be zero only if $\sigma_1 = -\sigma_4 = 0$, thus showing that all of the charge placed on the plates accumulates on the facing surfaces. (d) Show that the field in the region outside that bounded by the plates is identically zero. The arrangement in this problem is called a *parallel plate capacitor.*

4-3
The Restricted Faraday Law

The second law of electrostatics that follows directly from Coulomb's law and the principle of superposition is a restricted form of what we shall later call Faraday's law; in essence it states that the force field $\mathbf{F}_q(\mathbf{r}) = q\mathbf{E}(\mathbf{r})$ corresponding to the electro*static* field $\mathbf{E}(\mathbf{r})$ is conservative. This law is expressed analytically in the integral form

$$\oint \mathbf{F}_q(\mathbf{r})\cdot d\mathbf{r} = 0 \Longrightarrow \oint \mathbf{E}(\mathbf{r})\cdot d\mathbf{r} = 0 \tag{4-30}$$

where the (*closed*) path of integration is arbitrary [see Eq. (0-35)], or in the differential form

$$\nabla \times \mathbf{F}_q = 0 \Longrightarrow \nabla \times \mathbf{E} = 0 \tag{4-31}$$

[See Eq. (0-53).] Since the validity of Eq. (4-30) [and hence of Eq. (4-31)] for an arbitrary static field follows by superposition from its validity for the static field of a point charge, we need only establish Eq. (4-30) for a point charge. Let the charge have strength q' and be located at $\mathbf{r}'$. Then its field is given by Eq. (4-5) and we find that

$$\oint \mathbf{E}\cdot d\mathbf{r} = \frac{q'}{4\pi\epsilon_0}\oint \frac{\mathbf{r}-\mathbf{r}'}{|\mathbf{r}-\mathbf{r}'|^3}\cdot d\mathbf{r}$$

$$= \frac{q'}{4\pi\epsilon_0}\oint \frac{(x-x')\,dx + (y-y')\,dy + (z-z')\,dz}{[(x-x')^2 + (y-y')^2 + (z-z')^2]^{3/2}}$$

$$= -\frac{q'}{4\pi\epsilon_0}\oint\left(dx\frac{\partial}{\partial x} + dy\frac{\partial}{\partial y} + dz\frac{\partial}{\partial z}\right)\frac{1}{[\cdots]^{1/2}}$$

$$= -\frac{q'}{4\pi\epsilon_0}\oint d\left(\frac{1}{|\mathbf{r} - \mathbf{r}'|}\right)$$

$$= -\frac{q'}{4\pi\epsilon_0}\frac{1}{|\mathbf{r} - \mathbf{r}'|}\Bigg|_{\text{starting }\mathbf{r}}^{\text{finishing }\mathbf{r}} = 0 \tag{4-32}$$

the final form following because the path is closed and the starting and finishing points are the same point. Equations (4-30) and (4-31) are therefore established for the general electrostatic field. In contrast to Gauss's law, this restricted form of Faraday's law is valid only for static electric fields. In Chapter 6, we shall look to experiment to determine the appropriate modifications when the electric field ceases to be static. The resulting generalization, of course, is then more than merely a consequence of Coulomb's law.

4-4 The Electrostatic Potential

As discussed in Section 0-3, every conservative force field $\mathbf{F}_q(\mathbf{r})$ has associated with it a scalar potential energy field $U(\mathbf{r})$ given by Eq. (0-40), which assumes the form

$$U(\mathbf{r}) = U(\mathbf{r}_0) - \int_{\mathbf{r}_0}^{\mathbf{r}} q\mathbf{E}(\mathbf{r})\cdot d\mathbf{r} \tag{4-33}$$

for the electrostatic force field $\mathbf{F}_q = q\mathbf{E}$. Here, $\mathbf{r}_0$ is an arbitrary reference point and $U(\mathbf{r}_0)$ may be assigned an arbitrary value; both are commonly chosen to simplify the form of $U(\mathbf{r})$. Further, by Eq. (0-48), the force field can be recovered from the potential energy field by evaluating the negative gradient,

$$\mathbf{F}_q(r) = q\mathbf{E}(\mathbf{r}) = -\nabla U(\mathbf{r}) \tag{4-34}$$

and by Eq. (0-39) the potential energy field appears in the expression

$$\tfrac{1}{2}mv^2 + U(\mathbf{r}) = \text{constant} \tag{4-35}$$

for the conservation of the mechanical energy of a particle of mass m moving in the given force field with velocity $\mathbf{v}$. It is customary to eliminate the test charge q by dividing Eqs. (4-33) and (4-34) by q and introducing the *electrostatic potential field* $V(\mathbf{r})$ as the potential energy *per unit charge*, viz.,

$$V(\mathbf{r}) = \frac{U(\mathbf{r})}{q} \tag{4-36}$$

Equation (4-33) then becomes

$$V(\mathbf{r}) = V(\mathbf{r}_0) - \int_{\mathbf{r}_0}^{\mathbf{r}} \mathbf{E}(\mathbf{r})\cdot d\mathbf{r} \tag{4-37}$$

and permits calculation of $V(\mathbf{r})$ if $\mathbf{E}(\mathbf{r})$ is known. Determination of $\mathbf{E}(\mathbf{r})$ from knowledge of $V(\mathbf{r})$ involves Eq. (4-34), which assumes the form

$$\mathbf{E}(\mathbf{r}) = -\nabla V(\mathbf{r}) \tag{4-38}$$

Finally, the conservation law, Eq. (4-35), becomes

$$\tfrac{1}{2}mv^2 + qV(\mathbf{r}) = \text{constant} \tag{4-39}$$

(*Warning:* The factor of q difference between the electrostatic *potential* and the electrostatic *potential energy* is easily forgotten.) From Eq. (4-36), we infer that the mks unit of potential is the joule/coulomb (J/C), a combination given the name *volt* (V). We then conclude from Eq. (4-38) that the mks unit of electric field, previously identified as the N/C, must also be expressible as the V/m; the second name is the more common.

We have now shown the existence of a potential function for an electrostatic field and we have obtained expressions for finding the potential from the field [Eq. (4-37)] and for finding the field from the potential [Eq. (4-38)]. We next obtain an expression for determining the potential directly from the source distribution. Let the source first be a point charge of strength q at the origin. If the reference point $\mathbf{r}_0$ is taken at infinity and $V(\mathbf{r}_0)$ is set equal to zero, then combining Eq. (4-5) (with $q' \longrightarrow q$, $\mathbf{r}' = 0$) with Eqs. (4-37) and (0-30) gives

$$\begin{aligned} V(\mathbf{r}) &= -\int_\infty^{\mathbf{r}} \frac{q}{4\pi\epsilon_0}\frac{\hat{\mathbf{r}}}{r^2}\cdot(\hat{\mathbf{r}}\,dr + \hat{\boldsymbol{\theta}}r\,d\theta + \hat{\boldsymbol{\phi}}r\sin\theta\,d\phi) \\ &= -\int_\infty^{r} \frac{q}{4\pi\epsilon_0}\frac{dr}{r^2} = \frac{q}{4\pi\epsilon_0 r} \end{aligned} \tag{4-40}$$

Translation of the coordinate system so that the point charge is at $\mathbf{r}_0$ and the field point is at $\mathbf{r}$ merely replaces r in Eq. (4-40) with $|\mathbf{r} - \mathbf{r}_0|$. Thus the potential established by a point charge q located at $\mathbf{r}_0$ is given by

$$V(\mathbf{r}) = \frac{q}{4\pi\epsilon_0}\frac{1}{|\mathbf{r} - \mathbf{r}_0|} \tag{4-41}$$

Finally, by superposition we obtain the expression

$$V(\mathbf{r}) = \frac{1}{4\pi\epsilon_0}\sum_i \frac{q_i}{|\mathbf{r} - \mathbf{r}_i|} \tag{4-42}$$

for the potential established by an array of point charges q_i at $\mathbf{r}_i$. Thus, for example, the potential established by the dipole shown in Fig. 4-6 is given by

$$V(\mathbf{r}) = \frac{q}{4\pi\epsilon_0}\left[\frac{1}{|\mathbf{r} - \frac{1}{2}\mathbf{a}|} - \frac{1}{|\mathbf{r} + \frac{1}{2}\mathbf{a}|}\right] \tag{4-43}$$

which reduces to the simpler form

$$V(\mathbf{r}) = \frac{\mathbf{p}\cdot\mathbf{r}}{4\pi\epsilon_0 r^3} = \frac{p\cos\theta}{4\pi\epsilon_0 r^2}, \quad \mathbf{p} = q\mathbf{a} \tag{4-44}$$

in the limit $r \gg a$ (P4-23) and varies as r^{-2}, contrasting with the variation of the potential of a monopole with r^{-1}.

We could relate the potential established by a distributed source to the source itself by applying to Eq. (4-42) an argument such as that leading to Eq. (4-4). To set the stage for the corresponding magnetic development, however, we shall here adopt a different approach. Note first the mathematical identity,

$$\frac{\mathbf{r} - \mathbf{r}'}{|\mathbf{r} - \mathbf{r}'|^3} = -\nabla \frac{1}{|\mathbf{r} - \mathbf{r}'|} \tag{4-45}$$

where the gradient involves derivatives with respect to the components of $\mathbf{r}$. Substitution of this identity into Eq. (4-4) leads ultimately to

$$\mathbf{E}(\mathbf{r}) = -\nabla\left(\frac{1}{4\pi\epsilon_0} \int \frac{dq'}{|\mathbf{r} - \mathbf{r}'|}\right) \tag{4-46}$$

in which the gradient has been written in front of the integral because it does not involve differentiations with respect to any of the variables of integration. Comparison with Eq. (4-38) then results in the identification

$$V(\mathbf{r}) = \frac{1}{4\pi\epsilon_0} \int \frac{dq'}{|\mathbf{r} - \mathbf{r}'|} + \chi(\mathbf{r}) \tag{4-47}$$

where $\chi(\mathbf{r})$ is arbitrary except that its gradient must vanish if the negative gradient of Eq. (4-47) is to give the correct field, i.e.,

$$\nabla\chi = \frac{\partial \chi}{\partial x}\,\hat{\mathbf{i}} + \frac{\partial \chi}{\partial y}\,\hat{\mathbf{j}} + \frac{\partial \chi}{\partial z}\,\hat{\mathbf{k}} = 0 \tag{4-48}$$

Equation (4-48) is a vector equation and each component must be separately zero. Thus χ cannot depend on x, y, or z; at worst, it can be a constant and we have finally that

$$V(\mathbf{r}) = \frac{1}{4\pi\epsilon_0} \int \frac{dq'}{|\mathbf{r} - \mathbf{r}'|} + \text{constant} \tag{4-49}$$

where the integral extends over the charge distribution and dq' is expressed as appropriate to the distribution, e.g.,

$$V(\mathbf{r}) = \frac{1}{4\pi\epsilon_0} \int \frac{\rho(\mathbf{r}')}{|\mathbf{r} - \mathbf{r}'|}\, dv' + \text{constant} \tag{4-50}$$

for a volume distribution. Equations (4-49) and (4-50) not only provide for a calculation of V directly from known sources but also express the relationship between (a static) V and its sources even when the sources are not explicitly known.[2]

An extremely important equation for the electrostatic potential is obtained by substituting the equation $\mathbf{E} = -\nabla V$ into Gauss's law, $\nabla \cdot \mathbf{E} = \rho/\epsilon_0$, to

[2]The existence of V might also have been inferred from the equation $\nabla \times \mathbf{E} = 0$ by applying the theorem in item (3) of Section 2-5. This route to Eq. (4-38), however, does not provide an expression for calculating V directly from its sources.

TABLE 4-1 Interrelationships Among Static ρ, $\mathbf{E}$, and V

Given \ Find	ρ	$\mathbf{E}$	V
ρ	—	$\frac{1}{4\pi\epsilon_0}\int \frac{\mathbf{r}-\mathbf{r}'}{\lvert\mathbf{r}-\mathbf{r}'\rvert^3}\rho(\mathbf{r}')\,dv'$	$\frac{1}{4\pi\epsilon_0}\int \frac{\rho(\mathbf{r}')}{\lvert\mathbf{r}-\mathbf{r}'\rvert}dv'$
$\mathbf{E}$	$\epsilon_0\,\nabla\cdot\mathbf{E}$	—	$-\int \mathbf{E}\cdot d\mathbf{r}$
V	$-\epsilon_0\,\nabla^2 V$	$-\nabla V$	—

obtain

$$\nabla\cdot(\nabla V) = \nabla^2 V = -\frac{\rho}{\epsilon_0} \tag{4-51}$$

where the *Laplacian operator* ∇^2 is defined in Section 2-5. This equation can be used to determine the source of a field if the potential is known; it is called *Poisson's equation* if $\rho \neq 0$ and *Laplace's equation* if $\rho = 0$. One solution to Eq. (4-51) is, of course, expressed in Eq. (4-50), but this solution is not always useful. A detailed examination of this equation is postponed to Chapter 8.

The interrelationships we have now developed among the charge density, the electrostatic field, and the electrostatic potential are summarized in Table 4-1.

In addition to representing an electric field graphically by displaying its field lines, we can also represent the field by showing its *equipotential contours*, which are curves or surfaces at every point of which the electrostatic potential has the *same* value. The equipotential contours for several selected fields are shown by the light curves in Figs. 4-5 and 4-8. The field itself is strong where the contours are close together [field = rate of change of potential; see Eq. (4-38)]. Further, the field is at every point perpendicular to the equipotential contours, a property that we prove by evaluating the potential difference $V(\mathbf{r}_a + d\mathbf{r}) - V(\mathbf{r}_a)$ between two infinitesimally separated points $\mathbf{r}_a + d\mathbf{r}$ and $\mathbf{r}_a$ lying in the *same* equipotential contour. By hypothesis, this potential difference is zero, but it is also calculable directly from Eq. (4-37); we find that

$$0 = -\int_{\mathbf{r}_a}^{\mathbf{r}_a+d\mathbf{r}} \mathbf{E}\cdot d\mathbf{r} = -\mathbf{E}(\mathbf{r}_a)\cdot d\mathbf{r} \tag{4-52}$$

Thus, for the particular displacement assumed, the electric field is perpendicular to $d\mathbf{r}$. But $d\mathbf{r}$ was any displacement lying in an equipotential contour and the electric field is therefore perpendicular to the equipotential contour. Q.E.D.

That the electric field at the surface of a conductor is perpendicular to that surface now leads us to suspect that the surface of a conductor must be an equipotential. Indeed, since there is no field anywhere within the body of a conductor in a static field, the potential difference between *any* two points in such a conductor—not just between two points on the surface—is zero and the *entire volume* of space occupied by a conductor in a static field is therefore an equipotential region. Thus, we can speak unambiguously of *the* potential of a conductor. Now, suppose that *two* conductors, one carrying charge Q and the other carrying charge $-Q$, are placed near one another in a region previously free of fields. Once the static fields produced by the charges on the conductors have been established, the charge on each conductor is distributed so that each conductor is at a definite potential. Any device of the sort described is called a *capacitor*, and the ratio of the charge Q to the *difference* ΔV in potential between the two conductors is called the *capacitance* C of the arrangement,

$$C = \frac{Q}{\Delta V} \tag{4-53}$$

In most cases, ΔV is proportional to Q, and C is therefore independent of Q; C is essentially a geometric quantity, determined largely by the size, shape, and separation of the two conductors. The mks unit of capacitance is the coulomb per volt (C/V), called the *farad* (F), which turns out to be inconveniently large; the microfarad (μF; 10^{-6} F) and even the picofarad (pF; 10^{-12} F) occur more commonly.

PROBLEMS

P4-17. (a) Show that the electrostatic potential function associated with a *constant* electric field $\mathbf{E}$ is given by $-\mathbf{E}\cdot\mathbf{r}$, where the reference point of zero potential is taken at the coordinate origin. (b) Show that the potential *energy* of a dipole of moment $\mathbf{p}$ placed in a constant electric field $\mathbf{E}$ is given by $-\mathbf{p}\cdot\mathbf{E}$.

P4-18. (a) Substituting the field obtained in Eq. (4-25) into Eq. (4-37), show that the electrostatic potential at a distance $\imath$ from the axis of an infinite line charge is given by

$$V(\imath) = -\frac{\lambda}{2\pi\epsilon_0}\,\ell n\,\frac{\imath}{a}$$

if the potential is taken to be zero at $\imath = a$. Why is the reference point not taken either at ∞ or at the axis? (b) Recover the field from the potential by applying Eq. (4-38).

P4-19. Starting with Eq. (4-26), show that the charge density σ at a point on the surface of a conductor in a static field described by the potential V is given by $\sigma = -\epsilon_0\,\nabla V\cdot\hat{\mathbf{n}}$, where $\hat{\mathbf{n}}$ is a unit outward normal to the conductor and the gradient is evaluated at the point where the charge density is desired.

P4-20. A nonrelativistic alpha particle of mass m and charge $2q$ with initial speed v experiences a head-on collision with a *heavy* nucleus of atomic number Z, and hence of charge Zq. Use conservation of energy to determine the distance of closest approach to the nucleus.

P4-21. Two spherical conducting raindrops of radius a are both charged to the same potential V relative to a reference point at infinity. Determine the potential of the single drop that results when the two drops are allowed to merge.

P4-22. What is the highest potential to which a spherical conductor of radius 20 cm can be charged if air becomes conducting when a field exceeding 3×10^6 V/m is produced?

P4-23. Derive Eq. (4-44) for the potential of a dipole by expanding Eq. (4-43) in powers of a/r and then calculate the field of the dipole. Compare your result with Eq. (4-11). *Hint:* Use the gradient in spherical coordinates, Eq. (0-57).

P4-24. Determine the electrostatic potential of a linear quadrupole [Fig. 4-7(a)] having its axis along the z axis when the quadrupole is observed from a distance large compared to its dimensions, and then find the field. Let the separation of adjacent charges be a. The coordinates in Fig. 4-6 with the center of the quadrupole at the origin are also convenient here.

P4-25. Consider a uniformly charged plane circular disc of radius a lying in the x-y plane with its center at the origin. (a) Write an integral (multiple if necessary) giving the electrostatic potential at the point $\mathbf{r}$. (b) Determine the potential $V(z)$ at the point $(0, 0, z)$ and sketch a graph of $V(z)$ versus z. (c) Obtain approximate expressions for the potential $V(z)$ in the two regions $z \gg a$ and $z \ll a$ and interpret the two limiting expressions physically. (d) Calculate the z-component of the electric field at the point $(0, 0, z)$ and sketch a graph of this component as a function of z. (e) Can you calculate the x- and y-components of the electric field at $(0, 0, z)$ from the potential obtained in part (b)? If so, do it; if not, explain why not. *Optional:* By applying the integral form of Gauss's law to a small cylindrical pillbox centered on the axis of the disc and a distance z above its plane, show that the radial component of the electric field at points close to the axis is given by

$$E_{\imath}(\imath, z) \approx -\frac{\imath}{2} \frac{\partial E_z}{\partial z}\bigg|_{\imath=0}$$

where $\imath$ is the distance of the point from the axis. Can you obtain this same result by manipulating with the differential form of Gauss's law?

P4-26. Let a portion of the z axis in the region $-a \leq z \leq a$ carry a charge Q uniformly distributed along its length with linear density λ. (a) Set up an integral giving the electrostatic potential at any point in space. (b) Evaluate the potential at points on the positive z axis for which $z > a$. (c) Show that the result obtained in part (b) approaches $Q/4\pi\epsilon_0 z$ in the limit $z \gg a$. Note the power series expansion $\ell n\,(1 + x) = x - \frac{1}{2}x^2 + \frac{1}{3}x^3 + \cdots$. (d) Explain how you could have predicted the limiting value in part (c) without actually evaluating the integral in part (b).

P4-27. Determine the electric field and the source distribution corresponding to the *Yukawa potential*, $V(\mathbf{r}) = Qe^{-\alpha r}/4\pi\epsilon_0 r$, where α is a constant.

P4-28. An infinite conducting plate in the plane $z = 0$ is maintained at potential zero and a similar plate in the plane $z = d$ is maintained at potential V_0. Assume that there is no charge in the region between the plates. Because the plates are invariant to arbitrary translation parallel to the x-y plane, the potential can be a function only of z, $V = V(z)$, and Laplace's equation reduces to $d^2V/dz^2 = 0$. By solving Laplace's equation and imposing the given values of V at the two plates, find $V(z)$ for $0 < z < d$. Then find the field between the plates and the charge density on each plate. *Hint:* See Eq. (4-26).

P4-29. Describe the equipotential contours corresponding to the three-dimensional field of a point charge.

P4-30. Obtain a carefully drawn sketch of the equipotentials established in the y-z plane by a point dipole having dipole moment $\mathbf{p} = p\hat{\mathbf{k}}$. (Compare P0-16.)

P4-31. The equipotential curves established by a particular two-dimensional charge distribution are shown in Fig. P4-31. Determine both the magnitude and the direction of the electric field at point A.

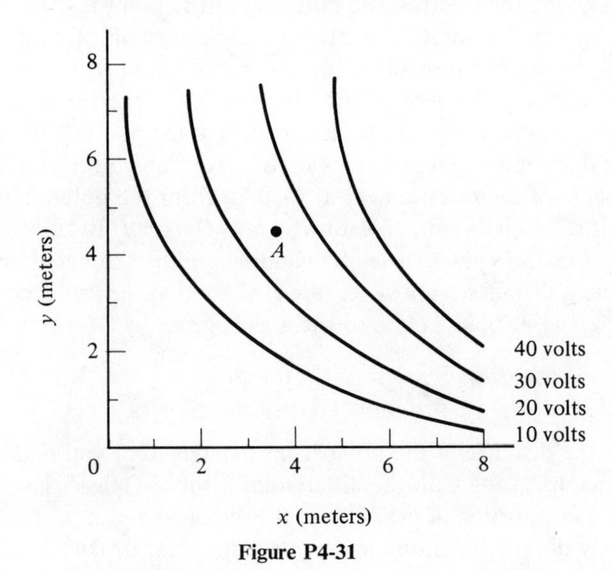

Figure P4-31

P4-32. A potential difference is applied between the two conducting plates shown in Fig. P4-32. Sketch the electric field lines and the equipotential lines in the region between the plates.

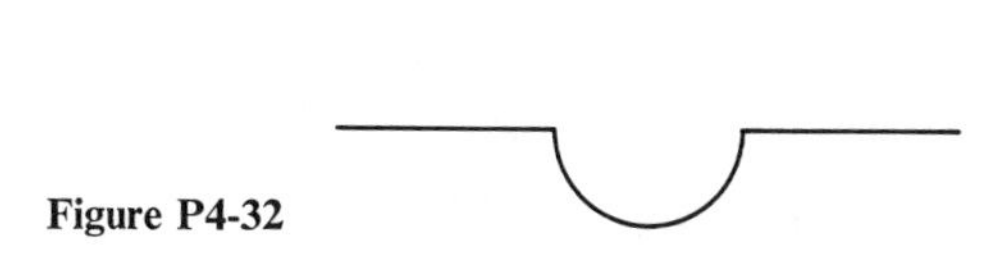

Figure P4-32

P4-33. Accepting that the field between two equally but oppositely charged parallel conducting plates is given by $(\sigma/\epsilon_0)\hat{\mathbf{n}}$, where σ is the charge density on the plates and $\hat{\mathbf{n}}$ is a unit vector perpendicular to the plates (P4-16), show that the capacitance of this parallel plate capacitor is given by $C = \epsilon_0 A/d$, where A is the surface area of the plates and d their separation. Ignore fringing at the edges of the plates.

P4-34. Show that the capacitance of a capacitor consisting of two coaxial cylindrical shells of length L and radii a and b, $a < b$, is given by $C = 2\pi\epsilon_0 L/\ \ell n\ (b/a)$. Ignore fringing at the ends of the capacitor.

P4-35. (a) Show that the capacitance of a capacitor consisting of two concentric spherical shells of radii a and b, $a < b$, is given by $C = 4\pi\epsilon_0 ab/(b - a)$. (b) The capacitance of an isolated sphere is defined by allowing $b \longrightarrow \infty$. What (numerically) is the radius of an isolated sphere having a capacitance of 1 F? Compare your result with the radius of the earth, $R_e = 6.37 \times 10^6$ m. (The distance from the earth to the moon is about $60R_e$.)

4-5 Energy in the Electrostatic Field

Because it takes work to assemble a distribution of charges from the state in which all elementary charges composing the distribution are infinitely remote from one another, the distribution (once assembled) possesses the capacity to do useful work. i.e., has a potential energy. Further, since the electrostatic forces against which the agent assembling the distribution does work are conservative, the amount of recoverable energy stored in the distribution (i.e., the potential energy) is exactly equal to the amount of work needed to assemble the distribution in the first place. Consider, for example, the buildup of an assembly of point charges q_i located at positions $\mathbf{r}_i$ by bringing the charges in one at a time to their final positions. The first charge can be brought from infinity to its final position $\mathbf{r}_1$ with no work, since there are as yet no charges in the distribution. To bring in the second charge, work must be done against the forces exerted by the first charge on the second. This work is equal to the potential energy of the second charge when it finally reaches the position $\mathbf{r}_2$. Accepting the electrostatic potential given by Eq. (4-41), we find that, once the first two charges have been positioned, the potential energy of the system is

$$\Delta W_2 = q_2\left[\frac{q_1}{4\pi\epsilon_0}\frac{1}{|\mathbf{r}_1 - \mathbf{r}_2|}\right] = \frac{q_1 q_2}{4\pi\epsilon_0}\frac{1}{|\mathbf{r}_1 - \mathbf{r}_2|} \tag{4-54}$$

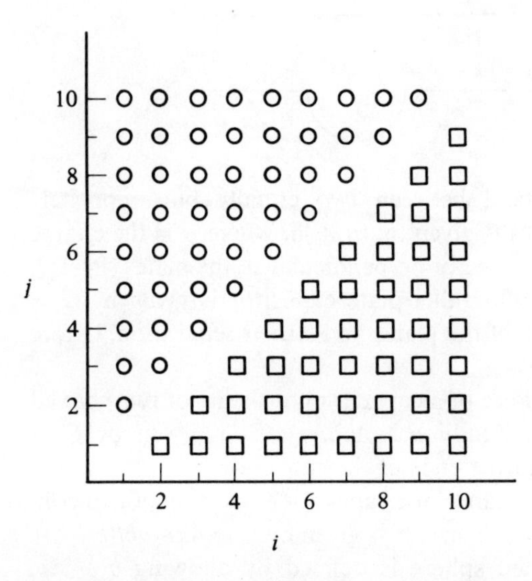

Fig. 4-14. Points over which the sum giving the potential energy of a distribution of point charges extends.

To bring up the third charge, work must be done against both of the charges that are already in position. The work expended to bring in the third charge then is

$$\Delta W_3 = q_3\left[\frac{q_1}{4\pi\epsilon_0}\frac{1}{|\mathbf{r}_1 - \mathbf{r}_3|} + \frac{q_2}{4\pi\epsilon_0}\frac{1}{|\mathbf{r}_2 - \mathbf{r}_3|}\right] \tag{4-55}$$

and so on; the amount of work needed to bring in the jth charge is given by

$$\Delta W_j = q_j \sum_{i=1}^{j-1} \frac{q_i}{4\pi\epsilon_0}\frac{1}{|\mathbf{r}_i - \mathbf{r}_j|} \tag{4-56}$$

The total potential energy of the charge distribution then is

$$W = \sum_{j=2}^{N}\sum_{i=1}^{j-1} \frac{q_i q_j}{4\pi\epsilon_0}\frac{1}{|\mathbf{r}_i - \mathbf{r}_j|} \tag{4-57}$$

The double sum appearing in Eq. (4-57) can now be rewritten in a more useful form. If the two summation indices are regarded as locating points in a two-dimensional plane, the sum as it stands can be interpreted as a sum over the points marked with a circle in Fig. 4-14. Now, for every circled point with coordinates (i, j) there is a "squared" point with coordinates (j, i). Since the summand in Eq. (4-57) has the same value for (i, j) as it has for (j, i), the sum might alternatively be viewed as extending over the "squared" points instead of over the circled points. Or—what is more useful—the sum may be regarded as extending over all the marked points provided a factor of $\frac{1}{2}$ is inserted to correct for the resulting double counting. We thus find that Eq. (4-57) can be rewritten in the form

$$W = \frac{1}{2}\sum_{i,j=1}^{N}{}' \frac{q_i q_j}{4\pi\epsilon_0}\frac{1}{|\mathbf{r}_i - \mathbf{r}_j|} \tag{4-58}$$

where the prime means that the terms for which $i = j$ are to be omitted in the double sum. If we now recognize the sum over j in Eq. (4-58) as the potential established at point i by all of the other charges, we can rewrite Eq. (4-58) in the form

$$W = \frac{1}{2}\sum_{i=1}^{N} q_i V_i \tag{4-59}$$

Finally, if q_i in Eq. (4-59) is interpreted as an element of a broad charge distribution, it might more appropriately be denoted Δq_i. Then the sum over these elements becomes an integral as the size of the elements is diminished to zero. Thus, for a broad distribution, we find that the stored potential energy is given by

$$W = \frac{1}{2}\int V(\mathbf{r})\, dq \tag{4-60}$$

where dq would be written more explicitly in terms of a volume, surface, or linear charge density as appropriate to the distribution.

One is tempted to ask, *Where* is this energy stored? The most fruitful answer can be inferred from the following rewriting of Eq. (4-60). Assume a volume density of charge. Then, Eq. (4-60) becomes

$$W = \frac{1}{2}\int V\rho\, dv \tag{4-61}$$

Now, note that

$$\begin{aligned} \nabla\cdot(V\mathbf{E}) &= \nabla V\cdot\mathbf{E} + V\nabla\cdot\mathbf{E} \\ &= -\mathbf{E}\cdot\mathbf{E} + \frac{1}{\epsilon_0}V\rho \end{aligned} \tag{4-62}$$

where Gauss's law, Eq. (4-29), has been invoked. Solving for $V\rho$ and substituting the result into Eq. (4-61), we find that

$$W = \frac{1}{2}\epsilon_0\int \mathbf{E}\cdot\mathbf{E}\, dv + \frac{1}{2}\epsilon_0\oint V\mathbf{E}\cdot d\mathbf{S} \tag{4-63}$$

where the divergence theorem has been invoked in the second term on the right. Now, the volume integral extends over all space and the surface integral therefore extends over a very large surface at infinity. If the sources of the field are confined to some finite region of space, V and $\mathbf{E}$ fall off with increasing r at least as rapidly as r^{-1} and r^{-2}, respectively, but $d\mathbf{S}$ increases only as r^2. Thus, as r becomes large, $V\mathbf{E}\cdot d\mathbf{S}$ falls off as r^{-1}, and the second integral in Eq. (4-63) becomes negligible in the limit as the surface over which it extends becomes indefinitely large. We find, therefore, that

$$W = \int \left(\tfrac{1}{2}\epsilon_0\mathbf{E}\cdot\mathbf{E}\right) dv \tag{4-64}$$

the integral extending over all space. Because of its role in Eq. (4-64), it is appropriate to introduce an electric *energy density* u_E given by

$$u_E = \tfrac{1}{2}\epsilon_0\mathbf{E}\cdot\mathbf{E} \tag{4-65}$$

and to regard the energy stored in a charge distribution as residing in the electric field itself. From one point of view, this association merely enables easy bookkeeping regarding the energy required to assemble a charge distribution. From another point of view, however, we have endowed the field with another "real" property and have thereby attached further reality to the field itself.

PROBLEMS

P4-36. Two positive and two negative charges, all of the same magnitude q, are placed at the vertices of a square of side a. If the positive charges are on adjacent vertices, determine the work required to assemble the distribution.

P4-37. (a) Suppose two initially uncharged conductors are charged to Q and $-Q$, respectively, by moving small amounts of charge dq slowly from one to the other. Let the potential difference between the conductors when the charges have reached the values q and $-q$ be $V(q)$. Show that the work done to produce the final distribution is given by $W = \int_0^Q V(q)\,dq$. (b) Suppose the capacitance of the arrangement is independent of the charge. Show that $W = Q^2/2C$. (c) Accepting the capacitance and the field given in P4-33, express W for a parallel plate capacitor in terms of the field within it and show that $W =$ (energy density) $\times$ (volume between plates).

P4-38. Starting with Eq. (4-60), show that the energy stored in a system of conductors, the ith one being at potential V_i and carrying charge q_i, is given by $W = \frac{1}{2}\sum q_i V_i$.

P4-39. (a) By integrating the energy density of the electric field over all space, determine the energy required to distribute a total charge Q uniformly over the surface of a spherical shell of radius a. (b) Assuming that the charge on an electron is uniformly distributed over a spherical shell and further assuming that the relativistic rest energy of the electron has an electromagnetic origin, estimate (*numerically*) the radius of the electron.

P4-40. (a) Show that the energy of an arbitrary charge distribution described by a charge density $\rho(\mathbf{r})$ is given by $\int \rho V\,dv$ when the charge distribution is placed in an *external* potential V. (b) Show that, when V corresponds to a uniform field $\mathbf{E}$, this energy reduces to $-\mathbf{p}\cdot\mathbf{E}$, where $\mathbf{p}$ is the dipole moment of the distribution (See P4-17.)

4-6 The Multipole Expansion of the Electrostatic Potential

Exact evaluation of Eq. (4-50) is possible only for a very few charge distributions. Fortunately, we are very often interested in the potential only in regions remote from its source. In those regions, the potential is given approximately (but quite accurately) by the first contributing term in a binomial expansion, as we saw, for example, in our treatment of the electric dipole. If the subsequent terms in the binomial expansion are retained, a series expressing the potential as an infinite sum is obtained. As the observation point moves closer to the source distribution, more and more of these terms must be preserved in order to obtain an accurate approximation for the potential. This infinite sum for the potential is called the *multipole expansion* of the potential and the individual terms in the sum are the monopole, dipole, quadrupole, etc., contributions. The different terms are distinguished from one another in part by their dependence on the distance r from the source to the observation point; each term in the series has one more power of r in the denominator than its predecessor. Successive terms also have progressively more involved dependences on the angles specifying the direction of the observation point relative to the source.

A more detailed development of the multipole expansion is based on a binomial expansion of the quantity $1/|\mathbf{r} - \mathbf{r}'|$. Using the binomial theorem, we find that

$$\begin{aligned}\frac{1}{|\mathbf{r}-\mathbf{r}'|} &= \frac{1}{[r^2 - 2\mathbf{r}\cdot\mathbf{r}' + (r')^2]^{1/2}} \\ &= \frac{1}{r}\left[1 - 2\hat{\mathbf{r}}\cdot\frac{\mathbf{r}'}{r} + \left(\frac{r'}{r}\right)^2\right]^{-1/2} \\ &= \frac{1}{r}\left[1 + \hat{\mathbf{r}}\cdot\frac{\mathbf{r}'}{r} + \frac{3(\mathbf{r}'\cdot\hat{\mathbf{r}})^2 - (r')^2}{2r^2} + \cdots\right] \end{aligned} \tag{4-66}$$

A more convenient alternative expression of this expansion is obtained if the components of $\mathbf{r}$ and $\mathbf{r}'$ are denoted by x_1, x_2, x_3 and x'_1, x'_2, x'_3 rather than by x, y, z and x', y', z'. We then find from Eq. (4-66) that

$$\frac{1}{|\mathbf{r}-\mathbf{r}'|} = \frac{1}{r} + \frac{\hat{\mathbf{r}}\cdot\mathbf{r}'}{r^2} + \frac{3(\sum_i x'_i x_i)^2 - (r')^2 \sum_j x_j^2}{2r^5} + \cdots \tag{4-67}$$

The object now is to separate the primed and unprimed coordinates. Let the numerator in the third term be written as follows:

$$\begin{aligned} 3(\sum_i x'_i x_i)^2 - (r')^2 \sum_j x_j^2 &= 3 \sum_{i,j} x'_i x_i x'_j x_j - (r')^2 \sum_{i,j} x_i\, \delta_{ij} x_j \\ &= \sum_{i,j} x_i x_j [3x'_i x'_j - \delta_{ij}(r')^2] \end{aligned} \tag{4-68}$$

where the Kronecker delta, δ_{ij}, is unity when $i = j$ and zero when $i \neq j$. We thus find that

$$\frac{1}{|\mathbf{r}-\mathbf{r}'|} = \frac{1}{r} + \frac{\mathbf{r}\cdot\mathbf{r}'}{r^3} + \frac{1}{2r^5}\sum_{i,j} x_i x_j [3x'_i x'_j - \delta_{ij}(r')^2] + \cdots \tag{4-69}$$

and in each term the source coordinates are explicitly separated from the observation coordinates.

The multipole expansion for the potential is now obtained by substituting Eq. (4-69) into Eq. (4-50); writing each term separately, we find that

$$V(\mathbf{r}) = \frac{\int dq'}{4\pi\epsilon_0 r} + \frac{\mathbf{r} \cdot \int \mathbf{r}'\, dq'}{4\pi\epsilon_0 r^3} + \frac{\sum_{i,j} x_i x_j \int [3x_i' x_j' - \delta_{ij}(r')^2]\, dq'}{8\pi\epsilon_0 r^5} + \cdots \tag{4-70}$$

All integrals extend over the source distribution. For a given source, these integrals are constants characterizing the distribution; *they do not depend on the observation point.*

Each of the integrals in Eq. (4-70) has a physical interpretation. The first, for example, is given by

$$\int dq' = Q \tag{4-71}$$

where Q is the net charge in the distribution and the first term in Eq. (4-70) is the potential of a point charge. It is the most significant contribution to $V(\mathbf{r})$ when r is very large. Comparison of the second term with Eq. (4-44) leads to the identification of a dipole moment

$$\int \mathbf{r}'\, dq' = \mathbf{p} \tag{4-72}$$

characterizing an arbitrary charge distribution. If the total charge Q happens to be zero, the most significant contribution to the potential at large distances is the dipole contribution. The third term is the quadrupole contribution to the electrostatic potential and the nine numbers

$$Q_{ij} = \int [3x_i' x_j' - \delta_{ij}(r')^2]\, dq' \tag{4-73}$$

are the elements of the *quadrupole moment tensor*, which is clearly symmetric, since Q_{ij} necessarily equals Q_{ji}. (Why?) For specific charge distributions, some elements may be zero and furthermore some components other than those that are necessarily equal may happen accidentally to be equal. Although Q, $\mathbf{p}$, Q_{ij}, and higher moments do not depend in any way on the observation point, they are moments of the charge distribution *relative to a particular coordinate system* and are not all necessarily invariant to changes in that coordinate system. (See P3-18.) Usually, however, a most natural coordinate system is apparent from the charge distribution and the selection of that specific coordinate system is left understood.

In terms of the monopole, dipole, and quadrupole moments, the first three terms of the multipole expansion of the potential are

$$V(\mathbf{r}) = \frac{Q}{4\pi\epsilon_0 r} + \frac{\mathbf{r}\cdot\mathbf{p}}{4\pi\epsilon_0 r^3} + \frac{\sum_{i,j} x_i Q_{ij} x_j}{8\pi\epsilon_0 r^5} + \cdots \tag{4-74}$$

Every subsequent term in this expansion can also be written in terms of the coordinates of the *observation* point and a set of numbers determined *solely* by the charge distribution. We shall not carry the expansion beyond this point.

A corresponding expansion for the electric field of an arbitrary charge distribution is most easily obtained by computing the gradient of Eq. (4-74).

PROBLEMS

P4-41. Consider a linear quadrupole consisting of a point charge of strength $2q$ at the origin and point charges of strength $-q$ at $(0, 0, a)$ and $(0, 0, -a)$. Show that the monopole and dipole moments of this distribution are zero, calculate the elements of the quadrupole moment tensor, and write out the electrostatic potential as a function of the spherical polar coordinates of the observation point. Finally, calculate the field of this quadrupole by evaluating the gradient of the potential. *Hint:* Let the integrals become appropriate sums over the point charges in the quadrupole.

P4-42. (a) Calculate the monopole, dipole, and quadrupole moments of the ring in Fig. P4-7, which carries a linear charge density given by $\lambda(\phi) = \lambda_0(1 + \sin\phi)$. (b) Write out the first three terms in the multipole expansion of the electrostatic potential of this ring in terms of the spherical coordinates of the observation point.

P4-43. A spherically symmetric charge distribution is described by a charge density that is a function only of the spherical radial coordinate. Show that the dipole and quadrupole moments of such a distribution are zero.

P4-44. Consider a charge distribution that is invariant to rotation about the z axis. Such a distribution is described by a charge density ρ that depends only on the spherical coordinates r, θ: $\rho = \rho(r, \theta)$. (a) Obtain a two-dimensional integral giving the total charge in the distribution. (b) Show that the dipole moment of this distribution has only a z-component and obtain a two-dimensional integral for this component. (c) Show that the components of the electric quadrupole moment satisfy $Q_{12} = Q_{21} = Q_{23} = Q_{32} = Q_{31} = Q_{13} = 0$ and $Q_{11} = Q_{22} = -\frac{1}{2}Q_{33}$, and obtain an integral giving Q_{33}. (d) Write out the electrostatic potential (through the quadrupole term), express the result in terms of the spherical coordinates of the observation point, and note the appearance of the Legendre polynomials. (See Table 8-1 with $x = \cos\theta$.) (e) The electron cloud in one of the excited states of the hydrogen atom is described by the charge density

$$\rho(\mathbf{r}) = -\frac{q}{64\pi a_0^5} r^2 e^{-r/a_0} \sin^2\theta$$

where q is the magnitude of the electronic charge and a_0 is the Bohr radius. Calculate the monopole, dipole, and quadrupole moments for this state. Don't be afraid of integral tables and don't overlook the contributions of the nucleus.

P4-45. Consider a line distribution of charge lying along the z axis and carrying linear charge density $\lambda(z)$. Further, suppose $\lambda(z) = 0$ except in some finite range of z near $z = 0$. Write an integral for the potential at an arbitrary point in space and then expand the integrand to show that

$$V(\mathbf{r}) = \frac{1}{4\pi\epsilon_0} \sum_{n=0}^{\infty} \frac{P_n(\cos\theta)}{r^{n+1}} \int (z')^n \lambda(z')\, dz'$$

where the integral in each term can be interpreted as a multipole moment. Use spherical coordinates for the field point and note the generating function for the Legendre polynomials $P_n(t)$ in PB-5 in Appendix B.

P4-46. As shown in P4-40(a), the energy W of a charge distribution ρ placed in an *external* potential V is given by $W = \int \rho V\, dv$. Assuming V is established entirely by charges outside the region of integration, expand V in a (three-dimensional) Taylor series about the origin to show that

$$W = QV(0) - \mathbf{p}\cdot\mathbf{E}(0) - \frac{1}{6}\sum_{i,j} Q_{ij} \left.\frac{\partial E_j}{\partial x_i}\right|_0 + \cdots$$

where Q, $\mathbf{p}$, and Q_{ij} are the monopole, dipole, and quadrupole moments of the distribution. *Hint:* For the *external* field, $\nabla\cdot\mathbf{E} = 0$ throughout the volume of integration. See J. D. Jackson, *Classical Electrodynamics* (John Wiley & Sons, Inc., New York, 1962), p. 101.

Supplementary Problems

P4-47. Let the electric field established at $\mathbf{r}$ by a point charge q at $\mathbf{r}'$ be written as $q\mathscr{E}(\mathbf{r}, \mathbf{r}')$. (a) Write an analytic expression for $\mathscr{E}$ and interpret $\mathscr{E}$ physically. (b) A charge q is located at $\mathbf{r}_0 + \frac{1}{2}\mathbf{a}$ and a charge $-q$ is located at $\mathbf{r}_0 - \frac{1}{2}\mathbf{a}$. Write a formal expression for the field of the resulting dipole in terms of $\mathscr{E}$ and, using a Taylor expansion about $\mathbf{r}_0$, show that

$$\mathbf{E}_{\text{dipole}}(\mathbf{r}) = (\mathbf{p}\cdot\nabla_0)\mathscr{E}(\mathbf{r}, \mathbf{r}_0)$$

where $\mathbf{p} = q\mathbf{a}$ and ∇_0 acts on the components of $\mathbf{r}_0$. In effect the field of the dipole can be obtained as the derivative of the field of a monopole. (c) Substitute the explicit analytic expression for $\mathscr{E}$ into the result of part (b) and derive an expression for the field of a dipole in terms of $\mathbf{p}$, $\mathbf{r}$, and $\mathbf{r}_0$. *Note:* A discussion of this property of the dipole field may be found, for example, in R. P. Feynman, R. B. Leighton, and M. Sands, *The Feynman Lectures on Physics* (Addison-Wesley Publishing Company, Inc., Reading, Mass. 1964), Volume II, Section 6-4.

P4-48. The hydrogen atom in its ground state consists of a positively charged (point) nucleus carrying charge q surrounded by a (spherically symmetric) electron cloud described by the volume charge density

$$\rho(\mathbf{r}) = -\frac{q}{\pi a_0^3} e^{-2r/a_0}$$

where the coordinate origin is taken at the nucleus and a_0 is the Bohr radius.

Use symmetry and Gauss's law to find the electric field as a function of **r**. Don't be afraid of integral tables.

P4-49. A total charge Q is uniformly distributed throughout a sphere of radius a with its center at the origin. Let the (constant) charge density be ρ. (a) Use symmetry and Gauss's law to determine the electric field at all points in space. Make sure the arguments supporting each step of the calculation are clearly presented. Do not overlook the region $r < a$. (b) Sketch a graph of the radial component of the field as a function of r. Sketch also the magnitude of the field for a point charge of strength Q and compare the two fields. (c) Taking the reference point at infinity, calculate and sketch a graph of the electrostatic potential as a function of r. (d) Suppose a *point* charge of strength $-Q$ and mass m is released from rest at a point inside the distribution described in this problem. Ignoring collisions of this charge with the charges composing the distribution, describe its subsequent motion and determine its frequency of oscillation. Evaluate this frequency explicitly for $Q = 1.6 \times 10^{-19}$ C, $a_0 = 0.5 \times 10^{-10}$ m and $m = 9.1 \times 10^{-31}$ kg and compare the result with typical frequencies of light. The system described by these numerical values is a simplified version of the now-discarded "plum pudding" model of the hydrogen atom.

P4-50. Consider a uniformly charged infinite plane sheet lying in the x-y plane. Set up Eq. (4-49) for the potential of this distribution and explain physically why this approach does not yield a useful evaluation of the potential. Obtain and sketch a graph of the potential as a function of z by some alternative means.

P4-51. Consider two infinite plane sheets, one lying in the plane $z = 0$ and carrying a constant surface charge density σ and the other lying in the plane $z = d$ with surface charge density $-\sigma$. Using the results in Eq. (4-13), sketch graphs of the z-component of the electric field established by each plane separately and by the combination. Then, taking the reference point at the point $z = 0$, determine the electrostatic potential and sketch a graph. Finally, show that, if d becomes vanishingly small and $\sigma \longrightarrow \infty$ so that σd remains constant, a macroscopic observer walking from one side of the arrangement to the other experiences a discontinuity in the potential but no discontinuity in the field.

P4-52. Consider two dipoles, one of moment **p** at the origin and the other of moment **p**′ at **R**. (a) Show that the potential energy U of the dipole at **R** in the field of the dipole at the origin is given by

$$U = \frac{\mathbf{p}\cdot\mathbf{p}' - 3(\hat{\mathbf{R}}\cdot\mathbf{p})(\hat{\mathbf{R}}\cdot\mathbf{p}')}{4\pi\epsilon_0 R^3}$$

Hint: Treat the second dipole as two closely spaced charges and use Eq. (4-44) to obtain the potential energy of each charge in the field of the dipole at the origin. (b) Examine the characteristics of this dipole-dipole interaction. For example, how does U depend on the relative orientation of the two dipoles? Are there any configurations of static equilibrium? If so, are they stable or unstable? What forces and torques exist on each dipole? Use a desk calculator or computer if it seems appropriate.

P4-53. By integrating the energy density in the field of two point charges, derive an expression for the potential energy of one charge in the field of the other. *Hints:* (1) Let the fields of each charge separately be $\mathbf{E}_1$ and $\mathbf{E}_2$. The energy density is then related to

$$(\mathbf{E}_1 + \mathbf{E}_2)\cdot(\mathbf{E}_1 + \mathbf{E}_2) = \mathbf{E}_1\cdot\mathbf{E}_1 + 2\mathbf{E}_1\cdot\mathbf{E}_2 + \mathbf{E}_2\cdot\mathbf{E}_2$$

Interpret the divergent integrals arising from the first and third terms as electron self-energies and look more closely at the second term, which gives rise to a convergent integral. (2) Let one charge be located at the origin and the other on the z axis, and use spherical coordinates.

P4-54. Determine the force of attraction between the plates of a parallel plate capacitor (P4-16 and P4-33) in terms of the area A and the separation d of the plates and the applied potential ΔV. *Hint:* How much work is required to effect a small virtual displacement of one of the plates? Assume first that the charge on the plates remains constant during the virtual displacement but then show that the same force is found if the potential difference between the plates is assumed constant instead. *Note:* In the Kelvin absolute electrometer, a measurement of the force between the plates of a parallel plate capacitor is used to determine the potential difference between the plates.

P4-55. A charged ring is oriented as in Fig. P4-7 but carries a charge Q *uniformly* distributed over its circumference. (a) Show that the potential $V(\imath, \phi, z)$ established by this ring at the point $(\imath, \phi, z)$ in cylindrical coordinates is given by

$$\mathcal{V}(\imath, z) = \frac{V(\imath, \phi, z)}{Q/4\pi\epsilon_0 a} = \frac{1}{2\pi}\int_0^{2\pi} \frac{d\phi'}{[1 - 2(\imath/a)\cos\phi' + (\imath^2/a^2) + (z^2/a^2)]^{1/2}}$$

which, as expected from the rotational invariance of the distribution, is independent of ϕ. (b) Evaluate the integral analytically for a field point on the z axis, $\imath = 0$, showing that

$$\mathcal{V}(0, z) = \frac{1}{\sqrt{1 + (z^2/a^2)}} \Longrightarrow V(0, \phi, z) = \frac{Q}{4\pi\epsilon_0 a}\frac{1}{\sqrt{1 + (z^2/a^2)}}$$

Then, using a desk calculator or computer to obtain the necessary points, plot a careful graph of $\mathcal{V}(0, z)$ versus z/a. *Optional:* (1) Look up Simpson's rule (or some other technique for numerical integration) in an available book on calculus or numerical analysis, write a computer program to evaluate $\mathcal{V}(\imath, z)$ numerically for values of $\imath/a$ and z/a supplied as input data, test the adequacy of your program by comparing what it produces for $\imath/a = 0$ with the results of part (b), and then use your program to examine the behavior of

$\mathcal{V}(\imath, z)$ as a function of $\imath/a$ for representative, fixed values of z/a (including $z/a = 0$). (2) Sketch a few of the equipotential contours of this field.

P4-56. Suppose that equations or integrals from which the x- and y-components of a particular electric field **E** in the x-y plane can be calculated are available. We can then trace the field lines as follows. Starting at some point (x_1, y_1), we move a distance d in the direction of the (calculable) field at (x_1, y_1). If d is not too large, the point (x_2, y_2) that we reach lies approximately on

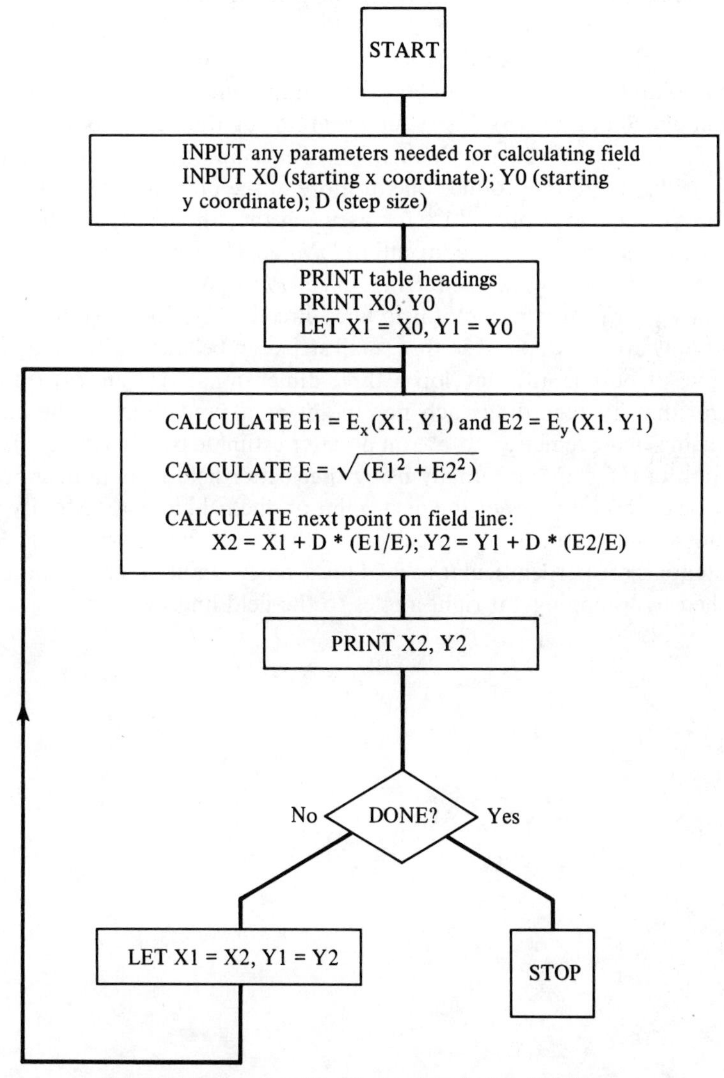

Fig. P4-56. An algorithm for tracing electric field lines.

the field line through (x_1, y_1). But the coordinates of the two points are related by

$$x_2 = x_1 + d \cos \theta = x_1 + d \frac{E_x(x_1, y_1)}{|\mathbf{E}(x_1, y_1)|}$$

$$y_2 = y_1 + d \sin \theta = y_1 + d \frac{E_y(x_1, y_1)}{|\mathbf{E}(x_1, y_1 |}$$

where θ is the angle between the field at (x_1, y_1) and the x axis. (Why?) Once the first step has been made, we replace the starting point with the new point and repeat the calculation. The algorithm is summarized in Fig. P4-56; its output is a table of the coordinates of the points on the field line through the starting point. (a) Write a computer program to implement this algorithm on an available computer. (b) Test your program for the field produced by an array of a few point charges. You might, for example, generate points on several field lines and reproduce one or more of the plots in Figs. 4-4, 4-5, or 4-8. Can you invent a procedure for assuring that the density of field lines conforms to the established convention? *Note:* This method has been discussed, for example, by J. R. Merrill, *Am. J. Phys.* **39**, 791 (1971). *Optional:* (1) Modify your program to plot field lines directly. (2) Modify your program to print only every Nth point so that small steps can be taken without generating reams of output. (3) Develop a three-dimensional version. (4) Improve the algorithm by regarding each new point as a prediction of the correct value, using that prediction to obtain a better estimate of the average field in the region of the step, and finally using that better field to obtain improved values for the coordinates of the next point on the field line. (5) Modify your program so that it traces equipotentials rather than field lines. *Hint:* Since equipotentials are perpendicular to field lines, a new point on an equipotential is reached by "stepping" at right angles to the field line.

5

The Magnetic Induction Field Produced by Steady Currents

Two apparently distinct sources of magnetic induction fields are known: permanent magnets and electric currents. A direct treatment of the field produced by permanent magnets involves introducing the concept of a point magnetic pole, realized physically by one end of a long slender bar magnet whose second end has a negligible effect in the region near the first end. Two types of poles—plus and minus—can be identified; the interactions between these poles can be expressed by a law similar to Coulomb's law; the magnetic induction field is defined as the force on a unit positive test pole; the principle of superposition can be applied to obtain expressions for the magnetic induction field established by a distribution of magnetic poles in space; and a magnetic scalar potential, whose gradient gives the negative of the magnetic induction field, can be introduced. Approached this way, the magnetic induction field of permanent magnets has a description very similar to that of the electric field of fixed charges.

This approach to permanent magnetism, however, has at least three disadvantages. First, there are experimental difficulties in isolating individual magnetic poles. Even if the second end of the slender bar magnet is far away, the two ends are still intimately related, for the remote end cannot simply be broken off and discarded. Experimentally, a magnet broken in two produces two magnets, each with the usual positive pole at one end and negative pole at the other end. In effect, magnetic dipoles exist in nature, and consequently all higher magnetic multipoles can be found, but the magnetic monopole seems not to exist. Second, the pole formalism is simple only for determining the fields in the region *outside of* permanent magnets. Finally, the pole formalism cannot be easily extended to include the magnetic induction fields produced by electric currents.

That electric currents give rise to magnetic induction fields is a comparatively recent discovery (Oersted, c. 1820). Since Oersted's time, it has been found that all magnetic induction fields, including those of permanent magnets, can be traced to currents, and we therefore adopt the more general approach in which currents are viewed as the primary source of magnetic induction fields. In this chapter, we shall explore the relationships between *steady* (i.e., time-independent) currents and the resulting *static* magnetic induction fields, paralleling the development as closely as possible after that of the static electric field in Chapter 4. The fields produced by time-dependent currents will be treated in Chapter 6.

5-1 The Law of Biot-Savart

The starting point for relating the static magnetic induction field to its sources is the law for the force per unit length between two parallel current-carrying wires (Section 1-4). In essence, we seek to reformulate this law so that the result gives the force of interaction between two short segments or *elements* of the wires. Unfortunately, two such elements cannot be isolated physically without destroying the current in the circuits. Thus, the forces of interaction between two isolated current elements cannot be examined *experimentally*. The best that we can do is *postulate* a law for that elementary interaction, choosing a form that (hopefully) gives experimentally verifiable results when it is integrated over all the elements making up a physically constructible circuit. Let us therefore suppose that there exists between two small parallel elements—one in each of two wires—a force given in *magnitude* by

$$dF = K\frac{(I\,d\ell)(I'\,d\ell')}{(r'')^2}\sin\theta \tag{5-1}$$

where K is a constant and the other symbols are defined in Fig. 5-1. To show that this assumed elementary force gives the correct value for the force per unit length between two parallel wires, we integrate Eq. (5-1) over the "primed" wire, obtaining

$$F_{II'} = KII'\int_{-\infty}^{\infty}\frac{\sin\theta}{(r'')^2}\,d\ell' \tag{5-2}$$

for the force *per unit length* on the "unprimed" wire. From the geometry of Fig. 5-1, we find that $r'' = \sqrt{s^2 + (\ell - \ell')^2}$ and that $\sin\theta = s/r''$, and Eq. (5-2) then becomes

$$F_{II'} = KII's\int_{-\infty}^{\infty}\frac{d\ell'}{[s^2 + (\ell - \ell')^2]^{3/2}} = 2K\frac{II'}{s} \tag{5-3}$$

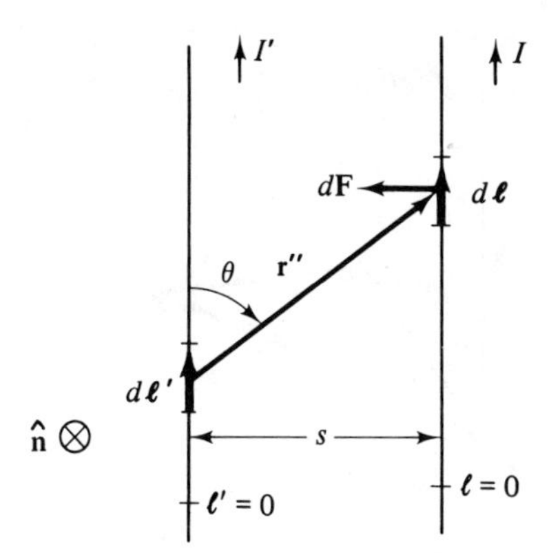

Fig. 5-1. Coordinates and vectors used in expressing the magnetic force between two parallel wires.

which is in agreement with Eq. (1-7) provided we set $K = \frac{1}{2}k_2$. In particular, in mks units, we set $K = \mu_0/4\pi$ and Eq. (5-1) for the assumed interaction between current elements becomes

$$dF = \frac{\mu_0}{4\pi} \frac{(I\, d\ell)(I'\, d\ell')}{(r'')^2} \sin\theta \tag{5-4}$$

We now press our assumed reformulation of Eq. (1-7) still further by writing Eq. (5-4) in a vector form, thereby including the direction of the force of interaction. The presence of the factor $\sin\theta$ in Eq. (5-4) suggests introducing the cross product

$$I'\, d\boldsymbol{\ell}' \times \mathbf{r}'' = I'\, d\ell' r'' \sin\theta \hat{\mathbf{n}} \tag{5-5}$$

where $\hat{\mathbf{n}}$ is a unit vector directed into the page in Fig. 5-1. In addition, the force between the two elements in Fig. 5-1 is an attractive force if both currents are positive, and the cross product $I\, d\boldsymbol{\ell} \times \hat{\mathbf{n}}$ then has the proper direction for the force $d\mathbf{F}$ on the element $I\, d\boldsymbol{\ell}$. We are thus led to *assume* the vector form

$$d\mathbf{F} = \frac{\mu_0}{4\pi} \frac{I\, d\boldsymbol{\ell} \times (I'\, d\boldsymbol{\ell}' \times \mathbf{r}'')}{r''^3} \tag{5-6}$$

for Eq. (5-4). Because $\hat{\mathbf{n}}$ and $d\boldsymbol{\ell}$ are perpendicular, the right-hand side of Eq. (5-6) has the correct magnitude; by its construction, it also has the correct direction.

At this point, we are confident only that Eq. (5-6) gives the correct force between two infinite parallel wires. We cannot even be confident that Eq. (5-6) is the *only* vector expression that reduces correctly in that special case. Further, because of the experimental impossibility of isolating individual current elements, we cannot subject Eq. (5-6) for the force between *current* elements to the sort of direct experimental test that can be given to Coulomb's law for the force between *charge* elements. Nonetheless, we now adopt Eq. (5-6) for the force of interaction between two current elements *arbitrarily* positioned and *arbitrarily* oriented in space, fully realizing (1) that this law has been *postulated* and (2) that its experimental support is found by exam-

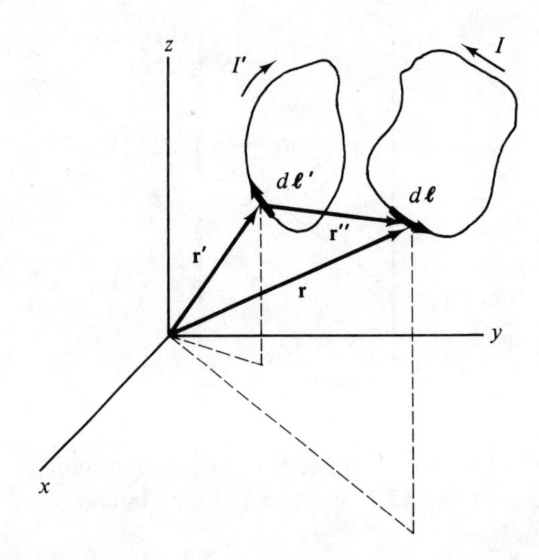

Fig. 5-2. Coordinates and vectors used in expressing the magnetic force between two arbitrary current loops.

ining how satisfactorily its *integral* predicts the forces between physical circuits.

An expression for determining the magnetic induction field from knowledge of its sources emerges if we integrate Eq. (5-6) over a portion Γ of the unprimed circuit in Fig. 5-2 and over the entire primed circuit. We then find that the force $\mathbf{F}$ on Γ is given by

$$\mathbf{F} = \int_{\Gamma} I\, d\boldsymbol{\ell} \times \left[\frac{\mu_0}{4\pi} \oint \frac{I'\, d\boldsymbol{\ell}' \times (\mathbf{r} - \mathbf{r}')}{|\mathbf{r} - \mathbf{r}'|^3} \right] \tag{5-7}$$

where $\mathbf{r}''$ has been replaced by its equivalent $\mathbf{r} - \mathbf{r}'$ (Fig. 5-2). Comparison of Eq. (5-7) with Eq. (3-25) now leads us to interpret the primed circuit as the source of a magnetic induction field given by

$$\mathbf{B}(\mathbf{r}) = \frac{\mu_0}{4\pi} \oint \frac{I'\, d\boldsymbol{\ell}' \times (\mathbf{r} - \mathbf{r}')}{|\mathbf{r} - \mathbf{r}'|^3} \tag{5-8}$$

Although it is very much an abstraction, it is convenient at times to suppose that the differential element $I'\, d\boldsymbol{\ell}'$ at $\mathbf{r}'$ makes a differential contribution

$$d\mathbf{B} = \frac{\mu_0}{4\pi} \frac{I'\, d\boldsymbol{\ell}' \times (\mathbf{r} - \mathbf{r}')}{|\mathbf{r} - \mathbf{r}'|^3} \tag{5-9}$$

to the field at $\mathbf{r}$. Equivalently, when the currents in a distribution not guided by wires are *steady* (i.e., when $\boldsymbol{\nabla} \cdot \mathbf{J} = 0$ [Eq. (2-47)], which implies that the lines of $\mathbf{J}$ close on themselves [Section 2-3], which in turn implies that the current distribution can be regarded as a superposition of current loops), then Eqs. (5-8) and (5-9) can be written in terms of current densities by making the replacement $I'\, d\boldsymbol{\ell}' \rightarrow \mathbf{J}(\mathbf{r}')\, dv'$ (P2-7). Thus, we obtain from Eq.(5-8), for example, that the magnetic induction established by a steady current distribution having current density $\mathbf{J}(\mathbf{r})$ is given by

$$\mathbf{B}(\mathbf{r}) = \frac{\mu_0}{4\pi}\int \frac{\mathbf{J}(\mathbf{r}') \times (\mathbf{r} - \mathbf{r}')}{|\mathbf{r} - \mathbf{r}'|^3}\, dv' \tag{5-10}$$

where, as usual, $\mathbf{r}$ and $\mathbf{r}'$ represent the field and source points, respectively. Equations (5-8)–(5-10) are all forms of the *Biot-Savart law,* which not only determines the magnetic induction field of given steady sources but also relates the field to the sources even when neither is known explicitly. Equation (5-10) should be compared with the expression for the electric field obtained by setting $dq' = \rho(\mathbf{r}')\, dv'$ in Eq. (4-7). Equation (5-10) is more complicated because the sources of the magnetic induction field (currents) are vectors, while the sources of the electric field (charges) are scalars.

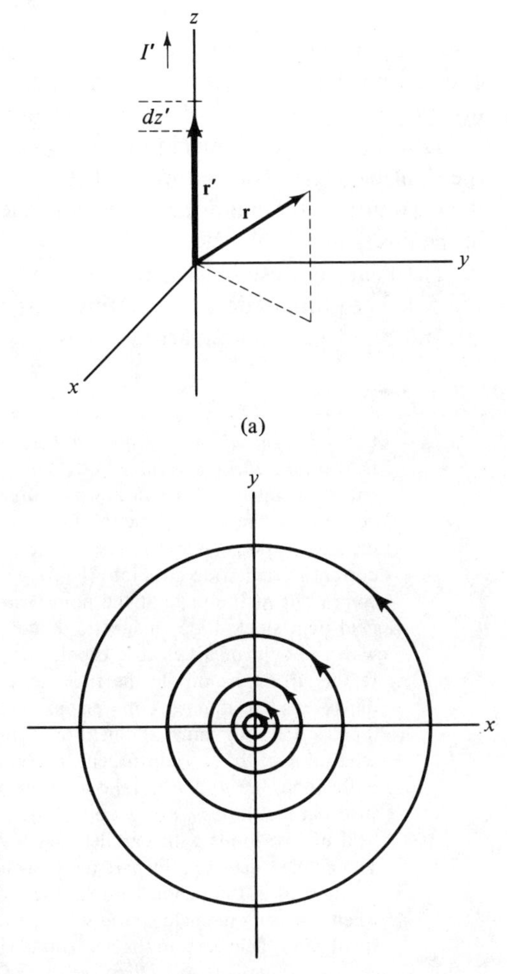

Fig. 5-3. The magnetic induction field of a long, straight, current-carrying wire: (a) a convenient choice of coordinates and (b) the field lines in a plane perpendicular to the wire. In part (b), the current comes out of the page at the origin.

We shall illustrate the Biot-Savart law by calculating the magnetic induction of a long straight wire carrying a current I' along the z axis toward $z = +\infty$ [Fig. 5-3(a)]. In cylindrical coordinates, we have that

$$\mathbf{r} = \imath\hat{\boldsymbol{\imath}} + z\hat{\mathbf{k}} \qquad \mathbf{r}' = z'\hat{\mathbf{k}} \qquad d\boldsymbol{\ell}' = dz'\hat{\mathbf{k}}$$

$$\mathbf{r} - \mathbf{r}' = \imath\hat{\boldsymbol{\imath}} + (z - z')\hat{\mathbf{k}}$$

$$|\mathbf{r} - \mathbf{r}'| = [\imath^2 + (z - z')^2]^{1/2}$$

$$I'\,d\boldsymbol{\ell}' \times (\mathbf{r} - \mathbf{r}') = I'\imath\,dz'\hat{\boldsymbol{\phi}}$$

Substitution of these quantities into Eq. (5-8) then gives

$$\mathbf{B}(\mathbf{r}) = \frac{\mu_0}{4\pi}\int_{-\infty}^{\infty}\frac{I'\imath\,dz'}{[\imath^2 + (z - z')^2]^{3/2}}\hat{\boldsymbol{\phi}} = \frac{\mu_0 I'}{2\pi\imath}\hat{\boldsymbol{\phi}} \tag{5-11}$$

Thus, the magnetic induction of a long wire is directed everywhere in the $\hat{\boldsymbol{\phi}}$ direction, i.e., in circles centered on the wire, and varies in magnitude inversely as the distance of the observation point from the wire. The field lines for this field are shown in Fig. 5-3(b). Equation (5-11), of course, includes the right-hand rule: If the thumb of the right hand points in the direction of the current and the palm faces the wire, the fingers curl around the current in the direction of the field.

The field lines established by a few other selected sources are shown in Fig. 5-4. The primary data for plotting these fields were obtained by numerical integration of the Biot-Savart law (P5-31).

Fig. 5-4. (*Shown on the following 2 pages.*) The magnetic induction field produced by several selected current distributions. Each distribution consists of one or more *circular* current loops, all carrying currents of the same strength. The plane of each loop is perpendicular to the page and intersects the page in the dashed line; the current in each loop goes into the page at the point marked $\times$ and comes out of the page at the point marked $\cdot$. Part (a) shows the field of a single loop (magnetic dipole) and should be compared with the field of an electric dipole shown in Fig. 4-5. In regions remote from the dipole the two fields are very similar; marked differences are apparent in regions near to and "inside of" the dipoles. Part (b) shows the field of a Helmholtz coil, in which the separation of the two constituent current loops is equal to the radius of the loops (P5-4). Part (c) shows the field of a magnetic quadrupole (two parallel loops with *opposite* currents) and part (d) shows the field of two loops with parallel axes but lying in the same plane. The dashed field line in part (d) is included to show more detail of the field in the region between the loops but should be omitted when assessing the field strength from the spacing of the field lines. In all cases, field lines in the immediate vicinity of each wire are not shown because of their closeness to one another.

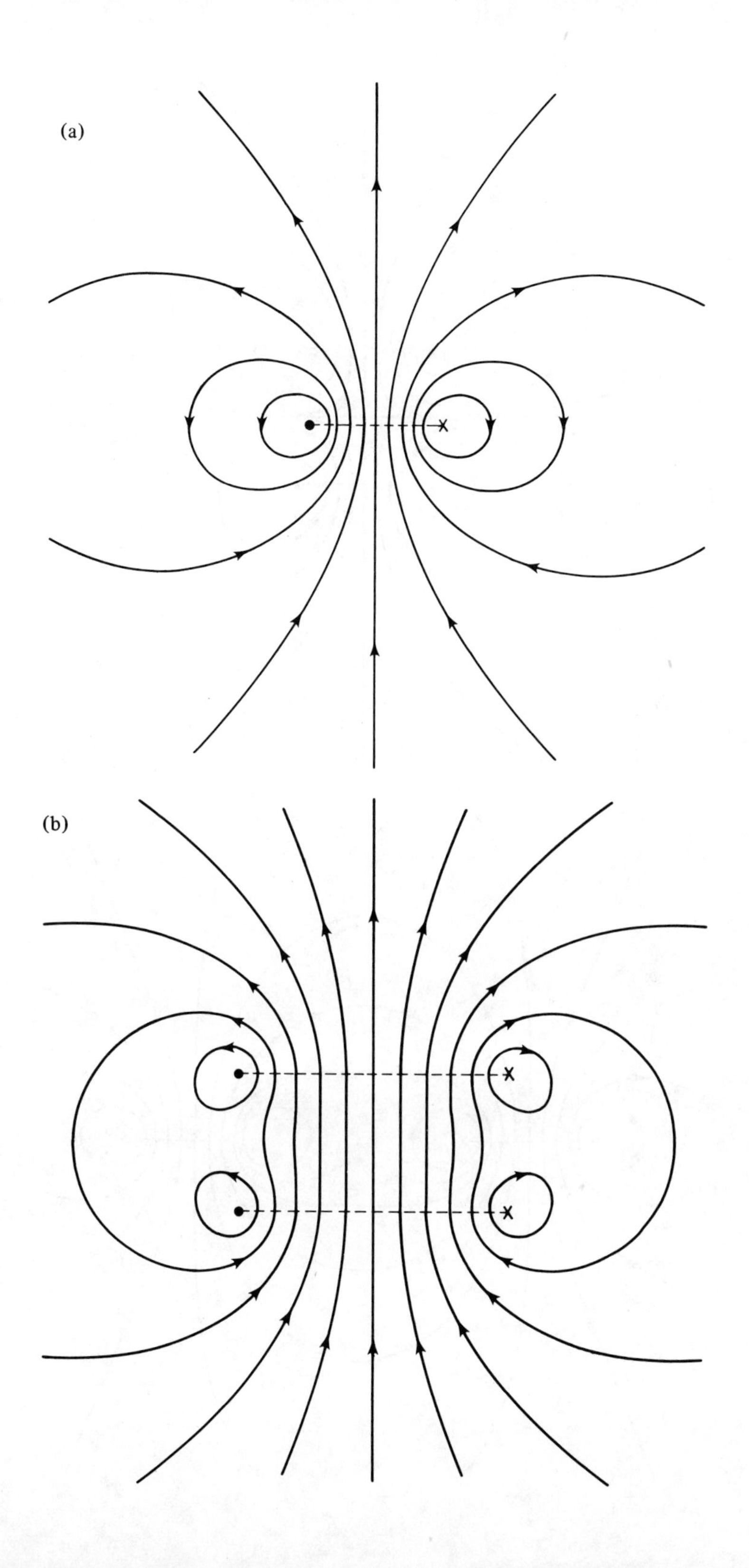

(a)
(b)

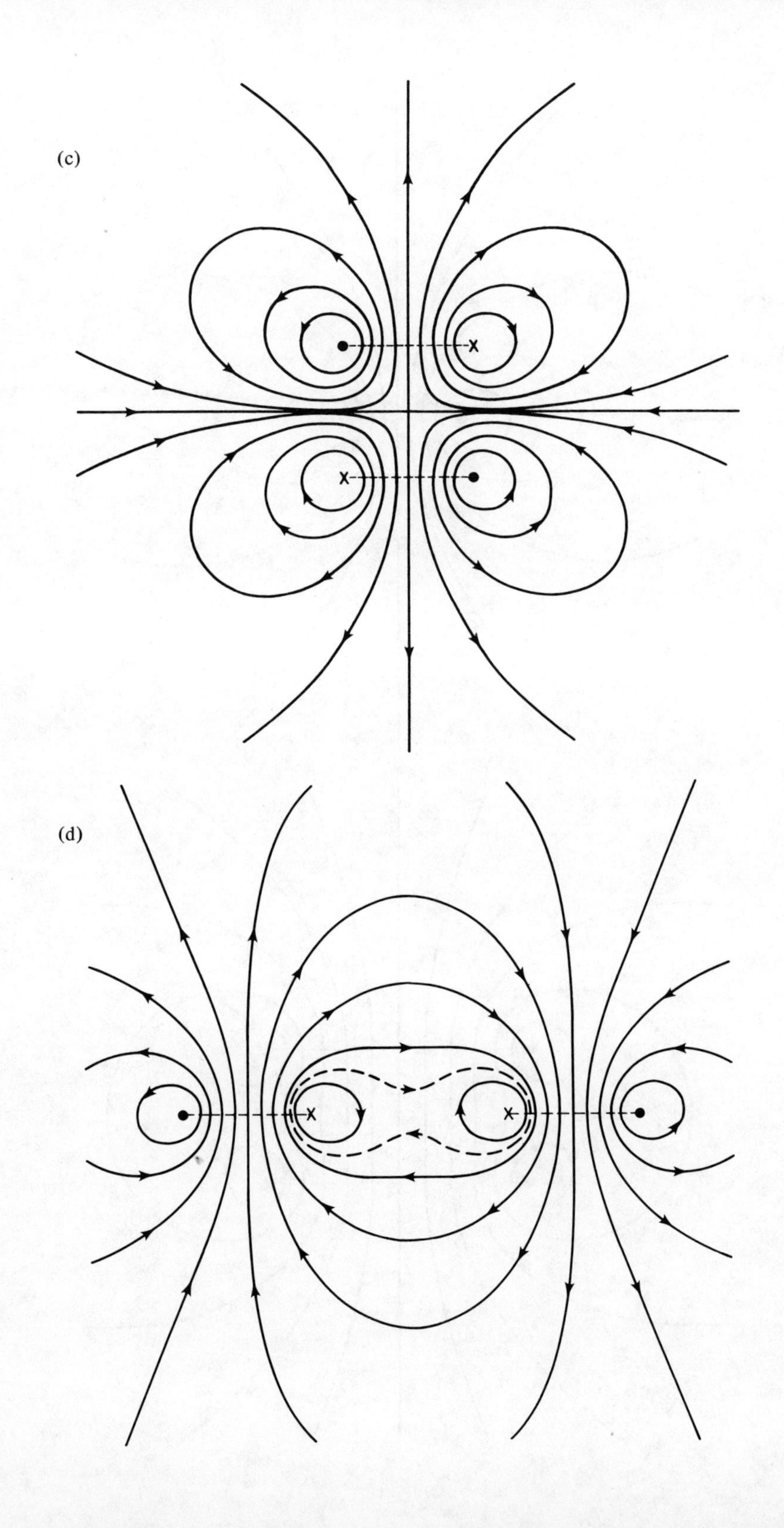
(c)
(d)

PROBLEMS

P5-1. Describe a charge distribution having zero charge density but nonzero current density, thereby showing how it is possible for magnetic induction fields to exist in the absence of electric fields.

P5-2. A rigid object within which the charge density is nonzero moves with velocity $\mathbf{v}$ but does not rotate. Show that the magnetic induction $\mathbf{B}$ and the electric field $\mathbf{E}$ established by this distribution are related by $\mathbf{B} = \mu_0\epsilon_0\mathbf{v} \times \mathbf{E}$ *Hint*: Use Eq. (5-10) and note that $\mathbf{J} = \rho\mathbf{v}$. (Strictly, this result is valid only if $v \ll$ speed of light.)

P5-3. A rectangular loop carrying current I' is placed near a long straight wire lying in the plane of the loop and carrying current I (Fig. P5-3). Find the net force on the loop. Is the loop attracted to or repelled from the straight wire?

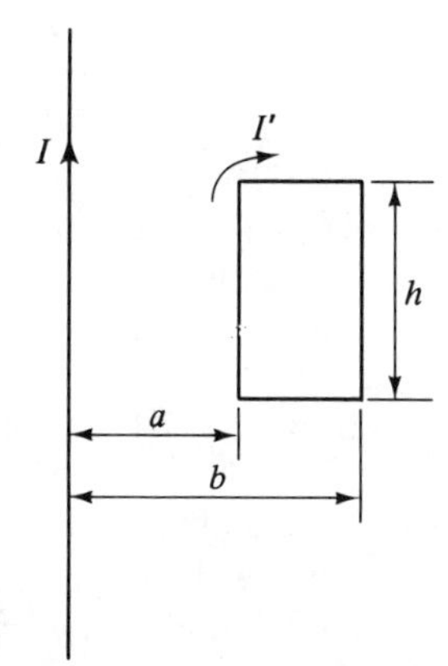

Figure P5-3

P5-4. (a) A circular current loop of radius a lies in the x-y plane with its center at the origin. Using the Biot-Savart law, show that the magnetic induction at a point on the z axis is given by

$$\mathbf{B}(0, 0, z) = \frac{\mu_0}{2\pi} \frac{\mathbf{m}}{(z^2 + a^2)^{3/2}}$$

where $\mathbf{m}$ is the magnetic dipole moment of the loop [Eq. (3-35)]. Obtain a careful graph of B_z/B_0 versus z/a, where B_0 is the field at $z = 0$. (b) The configuration known as a Helmholtz coil consists of two identical loops of radius a located with their centers on the z axis and with their planes in the planes $z = -\frac{1}{2}a$ and $z = \frac{1}{2}a$, each carrying the *same* current (magnitude *and* direction). Find an expression for the field produced on the z axis by such an arrangement; plot a careful graph of B_z/B_0 versus z/a, where B_0 is the field at $z = 0$; and determine how far from the origin one can go along the z axis before the field has fallen to 90% of its value at the origin. (c) Compare the two graphs and comment on the differences. *Hints*: (1) The contribution of each loop in part (b) can be obtained by a simple transformation of the result in part (a). (2) Use a desk calculator or a computer.

P5-5. The circulation of an electron in an atomic orbit constitutes a current loop. If the orbit has radius a and the electron (charge $-q_e$) circulates with speed v, determine the resulting current, find the magnetic dipole moment [Eq. (3-35)] of the orbiting electron, and find the magnetic induction field at the center of the orbit. Using values for the hydrogen atom, estimate the electron magnetic moment and the field at the nucleus *numerically*. *Hint*: Use the result of P5-4(a). Part of the hyperfine interaction between a nucleus and its orbital electrons is mediated by this magnetic induction field.

P5-6. A circular current loop of radius a lies in the x-y plane with its center at the origin and carries a current I' counterclockwise as viewed from a point on the positive z axis. (a) Starting with the Biot-Savart law, show that

$$\mathbf{B}(x, 0, z) = \frac{\mu_0 I' a}{2\pi} \int_0^{\pi} \frac{z \cos\phi' \hat{\mathbf{i}} + (a - x\cos\phi')\hat{\mathbf{k}}}{[x^2 + z^2 + a^2 - 2ax\cos\phi']^{3/2}}\, d\phi'$$

where ϕ' is the angle between the x axis and a point on the loop. (b) Verify that this result reduces correctly to the result in P5-4 when $x = 0$. (c) Expand the integrand in powers of a/r, where $r^2 = x^2 + z^2$, and show that, when the observation point is remote from the loop,

$$\mathbf{B}(\mathbf{r}) = \frac{\mu_0}{4\pi r^3}[3(\mathbf{m}\cdot\hat{\mathbf{r}})\hat{\mathbf{r}} - \mathbf{m}]$$

where $\mathbf{m}$ is the magnetic dipole moment [Eq. (3-35)] of the current loop and $\hat{\mathbf{r}}$ is a unit vector directed toward the point of observation from the center of the loop. *Hint:* Let θ be the angle between the z axis and the direction of observation. Then $z = r\cos\theta$ and $x = r\sin\theta$.

5-2 The Magnetic Flux Law

Just as Coulomb's law gave rise to Gauss's law and the restricted Faraday law for static electric fields, the Biot-Savart law gives rise to the magnetic flux law and Ampere's circuital law for static magnetic induction fields. Statement of the flux law, which is the topic of this section, involves the concept of the flux $\mathbf{\Phi}_m$ of the magnetic induction across a surface Σ, where, by definition,

$$\mathbf{\Phi}_m = \int_{\Sigma} \mathbf{B}\cdot d\mathbf{S} \tag{5-12}$$

Statements similar to those made following Eq. (4-14) apply also to Eq. (5-12). The mks unit of magnetic flux is the *weber* (Wb), and the older name—Wb/m²—for the mks unit of $\mathbf{B}$ is thus consistent with Eq. (5-12).

Seeking a statement analogous to Gauss's law, we shall now examine the flux of a static magnetic induction out of an *arbitrary closed* surface. Using the divergence theorem, we can write this flux as a volume integral, i.e.,

$$\oint_{\Sigma} \mathbf{B} \cdot d\mathbf{S} = \int_{V} \nabla \cdot \mathbf{B} \, dv \tag{5-13}$$

where V is the volume bounded by Σ. With $\mathbf{B}$ given by Eq. (5-10), however, $\nabla \cdot \mathbf{B}$ can be evaluated; we find that

$$\begin{aligned} \nabla \cdot \mathbf{B} &= \frac{\mu_0}{4\pi} \nabla \cdot \int \mathbf{J}(\mathbf{r}') \times \frac{\mathbf{r} - \mathbf{r}'}{|\mathbf{r} - \mathbf{r}'|^3} \, dv' \\ &= -\frac{\mu_0}{4\pi} \int \mathbf{J}(\mathbf{r}') \cdot \left[\nabla \times \frac{\mathbf{r} - \mathbf{r}'}{|\mathbf{r} - \mathbf{r}'|^3} \right] dv' \\ &= \frac{\mu_0}{4\pi} \int \mathbf{J}(\mathbf{r}') \cdot \left[\nabla \times \nabla \left(\frac{1}{|\mathbf{r} - \mathbf{r}'|} \right) \right] dv' \end{aligned} \tag{5-14}$$

where the second equation is obtained from the first by (1) reversing the order of differentiation with respect to the components of $\mathbf{r}$ and integration on the components of $\mathbf{r}'$ and (2) applying Eq. (C-12), and the third equation is obtained from the second by using Eq. (4-45). Since, however, the curl of any gradient is zero [Eq. (C-17)], we find from Eq. (5-14) that

$$\nabla \cdot \mathbf{B} = 0 \tag{5-15}$$

and then from Eq. (5-13) that

$$\oint_{\Sigma} \mathbf{B} \cdot d\mathbf{S} = 0 \tag{5-16}$$

where Σ is *arbitrary*. Equations (5-15) and (5-16) are equivalent (P5-8); they state the *magnetic flux law* in differential and integral form, respectively. When $\mathbf{B}$ is static, this law is a direct consequence of the Biot-Savart law. In fact, the flux law carries over to nonstatic fields without change of form. That extension, however, can be made only on the basis of experimental observations beyond those supporting the Biot-Savart law. In that extension, therefore, the magnetic flux law becomes something more than a mere restatement of a portion of the Biot-Savart law. A fuller treatment of time-dependent fields is postponed until Chapter 6.

Equation (5-16) can be given an illuminating physical interpretation. The electric analog of Eq. (5-16) is Gauss's law,

$$\oint_{\Sigma} \mathbf{E} \cdot d\mathbf{S} = \frac{1}{\epsilon_0} \int_{V} \rho \, dv \tag{5-17}$$

That is, the net flux of $\mathbf{E}$ out of a closed surface is proportional to the sum of all the point electric charges inside the surface. To date, however, no one has succeeded in isolating a corresponding magnetic point charge. If we were not aware of *electric* charge, we would put a zero on the right-hand side of

Eq. (5-17). Since in effect we are not aware of *magnetic* charge, we put a zero on the right-hand side of the magnetic analog of Eq. (5-17), i.e., of Eq. (5-16). Thus, Eq. (5-16) can be viewed as a mathematical statement that magnetic monopoles are not needed to account for any known experimental results. Equivalently, Eq. (5-15) requires the lines of the magnetic induction to close on themselves (Section 2-3), starting or stopping of these lines occurring only on (the unneeded) magnetic charges. Should someone someday successfully isolate a magnetic charge, Eqs. (5-15) and (5-16) would require modification to incorporate these new results.

PROBLEMS

P5-7. Calculate the flux of the magnetic induction produced by the long straight wire across the plane surface bounded by the rectangular loop in Fig. P5-3.

P5-8. Show that Eq. (5-15) can be derived from Eq. (5-16) and vice versa, thus verifying the equivalence of these two expressions. *Hint*: To obtain Eq. (5-15) from Eq. (5-16), let $\boldsymbol{\Sigma}$ become small and apply Eq. (2-25).

5-3 Ampere's Circuital Law

The second basic magnetostatic law that follows as a consequence of the Biot-Savart law involves the line integral of the magnetic induction field about an *arbitrary closed* path $\boldsymbol{\Gamma}$. Using Stokes' theorem, we can write this line integral as a surface integral, i.e,

$$\oint_{\Gamma} \mathbf{B}\cdot d\boldsymbol{\ell} = \int_{\Sigma} \nabla \times \mathbf{B}\cdot d\mathbf{S} \tag{5-18}$$

where $\boldsymbol{\Gamma}$ bounds the (open) surface $\boldsymbol{\Sigma}$. With $\mathbf{B}$ given by Eq. (5-10), however, $\nabla \times \mathbf{B}$ can be evaluated; we find that

$$\begin{aligned}\nabla \times \mathbf{B} &= \frac{\mu_0}{4\pi}\nabla \times \int \mathbf{J}(\mathbf{r}') \times \frac{(\mathbf{r}-\mathbf{r}')}{|\mathbf{r}-\mathbf{r}'|^3}\,dv' \\ &= -\frac{\mu_0}{4\pi}\int \mathbf{J}(\mathbf{r}')\left(\nabla' \cdot \frac{(\mathbf{r}-\mathbf{r}')}{|\mathbf{r}-\mathbf{r}'|^3}\right)dv' \\ &\quad + \frac{\mu_0}{4\pi}\int [\mathbf{J}(\mathbf{r}')\cdot\nabla']\frac{(\mathbf{r}-\mathbf{r}')}{|\mathbf{r}-\mathbf{r}'|^3}\,dv'\end{aligned} \tag{5-19}$$

where ∇ has been taken under the integral sign, the resulting triple product has been expanded using Eq. (C-16), and finally the operator ∇, which at this point acts only on functions of the combination $\mathbf{r} - \mathbf{r}'$, has been replaced by the operator $-\nabla'$, where ∇' differentiates with respect to the components of $\mathbf{r}'$. Next, we use Eq. (C-26) to rewrite the second integral in Eq. (5-19), finding that

$$\nabla \times \mathbf{B} = -\frac{\mu_0}{4\pi} \int \mathbf{J}(\mathbf{r}') \left(\nabla' \cdot \frac{(\mathbf{r} - \mathbf{r}')}{|\mathbf{r} - \mathbf{r}'|^3} \right) dv'$$
$$- \frac{\mu_0}{4\pi} \int \frac{\mathbf{r} - \mathbf{r}'}{|\mathbf{r} - \mathbf{r}'|^3} \nabla' \cdot \mathbf{J}(\mathbf{r}')\, dv'$$
$$+ \frac{\mu_0}{4\pi} \oint \frac{\mathbf{r} - \mathbf{r}'}{|\mathbf{r} - \mathbf{r}'|^3} \mathbf{J}(\mathbf{r}') \cdot d\mathbf{S}' \tag{5-20}$$

Here the third integral extends over the surface bounding the region in which $\mathbf{J}(\mathbf{r}')$ differs from zero. On this surface, however, $\mathbf{J}(\mathbf{r}') \cdot d\mathbf{S}'$—essentially the normal component of $\mathbf{J}(\mathbf{r}')$—is necessarily zero, for otherwise current would be flowing out of the region. Thus, the third integral in Eq. (5-20) has the value zero.[1] But for the *steady currents* to which the Biot-Savart law applies, $\nabla \cdot \mathbf{J}$ is also zero [Eq. (2-47)] and the second integral in Eq. (5-20) makes no contribution. Finally, differentiation shows that $\nabla' \cdot [(\mathbf{r} - \mathbf{r}')/|\mathbf{r} - \mathbf{r}'|^3] = 0$ except when $\mathbf{r}' = \mathbf{r}$. Thus, the remaining integral in Eq. (5-20) can be reduced to an integral over a small volume $\Delta V'$ surrounding the point $\mathbf{r}$, i.e.,

$$\nabla \times \mathbf{B} = -\frac{\mu_0}{4\pi} \int_{\Delta V'} \mathbf{J}(\mathbf{r}') \left(\nabla' \cdot \frac{\mathbf{r} - \mathbf{r}'}{|\mathbf{r} - \mathbf{r}'|^3} \right) dv' \tag{5-21}$$

Now if we choose $\Delta V'$ sufficiently small that $\mathbf{J}(\mathbf{r}')$ varies insignificantly throughout $\Delta V'$, then $\mathbf{J}(\mathbf{r}')$ can be evaluated at the central point $\mathbf{r}$ and taken outside the integral. The remaining integral can then be rewritten as a surface integral by using the divergence theorem; we find that

$$\nabla \times \mathbf{B} = +\frac{\mu_0 \mathbf{J}}{4\pi} \oint \frac{\mathbf{r}' - \mathbf{r}}{|\mathbf{r}' - \mathbf{r}|^3} \cdot d\mathbf{S}' \tag{5-22}$$

where $\mathbf{J}(\mathbf{r})$ has been written simply $\mathbf{J}$ and the minus sign in front of Eq. (5-21) has been absorbed by reversing the terms in the numerator of the integrand. Now, the replacements $\mathbf{r}' \rightarrow \mathbf{r}$, $\mathbf{r} \rightarrow \mathbf{r}_0$, $d\mathbf{S}' \rightarrow d\mathbf{S}$ in Eq. (5-22) convert the integral into the integral evaluated at Eq. (2-37). Since $\mathbf{r}$ lies *inside* the surface of integration, the integral in Eq. (5-22) has the value 4π and we finally find that

$$\nabla \times \mathbf{B} = \mu_0 \mathbf{J} \tag{5-23}$$

and then from Eq. (5-18) that

$$\oint_\Gamma \mathbf{B} \cdot d\boldsymbol{\ell} = \mu_0 \int_\Sigma \mathbf{J} \cdot d\mathbf{S} \tag{5-24}$$

where the surface Σ is *any* surface bounded by the curve Γ, Γ is *arbitrary*, and the usual right-hand rule relates the direction in which Γ is traversed to the direction assigned to $d\mathbf{S}$. Equations (5-23) and (5-24) are equivalent (Why?

[1] In fact, some current distributions extending to infinity can be permitted without changing this conclusion. In particular, currents such as those carried from minus infinity to plus infinity by infinitely long wires are allowed. See P5-9.

Compare P5-8); they state Ampere's circuital law in differential and integral form, respectively. In words, Eq. (5-24) states that the line integral of **B** about a *closed* path is equal to μ_0 times the current flowing across *any* surface bounded by that path. In contrast to the magnetic flux law, Ampere's circuital law is valid only for static magnetic induction fields. We shall develop the necessary modifications in Chapter 6, noting now only that this later generalization of Ampere's law rests on additional experimental observations and is therefore more than merely a consequence of the Biot-Savart law.

As an example of the application of both the circuital law and the flux law, we shall evaluate the magnetic induction established by an infinitely long, cylindrical wire of radius a having its axis coincident with the z axis and carrying a total current I uniformly distributed over its cross section with current density $(I/\pi a^2)\hat{\mathbf{k}}$ (Fig. 5-5). As with the line charge treated in Section 4-2, we here apply symmetry to simplify the most general field at the point **r**, viz.,

$$\mathbf{B}(\mathbf{r}) = B_{\imath}(\imath, \phi, z)\hat{\boldsymbol{\imath}} + B_{\phi}(\imath, \phi, z)\hat{\boldsymbol{\phi}} + B_z(\imath, \phi, z)\hat{\mathbf{k}} \tag{5-25}$$

in cylindrical coordinates. The present distribution shares invariance to rotation about the z axis and invariance to translation along the z axis with the

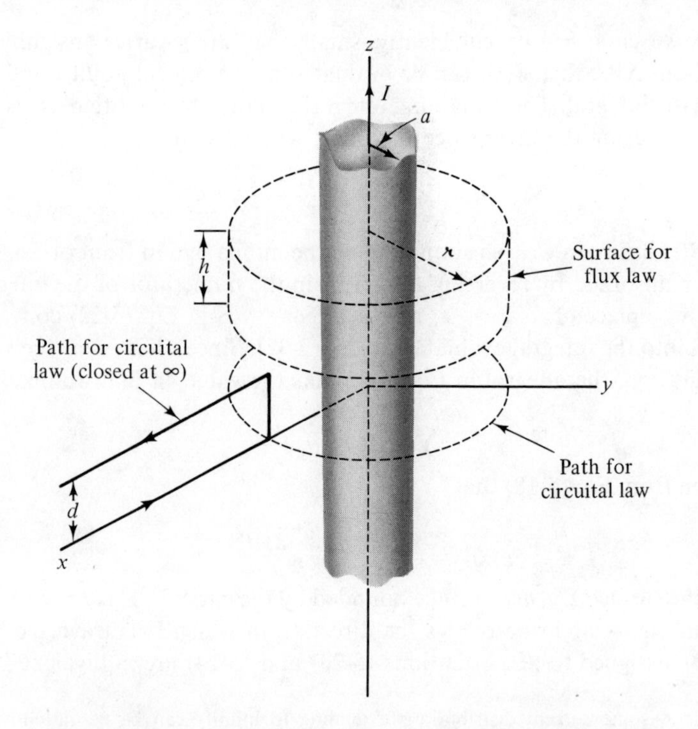

Fig. 5-5. Surfaces and paths for the several integrals involved in determining the magnetic induction field of a long straight wire of non-zero radius.

line charge, and the three components in Eq. (5-25) must therefore be independent of ϕ and z. But the current distribution is not invariant to reflection in the x-y plane, so the remaining conclusions concerning the symmetries of the field of a line charge do not apply to the current. Symmetry supports the reduction

$$\mathbf{B}(\mathbf{r}) = B_{\mathscr{r}}(\mathscr{r})\hat{\boldsymbol{\mathscr{r}}} + B_\phi(\mathscr{r})\hat{\boldsymbol{\phi}} + B_z(\mathscr{r})\hat{\mathbf{k}} \tag{5-26}$$

which is applicable whether $\mathscr{r} < a$ *or* $\mathscr{r} > a$, but (as presented here) goes no further. Additional information about these components, however, can be obtained by applying the two basic laws to suitable surfaces or curves. Apply the flux law, for example, to the surface bounded by a cylinder of radius $\mathscr{r}$ and height h and two planes, as shown in Fig. 5-5. Only B_z contributes to the flux across the upper and lower plane surfaces, but B_z is independent of z and the flux entering at the lower surface is therefore exactly equal to the flux leaving at the upper surface; the two surfaces together make no net contribution to $\oint \mathbf{B}\cdot d\mathbf{S}$. The cylindrical portion of the surface, however, contributes an amount $2\pi\mathscr{r}hB_{\mathscr{r}}(\mathscr{r})$, since the component of $\mathbf{B}$ normal to the cylindrical surface is everywhere $B_{\mathscr{r}}$ and $B_{\mathscr{r}}$ has the same value at every point on the surface. Thus,

$$\oint \mathbf{B}\cdot d\mathbf{S} = 2\pi\mathscr{r}hB_{\mathscr{r}}(\mathscr{r}) \tag{5-27}$$

for this surface. Since the flux law requires $\oint \mathbf{B}\cdot d\mathbf{S} = 0$, we conclude that $B_{\mathscr{r}} = 0$ and hence that

$$\mathbf{B}(\mathbf{r}) = B_\phi(\mathscr{r})\hat{\boldsymbol{\phi}} + B_z(\mathscr{r})\hat{\mathbf{k}} \tag{5-28}$$

Now, apply the circuital law to a path coming in along the x axis to $x = \mathscr{r}$, going up parallel to the z axis to $x = \mathscr{r}$, $z = d$, and then returning to infinity along the line $z = d$ in the x-z plane. Finally, regard the path to be closed at infinity. No current crosses the surface bounded by this path, *even if the path extends to a point inside the wire*. Thus, the circuital law requires that $\oint \mathbf{B}\cdot d\boldsymbol{\ell} = 0$. Assuming that $|\mathbf{B}| \longrightarrow 0$ at large distances from the wire[2] and taking $\mathbf{B}$ as in Eq. (5-28), we determine for this path that

$$\oint \mathbf{B}\cdot d\boldsymbol{\ell} = B_z(\mathscr{r})\, d \tag{5-29}$$

which will be zero only if $B_z = 0$. Thus, $\mathbf{B}$ is finally reduced to

$$\mathbf{B}(\mathbf{r}) = B_\phi(\mathscr{r})\hat{\boldsymbol{\phi}} \tag{5-30}$$

To determine B_ϕ more explicitly, we apply the circuital law again, this time to a circle of radius $\mathscr{r}$ lying in the x-y plane with its center at the origin. On

[2]This statement follows from the Biot-Savart law, but it also is part of our general expectation about the relationship between any field and its source, namely that the effect of a (localized) source must become less strong as the observation point becomes more remote. The difficulty in applying this general expectation to specific cases lies in deciding whether a specific source is *sufficiently* localized. For the infinite wire in the present context, this assumption proves to be realistic. It would not apply to the electric field of a two-dimensionally infinite sheet [Eq. (4-13)].

this circle $d\boldsymbol{\ell} = r\, d\phi\, \hat{\boldsymbol{\phi}}$ and, with attention to the proper relationship between the direction of traversal of the path and the direction assigned the surface, $d\mathbf{S} = r'\, dr'\, d\phi'\, \hat{\mathbf{k}}$. Thus, for this path,

$$\oint \mathbf{B}\cdot d\boldsymbol{\ell} = \int_0^{2\pi} B_\phi(r)\hat{\boldsymbol{\phi}}\cdot r\, d\phi\, \hat{\boldsymbol{\phi}} = 2\pi r B_\phi(r) \tag{5-31}$$

and

$$\int \mathbf{J}\cdot d\mathbf{S} = \int_0^{r} r'\, dr' \int_0^{2\pi} d\phi' \frac{I}{\pi a^2} = \begin{cases} I\left(\dfrac{r}{a}\right)^2, & r < a \\ I, & r > a \end{cases} \tag{5-32}$$

and the circuital law gives

$$B_\phi(r) = \begin{cases} \dfrac{\mu_0 I}{2\pi a}\left(\dfrac{r}{a}\right), & r < a \\ \dfrac{\mu_0 I}{2\pi a}\left(\dfrac{a}{r}\right), & r > a \end{cases} \tag{5-33}$$

A graph of $B_\phi(r)$ versus r, plotted in the dimensionless form $B_\phi/(\mu_0 I/2\pi a)$ versus r/a, is shown in Fig. 5-6. Note that this result in the region outside the wire is identical with Eq. (5-11) for a wire of zero radius.

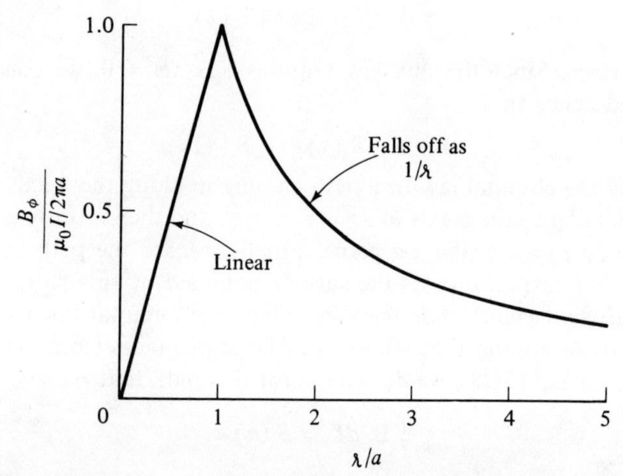

Fig. 5-6. B_ϕ versus r for the magnetic induction field of a long straight wire of non-zero radius.

PROBLEMS

P5-9. Let the source distribution for Eq. (5-20) be an infinitely long straight wire carrying current I along the z axis from $z = -\infty$ to $z = +\infty$. Regard the surface integral in Eq. (5-20) to extend over a cylindrical surface of finite length surrounding the wire and then let the ends of the cylinder recede to $\pm\infty$. Show that the surface integral approaches the value zero in that limit.

P5-10. A coaxial cable consists of a center wire of radius a and a coaxial cylindrical shell of radius b, $b > a$. Suppose that the center wire of a long straight coaxial cable carries a current I along the z axis toward $z = +\infty$ and that the shell carries the same current in the opposite direction. Assuming that the current in the wire is uniformly distributed over the wire, find the magnetic induction field as a function of distance $\imath$ from the axis of the cable and draw a graph of the azimuthal component of the field versus $\imath$.

P5-11. A simple solenoid is made by winding a wire many times around a (nonmagnetic) cylindrical core, the resulting coil looking like a very tight coil spring. Consider an (idealized) infinitely long solenoid of circular cross section (radius a) wound with n turns of wire per unit length, the wire carrying current I. Let the coil be wound tightly enough so that effects arising from the pitch of the windings can be ignored. (a) Use the flux law and symmetry to show that the resulting field has no (cylindrical) radial component. (b) Use the circuital law and symmetry to show that the resulting field has no azimuthal component. (c) Use the circuital law to show that the axial component of the resulting field is zero outside and $\mu_0 nI$ inside the solenoid and argue that the field is uniform across the cross section of the solenoid. (d) From the sign conventions built into the basic laws, verify the right-hand rule relating the direction of the current around the solenoid to the direction of the field. (Fingers around the solenoid in the direction of the current; thumb in the direction of the field.)

P5-12. A solenoid (P5-11) of *finite* length and radius a is wound with a total of N turns of a wire carrying current I and is then bent so that its two ends meet, forming a toroid (doughnut) with central radius b, $b > a$. Let the axis of the toroid coincide with the z axis and let its median plane lie in the x-y plane. Show that the magnetic induction inside the toroid is given by $\mathbf{B} = (\mu_0 NI/2\pi\imath)\hat{\boldsymbol{\phi}}$ and show that the flux Φ_m of this field across the plane surface bounded by a single turn of the coil is given by

$$\Phi_m = \frac{\mu_0 NIa}{\pi} \int_{-1}^{1} \frac{\sqrt{1-\xi^2}}{\xi + (b/a)}\, d\xi$$

Optional: Use numerical integration (e.g., Simpson's rule) on a computer to explore Φ_m as a function of b/a.

5-4
The Magnetic Vector Potential

An expression for a magnetic analog of the electrostatic potential can be derived by applying an argument similar to that following Eq. (4-45) to the Biot-Savart law. Substituting the identity, Eq. (4-45), into Eq. (5-8), for example, we obtain

$$\mathbf{B}(\mathbf{r}) = \frac{\mu_0}{4\pi} \oint \left(\nabla \frac{1}{|\mathbf{r}-\mathbf{r}'|}\right) \times I'\, d\boldsymbol{\ell}' \tag{5-34}$$

Since $I'\,d\boldsymbol{\ell}'$ does not depend on $\mathbf{r}$, however, it is constant in so far as the derivatives in ∇ are concerned. Using Eq. (C-8) with $\Phi = 1/|\mathbf{r} - \mathbf{r}'|$ and $\mathbf{Q} = I'\,d\boldsymbol{\ell}'$, we can therefore write the integrand in Eq. (5-34) as $\nabla \times (\Phi\mathbf{Q})$. Finally, since ∇ does not act on the variables of integration, we can interchange the order of differentiation and integration to find that

$$\mathbf{B}(\mathbf{r}) = \nabla \times \left(\frac{\mu_0}{4\pi} \oint \frac{I'\,d\boldsymbol{\ell}'}{|\mathbf{r} - \mathbf{r}'|}\right) \tag{5-35}$$

A similar argument applied to Eq. (5-10) gives

$$\mathbf{B}(\mathbf{r}) = \nabla \times \left(\frac{\mu_0}{4\pi} \int \frac{\mathbf{J}(\mathbf{r}')}{|\mathbf{r} - \mathbf{r}'|}\,dv'\right) \tag{5-36}$$

for a source distribution described by the current density $\mathbf{J}(\mathbf{r})$. From either form, we infer that the magnetic induction field can be derived from the curl of another vector field,

$$\mathbf{B} = \nabla \times \mathbf{A} \tag{5-37}$$

where $\mathbf{A}$ is called the (*magnetic*) *vector potential.* From Eqs. (5-35) and (5-36), however, we can conclude only that

$$\mathbf{A}(\mathbf{r}) = \frac{\mu_0}{4\pi} \oint \frac{I'\,d\boldsymbol{\ell}'}{|\mathbf{r} - \mathbf{r}'|} + \mathbf{W}(\mathbf{r}) \tag{5-38}$$

$$= \frac{\mu_0}{4\pi} \int \frac{\mathbf{J}(\mathbf{r}')}{|\mathbf{r} - \mathbf{r}'|}\,dv' + \mathbf{W}(\mathbf{r}) \tag{5-39}$$

where $\mathbf{W}(\mathbf{r})$ is an *arbitrary* vector field satisfying $\nabla \times \mathbf{W} = 0$—a restriction that is considerably less severe than the corresponding restriction on the function χ in Eq. (4-47). Since any field with zero curl can always be expressed as a gradient [item (3) in Section 2-5], the condition $\nabla \times \mathbf{W} = 0$ will be satisfied if (and only if) $\mathbf{W} = \nabla\lambda$ but imposes no constraint on the scalar field λ. Thus Eqs. (5-35)–(5-37) together define $\mathbf{A}$ no more explicitly than the equations

$$\mathbf{A}(\mathbf{r}) = \frac{\mu_0}{4\pi} \oint \frac{I'\,d\boldsymbol{\ell}'}{|\mathbf{r} - \mathbf{r}'|} + \nabla\lambda(\mathbf{r}) \tag{5-40}$$

$$= \frac{\mu_0}{4\pi} \int \frac{\mathbf{J}(\mathbf{r}')}{|\mathbf{r} - \mathbf{r}'|}\,dv' + \nabla\lambda(\mathbf{r}) \tag{5-41}$$

where λ is *arbitrary*; whatever the value of λ, $\nabla \times \mathbf{A}$ will give the correct $\mathbf{B}$-field. Equations (5-40) and (5-41) not only provide for a calculation of $\mathbf{A}$ directly from known sources but also express a relationship between (a static) $\mathbf{A}$ and its sources even when the sources are not explicitly known. Equation (5-41) should be compared with the electrostatic analog, Eq. (4-50).[3]

[3]The existence of $\mathbf{A}$ might also have been inferred from the flux law, $\nabla \cdot \mathbf{B} = 0$, by applying the theorem in item (4) of Section 2-5. Since the flux law is valid for time-dependent fields, this route to Eq. (5-37) demonstrates the existence of $\mathbf{A}$ more generally than the route adopted above. It does not, however, provide an expression for calculating $\mathbf{A}$ directly from its sources.

In consequence of Eqs. (5-40) and (5-41), the difference between two equivalent vector potentials $\mathbf{A}_1$ and $\mathbf{A}_2$, distinguished by the choices λ_1 and λ_2 for λ, is given by

$$\mathbf{A}_2 - \mathbf{A}_1 = \nabla(\lambda_2 - \lambda_1) \Longrightarrow \mathbf{A}_2 = \mathbf{A}_1 + \nabla\Lambda \tag{5-42}$$

where $\Lambda = \lambda_2 - \lambda_1$ is arbitrary, since λ_2 and λ_1 are themselves independently arbitrary. This transformation from one vector potential to a second equivalent vector potential is called a *gauge transformation*, and the fact that addition of an arbitrary gradient to the vector potential does not change the corresponding **B**-field is referred to as the *gauge invariance* of **B**; each vector potential by itself is said to be *in a* (particular) *gauge*. Now the vector potential can be chosen not only so that $\mathbf{B} = \nabla \times \mathbf{A}$ but also so that $\nabla \cdot \mathbf{A}$ has any chosen value whatever. We prove this property by noting first that Eq. (5-42) gives

$$\nabla \cdot \mathbf{A}_2 = \nabla \cdot \mathbf{A}_1 + \nabla^2 \Lambda \tag{5-43}$$

Thus if we have a vector potential $\mathbf{A}_1$ in hand and $\nabla \cdot \mathbf{A}_1$ does not have the desired value, we first find Λ by solving

$$\nabla^2 \Lambda = -\nabla \cdot \mathbf{A}_1 + d \tag{5-44}$$

where $d = d(\mathbf{r})$ is the desired divergence, and then find an equivalent vector potential $\mathbf{A}_2$ from Eq. (5-42). By its construction, $\mathbf{A}_2$ will then have the desired divergence. In summary, gauge invariance of **B** corresponds to the freedom to select $\nabla \cdot \mathbf{A}$ arbitrarily. When treating static fields, $\nabla \cdot \mathbf{A}$ is commonly set equal to zero; we shall later see that other choices are more suitable for time-dependent fields.

A differential equation relating **A** and **J** when both are *static* can be obtained by substituting Eq. (5-37) into Eq. (5-23) and then using Eq. (C-19); we find that

$$\nabla \times (\nabla \times \mathbf{A}) = \nabla(\nabla \cdot \mathbf{A}) - \nabla^2 \mathbf{A} = \mu_0 \mathbf{J} \tag{5-45}$$

Thus, choosing the so-called *Coulomb gauge*, in which $\nabla \cdot \mathbf{A} = 0$, we find finally that

$$\nabla^2 \mathbf{A} = -\mu_0 \mathbf{J} \tag{5-46}$$

which should be compared to the electrostatic analog, Eq. (4-51).

The interrelationships we have now developed among the current density, the magnetic induction, and the vector potential are summarized in Table 5-1. The entries in this table should be compared with those in Table 4-1.

To illustrate the application of some of the relationships developed in this section, we shall calculate the magnetic induction produced by an arbitrary current loop at a point remote from the loop. We first evaluate the vector potential, which we take to be given by Eq. (5-40) with $\lambda = 0$. When $|\mathbf{r}| = r \gg |\mathbf{r}'| = r'$ for all $\mathbf{r}'$ within the region of integration (i.e., far from the loop), we can expand the denominator in powers of r'/r and keep only the

TABLE 5-1 Interrelationships Among Steady **J** and Static **B** and **A**

Given \ *Find*	**J**	**B**	**A**
J	—	$\frac{\mu_0}{4\pi}\int \frac{\mathbf{J}(\mathbf{r}') \times (\mathbf{r}-\mathbf{r}')}{\lvert\mathbf{r}-\mathbf{r}'\rvert^3}\,dv'$	$\frac{\mu_0}{4\pi}\int \frac{\mathbf{J}(\mathbf{r}')}{\lvert\mathbf{r}-\mathbf{r}'\rvert}\,dv'$
B	$\frac{1}{\mu_0}\nabla \times \mathbf{B}$	—	See P5-30
A	$-\frac{1}{\mu_0}\nabla^2 \mathbf{A}$[a]	$\nabla \times \mathbf{A}$	—

[a]provided **A** is in Coulomb gauge, for which $\nabla\cdot\mathbf{A} = 0$.

first two terms; we find that

$$\mathbf{A}(\mathbf{r}) = \frac{\mu_0 I'}{4\pi r}\oint d\mathbf{r}' + \frac{\mu_0 I'}{4\pi r^3}\oint (\mathbf{r}\cdot\mathbf{r}')\,d\mathbf{r}' + \cdots \tag{5-47}$$

where we have recognized that $d\boldsymbol{\ell}'$ in Eq. (5-40) in fact stands for an increment in $\mathbf{r}'$, which can equally well be denoted by $d\mathbf{r}'$. Now $\oint d\mathbf{r}'$ is a vector from the start of the path to the end and is zero for a *closed* path. Thus, only the second term in Eq. (5-47) contributes. This term can be evaluated by applying Eq. (C-20) followed by Eq. (C-15) with all derivatives acting on the primed variables and $\mathbf{r}$ being a constant; we find that

$$\begin{aligned}\oint (\mathbf{r}\cdot\mathbf{r}')\,d\mathbf{r}' &= \int d\mathbf{S}' \times \nabla'(\mathbf{r}\cdot\mathbf{r}') \\ &= \int d\mathbf{S}' \times [(\mathbf{r}\cdot\nabla')\mathbf{r}' + \mathbf{r} \times (\nabla' \times \mathbf{r}')] \\ &= \int d\mathbf{S}' \times \mathbf{r}\end{aligned} \tag{5-48}$$

the last form following because $(\mathbf{r}\cdot\nabla')\mathbf{r}' = \mathbf{r}$ and $\nabla' \times \mathbf{r}' = 0$. Substituting Eq. (5-48) into the second term of Eq. (5-47) and omitting the (vanishing) first term, we have that

$$\mathbf{A}(\mathbf{r}) = \frac{\mu_0}{4\pi r^3}\left(\int I'\,d\mathbf{S}'\right) \times \mathbf{r} = \frac{\mu_0}{4\pi}\frac{\mathbf{m}\times\mathbf{r}}{r^3} \tag{5-49}$$

where $\mathbf{m}$ is the dipole moment of the current loop, which has already appeared in a different context in Eq. (3-35). Direct evaluation of $\nabla \times \mathbf{A}$ then gives

$$\mathbf{B} = \frac{\mu_0}{4\pi r^3}[3(\mathbf{m}\cdot\hat{\mathbf{r}})\hat{\mathbf{r}} - \mathbf{m}] \tag{5-50}$$

$$= \frac{\mu_0 m}{4\pi r^3}[2\cos\theta\,\hat{\mathbf{r}} + \sin\theta\,\hat{\boldsymbol{\theta}}] \tag{5-51}$$

the second form applying in spherical coordinates when $\mathbf{m} = m\hat{\mathbf{k}}$ (P5-19). The *far* field of a magnetic dipole thus has exactly the same form as the *far* field of an electric dipole [compare Eqs. (5-50) and (5-51) with Eqs. (4-10) and (4-11)], but the similarity does not extend to the *near* fields.

Although the vector potential **A** is important to many theoretical developments, it is less useful in practical problems than the scalar potential V. Part of the utility of V, for example, arises because the electrostatic field **E** is often easily determined by first evaluating V from a *scalar* integral [Eq. (4-49)] over known sources and then evaluating $\mathbf{E} = -\nabla V$. The corresponding procedure for determining the magnetostatic field **B**—namely evaluate **A** from a *vector* integral [Eq. (5-40) or (5-41)] over known sources and then evaluate $\mathbf{B} = \nabla \times \mathbf{A}$—is only occasionally easier than determining **B** directly from the Biot-Savart law. Not only is the curl of a vector more difficult to evaluate than the gradient of a scalar, but the integrals for **A** are often more difficult than those for V, particularly because **A** must be known over a broader region of space than V. To be more specific, if we seek only $E_z(0, 0, z)$, then it is sufficient to know $V(0, 0, z)$ in order to obtain $E_z(0, 0, z) = -\partial V(0, 0, z)/\partial z$. If, however, we seek $B_z(0, 0, z)$ we must know $A_x(x, y, z)$ and $A_y(x, y, z)$, for only then can we calculate $(\nabla \times \mathbf{A})_z = \partial A_y/\partial x - \partial A_x/\partial y$ correctly at the point $(0, 0, z)$. Said another way, symmetries constraining **E** can be exploited to yield a useful V, but symmetries constraining **B** cannot be so easily exploited to yield a useful **A**.

A second part of the utility of V arises because its physical interpretation is very immediate and hence intuitive: qV gives the energy of a charge q in the field V; $\frac{1}{2}\int \rho V\, dv$ gives the energy required to assemble the distribution ρ that is the source of V. Although **A** has physical interpretations—it plays a role in determining the energy required to establish a current distribution; **A** and also V have effects on the phase of quantum wave functions—these interpretations are less immediate and less intuitive than those of V. It is therefore natural to think of V as a "real" field—whatever that means—and to think of **A** more as a mathematical tool, even though such a classification in fact misrepresents particularly the quantum significance of **A**.[4]

PROBLEMS

P5-13. A portion of a circuit carrying current I lies along the z axis in the region $a \leq z \leq b$. Let I flow from $z = a$ toward $z = b$. Use Eq. (5-40) with $\lambda = 0$ to calculate the contribution of this portion of the circuit to the vector potential at an arbitrary point **r**. Don't be afraid of integral tables. *Optional*: Let $a = -b$ and use a computer to obtain a graph of $A_z/(\mu_0 I/4\pi)$ versus $\imath/b$, where $\imath$ is the distance from the wire to an observation point in the x-y plane.

[4]For a discussion of the "reality" of **A**, see, for example, R. P. Feynman, R. B. Leighton, and M. Sands, *The Feynman Lectures on Physics* (Addison-Wesley Publishing Company, Inc., Reading, Mass., 1964), Volume II, Lecture 15.

P5-14. By setting up and evaluating Eq. (5-40) with $\lambda = 0$, find the vector potential established at a point on the axis and a distance z from the center of a circular loop of radius a carrying current I. How much of the magnetic induction field can be correctly found from the curl of this potential? Explain your answer fully.

P5-15. Show that $\mathbf{A}$ as given by Eq. (5-41) with $\lambda = 0$ has zero divergence. Assume that the currents are confined to a finite region of space. *Hint*: Remember that Eq. (5-41) applies only for steady currents, for which $\nabla \cdot \mathbf{J} = 0$.

P5-16. Let $\mathbf{B} = B_0\hat{\mathbf{k}}$ with B_0 constant. (a) Show that $\mathbf{A}_1 = \frac{1}{2}\mathbf{B} \times \mathbf{r}$ and $\mathbf{A}_2 = B_0 x\hat{\mathbf{j}}$ are both acceptable vector potentials. (b) Find a function Λ such that $\mathbf{A}_2 = \mathbf{A}_1 + \nabla\Lambda$. (c) Show that both $\mathbf{A}_1$ and $\mathbf{A}_2$ have zero divergence, thereby demonstrating that $\nabla \times \mathbf{A} = \mathbf{B}$ and $\nabla \cdot \mathbf{A} = 0$ together do not necessarily define $\mathbf{A}$ uniquely.

P5-17. Find a vector potential for the field of a long straight wire. *Hints*: (1) Assume $\mathbf{A} = A_z(\imath)\hat{\mathbf{k}}$ and require $\nabla \times \mathbf{A}$ to give the field in Eq. (5-11). (2) See Eq. (0-63).

P5-18. (a) Use Stokes' theorem to show that $\oint \mathbf{A} \cdot d\boldsymbol{\ell} = \int \mathbf{B} \cdot d\mathbf{S}$, where $\mathbf{B} = \nabla \times \mathbf{A}$. (b) An infinitely long solenoid of radius a with n turns per unit length of a wire carrying current I produces the field $\mathbf{B} = \mu_0 nI\hat{\mathbf{k}}$ when $\imath < a$ and $\mathbf{B} = 0$ when $\imath > a$ (P5-11). Here, $\hat{\mathbf{k}}$ is a unit vector parallel to the axis of the solenoid. Assume a vector potential having only a $\hat{\boldsymbol{\phi}}$-component, and use part (a) to find this vector potential. (c) Verify that $\nabla \times \mathbf{A} = \mathbf{B}$ both inside and outside the solenoid. See Eq. (0-63). (d) Ponder what it means for there to exist a vector potential *outside* the solenoid where there is *no* field.

P5-19. Derive Eqs. (5-50) and (5-51) for the field of a magnetic dipole by evaluating the curl of the corresponding vector potential, Eq. (5-49).

5-5
Energy in the Static Magnetic Induction Field

Paralleling our development of electrostatics, we should at this point discuss energy storage in a static magnetic induction field. Unfortunately, we can develop that topic most naturally only after we are able to calculate the work required to establish a current, and that calculation in turn requires the (general) Faraday law of induction, which we shall develop in Section 6-1. The best we can do at this point is obtain the correct result by a *plausibility* argument based on the analogous electrostatic expressions. Beginning with Eq. (4-61), we replace ρ with $\mathbf{J}$, V with $\mathbf{A}$, and the product ρV with the dot product $\mathbf{J} \cdot \mathbf{A}$. The result for the work W required to establish the *steady* current distribution $\mathbf{J}$ is

$$W = \tfrac{1}{2} \int \mathbf{J} \cdot \mathbf{A}\, dv \tag{5-52}$$

where $\mathbf{A}$ is the vector potential established by $\mathbf{J}$ and the integral extends over all space. Using Eqs. (5-23), (C-12), and (5-37) and the divergence theorem, we find alternatively that

$$
\begin{aligned}
W &= \frac{1}{2\mu_0}\int (\nabla \times \mathbf{B})\cdot\mathbf{A}\,dv \\
&= \frac{1}{2\mu_0}\int B^2\,dv - \frac{1}{2\mu_0}\oint_\Sigma (\mathbf{A}\times\mathbf{B})\cdot d\mathbf{S} \\
&= \frac{1}{2\mu_0}\int B^2\,dv
\end{aligned} \tag{5-53}
$$

where Σ is a large spherical surface that ultimately recedes to infinity, in which limit (if $\mathbf{J}$ is localized) $|\mathbf{A}| \propto r^{-2}$, $|\mathbf{B}| \propto r^{-3}$—see Eqs. (5-49) and (5-50)—and $|d\mathbf{S}| \propto r^2$. Thus, $\mathbf{A}\times\mathbf{B}\cdot d\mathbf{S} \propto r^{-3}$ and the above integral over Σ approaches zero. From Eq. (5-53) we infer that a magnetic energy density

$$
u_B = \frac{B^2}{2\mu_0} \tag{5-54}
$$

should be assigned to a region of space where the magnetic induction is $\mathbf{B}$. We stress, however, that Eq. (5-52), which is correct only for static fields, and Eq. (5-53), which turns out to be correct more generally, have not yet been rigorously derived.

PROBLEMS

P5-20. Let a toroidal coil be made by winding N turns on a core whose cross section is a rectangle of height h and width w and then bending the core into a "doughnut" with inside radius a and outside radius $a + w$. If the coil carries current I, show that the energy W stored in the coil is given by

$$
W = \frac{\mu_0}{4\pi}N^2I^2h\,\ell n\left(\frac{a+w}{a}\right)
$$

P5-21. The energy stored in a static magnetic induction field as given by Eq. (5-53) is obviously gauge-invariant. Show that Eq. (5-52) is also gauge-invariant, i.e., that $\frac{1}{2}\int \mathbf{J}\cdot\mathbf{A}_1\,dv = \frac{1}{2}\int \mathbf{J}\cdot\mathbf{A}_2\,dv$, where $\mathbf{A}_1$ and $\mathbf{A}_2$ are related by Eq. (5-42), and state any assumptions in your proof. *Hint*: Remember that $\nabla\cdot\mathbf{J} = 0$ for steady currents.

5-6 The Multipole Expansion of the Magnetic Vector Potential

A multipole expansion of the magnetic vector potential established by a current loop can be obtained by substituting Eq. (4-69) into Eq. (5-40) with $\lambda = 0$; we find that

$$
\mathbf{A}(\mathbf{r}) = \frac{\mu_0}{4\pi}\oint I'\,d\boldsymbol{\ell}'\left(\frac{1}{r} + \frac{\mathbf{r}\cdot\mathbf{r}'}{r^3} + \frac{1}{2r^5}\sum_{i,j} x_i x_j[3x_i'x_j' - \delta_{ij}(r')^2] + \cdots\right) \tag{5-55}
$$

The first two terms are identical with those in Eq. (5-47); Eq. (5-49) expresses their value in terms of the magnetic dipole moment

$$\mathbf{m} = \int I'\,d\mathbf{S}' = \tfrac{1}{2}\oint \mathbf{r}' \times I'\,d\boldsymbol{\ell}' \tag{5-56}$$

as defined in Eqs. (3-35) and (3-36). To express the third term more compactly, we introduce the nine-component quantity

$$\mathbf{Q}_{ij}^{(m)} = \oint [3x_i'x_j' - \delta_{ij}(r')^2]I'\,d\boldsymbol{\ell}' \tag{5-57}$$

each component of which is a *vector*. We then have from Eq. (5-55) that

$$\mathbf{A}(\mathbf{r}) = \frac{\mu_0}{4\pi r^3}\mathbf{m} \times \mathbf{r} + \frac{\mu_0}{8\pi r^5}\sum_{i,j} x_i\mathbf{Q}_{ij}^{(m)}x_j + \cdots \tag{5-58}$$

from which the corresponding field can be derived by evaluating $\nabla \times \mathbf{A}$. We make three further observations: (1) If the source consists of several current loops, the corresponding values of $\mathbf{m}$ and $\mathbf{Q}_{ij}^{(m)}$ are merely sums of the values for each loop separately (Why?); (2) when the currents are steady ($\nabla \cdot \mathbf{J} = 0$), a general current distribution can be thought of as a superposition of current loops and Eq. (5-58) applies with

$$\mathbf{m} = \tfrac{1}{2}\int \mathbf{r}' \times \mathbf{J}(\mathbf{r}')\,dv' \tag{5-59}$$

$$\mathbf{Q}_{ij}^{(m)} = \int [3x_i'x_j' - \delta_{ij}(r')^2]\mathbf{J}(\mathbf{r}')\,dv' \tag{5-60}$$

and (3) the quantity $\mathbf{Q}_{ij}^{(m)}$ introduced here is not the conventional magnetic quadrupole moment tensor, which is defined in terms of convenient (but fictitious) magnetic monopoles (see Section 11-4) and has nine *scalar* components.

PROBLEMS

P5-22. A current distribution consists of two parallel circular current loops of radius a, one located parallel to the x-y plane with its center at $(0, 0, -\frac{1}{2}b)$ and carrying current I clockwise as viewed from a point on the positive z axis and the other with its center at $(0, 0, \frac{1}{2}b)$ and carrying current I counterclockwise. (a) Show that this distribution has zero magnetic dipole moment. (b) Calculate the nine *vector* components of $\mathbf{Q}_{ij}^{(m)}$. (c) Write out the quadrupole term in the vector potential using spherical coordinates (Table 0-1) and calculate the corresponding magnetic induction. *Optional*: Evaluate this field by writing a superposition of two terms, each obtained from Eq. (5-50), and expanding in powers of b/r.

P5-23. Starting from Eq. (5-41), derive Eq. (5-58) in which $\mathbf{m}$ and $\mathbf{Q}_{ij}^{(m)}$ are defined by Eqs. (5-59) and (5-60).

Supplementary Problems

P5-24. When Γ is a closed path, Eq. (5-7) gives the force between two complete circuits. Show that this force satisfies Newton's third law. *Hint*: Use Eq. (C-1) and note that one of the resulting (closed) integrals involves an exact differential and hence has the value zero.

P5-25. (a) Using the result in P5-4, show that the field at $(0, 0, z)$ on the axis of a solenoid (P5-11) of radius a is given by

$$\mathbf{B}(0, 0, z) = \frac{\mu_0 n \mathbf{m}}{2\pi} \int_{-L/2}^{L/2} \frac{dz'}{[a^2 + (z - z')^2]^{3/2}}$$

where n is the number of turns per unit length, $\mathbf{m}$ is the magnetic moment of a *single* turn, L is the length of the solenoid, and the axis of the solenoid coincides with the z axis. (b) Calculate the field B_∞ when $L = \infty$ and reexpress the above equation in the dimensionless form $\mathbf{B}/B_\infty = \cdots$. (c) Using a computer, evaluate this integral numerically to study $B_z(0, 0, z)/B_\infty$ as a function of z/a for several values of L/a. For each case examined, determine how far from the center you can go before the field has changed by 1%. Under what circumstances, can the field at the center be calculated to 1% by assuming the solenoid to be infinitely long?

P5-26. In a region of space where $\mathbf{J}$ is zero, $\nabla \times \mathbf{B}$ is also zero and $\mathbf{B}$ can be derived from a *magnetic scalar potential* $V^{(m)}$, defined so that $\mathbf{B} = -\mu_0 \nabla V^{(m)}$. (a) Show that $V^{(m)}$ satisfies Laplace's equation, $\nabla^2 V^{(m)} = 0$. (b) Given the field and potential of an electric dipole, Eqs. (4-10) and (4-44), and the field of a magnetic dipole, Eq. (5-50), infer the corresponding magnetic scalar potential. (c) Find a magnetic scalar potential from which the field outside an infinitely long wire, Eq. (5-11), can be derived and explain why your result is not a single-valued function of position. (d) Find a magnetic scalar potential from which the field on the axis of a circular current loop as given in P5-4 can be derived.

P5-27. A magnetic dipole with moment $\mathbf{m} = m\hat{\mathbf{k}}$, $m > 0$, is located at the origin and produces a field given by Eq. (5-50). A circular loop of radius a and carrying current I' is placed with its center at $(0, 0, b)$ and its plane parallel to the x-y plane. (a) Calculate the force between the dipole and the current loop, first exactly and then in the limit $b \gg a$; express the force in terms of the dipole moment $\mathbf{m}' = m'\hat{\mathbf{k}}$ of the loop (and other relevant parameters); and note how this force depends on b and on the sign of m'. (b) Sketch a graph of the exact force versus b. For what value of b is the force a maximum? This dipole-dipole interaction, explored in part here and somewhat more fully for the electric case in P4-52, is important in determining many molecular, atomic, and nuclear properties.

P5-28. Since $\mathbf{B} = \nabla \times \mathbf{A}$, Stokes' theorem apparently combines with the flux law to give $\oint \mathbf{B} \cdot d\mathbf{S} = \oint \nabla \times \mathbf{A} \cdot d\mathbf{S} = \oint \mathbf{A} \cdot d\boldsymbol{\ell} = 0$. The final two parts of this equation, however, constitute the condition for deriving $\mathbf{A}$ from a scalar via $\mathbf{A} = \nabla\phi$. But then $\mathbf{B} = \nabla \times \nabla\phi = 0$ identically. All **B**-fields therefore vanish! Find the fallacy. This puzzler was presented by G. Arfken, *Am. J. Phys.* **27**, 526 (1959).

P5-29. The force on a volume V of a current distribution $\mathbf{J}$ placed in a magnetic induction $\mathbf{B}$ is given by $\int_V \mathbf{J} \times \mathbf{B}\, dv$ [Eq. (3-21)]. This force can also be expressed in terms of a *magnetic pressure*, $p_m(\mathbf{r})$. (a) Show that the force on V is given in terms of $p_m(\mathbf{r})$ by $-\oint_\Sigma p_m\, d\mathbf{S}$ where Σ bounds V and $d\mathbf{S}$ is an *outward* normal. (b) Use Eq. (C-21) to write the surface integral as a volume integral and infer that $\nabla p_m = -\mathbf{J} \times \mathbf{B}$. (c) Assuming that $\mathbf{J} = J(\imath)\hat{\mathbf{k}}$ and that $\mathbf{B} = B(\imath)\hat{\boldsymbol{\phi}}$ in cylindrical coordinates, use the circuital law to show that

$$B(\imath) = \frac{\mu_0}{\imath}\int_0^{\imath} \imath' J(\imath')\, d\imath'$$

(d) Show that

$$p_m(\imath) - p_m(\imath_0) = \frac{B^2(\imath)}{2\mu_0} - \frac{B^2(\imath_0)}{2\mu_0} + \frac{1}{\mu_0}\int_{\imath_0}^{\imath} \frac{B^2(\imath')}{\imath'}\, d\imath'$$

where $\imath_0$ is an arbitrary reference point. *Hint*: Differentiate the result in part (c) with respect to $\imath$, solve for $J(\imath)$ in terms of $B(\imath)$, evaluate $\mathbf{J} \times \mathbf{B}$, and set the result equal to $-\nabla p_m = -(dp_m/d\imath)\hat{\boldsymbol{\imath}}$. (e) Determine and sketch a graph of $p_m(\imath)$ as a function of $\imath$ for the field of a long wire of finite radius, Eq. (5-33). Choose $\imath_0 = 0$ and set $p_m(0) = 0$.

P5-30. One expression determining $\mathbf{A}$ directly from $\mathbf{B}$ is obtained by substituting $\mathbf{J} = \nabla \times \mathbf{B}/\mu_0$ into the expression determining $\mathbf{A}$ from $\mathbf{J}$. From this expression, show that

$$\mathbf{A}(\mathbf{r}) = \frac{1}{4\pi}\int \mathbf{B}(\mathbf{r}') \times \frac{\mathbf{r} - \mathbf{r}'}{|\mathbf{r} - \mathbf{r}'|^3}\, dv'$$

and state any assumptions made in your development. *Optional*: Seek an expression determining $\mathbf{A}$ from $\mathbf{B}$ that does not involve the function $1/|\mathbf{r} - \mathbf{r}'|$ or some power of it. Compare $-\int \mathbf{E}\cdot d\mathbf{r}$ for determining V from $\mathbf{E}$.

P5-31. The magnetic induction produced at the point $(x, y, 0)$ by a circular current loop of radius a lying in the y-z plane with its center at the origin and carrying a current i counterclockwise as viewed from a point on the positive x axis is given by

$$\mathbf{B}(x, y, 0) = \frac{\mu_0 ia}{2\pi}\int_0^{\pi} \frac{x\cos\phi'\hat{\mathbf{j}} + (a - y\cos\phi')\hat{\mathbf{i}}}{[x^2 + y^2 + a^2 - 2ay\cos\phi']^{3/2}}\, d\phi'$$

where ϕ' is the angle between the positive y axis and a point on the loop. (Compare P5-6.) (a) Express this integral in dimensionless form by introducing $X = x/a_0$, $Y = y/a_0$, $A = a/a_0$, $I = i/i_0$, and $\mathbf{b} = \mathbf{B}/B_0$, where a_0 and i_0 are convenient values for length and current and $B_0 = \mu_0 i_0/2a_0$ is the field at the center of a loop of radius a_0 carrying current i_0. (b) Write a program for an available computer to evaluate b_x and b_y numerically. Assume that the current I, the radius A, and the position X, Y are specified as input. (c) Test your program by evaluating **b** at several points on the x axis and comparing the results with the analytic result $\mathbf{b}(X, 0, 0) = IA^2\hat{\mathbf{i}}/(A^2 + X^2)^{3/2}$. (Compare P5-4.) (d) Explore the field at representative points in the x-y plane off the axis of the loop. For example, examine $\mathbf{b}(0, Y, 0)$ as a function of Y. *Note:* An alternative strategy for calculating the field of a loop numerically has been discussed by J. R. Merrill, *Am. J. Phys.* **39**, 791 (1971). *Optional*: (1) Modify your program to compute the field produced in the x-y plane by several loops, all of whose axes coincide with the x axis. Let the current, radius, and location on the x axis of each loop be specified as input. Use this program to study the field of the Helmholtz coil (two loops separated by their common radius and carrying the same current), the solenoid (many—say 10—closely spaced loops), and other configurations of your choosing. (2) Use one (or both) of the programs here written in a program similar to that of Fig. P4-56 to print out (or plot) points on the field lines of the magnetic induction in the x-y plane and study several configurations of loops. Can you invent a procedure to assure that field lines are drawn with the correct spacing? (3) Use one (or both) of these programs as part of a program based on Fig. 3-1 or 3-2 to print out (or plot) points on the trajectory of a particle moving in the field of one or more loops. (4) Write a program to compute the field established in the x-y plane by two loops of radius a located in the y-z plane with their centers at $(0, d, 0)$ and $(0, -d, 0)$, $d > a$, and study the resulting field by printing its values or by printing (or plotting) points on selected field lines.

P5-32. Show that the vector potential $\mathbf{A}(\mathbf{r})$ established by a current distribution described by a *surface* current density $\mathbf{j}(\mathbf{r})$ (see P2-25) is given by

$$\mathbf{A}(\mathbf{r}) = \frac{\mu_0}{4\pi} \int_\Sigma \frac{\mathbf{j}(\mathbf{r}')}{|\mathbf{r} - \mathbf{r}'|}\, dS'$$

where Σ is the surface in which the current flows and dS' is the (scalar) area of an element of the surface.

6

The Electromagnetic Field Produced by Time-Dependent Charge Distributions: Maxwell's Equations in Vacuum

In this chapter, we shall develop relationships among time-dependent charge and current distributions and the resulting time-dependent fields. We have already defined these more general fields, $\mathbf{E}(\mathbf{r}, t)$ and $\mathbf{B}(\mathbf{r}, t)$, in Chapter 3, essentially by the Lorentz force

$$\mathbf{F}_q = q(\mathbf{E} + \mathbf{v} \times \mathbf{B}) \tag{6-1}$$

on a charge q moving in the fields with velocity $\mathbf{v}$. Further, we have developed the equation of continuity

$$\oint \mathbf{J} \cdot d\mathbf{S} = -\int \frac{\partial \rho}{\partial t}\, dv \qquad \nabla \cdot \mathbf{J} = -\frac{\partial \rho}{\partial t} \tag{6-2}$$

in Chapter 2 in a fully time-dependent form. In Chapters 4 and 5, however, we restricted our attention to the *static* electric fields of *fixed* charges and to the *static* magnetic induction fields of *steady* currents, finding (1) that the *static* **E**-field has nonzero *divergence* and zero *curl* and satisfies Gauss's law

$$\oint \mathbf{E} \cdot d\mathbf{S} = \frac{1}{\epsilon_0}\int \rho\, dv \qquad \nabla \cdot \mathbf{E} = \frac{1}{\epsilon_0}\rho \tag{6-3}$$

and the restricted Faraday law

$$\oint \mathbf{E} \cdot d\boldsymbol{\ell} = 0 \qquad \nabla \times \mathbf{E} = 0 \tag{6-4}$$

(2) that the *static* **B**-field has zero *divergence* and nonzero *curl* and satisfies

the flux law

$$\oint \mathbf{B} \cdot d\mathbf{S} = 0 \qquad\qquad \nabla \cdot \mathbf{B} = 0 \tag{6-5}$$

and Ampere's circuital law

$$\oint \mathbf{B} \cdot d\boldsymbol{\ell} = \mu_0 \int \mathbf{J} \cdot d\mathbf{S} \qquad\qquad \nabla \times \mathbf{B} = \mu_0 \mathbf{J} \tag{6-6}$$

and (3) that the *static* **E**-field and the *static* **B**-field are entirely independent. (Neither field occurs in the equations satisfied by the other.) In the first three sections of this chapter, we shall examine the applicability of Eqs. (6-3)–(6-6) to time-dependent fields, obtaining *Maxwell's equations* for these fields and finding qualitatively (1) that the time-dependent **E**- and **B**-fields are interdependent and must therefore be thought of together as a single *electromagnetic field*, (2) that Eqs. (6-4) and (6-6) each require an additional term, and (3) that Eqs. (6-3) and (6-5) carry over to time-dependent fields with no change in form.[1] In the remaining sections of the chapter, we shall develop a few very general consequences of Maxwell's equations. Except for a few problems, actual solution of these equations is postponed to later chapters.

6-1 Electromagnetic Induction: Faraday's Law[2]

The additional term needed to extend Eq. (6-4) to time-dependent fields was discovered early in the nineteenth century on the basis of direct experimental evidence acquired independently by Faraday in England and by Henry in the United States. Qualitatively, the phenomenon of electromagnetic induction discovered by these men associates an induced *electric* field with a *changing magnetic* induction field. This connection and some of its consequences are the subjects of this section.

ELECTROMOTIVE FORCE

The basic law extending Eq. (6-4) to time dependent fields is stated in terms of a quantity known as the *electromotive force* (emf) or the *electromotance*. To define this quantity, we imagine the following physical situation: In some region of space R, let there exist general time-dependent electric and magnet-

[1]Absence of a change in *form* does not mean that there is no change in *substance*. These two equations look the same for both time-independent and time-dependent fields, but that identity in appearance can be accepted only after the more general equations have been subjected to experimental tests with time-dependent fields.

[2]Faraday's law is perhaps the most difficult of the basic laws of electromagnetism to introduce. Several of the common approaches are compared in P. J. Scanlon, R. N. Henrikson, and J. R. Allen, *Am. J. Phys.* **37**, 698 (1969). The reader may wish to compare the approach here adopted with that in R. P. Feynman, R. B. Leighton, and M. Sands, *The Feynman Lectures on Physics* (Addison-Wesley Publishing Company, Inc., Reading, Mass., 1964), Volume II, Lectures 16 and 17, with that in E. M. Purcell, *Electricity and Magnetism* (McGraw-Hill Book Company, New York, 1965), Chapter 7, or with that in other available texts on the subject.

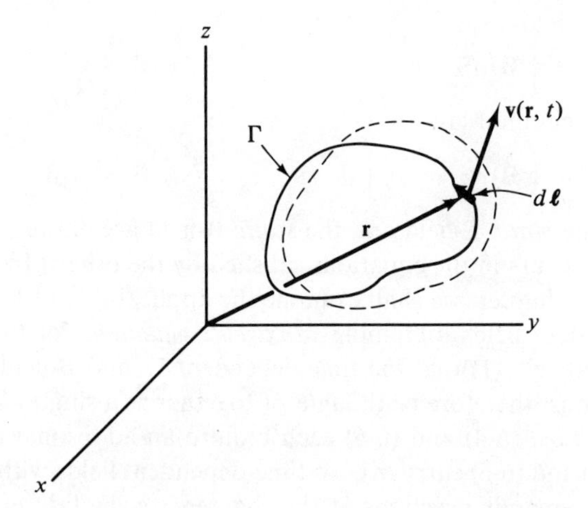

Fig. 6-1. Positions of an arbitrary moving path at times t (solid) and $t + \Delta t$ (dashed). The electric and magnetic induction fields in the vicinity of the path are not shown.

ic fields $\mathbf{E}(\mathbf{r}, t)$ and $\mathbf{B}(\mathbf{r}, t)$. Further, select a *closed* path $\mathbf{\Gamma}$—perhaps but not necessarily defined by a wire—lying wholly in R and let that path (or portions of it) be moving—perhaps distorting—in some arbitrary way so that at time t the point $\mathbf{r}$ on the path is moving with velocity $\mathbf{v}(\mathbf{r}, t)$. Finally, if there is a wire defining the path, include in $\mathbf{E}$ and $\mathbf{B}$ the contribution of any charge and current distributions on the wire and allow for the possibility of batteries in the wire, temperature differences, and perhaps other external influences. The positions of the path at times t and $t + \Delta t$ are shown in Fig. 6-1. Now, consider a test charge of strength q at an arbitrary point on the path and moving with the path. This test charge experiences a total force $\mathbf{f}_q(\mathbf{r}, t)$ consisting not only of electromagnetic forces but also of effective (nonelectromagnetic) forces from chemical effects in the batteries, thermal effects in the wire, etc. (The test charge is assumed to be at rest relative to its point on the path so that resistive forces arising from collisions of the test charge with the microscopic constituents of the wire do not arise.) The emf $\mathscr{E}(t)$ about the selected path at time t is now defined as the work done *by* $\mathbf{f}_q(\mathbf{r}, t)$ *on* a *unit* charge moved once around the path, i.e.,

$$\mathscr{E}(t) = \oint_\Gamma \frac{\mathbf{f}_q(\mathbf{r}, t)}{q} \cdot d\boldsymbol{\ell} \tag{6-7}$$

$$\begin{aligned} &= \oint_\Gamma [\mathbf{E}(\mathbf{r}, t) + \mathbf{v}(\mathbf{r}, t) \times \mathbf{B}(\mathbf{r}, t)] \cdot d\boldsymbol{\ell} \\ &\quad + \frac{1}{q} \oint_\Gamma \textbf{effective chemical forces} \cdot d\boldsymbol{\ell} \\ &\quad + \frac{1}{q} \oint_\Gamma \textbf{effective thermal forces} \cdot d\boldsymbol{\ell} + \cdots \end{aligned} \tag{6-8}$$

Although the definition of $\mathscr{E}$ is very similar to the integral giving a (static) potential difference ΔV about a closed path and both quantities are measured in volts, $\mathscr{E}$ and ΔV are physically very different; $\mathscr{E}$ is in general not equal to zero and hence arises from nonconservative force fields while ΔV is necessarily always zero and corresponds to a conservative force field.

Despite the abstractness of the definition in Eq. (6-7), $\mathscr{E}$ is a measurable quantity, for it is related in a simple way to the current produced in a conducting wire that follows the selected path.[3] Thus, the dependence of $\mathscr{E}$ on various parameters can be studied *experimentally* in order to formulate a law (or laws) determining $\mathscr{E}$ from these parameters. In the present context, we are interested in the results of such experimental studies when batteries, temperature differences, and the like are not present. We therefore omit the chemical and thermal emf's from Eq. (6-8) and denote the term of interest by $\mathscr{E}_{\mathrm{em}}$,

$$\mathscr{E}_{\mathrm{em}} = \oint_{\Gamma} (\mathbf{E} + \mathbf{v} \times \mathbf{B}) \cdot d\boldsymbol{\ell} \tag{6-9}$$

where the arguments $\mathbf{r}$ and t have been suppressed.

MOTIONAL EMF

Under some circumstances, our previous theoretical framework permits a rewriting of Eq. (6-9) in a useful alternative form. Consider first a conducting rod of length ℓ that is moving through a uniform magnetic induction field. Before the rod is set into motion, the mobile electrons in the rod are uniformly distributed and the electrons are neutralized by the immobile positive ions that make up the rest of the conductor. On no portion of the rod is there an excess charge of either sign. When the rod is set into motion in the external field, the charges—both positive and negative—experience forces, and the electrons, which are free to move, respond by drifting away from their positions in the stationary rod. A charge imbalance is therefore established in the rod. This imbalance in turn produces an electric field both inside and outside the rod, and the charges in the rod now experience forces from this induced electric field as well as from the original magnetic induction. Ultimately, an equilibrium charge distribution is reached when sufficient charge separation has occurred so that the electric and magnetic forces on the electrons exactly cancel. All (macroscopic) motion of the electrons then ceases.[4] The resulting equilibrium charge distribution is shown in Fig. 6-2.

Let us now evaluate the integral in Eq. (6-9) counterclockwise about the path shown in Fig. 6-3. We assume that, at every instant of time, the **E**-field throughout the region containing the path is the *static* field appropriate to the instantaneous position of the rod; i.e., we assume that the **E**-field follows

[3] $\mathscr{E} = IR$, where I is the current and R the resistance of the wire; see Chapter 9.

[4] The time required for this equilibrium to be established depends on properties of the conductor (see Chapter 9) but is not of concern here. For a perfect conductor, the equilibrium is established instantaneously.

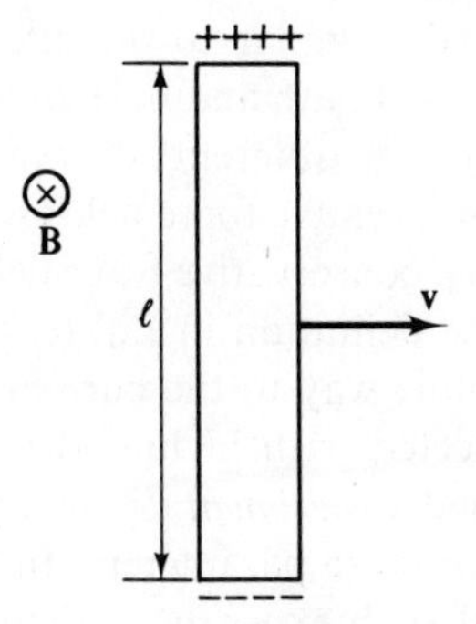

Fig. 6-2. A conducting rod moving at right angles to a constant magnetic induction field. Once a steady state has been reached, one end of the rod will be positively charged and the other end will be negatively charged.

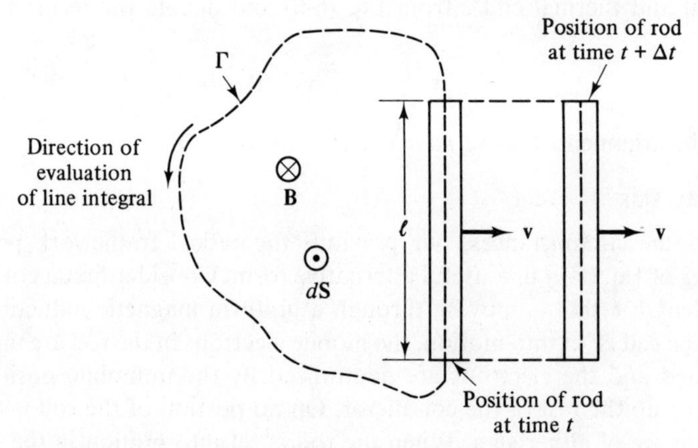

Fig. 6-3. An imagined path for the evaluation of a line integral. The portion of the path lying inside the moving rod moves with the rod; the remainder of the path is at rest.

changes in the position of the rod instantly. Since we shall later find that electromagnetic fields follow changes in their sources only after a delay determined by the speed of light and the distance from the source, our assumption here in effect requires that the speed of the rod be small compared to that of light and that the path not extend too far from the rod. When $\mathbf{E}$ can be regarded as static, however, $\oint \mathbf{E}\cdot d\boldsymbol{\ell} = 0$. Further, since $\mathbf{v}$ is zero everywhere except in the rod, Eq. (6-9), which now gives what is called a *motional emf*, reduces to

$$\mathscr{E}_{\text{em}}^{\text{mot}} = \int_{\text{lower end}}^{\text{upper end}} (\mathbf{v} \times \mathbf{B})\cdot d\boldsymbol{\ell} = vB\ell \qquad (6\text{-}10)$$

where the final evaluation follows because $\mathbf{v}$ and $\mathbf{B}$ are constant and mutually perpendicular at all points on the rod and $\mathbf{v} \times \mathbf{B}$ is parallel to $d\boldsymbol{\ell}$. Now, however, we note that

$$\frac{d\Phi_m(t)}{dt} = \lim_{\Delta t \to 0} \frac{\Phi_m(t + \Delta t) - \Phi_m(t)}{\Delta t}$$

$$= \lim_{\Delta t \to 0} \left(- \frac{(v\,\Delta t)\ell B}{\Delta t} \right) = -v\ell B \tag{6-11}$$

where $\Phi_m(t)$ is the flux of $\mathbf{B}$ across the surface bounded by the path $\mathbf{\Gamma}$ at time t. The difference $\Phi_m(t + \Delta t) - \Phi_m(t)$ is then just the flux across the area added to the path in time Δt and is *negative* because the added flux goes *into* the page in Fig. 6-3 while the positive direction defined by our counterclockwise traversal of the path is out of the page. Equations (6-10) and (6-11) together finally yield that

$$\mathscr{E}_{\text{em}}^{\text{mot}} = -\frac{d\Phi_m}{dt} \tag{6-12}$$

which by its development is valid only when (1) the rod moves slowly enough and (2) the change in flux comes about by motion of the rod in a *static* **B**-field. In words, the motional emf about the path in this situation is the *negative* rate of change of the magnetic flux across a surface bounded by the path.

The result in Eq. (6-12) for a *motional* emf can also be derived when the path moves in a more arbitrary way and the **B**-field is not uniform (but still static). We continue to assume that **E** is static, which means in particular that any conductors along the path must move slowly. With that assumption, $\oint \mathbf{E} \cdot d\boldsymbol{\ell} = 0$ for the more general path of Fig. 6-1, and Eq. (6-9) gives

$$\mathscr{E}_{\text{em}}^{\text{mot}} = \oint_{\Gamma} (\mathbf{v} \times \mathbf{B}) \cdot d\boldsymbol{\ell} \tag{6-13}$$

which is nonzero only if at least a portion of the path has a nonzero velocity. On the other hand, the rate of change of flux across a surface bounded by the path in Fig. 6-1 is given by

$$\frac{d\Phi_m}{dt} \approx \lim_{\Delta t \to 0} \frac{1}{\Delta t} \sum_i \mathbf{B}(\mathbf{r}_i) \cdot \Delta\mathbf{S}_i \tag{6-14}$$

where the path has been broken into segments $\Delta\boldsymbol{\ell}_i$, $\mathbf{r}_i$ is a point on the ith segment, and $\Delta\mathbf{S}_i$ is the area added when the ith segment moves a distance $\mathbf{v}(\mathbf{r}_i, t)\,\Delta t$ to its position at time $t + \Delta t$ (Fig. 6-4). Since $\mathbf{v}(\mathbf{r}_i, t)\,\Delta t \times \Delta\boldsymbol{\ell}_i$ agrees both in magnitude and direction with $\Delta\mathbf{S}_i$ (the direction of $\Delta\mathbf{S}_i$ must agree with the positive direction determined by applying the right-hand rule to the path at $t + \Delta t$), we can rewrite Eq. (6-14) as

$$\frac{d\Phi_m}{dt} \approx \lim_{\Delta t \to 0} \frac{1}{\Delta t} \sum_i \mathbf{B}(\mathbf{r}_i) \cdot [\mathbf{v}(\mathbf{r}_i, t)\,\Delta t \times \Delta\boldsymbol{\ell}_i] \tag{6-15}$$

$$\approx \sum_i \mathbf{B}(\mathbf{r}_i) \cdot [\mathbf{v}(\mathbf{r}_i, t) \times \Delta\boldsymbol{\ell}_i] \tag{6-16}$$

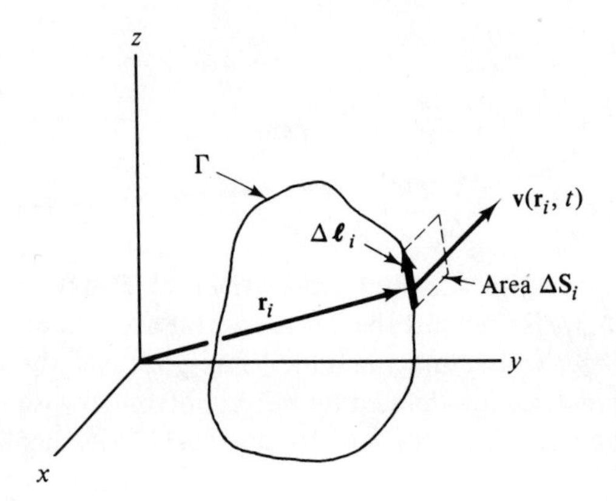

Fig. 6-4. Coordinates and vectors for a general evaluation of a motional emf.

In the limit as $|\Delta \boldsymbol{\ell}_i| \longrightarrow 0$ for all i, Eq. (6-16) becomes

$$\frac{d\Phi_m}{dt} = \oint_\Gamma \mathbf{B}\cdot(\mathbf{v} \times d\boldsymbol{\ell}) = -\oint_\Gamma (\mathbf{v} \times \mathbf{B})\cdot d\boldsymbol{\ell} \tag{6-17}$$

where the final form involves exchanging the dot and cross in the previous form (P0-7) and then reversing the order of the factors in the resulting cross product. By comparing Eqs. (6-13) and (6-17), however, we find that Eq. (6-12) in fact gives the motional emf under more general circumstances than its initial derivation implied, but even this more general development still restricts Eq. (6-12) to *motional* emf.

Although the minus sign in Eq. (6-12) has apparently been required merely to preserve certain sign conventions, the fact that a minus sign rather than a plus sign is needed expresses a physically significant aspect of motional emf's. The direction assigned to $d\boldsymbol{\ell}$ in evaluating $\mathscr{E}_{\text{em}}^{\text{mot}}$ defines the direction in which an induced current would flow if the path is replaced by a conducting wire and $\mathscr{E}_{\text{em}}^{\text{mot}} > 0$. (Why?) Under those circumstances, the induced current produces a magnetic induction field whose flux contributes *positively* to the flux across the surface bounded by the path. (Imagine a positive or counterclockwise current in the path of Fig. 6-3. What is the direction of the resulting **B**-field over the surface bounded by the path?) If, however, $\mathscr{E}_{\text{em}}^{\text{mot}} > 0$, Eq. (6-12) requires that $d\Phi_m/dt < 0$. Thus, a *negative* change in the flux induces a current whose magnetic induction contributes *positively* to the flux. The minus sign in Eq. (6-12) in effect states that *the induced emf brought about by a change in magnetic flux is always of such direction as to oppose the change in flux*; it is the mathematical expression of what is called *Lenz's law.*

FARADAY'S LAW OF ELECTROMAGNETIC INDUCTION

In the previous paragraphs, we have seen that an induced emf is generated if a path moves or distorts to encompass a different total magnetic flux. But the flux encompassed by a path also changes when the magnetic induction is time-dependent and the path itself remains stationary. These new circumstances occur, for example, when a current loop or bar magnet is moved toward or away from a fixed path in space. The observations of Faraday and Henry were that a time-dependent magnetic flux also induces an emf about a given path and, further, that Eq. (6-12), *derived* for *motional* emf, *actually applies* as well when the changing flux comes about through a time-dependent magnetic induction. *Experimental* evidence beyond any introduced earlier in this book thus supports the *Faraday law of electromagnetic induction*, which we can write either as

$$\mathscr{E}_{\text{em}} = -\frac{d\Phi_m}{dt} \tag{6-18}$$

or, using Eqs. (6-9) and (5-12), as

$$\oint_\Gamma (\mathbf{E} + \mathbf{v} \times \mathbf{B})\cdot d\boldsymbol{\ell} = -\frac{d}{dt}\int_\Sigma \mathbf{B}\cdot d\mathbf{S} \tag{6-19}$$

where $\boldsymbol{\Sigma}$ is any surface bounded by $\boldsymbol{\Gamma}$. The minus sign expressing Lenz's law is still present and the fields may now have a general dependence on time. Further, Eq. (6-19) incorporates both motional emf's and emf's induced about fixed paths by time-dependent **B**-fields. We stress again that Eqs. (6-18) and (6-19) cannot be accepted solely on the basis of their agreement with the results obtained for motional emf. The law of induction has other aspects and these must be subjected to careful *experimental* test before accepting the law. This law cannot be derived from principles presented earlier in this book; it is inherently an experimental law and takes its justification from the agreement of its predictions with experimental observation.

We now rewrite Eq. (6-19) in a form that compares more directly with Eq. (6-4). Since both **B** and the surface of integration in Eq. (6-19) may depend on time, the total time derivative of the flux in general has two contributions; symbolically,

$$\frac{d}{dt}\int_\Sigma \mathbf{B}\cdot d\mathbf{S} = \int_\Sigma \frac{\partial \mathbf{B}}{\partial t}\cdot d\mathbf{S} - \oint_\Gamma (\mathbf{v} \times \mathbf{B})\cdot d\boldsymbol{\ell} \tag{6-20}$$

where the first term expresses the rate of change of flux arising from the explicit time dependence of **B** and the second term—obtained directly from Eq. (6-17)—gives the rate of change of flux arising from the motion of the boundaries of the surface of integration. Substituting Eq. (6-20) into Eq. (6-19), we find finally that

$$\oint_\Gamma \mathbf{E}\cdot d\boldsymbol{\ell} = -\int_\Sigma \frac{\partial \mathbf{B}}{\partial t}\cdot d\mathbf{S} \tag{6-21}$$

terms involving $\mathbf{v} \times \mathbf{B}$ having canceled. This form of Faraday's law is the preferred form for theoretical purposes. It reduces more obviously than Eq. (6-19) to Eq. (6-4) when **B** is static and can thus be viewed as the generalization of Eq. (6-4) to time-dependent fields. In words, Eq. (6-21) states that a *time-dependent magnetic induction field is accompanied by an induced, nonconservative electric field.*

Of all the basic laws of electromagnetism, Faraday's law is perhaps of greatest practical import, for this law expresses the essential physics of transformers and electric generators which in turn are crucial to the technological society we take so much for granted. To examine these applications here, however, would take us far afield; we relegate some elementary aspects to the problems and the rest of that subject to other books.

PROBLEMS

P6-1. In our derivation of Eq. (6-12) for motional emf, we restricted both **E** and **B** to be static. Subsequently, we set $\oint \mathbf{E} \cdot d\boldsymbol{\ell} = 0$ and clearly utilized the assumption of a static **E**-field. Where did we utilize the assumption of a static **B**-field?

P6-2. Convince yourself that the expression substituted for $\Delta\mathbf{S}_i$ in going from Eq. (6-14) to Eq. (6-15) indeed has the correct direction for all possible values of $\mathbf{v}(\mathbf{r}_i, t)$, not just for the direction we happen to have drawn in Fig. 6-4.

P6-3. A rectangular conducting loop of mass m falls under gravity across the poles of a permanent magnet producing a constant field **B** out of the page (Fig. P6-3). Let the field extend over a region of width w. Find the terminal velocity v that the loop acquires before falling out of the region occupied by the field. *Hint*: Let the loop have resistance R and accept the relationship $\mathscr{E} = IR$ between the induced emf $\mathscr{E}$ and the current I in the loop.

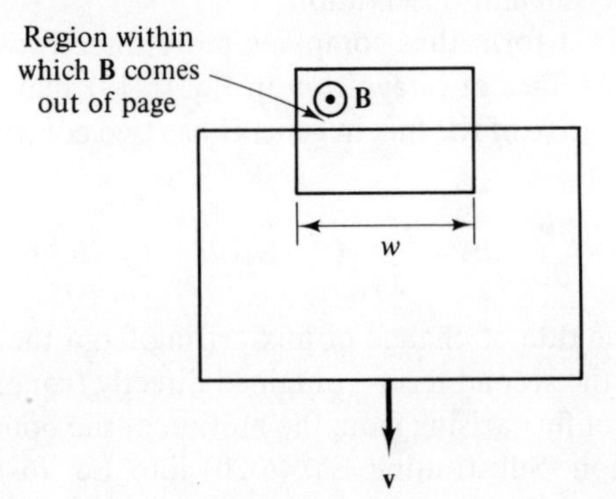

Figure P6-3

P6-4. A conducting disc of radius R rotates at constant angular speed ω about its axis, which is parallel to a constant magnetic induction **B**. Calculate the motional emf induced in a path running along a radius from the center to the rim of the disc and then closing along a stationary path lying outside the disc. This device is called a Faraday disc dynamo.

P6-5. A conducting circular loop of area A is arranged so that it can be rotated at constant angular speed ω about a diameter. (a) Let the axis of rotation be perpendicular to a constant magnetic induction **B**. Determine the induced emf as a function of time. What could you do to increase the maximum emf without changing ω? This device is a simple generator. (b) Let the orientation of the axis be adjustable and suppose that you can display the induced emf as a function of time on an oscilloscope. Describe a means by which this device could be used to measure both the magnitude and the direction of an unknown field. This device is now called a search coil.

P6-6. A magnetic dipole has moment $m(t)\hat{\mathbf{k}}$ and is located at the origin. Let the dipole moment be increasing ($dm/dt > 0$). Determine the direction of the emf induced about a circular path oriented as described in each of the following situations: (a) plane perpendicular to the z axis, the center at $(0, 0, b)$, $b > 0$; (b) in the x-y plane, the center at $(0, b, 0)$; and (c) plane perpendicular to the y axis, the center at $(0, b, 0)$.

P6-7. A long straight wire carries a current $I(t)$ as shown in Fig. P5-3. (a) Determine (exactly) the magnitude and direction of the emf induced about the rectangular loop shown in the figure. *Hint*: The flux across the loop was calculated in P5-7. (b) Suppose the current in the straight wire changes with time as illustrated in Fig. P6-7, where $I > 0$ means a current in the positive

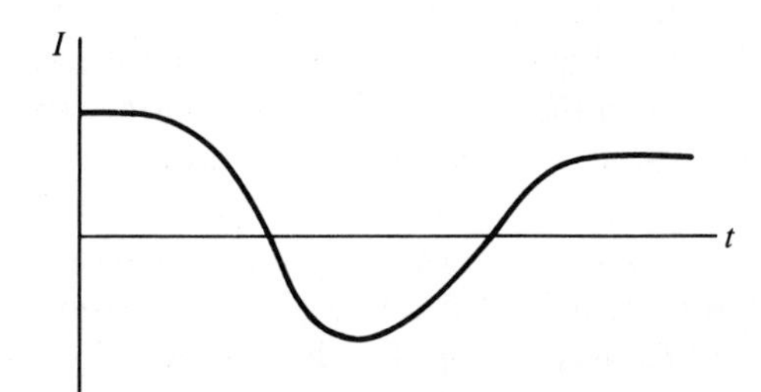

Figure P6-7

direction indicated in Fig. P5-3. Sketch a qualitative graph of the current in the rectangular loop as a function of time, making sure that the direction of that current is clear.

P6-8. A conducting ring is placed on top of a solenoid (P5-11) that has its axis vertical. When the current is turned on in the solenoid, the ring flies into the air. Explain this phenomenon for yourself and then read the discussion given by E. J. Churchill and J. D. Noble, *Am. J. Phys.* **39**, 285 (1971).

P6-9. A uniform, time-dependent magnetic induction $\mathbf{B} = B(t)\hat{\mathbf{k}}$ exists in space. Determine the emf induced about a fixed circular path of radius r lying in the x-y plane with its center at the origin and then, assuming that the induced electric field $\mathbf{E}$ is tangent to the path at each point (x, y) on the path, show that

$$\mathbf{E} = -\frac{r}{2}\frac{dB}{dt}\hat{\boldsymbol{\phi}} = \frac{1}{2}\frac{dB}{dt}(y\hat{\mathbf{i}} - x\hat{\mathbf{j}})$$

Characteristics of this field are further explored in P6-36.

P6-10. In a betatron an evacuated toroidal doughnut, inside of which moves a beam of electrons, is placed between the poles of a large electro-

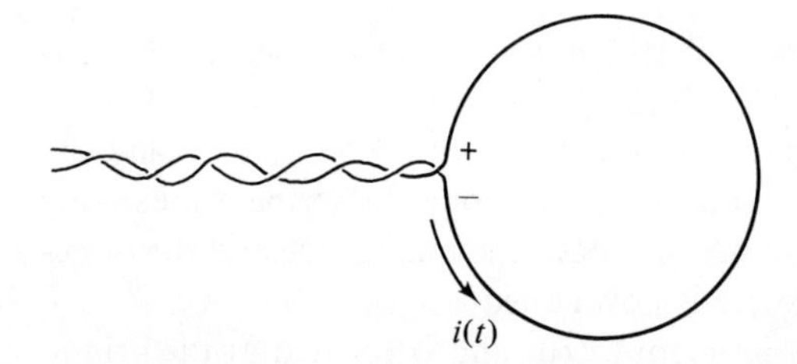

Fig. 6-5. Loop carrying a current $i(t)$ in a magnetic induction field. The field is established partly by the current in the loop and partly by other currents that are not shown. The signs indicate the direction of the induced emf when the magnetic flux increases out of the paper.

magnet. As the magnetic induction field is increased, an emf is induced about the doughnut and the electrons are accelerated. Simultaneously, the same changing field provides the centripetal force necessary to keep the electron in its circular orbit. Show that the field at the orbit must at all times be one-half the (space) average of the field over the surface bounded by the orbit if the electrons are to remain in an orbit of fixed radius.

ENERGY IN THE STATIC MAGNETIC INDUCTION FIELD

The law of induction now makes possible a rigorous justification of Eq. (5-52) for the energy required to establish a *steady* current distribution. We ask first for the work required to establish a steady current I in a closed circuit. Although the circuit can be of arbitrary shape, it is convenient to think of it as a simple loop, such as that shown in Fig. 6-5. Let the current in this loop at time t be $i(t)$, where $i(t)$ increases from 0 to I in some time interval $0 < t < T$. Further, let the (magnetic) flux $\Phi_m(t)$ across the loop originate in part from the current in the loop and in part from other changing currents that are external to the loop. Finally, adopt the (consistent) conventions that $i(t) > 0$ means a counterclockwise current and $\Phi_m(t) > 0$ means a flux out of the page. Figure 6-5 shows the direction of the induced emf $\mathscr{E}$ when $d\Phi/dt > 0$ (and hence $\mathscr{E} = -|\mathscr{E}|$). In the situation described, the current i is flowing *against* the induced emf and whatever agent is maintaining the current must therefore be doing *positive* work on the charges as they move around the loop. In particular, if a charge ΔQ is carried around the loop at a time when the emf is $\mathscr{E}$, the amount of work ΔW done on the charge is given by

$$\Delta W = -\mathscr{E}\,\Delta Q \tag{6-22}$$

(The minus sign must be inserted explicitly to assure that ΔW will be positive when ΔQ is positive and $\mathscr{E}$ is intrinsically negative.) If the charge transport represented by ΔQ takes place in a small time interval Δt, division of Eq. (6-22) by Δt and passage to the limit $\Delta t \to 0$ yields the expression

$$P(t) = \frac{dW}{dt} = -\mathscr{E}(t)i(t) \tag{6-23}$$

for the instantaneous power output $P(t)$ of the agent maintaining the current.

Thus, the total energy W required to establish the current I is given by

$$W = \int_0^T P\,dt = -\int_0^T \mathscr{E} i\,dt = \int_0^T i\frac{d\Phi_m}{dt}\,dt = \int_{\Phi_m(0)}^{\Phi_m(T)} i\,d\Phi_m \tag{6-24}$$

where the last two forms are obtained by using Faraday's law.

We next ask for the work required to establish steady currents *simultaneously* in each of several nearby loops. Let the current at time t in the rth loop be $i_r(t)$, where $i_r(0) = 0$ and $i_r(T) = I_r$, and let the flux across the rth loop be $\Phi_{mr}(t)$. Then, summing the work required for each loop separately, we obtain for the total work W_t the expression

$$W_t = \sum_r \int_0^T i_r \frac{d\Phi_{mr}}{dt}\,dt \tag{6-25}$$

Now, the flux across the rth loop is the sum of *independent* contributions from all of the loops, *including* the rth, i.e.,

$$\Phi_{mr} = \sum_s \int_{r\text{th loop}} (\mathbf{B} \text{ at } r\text{th loop due to } s\text{th loop})\cdot d\mathbf{S} \tag{6-26}$$

From the Biot-Savart law, however, the **B**-field at the rth loop due to the current i_s in the sth loop is directly proportional to i_s. This same proportionality therefore applies also to the contribution of the sth loop to the *flux* across the rth loop. Further, since we assume that all the loops are fixed in space, the entire time dependence of this contribution to the flux is contained in i_s. We can therefore display the time dependence of Φ_{mr} explicitly and fully by writing Eq. (6-26) in the form

$$\Phi_{mr}(t) = \sum_s M_{rs} i_s(t) \tag{6-27}$$

where M_{rs} is a *constant* determined by the shape, size, and relative positions of loops r and s but independent of the current in either loop; it is called the *mutual inductance* of the rth loop with respect to the sth loop when $r \neq s$ and the *self-inductance* of the rth loop when $r = s$, and $M_{rs} = M_{sr}$ (P6-13). The mks unit of inductance is the *henry* (H), which may be alternatively expressed as a Wb/A or as a V·sec/A. Now, Eq. (6-27) yields not only the more explicit definition

$$M_{rs} = \frac{\partial \Phi_{mr}}{\partial i_s} \tag{6-28}$$

for M_{rs} but also, on substitution into Eq. (6-25), yields the result

$$W_t = \sum_{r,s} M_{rs} \int_0^T i_r \frac{di_s}{dt}\,dt \tag{6-29}$$

Let us now build up the currents in such a way that at time t the current in each loop is some fixed fraction of its final value, i.e., so that $i_r(t) = \alpha(t) I_r$

with $\alpha(t)$ independent of r and $\alpha(0) = 0$, $\alpha(T) = 1$. Then, Eq. (6-29) gives

$$W_t = \sum_{r,s} M_{rs} \int_0^T \alpha(t) I_r \frac{d\alpha(t)}{dt} I_s \, dt$$

$$= \sum_{r,s} M_{rs} I_r I_s \int_0^1 \alpha \, d\alpha$$

$$= \tfrac{1}{2} \sum_{r,s} M_{rs} I_r I_s \tag{6-30}$$

$$= \tfrac{1}{2} \sum_{r} I_r \Phi_{mr} \tag{6-31}$$

where Eq. (6-31) follows from Eq. (6-30) on substitution of Eq. (6-27) with $t = T$; Φ_{mr} in Eq. (6-31) thus denotes the final steady flux across the rth loop.

Finally, we transform Eq. (6-31) into a form more suited for calculating the energy required to establish a current distribution described by the *steady* current density **J**. Such a distribution can be regarded as a superposition of many current loops. Thus, in terms of the vector potential **A**, we find from Eq. (6-31) that

$$W_t = \tfrac{1}{2} \sum_r I_r \int_{r\text{th loop}} \mathbf{B} \cdot d\mathbf{S}$$

$$= \tfrac{1}{2} \sum_r I_r \oint_{r\text{th loop}} \mathbf{A} \cdot d\boldsymbol{\ell} \qquad \text{(see P5-18)}$$

$$= \tfrac{1}{2} \sum_r \int_{\text{volume occupied by } r\text{th loop}} \mathbf{J} \cdot \mathbf{A} \, dv \qquad \text{(see P2-7)}$$

$$= \tfrac{1}{2} \int_{\text{entire distribution}} \mathbf{J} \cdot \mathbf{A} \, dv \tag{6-32}$$

and we have supplied the justification both for Eq. (5-52) and for the development in Section 5-5.

PROBLEMS

P6-11. Obtain an integral—probably multiple—for the self-inductance L of an isolated circular current loop of radius a. *Hint*: The field of the loop is given in sufficient generality in P5-6. *Optional*: Express the result so that the integral is a pure number and then evaluate the integral by whatever means you can.

P6-12. Calculate the mutual inductance M of the arrangement in Fig. P5-3.

P6-13. Show that the mutual inductance M_{12} of two current loops is given by

$$M_{12} = \frac{\mu_0}{4\pi} \oint_1 \oint_2 \frac{d\boldsymbol{\ell}_1 \cdot d\boldsymbol{\ell}_2}{|\mathbf{r}_1 - \mathbf{r}_2|}$$

and hence that $M_{12} = M_{21}$. *Hint*: See P5-18.

P6-14. In developing Eqs. (6-30) and (6-31) from Eq. (6-29), we selected a general but still not entirely arbitrary way to build up the currents to their final values. To what extent (if any) does the final result depend on that selection? Explain your answer fully.

P6-15. A coil is made by winding N turns of wire on a frame of nonmagnetic material. The frame consists of a portion of a hollow cylinder of inner radius a, outer radius b, and height h (Fig. P6-15). A current I flows through

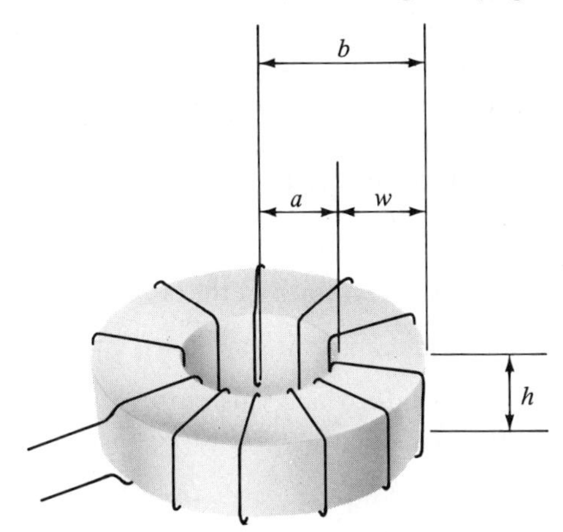

Figure P6-15

the wire. Assume that the magnetic induction **B** is confined to the region interior to this frame and that at every point in this region **B** is directed tangent to a circle through that point, which circle has its center on the axis of the frame. (a) Use the circuital law to determine the magnitude of the magnetic induction field at all points inside the frame. (b) Determine exactly the magnetic flux crossing a surface bounded by a single turn of the windings. (c) Show that the self-inductance of the arrangement is given by

$$L = \frac{\mu_0}{2\pi} N^2 h \, \ell n \frac{b}{a}$$

Note specifically the dependence on the *square* of N; doubling the number of turns quadruples the self-inductance. (d) Determine the limit of the self-inductance when $w \ll a$, where $w = b - a$ is the thickness of the frame, and note the appearance of the area wh of the cross section of the frame. *Hint*: When $\epsilon \ll 1$, $\ell n\,(1 + \epsilon) \approx \epsilon$. (e) The energy required to establish a current I in this coil was calculated in P5-20 by integrating the magnetic energy density over the volume of the coil. By a direct transformation of that result, show that this energy is given by $\frac{1}{2}LI^2$, in agreement with Eq. (6-30). (f) A second winding of N' turns of wire is now wound on the frame pictured in this problem. Calculate the mutual inductance M between the two coils. This arrangement of two coils is a crude transformer.

6-2
A Contradiction and its Resolution: Displacement Current

The second of the two terms needed to extend Eqs. (6-3)–(6-6) to time-dependent fields was predicted theoretically in the 1860's by Maxwell but was not confirmed experimentally until some 20 years later. Qualitatively, this term associates an induced *magnetic* induction field with a *changing electric* field, a connection that we shall explore in this section.

We shall begin by demonstrating that the basic equations in their present form [Eqs. (6-1), (6-2), (6-3), (6-5), (6-6), and (6-21)] are inconsistent. Specifically Eq. (6-6), whose divergence necessarily requires $\nabla \cdot \mathbf{J} = 0$ (Why?), contradicts Eq. (6-2) unless $\partial \rho / \partial t$ happens to be zero. To set the stage for the subsequent development, however, we elect to demonstrate this contradiction also by an argument based on the integral form of the laws. Consider, then, $\oint_{\Sigma} \mathbf{J} \cdot d\mathbf{S}$ over some *closed* surface Σ. Let an arbitrary *closed* curve $\boldsymbol{\Gamma}$ be scribed on Σ and denote the two portions into which this curve divides the surface by Σ_1 and Σ_2 (Fig. 6-6). With $d\mathbf{S}$ chosen always to be the *outward* normal on Σ, Eq. (6-6) now gives

$$\mu_0 \int_{\Sigma_1} \mathbf{J} \cdot d\mathbf{S} = \oint_{\Gamma_1} \mathbf{B} \cdot d\boldsymbol{\ell}; \qquad \mu_0 \int_{\Sigma_2} \mathbf{J} \cdot d\mathbf{S} = \oint_{\Gamma_2} \mathbf{B} \cdot d\boldsymbol{\ell} \tag{6-33}$$

where $\boldsymbol{\Gamma}_1$ and $\boldsymbol{\Gamma}_2$ are the curves bounding Σ_1 and Σ_2, respectively, *and each is traversed in the proper sense to agree with the conventions implicit in Eq. (6-6)*. Although $\boldsymbol{\Gamma}_1$ and $\boldsymbol{\Gamma}_2$ pass through the same points in space, *they are traversed in opposite directions*. Thus $\oint_{\Gamma_1} \mathbf{B} \cdot d\boldsymbol{\ell} = -\oint_{\Gamma_2} \mathbf{B} \cdot d\boldsymbol{\ell}$ and, on adding the two parts of Eq. (6-33), we find that Eq. (6-6) yields $\oint \mathbf{J} \cdot d\mathbf{S} = 0$, contradicting this time the integral form of Eq. (6-2) unless $\partial \rho / \partial t = 0$.

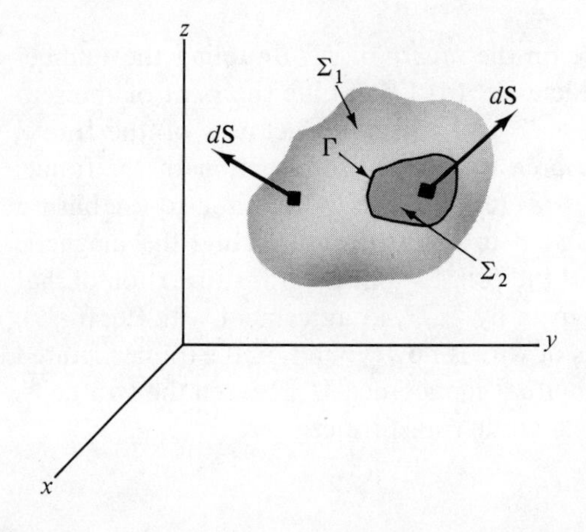

Fig. 6-6. A closed surface divided into two open surfaces by the curve $\boldsymbol{\Gamma}$.

A fruitful way to approach the contradiction between Eq. (6-2) and Eq. (6-6) is to examine symmetries that one might expect to be present in the equations. Basing our intuition on terms already present, we might expect a generalization of the basic equations to have the appearance

$$\oint \mathbf{E} \cdot d\mathbf{S} = \frac{1}{\epsilon_0} \int \rho \, dv \tag{6-34}$$

$$\oint \mathbf{B} \cdot d\mathbf{S} = \alpha \int \begin{pmatrix} \text{magnetic charge} \\ \text{density} \end{pmatrix} dv \tag{6-35}$$

$$\oint \mathbf{E} \cdot d\boldsymbol{\ell} = \beta \int \begin{pmatrix} \textbf{magnetic current} \\ \textbf{density} \end{pmatrix} \cdot d\mathbf{S} - \int \frac{\partial \mathbf{B}}{\partial t} \cdot d\mathbf{S} \tag{6-36}$$

$$\oint \mathbf{B} \cdot d\boldsymbol{\ell} = \mu_0 \int \mathbf{J} \cdot d\mathbf{S} + \gamma \int \frac{\partial \mathbf{E}}{\partial t} \cdot d\mathbf{S} \tag{6-37}$$

where α, β, and γ are as yet undetermined constants. We have already confirmed the correctness of the terms not multiplied by an undetermined constant. Of the remaining three terms, two—those multiplied by α and β—describe effects that would be attributed to isolated (perhaps microscopic) magnetic monopoles. To date, however, no experiments have given any suggestion that isolated magnetic monopoles can be found in nature. Although it would be an interesting exercise to keep these terms and explore their consequences (perhaps predicting an experiment that would detect magnetic monopoles), we nonetheless elect to set $\alpha = \beta = 0$, not so much because these terms are known to be incorrect as because no experiments to date require them for an adequate accounting. Should some future experiment reveal magnetic monopoles, the terms are easily reinstated.

With the proper choice of the constant γ, however, Eq. (6-37) no longer contradicts Eq. (6-2). By the same arguments invoked in the sentences following Eq. (6-33), we find from Eq. (6-37) that

$$\oint \mathbf{J} \cdot d\mathbf{S} + \frac{\gamma}{\mu_0} \frac{\partial}{\partial t} \oint \mathbf{E} \cdot d\mathbf{S} = 0 \tag{6-38}$$

Substituting from Eq. (6-34), however, we find that

$$\oint \mathbf{J} \cdot d\mathbf{S} + \frac{\gamma}{\mu_0 \epsilon_0} \frac{\partial}{\partial t} \int \rho \, dv = 0 \tag{6-39}$$

which agrees with Eq. (6-2) provided we set $\gamma = \mu_0 \epsilon_0$. Thus, if we replace Eq. (6-6) with

$$\oint \mathbf{B} \cdot d\boldsymbol{\ell} = \mu_0 \int \left(\mathbf{J} + \epsilon_0 \frac{\partial \mathbf{E}}{\partial t} \right) \cdot d\mathbf{S} \tag{6-40}$$

the contradiction is resolved. In effect we have identified in $\epsilon_0 \partial \mathbf{E}/\partial t$ a new current density, called the *displacement current density*, and we have predicted theoretically that this new current density must be added to the current density **J** arising from macroscopic charge transport. We shall call Eq. (6-40) the *generalized circuital law.*

PROBLEMS

P6-16. A current I flows along a wire toward one of two parallel conducting plates and away from the second (Fig. P6-16). As a result a charge $Q(t)$ accumulates on the first plate and a charge $-Q(t)$ on the second, and a time-dependent electric field is established between the plates. For this situation, the surface integral on the right in Eq. (6-40) may in particular be evaluated either over the mouth of the "sack" shown or over the sack itself. Further, the two surface integrals must have the same value. (Why?) (a) Neglect fringing and show that in fact the right-hand side of Eq. (6-40) indeed does have the same value for both surfaces. Assume that the bottom of the sack is parallel to the plates. (b) Qualitatively, what factor would preserve the identity of the two surface integrals if the sack intersected the plate and the displacement current across its bottom were therefore reduced?

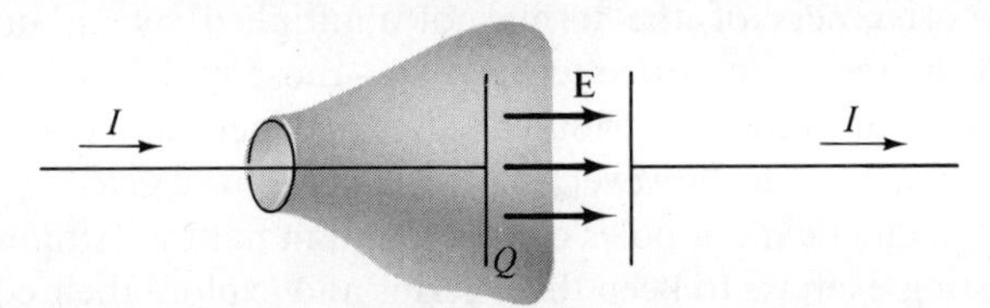

Figure P6-16

P6-17. A point sample of uranium located at the origin radiates (charged) alpha particles uniformly in all directions, thus generating a radial current density given by $\mathbf{J}(\mathbf{r}, t) = [I(t)/4\pi r^2]\hat{\mathbf{r}}$, where $I(t)$ is the current crossing a sphere of radius r with its center at the uranium. Let $Q(t)$ be the total charge on the uranium at time t. (a) Use the equation of continuity to show that $I = -dQ/dt$. (b) Use symmetry and the flux law to show that this current distribution produces *no* magnetic induction field. (c) Assuming that the electric field follows changes in the charge distribution instantaneously, use symmetry and Gauss's law to find $\mathbf{E}(\mathbf{r}, t)$. (d) Show explicitly that Eq. (6-37) is correct for this case only if $\gamma = \mu_0\epsilon_0$. *Hints*: (1) Apply Eq. (6-37) to a path bounding a portion of a spherical surface of radius r. (2) Note that the integral $\int \hat{\mathbf{r}}\cdot d\mathbf{S}/r^2$ need not be explicitly evaluated in order to obtain the desired conclusion. (e) Identify any approximations or tacit assumptions made in this problem.

6-3
Maxwell's Equations

With the addition of the term involving the displacement current, we have at last completed the development of the basic equations satisfied by the (time-dependent) electromagnetic field. Although these equations, called *Maxwell's equations*, have emerged from an examination of experimental evidence and thus are grounded in experiment, they have been tested in such a wide variety of situations that we tend to regard them theoretically as irrefutable postulates that require no further experimental verification. In this regard, we must be cautious, for phenomena presently unknown could well require modification of these equations in the future.

Maxwell's equations have two useful forms. In their *integral* form, which has been our primary concern so far, these equations are

(1) Gauss's law, Eq. (6-34):

$$\oint \mathbf{E}\cdot d\mathbf{S} = \frac{1}{\epsilon_0}\int \rho\, dv \tag{6-41}$$

(2) The magnetic flux law, Eq. (6-35) with $\alpha = 0$:

$$\oint \mathbf{B}\cdot d\mathbf{S} = 0 \tag{6-42}$$

(3) Faraday's law of induction, Eq. (6-36) with $\beta = 0$:

$$\oint \mathbf{E}\cdot d\boldsymbol{\ell} = -\int \frac{\partial \mathbf{B}}{\partial t}\cdot d\mathbf{S} \tag{6-43}$$

(4) The generalized circuital law, Eq. (6-37) with $\gamma = \mu_0\epsilon_0$:

$$\oint \mathbf{B}\cdot d\boldsymbol{\ell} = \mu_0\int \mathbf{J}\cdot d\mathbf{S} + \mu_0\epsilon_0\int \frac{\partial \mathbf{E}}{\partial t}\cdot d\mathbf{S} \tag{6-44}$$

Throughout these equations, the various surfaces and curves are arbitrary, subject only to the convention relating the direction of $d\boldsymbol{\ell}$ to that of $d\mathbf{S}$ in Eqs. (6-43) and (6-44) by the right-hand rule.

Despite the generality of Eqs. (6-41)–(6-44), they are not particularly suited to the direct calculation of fields except when the sources ρ and $\mathbf{J}$ are known and exhibit considerable symmetry. Further awkwardness arises because each integral involves a field, a charge density, or a current density at *every* point on some curve, surface, or volume. The key to a reformulation that eliminates these objections is to allow the arbitrary curves, surfaces, and volumes in Eqs. (6-41)–(6-44) to become indefinitely small. The argument transforming Eqs. (6-41) and (6-42) is therefore the same as that presented in Eqs. (4-27)–(4-29) except that we now must regard time as a parameter that remains fixed throughout the argument; we find that Eq. (6-41) becomes

$$\nabla\cdot\mathbf{E} = \frac{1}{\epsilon_0}\rho \quad \text{(Gauss's law)} \tag{6-45}$$

and that Eq. (6-42) becomes

$$\nabla\cdot\mathbf{B} = 0 \quad \text{(magnetic flux law)} \tag{6-46}$$

Equations (6-43) and (6-44) are transformed to differential form by applying the relationship in Eq. (0-61) and by regarding the integrand in the surface integral to be constant over the (now small) surface $\Delta\mathbf{S}$. From Eq. (6-43)

TABLE 6-1 Maxwell's Equations in mks Units

Vector form	Cartesian components
$\nabla \cdot \mathbf{E} = \dfrac{\rho}{\epsilon_0}$	$\dfrac{\partial E_x}{\partial x} + \dfrac{\partial E_y}{\partial y} + \dfrac{\partial E_z}{\partial z} = \dfrac{\rho}{\epsilon_0}$
$\nabla \cdot \mathbf{B} = 0$	$\dfrac{\partial B_x}{\partial x} + \dfrac{\partial B_y}{\partial y} + \dfrac{\partial B_z}{\partial z} = 0$
	$\dfrac{\partial E_z}{\partial y} - \dfrac{\partial E_y}{\partial z} = -\dfrac{\partial B_x}{\partial t}$
$\nabla \times \mathbf{E} = -\dfrac{\partial \mathbf{B}}{\partial t}$	$\dfrac{\partial E_x}{\partial z} - \dfrac{\partial E_z}{\partial x} = -\dfrac{\partial B_y}{\partial t}$
	$\dfrac{\partial E_y}{\partial x} - \dfrac{\partial E_x}{\partial y} = -\dfrac{\partial B_z}{\partial t}$
	$\dfrac{\partial B_z}{\partial y} - \dfrac{\partial B_y}{\partial z} = \mu_0 J_x + \mu_0\epsilon_0\dfrac{\partial E_x}{\partial t}$
$\nabla \times \mathbf{B} = \mu_0\mathbf{J} + \mu_0\epsilon_0\dfrac{\partial \mathbf{E}}{\partial t}$	$\dfrac{\partial B_x}{\partial z} - \dfrac{\partial B_z}{\partial x} = \mu_0 J_y + \mu_0\epsilon_0\dfrac{\partial E_y}{\partial t}$
	$\dfrac{\partial B_y}{\partial x} - \dfrac{\partial B_x}{\partial y} = \mu_0 J_z + \mu_0\epsilon_0\dfrac{\partial E_z}{\partial t}$

we find that $\nabla \times \mathbf{E}\cdot\Delta\mathbf{S} \approx -(\partial\mathbf{B}/\partial t)\cdot\Delta\mathbf{S}$. Since $\Delta\mathbf{S}$ is arbitrary, this expression in turn implies that

$$\nabla \times \mathbf{E} = -\frac{\partial \mathbf{B}}{\partial t} \quad \text{(Faraday's law)} \tag{6-47}$$

A similar argument applied to Eq. (6-44) yields

$$\nabla \times \mathbf{B} = \mu_0\mathbf{J} + \mu_0\epsilon_0\frac{\partial \mathbf{E}}{\partial t} \quad \text{(generalized circuital law)} \tag{6-48}$$

and we have completed the transformation of Eqs. (6-41)–(6-44) to their *differential* form. The resulting equations are summarized in Table 6-1 both in vector form, which is independent of any particular coordinate system, and in the form satisfied by the Cartesian components of the fields. In contrast to the integral form, the differential form of Maxwell's equations relates aspects of the fields at single points in space-time to the sources at that *same* point in space-time. The calculational difficulties associated with integrals over extended regions of space have therefore been replaced by whatever problems are associated with solving coupled partial differential equations. In particular, suitable boundary and initial conditions must be specified to supplement the differential equations. We shall see in subsequent chapters, however, not only that the necessary boundary and initial conditions are often easy to obtain but also that partial differential equations are often easily solved by at least one of several well-developed techniques, even when the sources ρ and $\mathbf{J}$ are not fully known until *after* the fields have been found.

In the remainder of this book, uses of the differential form of Maxwell's equations will predominate over uses of the integral form. These equations admit an immense variety of solutions, some of which we shall explore in the problems and in later chapters. In the remainder of this chapter, we shall treat energy and momentum in a general electromagnetic field and examine two useful reformulations of Maxwell's equations.

PROBLEMS

P6-18. Stokes' theorem and the divergence theorem can be used to convert Maxwell's equations to differential form. Stokes' theorem, for example, applied to Eq. (6-43), gives

$$\oint \mathbf{E}\cdot d\boldsymbol{\ell} = \int (\nabla \times \mathbf{E})\cdot d\mathbf{S} = -\int \frac{\partial \mathbf{B}}{\partial t}\cdot d\mathbf{S}$$

Since this equation must be correct for an *arbitrary* path, it must also be correct for an *arbitrary* surface, which can be the case only if the integrands are equal at every point in space, and we infer Eq. (6-47). Derive the rest of Maxwell's equations by similar arguments.

P6-19. Show that the solutions to Maxwell's equations satisfy the principle of superposition. *Hint*: Let the field $\mathbf{E}_1$, $\mathbf{B}_1$ satisfy the equations with sources ρ_1 and $\mathbf{J}_1$ and let $\mathbf{E}_2$, $\mathbf{B}_2$ satisfy the equations with sources ρ_2 and $\mathbf{J}_2$. Then show that the field $\mathbf{E}_1 + \mathbf{E}_2$, $\mathbf{B}_1 + \mathbf{B}_2$ satisfies the equations with sources $\rho_1 + \rho_2$ and $\mathbf{J}_1 + \mathbf{J}_2$.

P6-20. Let ρ and $\mathbf{J}$ be zero. Then show that Maxwell's equations are invariant to the transformation

$$\mathbf{E}' = \mathbf{E}\cos\theta + \frac{1}{\mu_0}\mathbf{B}\sin\theta$$

$$\mathbf{B}' = -\epsilon_0\mathbf{E}\sin\theta + \mathbf{B}\cos\theta$$

where θ is an arbitrary constant angle. Note particularly the form of this transformation when $\theta = \frac{1}{2}\pi$—in a sense, the fields $\mathbf{E}$ and $\mathbf{B}$ can be interchanged! *Hint*: Let $\mathbf{E}$, $\mathbf{B}$ satisfy Maxwell's equations and show that $\mathbf{E}'$, $\mathbf{B}'$ do also.

P6-21. Starting with Maxwell's equations, show that there exist in a region free of charges and currents no static solutions for $\mathbf{E}$ or $\mathbf{B}$ that depend on only one Cartesian coordinate.

P6-22. The electrostatic field in some region of space has everywhere the same direction, say the z direction. Show that $\mathbf{E}$ cannot depend on either x or y. If there is no charge in this region of space, show that $\mathbf{E}$ must be a constant field.

P6-23. Derive the equation of continuity in differential form directly from Maxwell's equations in differential form.

P6-24. Starting with the Lorentz force and Maxwell's equations in differential form, derive Coulomb's law.

P6-25. Starting with the differential form of Maxwell's equations, let $\mathbf{B} = B(t)\hat{\mathbf{k}}$ and derive the result in P6-9 for an associated electric field. Carefully enumerate any assumptions made. Is it possible for ρ and/or $\mathbf{J}$ to be zero? If not, what must they be?

6-4 Energy in the Electromagnetic Field

Maxwell's equations can be manipulated to obtain a relationship that can be interpreted as a statement of energy balance. We begin by subtracting the dot product of Eq. (6-47) with $\mathbf{B}$ from the dot product of Eq. (6-48) with $\mathbf{E}$, obtaining

$$\mathbf{E}\cdot(\nabla \times \mathbf{B}) - \mathbf{B}\cdot(\nabla \times \mathbf{E}) = \mu_0 \mathbf{J}\cdot\mathbf{E} + \mu_0 \frac{\partial}{\partial t}\left(\frac{\epsilon_0 E^2}{2} + \frac{B^2}{2\mu_0}\right) \tag{6-49}$$

We next invoke Eq. (C-12) to find the differential statement

$$-\nabla \cdot \left(\mathbf{E} \times \frac{\mathbf{B}}{\mu_0}\right) = \mathbf{J}\cdot\mathbf{E} + \frac{\partial}{\partial t}\left(\frac{\epsilon_0 E^2}{2} + \frac{B^2}{2\mu_0}\right) \tag{6-50}$$

which is equivalent to the integral statement

$$-\oint_\Sigma \left(\mathbf{E} \times \frac{\mathbf{B}}{\mu_0}\right) \cdot d\mathbf{S} = \int_V \mathbf{J}\cdot\mathbf{E}\, dv + \frac{\partial}{\partial t}\int_V \left(\frac{\epsilon_0 E^2}{2} + \frac{B^2}{2\mu_0}\right) dv \tag{6-51}$$

obtained by integrating Eq. (6-50) over a *fixed* volume V and using the divergence theorem to express the left-hand side as an integral over the surface Σ bounding V.

We now seek a physical interpretation of the terms in Eq. (6-51). In accordance with Eq. (3-39), $\int \mathbf{J}\cdot\mathbf{E}\, dv$ is the rate at which the field does work on the particles composing the current in V. We have also seen [Eqs. (4-64) and (5-53)] that the integral over V in the final term of Eq. (6-51) gives the energy stored in the fields when the fields are static. Consistent with these properties, we now simply assign an energy density

$$u_{\mathrm{EM}} = \frac{\epsilon_0 E^2}{2} + \frac{B^2}{2\mu_0} \tag{6-52}$$

to the time-dependent field as well. With that assignment, Eq. (6-51) has the semiverbal expression

$$\begin{aligned} -\oint_\Sigma \left(\mathbf{E} \times \frac{\mathbf{B}}{\mu_0}\right) \cdot d\mathbf{S} &= \begin{pmatrix}\text{rate at which mechanical energy}\\ \text{of particles increases}\end{pmatrix} \\ &\quad + \begin{pmatrix}\text{rate at which energy stored}\\ \text{in fields increases}\end{pmatrix} \\ &= \begin{pmatrix}\text{rate at which total energy}\\ \text{in } V \text{ increases}\end{pmatrix} \end{aligned} \tag{6-53}$$

Now, in the absence of explicit sources of energy within V, an increase in the total energy in V can come about only if energy flows into V over the surface

Σ. We are thus led to interpret the surface integral in Eq. (6-53) as the rate at which energy flows into V from the space outside V. Verbally, and without the minus sign, we thus have that

$$\oint_\Sigma \left(\mathbf{E} \times \frac{\mathbf{B}}{\mu_0}\right) \cdot d\mathbf{S} = \begin{pmatrix}\text{rate at which energy flows out of} \\ V \text{ across the bounding surface } \Sigma\end{pmatrix} \tag{6-54}$$

The vector

$$\mathbf{S} = \mathbf{E} \times \frac{\mathbf{B}}{\mu_0} \tag{6-55}$$

that appears in Eq. (6-54) is called the *Poynting vector*; its direction is the direction in which the electromagnetic fields **E** and **B** are transporting energy and its magnitude is the rate at which energy is transported across a surface of unit area oriented perpendicular to the direction of energy flow. It is often stated that the only suitable interpretation of Eq. (6-54) is that the *integral* of the Poynting vector over a closed surface Σ represents energy transported out of the volume bounded by Σ. In this view, the interpretation of **S** as itself representing a point-by-point energy flow is regarded as a convenient fiction leading to inconsistencies if taken too literally.[5] Whether **S** does or does not represent a point-by-point energy flow remains a matter for debate.[6]

Although we cannot be sure either that the interpretations given in the previous paragraphs are the only possible interpretations or that Eqs. (6-50) and (6-51) are the only expressions that are both consistent with Maxwell's equations and interpretable as expressions of energy balance,[7] these equations and the above interpretation are pleasing in their simplicity and experimentally adequate to all tests made of them to date. We therefore accept the interpretation not so much because of any inevitability we might like it to have as because of its suitability. In accepting that interpretation, we not only have allowed for general time-dependent fields to *store* energy but also have attributed to these fields a capacity to *transport* energy through space.

PROBLEM

P6-26. A straight cylindrical metal wire of radius b carries a current I along the z axis in response to the application of an electric field $\mathbf{E} = E\hat{\mathbf{k}}$ to points inside the wire. Determine the direction and magnitude of the Poynting vector at the surface of the wire, integrate the normal component of the

[5]See, for example, J. R. Reitz and F. J. Milford, *Foundations of Electromagnetic Theory* (Addison-Wesley Publishing Company, Inc., Reading, Mass., 1967), Second Edition, p. 299; D. R. Corson and P. Lorrain, *Introduction to Electromagnetic Fields and Waves* (W. H. Freeman and Company, Publishers, San Francisco, 1962), pp. 321ff.; and other books on electricity and magnetism.

[6]See W. H. Furry, *Am. J. Phys.* **37**, 621 (1969) and the references given there.

[7]See R. P. Feynman, R. B. Leighton, and M. Sands, *The Feynman Lectures on Physics* (Addison-Wesley Publishing Co., Reading, Massachusetts, 1964) Vol. II, Lecture 27.

Poynting vector over a segment of the wire of length L, and compare your result with the Joule heat produced in this segment. Now, ponder what it means for the energy dissipated in the wire to enter from the space outside the wire rather than to propagate along the wire. *Hint*: The Joule heat in a wire is given by IV, where V is the potential difference between the two ends of the wire.

6-5 Momentum in the Electromagnetic Field

Maxwell's equations can also be manipulated to obtain a relationship that can be interpreted as a statement of momentum balance. Suppose that we have a system of charges and currents distributed in space and that the electromagnetic field established *by this system* is **E**, **B**. The fields, of course, satisfy Maxwell's equations. The force **F** on the system of charges and currents, which is also the time rate of change of the (mechanical) momentum **P** of the system, is then given by

$$\mathbf{F} = \frac{d\mathbf{P}}{dt} = \int (\rho\mathbf{E} + \mathbf{J} \times \mathbf{B})\, dv \tag{6-56}$$

where the integral extends over the volume containing the distribution, which is equivalent to extending over all space, since ρ and **J** are zero outside the distribution. We now use Maxwell's equations to rewrite Eq. (6-56). First, substitute for ρ and **J** from Eqs. (6-45) and (6-48) to find that

$$\frac{d\mathbf{P}}{dt} = \int \left[(\epsilon_0 \nabla \cdot \mathbf{E})\mathbf{E} + \frac{1}{\mu_0}(\nabla \times \mathbf{B}) \times \mathbf{B} - \epsilon_0 \frac{\partial \mathbf{E}}{\partial t} \times \mathbf{B}\right] dv \tag{6-57}$$

Next, we seek an expression in which the time derivative under the integral can be removed from the integral. We write

$$\begin{aligned}\int \left(\frac{\partial \mathbf{E}}{\partial t} \times \mathbf{B}\right) dv &= \int \frac{\partial}{\partial t}(\mathbf{E} \times \mathbf{B})\, dv - \int \left(\mathbf{E} \times \frac{\partial \mathbf{B}}{\partial t}\right) dv \\ &= \frac{d}{dt}\int \mathbf{E} \times \mathbf{B}\, dv + \int [\mathbf{E} \times (\nabla \times \mathbf{E})]\, dv \end{aligned} \tag{6-58}$$

where the second form follows partly on substitution from Eq. (6-47) and partly because the integrals extend over fixed volumes so that $\partial/\partial t$ can be taken outside the integral and then written as a total derivative. Then we substitute Eq. (6-58) into Eq. (6-57) and rearrange the terms to obtain that

$$\begin{aligned}\frac{d}{dt}\left(\mathbf{P} + \int \epsilon_0 \mathbf{E} \times \mathbf{B}\, dv\right) = \int \Big[&\epsilon_0(\nabla \cdot \mathbf{E})\mathbf{E} - \epsilon_0 \mathbf{E} \times (\nabla \times \mathbf{E}) \\ &- \frac{1}{\mu_0}\mathbf{B} \times (\nabla \times \mathbf{B})\Big]\, dv \end{aligned} \tag{6-59}$$

Finally, because the integral under the time derivative appears on the same footing as the mechanical momentum **P**, it is appropriate to make the identification

$$\begin{pmatrix}\textbf{momentum in}\\ \text{the field } \mathbf{E}, \mathbf{B}\end{pmatrix} = \int \epsilon_0 \mathbf{E} \times \mathbf{B}\, dv \tag{6-60}$$

and to interpret the quantity

$$\mathcal{G} = \epsilon_0 \mathbf{E} \times \mathbf{B} \tag{6-61}$$

as a momentum density associated with the electromagnetic field. From Eq. (6-55) it is apparent that $\mathcal{G}$ and $\mathbf{S}$ are here related by

$$\mathcal{G} = \epsilon_0 \mu_0 \mathbf{S} \tag{6-62}$$

but this relationship applies *only* to fields in vacuum. When matter is introduced (Chapter 12), Maxwell's equations assume a slightly different form and a significantly different degree of difficulty. The momentum density and the Poynting vector are then not even determined from the same pair of fields.

6-6 A Reformulation: Maxwell's Equations for the Potentials

The existence of scalar and vector potentials for the general time-dependent field and also equations satisfied by these potentials can be inferred *directly* from Maxwell's equations *without* reference to previous developments. We first note that the flux law, Eq. (6-46), implies the existence of a vector potential **A** in terms of which

$$\mathbf{B} = \nabla \times \mathbf{A} \tag{6-63}$$

[See item (4) in Section 2-5.] Since mixed second partial derivatives can be evaluated in either order, Faraday's law, Eq. (6-47), then becomes

$$\nabla \times \left(\mathbf{E} + \frac{\partial \mathbf{A}}{\partial t}\right) = 0 \tag{6-64}$$

which implies the existence of a scalar potential V, in terms of which

$$\mathbf{E} + \frac{\partial \mathbf{A}}{\partial t} = -\nabla V \Longrightarrow \mathbf{E} = -\nabla V - \frac{\partial \mathbf{A}}{\partial t} \tag{6-65}$$

[See item (3) in Section 2-5.] Note that Eqs. (6-63) and (6-65) reduce correctly to their static counterparts, Eqs. (5-37) and (4-38), when the potentials are static. The most interesting aspect of the time-dependent expressions is the appearance of **A** in the equation giving **E**.

The phenomenon of gauge invariance carries over from static to time-dependent potentials. Certainly, **B** as given by Eq. (6-63) is unchanged if an arbitrary gradient is added to **A**; i.e, **B** is invariant to the transformation

$$\mathbf{A}_2 = \mathbf{A}_1 + \nabla \Lambda \tag{6-66}$$

where Λ now may have both spatial and temporal dependence. The electric field $\mathbf{E}$ as given by Eq. (6-65) will be invariant to this change in $\mathbf{A}$, however, only if the transformation in $\mathbf{A}$ is accompanied by a transformation in V chosen so that

$$-\nabla V_1 - \frac{\partial \mathbf{A}_1}{\partial t} = -\nabla V_2 - \frac{\partial \mathbf{A}_2}{\partial t} \tag{6-67}$$

Substituting for $\mathbf{A}_2$ from Eq. (6-66), we find that

$$-\nabla V_1 = -\nabla V_2 - \nabla \frac{\partial \Lambda}{\partial t} \Longrightarrow V_2 = V_1 - \frac{\partial \Lambda}{\partial t} \tag{6-68}$$

except for a possible arbitrary constant. Thus, the time-dependent electromagnetic field $\mathbf{E}$, $\mathbf{B}$ is unchanged when the vector and scalar potentials are simultaneously transformed by Eqs. (6-66) and (6-68), where Λ is arbitrary, and this transformation is therefore a natural generalization of the gauge transformation introduced in Section 5-4. Here as there, one consequence of the gauge invariance of the fields is the arbitrariness of $\nabla \cdot \mathbf{A}$, which can be given any value that we find convenient (P6-27).

Two of Maxwell's equations—the homogeneous pair, Eqs. (6-46) and (6-47)—are automatically satisfied when the fields are expressed in terms of the potentials. (Why?) The remaining two—the inhomogeneous pair, Eqs. (6-45) and (6-48)—generate differential equations whose solutions determine the potentials. Combined with Eq. (6-65), for example, Eq. (6-45) yields

$$-\nabla^2 V - \frac{\partial}{\partial t} \nabla \cdot \mathbf{A} = \frac{\rho}{\epsilon_0} \tag{6-69}$$

where the Laplacian, $\nabla^2 = \nabla \cdot \nabla$, is introduced initially in Section 2-5. Similarly, substituting Eqs. (6-63) and (6-65) into Eq. (6-48) and using Eq. (C-19) to reexpress $\nabla \times (\nabla \times \mathbf{A})$, we find after rearranging terms that

$$\left(\nabla^2 - \mu_0 \epsilon_0 \frac{\partial^2}{\partial t^2}\right) \mathbf{A} = -\mu_0 \mathbf{J} + \nabla\left(\nabla \cdot \mathbf{A} + \mu_0 \epsilon_0 \frac{\partial V}{\partial t}\right) \tag{6-70}$$

Now, we can simplify Eq. (6-70) by imposing the so-called *Lorentz condition*

$$\nabla \cdot \mathbf{A} + \mu_0 \epsilon_0 \frac{\partial V}{\partial t} = 0 \tag{6-71}$$

on the potentials, thereby stipulating the value of $\nabla \cdot \mathbf{A}$ and selecting what is called a *Lorentz gauge* for the potentials. In this gauge, Eq. (6-70) becomes

$$\left(\nabla^2 - \mu_0 \epsilon_0 \frac{\partial^2}{\partial t^2}\right) \mathbf{A} = -\mu_0 \mathbf{J} \tag{6-72}$$

Further, on substituting $\nabla \cdot \mathbf{A}$ from Eq. (6-71), we find from Eq. (6-69) that

$$\left(\nabla^2 - \mu_0 \epsilon_0 \frac{\partial^2}{\partial t^2}\right) V = -\frac{\rho}{\epsilon_0} \tag{6-73}$$

The similarity of these two equations is part of the reason for imposing the Lorentz condition. In Lorentz gauge, the time-dependent potentials both satisfy the so-called *inhomogeneous wave equation*, which we shall examine more fully in Chapter 14. When **A** and V are time-independent, Eqs. (6-72) and (6-73) reduce to Poisson's equation,

$$\nabla^2\mathbf{A} = -\mu_0\mathbf{J} \qquad \nabla^2 V = -\frac{\rho}{\epsilon_0} \tag{6-74}$$

or, when **J** and ρ are zero, to Laplace's equation,

$$\nabla^2\mathbf{A} = 0 \qquad \nabla^2 V = 0 \tag{6-75}$$

both of which we shall study more fully in Chapter 8.

PROBLEMS

P6-27. Show that the arbitrariness of the gauge function relating two equivalent magnetic vector potentials means that the divergence of the vector potential is also arbitrary. *Hint*: See Section 5-4.

P6-28. Let $\mathbf{A}_1$ and V_1 be potentials in Lorentz gauge. Further, let the potentials $\mathbf{A}_2$ and V_2 be obtained from $\mathbf{A}_1$ and V_1 by a gauge transformation. Show that $\mathbf{A}_2$ and V_2 are also potentials in Lorentz gauge provided only that the gauge function satisfies the homogeneous wave equation. Comment on the uniqueness of the Lorentz gauge potentials.

P6-29. (a) Using the Poisson equation in spherical coordinates, find the electrostatic potential established by the charge distribution $\rho(r) = \rho_0$, $r < a$; $\rho(r) = 0$, $r > a$. (b) Find the electric field. *Hints*: (1) Symmetry rules out a dependence on the coordinates θ and ϕ. (Why?) (2) Solve the problem in the two domains $r < a$ and $r > a$. The potential and its first derivative must be continuous everywhere, in particular at $r = a$, and the potential must be finite everywhere, in particular at $r = 0$. (Why?) (3) For definiteness, set the arbitrary constant so that $V(\infty) = 0$.

P6-30. An infinite conducting plate in the plane $z = 0$ is maintained at potential zero and a similar plate in the plane $z = d$ is maintained at potential V_0. There is no charge between the plates. (a) Use symmetry to show that the potential can depend only on z. (b) Solve the relevant form of Laplace's equation and impose the conditions at $z = 0$ and $z = d$ to show that $V = V_0 z/d$. (c) Find the electric field in the region between the plates. (d) Find the charge density on each plate. *Hint*: See Eq. (4-26).

P6-31. Following the pattern illustrated more fully in P6-30, use Laplace's equation and suitable boundary conditions to find the potential, field, and all relevant charge densities in the region between two infinitely long coaxial conducting cylinders, the inner of radius a being at potential V_a and the outer of radius b being at potential V_b.

P6-32. Following the pattern illustrated more fully in P6-30, use Laplace's equation and suitable boundary conditions to find the potential, field, and all relevant charge densities in the region between two concentric spherical conducting shells, the inner shell of radius a being at potential V_a and the outer of radius b being at potential V_b. Express the field in terms of the charge on the inner sphere and comment.

6-7
Another Reformulation: Decoupling the Equations for the Fields

Maxwell's equations as summarized in Section 6-3 are first-order equations and the equations for **E** and for **B** are coupled when the fields are time-dependent. At the expense of generating second-order equations, however, we can find *apparently* uncoupled equations for **E** and for **B**. For example, we can evaluate the curl of Eq. (6-47) and use Eq. (C-19) to expand $\nabla \times (\nabla \times \mathbf{E})$, finding that

$$\nabla(\nabla \cdot \mathbf{E}) - \nabla^2 \mathbf{E} = -\frac{\partial}{\partial t}(\nabla \times \mathbf{B}) \tag{6-76}$$

We next substitute from Eqs. (6-45) and (6-48) and rearrange the terms to obtain the equation

$$\left(\nabla^2 - \mu_0 \epsilon_0 \frac{\partial^2}{\partial t^2}\right)\mathbf{E} = \frac{1}{\epsilon_0}\nabla\rho + \mu_0 \frac{\partial \mathbf{J}}{\partial t} \tag{6-77}$$

A similar argument, beginning with the curl of Eq. (6-48), yields

$$\left(\nabla^2 - \mu_0 \epsilon_0 \frac{\partial^2}{\partial t^2}\right)\mathbf{B} = -\mu_0 \nabla \times \mathbf{J} \tag{6-78}$$

Neither of these equations is particularly simple except when ρ and **J** are zero. Further, the act of differentiating Maxwell's equations in general introduces extraneous solutions—just as squaring an algebraic equation introduces extraneous solutions—and any solution obtained for Eqs. (6-77) and (6-78) must always be verified by substitution into the original first-order Maxwell equations. In that substitution, additional constraints relating constants and sometimes functions in the solution for **E** to similar entities in the solution for **B** will usually emerge. Thus the apparent separation of **E** from **B** in Eqs. (6-77) and (6-78) is illusory, as it must be since **E** and **B** are inextricably intertwined when they depend on time. Even so, we shall find that Eqs. (6-77) and (6-78) often can be exploited to advantage in solving some problems.

PROBLEM

P6-33. Evaluate the curl of Eq. (6-48) and derive Eq. (6-78).

Supplementary Problems

P6-34. In a sense, Maxwell's equations are not entirely independent of one another. Show from Faraday's law that $\partial(\nabla \cdot \mathbf{B})/\partial t = 0$ and hence show that, if $\nabla \cdot \mathbf{B}$ ever was zero at some time in the past, it must still be zero. Can you find any other interconnections of this sort?

P6-35. Imagine magnetic monopoles to be found. Maxwell's equations would then have the form

$$\nabla \cdot \mathbf{E} = \frac{1}{\epsilon_0}\rho \qquad \nabla \times \mathbf{E} = -\frac{\partial \mathbf{B}}{\partial t} + \beta \mathbf{J}_m$$

$$\nabla \cdot \mathbf{B} = \alpha \rho_m \qquad \nabla \times \mathbf{B} = \mu_0 \epsilon_0 \frac{\partial \mathbf{E}}{\partial t} + \mu_0 \mathbf{J}$$

where α and β are constants and ρ_m and $\mathbf{J}_m$ are the density of magnetic poles and the magnetic pole current density. (a) Derive an equation of continuity involving ρ_m and $\mathbf{J}_m$, and note that it is entirely separate from the equation involving ρ and $\mathbf{J}$. (b) Find an equation analogous to the energy equation, Eq. (6-50), and interpret its terms physically. (c) Do there exist potentials from which these fields can be derived? Why or why not? If they exist, express the equations in terms of them.

P6-36. The uniform but time-dependent field $\mathbf{B} = B(t)\hat{\mathbf{k}}$ is invariant to arbitrary translation and hence does not define any origin. Thus, we might equally well have centered the loop in P6-9 at (x_0, y_0) rather than at the origin, and we would then have obtained

$$\begin{aligned}\mathbf{E}(\mathbf{r}) &= \frac{1}{2}\frac{dB}{dt}[(y - y_0)\hat{\mathbf{i}} - (x - x_0)\hat{\mathbf{j}}] \\ &= \frac{1}{2}\frac{dB}{dt}(\mathbf{r} - \mathbf{r}_0) \times \hat{\mathbf{k}}\end{aligned}$$

Show explicitly that $\oint_\Gamma \mathbf{E} \cdot d\boldsymbol{\ell} = -d\Phi_m/dt$, where Γ is *any* path in the x-y plane—not necessarily a circle centered at (x_0, y_0). In particular, since Γ may be a circle centered at the origin, this problem shows that Faraday's law can be satisfied for that path *without* having $\mathbf{E}$ everywhere tangent to the path. *Optional*: Use analog or digital methods (Section 3-2) to explore the trajectories of particles moving in the electromagnetic field of this problem. [See K. Shen, E. D. Alton, and H. C. S. Hsuan, *Am. J. Phys.* **38**, 1133 (1970), and D. M. Cook, *Am. J. Phys.* **40**, 210 (1972).]

P6-37. (a) Show that

$$[(\nabla \cdot \mathbf{Q})\mathbf{Q} - \mathbf{Q} \times (\nabla \times \mathbf{Q})]_i = \sum_j \frac{\partial R_{ij}}{\partial x_j}$$

where

$$R_{ij} = Q_i Q_j - \tfrac{1}{2} Q^2 \delta_{ij}$$

and $\delta_{ij} = 1$ if $i = j$ and is zero otherwise. *Hint*: The notation of P0-28 may be useful. (b) Show that the ith component of the right-hand side of Eq. (6-59) can be expressed as a surface integral

$$\sum_j \oint T_{ij}\, dS_j$$

where dS_j is a component of the vector $d\mathbf{S}$, provided T_{ij} is suitably identified. T_{ij} is known as the *Maxwell stress tensor*, the terminology hanging over from the days when the electromagnetic field was viewed as a distortion or stress in an elastic, mechanical ether.

P6-38. Show that in a region free of charges and currents Maxwell's equations are all satisfied if the fields are derived from a single potential **Q** by

$$\mathbf{E} = -\nabla \times \frac{\partial \mathbf{Q}}{\partial t} \qquad \mathbf{B} = \nabla \times (\nabla \times \mathbf{Q})$$

and **Q** satisfies

$$\left(\nabla^2 - \mu_0 \epsilon_0 \frac{\partial^2}{\partial t^2}\right)\mathbf{Q} = 0$$

P6-39. Let the plate at $z = 0$ in P6-30 be the cathode of a simple vacuum diode and the plate at $z = d$ be the anode. The cathode emits electrons into the region between the plates and the Poisson equation $d^2V/dz^2 = -\rho(z)/\epsilon_0$, where $\rho(z)$ is the charge density at the coordinate z between the plates, replaces the Laplace equation. We still require $V(0) = 0$ and $V(d) = V_0$, but the basic equation cannot be solved until more is known about $\rho(z)$. Assume that the electrons start from rest at the cathode. (a) Show that $\rho(z) = J(z)/v(z) = J(z)\sqrt{m/2eV(z)}$, where $J(z)$ and $v(z)$ are the current density and electron velocity at the coordinate z between the plates and $-e$ and m are the charge and mass of the electron. (b) Assuming that charge is conserved and that a steady state has been reached between the plates, argue that $J(z)$ in fact cannot depend on z and hence obtain an equation for $V(z)$ that involves only one unknown function. (c) Show that for steady current flow, $dV/dz = 0$ at $z = 0$. (d) Solve the equation obtained in part (b) subject to all boundary conditions and derive Child's law that the current in this simple diode is proportional to the three-halves power of the anode to cathode potential. (e) Comment on the physical limitations of Child's law and sketch a graph of the expected current-voltage characteristics of this simple diode, taking into account whatever limitations you have pointed out. Can you infer the meaning of the phrase *space-charge-limited* from knowledge that the phrase is used to describe the current in the region to which Child's law applies? This derivation is presented in some detail in Appendix 8 of K. R. Spangenberg, *Fundamentals of Electron Devices* (McGraw-Hill Book Company, New York, 1957). A treatment of the analogous problem in cylindrical coordinates is given by Ll. G. Chambers, *Am. J. Phys.* **36**, 911 (1968).

Interlude: A Change of View

Although we must yet consider numerous applications and at least one important generalization, the essential theory of the electromagnetic field is now complete.[1] At this point in this book a marked change in point of view occurs. Initially the electric and magnetic induction fields were introduced in Chapter 3 to separate the interaction between a source distribution and a test distribution into two parts, and our initial motivation for examining the fields lay in a desire to treat the general interaction between two arbitrary distributions. Within this limited context, the fields were merely a useful contrivance invented specifically to simplify the study of that interaction. As the properties of the fields were further articulated in Chapters 4–6, however, the fields gradually acquired a broader significance. We assigned an energy content and attributed to the fields a capacity to transport this energy through space, and we assigned a momentum content and attributed to the fields a capacity to transport this momentum through space. With each additional attribute, the fields became more and more real *in themselves* and less and less a mere contrivance introduced for the sake of a limited initial concern. In the rest of this book, our interest remains focused on examining the properties of the fields in various circumstances and on determining the fields from their sources, but the motivation underlying that focus has changed

[1]It is suggested that the reader hold Chapters 1–6 between his thumb and index finger and note the compactness of that theory.

from a desire to study the fields so that we can determine forces of interaction to a desire to study the fields because they have become real physical entities in their own right. Before proceeding to the remainder of this book, the reader may find it valuable to review the evolution of this change in viewpoint by studying again the upper portion of the flow chart following the Preface and by working the following problem.

PROBLEM

Starting with Coulomb's law and the analogous expression for the force per unit length on one of two parallel, current-carrying wires, repeat the essentials of the development in Chapters 3–6 to obtain the differential form of Maxwell's equations in *Gaussian* units, namely

$$\nabla \cdot \mathbf{E} = 4\pi\rho \qquad \nabla \cdot \mathbf{B} = 0$$

$$\nabla \times \mathbf{E} = -\frac{1}{c}\frac{\partial \mathbf{B}}{\partial t} \qquad \nabla \times \mathbf{B} = \frac{4\pi}{c}\mathbf{J} + \frac{1}{c}\frac{\partial \mathbf{E}}{\partial t}$$

Along the way, obtain also expressions in Gaussian units for (a) the electrostatic field established by a given charge distribution, (b) the electrostatic potential established by a given charge distribution, (c) the law of Biot-Savart, and (d) the magnetic vector potential established by a given steady current distribution. Then, arguing from Maxwell's equations, find expressions in Gaussian units for (e) the energy density in an electromagnetic field, (f) the Poynting vector, (g) the momentum density in an electromagnetic field, (h) Poisson's equation for the static scalar potential and the analogous equation for the static vector potential, and (i) the wave equation for the time-dependent electric field in vacuum. In writing a solution to this exercise, stress particularly the logical development. Make it clear where definitions are made, where reference is made to experimental results, where mathematical identities are used, etc. Describe each calculation briefly, but abbreviate the presentation of mathematical details as much as is consistent with clarity and use theorems such as the divergence theorem without proof. *Note:* The reader who is particularly interested in electromagnetic units is referred to several articles by R. T. Birge [*Am. J. Phys.* **2**, 41 (1934); **3**, 102 (1935); and **3**, 171 (1935)] and to the Appendix on Units and Dimensions in J. D. Jackson, *Classical Electrodynamics* (John Wiley & Sons, Inc., New York, 1962).

7

Plane Electromagnetic Waves in Vacuum

In a region free of charges and currents ($\rho = 0$ and $\mathbf{J} = 0$), Maxwell's equations reduce to

$$\nabla \cdot \mathbf{E} = 0 \qquad \nabla \times \mathbf{E} = -\frac{\partial \mathbf{B}}{\partial t} \tag{7-1), (7-2}$$

$$\nabla \cdot \mathbf{B} = 0 \qquad \nabla \times \mathbf{B} = \mu_0 \epsilon_0 \frac{\partial \mathbf{E}}{\partial t} \tag{7-3), (7-4}$$

The solutions to these equations can be at least partially classified by examining their dependence on the three Cartesian coordinates (x, y, z) and on the time t. Constant fields, which depend on none of these variables, obviously satisfy Eqs. (7-1)–(7-4), for every term in those equations contains a derivative. There are no solutions depending only on one of the four variables (x, y, z, t). If, for example, we assume a solution in which $\mathbf{E}$ and $\mathbf{B}$ both depend only on t, we cannot satisfy either Eq. (7-2) or (7-4), for the left-hand side would be zero and the right-hand side would be nonzero. That no solutions depending only on x or on y or on z can exist was proved in P6-21. Solutions depending on two or more of the four variables (x, y, z, t) are less trivial. We postpone a detailed examination of some of these solutions to Chapters 8 and 14, confining our attention in this chapter to solutions that depend on t and essentially on one Cartesian coordinate, which we initially take to be z. Further, we shall examine only some of the simpler solutions, for Eqs. (7-1)–(7-4) are linear and homogeneous and we can therefore superpose any number of simpler solutions to generate more complicated solutions.

7-1 Elementary Fields Depending on z and t; Plane Electromagnetic Waves

The essential features of solutions depending on $(\mathbf{r}, t)$ can be illustrated by considering solutions depending on (z, t). Suppose then that

$$\begin{aligned} \mathbf{E} &= E_x(z,t)\hat{\mathbf{i}} + E_y(z, t)\hat{\mathbf{j}} + E_z(z, t)\hat{\mathbf{k}} \\ \mathbf{B} &= B_x(z, t)\hat{\mathbf{i}} + B_y(z, t)\hat{\mathbf{j}} + B_z(z, t)\hat{\mathbf{k}} \end{aligned} \tag{7-5}$$

where hereafter we shall leave the arguments (z, t) understood. For this electromagnetic field, Eqs. (7-1)–(7-4) reduce to

$$\frac{\partial E_z}{\partial z} = 0; \qquad \frac{\partial E_z}{\partial t} = 0 \Longrightarrow E_z = \text{constant} \tag{7-6}$$

$$\frac{\partial B_z}{\partial z} = 0; \qquad \frac{\partial B_z}{\partial t} = 0 \Longrightarrow B_z = \text{constant} \tag{7-7}$$

$$\frac{\partial E_y}{\partial z} = \frac{\partial B_x}{\partial t}; \qquad \frac{\partial B_x}{\partial z} = \mu_0 \epsilon_0 \frac{\partial E_y}{\partial t} \tag{7-8}$$

$$\frac{\partial E_x}{\partial z} = -\frac{\partial B_y}{\partial t}; \quad \frac{\partial B_y}{\partial z} = -\mu_0 \epsilon_0 \frac{\partial E_x}{\partial t} \tag{7-9}$$

(See Table 6-1.) Since by superposition we can readily add a constant field to our solution at any time, we can take E_z and B_z to be zero without serious loss.

Equations (7-8) and (7-9) are more interesting than Eqs. (7-6) and (7-7), although even here there is a separation: Equation (7-8) involves E_y and B_x while Eq. (7-9) involves E_x and B_y. Apart from an arbitrary constant field, the most general solution in the form of Eq. (7-5) therefore consists of two parts, the first a solution of Eq. (7-8) having only E_y and B_x nonzero and the second a solution of Eq. (7-9) having only E_x and B_y nonzero. We shall consider in detail only the second part, for which

$$\mathbf{E} = E_x(z, t)\hat{\mathbf{i}} \qquad \mathbf{B} = B_y(z, t)\hat{\mathbf{j}} \tag{7-10}$$

and E_x and B_y satisfy Eq. (7-9). If the first member of Eq. (7-9) is differentiated with respect to z and the second with respect to t and then the equality of the two mixed second partial derivatives is recognized, we find that

$$\frac{\partial^2 E_x}{\partial z^2} = \mu_0 \epsilon_0 \frac{\partial^2 E_x}{\partial t^2} \tag{7-11}$$

By a similar argument, we find also that

$$\frac{\partial^2 B_y}{\partial z^2} = \mu_0 \epsilon_0 \frac{\partial^2 B_y}{\partial t^2} \tag{7-12}$$

Thus, the two components E_x and B_y both satisfy the one-dimensional wave equation. Since our initial objective is not to find the *most general* solution but merely to find *a* solution, let us try the function

$$E_x(z, t) = E_{x0} \cos(\kappa z - \omega t + \phi) \tag{7-13}$$

where the *amplitude* E_{x0}, the *wave number* κ (in m^{-1}), the *angular frequency* ω (in radian/sec), and the *phase* ϕ are constants to be determined. By direct substitution, we find that this function satisfies Eq. (7-11) only if κ and ω are related by

$$\kappa = \omega\sqrt{\mu_0\epsilon_0} \tag{7-14}$$

but that E_{x0}, ϕ, and either κ or ω can be chosen arbitrarily. The function $B_y(z, t)$ that must accompany Eq. (7-13) is now determined by substituting Eq. (7-13) into the first member of Eq. (7-9); we find that

$$\frac{\partial B_y}{\partial t} = \kappa E_{x0} \sin(\kappa z - \omega t + \phi) \tag{7-15}$$

which, on integration with respect to t, gives

$$B_y(z, t) = \frac{\kappa E_{x0}}{\omega} \cos(\kappa z - \omega t + \phi) + g(z) \tag{7-16}$$

where $g(z)$ is an arbitrary function of z (but cannot depend on t). Requiring that Eqs. (7-16) and (7-13) satisfy the second member of Eq. (7-9) gives $dg/dz = 0$, whence $g(z)$ must in fact be constant. Since we can always add a constant solution by superposition, we take $g = 0$. One solution of Eq. (7-9) therefore is

$$\begin{aligned} \mathbf{E} &= E_{x0} \cos(\kappa z - \omega t + \phi)\hat{\mathbf{i}} \\ \mathbf{B} &= \frac{\kappa}{\omega} E_{x0} \cos(\kappa z - \omega t + \phi)\hat{\mathbf{j}} = \frac{1}{\omega}\boldsymbol{\kappa} \times \mathbf{E} \end{aligned} \tag{7-17}$$

where $\boldsymbol{\kappa} = \kappa\hat{\mathbf{k}}$ and is known as the *propagation vector*. The two parts of Eq. (7-17) are inseparable: If the **E**-field is present, the **B**-field must be present, and vice versa. Note, in particular, that a **B**-field in the y direction is associated with an **E**-field in the x direction.

More general solutions to the original equations can be obtained by superposing Eq. (7-17) with other solutions of the same form but having different values for E_{x0}, ϕ, and κ or by superposing Eq. (7-17) with solutions to Eq. (7-8), which are found in P7-2 to have the form

$$\begin{aligned} \mathbf{E} &= E_{y0} \cos(Kz - \Omega t + \Phi)\hat{\mathbf{j}} \\ \mathbf{B} &= -\frac{K}{\Omega} E_{y0} \cos(Kz - \Omega t + \Phi)\hat{\mathbf{i}} = \frac{1}{\Omega}\mathbf{K} \times \mathbf{E} \end{aligned} \tag{7-18}$$

where $\mathbf{K} = K\hat{\mathbf{k}}$ and $K = \Omega\sqrt{\mu_0\epsilon_0}$. In Eq. (7-18), E_{y0}, Φ, and K are arbitrary and independent of E_{x0}, ϕ, and κ in Eq. (7-17). We shall treat these superpositions more fully in Sections 7-3 and 7-4.

We next examine the general properties of the solution expressed in Eq. (7-17). Graphs of E_x versus z for the three times $t = 0$, $t = \pi/2\omega$, and $t = \pi/\omega$ are shown in Fig. 7-1. Apparently, E_x can be described as a sinusoidal

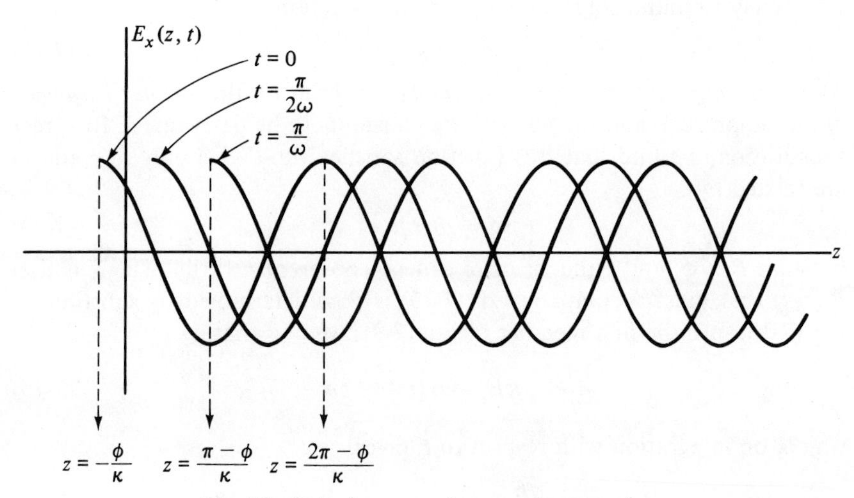

Fig. 7-1. $E_x(z, t)$ versus z for several values of t.

wave propagating in the positive z direction, a suggestion that is confirmed by the appearance of the combination $z - (\omega/\kappa)t$ when E_x is written in the form

$$E_x(z, t) = E_{x0} \cos\left[\kappa\left(z - \frac{\omega}{\kappa}t\right) + \phi\right] \tag{7-19}$$

From this form, we conclude that the wave propagates with the (constant) speed

$$c = \frac{\omega}{\kappa} = \frac{1}{\sqrt{\mu_0 \epsilon_0}} \tag{7-20}$$

[see Eq. (7-14) and compare also P1-8], which—with sufficient knowledge of the properties of the wave equation—we could have inferred directly from the coefficient $\mu_0\epsilon_0$ in Eqs. (7-11) and (7-12). Further, because of the periodicity of the cosine function, the value of cos $(\cdots)$ in Eq. (7-19) is not changed if κz is changed by 2π or (equivalently) if z is changed by

$$\lambda = \frac{2\pi}{\kappa} \tag{7-21}$$

Thus, $E_x\,(z, t)$ is the same at $z + \lambda$ as it is at z, where z is arbitrary. The parameter λ is called the *wavelength* of the wave and it measures the spatial separation between corresponding points on successive cycles of the wave at a given time; the mks unit of wavelength is the meter. Similarly (P7-3), the *period* T (in seconds), which is the reciprocal of the (circular) *frequency* ν (in cycles per second, called a hertz, Hz), is given by

$$T = \frac{1}{\nu} = \frac{2\pi}{\omega} \tag{7-22}$$

and measures the temporal separation between successive occurrences of

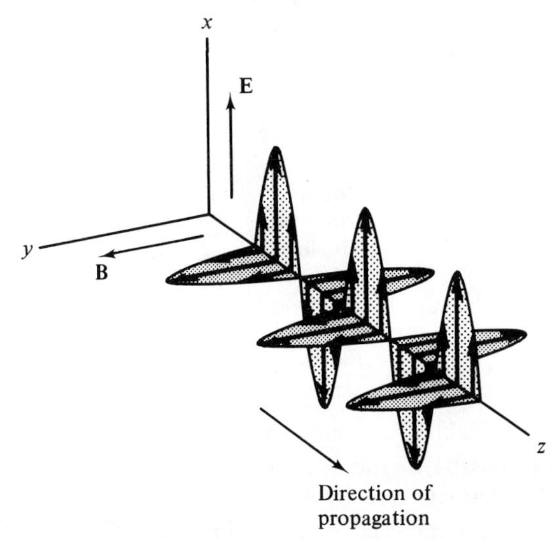

Fig. 7-2. Fields in a sinusoidal electromagnetic wave. The **E**-Field is parallel to the x axis and to the x-z plane and the **B**-Field is parallel to the y axis and to the y-z plane. The fields extend to infinity in all directions.

(say) a maximum of the field at a *fixed* point in space. The magnetic induction field **B** associated with the electric field **E** in Eq. (7-17) has the same analytic form as **E** but is directed along the y axis. In combination, the two fields constitute an electromagnetic wave, and the fields in this wave are shown at a particular instant of time in Fig. 7-2.

In brief, Maxwell's equations predict the existence of electromagnetic waves in empty space ($\rho = 0, \mathbf{J} = 0$). The speed of propagation of these waves is independent of frequency and, as given by Eq. (7-20), has the numerical value

$$c = \left(\left[(8.85418 \pm 0.00002) \times 10^{-12} \frac{C^2}{N \cdot m^2}\right]\left[4\pi \times 10^{-7} \frac{N}{A^2}\right]\right)^{-1/2}$$
$$= (2.99793 \pm 0.00001) \times 10^8 \text{ m/sec} \tag{7-23}$$

which, within experimental uncertainty, is the speed of light! (See Table 1-1.) This unexpected occurrence cannot be explained as a pure coincidence; rather we conclude that light is an electromagnetic wave. This prediction, *which united two previously distinct areas of physics* (optics and electromagnetism), was first made by Maxwell in 1861. It was not until the 1880's that electromagnetic waves at frequencies outside of the visible spectrum were first detected (by Hertz).

All of the solutions obtained in this section are referred to as *plane* waves because at every instant of time the surfaces over which the electric field has the same value are planes, in this case, perpendicular to the z axis. (The

TABLE 7-1 Names in Common Use for Regions of the Electromagnetic Spectrum

Wavelength, λ (*m*)	*Name*	*Frequency*, ν (*Hz*)
$< \approx 10^{-10}$	Gamma rays	$> \approx 3 \times 10^{18}$
$< \approx 10^{-8}$	X-rays	$> \approx 3 \times 10^{16}$
4×10^{-7} — $\approx 10^{-9}$	Ultraviolet	7×10^{14} — $\approx 3 \times 10^{17}$
7×10^{-7} — 4×10^{-7}	Visible	4×10^{14} — 7×10^{14}
$\approx 10^{-4}$ — 7×10^{-7}	Infrared	$\approx 3 \times 10^{12}$ — 4×10^{14}
$\approx 10^{0}$ — $\approx 10^{-4}$	Microwaves	$\approx 3 \times 10^{8}$ — $\approx 3 \times 10^{12}$
$\approx 10^{0}$	Television	$\approx 3 \times 10^{8}$
$\approx 10^{2}$	Radio	$\approx 3 \times 10^{6}$

magnetic induction field is, of course, also constant over these same planes.) The solutions discussed in this section are also said to be *monochromatic* because they are characterized by a definite wave number, which means in turn that they are characterized by a definite wavelength and hence by a single, well-defined *color*, where the term *color* may refer to wavelengths *outside* the visible portion of the electromagnetic spectrum. Names commonly applied to various regions of this spectrum are summarized in Table 7-1.

PROBLEMS

P7-1. Introduce the variable transformations $\xi = z + at, \eta = z - at$ into the scalar wave equation

$$\frac{\partial^2 u}{\partial z^2} = \frac{1}{a^2}\frac{\partial^2 u}{\partial t^2}$$

and from the result show that $u(z, t) = f(z - at) + g(z + at)$, where f and g are arbitrary functions. Describe qualitatively the essential features of each term in this general solution.

P7-2. Obtain the fields in Eq. (7-18) by applying the arguments in Section 7-1 to Eq. (7-8).

P7-3. Present an argument based on the periodicity of the cosine function to derive Eq. (7-22) relating T and ω.

P7-4. Show that the relationship $\omega = \kappa c$ is equivalent to the more familiar statement $\lambda\nu = c$, where λ is the wavelength and ν the frequency (in Hz) of the wave. Numerically, what is ν for visible light, $\lambda \approx 5000$ Å? $\cdots$ for microwaves, $\lambda \approx 10$ cm?

P7-5. Use Maxwell's equations to determine the magnetic induction associated with the electric field

$$\mathbf{E} = E_0[\hat{\mathbf{i}} \cos(\kappa y - \omega t) + \hat{\mathbf{k}} \sin(\kappa y - \omega t)]$$

where E_0 is a constant.

P7-6. Find the magnetic induction field associated with an electric field of the form $\mathbf{E} = f(z - ct)\hat{\mathbf{i}}$.

7-2
Energy and Momentum in Plane Waves

According to the interpretations in Sections 6-4 and 6-5, the electromagnetic field given by Eq. (7-17) carries both energy and momentum. We shall consider the energy first. For the field in Eq. (7-17), the Poynting vector, which gives the rate at which energy is transported by the fields, is given by

$$\mathbf{S} = \frac{1}{\mu_0}\mathbf{E} \times \mathbf{B} = \frac{\kappa E_{x0}^2}{\mu_0 \omega} \cos^2 (\kappa z - \omega t + \phi)\hat{\mathbf{k}} \qquad (7\text{-}24)$$

[See Eq. (6-55).] As expected, $\mathbf{S}$ is parallel to the direction of propagation of the wave. In addition, $\mathbf{S}$ fluctuates both in space and in time. Since the period of microwaves and of waves of shorter wavelength is macroscopically small ($< \approx 10^{-8}$ sec), at least in these regions of the electromagnetic spectrum typical macroscopic measurements, which extend over many, many cycles of the wave, are sensitive not to the rapid fluctuations in the rate of energy transport but to the average rate of energy transport over many cycles. Although a macroscopic time interval may not contain exactly an integral number of cycles, a fraction of a cycle added to a large number of complete cycles is insignificant, and we can average over an integral number of cycles. Because of the periodicity of Eq. (7-24), however, that average is equivalent to the average over a single cycle. Thus, we find at a fixed point in space (fixed z) that

$$\begin{aligned}\langle \mathbf{S} \rangle &= \frac{\kappa E_{x0}^2}{\mu_0 \omega} \langle \cos^2 (\kappa z - \omega t + \phi)\rangle \hat{\mathbf{k}} \\ &= \frac{\kappa E_{x0}^2}{2\mu_0 \omega}\hat{\mathbf{k}}\end{aligned} \qquad (7\text{-}25)$$

where $\langle f(t) \rangle$ denotes the average of $f(t)$ over a single cycle,

$$\langle f(t) \rangle = \frac{1}{T}\int_0^T f(t)\, dt; \qquad T = \frac{2\pi}{\omega} \qquad (7\text{-}26)$$

and we have recognized that $\langle \cos^2 (\cdots)\rangle = \frac{1}{2}$ (P7-7). Further, as shown in P7-8, the *average* rate at which energy is transported through space by this plane wave is related to the *average* energy density in the fields in such a way as to justify thinking of the energy as moving rigidly through space with the wave.

The *momentum* transported by the plane wave in Eq. (7-17) leads to another interesting result. From Eq. (6-61), we find that the momentum density in this plane wave is

$$\mathbf{\mathcal{G}} = \epsilon_0 \mathbf{E} \times \mathbf{B} = \frac{\epsilon_0 \kappa E_{x0}^2}{\omega} \cos^2 (\kappa z - \omega t + \phi)\hat{\mathbf{k}} \qquad (7\text{-}27)$$

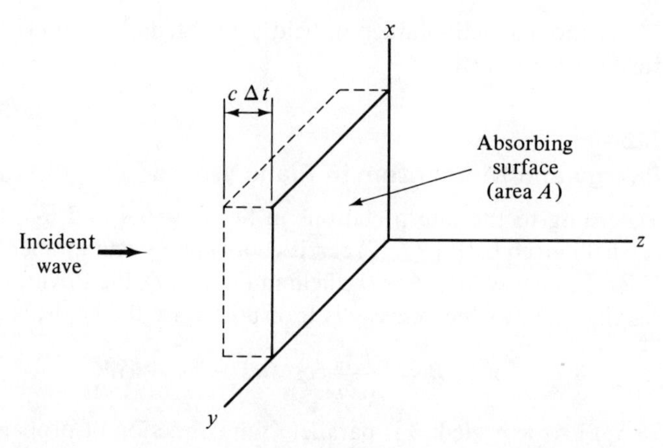

Fig. 7-3. A plane wave incident normally on an absorbing surface.

Suppose this plane wave is incident on an absorbing surface located at $z = 0$ (Fig. 7-3). Since this surface absorbs momentum from the wave, it experiences a force $\mathbf{F}$ given by

$$\mathbf{F} = \frac{\mathbf{\Delta(momentum)}}{\Delta t} \tag{7-28}$$

where Δt is the time interval over which the absorption occurs. But the change in momentum of the electromagnetic field is equal to the amount of momentum in the volume of depth $c\,\Delta t$, since in time Δt the radiation in that volume is absorbed. If Δt is small enough, the momentum density at the surface can be assumed to be applicable to the entire volume. Thus, in magnitude,

$$F = \frac{1}{\Delta t}\left[\frac{\epsilon_0 \kappa E_{x0}^2}{\omega}\cos^2(\omega t - \phi)\right][Ac\,\Delta t] \tag{7-29}$$

where A is the area of the surface and we have set $z = 0$. Defining the *radiation pressure* p_r on the surface by F/A and setting $\kappa c/\omega = 1$ [Eq. (7-20)], we find that

$$p_r = \epsilon_0 E_{x0}^2 \cos^2(\omega t - \phi) \tag{7-30}$$

On a macroscopic time scale, however, this instantaneous pressure fluctuates rapidly throughout a large portion of the electromagnetic spectrum. Thus, we can again connect macroscopic physical observations with an average over one cycle, and we find the expression

$$\langle p_r \rangle = \tfrac{1}{2}\epsilon_0 E_{x0}^2 \tag{7-31}$$

for the observed radiation pressure. (Remember that $\langle \cos^2(\cdots)\rangle = \frac{1}{2}$.) For a field $E_{x0} = 1$ V/m, this pressure is 4.4×10^{-12} N/m^2 or about 10^{-17} atmospheres.

PROBLEMS

P7-7. Show by direct integration, as in Eq. (7-26), that $\langle \cos^2 (\kappa z - \omega t + \phi) \rangle = \frac{1}{2}$ when κ, z, ω, and ϕ are constants.

P7-8. Consider the wave in Eq. (7-17). (a) Show that $\langle u_E \rangle = \langle u_B \rangle$, where u_E and u_B are the electric and magnetic energy densities, respectively. (b) Show that $|\langle \mathbf{S} \rangle| = c[\langle u_E \rangle + \langle u_B \rangle]$, and interpret this equation physically. (c) Express the radiation pressure, Eq. (7-31), in terms of the average energy density $\langle u_{EM} \rangle = \langle u_E \rangle + \langle u_B \rangle$.

P7-9. The light from the sun delivers about 1300 W/m^2 to the surface of the earth. Assume the radiation to be a plane wave described by Eq. (7-17) and let it be incident normally on the earth's surface. (a) Calculate the amplitude of the electric field in this wave and the resulting radiation pressure on the earth's surface. *Hint:* Neither of these quantities depends specifically on the wavelength of the light; see Eq. (7-20). (b) The energy of a single photon of frequency ν is $h\nu$, where h is Planck's constant. Assume sunlight to have the wavelength 5500 Å and calculate the number of photons incident per second on a square meter of the earth's surface.

P7-10. Let the radiation impinging at normal incidence on a wall be reflected rather than absorbed. Determine the radiation pressure on the wall.

P7-11. (a) Determine the radiation pressure experienced by a perfectly reflecting plane surface bathed in monochromatic light whose incident direction makes an angle θ with the normal to the surface. Express your answer in terms of the average energy density $\langle u_{EM} \rangle$ in the incident wave. (b) Determine the pressure experienced by the same surface bathed in monochromatic light incident isotropically from all directions on one side of the surface. *Hint:* See F. K. Richtmyer, E. H. Kennard, and T. Lauritsen, *Introduction to Modern Physics* (McGraw-Hill Book Company, New York, 1955), Fifth Edition, Chapter 4.

7-3
Superposition of Waves of the Same Frequency: Polarization and Interference

The electromagnetic wave is a *transverse* wave, which means that the oscillating electric and magnetic fields composing the wave are perpendicular to the direction of propagation. Since there are two linearly independent directions perpendicular to a given direction, there are two distinct plane waves with the same direction of propagation. These two waves are said to be in different states of *polarization*, the type of polarization being conventionally classified by the behavior of the *electric* field in the wave. A *linearly* polarized wave, for example, has an electric field that is always directed parallel to some fixed line in space. The wave in Eq. (7-17) is therefore linearly polarized in the x direction.

More general states of polarization can be constructed by superposing two waves having the *same* frequency [and hence the *same* wave number; see Eq. (7-14)] but polarized at right angles to one another. The resulting electric field, which determines the state of polarization, is obtained by setting K and Ω in Eq. (7-18) equal to κ and ω and then adding the fields to those in Eq. (7-17); we find, for example, that

$$\mathbf{E} = E_{x0} \cos(\kappa z - \omega t)\hat{\mathbf{i}} + E_{y0} \cos(\kappa z - \omega t + \Phi)\hat{\mathbf{j}} \tag{7-32}$$

where, for a simpler discussion, ϕ has been taken equal to zero. (Only the phase *difference* between the x- and y-components is important.) To determine the state of polarization of the wave in Eq. (7-32), we examine the field in the plane $z = 0$, where

$$\mathbf{E} = E_{x0} \cos \omega t\,\hat{\mathbf{i}} + E_{y0} \cos(\omega t - \Phi)\hat{\mathbf{j}} \tag{7-33}$$

As a first example, let $\Phi = 0$. The two components of the wave are then said to be *in phase*, and

$$\mathbf{E} = (E_{x0}\hat{\mathbf{i}} + E_{y0}\hat{\mathbf{j}}) \cos \omega t \tag{7-34}$$

Thus, as viewed from a point on the positive z axis, the tip of the electric field vector in the x-y plane moves back and forth along a line having the direction of the *constant* vector $E_{x0}\hat{\mathbf{i}} + E_{y0}\hat{\mathbf{j}}$. This line makes an angle θ satisfying

$$\tan \theta = \frac{E_{y0}}{E_{x0}} \tag{7-35}$$

with the x axis (Fig. 7-4). The wave described by Eq. (7-32) when $\Phi = 0$ is a *linearly polarized* plane wave, polarized in the direction shown. When $\Phi = \pi$, the two components of the wave are 180° *out of phase*, and the composite wave is also linearly polarized, but in the direction making an angle $-\theta$ with the x axis (P7-12).

A more interesting case occurs when the two components of Eq. (7-32)

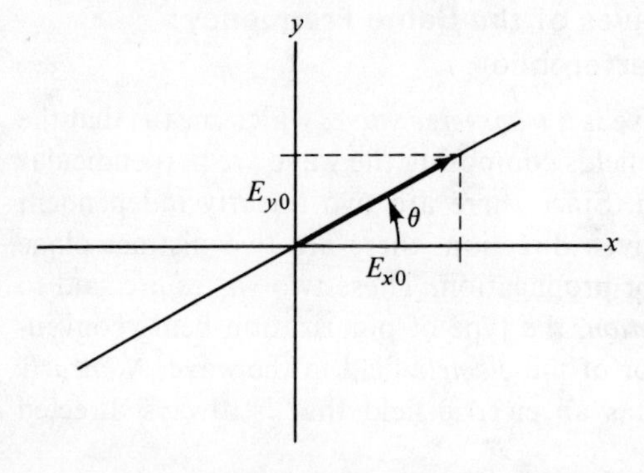

Fig. 7-4. Direction of the **E**-field in a plane wave linearly polarized at an angle θ to the x axis. The **E**-field oscillates in magnitude but is always directed one way or the other along the line shown. Its maximum amplitude is given by the vector $E_{x0}\hat{\mathbf{i}} + E_{y0}\hat{\mathbf{j}}$.

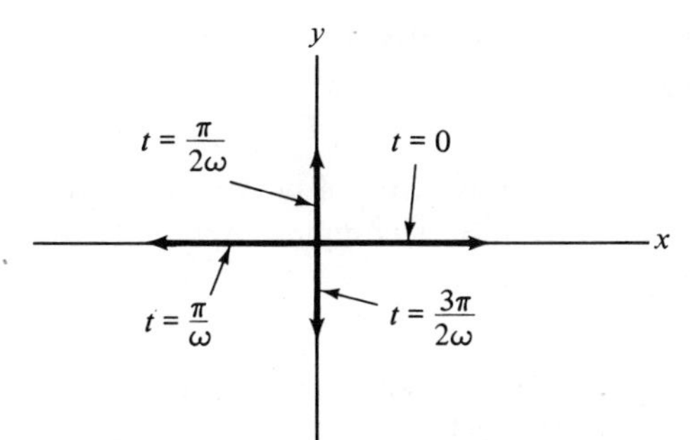

Fig. 7-5. Successive values of **E** for the elliptically polarized wave of Eq. (7-36). The vectors along the x axis have magnitude E_{x0} and those along the y axis have magnitude E_{y0}.

are 90° out of phase, $\Phi = \frac{1}{2}\pi$. In this case,

$$\mathbf{E} = E_{x0} \cos \omega t \hat{\mathbf{i}} + E_{y0} \sin \omega t \hat{\mathbf{j}} \tag{7-36}$$

More specifically the electric field is given at successive times by

$$\begin{aligned} \mathbf{E}(0) &= E_{x0} \hat{\mathbf{i}} \\ \mathbf{E}\left(\frac{\pi}{2\omega}\right) &= E_{y0} \hat{\mathbf{j}} \\ \mathbf{E}\left(\frac{\pi}{\omega}\right) &= -E_{x0} \hat{\mathbf{i}} \\ \mathbf{E}\left(\frac{3\pi}{2\omega}\right) &= -E_{y0} \hat{\mathbf{j}} \end{aligned} \tag{7-37}$$

From the successive positions of **E** shown in Fig. 7-5 we conclude that the **E**-vector rotates counterclockwise as seen from a point on the positive z axis. If we eliminate t from the two equations, $E_x(t) = E_{x0} \cos \omega t$ and $E_y(t) = E_{y0} \sin \omega t$, we find that

$$\left(\frac{E_x}{E_{x0}}\right)^2 + \left(\frac{E_y}{E_{y0}}\right)^2 = 1 \tag{7-38}$$

which is the equation of an ellipse with semiaxes E_{x0} and E_{y0}. Thus, the tip of the **E**-vector in this case traces out an ellipse and the wave is said to be *left elliptically polarized*, the adjective *left* designating the counterclockwise direction of rotation. Sometimes the phrase *positive helicity* is used to describe this direction of the polarization. By similar arguments (P7-13) one can show that when $\Phi = 3\pi/2$ the electric field rotates clockwise, its tip again tracing out an ellipse; the result is a *right elliptically polarized* wave, sometimes said to have *negative helicity*.

If $E_{x0} = E_{y0}$, then Eq. (7-38) is the equation of a circle and we have a special case of elliptic polarization called *circular polarization*. A *left* circularly polarized wave has an electric field given in the plane $z = 0$ by

$$\mathbf{E} = E_{x0}(\cos \omega t \hat{\mathbf{i}} + \sin \omega t \hat{\mathbf{j}}) \tag{7-39}$$

and a *right* circularly polarized wave is characterized by an electric field given by

$$\mathbf{E} = E_{x0}(\cos \omega t\hat{\mathbf{i}} - \sin \omega t\hat{\mathbf{j}}) \tag{7-40}$$

The various states of polarization are more easily related to one another if we introduce a complex representation for the fields.[1] Recognizing the Euler identity

$$e^{i\theta} = \cos\theta + i \sin\theta \tag{7-41}$$

[Eq. (D-17)], we can then replace the fields in Eq. (7-17), for example, with the complex fields

$$\boldsymbol{\mathcal{E}} = \mathcal{E}_{x0}e^{i(\kappa z - \omega t)}\hat{\mathbf{i}}, \qquad \boldsymbol{\mathcal{B}} = \frac{\kappa}{\omega}\mathcal{E}_{x0}e^{i(\kappa z - \omega t)}\hat{\mathbf{j}} \tag{7-42}$$

where the complex *amplitude* $\mathcal{E}_{x0}$ is defined by

$$\mathcal{E}_{x0} = E_{x0}e^{i\phi} \tag{7-43}$$

Whenever we need to, we can recover the physical fields **E** and **B** by taking the real part of the complex fields, i.e.,

$$\mathbf{E} = \operatorname{Re} \boldsymbol{\mathcal{E}}, \qquad \mathbf{B} = \operatorname{Re} \boldsymbol{\mathcal{B}} \tag{7-44}$$

A similar pair of complex fields can be introduced to represent the physical fields in Eq. (7-18). Since the complex fields corresponding to two or more physical fields can be added to give the complex field corresponding to the superposition of the physical fields [$\operatorname{Re}(z_1 + z_2) = \operatorname{Re} z_1 + \operatorname{Re} z_2$], we then find that the complex electric fields for the basic states of polarization have the form

$$\begin{aligned}
&\text{linearly polarized, } x \text{ direction:} && \boldsymbol{\mathcal{E}} = E_0\hat{\mathbf{i}}e^{i(\kappa z - \omega t)} \\
&\text{linearly polarized, } y \text{ direction:} && \boldsymbol{\mathcal{E}} = E_0\hat{\mathbf{j}}e^{i(\kappa z - \omega t)} \\
&\text{right circularly polarized:} && \boldsymbol{\mathcal{E}} = E_0(\hat{\mathbf{i}} - i\hat{\mathbf{j}})e^{i(\kappa z - \omega t)} \\
&\text{left circularly polarized:} && \boldsymbol{\mathcal{E}} = E_0(\hat{\mathbf{i}} + i\hat{\mathbf{j}})e^{i(\kappa z - \omega t)}
\end{aligned} \tag{7-45}$$

The complex fields in Eq. (7-42) also facilitate examining the effect of superposing two or more waves having the *same* frequency and the *same* polarization but different amplitudes and phases. Let the sth such wave be represented by the complex field

$$\begin{aligned}
\boldsymbol{\mathcal{E}}^{(s)} &= E_0^{(s)}e^{i\phi_s}e^{i(\kappa z - \omega t)}\hat{\mathbf{i}} \\
\boldsymbol{\mathcal{B}}^{(s)} &= \frac{\kappa}{\omega}E_0^{(s)}e^{i\phi_s}e^{i(\kappa z - \omega t)}\hat{\mathbf{j}}
\end{aligned} \tag{7-46}$$

where $E_0^{(s)}$ is a *real* number. The complex fields $\boldsymbol{\mathcal{E}}$, $\boldsymbol{\mathcal{B}}$ representing the superposition then are

$$\boldsymbol{\mathcal{E}} = \mathcal{E}_0e^{i(\kappa z - \omega t)}\hat{\mathbf{i}}, \qquad \boldsymbol{\mathcal{B}} = \frac{\kappa}{\omega}\mathcal{E}_0e^{i(\kappa z - \omega t)}\hat{\mathbf{j}} \tag{7-47}$$

[1]The properties of complex numbers are summarized in Appendix D.

where

$$\mathcal{E}_0 = \sum_s E_0^{(s)} e^{i\phi_s} \tag{7-48}$$

We ask now for the rate at which energy is transported by this combined wave. The answer is given by the time average of the Poynting vector (Why?), but we must evaluate this quantity carefully. Its definition involves a *product* of the physical fields, and we cannot multiply the complex fields and then extract a real part; we must take the real parts first [$\text{Re}\,(z_1 z_2) \neq (\text{Re}\,z_1)(\text{Re}\,z_2)$]. Thus, we must proceed as follows:

$$\begin{aligned}\langle \mathbf{S} \rangle &= \frac{1}{\mu_0}\langle \mathbf{E} \times \mathbf{B} \rangle = \frac{1}{\mu_0}\langle (\text{Re}\,\boldsymbol{\mathcal{E}}) \times (\text{Re}\,\boldsymbol{\mathcal{B}}) \rangle \\ &= \frac{1}{4\mu_0}\langle (\boldsymbol{\mathcal{E}} + \boldsymbol{\mathcal{E}}^*) \times (\boldsymbol{\mathcal{B}} + \boldsymbol{\mathcal{B}}^*) \rangle \quad \text{[see Eq. (D-10)]} \\ &= \frac{1}{4\mu_0}[\langle \boldsymbol{\mathcal{E}} \times \boldsymbol{\mathcal{B}} \rangle + \langle \boldsymbol{\mathcal{E}} \times \boldsymbol{\mathcal{B}}^* \rangle + \langle \boldsymbol{\mathcal{E}}^* \times \boldsymbol{\mathcal{B}} \rangle + \langle \boldsymbol{\mathcal{E}}^* \times \boldsymbol{\mathcal{B}}^* \rangle]\end{aligned} \tag{7-49}$$

Now with $\boldsymbol{\mathcal{E}}$ and $\boldsymbol{\mathcal{B}}$ in the form of Eq. (7-47), the time dependence of $\boldsymbol{\mathcal{E}} \times \boldsymbol{\mathcal{B}}$, lies in a factor $e^{-2i\omega t}$ and $\langle \boldsymbol{\mathcal{E}} \times \boldsymbol{\mathcal{B}} \rangle = 0$. Similarly, $\boldsymbol{\mathcal{E}}^* \times \boldsymbol{\mathcal{B}}^*$ involves a factor $e^{2i\omega t}$ and $\langle \boldsymbol{\mathcal{E}}^* \times \boldsymbol{\mathcal{B}}^* \rangle = 0$. The terms $\boldsymbol{\mathcal{E}} \times \boldsymbol{\mathcal{B}}^*$ and $\boldsymbol{\mathcal{E}}^* \times \boldsymbol{\mathcal{B}}$, however, contain no explicit time dependence. Each is therefore equal to its own time average and we have from Eq. (7-49) that

$$\begin{aligned}\langle \mathbf{S} \rangle &= \frac{1}{4\mu_0}[\boldsymbol{\mathcal{E}} \times \boldsymbol{\mathcal{B}}^* + \boldsymbol{\mathcal{E}}^* \times \boldsymbol{\mathcal{B}}] \\ &= \frac{1}{2\mu_0}\,\text{Re}\,(\boldsymbol{\mathcal{E}} \times \boldsymbol{\mathcal{B}}^*) = \frac{1}{2\mu_0}\,\text{Re}\,(\boldsymbol{\mathcal{E}}^* \times \boldsymbol{\mathcal{B}})\end{aligned} \tag{7-50}$$

[See Eq. (D-10)]. Finally, we obtain

$$\langle \mathbf{S} \rangle = \frac{\kappa}{2\mu_0\omega}|\mathcal{E}_0|^2\hat{\mathbf{k}} \tag{7-51}$$

$$= \frac{\kappa}{2\mu_0\omega}\left|\sum_s E_0^{(s)} e^{i\phi_s}\right|^2\hat{\mathbf{k}} \tag{7-52}$$

for the average Poynting vector of the fields in Eq. (7-47). Its magnitude depends on the amplitudes and the phases of the component waves but is *not* just the sum of independent contributions from these components—that would be given by $(\kappa/2\mu_0\omega)\sum_s [E_0^{(s)}]^2$ (Why?)—so the individual components interact with one another to determine the rate at which energy is transported by the total wave. This phenomenon of *interference* between two or more waves that are simultaneously present is a characteristic property of waves and is responsible for the useful functioning of such diverse apparatus as grating spectrometers and arrays of antennas for radio and television broadcasting.

As a more specific example of the consequences of Eq. (7-52), suppose there are N waves, all having the *same* amplitude $E_0^{(s)} = E_0$. Further, let the

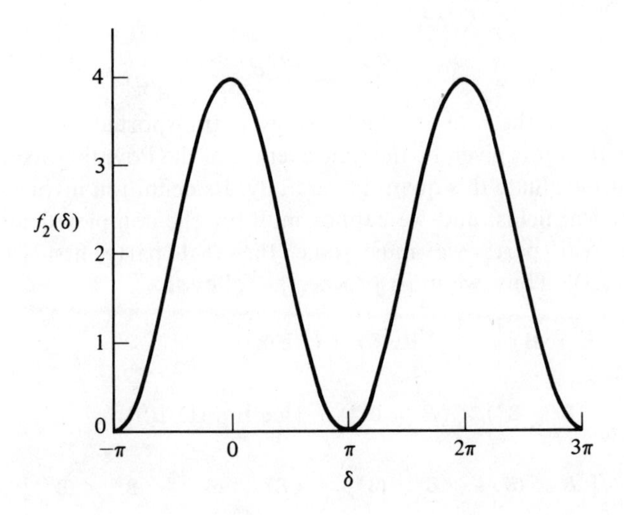

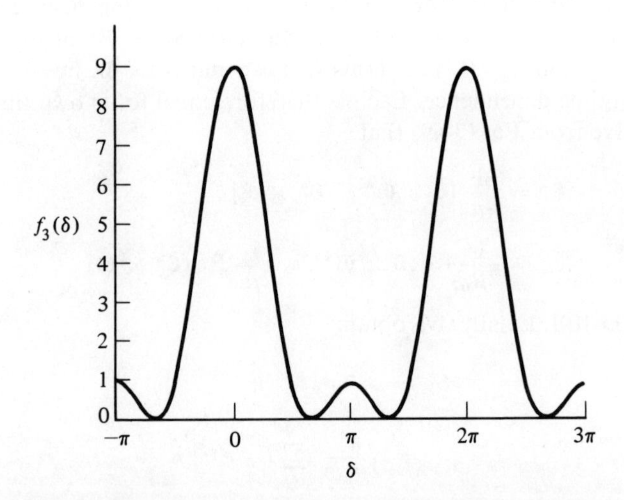

Fig. 7-6. Graphs of $f_N(\delta)$ versus δ for selected values of N (*above, and on opposite page*). Note particularly that the vertical scales are different for each graph.

phase of the sth wave differ from that of the $(s-1)$st wave by an amount δ that is independent of s, so that $\phi_s = (s-1)\delta$ if we set $\phi_1 = 0$. In this case, Eq. (7-52) becomes

$$\langle \mathbf{S} \rangle = \frac{\kappa E_0^2}{2\mu_0 \omega} f_N(\delta)\hat{\mathbf{k}} \tag{7-53}$$

where $f_N(\delta)$ is defined by

$$f_N(\delta) = \left| \sum_{s=1}^{N} e^{i(s-1)\delta} \right|^2 \tag{7-54}$$

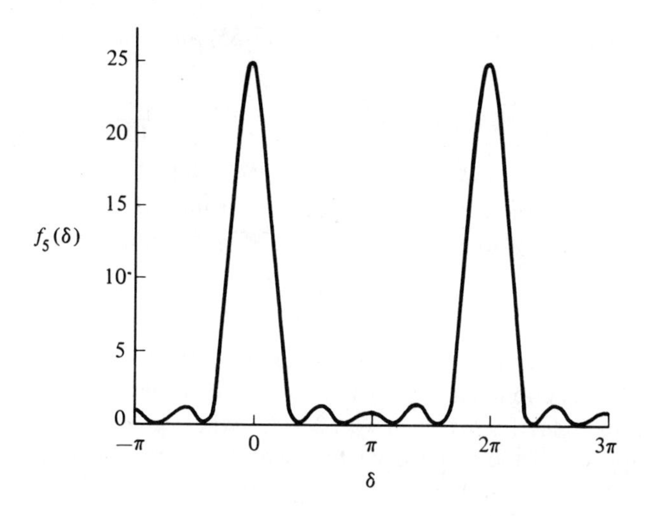

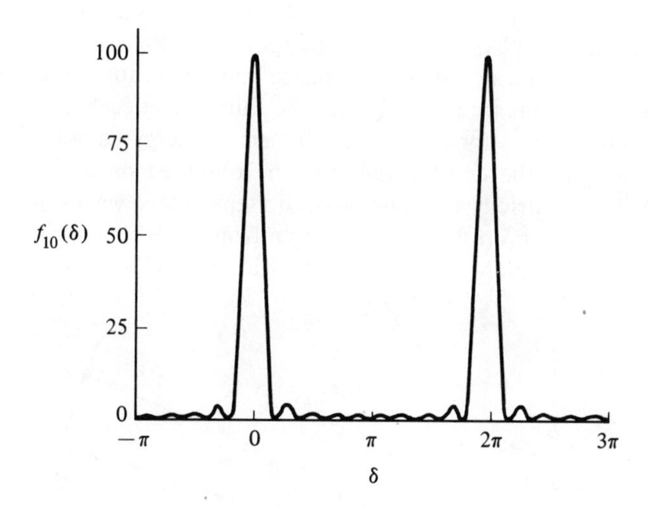

Apart from a constant factor, $f_N(\delta)$ gives the rate at which this composite wave transports energy, expressing that rate as a function of the phase difference δ between consecutive individual waves. The sum in Eq. (7-54), however, can by evaluated analytically. Since

$$\sum_{s=1}^{N} e^{i(s-1)\delta} = 1 + e^{i\delta} + \cdots + e^{i(N-1)\delta} \tag{7-55}$$

we find that

$$e^{i\delta} \sum_{s=1}^{N} e^{i(s-1)\delta} = e^{i\delta} + \cdots + e^{iN\delta} \tag{7-56}$$

Thus, on subtracting Eq. (7-56) from Eq. (7-55) and rearranging the result, we obtain

$$\sum_{s=1}^{N} e^{i(s-1)\delta} = \frac{1 - e^{iN\delta}}{1 - e^{i\delta}} = e^{i(N-1)\delta/2}\,\frac{\sin(N\delta/2)}{\sin(\delta/2)} \tag{7-57}$$

where the last form follows after factoring $e^{iN\delta/2}$ and $e^{i\delta/2}$ out of the numerator and denominator, respectively, and then using the results of PD-12 in Appendix D. Finally, on substituting Eq. (7-57) into Eq. (7-54), we find that

$$f_N(\delta) = \frac{\sin^2(N\delta/2)}{\sin^2(\delta/2)} \tag{7-58}$$

Graphs of this function for several values of N are shown in Fig. 7-6. For every N the function has *principal maxima* of height N^2 when $\delta = 2n\pi$, where n is an integer, *zeros* when $\delta = 2m\pi/N$, where m is an integer not equal to a multiple of N, and *secondary maxima* at $N - 2$ points between adjacent principal maxima. As N becomes larger, the principal maxima become narrower (P7-17) and the secondary maxima become less significant.

To make the example of the previous paragraph even more meaningful, let us place it in a physical context. Suppose we have a linear array of N equally spaced sources oscillating in phase and all emitting electromagnetic radiation of the same frequency (Fig. 7-7). Further, let each source have the same amplitude and suppose each source emits energy equally in all directions. Finally, let the emitted radiation be observed on a *distant* viewing screen. Although strictly the sources do not emit plane waves, the radiation reaching the point P on the viewing screen from each source can be treated approximately as a plane wave propagating along the line from the source to

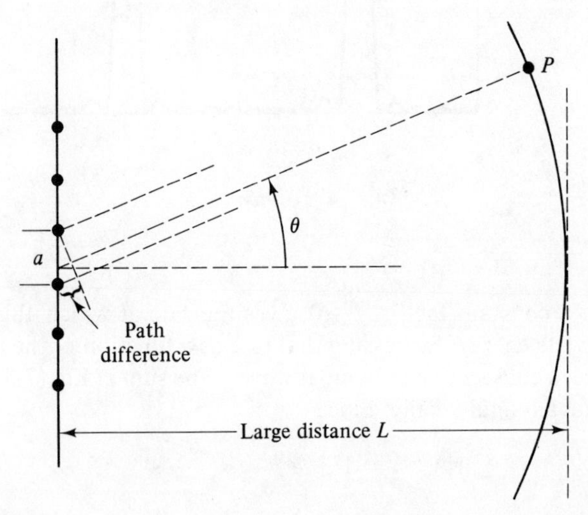

Fig. 7-7. A linear array of sources radiating to a distant viewing screen.

P. Because of the assumed distance of the viewing screen from the sources, however, lines from all of the sources to P are nearly parallel and all waves reaching P are therefore coming from essentially the same direction. Further, these waves have all traveled essentially the same distance from their respective sources, so any attenuation in amplitude will affect all of them nearly equally and the separate waves as they arrive at P will have nearly the same amplitude. Indeed, again because of the distance from the sources to the viewing screen, the amplitude of the individual waves is nearly independent of the position of P on the screen. Finally, the phase difference between the waves arriving at P from adjacent sources is determined by the difference in length between the two paths from P to the sources, a difference of one wavelength λ corresponding to a phase difference of 2π. (Why?) Since λ is typically small, we cannot here ignore the small differences in these paths as we have done earlier in the paragraph. When a line from the sources—any source will do (Why?)—to P makes an angle θ with a normal to the line of sources, the path difference for waves from adjacent sources is $a \sin \delta$, where a is the separation of the sources. This path difference corresponds to a phase difference δ given by

$$\delta = 2\pi\left(\frac{a \sin \theta}{\lambda}\right) = \kappa a \sin \theta \qquad (7\text{-}59)$$

which is the same for any two adjacent sources. Thus, with the various approximations made in this paragraph, the situation at P in Fig. 7-7 exactly duplicates the situation to which Eq. (7-58) applies, provided we identify δ as in Eq. (7-59), and, apart from the distortion of the horizontal scale caused by the nonlinear relationship between δ and θ, the graphs in Fig. 7-6 show the intensity of the radiation received at P as a function of the position of P on the viewing screen. Some additional aspects of these radiation patterns are explored in P7-18, P7-19, and P7-33.

PROBLEMS

P7-12. Show that when $\Phi = \pi$, Eq. (7-32) represents the electric field in a linearly polarized wave and determine the angle α between the direction of polarization and the direction of increasing x.

P7-13. Show that when $\Phi = 3\pi/2$, Eq. (7-32) represents the electric field in a right elliptically polarized wave.

P7-14. Show that for general Φ, Eq. (7-32) represents the electric field in an elliptically polarized wave and determine the angles that the axes of the ellipse make with the direction of increasing x.

P7-15. Show that the *imaginary* part of Eq. (7-42) satisfies Maxwell's equations with $\rho = 0$ and $\mathbf{J} = 0$. Thus, we can take either the real or the imaginary part of the complex field as a physical field.

P7-16. Verify the third and fourth members of Eq. (7-45).

P7-17. (a) Defining the width Δ of the principal maximum as the distance (in δ) from the peak to the nearest zero, find Δ in terms of N for the pattern given by Eq. (7-58). (b) *Estimate* the height of the secondary maximum immediately adjacent to the principal maximum and compare this height to that of the principal maximum. *Hint:* (1) *Approximately* for what value of δ does this secondary maximum occur? (2) Assume N fairly large.

P7-18. (a) Show that $f_2(\delta) = 4\cos^2(\delta/2) = 2(1 + \cos\delta)$. (b) With δ as given by Eq. (7-59), draw polar graphs of f_2 as a function of θ for $a/\lambda = .5, 1, 2$, and other values if necessary. Consider only $-\frac{1}{2}\pi \leq \theta \leq \frac{1}{2}\pi$. (c) Describe in words the effect of changes in a/λ on the pattern.

P7-19. (a) Show that $f_4(\delta) = 8(1 + \cos\delta)\cos^2\delta$. (b) Obtain a careful graph of f_4 versus δ, find the positions of the two secondary maxima to three significant figures, and determine the height of these maxima. *Hint:* Use a desk calculator or computer. (c) With δ as given by Eq. (7-59), draw polar graphs of f_4 as a function of θ for selected values of a/λ and describe in words the effect of changes in a/λ on the pattern. Consider only $-\frac{1}{2}\pi \leq \theta \leq \frac{1}{2}\pi$. *Hint:* Again, use a desk calculator or computer.

P7-20. Suppose a very large number N of sources contribute waves whose amplitudes at the observation point fall off so that $E_0^{(s)} = r^{s-1}E_0^{(1)}$, where $0 < r < 1$, and whose phases satisfy $\phi_s = (s-1)\delta$. Show from Eq. (7-52) that the rate of energy transport is proportional to

$$f'_N(\delta) = \left|\frac{1 - r^N e^{iN\delta}}{1 - re^{i\delta}}\right|^2 \xrightarrow[N\to\infty]{} \frac{1}{1 - 2r\cos\delta + r^2}$$

and examine the latter form as a function of δ for various r. Can you fit this model into a physical context? *Hint:* Use a computer.

7-4 Superposition of Waves of Different Frequencies; Spectral Decomposition

The complex fields also facilitate a study of superposition when the frequencies of the superposed waves are different. We shall consider here only the superposition of waves having the *same linear* polarization. Since the solutions expressed in Eq. (7-42) are valid for any $\mathcal{E}_{x0}$ and for any κ, provided $\omega = \kappa c$, we can build more general solutions by superposing these simpler solutions for a discrete *spectrum* of values $\{\mathcal{E}_{x0}(\kappa_i), \kappa_i, \omega_i; i = 1, 2, \ldots\}$, or, more usefully, by superposing simpler solutions for a continuous spectrum of κ by integration, specifically

$$\mathcal{E}(z, t) = \hat{\mathbf{i}} \int_0^\infty \mathcal{E}_{x0}(\kappa) e^{i(\kappa z - \omega t)}\, d\kappa \tag{7-60}$$

$$\mathcal{B}(z, t) = \frac{1}{c}\hat{\mathbf{j}} \int_0^\infty \mathcal{E}_{x0}(\kappa) e^{i(\kappa z - \omega t)}\, d\kappa \tag{7-61}$$

where ω is a function of κ ($\omega = \kappa c$) and $\mathcal{E}_{x0}(\kappa)$ is an arbitrary (complex) function of the (real) variable κ. We shall need also expressions for the physical fields. For the physical **E**-field, we obtain

$$\begin{aligned}\mathbf{E} &= \text{Re}\,\boldsymbol{\mathcal{E}} = \tfrac{1}{2}(\boldsymbol{\mathcal{E}} + \boldsymbol{\mathcal{E}}^*)\\ &= \tfrac{1}{2}\hat{\mathbf{i}}\int_0^\infty [\mathcal{E}_{x0}(\kappa)e^{i(\kappa z-\omega t)} + \mathcal{E}_{x0}^*(\kappa)e^{-i(\kappa z-\omega t)}]\,d\kappa\\ &= \tfrac{1}{2}\hat{\mathbf{i}}\left[\int_0^\infty \mathcal{E}_{x0}(\kappa)e^{i(\kappa z-\omega t)}\,d\kappa + \int_{-\infty}^0 \mathcal{E}_{x0}^*(-\kappa)e^{i(\kappa z-\omega t)}\,d\kappa\right]\\ &= \hat{\mathbf{i}}\int_{-\infty}^\infty A(\kappa)e^{i(\kappa z-\omega t)}\,\frac{d\kappa}{2\pi}\end{aligned} \tag{7-62}$$

where the *spectral function* $A(\kappa)$, which relates to how much of a particular wave number κ is present in the general field, is defined by

$$A(\kappa) = \begin{cases}\pi\mathcal{E}_{x0}(\kappa), & \kappa > 0\\ \pi\mathcal{E}_{x0}^*(-\kappa), & \kappa < 0\end{cases} \tag{7-63}$$

This extension of κ into the physically meaningless region $\kappa < 0$ should be noted. Mathematically this extension is useful, but it cannot be allowed to confuse any subsequent physical interpretations. A similar argument applied to Eq. (7-61) gives

$$\mathbf{B}(z, t) = \frac{1}{c}\hat{\mathbf{j}}\int_{-\infty}^\infty A(\kappa)e^{i(\kappa z-\omega t)}\,\frac{d\kappa}{2\pi} \tag{7-64}$$

for the physical **B**-field. The property

$$A(\kappa)^* = A(-\kappa) \tag{7-65}$$

which follows directly from Eq. (7-63), assures that both $\mathbf{E}(z, t)$ and $\mathbf{B}(z, t)$ as given by Eqs. (7-62) and (7-64) will in fact be real (P7-21). Further, the spectral function corresponding to a plane wave whose **E**-field is *known* at some initial time $t = 0$ can now be determined by Fourier inversion of Eq. (7-62); we find that

$$A(\kappa) = \int_{-\infty}^\infty E_x(z, 0)e^{-i\kappa z}\,dz \tag{7-66}$$

(See Appendix D.) Since $A(\kappa)$ in Eq. (7-66) is well-defined even if $E_x(z, 0) = 0$ except in some small range of z, we have now made it possible to express general plane waves that may be nonzero only in some finite range of z. Further, we have found expressions for calculating the physical fields from the spectral function [Eqs. (7-62) and (7-64)] and also for calculating the spectral function from the fields [Eq. (7-66)]. The formalism for describing these general superpositions is thus complete.

We shall now illustrate the use of this formalism to calculate the spectral distribution of *energy* in this general, linearly polarized, plane wave. Let us first calculate the total energy W transported across a unit surface oriented with its plane perpendicular to the direction of propagation; we find

$$W = \int_{-\infty}^{\infty} |\mathbf{S}|\, dt = \int_{-\infty}^{\infty} \left| \frac{1}{\mu_0}(\mathbf{E} \times \mathbf{B}) \right| dt$$

$$= \int_{-\infty}^{\infty} dt \left(\frac{1}{\mu_0} \int_{-\infty}^{\infty} \frac{d\kappa}{2\pi} A(\kappa) e^{i(\kappa z - \omega t)} \right) \left(\frac{1}{c} \int_{-\infty}^{\infty} A(\kappa') e^{i(\kappa' z - \omega' t)} \frac{d\kappa'}{2\pi} \right) \qquad (7\text{-}67)$$

Now, for the sake of the final result, let us change the sign of κ' and recognize Eq. (7-65) to obtain

$$W = \frac{1}{\mu_0 c} \int_{-\infty}^{\infty} dt \int_{-\infty}^{\infty} \frac{d\kappa}{2\pi} \int_{-\infty}^{\infty} \frac{d\kappa'}{2\pi} A(\kappa) A(\kappa')^* e^{i[(\kappa - \kappa')z - (\omega - \omega')t]} \qquad (7\text{-}68)$$

For simplicity in the manipulation, let the origin of the coordinate system be taken in the surface across which the energy flux is computed. Then $z = 0$ and, since $\omega = \kappa c$ and $\omega' = \kappa' c$, Eq. (7-68) becomes

$$W = \frac{1}{\mu_0 c^2} \int_{-\infty}^{\infty} d\xi \int_{-\infty}^{\infty} \frac{d\kappa}{2\pi} \int_{-\infty}^{\infty} \frac{d\kappa'}{2\pi} A(\kappa) A(\kappa')^* e^{-i(\kappa - \kappa')\xi} \qquad (7\text{-}69)$$

where $\xi = ct$. Now, let us write this expression in a form to which the Fourier integral theorem (Appendix D) can be readily applied:

$$W = \frac{1}{\mu_0 c^2} \int_{-\infty}^{\infty} \frac{d\kappa}{2\pi} A(\kappa) \left[\int_{-\infty}^{\infty} d\xi e^{-i\kappa\xi} \left(\int_{-\infty}^{\infty} \frac{d\kappa'}{2\pi} A(\kappa')^* e^{i\kappa'\xi} \right) \right] \qquad (7\text{-}70)$$

In this form, the innermost integral can be interpreted as giving that function of ξ whose Fourier transform is $A(\kappa')^*$. But then the middle integral expresses the Fourier transform of that function. Together, the integrals on κ' and ξ merely produce $A(\kappa)^*$! [See Eqs. (D-30) and (D-31).] Thus,

$$W = \frac{1}{\mu_0 c^2} \int_{-\infty}^{\infty} |A(\kappa)|^2 \frac{d\kappa}{2\pi} \qquad (7\text{-}71)$$

To extract a physical interpretation, however, we must remove all reference to the unphysical negative wave numbers. From Eq. (7-65) it follows that $|A(\kappa)|^2 = |A(-\kappa)|^2$ (Why?); thus, $|A(\kappa)|^2$ is an *even* function of κ and Eq. (7-71) may be written alternatively as

$$W = \frac{1}{\pi \mu_0 c^2} \int_0^{\infty} |A(\kappa)|^2 \, d\kappa \qquad (7\text{-}72)$$

from which it is now natural to interpret the integrand as giving the distribution $u(\kappa)$ of energy over all wave numbers,

$$u(\kappa) = \frac{1}{\pi \mu_0 c^2} |A(\kappa)|^2 \qquad (7\text{-}73)$$

In essence, the function $u(\kappa)$ predicts the reading of a spectral analyzer at wave number κ when a wave defined by the spectral function $A(\kappa)$ is incident on the instrument. The transformation of Eq. (7-73) into expressions for the distribution of energy over frequency ω and over wavelength λ is the topic of P7-23.

PROBLEMS

P7-21. Verify that the condition $A(\kappa)^* = A(-\kappa)$ assures that the fields given by Eqs. (7-62) and (7-64) will be real.

P7-22. (a) Suppose $\mathbf{E}(0, t)$ is known. Find an expression for $A(\kappa)$. (b) Let $\mathbf{E}(0, t) = E_0\hat{\mathbf{i}}$, $-T \leq t \leq T$, and $\mathbf{E}(0, t) = 0$ outside that time interval. Find $A(\kappa)$ and $u(\kappa)$ and sketch a graph of the latter as a function of κ. (c) Show that the width of $A(\kappa)$ in κ is inversely proportional to T. This property is very closely related to the Heisenberg uncertainty relations in quantum mechanics. *Optional:* Find and describe the **B**-field associated with the **E**-field in part (b). *Hint:* In terms of the signum function sgn(α) defined by $\text{sgn}(\alpha) = 1$, $\alpha > 0$; $= 0$, $\alpha = 0$; $= -1$, $\alpha < 0$, note the integral

$$\int_{-\infty}^{\infty} \frac{\sin \alpha x}{x}\, dx = \pi\, \text{sgn}(\alpha)$$

P7-23. Starting with Eq. (7-72), obtain an expression giving the distribution of energy (a) in frequency ω and (b) in wavelength λ.

7-5 Plane Waves in Three Dimensions

A three-dimensional plane wave solution to Maxwell's equations can be inferred from the one-dimensional solution already obtained. Instead of a coordinate z, which measures the displacement of the general point $\mathbf{r}$ from the x-y plane, we should have a coordinate measuring the displacement of the point $\mathbf{r}$ from some more general plane whose normal vector is, say, $\hat{\mathbf{n}}$. Thus, we expect that z, which is equivalent to $\hat{\mathbf{k}}\cdot\mathbf{r}$, should be replaced by $\hat{\mathbf{n}}\cdot\mathbf{r}$, and further that the vector $\hat{\mathbf{n}}$ (as did $\hat{\mathbf{k}}$ before) should coincide with the direction of propagation of the wave. We are thus led to guess a more general solution of the form

$$\begin{aligned} \mathcal{E}(\mathbf{r}, t) &= \mathbf{E}_0 e^{i(\boldsymbol{\kappa}\cdot\mathbf{r}-\omega t)} \\ \mathcal{B}(r, t) &= \mathbf{B}_0 e^{i(\boldsymbol{\kappa}\cdot\mathbf{r}-\omega t)} \end{aligned} \tag{7-74}$$

where $\mathbf{E}_0$ and $\mathbf{B}_0$ will be assumed real and $\boldsymbol{\kappa} = \kappa\hat{\mathbf{n}}$. It is shown in P7-25 that the fields in Eq. (7-74) in fact do satisfy Maxwell's equations when $\rho = 0$ and $\mathbf{J} = 0$ provided that

$$\begin{aligned} \boldsymbol{\kappa}\cdot\mathbf{E}_0 &= 0 \Longrightarrow \boldsymbol{\kappa} \perp \mathbf{E}_0 \\ \boldsymbol{\kappa}\cdot\mathbf{B}_0 &= 0 \Longrightarrow \boldsymbol{\kappa} \perp \mathbf{B}_0 \\ \omega &= \kappa c \\ \mathbf{B}_0 &= \frac{1}{\omega}\boldsymbol{\kappa} \times \mathbf{E}_0 \end{aligned} \tag{7-75}$$

Furthermore, the time average of the Poynting vector is given by

$$\langle \mathbf{S} \rangle = \frac{E_0^2}{2\mu_0\omega}\boldsymbol{\kappa} \tag{7-76}$$

confirming our intuition that $\boldsymbol{\kappa}$ gives the direction of propagation of the wave. Consistent with Eq. (7-75), we can think of the plane wave propagating in three dimensions to be defined as follows:

(1) Specify $\boldsymbol{\kappa}$, i.e., the direction of propagation and the wave number, *arbitrarily*.
(2) Specify $\mathbf{E}_0$ *arbitrarily*, subject only to the constraint that $\mathbf{E}_0$ be perpendicular to $\boldsymbol{\kappa}$.
(3) $\mathbf{B}_0$ and ω are then determined by Eq. (7-75).

Plane waves in three dimensions are discussed in much greater detail in Chapter 13.

PROBLEMS

P7-24. Obtain expressions for the (complex) electric and magnetic induction fields in a plane wave that is (a) linearly polarized in the y direction and propagating in the positive x direction, (b) linearly polarized in the z direction and propagating in a direction parallel to the x-y plane at an angle θ to the positive x axis, and (c) right circularly polarized and propagating in the positive x direction.

P7-25. Because Maxwell's equations are linear in the fields, the entire complex field can be required to satisfy the equations with the assurance that the real part alone then necessarily satisfies the equations by itself. (a) Substitute Eq. (7-74) into Maxwell's equations with $\rho = 0$ and $\mathbf{J} = 0$ and derive the conditions expressed in Eq. (7-75). (b) Obtain Eq. (7-76) for the time-averaged Poynting vector of the fields in Eq. (7-74). *Caution:* The real parts of the complex fields must be taken *before* evaluating the cross product $\mathbf{E} \times \mathbf{B}$. (Why?)

P7-26. For the wave in Eq. (7-74), show that

$$\langle \mathbf{S} \rangle = \frac{1}{2\mu_0} \operatorname{Re}\,(\boldsymbol{\mathcal{E}}^* \times \boldsymbol{\mathcal{B}}) = \frac{1}{2\mu_0} \operatorname{Re}\,(\boldsymbol{\mathcal{E}} \times \boldsymbol{\mathcal{B}}^*)$$

$$\langle u_{\mathrm{E}} \rangle = \frac{\epsilon_0}{4} \operatorname{Re}\,(\boldsymbol{\mathcal{E}}^* \cdot \boldsymbol{\mathcal{E}}) \qquad \langle u_{\mathrm{B}} \rangle = \frac{1}{4\mu_0} \operatorname{Re}\,(\boldsymbol{\mathcal{B}}^* \cdot \boldsymbol{\mathcal{B}})$$

where $\mathbf{S}$, u_{E}, and u_{B} are the Poynting vector, the electric energy density, and the magnetic energy density, respectively.

Supplementary Problems

P7-27. Find the most general solution to the scalar wave equation

$$\nabla^2 u = \frac{1}{a^2} \frac{\partial^2 u}{\partial t^2}$$

in spherical coordinates if the solution is known to depend only on r and t.

Hints: (1) Use ∇^2 as given by Eq. (2-53). (2) Make the variable transformation $u(r, t) = f(r, t)/r$. (3) See P7-1.

P7-28. Show that the wave equation

$$\frac{\partial^2 u}{\partial z^2} = \frac{1}{c^2}\frac{\partial^2 u}{\partial t^2}$$

is invariant to the Lorentz transformation $z' = \gamma(z - vt)$, $t' = \gamma[t - (v/c^2)z]$, where $\gamma = \sqrt{1 - (v/c)^2}$ and v is the speed of the primed coordinate system relative to the unprimed system. That is, show that a function $u(z, t)$ satisfying the above equation also satisfies

$$\frac{\partial^2 u}{\partial z'^2} = \frac{1}{c^2}\frac{\partial^2 u}{\partial t'^2}$$

where z' and t' are determined from z and t by the Lorentz transformation.

P7-29. For the plane wave in Eq. (7-17), find a suitable *Lorentz gauge* vector potential **A** when the scalar potential is taken to be zero.

P7-30. In experiments on nuclear magnetic resonance (nmr), the sample is often placed in a region free of charges and currents and subjected to a magnetic induction field of the form

$$\mathbf{B}(t) = B_0\hat{\mathbf{k}} + b\hat{\mathbf{i}}\cos\omega_0 t - b\hat{\mathbf{j}}\sin\omega_0 t$$

where B_0, b, and ω_0 are constants. (a) Describe this field in words. (b) Show that there exists no electric field **E** that can be added to **B** so as to produce a pair of fields satisfying Maxwell's equations and explain how the physicist doing experimental nmr obtains the specified field.

P7-31. Calculate the force experienced by a perfectly reflecting sphere of radius a placed in the path of a linearly polarized, plane, monochromatic electromagnetic wave. *Hints:* (1) Recall that the angle of incidence equals the angle of reflection, where both are measured with respect to the normal to the surface at the point of incidence. (2) See P7-11.

P7-32. If the common factors in Eq. (7-45) are ignored, we can make the correspondences

$\begin{pmatrix}1\\0\end{pmatrix} \Longleftrightarrow \hat{\mathbf{i}} \Longleftrightarrow$ wave linearly polarized in the x direction and propagating in the z direction

$\begin{pmatrix}0\\1\end{pmatrix} \Longleftrightarrow \hat{\mathbf{j}} \Longleftrightarrow$ wave linearly polarized in the y direction and propagating in the z direction

$\begin{pmatrix}1\\-i\end{pmatrix} \Longleftrightarrow \hat{\mathbf{i}} - i\hat{\mathbf{j}} \Longleftrightarrow$ right circularly polarized wave propagating in the z direction

$\begin{pmatrix}1\\i\end{pmatrix} \Longleftrightarrow \hat{\mathbf{i}} + i\hat{\mathbf{j}} \Longleftrightarrow$ left circularly polarized wave propagating in the z direction

These two-dimensional vectors, and others like them, are called *Jones vectors.* Within this representation, many optical devices can be put in correspondence with a two-by-two *Jones matrix.* The Jones vector corresponding to the output of such a device is then obtained by multiplying the Jones vector corresponding to the input by the Jones matrix of the device. (a) Show that the Jones vector representing a wave linearly polarized in the x direction is a linear combination of the Jones vectors representing the two states of circular polarization. (b) Find the Jones vector representing light linearly polarized at an angle α to the x axis. (c) An optical device is represented by the Jones matrix

$$\begin{pmatrix} \cos\phi & \sin\phi \\ -\sin\phi & \cos\phi \end{pmatrix}$$

Describe the effect of this device on an incident plane wave linearly polarized in a direction making an angle θ with respect to the positive x axis. This so-called *Jones calculus* is developed in some detail in G. R. Fowles, *Introduction to Modern Optics* (Holt, Rinehart and Winston, Inc., New York, 1968), Chapter 2.

P7-33. Relax the assumption that the sources in Fig. 7-7 are equally spaced and in phase, letting the sth source be located more generally at y_s on the y axis. Further, let the sth source have an intrinsic phase α_s. (a) Using the same approximations as in the text, show that the phase of the sth contribution to the amplitude at P is given by $\phi_s = \kappa y_s \sin\theta + \alpha_s$, where the phase reference is a (possibly imaginary) source at $y = 0$. (b) Write a program that accepts the number of sources N, the emitted wavelength λ, the positions y_s, the intrinsic phases α_s, and the amplitudes $E_0^{(s)}$ as input and then computes and prints out the square magnitude in Eq. (7-52) for a specified range of θ. (c) Use this program on an available computer to obtain a polar graph of intensity versus angle for the combination of two sources for which $E_0^{(1)} = 1$, $y_1 = \lambda$, $\alpha_1 = 0$ and $E_0^{(2)} = 1$, $y_2 = -\lambda$, $\alpha_2 = \frac{1}{2}\pi$. *Optional:* (1) Examine other arrays of sources of your choice. (2) Modify the technique to allow for positioning the sources in a plane but not necessarily on a line and use the resulting program to examine intensity versus angle for several such arrays.

P7-34. In our discussion of interference, we tacitly assumed that the interfering waves were *coherent*, which means that the phase difference between any two individual waves remained constant so that at a particular point the two waves were, for example, *always* in phase or *always* 147° out of phase or something similar. If the phase difference between any two individual waves changes rapidly and randomly on the time scale of a macroscopic observation, any interference will be washed out because the waves are as much in phase as out of phase during the observation. In the latter case, which typically arises when the waves are derived from independent sources rather than by splitting the wave from a single source, the waves are said to be *incoherent*. Mathematically, the superposition of incoherent waves can be represented by averaging Eq. (7-52) over assumed random and independent fluctuations of all ϕ_s. Recognizing that such an average of $e^{i\alpha}$, where α varies randomly, is zero, show that Eq. (7-52) reduces to the sum of independent contributions from each wave when the waves are incoherent.

P7-35. In this problem, we shall explore a simple numerical technique for solving the wave equation

$$\frac{\partial^2 u}{\partial z^2} = \frac{1}{a^2}\frac{\partial^2 u}{\partial t^2}$$

subject to the conditions $u(0, t) = u(L, t) = 0$, $u(z, 0) = f(z)$, and $u_t(z, 0) = 0$, where a is a constant and $f(z)$ is a known function of z. [Think of a string stretched between $z = 0$ and $z = L$ and released from rest with an initial displacement given by $f(z)$.] We begin by superposing a rectangular grid on the domain $0 \leq z \leq L, 0 \leq t < \infty$ of the problem (Fig. P7-35). Let the lines of the grid be separated by Δz and Δt in the two coordinate directions, where both Δz and Δt can be chosen as small as necessary. We now interpret solving this problem numerically to mean obtaining values for u at all intersections on this grid. To facilitate writing subsequent equations, let $z_i = i\,\Delta z$, $i = 0, 1, 2, \cdots$; $t_j = j\,\Delta t$, $j = 0, 1, 2, \cdots$; and $u_{i,j} = u(z_i, t_j)$. (a) Approximate the derivatives in the above differential equation and show that

$$u_{i,j+1} = \alpha(u_{i+1,j} + u_{i-1,j}) + 2(1 - \alpha)u_{i,j} - u_{i,j-1} \tag{1}$$

where $\alpha = a^2(\Delta t)^2/(\Delta z)^2$. *Hint*: Show first that $d^2f(\xi)/d\xi^2 \approx [f(\xi + \Delta\xi) - 2f(\xi) + f(\xi - \Delta\xi)]/(\Delta\xi)^2$. (b) Equation (1) permits a direct calculation of u at points on the line $t = t_{j+1}$ provided values are known on the *two* previous lines $t = t_j$ and $t = t_{j-1}$. (Why?) To get started, we therefore need the values $u_{i,0}$ and $u_{i,1}$ for all i. The initial condition on u gives $u_{i,0} = f(z_i)$. We obtain

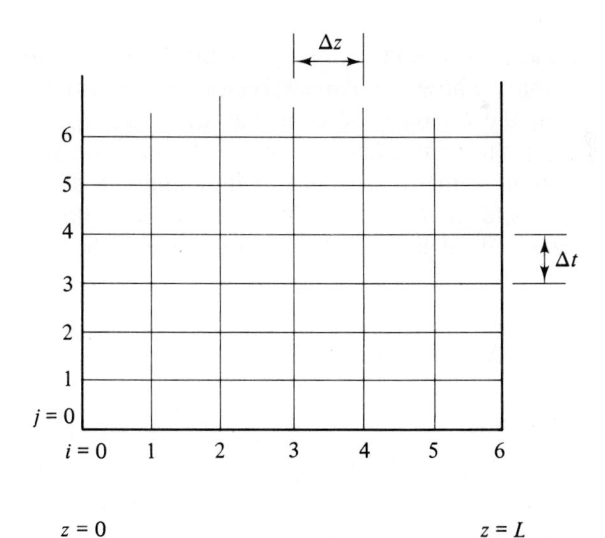

Figure P7-35

$u_{i,1}$ by introducing temporarily the values $u_{i,-1}$. Then $u_t(z_i, 0) \approx (u_{i,1} - u_{i,-1})/(2\,\Delta t)$. Thus, the initial condition on u_t requires that $u_{i,1} = u_{i,-1}$. Now, set $j = 0$ in Eq. (1) and show that

$$u_{i,1} = \tfrac{1}{2}[\alpha(u_{i+1,0} + u_{i-1,0}) + 2(1 - \alpha)u_{i,0}] \tag{2}$$

Thus, we calculate $u_{i,0}$ from the initial condition on u, then calculate $u_{i,1}$ from Eq. (2), and then use Eq. (1) to obtain $u_{i,2}, u_{i,3}, \cdots$ in succession. (c) Write a program to implement this algorithm on an available computer. Assume that L, a, Δz, and Δt are provided as input. (d) Use your program to find the solution at several successive times when $L = 10$ cm, $a = .5$ cm/sec, $\Delta z = 1$ cm, $\Delta t = 1$ sec, and in consistent units $f(z) = z(10 - z)/25$. Try also some other cases of your choosing. Note, however, that this method is unstable (and hence unsatisfactory) unless $\alpha \leq 1$. *Optional:* (1) Modify your program to print $u_{i,j}$ only for every Nth value of j, thus permitting smaller time steps without generating excessive output. (2) Develop a means to extend this method to cases where $u_t(z, 0) = g(z) \neq 0$.

P7-36. The linearly polarized electromagnetic wave given by Eq. (7-42) is incident on an electron that, in the absence of the external field, undergoes damped harmonic motion about a nominal equilibrium position, say the origin. Let the spring constant be k and the damping constant be b. (a) Show that the force arising on the electron from the **B**-field in the wave is smaller than the force from the **E**-field by a factor v/c, where v is the electron speed, and hence argue that, when $v/c \ll 1$, the equation of motion of the electron is

$$m\frac{d^2x}{dt^2} + b\frac{dx}{dt} + kx = -qE_{x0}e^{-i\omega t} \tag{1}$$

where m and $-q$ are the mass of and the charge on the electron, respectively, and $\mathcal{E}_{x0} \longrightarrow E_{x0}$ has been taken to be real. Note that we have tacitly introduced a *complex* position x whose *real* part expresses the physical displacement of the electron along the x axis. (b) Assume that $x(t) = x_0 \exp(-i\omega t)$ and find a solution of Eq. (1) for x as a function of t. (c) Discuss this solution, paying particular attention to the dependence of its amplitude and phase on the frequency of the incident wave. *Note:* The calculation in this problem is the first step in a classical calculation of the scattering of light by a bound electron.

8

Potential Theory

In Eq. (6-75), we found that, in a region where the charge and current densities are zero, both the scalar electro*static* potential and the components of the vector magneto*static* potential satisfy Laplace's equation,

$$\nabla^2 V = 0 \tag{8-1}$$

where the Laplacian operator ∇^2 in different coordinate systems is given in Eqs. (2-51)–(2-53). The body of knowledge relating to methods for solving Eq. (8-1) and to properties of the resulting solutions is known as *potential theory* and is the topic of this chapter. We shall illustrate the general theory exclusively with problems in electrostatics, but the theory itself can be applied wherever Laplace's equation appears. Such diverse areas of physics as steady-state heat flow, static deflections of elastic membranes, and irrotational fluid flow can thus be studied by the methods here treated. In Sections 8-1 and 8-2, we shall discuss how problems involving Laplace's equation must be formulated in order to be unambiguous and completely stated. In the next six sections, we shall examine various direct and indirect methods for finding solutions to these problems. Finally, we shall devote a section to Poisson's equation, obtained by replacing the right-hand side of Eq. (8-1) with a nonzero inhomogeneity. We shall leave a consideration of the equations satisfied by the time-dependent potentials to Chapter 14.

8-1
Boundary Conditions

No problem involving a differential equation is fully stated unless the equation is supplemented with conditions that stipulate the behavior of the solution at the boundaries of the region in which a solution is sought. One common boundary condition imposed on the electrostatic potential specifies the value of the potential at all points on the boundary and is called a *Dirichlet boundary condition.* If the boundary happens to be defined by the surface of a conductor (which is an equipotential surface when the field is static), then the potential must assume a *constant* value on the boundary. Sometimes it is sufficient to insist merely that the conductor be an equipotential, leaving the specific value of its potential to be determined as part of the problem.

A different boundary condition is imposed on a solution to Laplace's equation when the charge density σ at points on a conducting boundary is known. We found in P4-19 that this charge density is given by

$$\sigma = -\epsilon_0 \nabla V \cdot \hat{\mathbf{n}} = -\epsilon_0 \frac{\partial V}{\partial n} \tag{8-2}$$

where $\hat{\mathbf{n}}$ is a unit vector normal to the boundary and directed *into* the volume in which the solution to Laplace's equation is sought, and $\partial V/\partial n = \hat{\mathbf{n}} \cdot \nabla V$ is the *normal derivative* of V. Equation (8-2) is meaningful only when evaluated at points on the boundary. When σ is known initially, Eq. (8-2) determines values that must be imposed on the normal derivative of the solution at the boundary. Conditions imposed on the normal derivative are called *Neumann boundary conditions.* In reverse, Eq. (8-2) can be used to *determine* charge densities on a conducting surface if the potential has been found by other means.

Other types of boundary conditions are more involved. *Mixed boundary conditions* specify values on the boundary for the combination $aV + b\,\partial V/\partial n$, where a and b do not depend on V but may vary from point to point on the boundary. *Asymptotic boundary conditions* require V or perhaps something determined from V to approach some given limiting form as the field point is moved to infinity. In most cases V and its first derivatives must be everywhere finite, and this requirement sometimes plays the role of a boundary condition. Finally, when we extend our formalism to include matter (Chapters 9–13), *internal* boundaries appear at the interface between two different kinds of matter (or between matter and vacuum), and we shall need to develop boundary conditions relating the fields and potentials on one side of the interface to those on the other side of the interface.

We shall define two additional terms: (1) A *boundary value problem* is any problem whose statement involves a differential equation and associated boundary conditions, and (2) a *homogeneous* boundary condition is a boundary condition that requires some linear combination of V and $\partial V/\partial n$, e.g., V, $\partial V/\partial n$, or $aV + b\,\partial V/\partial n$, to be (or approach) *zero* on the boundary.

8-2 Superposition and Uniqueness

In this section we shall describe two important general properties of solutions to Laplace's equation. The first property is stated in the *theorem of superposition*, whose proof we leave to P8-1: *If V_1 and V_2 satisfy Laplace's equation, then $V = aV_1 + bV_2$, where a and b are constants, also satisfies the equation.* At a point P on the boundary, $V(P) = aV_1(P) + bV_2(P)$. Thus, V will not in general satisfy the same boundary conditions as V_1 and V_2. A sufficient (but not necessary) condition to assure that V will satisfy the same boundary conditions as V_1 and V_2 is that *homogeneous* conditions of the *same* type be imposed on V_1 and V_2.

A second important property of solutions to Laplace's equation is contained in the *uniqueness theorem*: *The solution to Laplace's equation subject to Dirichlet boundary conditions is unique*; *the solution to Laplace's equation subject to Neumann boundary conditions is unique to within an additive constant.* To prove this theorem, we assume two solutions V_1 and V_2 to a particular boundary value problem and then show that at worst the solutions can differ by an additive constant. Suppose then that

$$\nabla^2 V_1 = 0 \qquad \nabla^2 V_2 = 0 \tag{8-3}$$

is some region R and that V_1 and V_2 both satisfy *either* the Dirichlet conditions

$$V_1 = f(\mathbf{r}), \qquad V_2 = f(\mathbf{r}); \quad f \text{ given} \tag{8-4}$$

or the Neumann conditions

$$\nabla V_1 \cdot \hat{\mathbf{n}} = g(\mathbf{r}), \qquad \nabla V_2 \cdot \hat{\mathbf{n}} = g(\mathbf{r}); \quad g \text{ given} \tag{8-5}$$

at points $\mathbf{r}$ *on the boundary* of R. Then the function $V = V_1 - V_2$ satisfies

$$\nabla^2 V = 0 \tag{8-6}$$

subject either to

$$V = 0 \tag{8-7}$$

or to

$$\nabla V \cdot \hat{\mathbf{n}} = 0 \tag{8-8}$$

on the boundary. From Eq. (8-6) and Eq. (C-11), we now find that

$$\nabla \cdot (V \nabla V) = |\nabla V|^2 + V \nabla^2 V = |\nabla V|^2 \tag{8-9}$$

Thus, upon integrating Eq. (8-9) over the volume of the region R and using the divergence theorem, we obtain finally that

$$\oint V \nabla V \cdot d\mathbf{S} = -\oint V \nabla V \cdot \hat{\mathbf{n}}\, |d\mathbf{S}| = \int |\nabla V|^2\, dv \tag{8-10}$$

where the minus sign appears because $d\mathbf{S}$ has the direction of the *outward* normal to the surface bounding R while $\hat{\mathbf{n}}$ in Eq. (8-8) was tacitly taken to point *inward*. Whether Eq. (8-7) or Eq. (8-8) applies, however, the surface integral in Eq. (8-10) is zero, and we find that

$$\int |\nabla V|^2 \, dv = 0 \tag{8-11}$$

Since the integrand in Eq. (8-11) is necessarily nonnegative, the integral can be zero only if

$$\nabla V = 0 \tag{8-12}$$

at *all* points in R. Finally, we conclude from Eq. (8-12) that

$$V = \text{constant} \Longrightarrow V_1 = V_2 + \text{constant} \tag{8-13}$$

since $V = V_1 - V_2$. [See Eq. (4-48) and the associated text.] Equation (8-13) establishes the uniqueness theorem for Neumann conditions; the theorem is established for Dirichlet conditions by noting that the condition $V_1 = V_2$ *on the boundary* requires the constant in Eq. (8-13) to be zero.

The assurance that the solution to a completely stated problem is unique is especially important to some methods of solving electrostatic problems. Sometimes we are forced to *guess* a solution and then *show* that this solution satisfies Laplace's equation and the associated boundary conditions. Without uniqueness, we could never be sure that such a guessed solution was the *only* solution. Furthermore, the uniqueness theorem helps us to understand the nature of the boundary conditions that must specified in order that a problem in electrostatics be well stated.

PROBLEMS

P8-1. Prove the theorem of superposition. *Hint*: Assume V_1 and V_2 satisfy Laplace's equation and show that $aV_1 + bV_2$, where a and b are constants, also satisfies the equation.

P8-2. (a) Prove that the solution to Laplace's equation is unique if the solution is required to have a specified value over part of the boundary and a specified normal derivative over the remainder of the boundary. (b) A solution to the two-dimensional Laplace equation

$$\nabla^2 V = \frac{\partial^2 V}{\partial x^2} + \frac{\partial^2 V}{\partial y^2} = 0$$

is desired in the interior of the square $|x| \leq a$, $|y| \leq a$. On the boundary, the solution is required to assume the values shown in Fig. P8-2(a). (1) Show that the solution is invariant to reflection in the x axis and also in the y axis. (2) Show that the solution to the problem in Fig. P8-2(b) is identical with the solution in the first quadrant to the problem in Fig. P8-2(a).

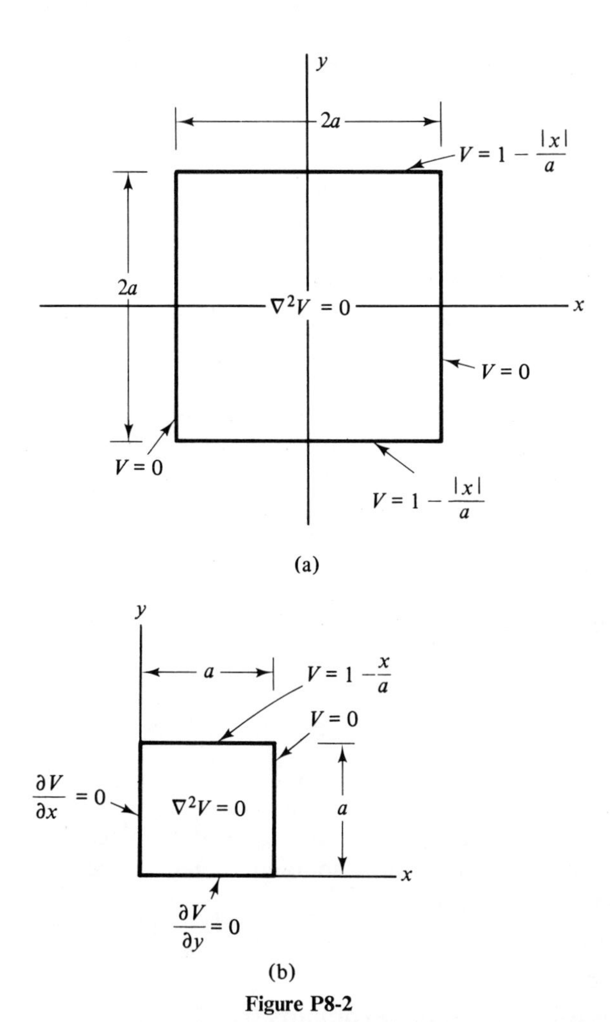

Figure P8-2

P8-3. In the presence of any *static* external field the entire volume occupied by a solid conductor of arbitrary shape is an equipotential (Section 4-4), and the (constant) potential V in this region then satisfies $\nabla^2 V = 0$ subject to the requirement that V be constant on the surface of the conductor. Show that this same mathematical statement applies if the solid conductor is replaced by a thin conducting shell enclosing the same region of space and then use the uniqueness theorem in an argument showing that the region enclosed by the *shell* is field-free. The phenomenon of this problem makes possible the *electrostatic shielding* of a region of space by enclosing it in a conducting shell. Indeed, a conducting shell made of screen wire is often adequate.

P8-4. Show that maximum and minimum values of the solution to $\nabla^2 V = 0$ can occur only on the boundaries of the region R to which the solution applies. *Hint:* Assume a maximum or minimum in V at the point P in the interior of R. What would the electric field look like in the vicinity of P? Now review P4-13.

8-3
One-Dimensional Problems

For some problems, symmetry arguments (or perhaps other arguments) can demonstrate that the electrostatic potential depends only on one of the three coordinates in the relevant coordinate system. Such a problem is essentially a one-dimensional problem, and stipulation of the boundary conditions reduces, for example, to specification of the value of V at two values of the one coordinate on which V depends. A potential depending only on the Cartesian coordinate x satisfies

$$\frac{d^2V}{dx^2} = 0 \Longrightarrow V = ax + b \tag{8-14}$$

while a potential depending only on the cylindrical variable $\imath$ satisfies

$$\frac{1}{\imath}\frac{d}{d\imath}\left(\imath\frac{dV}{d\imath}\right) = 0 \Longrightarrow V = a\,\ell n\,\imath + b \tag{8-15}$$

and a potential depending only on the spherical coordinate r satisfies

$$\frac{1}{r^2}\frac{d}{dr}\left(r^2\frac{dV}{dr}\right) = 0 \Longrightarrow V = \frac{a}{r} + b \tag{8-16}$$

In each solution, the constants a and b are determined by the boundary conditions. Verification of these solutions is left to the reader. (See also P6-30, P6-31, and P6-32.)

8-4
Two-Dimensional Problems by Separation of Variables

Two-dimensional problems, in which dependence on only one of the three spatial variables can be ruled out ab initio, are more interesting. The most common of these are problems in Cartesian coordinates that involve only x and y, problems in cylindrical coordinates that involve only $\imath$ and ϕ, problems in cylindrical coordinates that involve only $\imath$ and z, and problems in spherical coordinates that involve only r and θ. The treatment in this section is illustrative rather than exhaustive.

In Cartesian coordinates, Laplace's equation in the two variables x and y assumes the form

$$\frac{\partial^2 V}{\partial x^2} + \frac{\partial^2 V}{\partial y^2} = 0 \tag{8-17}$$

[See Eq. (2-51).] Without having any initial assurance that it will work, we adopt the method of *separation of variables* and try a solution of the form

$$V(x, y) = X(x)Y(y) \tag{8-18}$$

i.e., a product of functions of each variable separately. If the function in Eq. (8-18) is to satisfy Eq. (8-17), the functions $X(x)$ and $Y(y)$ must satisfy the condition

$$\frac{1}{X}\frac{d^2X}{dx^2} = -\frac{1}{Y}\frac{d^2Y}{dy^2} \tag{8-19}$$

obtained by substituting Eq. (8-18) into Eq. (8-17). In Eq. (8-19), however, the two independent variables x and y have been separated, each side of Eq. (8-19) depending on only one of these variables. Since variation of x, for example, cannot change the value of the side depending on y, the side depending on x must in fact be a constant. A similar conclusion applies to the side depending on y. Thus, we can extract the two *ordinary* differential equations

$$\frac{1}{X}\frac{d^2X}{dx^2} = k \qquad \frac{1}{Y}\frac{d^2Y}{dy^2} = -k \tag{8-20}$$

$$\Longrightarrow \frac{d^2X}{dx^2} - kX = 0 \qquad \frac{d^2Y}{dy^2} + kY = 0 \tag{8-21}$$

from Eq. (8-19). These two equations are coupled only because the *same separation constant* k appears in both equations. The solution of Eq. (8-21) is now immediate, viz.,

$$X = Ae^{\lambda x} + Be^{-\lambda x}, \qquad Y = A' \sin \lambda y + B' \cos \lambda y \tag{8-22}$$

where $\lambda = \sqrt{k}$. Finally, on substitution of Eq. (8-22) into Eq. (8-18), we find that

$$V_\lambda(x, y) = (Ae^{\lambda x} + Be^{-\lambda x})(A' \sin \lambda y + B' \cos \lambda y) \tag{8-23}$$

where the subscript λ has been added as a reminder of the dependence on a parameter that can assume any value whatever (including complex values) without destroying the basic property that Eq. (8-23) is a solution of Eq. (8-17). Our tentative assumption of a solution in product form has thus yielded not just a single solution but an entire family of solutions.

A more explicit evaluation of the integration constants A, A', B and B' and of the separation constant λ appearing in Eq. (8-23) requires a more specific problem, for these constants are determined by the boundary conditions. Sometimes the boundary conditions can be satisfied by a single value of λ. More often several values of λ are consistent with a portion of the boundary conditions and several "product" solutions corresponding to different values of λ must be superposed to find a solution fitting the rest of the boundary conditions. Suppose, for example, that a solution to Laplace's equation is sought in the region interior to the rectangle of Fig. 8-1 and that V is to assume the indicated values on the edges of this rectangle. Consider the three homogeneous boundary conditions first; we find that $V_\lambda(x, 0) = 0$,

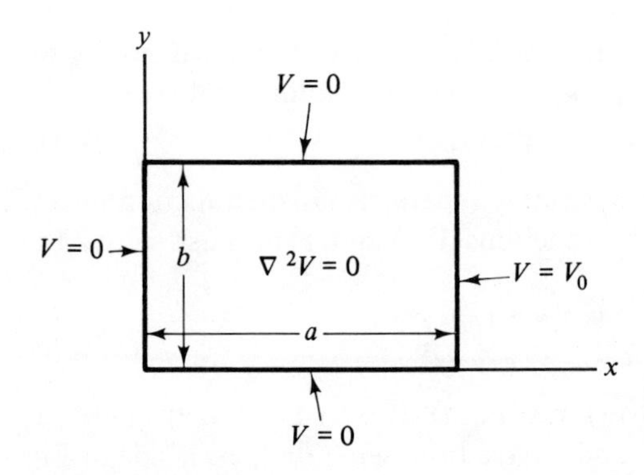

Fig. 8-1. Region in which a solution to Laplace's equation is sought. The boundary conditions to be imposed on the solution are shown.

which implies $B' = 0$, and that $V_\lambda(0, y) = 0$, which implies $A = -B$. Thus, Eq. (8-23) reduces to[1]

$$V_\lambda(x, y) = A'' \sinh \lambda x \sin \lambda y \tag{8-24}$$

The remaining homogeneous boundary condition requires that $V_\lambda(x, b) = 0$, which requires in turn that $\sin \lambda b = 0$ and hence that $\lambda b = n\pi$, with n a positive integer. We have therefore found a countable infinity of solutions

$$V_n(x, y) = A_n \sinh \frac{n\pi x}{b} \sin \frac{n\pi y}{b}, \quad n = 1, 2, \cdots \tag{8-25}$$

all of which are consistent with the three homogeneous boundary conditions. A more general solution $V(x, y)$ is now obtained by superposing these simpler solutions; we find that

$$V(x, y) = \sum_{n=1}^{\infty} A_n \sinh \frac{n\pi x}{b} \sin \frac{n\pi y}{b} \tag{8-26}$$

which clearly satisfies the three homogeneous boundary conditions and which will satisfy the final (inhomogeneous) boundary condition if

$$V(a, y) = V_0 = \sum_{n=1}^{\infty} A_n \sinh \frac{n\pi a}{b} \sin \frac{n\pi y}{b} \tag{8-27}$$

This series, however, is the Fourier sine series expansion for the constant V_0 over the interval $0 < y < b$. (See PD-15 in Appendix D.) Hence, the constant $A_n \sinh (n\pi a/b)$ must match the coefficients in that Fourier series. Once these constants have been found (P8-5), Eq. (8-26) expresses a solution to Laplace's equation subject to the given boundary conditions. By uniqueness, this series therefore expresses *the* solution to the problem.

For solutions that depend only on $\imath$ and ϕ in cylindrical coordinates, Laplace's equation assumes the form

$$\frac{1}{\imath} \frac{\partial}{\partial \imath}\left(\imath \frac{\partial V}{\partial \imath}\right) + \frac{1}{\imath^2} \frac{\partial^2 V}{\partial \phi^2} = 0 \tag{8-28}$$

[See Eq. (2-52).] If we employ the method of separation of variables and fur-

[1]The function $\sinh \lambda x$ is defined to be $\frac{1}{2}(e^{\lambda x} - e^{-\lambda x})$. We may also need $\cosh \lambda x = \frac{1}{2}(e^{\lambda x} + e^{-\lambda x})$.

ther insist that the solutions be periodic in ϕ with period 2π, we find ultimately—see P8-7—that the most general solution can be expressed as the infinite series

$$V(r, \phi) = a_0 \ln r + b_0 + \sum_{n=1}^{\infty} r^n(a_n \cos n\phi + b_n \sin n\phi) + \sum_{n=1}^{\infty} r^{-n}(c_n \cos n\phi + d_n \sin n\phi) \quad (8\text{-}29)$$

where the constants a_0, b_0, a_n, b_n, c_n, and d_n are determined by the boundary conditions. The requirement of periodicity imposed on this solution restricts its utility to problems in which the full angular range $0 \leq \phi \leq 2\pi$ is assumed; modifications will be necessary if the solution need not have the periodicity required of Eq. (8-29). Further, if the origin happens to be in the domain of the problem, the constants c_n and d_n for all n and also the constant a_0 must be zero to suppress terms that diverge at $r = 0$. Note finally that Eq. (8-29) with all constants except a_0 equal to zero reduces to the potential of a line charge (P4-18).

For solutions that depend only on r and z in cylindrical coordinates, Laplace's equation assumes the form

$$\frac{1}{r}\frac{\partial}{\partial r}\left(r\frac{\partial V}{\partial r}\right) + \frac{\partial^2 V}{\partial z^2} = 0 \quad (8\text{-}30)$$

[see Eq. (2-52)] and, again using the method of separation of variables, we find ultimately that a basic product solution has the form

$$V_\lambda(r, z) = [Ae^{\lambda z} + Be^{-\lambda z}][CJ_0(\lambda r) + DY_0(\lambda r)] \quad (8\text{-}31)$$

where λ is a separation constant that may assume any value (including complex values); A, B, C, and D are integration constants; and $J_0(x)$ and $Y_0(x)$ are two conventionally defined, linearly independent solutions to the *zeroth order Bessel equation,*

$$x\frac{d^2y}{dx^2} + \frac{dy}{dx} + xy = 0 \quad (8\text{-}32)$$

The functions $J_0(x)$ and $Y_0(x)$ are called the *zeroth-order Bessel* and *Neumann functions*, respectively, and are shown graphically in Fig. 8-2 for real values of x. In most problems, V must approach zero as $z \longrightarrow \infty$ so the increasing exponential in Eq. (8-31) must be suppressed. Further, if the origin $r = 0$ is included in the domain of the problem, the Neumann function Y_0 must be eliminated because it diverges at $r = 0$, thereby giving an unphysical potential. Under these additional conditions, the most general acceptable solution in product form is

$$V_\lambda(r, z) = C'e^{-\lambda|z|}J_0(\lambda r) \quad (8\text{-}33)$$

A possible superposition of these solution is then expressed by the integral

$$V(r, z) = \int_0^\infty C'(\lambda)e^{-\lambda|z|}J_0(\lambda r)\, d\lambda \quad (8\text{-}34)$$

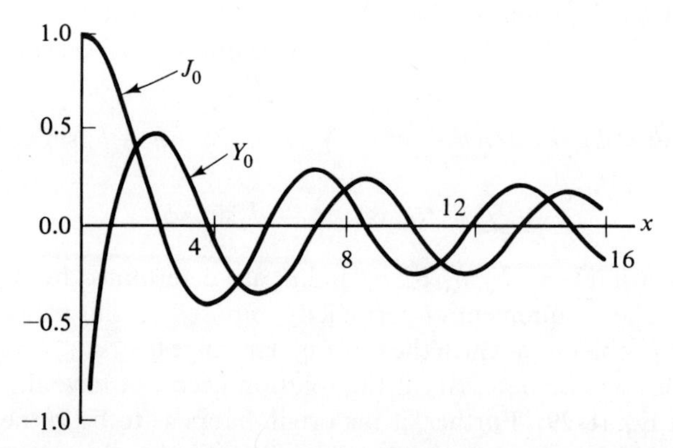

Fig. 8-2. $J_0(x)$ and $Y_0(x)$ versus x. Both functions are oscillatory with decaying amplitude as x increases. The first few zeroes of $J_0(x)$ occur at $x = 2.405$, 5.520, 8.654, 11.792, and 14.931.

where $C'(\lambda)$ would be determined by additional boundary conditions. Equation (8-31) is derived in P8-10, and one application of Eq. (8-34) is explored in P8-12.

Consider now solutions depending on the spherical variables r and θ. The relevant form of Laplace's equation is

$$\frac{1}{r^2}\frac{\partial}{\partial r}\left(r^2\frac{\partial V}{\partial r}\right) + \frac{1}{r^2 \sin\theta}\frac{\partial}{\partial\theta}\left(\sin\theta\frac{\partial V}{\partial\theta}\right) = 0 \tag{8-35}$$

[see Eq. (2-53)] and separation of variables by writing $V(r, \theta) = R(r)\Theta(\theta)$ leads ultimately to the two equations

$$\frac{d}{dr}\left(r^2\frac{dR}{dr}\right) - kR = 0 \tag{8-36}$$

$$\frac{1}{\sin\theta}\frac{d}{d\theta}\left(\sin\theta\frac{d\Theta}{d\theta}\right) + k\Theta = 0 \tag{8-37}$$

(See P8-13.) Consider Eq. (8-37) first. The variable transformation $x = \cos\theta$ reduces this equation to the form

$$(1 - x^2)\frac{d^2\Theta}{dx^2} - 2x\frac{d\Theta}{dx} + k\Theta = 0 \tag{8-38}$$

which is *Legendre's equation*. As with every second-order linear differential equation, Legendre's equation has two linearly independent solutions. For Eq. (8-38), however, it happens that both of these solutions diverge at $x = \pm 1$ ($\theta = 0, \pi$) unless k has one of the values given by $n(n + 1)$, where n is a nonnegative integer. Even when k has one of the specified values, one of the two solutions still diverges at $x = \pm 1$ ($\theta = 0, \pi$); those solutions that do not diverge are the *Legendre polynomials*, conventionally denoted $P_n(x)$. (The first few of these polynomials are plotted in Fig. 8-3 and several of their properties are summarized in Table 8-1.) Thus, whenever the values $\theta = 0$ and/or

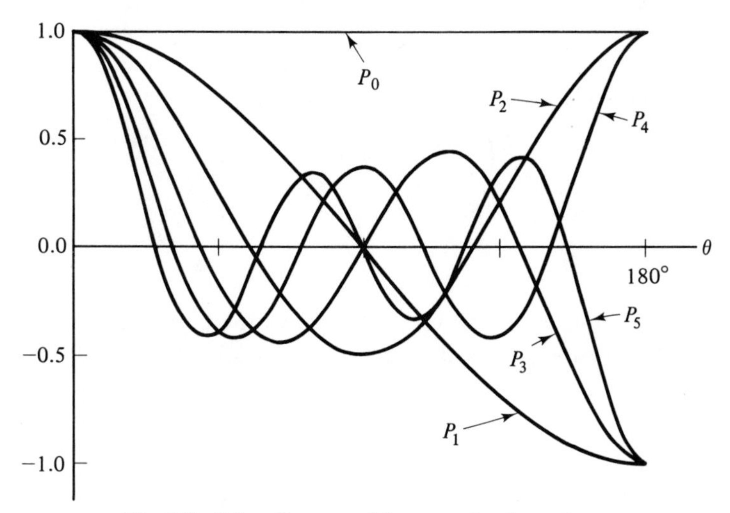

Fig. 8-3. $P_n(\cos\theta)$ versus θ for several values of n.

TABLE 8-1 Legendre Polynomials and Their Properties

$$P_0(x) = 1 \qquad P_1(x) = x \qquad P_2(x) = \tfrac{1}{2}(3x^2 - 1)$$

$$P_3(x) = \tfrac{1}{2}(5x^3 - 3x) \qquad P_4(x) = \tfrac{1}{8}(35x^4 - 30x^2 + 3)$$

$$P_5(x) = \tfrac{1}{8}(63x^5 - 70x^3 + 15x)$$

$$\frac{1}{\sqrt{1 - 2tx + t^2}} = \sum_{n=0}^{\infty} P_n(x)t^n$$

$$P_n(-x) = (-1)^n P_n(x)$$

$$(n + 1)P_{n+1}(x) - (2n + 1)xP_n(x) + nP_{n-1}(x) = 0$$

$$(1 - x^2)\frac{d^2P_n}{dx^2} - 2x\frac{dP_n}{dx} + n(n + 1)P_n = 0$$

$$(1 - x^2)\frac{dP_n}{dx} = n(P_{n-1} - xP_n) = (n + 1)(xP_n - P_{n+1})$$

$$\int_{-1}^{1} P_n(x)P_m(x)\,dx = \int_0^{\pi} P_n(\cos\theta)P_m(\cos\theta)\sin\theta\,d\theta = \frac{2}{2n + 1}\delta_{nm}$$

$$P_n(1) = 1 \qquad P_n(-1) = (-1)^n$$

$$P_1(x) = 0 \text{ at } \begin{Bmatrix} x = & 0.000 \\ \theta = & 90.0^\circ \end{Bmatrix}$$

$$P_2(x) = 0 \text{ at } \begin{Bmatrix} x = & 0.577 & -0.577 \\ \theta = & 54.7^\circ & 125.3^\circ \end{Bmatrix}$$

$$P_3(x) = 0 \text{ at } \begin{Bmatrix} x = & 0.774 & 0.000 & -0.774 \\ \theta = & 39.2^\circ & 90.0^\circ & 140.8^\circ \end{Bmatrix}$$

$$P_4(x) = 0 \text{ at } \begin{Bmatrix} x = & 0.861 & 0.340 & -0.340 & -0.861 \\ \theta = & 30.6^\circ & 70.1^\circ & 109.9^\circ & 149.4^\circ \end{Bmatrix}$$

$$P_5(x) = 0 \text{ at } \begin{Bmatrix} x = & 0.906 & 0.538 & 0.000 & -0.538 & -0.906 \\ \theta = & 25.0^\circ & 57.4^\circ & 90.0^\circ & 122.6^\circ & 155.0^\circ \end{Bmatrix}$$

π are included in the domain of the physical problem, we conclude (1) that the only solutions to Eq. (8-37) that are finite everywhere in the domain of the problem are the Legendre polynomials and (2) that the physically meaningful values of the separation constant k are given in terms of a nonnegative integer n by $n(n+1)$. The radial equation then becomes

$$\frac{d}{dr}\left(r^2\frac{dR_n}{dr}\right) - n(n+1)R_n = 0 \tag{8-39}$$

which has the solution

$$R_n = a_n r^n + \frac{b_n}{r^{n+1}} \tag{8-40}$$

(P8-13). Combining each solution for R_n with the corresponding Legendre polynomial and superposing these products for all possible values of n, we find the sum

$$V(r, \theta) = \sum_{n=0}^{\infty}\left(a_n r^n + \frac{b_n}{r^{n+1}}\right)P_n(\cos\theta) \tag{8-41}$$

which expresses a general solution to Laplace's equation for the conditions assumed in this paragraph. Modifications will be necessary if the values $\theta = 0$ and π are actually excluded from the domain of the problem. Further, if the origin happens to be in the domain of the problem, b_n must be zero for all n to suppress terms that diverge at $r = 0$.

Equation (8-41) reduces to several familiar results if the constants are given particular values. The constant potential is obtained by setting all constants except a_0 equal to zero; with all constants except b_0 equal to zero and $b_0 = Q/4\pi\epsilon_0$, Eq. (8-41) reduces to the potential of a point charge; setting $b_1 = p/4\pi\epsilon_0$ and all other constants equal to zero produces the potential of a point dipole with dipole moment $\mathbf{p} = p\hat{\mathbf{k}}$; the potential of a linear quadrupole emerges if all constants except b_2 are zero; the potential of a uniform field emerges if only a_1 is nonzero; and so on.

As an example of the use of Eq. (8-41) to determine the electrostatic potential under given circumstances, let us determine the potential established when an *uncharged, conducting sphere* is placed in a *uniform* external electric field. If we take the direction of the electric field to define the polar axis so that

$$\mathbf{E} = E_0\hat{\mathbf{k}} \tag{8-42}$$

and further if we place the origin at the center of the sphere, the entire problem is invariant to rotation about the z axis and the potential therefore cannot depend on the spherical coordinate ϕ. Several conditions must be satisfied by an acceptable solution: (1) In the region far from the sphere, the sphere will have little effect on the potential and hence

$$V(r, \theta) \xrightarrow[r\to\infty]{} -\mathbf{E}\cdot\mathbf{r} = -E_0 r\cos\theta = -E_0 r P_1(\cos\theta) \tag{8-43}$$

which is the potential of a uniform field (P4-17); (2) the potential *on the surface* of the sphere must be constant; (3) the sphere must be uncharged; (4) there can be no dependence on ϕ; (5) the solution must be finite everywhere, in particular at $\theta = 0$ and at $\theta = \pi$; and (6) the potential in the region exterior to the sphere must satisfy Laplace's equation. Conditions (4)–(6) restrict the most general potential in the region outside the sphere to be given by Eq. (8-41). Equation (8-41) will match Eq. (8-43) at large r, however, only if $a_1 = -E_0$ and the rest of the a_n's (including a_0) are zero. Thus, the asymptotic boundary condition reduces the most general solution to

$$V(r, \theta) = \frac{b_0}{r} + \left(\frac{b_1}{r^2} - E_0 r\right)P_1(\cos\theta) + \sum_{n=2}^{\infty} \frac{b_n}{r^{n+1}} P_n(\cos\theta) \tag{8-44}$$

where the terms multiplied by b_0 and b_1 have been written explicitly. If V is to be constant on the sphere, say $r = a$, we must then require that

$$V(a, \theta) = \frac{b_0}{a} + \left(\frac{b_1}{a^2} - E_0 a\right)P_1(\cos\theta) + \sum_{n=2}^{\infty} \frac{b_n}{a^{n+1}} P_n(\cos\theta) \tag{8-45}$$

be independent of θ, which in turn requires that $b_1 = E_0 a^3$ and $b_n = 0$ for $n \geq 2$. (Why? *Hint*: The P_n's are linearly independent.) Thus, Eq. (8-44) reduces even further for this problem to

$$V(r, \theta) = \frac{b_0}{r} + E_0 r\left(\frac{a^3}{r^3} - 1\right)\cos\theta \tag{8-46}$$

The condition that the conducting sphere be uncharged now determines the remaining constant b_0. From Eq. (8-2), we find that the charge density σ at points on the surface of the sphere is given by

$$\sigma(\theta) = -\epsilon_0 \nabla V \cdot \hat{\mathbf{r}}\Big|_{r=a} = -\epsilon_0 \frac{\partial V}{\partial r}\Big|_{r=a}$$
$$= \epsilon_0\left(\frac{b_0}{a^2} + 3E_0\cos\theta\right) \tag{8-47}$$

The total charge Q on the sphere is then given by integrating σ over the sphere,

$$Q = \int_{\text{sphere}} \sigma\, dS = \int_0^{2\pi}\int_0^{\pi} \epsilon_0\left(\frac{b_0}{a^2} + 3E_0\cos\theta\right)a^2\sin\theta\, d\theta\, d\phi$$
$$= 4\pi\epsilon_0 b_0 \tag{8-48}$$

and will be zero only if $b_0 = 0$. Thus, Eq. (8-47) reduces to

$$\sigma(\theta) = 3\epsilon_0 E_0 \cos\theta \tag{8-49}$$

and Eq. (8-46) becomes

$$V(r, \theta) = -E_0 r\left(1 - \frac{a^3}{r^3}\right)\cos\theta \tag{8-50}$$

Further properties of this field are explored in P8-15.

The method of separation of variables can also be applied to solve problems in three dimensions, but the mathematics of such problems in most coordinate systems is considerably more involved than that for two-dimensional problems. A few examples may be found in P8-37, P8-38, and P8-39.

PROBLEMS

P8-5. Complete the solution of the problem in Fig. 8-1 by finding the coefficients A_n in Eq. (8-27). *Hint*: See PD-15 in Appendix D.

P8-6. Find the electrostatic potential in the interior of a square region of side a if two adjacent sides are at zero potential and the other two sides are maintained at a potential V_0. *Hint*: Regard the given problem as a superposition of two problems, each of which has homogeneous boundary conditions on three sides of the square.

P8-7. Find the general solution, Eq. (8-29), for a potential depending on the cylindrical coordinates r and ϕ by applying separation of variables to Eq. (8-28).

P8-8. An infinitely long, uncharged, conducting cylinder of radius a is placed in a uniform external electric field directed perpendicular to the axis of the cylinder. Find the resulting electrostatic potential and determine the charge density on the surface of the cylinder. *Hint*: Let the axis of the cylinder coincide with the z axis and let the field be in the x direction.

P8-9. Determine the electrostatic potential in the region interior to a 90° conducting wedge if the two straight sides are at zero potential and the 90° arc at radius a is maintained at a constant potential V_0.

P8-10. Find the product form, Eq. (8-31), for a potential depending on the cylindrical coordinates r and z by applying separation of variables to Eq. (8-30).

P8-11. Assume a power series solution of the form $\sum_{n=0}^{\infty} \alpha_n x^n$ to the zeroth-order Bessel equation, Eq. (8-32), find the conditions imposed on the α_n's by Eq. (8-32), and find a solution if $y(0)$ is required to be one. The result is a power series representation of $J_0(x)$. Can you explain why this method does not yield *two* linearly independent solutions to the zeroth-order Bessel equation?

P8-12. A circular conducting disc of radius a lies in the x-y plane with its center at the origin and is charged to a potential V_0. Find the potential at all points in space, find and sketch the charge distribution on the disc as a function of r, and find the total charge Q on the disc. *Hints*: (1) Use Eq. (8-34) and employ the identity

$$\int_0^\infty J_0(\alpha x)\frac{\sin \beta x}{x}\,dx = \begin{cases} \dfrac{\pi}{2}, & \alpha < \beta \\[2ex] \sin^{-1}\dfrac{\beta}{\alpha}, & \alpha > \beta \end{cases}$$

Note the amazingly simple value for this apparently complicated integral.

In particular, the integral is *constant* for $\alpha < \beta$. (2) The similar identity

$$\int_0^\infty J_0(\alpha x)\sin\beta x\,dx = \begin{cases} \dfrac{1}{\sqrt{\beta^2-\alpha^2}}, & \alpha < \beta \\ 0, & \alpha > \beta \end{cases}$$

may be useful. See F. Bowman, *Introduction to Bessel Functions* (Dover Publications, Inc., New York, 1958), pp. 60–63.

P8-13. (a) Find the two equations, Eq. (8-36) and (8-37), into which Laplace's equation for a potential depending on the spherical coordinates r and θ separates when a solution in product form is assumed. (b) Accepting the separation constant $k = n(n+1)$ as determined from the angular equation, solve the radial equation to find Eq. (8-40). *Hint*: Guess a solution R_n of the form r^p. Why might you expect this guess to work? *Optional*: Assume a power series solution of the form $\sum_{n=0}^{\infty} \alpha_n x^{n+c}$ to Eq. (8-38); find the conditions that Eq. (8-38) imposes on c and the α_n's; find power series representing two basic solutions Θ_{odd} and Θ_{even} defined by $\Theta_{\text{odd}}(0) = 0$, $d\Theta_{\text{odd}}(0)/dx = 1$ and $\Theta_{\text{even}}(0) = 1$, $d\Theta_{\text{even}}(0)/dx = 0$; show that these series in general diverge at $x = \pm 1$ but that at least one of them terminates (and hence converges) when $k = n(n+1)$; and finally show that the terminating series is proportional to the corresponding Legendre polynomial.

P8-14. Use the recurrence relation in Table 8-1 to obtain $P_5(x)$ from $P_3(x)$ and $P_4(x)$.

P8-15. (a) Show that the potential in Eq. (8-50) is the potential of a dipole plus that of the uniform field and infer the dipole moment to assign to the conducting sphere. (b) Calculate the dipole moment of the sphere directly from the charge distribution of Eq. (8-49) and compare your result with that obtained in part (a). (c) Determine the amount of charge Q on the positively charged hemisphere of this sphere and calculate the separation s between two charges of this magnitude, one positive and the other negative, in order that the combination will have the same dipole moment as the actual sphere. (d) Calculate the field from the potential and show explicitly that the field is perpendicular to the sphere at points on the surface of the sphere. (e) Sketch a graph showing the field lines and equipotential surfaces.

P8-16. Find the potential established when a conducting sphere is placed in a uniform electric field if the sphere carries charge Q. Find also the charge density on the surface of the sphere.

P8-17. Consider a circular ring of radius a carrying charge Q distributed uniformly about its perimeter and lying in the x-y plane with its center at the origin. In spherical coordinates, the potential established by this ring is independent of ϕ, $V = V(r, \theta)$ (Why?), and in particular on the axis ($\theta = 0$), $V(r, 0) = Q/(4\pi\epsilon_0\sqrt{a^2 + r^2})$ [P4-55(b)]. Further, except at points on the ring, $\nabla^2 V$ is zero. (Why?) (a) Find the first four terms in the binomial expansion of $V(r, 0)$ in powers of a/r. (b) Show that Eq. (8-41) applies and find the values of the coefficients a_n and b_n by requiring Eq. (8-41) to agree with part (a) at points on the z axis. The resulting series is a multipole expansion of the potential established by the ring and is valid both on and off the z axis. Similar techniques can be used to find off-axis values for the potential established by a uniformly charged circular disc (P4-25) and for the magnetic *scalar* potential (P5-26) of a circular current loop, discussed by D. R. Corson and P. Lorrain, *Introduction to Electromagnetic Fields and Waves* (W. H. Freeman and Company, San Francisco, 1962), pp. 207–209.

8-5 Two-Dimensional Problems Using Complex Variables[2]

Another method for solving two-dimensional potential problems exploits the properties of what are called analytic functions of a complex variable, $z = x + iy$.[3] Any function $f(z)$ of the variable z is also a (complex) function of the *real* variables x and y and can itself be separated into real and imaginary parts, i.e.,

$$f(z) = V_1(x, y) + iV_2(x, y) \tag{8-51}$$

where V_1 and V_2 are both *real* functions of the *real* variables x and y. Thus, for example,

$$f(z) = z^2 = (x + iy)^2 = x^2 - y^2 + 2xyi$$
$$\Longrightarrow V_1(x, y) = x^2 - y^2; \qquad V_2(x, y) = 2xy \tag{8-52}$$

or

$$f(z) = \ell n\, z = \ell n\,(re^{i\phi}) = \ell n\, r + i\phi$$
$$= \ell n \sqrt{x^2 + y^2} + i \tan^{-1}\frac{y}{x}$$
$$\Longrightarrow V_1(x, y) = \ell n \sqrt{x^2 + y^2}; \qquad V_2(x, y) = \tan^{-1}\frac{y}{x} \tag{8-53}$$

and so on. The function $f(z)$ is said to be *analytic* if V_1 and V_2 satisfy the *Cauchy-Riemann conditions,*

$$\frac{\partial V_1}{\partial x} = \frac{\partial V_2}{\partial y} \qquad \frac{\partial V_1}{\partial y} = -\frac{\partial V_2}{\partial x} \tag{8-54}$$

Most reasonable functions, including z^n for any (positive or negative) integer n, $\sin z$, $\ell n\, z$, e^z, products of analytic functions, sums of analytic functions, and analytic functions of analytic functions in fact *are* analytic, except possibly at isolated points or along particular lines.

Analytic functions are important to two-dimensional potential theory because both the real and imaginary parts of an analytic function *automati-*

[2] The properties of complex numbers are summarized in Appendix D.
[3] Do not confuse the complex variable z with the Cartesian coordinate z.

cally satisfy Laplace's equation in two dimensions. We need merely differentiate one member of Eq. (8-54) with respect to x and the other with respect to y, obtaining

$$\frac{\partial^2 V_1}{\partial x^2} = \frac{\partial^2 V_2}{\partial x\,\partial y} \qquad \frac{\partial^2 V_1}{\partial y^2} = -\frac{\partial^2 V_2}{\partial y\,\partial x} \tag{8-55}$$

and then recognize the equality of the two mixed second partial derivatives to find that

$$\frac{\partial^2 V_1}{\partial x^2} + \frac{\partial^2 V_1}{\partial y^2} = 0 \tag{8-56}$$

The proof for V_2 is similar. Thus, every analytic function contains *two* solutions to Laplace's equation. Furthermore, again because of the Cauchy-Riemann conditions,

$$\begin{aligned}\nabla V_1 \cdot \nabla V_2 &= \left(\frac{\partial V_1}{\partial x}\hat{\mathbf{i}} + \frac{\partial V_1}{\partial y}\hat{\mathbf{j}}\right) \cdot \left(\frac{\partial V_2}{\partial x}\hat{\mathbf{i}} + \frac{\partial V_2}{\partial y}\hat{\mathbf{j}}\right) \\ &= \frac{\partial V_1}{\partial x}\frac{\partial V_2}{\partial x} + \frac{\partial V_1}{\partial y}\frac{\partial V_2}{\partial y} = 0 \\ &\Longrightarrow \nabla V_1 \perp \nabla V_2 \end{aligned} \tag{8-57}$$

Since the equipotentials of V_2 are perpendicular to $\mathbf{E}_2 = -\nabla V_2$ (Section 4-4), we conclude from Eq. (8-57) that these equipotentials are also tangent to $\mathbf{E}_1 = -\nabla V_1$. Similarly, the equipotentials of V_1 are tangent to $\mathbf{E}_2 = -\nabla V_2$. Thus, not only are V_1 and V_2 solutions to Laplace's equation in two dimensions, but also they are conjugate solutions in the sense that the field lines of one potential coincide with the equipotentials of the other. It is indeed unfortunate that these elegant properties in two dimensions cannot be extended to three dimensions.

We shall now illustrate one way in which the above properties can be used effectively.[4] Suppose $f(z) = z^2$. The resulting two potentials are then given by Eq. (8-52), and, in particular, the equipotential curves for these two potentials satisfy

$$x^2 - y^2 = \alpha \qquad 2xy = \beta \tag{8-58}$$

respectively. Here α and β are constants. In both cases, the equipotentials are hyperbolas. In the first case each hyperbola is bisected by the x or y axis, and in the second case each has the axes as asymptotes; the equipotentials for selected values of α and β are shown in Fig. 8-4(a). Once the equipotentials for a given analytic function are known, we can associate one or more physical situations with the potentials. If, for example, the regions exterior to the hyperbolas for $\alpha = \pm 4$ are occupied by conductors maintain-

[4]The treatment in this paragraph is patterned after a similar treatment in *The Feynman Lectures on Physics* by R. P. Feynman, R. B. Leighton, and M. Sands (Addison-Wesley Publishing Company, Inc., Reading, Mass., 1964), Volume II, Lecture 7, and is used here by permission of Addison-Wesley Publishing Company, Inc.

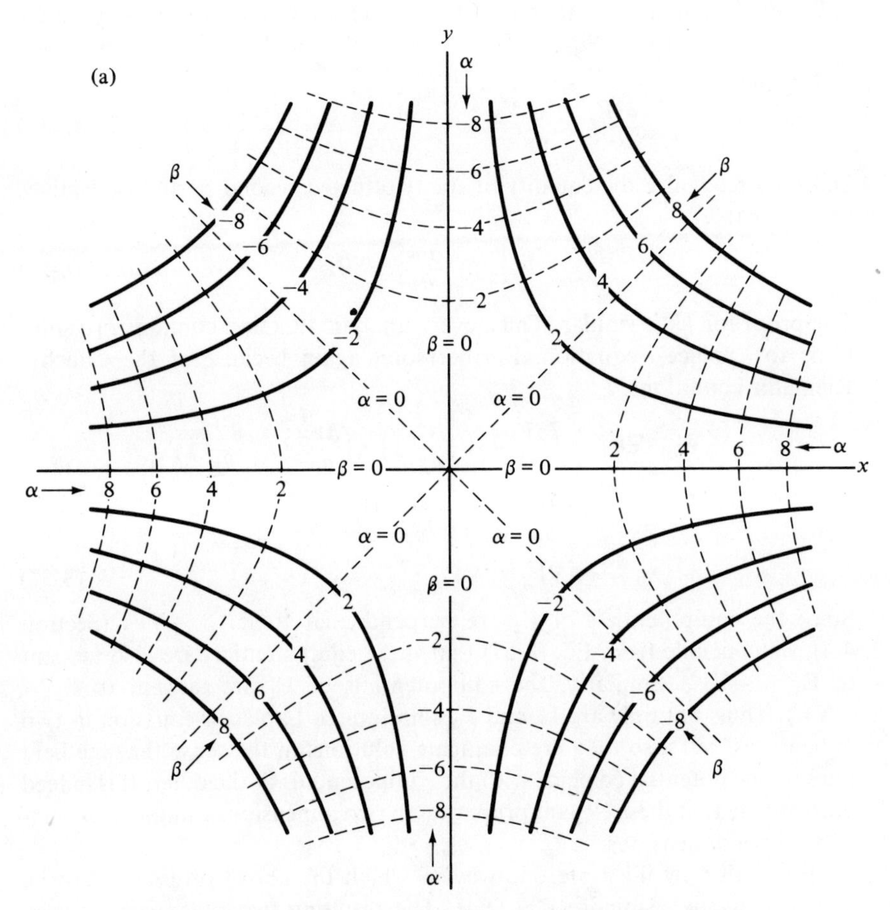

Fig. 8-4. Solutions to Laplace's equation obtained from the analytic function $f(z) = z^2$. Part (a) shows the two sets of equipotentials, with those for α (β) = constant being shown dashed (solid); part (b) (*opposite page*) shows the fields and equipotentials in a quadrupole lens.

ed at the potentials given by α, then the hyperbolas for values of α satisfying $|\alpha| \leq 4$ are the equipotentials in the region R between the conductors, and the hyperbolas for various values of β give the field lines in R [Fig. 8-4(b)]. The resulting field $\mathbf{E}_1$ in R is given by

$$\mathbf{E}_1 = -\nabla V_1 = -2x\hat{\mathbf{i}} + 2y\hat{\mathbf{j}} \tag{8-59}$$

and is linear in both coordinates; it is used in the so-called *quadrupole lens* for focusing beams of charged particles. Because the x and y axes, which correspond to $\beta = 0$, are at zero potential, the first quadrant in Fig. 8-4(a) shows the field lines (values of α) and the equipotentials (values of β) in a

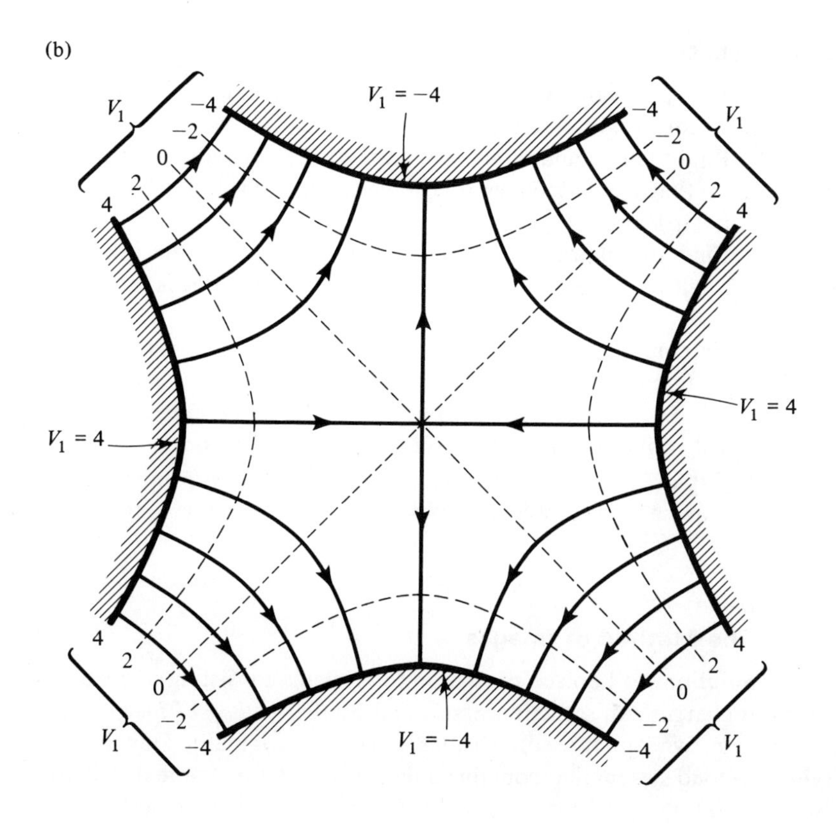

90° conducting wedge maintained at zero potential. It is left as an exercise for the reader to find still other physical problems described by this set of equipotentials.

The approach described in the above paragraphs in a sense attacks problems involving Laplace's equation by *first* finding a solution and *then* seeking the problem that it solves. We could assemble a table of such solutions simply by exploring several analytic functions. We would find that $f(z) = \ell n\, z$ relates physically to an infinite line charge, $f(z) = \sqrt{z}$ to a semiinfinite conducting plate, and so on. The method, however, is certainly not a general method and is made only a little more general by the theory of conformal mapping, which is beyond the scope of this book. (See P8-40.) Thus, although this technique is powerful, perhaps even beautiful, when it works, it is not useful for all problems or even for all two-dimensional problems.

PROBLEMS

P8-18. Verify that the two examples in Eqs. (8-52) and (8-53) satisfy the Cauchy-Riemann conditions, Eq. (8-54).

P8-19. Find the real and imaginary parts of z^n in cylindrical coordinates and recognize that you have generated some of the terms in Eq. (8-29). *Hint*: Write z in polar form.

P8-20. Find the charge density on the conductors in the quadrupole lens of Fig. 8-4(b) at the points where the conducting surfaces intersect the axes.

P8-21. Explore the equipotentials and field lines arising from $f(z) = \ell n\, z$. Include an evaluation of the field corresponding to V_1 and compare the result with Eq. (4-25) for the field of a line charge. Describe at least one other physical problem that would give rise to these fields.

P8-22. Explore the equipotentials and field lines arising from $f(z) = \sqrt{z}$ and describe at least one physical problem that would give rise to these fields.

8-6 The Method of Images

Solutions to Laplace's equation for systems consisting of conductors and point charges can sometimes be found by the *method of images*, which works in one, two, and three dimensions. In essence this method involves identifying a second system that contains only *simple*, *known* charge distributions (usually point charges) but that establishes equipotentials on surfaces coinciding with the conducting surfaces in the first system. The simpler system is chosen so that its potential can be easily written down. Because of the way the simpler distribution is chosen, however, its potential also satisfies the boundary conditions appropriate to the *original* system. Since solutions to Laplace's equation under those conditions are unique, the potential of the simpler system is therefore *everywhere* identical to that of the original system, at least in the region outside the conductors (which is the only region of interest anyway), and we have an immediate solution to the original problem. In general, however, the method is successful only if we happen to *know* a simpler system that establishes the necessary equipotentials. There is no systematic way to *find* the simpler system corresponding to an arbitrarily given set of equipotentials. The procedure is therefore much like that in the previous section: We compile a table of the equipotentials established by various simple charge distributions and then refer to this table whenever we think

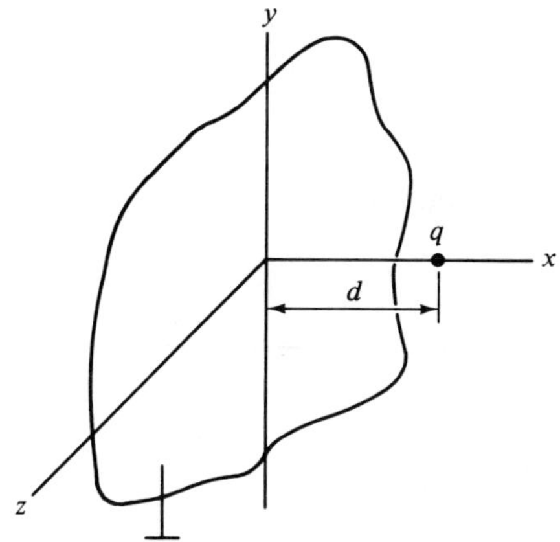

Fig. 8-5. A point charge in front of a grounded, conducting plate.

the method of images might be successful. In the remainder of this section, we shall develop a few entries for that table.

Suppose first that the potential established by a point charge q placed in front of an infinite plane conducting sheet is desired (Fig. 8-5). Let the conductor be grounded (i.e., at zero potential). We therefore seek a solution to Laplace's equation subject to the requirements (1) that the potential be zero in the y-z plane, (2) that there be a point charge q at $(d, 0, 0)$, and (3) that the potential go to zero as $x \longrightarrow \infty$. The domain of the problem is that region for which $x > 0$, exclusive of a small volume surrounding the point charge. Now the charge distribution shown in Fig. 8-6 has the right characteristics in the region $x > 0$ and has a zero potential surface in the y-z plane. (Why?) By the uniqueness theorem, the potential for $x > 0$ in Fig. 8-6 is then the same as that in Fig. 8-5 and the potential for the point charge in front of the grounded conductor is therefore given by

$$V(x, y, z) = \frac{q}{4\pi\epsilon_0}\left[\frac{1}{\sqrt{(x-d)^2 + y^2 + z^2}} - \frac{1}{\sqrt{(x+d)^2 + y^2 + z^2}}\right] \tag{8-60}$$

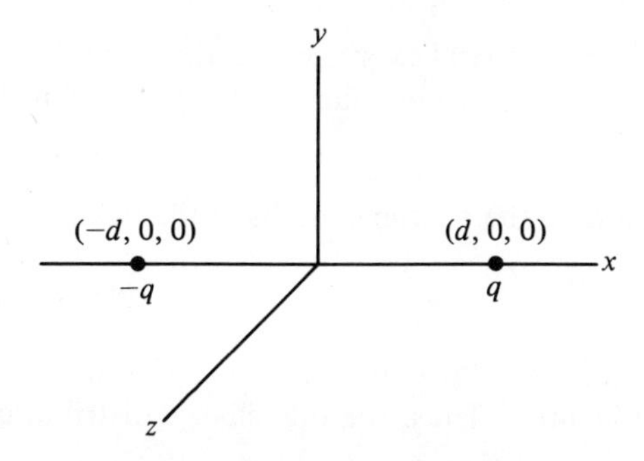

Fig. 8-6. A distribution of point charges equivalent to the charge and plate in Fig. 8-5. The equivalence is valid only in the region $x > 0$.

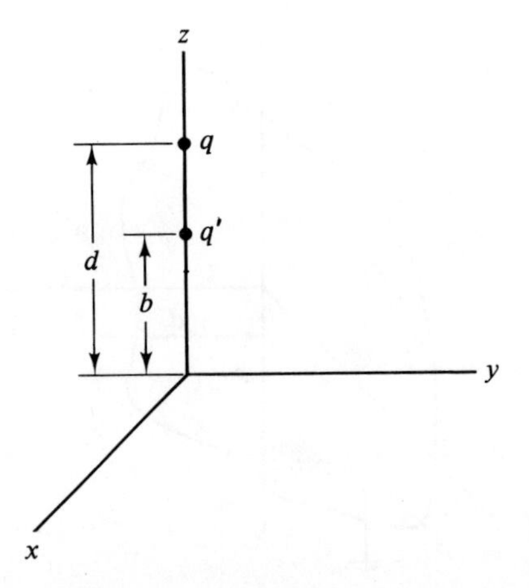

Fig. 8-7. Coordinates for evaluating the potential established by two point charges.

where (x, y, z) are the coordinates of the observation point. The first entry in our table of images then is the image of a point charge in an *infinite*, *plane* conductor at zero potential, which image is a charge of opposite sign located behind the conductor the same distance as the original charge is in front of the conductor. Here as always, image charges will turn out to be located *inside* conductors and these added charges are therefore always in a region of space that is not part of the region of interest.

A second useful set of point charges is shown in Fig. 8-7. The potential established by this distribution at a point (r, θ, ϕ) in spherical polar coordinates is

$$V(r, \theta, \phi) = \frac{q'}{4\pi\epsilon_0}\frac{1}{|\mathbf{r} - b\hat{\mathbf{k}}|} + \frac{q}{4\pi\epsilon_0}\frac{1}{|\mathbf{r} - d\hat{\mathbf{k}}|}$$

$$= \frac{q'}{4\pi\epsilon_0}\frac{1}{\sqrt{r^2 + b^2 - 2rb\cos\theta}} + \frac{q}{4\pi\epsilon_0}\frac{1}{\sqrt{r^2 + d^2 - 2rd\cos\theta}} \tag{8-61}$$

The surface on which $V(r, \theta, \phi) = 0$ therefore satisfies the condition

$$-\frac{q'}{q} = \sqrt{\frac{r^2 + b^2 - 2rb\cos\theta}{r^2 + d^2 - 2rd\cos\theta}} \tag{8-62}$$

which in general expresses a rather complicated relationship between r and θ. If, however, we introduce a length a defined by

$$bd = a^2 \tag{8-63}$$

and take the charge q' to have the value

$$q' = -q\frac{a}{d} \tag{8-64}$$

we find that the equation $r = a$ of a sphere of radius a reduces Eq. (8-62) to an identity. Thus, for this charge distribution (and actually for any pair of

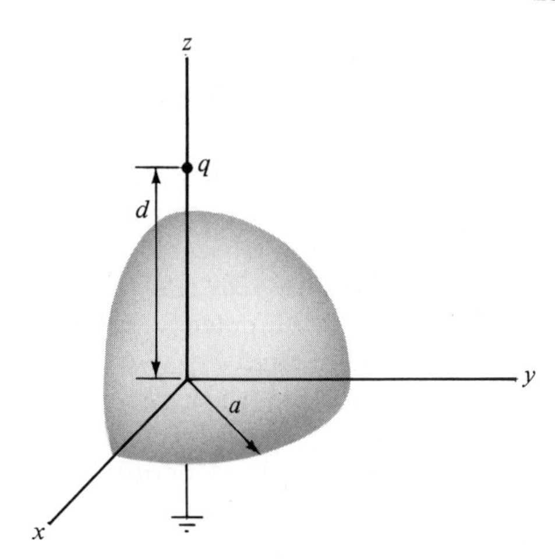

Fig. 8-8. A point charge outside a grounded, conducting sphere.

point charges of opposite sign; P8-30) there exists a *spherical* equipotential surface on which the potential is zero. Further, the potential can be changed to any other constant value by placing a third charge at the center of the sphere. (Why?) This second entry for our table of images can be used, for example, to find the potential established when a point charge q is placed a distance d from the center of a grounded conducting sphere of radius a (Fig. 8-8). In the region outside the conducting sphere, the potential is the same as that established by q and its image q' located at $(0, 0, b)$, where q' is given by Eq. (8-64) and b is given by Eq. (8-63). (See P8-27.)

A third useful entry for our table of images is the potential of two infinitely long, oppositely charged, parallel line charges. The equipotential surfaces are cylinders (P8-31). Thus, some potential problems involving line charges outside of conducting cylinders are readily solved by the method of images.

PROBLEMS

P8-23. (a) Determine the charge density $\sigma(y, z)$ on the plate in Fig. 8-5. (b) Calculate the total charge Q on the plate by direct integration over the plate and compare the result with the image charge. (c) Calculate the force **F** experienced by the charge q in Fig. 8-5 by integration over the plate and observe how the answer could have been more easily obtained using the image charge.

P8-24. Determine the image distribution for an arbitrary static charge distribution placed in front of an infinite, grounded, conducting plate. *Hint*: Consider first a distribution consisting of two point charges in front of the plate.

P8-25. Let a grounded conductor occupy all of space except the region for which $x > 0$, $y > 0$. Find the potential established in that region when a point charge q is placed at coordinates $(a, b, 0)$.

P8-26. A point charge q is placed midway between two infinite, parallel, plane conducting plates, both of which are grounded. Let the the plates be separated by a distance a. Use the method of images to find an infinite series for the potential in the region between the plates and examine carefully whether this series converges. Interesting discussions and several references may be found in papers by J. J. G. Scanio, *Am. J. Phys.*, **41**, 415 (1973) and by B. G. Dick, *Am. J. Phys.*, **41**, 1289 (1973).

P8-27. (a) Find the potential established by the system in Fig. 8-8 by eliminating b and q' from Eq. (8-61). (b) Show explicitly that $V = 0$ when $r = a$. (c) Find the charge density $\sigma(\theta)$ on the surface of the conducting sphere, and show by direct integration that the total charge on the sphere is equal to q'. Sketch a graph of $\sigma(\theta)$ versus θ. *Optional*: Determine the force of attraction between the charge q and the sphere by two different methods.

P8-28. A point charge q is placed a distance b from the center of a spherical cavity of radius a in a grounded conductor, with $b < a$. Determine the potential in the cavity and the charge density on its walls.

P8-29. Apply the method of images to find the charge density on an uncharged conducting sphere placed in a uniform electric field. *Hint*: Generate a uniform field by imagining the sphere between two point charges q and $-q$, each placed a distance L from the center of the sphere on an extension of a single diameter. Then, let $L \longrightarrow \infty$, keeping q/L^2 constant.

P8-30. Two point charges q and q', with $q/q' < 0$, are placed a distance s apart. Describe a way to determine which charge lies *inside* the resulting *spherical* equipotential and find the location of the center and the radius of that equipotential.

P8-31. Two infinite line charges lie in the x-y plane parallel to the x axis. The line at $y = d$ carries a linear charge density λ and the line at $y = -d$ carries a linear charge density $-\lambda$. Find an equation for the equipotential curves in the y-z plane, show that these curves are circles centered on the y axis. *Hint*: The potential of a single line charge is given in P4-18. *Optional*: Sketch a graph showing these equipotentials for several (positive *and* negative) values of the potential.

8-7 Numerical Solution of Laplace's Equation[5]

We have now explored several techniques for obtaining *analytic* solutions to Laplace's equation. None of these techniques, however, is perfectly general. Separation of variables works only if the bounding surfaces

[5]Much of the material in this section appears also in one of the author's contributions to *Computer-Oriented Physics Problems* edited by J. W. Robson (Commission on College Physics, College Park, Md., 1971) and is used here in slightly revised form by permission of the Commission on College Physics. A more detailed treatment may be found, for example, in B. Carnahan, H. A. Luther, and J. O. Wilkes, *Applied Numerical Methods* (John Wiley & Sons, Inc., New York, 1969), Chapter 7.

coincide with the natural surfaces in a coordinate system in which Laplace's equation is separable and, even when it does work, the method usually yields an infinite series whose physical meaning is difficult to visualize; the method of complex variables works only in two dimensions and then only if we happen (essentially accidentally) to find the right analytic function; and the method of images works only when we know a priori a simple charge distribution that establishes the correct equipotentials. In this section, we shall discuss a simple method to avoid these restrictions by solving Laplace's equation *numerically*.

We shall first derive a basic property of solutions to Laplace's equation. Assume $V(x, y)$ satisfies $\nabla^2 V = 0$. Now, from a Taylor series expansion (Appendix B), we find that

$$V(x + d, y) = V + d\frac{\partial V}{\partial x} + \frac{1}{2}d^2\frac{\partial^2 V}{\partial x^2} + \frac{1}{6}d^3\frac{\partial^3 V}{\partial x^3} + O(d^4) \quad (8\text{-}65)$$

$$V(x - d, y) = V - d\frac{\partial V}{\partial x} + \frac{1}{2}d^2\frac{\partial^2 V}{\partial x^2} - \frac{1}{6}d^3\frac{\partial^3 V}{\partial x^3} + O(d^4) \quad (8\text{-}66)$$

where functions with unspecified arguments are evaluated at the point (x, y) and $O(d^4)$ stands symbolically for terms of order d^4 and higher that have been omitted. Adding Eqs. (8-65) and (8-66), we find that

$$V(x + d, y) + V(x - d, y) = 2V + d^2\frac{\partial^2 V}{\partial x^2} + O(d^4) \quad (8\text{-}67)$$

Similarly,

$$V(x, y + d) + V(x, y - d) = 2V + d^2\frac{\partial^2 V}{\partial y^2} + O(d^4) \quad (8\text{-}68)$$

Finally, adding Eqs. (8-67) and (8-68), dividing the result by 4, and setting $\partial^2 V/\partial x^2 + \partial^2 V/\partial y^2 = 0$ (V satisfies $\nabla^2 V = 0$ by hypothesis), we find that

$$\begin{aligned} V(x, y) = \tfrac{1}{4}[&V(x + d, y) + V(x - d, y) \\ &+ V(x, y + d) + V(x, y - d)] + O(d^4) \end{aligned} \quad (8\text{-}69)$$

To within terms of order d^4, the solution to Laplace's equation at a particular point (x, y) is therefore given by the average of the solutions at four neighboring points, each displaced from the point (x, y) in a direction parallel to one of the coordinate axes by a distance d; i.e., the solution at point P in Fig. 8-9 is the average of the solutions at the four points P_1, P_2, P_3, and P_4.

To illustrate a simple procedure by which Eq. (8-69) can be used to obtain approximate solutions to Laplace's equation, we consider again the problem in Fig. 8-1, taking $a = b$ and $V_0 = 100$ in whatever units are appropriate. Let the region be divided by a square grid having four points in the interior of the region of interest (Fig. 8-10). An approximate solution to Laplace's equation is then obtained by finding values satisfying Eq. (8-69) at each interior point. The unknowns are represented by question marks in Table 8-2(a). (Fortunately, the procedure by which these unknowns are found does not involve the values at the four corners, two of which are ambiguous.) We

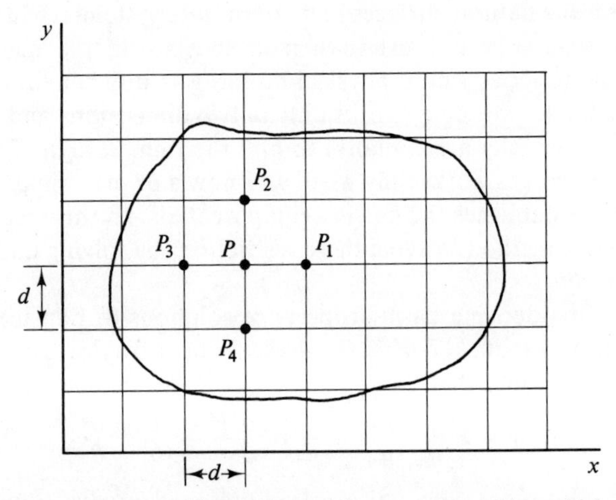

Fig. 8-9. A square grid superimposed on a region within which a solution to Laplace's equation is sought.

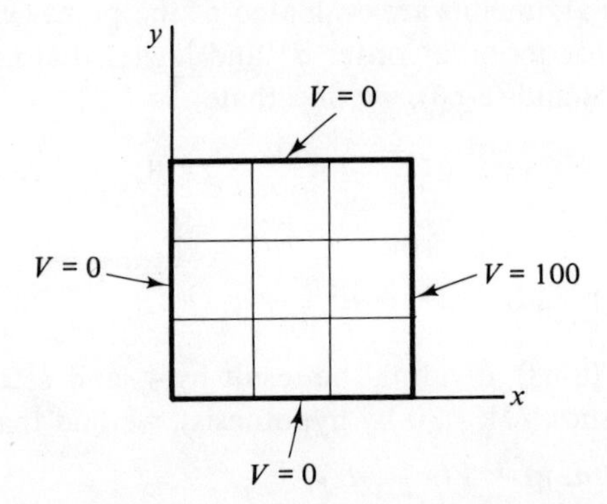

Fig. 8-10. A grid of four interior points superimposed on a square region in the x-y plane.

start by *guessing* the values at the interior points, as, for example, in Table 8-2(b). This guess is then improved by progressing systematically through the grid of *interior* points, replacing the value at each point with the average of the values at the four nearest neighboring points. First the 30 in the upper left corner of Table 8-2(b) is replaced by $(0 + 0 + 50 + 30)/4 = 20$ [Table 8-2(c)]; next the 50 in the upper right corner of Table 8-2(c) is replaced by $(20 + 0 + 100 + 50)/4 = 42.5$ [Table 8-2(d)]; then the 30 in the lower left corner of Table 8-2(d) is replaced by $(0 + 20 + 50 + 0)/4 = 17.5$ [Table 8-2(e)]; and finally the 50 in the lower right corner of Table 8-2(e) is replaced by $(17.5 + 42.5 + 100 + 0)/4 = 40$, yielding Table 8-2(f) and completing the first iteration of a process that can be repeated indefinitely. Table 8-2(g) shows the approximate solution obtained by repeating this procedure starting with the numbers in Table 8-2(f), and Table 8-2(h) shows the approximate solution resulting after the tenth successive application of this procedure. It is readily verified that the solution in Table 8-2(h) satisfies Eq. (8-69) exactly

TABLE 8-2 Successive Steps in the Numerical Solution of Laplace's Equation for the Problem in Fig. 8-10

(a) The unknowns. Each entry corresponds to an intersection of two lines of the grid in Fig. 8-10. Note the ambiguity at the two right-hand corners.

0	0	0	0,100
0	?	?	100
0	?	?	100
0	0	0	0,100

(b) An initial guess.

0	0	0	0,100
0	30	50	100
0	30	50	100
0	0	0	0,100

(c) The first step toward an improved approximation.

0	0	0	0,100
0	20	50	100
0	30	50	100
0	0	0	0,100

(d) The second step toward an improved approximation.

0	0	0	0,100
0	20	42.5	100
0	30	50	100
0	0	0	0,100

(e) The third step toward an improved approximation.

0	0	0	0,100
0	20	42.5	100
0	17.5	50	100
0	0	0	0,100

(f) The final step toward an improved approximation (completion of the first iteration).

0	0	0	0,100
0	20	42.5	100
0	17.5	40	100
0	0	0	0,100

(g) Approximate solution after the second iteration.

0	0	0	0,100
0	15.000	38.750	100
0	13.750	38.125	100
0	0	0	0,100

(h) Approximate solution after the tenth iteration.

0	0	0	0,100
0	12.500	37.500	100
0	12.500	37.500	100
0	0	0	0,100

at all points; continued iterations merely regenerate this solution. Note that at each calculation the most recently obtained value of the solution is used in the right-hand side of Eq. (8-69) and also that a poor initial guess requires a large number of iterations to achieve a desired accuracy but does not undermine the ultimate convergence of the process. The results of solving this

TABLE 8-3 Numerical Solution of the Example in the Text on a Grid with 225 Interior Points

0.0	0.0	0.0	0.0	0.0	0.0	0.0	0.0	0.0	0.0	0.0	0.0	0.0	0.0	0.0	0.0	xxxxx
0.0	0.4	0.9	1.4	1.9	2.5	3.3	4.1	5.2	6.6	8.4	10.8	14.2	19.7	29.4	49.6	100.0
0.0	0.8	1.7	2.7	3.7	4.9	6.4	8.1	10.2	12.8	16.1	20.5	26.5	35.2	48.3	68.9	100.0
0.0	1.2	2.5	3.9	5.4	7.2	9.2	11.6	14.6	18.2	22.7	28.6	36.1	46.2	59.7	77.7	100.0
0.0	1.6	3.2	4.9	6.9	9.1	11.6	14.6	18.3	22.7	28.1	34.8	43.2	53.7	66.7	82.3	100.0
0.0	1.8	3.7	5.8	8.0	10.6	13.6	17.1	21.2	26.2	32.2	39.5	48.2	58.7	70.9	84.9	100.0
0.0	2.0	4.1	6.4	8.9	11.7	15.0	18.8	23.3	28.7	35.1	42.6	51.5	61.8	73.5	86.4	100.0
0.0	2.2	4.4	6.8	9.4	12.4	15.9	19.9	24.6	30.2	36.7	44.4	53.3	63.5	74.9	87.2	100.0
0.0	2.2	4.5	6.9	9.6	12.6	16.2	20.2	25.0	30.6	37.2	45.0	53.9	64.0	75.3	87.4	100.0
0.0	2.2	4.4	6.8	9.4	12.4	15.9	19.9	24.6	30.2	36.7	44.4	53.3	63.5	74.9	87.2	100.0
0.0	2.0	4.1	6.4	8.9	11.7	15.0	18.8	23.3	28.7	35.1	42.6	51.5	61.8	73.5	86.4	100.0
0.0	1.8	3.7	5.8	8.0	10.6	13.6	17.1	21.2	26.2	32.2	39.5	48.2	58.7	70.9	84.9	100.0
0.0	1.6	3.2	4.9	6.9	9.1	11.6	14.6	18.3	22.7	28.1	34.8	43.2	53.7	66.7	82.3	100.0
0.0	1.2	2.5	3.9	5.4	7.2	9.2	11.6	14.6	18.2	22.7	28.6	36.1	46.2	59.7	77.7	100.0
0.0	0.8	1.7	2.7	3.7	4.9	6.4	8.1	10.2	12.8	16.1	20.5	26.5	35.2	48.3	68.9	100.0
0.0	0.4	0.9	1.4	1.9	2.5	3.3	4.1	5.2	6.6	8.4	10.8	14.2	19.7	29.4	49.6	100.0
0.0	0.0	0.0	0.0	0.0	0.0	0.0	0.0	0.0	0.0	0.0	0.0	0.0	0.0	0.0	0.0	xxxxx

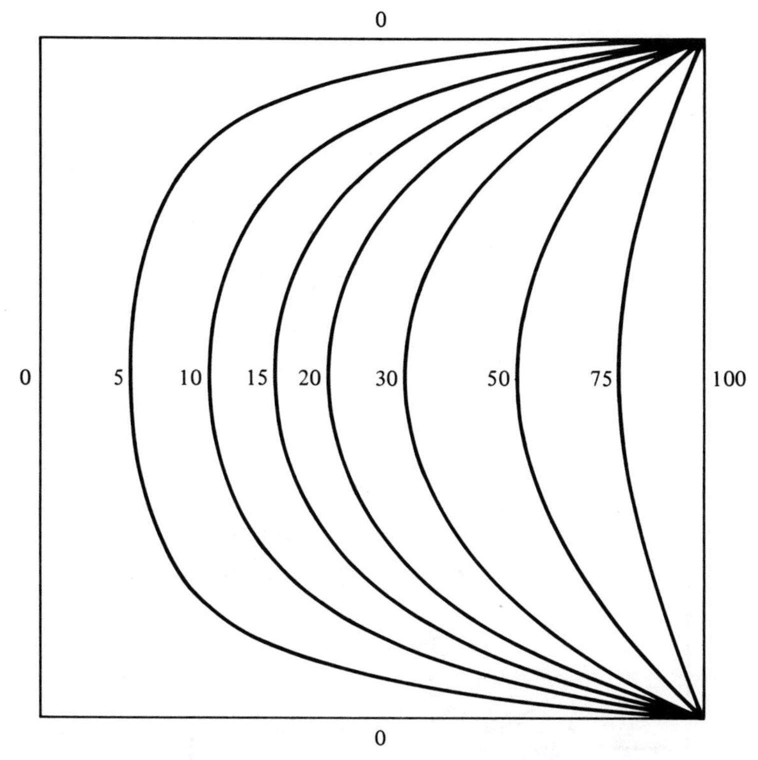

Fig. 8-11. Equipotentials obtained from a numerical solution of the example in the text. Each equipotential is labeled with the corresponding value of V.

same problem numerically on a finer grid are shown in Table 8-3 and in Fig. 8-11. (See P8-33.)

Although we have here considered only a *two-dimensional* problem in a *simple* geometry subject to *Dirichlet* boundary conditions, the method illustrated is not so restricted. It can be easily extended to three dimensions, and it is confined neither to Dirichlet boundary conditions nor to simple geometries. Certainly, application of the technique to these more general situations is more involved, but the basic idea remains unaltered. This simple method yields readily interpreted solutions for any unambiguously stated problem involving Laplace's equation. Indeed, numerical methods are the only methods that can be counted on to work for *all* problems.

PROBLEMS

P8-32. Verify the results given in Table 8-2(g) for the approximate solution to the example in the text after the second iteration.

P8-33. A simple algorithm for solving Laplace's equation numerically in a rectangular region is shown in Fig. P8-33(a). (a) Decide on an explicit

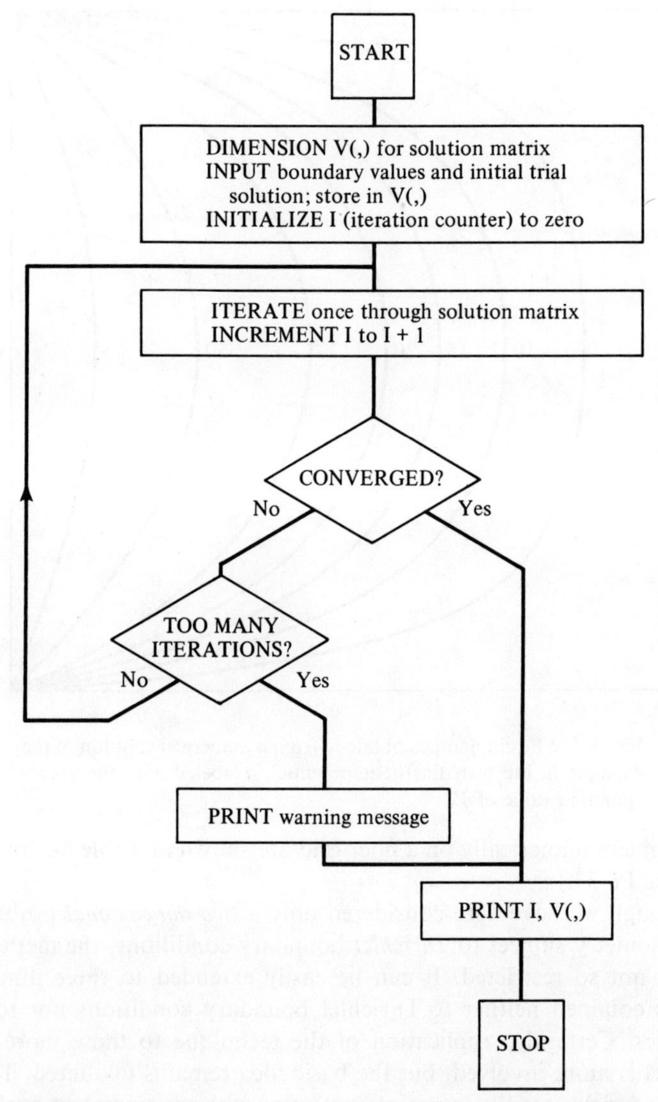

Fig. P8-33a. An algorithm for solving Laplace's equation numerically.

criterion for determining when the current approximation has converged adequately to the true solution. A possible criterion would terminate the iterations when the value assigned to each interior point in the grid differs by no more than some predetermined amount ϵ from the average of the values at the four neighboring points. (b) Translate the algorithm in Fig. P8-33,

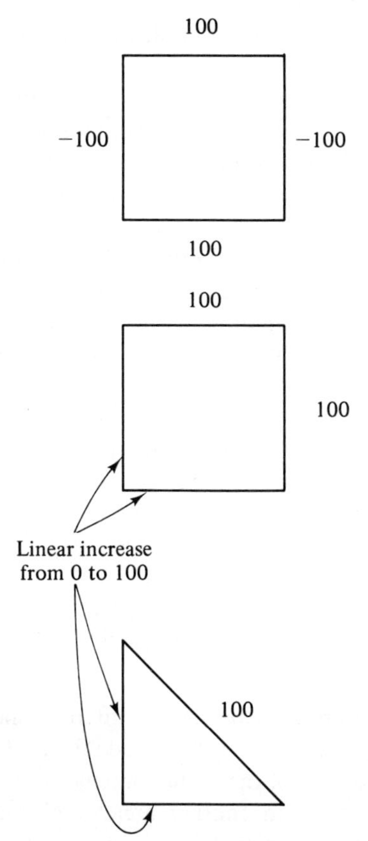

Fig. P8-33b. Suggested additional boundary conditions.

including your convergence criterion, into a computer program and execute your program to find an approximate solution to Laplace's equation when the boundary conditions are those in Fig. 8-10 and the grid has 25 interior points. Use the results in Table 8-2(h) as a basis for determining a judicious initial guess for the solution on this finer grid. (c) Find and sketch a graph of the charge density on the surface of the conducting boundary if the square has a side of 10 cm and V is expressed in volts. *Hint*: This charge density is given by the normal derivative of V [Eq. (8-2)]. Evaluate the derivative numerically, noting, for example, that $df/dx \approx [f(x + a) - f(x)]/a$, or (for greater accuracy) that $df/dx \approx [-f(x + 2a) + 4f(x + a) - 3f(x)]/2a$. *Optional*: (1) Go on to a grid of 121 interior points. (2) Incorporate calculation of the charge density in your program. (3) Modify your program to start with a grid of 1 interior point and then move automatically in succession to grids with 9, 49, 225, . . . interior points using the solution on each grid to obtain the initial guess for the solution on the next grid. (4) Find solutions for other boundary conditions, such as those shown in Fig. P8-33(b).

P8-34. (a) Develop a numerical method for finding a solution to the problem in Fig. P8-2(b). Take $a = 1$. *Hint*: On the axes where Neumann conditions are imposed, Eq. (8-69) cannot be used directly. We circumvent this problem at points on the y axis, for example, by imagining an extension of the domain into the region $x < 0$ and noting that the Neumann condition then requires that $[V(d, y) - V(-d, y)]/2d$ be zero, or $V(d, y) = V(-d, y)$. Thus, when Eq. (8-69) is evaluated at $x = 0$, the term in $V(-d, y)$ that arises can be replaced with $V(d, y)$ and the result involves only points lying within or on the boundaries of the problem. Similar arguments apply to the Neumann condition on the x axis and to the origin (which must be treated separately). Thus, there are four different classifications for grid points (interior, y axis, x axis, origin), each of which must be treated differently in the iterative "march" through the grid. (b) Either by hand or by machine, explore the solution to the problem in Fig. P8-2(b), find the equipotentials, and determine the charge densities on the conducting boundaries. *Hint*: See the hint in P8-33(c).

8-8 Solution of Laplace's Equation by Experiment: The Method of Analogy

Because Laplace's equation describes many different phenomena in many different areas of physics by *mathematically* identical statements, a solution to Laplace's equation for one of these phenomena can be taken over intact to the analogous phenomenon in another area of physics. In some cases, the physical situation corresponding to a given mathematical statement as applied to one area of physics can be easily constructed experimentally. The solution to Laplace's equation for this situation, and hence for analogous situations in other areas of physics as well, can then be *measured*. The most common example of this *method of analogy* in electrostatics is the (two-dimensional) field tray, in which electrodes having the desired shapes and maintained at the desired potentials are placed in a thin layer of a conducting solution. Because both are described mathematically by the same boundary value problem, the electrostatic potential in the field tray is identical to the potential established by a two-dimensional system of conductors located in free space but having the same geometry as those placed in the solution. The potential in the field tray, however, is the more easily measured potential. Deflections of elastic membranes, steady-state temperature distributions, flow patterns in fluids when the velocity field is steady and "curlless", and many other physical situations can be used to obtain solutions to Laplace's equation for analogous electrostatic problems, but a detailed development of these techniques would lead us away from the primary objectives of this book.

8-9
Poisson's Equation

When charge is present in the region of interest, the electro*static* potential satisfies Poisson's equation,

$$\nabla^2 V = -\frac{\rho}{\epsilon_0} \tag{8-70}$$

[See Eq. (6-74).] If the charge density ρ is *known*, then the general solution has the form

$$V(\mathbf{r}) = \begin{pmatrix}\text{any solution} \\ \text{to } \nabla^2 V = 0\end{pmatrix} + \frac{1}{4\pi\epsilon_0}\int \frac{\rho(\mathbf{r}')}{|\mathbf{r}-\mathbf{r}'|}\,dv' \tag{8-71}$$

where the added solution to the homogeneous equation provides the freedom to satisfy boundary conditions that may not be satisfied by the integral alone. [Compare Eq. (4-50) and the paragraph following Eq. (4-50).] Thus, if ρ is known, Eq. (8-70) is reducible to Laplace's equation and its solution can be obtained by any of the methods already described. Alternatively, if ρ is known, the single term $\rho(x, y)d^2/4\epsilon_0$ added to Eq. (8-69) provides a starting point for a direct numerical solution of Poisson's equation in two dimensions (P8-36).

When the charge density is not known *explicitly*, we need another equation relating V and ρ before Eq. (8-70) can be solved. The derivation of Child's law (P6-39) included finding such an equation. We conclude this section with another example by applying Eq. (8-70) to determine the character of the electron cloud attracted to a positive test charge q placed in an initially neutral ionized gas or plasma (which contains mobile positive and negative charges). Choose a coordinate system with its origin at the charge q. Then the equilibrium charge density ρ and the electrostatic potential V in the plasma can depend only on the spherical coordinate r, and Poisson's equation reduces in spherical coordinates to

$$\frac{1}{r^2}\frac{d}{dr}\left(r^2\frac{dV}{dr}\right) = -\frac{\rho}{\epsilon_0} = -\frac{q_e}{\epsilon_0}(n_+ - n_-) \tag{8-72}$$

where $n_-(r)$ is the density of electrons in the plasma, $n_+(r)$ is the density of ions (assumed singly ionized), and q_e is the charge on the ions. [See Eq. (2-10).] The particle densities n_+ and n_-, however, can be related to the potential by using a theorem from statistical mechanics: If the density of particles is n_0 in a region where the potential energy is U_0, then the density n in a region where the potential energy is U is given by

$$n = n_0 e^{-(U-U_0)/kT} \tag{8-73}$$

where k is Boltzmann's constant and T is the (absolute) temperature. To apply Eq. (8-73) to the plasma, we take the reference point at $r = \infty$, where we assume the electrostatic potential $V(r)$ has the value zero. Further, we assume that the plasma remains electrically neutral in the region remote from the test charge so that $n_+(\infty) = n_-(\infty)$, and we symbolize both by n_0, which is the density of particles of either charge in the undisturbed plasma. Remembering finally that potential *energy* is potential *times* charge, we find from Eq. (8-73) that

$$n_+ = n_0 e^{-q_e V/kT}; \qquad n_- = n_0 e^{q_e V/kT} \tag{8-74}$$

and hence from Eq. (8-72) that V by itself satisfies

$$\frac{1}{r^2}\frac{d}{dr}\left(r^2\frac{dV}{dr}\right) = -\frac{q_e n_0}{\epsilon_0}[e^{-q_e V/kT} - e^{q_e V/kT}] \tag{8-75}$$

Actually solving this equation to find V is left to P8-35. When $q_e V/kT \ll 1$ (high temperature), the result is

$$V = \frac{q}{4\pi\epsilon_0 r}e^{-r/D}; \qquad D^2 = \frac{\epsilon_0 kT}{2n_0 q_e^2} \tag{8-76}$$

where the length D, called the *Debye shielding length*, measures the scale of the exponential decay of this *shielded Coulomb potential* (sometimes called the *Yukawa potential*). For $n_0 = 10^{21}$ particles/m³ and $T = 10^8$ °K, which are typical of plasmas produced in controlled thermonuclear research, the Debye length is about 2×10^{-4} m. In a plasma described by these parameters, the influence of an extra charge therefore does not extend very far from the position of the charge.

PROBLEMS

P8-35. Solve Eq. (8-75) when $q_e V/kT \ll 1$ subject to the requirements that $V(r) \longrightarrow 0$ as $r \longrightarrow \infty$ and that $V(r) \longrightarrow q/4\pi\epsilon_0 r$ as $r \longrightarrow 0$. *Hints*: (1) $e^\alpha \approx 1 + \alpha$ when α is small. (2) Introduce a new dependent variable $V' = rV$.

P8-36. Repeat the development of Eq. (8-69) when $\nabla^2 V = -\rho/\epsilon_0$, describe a numerical algorithm similar to that in Section 8-7 for solving Poisson's equation, and then apply this algorithm (either by hand or by computer) to solve the equation $\nabla^2 V = -2$ in the interior of a square 10 cm on a side when V is required to be zero everywhere on the boundary. Physically, this problem relates more directly to the twisting of a square shaft under torsion than to a problem in electrostatics.

Supplementary Problems

P8-37. A rectangular parallelopiped is bounded by the coordinate planes and the planes $x = a$, $y = b$, and $z = c$. All surfaces of this parallelopiped are maintained at zero potential except the surface at $z = c$ and that surface is maintained at a constant potential V_0. Apply the method of separation of variables to find the potential in the interior of the parallelopiped.

P8-38. Apply the method of separation of variables to Laplace's equation in cylindrical coordinates when all three coordinates appear, solve the resulting equations, and write the most general solution subject only to the requirements that the origin be in the domain of the problem and the angle ϕ assume its full range. *Hints*: (1) Seek exponential solutions in z and trigonometric solutions in ϕ. (2) The equation $x^2y'' + xy' + (x^2 - n^2)y = 0$, which reduces to Eq. (8-32) when $n = 0$, is the *nth order Bessel equation*. The only solution that is finite for all values of x including the origin is called the *nth order Bessel function*, $J_n(x)$, and is tabulated, for example, in E. Jahnke and F. Emde, *Tables of Functions* (Dover Publications, Inc., New York, 1945).

P8-39. Apply the method of separation of variables to Laplace's equation in spherical coordinates when all three coordinates appear, solve the resulting equations, and write the most general solution subject only to the requirements that $\theta = 0$ and π be in the domain of the problem and that the angle ϕ assume its full range. *Hints*: (1) Seek trigonometric solutions in ϕ. (2) Set $x = \cos\theta$ in the equation for the θ dependence. The resulting equation, $(1 - x^2)y'' - 2xy' + [k - m^2/(1 - x^2)]y = 0$, which reduces to Eq. (8-38) when $m = 0$, is the *associated Legendre equation*. Here m is an *integer* that arises from solution of the equation for the ϕ dependence and k is another separation constant. Solutions finite for all x in the range $|x| \leq 1$ can be found only when k has one of the values $n(n + 1)$, where n is a nonnegative integer. One such solution exists for each value of n; it is called an *associated Legendre function*, $P_n^m(x)$, and is tabulated, for example, in E. Jahnke and F. Emde, *op. cit.*, P8-38.

P8-40. A *conformal mapping* from the complex z plane to the complex w plane associates points in the two planes by the transformation $w = f(z)$, where f is an analytic function of z. For some two-dimensional problems, proper selection of this transformation can convert a complicated geometry in the z plane to a simpler geometry in the w plane. A solution to Laplace's equation can then be found in the simpler geometry of the w plane and then transformed back to the z plane to yield a solution to the original problem. This problem illustrates this very powerful but also rather specialized technique. Suppose a solution to Laplace's equation in the half-plane $y > 0$ is desired when a conducting plate at zero potential lies along the positive x axis and a conducting plate at 100 V lies along the negative x axis. (a) Show that the transformation $w = \ell n\, z$ [Eq. (8-53)] transforms the region of interest in the z plane to the region $-\infty < u < \infty$, $0 < v < \pi$ in the w plane, where $u = \text{Re}\, w$ and $v = \text{Im}\, w$. Further, show that the positive x axis is transformed to the line $v = 0$ and the negative x axis to the line $v = \pi$. (b) Show that the desired solution in the w plane is the imaginary part of the analytic function $q(w) = 100\, w/\pi$. (c) Since an analytic function of an analytic function is analytic, the solution $V(x, y)$ in the z plane is then $\text{Im}\, q[w(z)]$. Show that $V(x, y) = (100/\pi) \tan^{-1}(y/x)$, sketch the equipotentials and field lines, and determine the charge density on the conducting plates. *Optional*: Show that the transformation $z' = [(i - z)/(i + z)]$ transforms the region $y > 0$ in the z plane to the interior of the unit circle $(x')^2 + (y')^2 = 1$ in the z' plane and find a solution to Laplace's equation in the interior of a unit circle when the bounding arc $0 < \phi' < \pi$ is kept at zero potential and the arc $\pi < \phi' < 2\pi$ is kept at 100 V. Sketch the equipotentials and field lines.

P8-41. Find an integral giving the solution $V(x, y)$ to Laplace's equation in the two-dimensional domain $-\infty < x < \infty$, $0 < y < \infty$ when V and its derivatives approach zero as $|x| \rightarrow \infty$ and as $y \rightarrow \infty$, and $V(x, 0) = 0$ except in $|x| \leq 1$, where $V(x, 0) = 1$. *Hint*: Take the Fourier transform (Appendix D) of Laplace's equation with respect to x (i.e., multiply by $e^{-i\kappa x}$ and integrate on x), note that the transform of $\partial^2 V/\partial x^2$ is $-\kappa^2 \tilde{V}(\kappa, y)$ (Why?), solve the resulting ordinary differential equation for $\tilde{V}(\kappa, y)$ but let the constants of integration depend on κ, and then invert the transform to return to $V(x, y)$, determining any constants by the boundary conditions.

P8-42. Obtain the integral in Eq. (8-71) by taking a three-dimensional Fourier transform (Appendix D) of Poisson's equation, solving for $\tilde{V}(\boldsymbol{\kappa})$, and then inverting the transform. *Hints*: (1) The transform of $\partial^2 V/\partial x^2$ is $-\kappa_x^2 \tilde{V}$. (Why?) (2) After writing the formal inversion integral, reexpress $\tilde{\rho}(\kappa)$ as a Fourier transform of $\rho(\mathbf{r})$ and interchange orders of integration to obtain

$$V(\mathbf{r}) = \frac{1}{8\pi^3 \epsilon_0} \int \rho(\mathbf{r}') \left[\int \frac{e^{i\boldsymbol{\kappa}\cdot(\mathbf{r}-\mathbf{r}')}}{\kappa^2}\, d\kappa_x\, d\kappa_y\, d\kappa_z \right] dv'$$

(3) Evaluate the integrals on the components of $\boldsymbol{\kappa}$ by introducing spherical coordinates in $\boldsymbol{\kappa}$-space, taking the polar axis in the direction of $\mathbf{r} - \mathbf{r}'$. (4) Don't be afraid of integral tables.

P8-43. (a) Prove Green's theorem, Eq. (C-27). (b) Show that

$$\int \Phi(\mathbf{r})\, \nabla^2 \left(\frac{1}{|\mathbf{r} - \mathbf{r}'|} \right) dv = -4\pi \Phi(\mathbf{r}')$$

Hint: $\nabla^2[1/|\mathbf{r} - \mathbf{r}'|] = 0$ everywhere except where $|\mathbf{r} - \mathbf{r}'| = 0$. (c) Apply these two results to show that a function V that satisfies Laplace's equation also satifies the *integral* equation

$$V(\mathbf{r}) = \frac{1}{4\pi} \oint_{\Sigma} \left[\frac{\nabla' V(\mathbf{r}')}{|\mathbf{r} - \mathbf{r}'|} - V(\mathbf{r}')\, \nabla' \left(\frac{1}{|\mathbf{r} - \mathbf{r}'|} \right) \right] \cdot d\mathbf{S}'$$

where Σ is any closed surface enclosing the point $\mathbf{r}$. This identity relates the solution to Laplace's equation at a *single point* $\mathbf{r}$ to the solution *and* its normal derivative *over a surface* enclosing $\mathbf{r}$. Similar integral equations are important to the Kirchhoff theory of diffraction.

9

Properties of Matter I: Conduction

Whether they be in the solid, liquid, gas, or plasma phase, macroscopic samples of matter exhibit several different responses when placed in external electric and magnetic induction fields. In Chapters 9, 10 and 11 we shall examine (electric) *conduction*, *dielectric polarization*, and *magnetization*[1] when the external fields are static, and in Chapter 12 we shall extend that treatment to include time-dependent external fields. Each of these responses can be approached either from a macroscopic or a microscopic viewpoint. In the macroscopic approach, we *accept* the properties of matter as exhibited in the macroscopic world, introduce appropriate descriptive concepts analogous to the charge and current densities of Chapter 2, and seek *experimentally determined* relationships among these descriptive concepts and the external fields. In the microscopic approach, we *postulate* a microscopic structure for a particular type of matter and seek to *understand* the observed macroscopic properties by relating them to the presumably simpler behavior of the assumed microscopic constituents. Although we shall not ignore either approach altogether, we shall place greater emphasis on the macroscopic approach.

Qualitatively and microscopically, a *conducting material* is a sample of matter that contains microscopic charged particles or *charge carriers* that are free to move macroscopic distances through the material. *Steady-state con-*

[1] A fourth possible phenomenon—*magnetic* conduction—will be of interest only after free magnetic monopoles have been found, for only then will macroscopic magnetic currents, i.e., a flow of magnetic monopoles, be physically realizable.

duction is the response observed when a conducting material is placed in a *static* electric field, e.g., when the two poles of a battery are connected to opposite ends of a metallic or semiconducting rod or when an ionized gas is placed between two parallel plates maintained at different potentials. Under these circumstances, the motion of the individual carriers has two components. Whether a field is present or not, these carriers move about erratically, collide frequently with one another and with any relatively immobile particles (e.g., "fixed" ions in the crystal lattice of the conducting material) that may be present, and experience random changes in velocity with every collision. Since this *random component* of the velocity of each particle averages to zero over time intervals that are macroscopically short but microscopically long, it results in no net macroscopic migration of charge through the conducting material. When the conducting material is placed in an external field, however, its constituent charges experience nonrandom forces and—in most materials—the charge carriers will be accelerated parallel or antiparallel to the field, depending on the sign of their charge. Further, the frequent collisions, which continue to occur even in the presence of the field, prevent the carriers from accelerating indefinitely. Once any transient effects resulting when the field is first turned on have decayed away (and they do so quite rapidly in "good" conducting materials; see P9-5), a balance between steady acceleration and frequent abrupt changes in velocity is achieved, and the overall microscopic effect of the field is the addition of a *steady* (nonrandom) *drift velocity* to the total velocity of each carrier. In the macroscopic world, a steady drift velocity of the carriers produces a steady macroscopic migration of charge (i.e., a steady current) through the conducting material, which is then said to be *conducting*.

9-1 Macroscopic Description: Conductivity and Ohm's Law

In the qualitative terms of the previous paragraphs, the macroscopic response of a conductor to an electric field is the appearance of a macroscopic current. Crudely, we can view the electric field as a cause and the current as its effect. Now a quantitative description of any phenomenon is built on quantitative definitions for concepts measuring its cause(s) and its effect(s). In the case of conduction, the necessary concepts have already been defined: With $\mathbf{r}$ a point in the conductor, we take the macroscopic electric field $\mathbf{E}(\mathbf{r})$ to measure the cause and the macroscopic current density $\mathbf{J}(\mathbf{r})$ to measure the effect. Conceptually, we then think of $\mathbf{J}$ as some function of $\mathbf{E}$,

$$\mathbf{J} = \mathbf{J}(\mathbf{E}) \tag{9-1}$$

but we must look to experiment to determine whether Eq. (9-1) applies to a particular sample of conducting material at all and, if it does apply, to determine its specific functional form *for that sample*. Equation (9-1), sometimes called a *constitutive relation*, is the point at which the *empirical* properties of matter find their way into the theoretical framework, and, in the *macroscopic* approach, the empirically dictated form of Eq. (9-1) is accepted without deeper question. That form, of course, may be extremely complicated. For most materials, however, **J** is parallel to **E** and Eq. (9-1) assumes the simpler form

$$\mathbf{J} = g(E)\mathbf{E}, \quad \mathbf{J} \parallel \mathbf{E} \tag{9-2}$$

where $g(E)$ remains to be determined experimentally for each material. Further, for a very large fraction of the materials to which Eq. (9-2) applies, the function $g(E)$ is actually independent of E, at least for the normal range of field strengths. For these materials, which are called *linear* or *ohmic* materials, Eq. (9-2) reduces to

$$\mathbf{J} = g\mathbf{E} \quad \text{(ohmic conductors)} \tag{9-3}$$

where the *conductivity* g is measured in the mks unit A/V·m, a unit we shall later call the mho/m.[2] For ohmic conductors, the value of g or equivalently the value of the *resistivity* $\eta = 1/g$ fully characterizes the extent to which the material conducts in response to a static electric field.[3] *Experimentally determined* values of η for several common materials are shown in Table 9-1. Metals and other "good" *conductors* characteristically have very high conductivities (small resistivities) that are only weakly dependent on temperature except at extremes of temperature. *Semiconductors* typically have much lower conductivities (higher resistivities) that are often more strongly dependent on temperature. *Insulators* have such small conductivities (high resistivities) that it is not appropriate to think of them as conducting materials at all. The limiting values $g \longrightarrow \infty$ ($\eta \longrightarrow 0$) and $g \longrightarrow 0$ ($\eta \longrightarrow \infty$) identify the idealized cases of the perfect conductor and the perfect insulator, respectively.

We shall now reexpress the equation $\mathbf{J} = g\mathbf{E}$ between current density and electric field in an ohmic conductor in a form more suited to the description of currents confined to wires. Assume a wire with cross-sectional area S and length ℓ and let a potential difference ΔV be maintained between its ends (Fig. 9-1). When the potential difference is first applied, transient currents will develop in the wire. Ultimately, however, these transients will establish a charge distribution along the surfaces of the wire, which charges in turn produce the fields necessary to "guide" the final steady-state current around any bends in the wire.[4] Steady-state conduction is established when

[2]The conductivity is defined by Eq. (9-3) in all systems of units with which the author is familiar.

[3]The symbols σ and ρ are often used for conductivity and resistivity, respectively, but that notation is subject to confusion with surface and volume charge densities. We use instead g and η, which follows the notation used by J. R. Reitz, and F. J. Milford, *Foundations of Electromagnetic Theory* (Addison-Wesley Publishing Company, Inc., Reading, Mass., 1967).

[4]See, for example, S. Parker, *Am. J. Phys.* **38**, 720 (1970) and the references given there.

TABLE 9-1 Resistivity η and Temperature Coefficient of Resistivity $d\eta/\eta\, dT$ for Selected Materials at Room Temperature

In this table, T represents temperature. The materials divide naturally into three groups: the *conductors* with resistivities on the order of 10^{-6} to 10^{-8} ohm·m, the *semiconductors* with resistivities on the order of 10^{2} to 10^{-4} ohm·m, and the *insulators* with resistivities greater than about 10^{8} ohm·m.

Material	*Resistivity,* η*(ohm·m)*	*Temperature Coefficient,* $\frac{1}{\eta}\frac{d\eta}{dT}[(°C)^{-1}]$	*Material*	*Resistivity,* η*(ohm·m)*
Aluminum[a]	2.824×10^{-8}	0.0039	Germanium[b]	
Brass[a]	7×10^{-8}	0.002	$<10^{13}$ cm^{-3} impurity	0.40
Constantan[a]	49×10^{-8}	10^{-5}	$\sim 4 \times 10^{14}$ cm^{-3} In; *p* type	0.044
Copper (annealed)[a]	1.724×10^{-8}	0.0039	$\sim 2 \times 10^{16}$ cm^{-3} In; *p* type	0.0017
Gold[a]	2.44×10^{-8}	0.0034	$\sim 7 \times 10^{14}$ cm^{-3} Sb; *n* type	0.0027
Iron (99.98%)[a]	10×10^{-8}	0.005	$\sim 5 \times 10^{16}$ cm^{-3} Sb; *n* type	5.1×10^{-4}
Lead[a]	22×10^{-8}	0.0039	Bakelite[c]	10^{8}–10^{14} [d]
Mercury[a]	95.783×10^{-8}	8.9×10^{-4}	Beeswax[c]	$\sim 10^{12}$
Nichrome[a]	100×10^{-8}	4×10^{-4}	Glass[c]	10^{9}–10^{14} [d]
Nickel[a]	7.8×10^{-8}	0.006	Glyptal[c]	10^{14}
Silver[a]	1.59×10^{-8}	0.0038	Paraffin (special)[c]	$>5 \times 10^{16}$
Tin[a]	11.5×10^{-8}	0.0042	Porcelain (unglazed)[c]	3×10^{12}
Tungsten[a]	5.6×10^{-8}	0.0045	Quartz (fused)[c]	$>5 \times 10^{16}$
			Shellac[c]	$\sim 10^{14}$
			Wood[c]	10^{8}–10^{11}

[a] *Handbook of Chemistry and Physics* (Chemical Rubber Publishing Company, Cleveland, 1965), Forty-sixth Edition, p. E-66. Used by permission of the Chemical Rubber Company.

[b] Computed after the method of P9-7 from data on Hall coefficients and conductivities given by H. Fritzsche, *Phys. Rev.* **99**, 406 (1955).

[c] *Handbook of Chemistry and Physics* (Chemical Rubber Publishing Company, Cleveland, 1961), Forty-third Edition, pp. 2566–2567. Used by permission of the Chemical Rubber Company.

[d] Depending on composition.

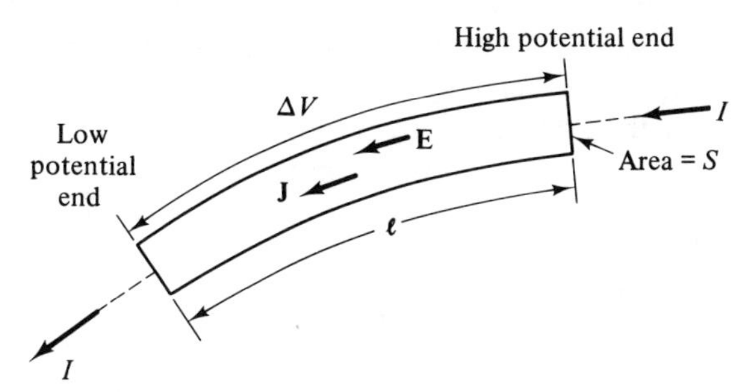

Fig. 9-1. A wire carrying a steady current.

the surface charges have been so adjusted that the current density **J** and the electric field **E** inside the wire are everywhere tangent to the wire. If we now assume that the current is uniformly distributed over the cross section of the wire, then $J = |\mathbf{J}|$ is constant throughout the wire. Further, in consequence of Eq. (9-3), $E = |\mathbf{E}|$ is also constant throughout the wire and the potential difference between the ends of the wire is

$$\Delta V = -\int_{\text{low end}}^{\text{high end}} \mathbf{E} \cdot d\boldsymbol{\ell} = E\ell \tag{9-4}$$

where the path follows the wire and $d\boldsymbol{\ell}$ is directed from the low to the high end. Further, the current flowing in the wire is given by

$$I = \int_{\Sigma} \mathbf{J} \cdot d\mathbf{S} = JS \tag{9-5}$$

where Σ is any surface intersecting the wire. Since the material of the wire has been assumed ohmic, however, $\mathbf{J} = g\mathbf{E}$, and we find from Eqs. (9-4) and (9-5) that

$$\begin{matrix} \Delta V = E\ell \\ I = gES \end{matrix} \Longrightarrow \frac{\Delta V}{I} = \frac{\ell}{gS} = \frac{\eta\ell}{S} = R \tag{9-6}$$

or that

$$\Delta V = IR \tag{9-7}$$

where the *resistance* R is a constant determined by the dimensions and conductivity of the wire; the mks unit of resistance is the *ohm* (Ω) and the mks unit of a reciprocal resistance, called a *conductance*, is the mho. A device whose electrical characteristics are described by Eq. (9-7) is called a (linear) *resistor*, and Eq. (9-7) itself is the familiar form of *Ohm's law*, which relates the potential difference ΔV across a (linear) resistor of resistance R to the current I through the resistor. Equation (9-3), however, is also referred to as Ohm's law. A paragraph discussing the limitations of Ohm's law is included in Section 9-2.

Finally, it is of interest to calculate the power dissipated in a conductor carrying a current. Clearly, power must be dissipated, for the external electric field is constantly doing work on the charges, but overall their kinetic energy remains unchanged. The collisions must convert energy to heat, called here *Joule heat*, at a rate exactly equal to the rate at which the external field supplies energy to the system. Thus, we can calculate the rate of energy dissipation in the conductor by calculating instead the rate P at which the external field does work on the charge carriers. That rate, however, is given by Eq. (3-39), and we have that

$$P = \int \mathbf{J}\cdot\mathbf{E}\, dv \tag{9-8}$$

where in mks units P is expressed in watts (W). The integral in Eq. (9-8) extends over the entire volume of the conductor. If the current happens to be in a wire, the replacement $\mathbf{J}\, dv \rightarrow I\, d\boldsymbol{\ell}$ yields the more familiar result

$$P = I \int_{\text{high end}}^{\text{low end}} \mathbf{E}\cdot d\boldsymbol{\ell} = I\,\Delta V = I^2 R = \frac{(\Delta V)^2}{R} \tag{9-9}$$

where the last two forms follow on substitution from Eq. (9-7).

9-2 Microscopic Description: Carrier Mobility and Collision Times

The basic microscopic response of a conductor to its presence in a static electric field is a steady drift velocity superimposed on the random motion of the individual charged particles making up the conductor. The classical microscopic description of conduction therefore involves relating particle drift velocities, which we view as the effects, to the applied electric field, which we continue to view as the cause. Since several different types of charge carrier may contribute to the total transport of charge, we allow for several different drift velocities. Thus, very symbolically, we expect the drift velocity $\mathbf{v}^{(a)}$ of carriers of type a to be some function of the applied electric field,

$$\mathbf{v}^{(a)} = \mathbf{v}^{(a)}(\mathbf{E}) \tag{9-10}$$

A theoretical prediction of the functional form of Eq. (9-10) for the carriers in a particular conductor can be made only after some further assumptions about the nature of the conductor are made. Since Eq. (9-10) is a statement about the behavior of *individual* microscopic particles, calculation of its specific form is intrinsically a quantum mechanical problem, but a brief quantum mechanical treatment would require a background beyond that assumed in this book. (See the references at the end of the chapter.) Instead we present a simple classical theory. Suppose the carriers in a conductor experience forces solely (or at least dominantly) from the external field except during collisions, and assume that the acceleration $\mathbf{a}^{(a)}$ of carriers of type a is constant between collisions. Then at a particular instant of time, the nonrandom component of the velocity $\mathbf{v}_i^{(a)}$ of the ith carrier of type a is

$$\mathbf{v}_i^{(a)} = \mathbf{a}^{(a)}\tau_i^{(a)} = \frac{q_a\tau_i^{(a)}}{m_a}\mathbf{E} \tag{9-11}$$

where q_a and m_a are the charge and mass of carriers of type a, $\tau_i^{(a)}$ is the time since the last collision of the ith particle of type a, and Newton's second law has been used to write $\mathbf{a}^{(a)}$ as $q_a\mathbf{E}/m_a$. The *average* drift velocity $\mathbf{v}^{(a)}$ for carriers of type a is then given by

$$\mathbf{v}^{(a)} = \langle\mathbf{v}_i^{(a)}\rangle = \frac{q_a\langle\tau_i^{(a)}\rangle}{m_a}\mathbf{E} = \frac{q_a\tau^{(a)}}{m_a}\mathbf{E} \tag{9-12}$$

where the *collision time* $\tau^{(a)} = \langle\tau_i^{(a)}\rangle$ physically is the average time between collisions. (Why? *Hint*: Collisions do not occur at regular intervals.) The quantity

$$\mu_a = \frac{q_a\tau^{(a)}}{m_a} \tag{9-13}$$

is called the *mobility* of carriers of type a and will be positive or negative, depending on whether the carrier is positively or negatively charged. In terms of the mobility, Eq. (9-12) becomes

$$\mathbf{v}^{(a)} = \mu_a\mathbf{E} \tag{9-14}$$

from which we see that μ_a is the average drift speed per unit field and must have the dimensions of velocity per electric field, (m/sec)/(V/m) in mks units. On the basis of the simple assumptions in this paragraph, we therefore predict that $\mathbf{v}^{(a)}$ in Eq. (9-10) should be proportional to $\mathbf{E}$ and that the constant of proportionality μ_a is determined by the intrinsic properties of the carriers and by the nature and frequency of the collisions experienced by the carriers as they migrate through the conductor. Except at extremes of temperature and field strength, the quantum mechanical description of conduction in many materials and particularly in metals leads to the same general conclusions, and the carrier mobilities are important (and frequently measured) parameters describing these materials.

Given Eq. (9-14), we can now *derive* Ohm's law. We need merely introduce the density of carriers of type a, $n^{(a)}$, and then substitute Eq. (9-14) into Eq. (2-18) to find that the macroscopic current density $\mathbf{J}$ is given by

$$\mathbf{J} = \sum_a q_a n^{(a)}\mathbf{v}^{(a)} = \left(\sum_a q_a n^{(a)}\mu_a\right)\mathbf{E} \tag{9-15}$$

which is Ohm's law, Eq. (9-3), provided we identify the coefficient of $\mathbf{E}$ with the conductivity. This classical microscopic description of conduction has thus yielded not only a derivation of Ohm's law but also the expression

$$g = \sum_a q_a n^{(a)}\mu_a = \sum_a \frac{(q_a)^2 n^{(a)}\tau^{(a)}}{m_a} \tag{9-16}$$

relating a macroscopic conductivity to microscopic mobilities and collision times. Particularly when only one type of carrier contributes to conduction, measurements of g are therefore effectively measurements of the product $qn\mu$ for the carrier involved. If q is assumed and n is known, say from Hall effect measurements (P9-7), measurements of the macroscopic conductivity yield values for the microscopic mobility. Note that g is necessarily positive even though the mobilities may have either sign.

Ohm's law and its microscopic equivalent, Eq. (9-14), are, of course, not universal laws of physics. There are circumstances that these laws do not describe adequately, and the microscopic development leading to Eq. (9-14) suggests areas in which we might expect Ohm's law to break down. We have, for example, tacitly assumed that the external field does not significantly influence the collision times, so that $\tau^{(a)}$ is independent of the field. Equivalently, we have required that the drift speeds be small compared to thermal speeds, which requirement in turn limits the strength of the field so that high drift speeds are not produced (P9-6). We also tacitly assumed the density of carriers to be independent of the strength of the applied field. But sufficiently strong fields may ionize more of the neutral particles present (or further ionize multielectron atoms that are already partly ionized), thus increasing the density of carriers and destroying the validity of Ohm's law. Third, we have assumed the conducting material to be *isotropic* (the same in all directions) and the resulting conductivity is a scalar. But some few materials are nonisotropic and many materials can be made nonisotropic, for example, by the application of a uniform external magnetic induction field. These materials conduct differently for electric fields applied in different directions and the resulting current density may not even be parallel to the applied field. Salvaging Ohm's law for these circumstances then requires allowing the conductivity to become a nine-component entity called the *conductivity tensor* (P9-8). Still further, we have assumed the conductor to be *homogeneous* (the same at all interior points). Nonohmic devices, such as transistors and solid-state rectifiers, can be made by assembling a conducting sample whose characteristics (conductivity, carrier concentration, etc.) change either slowly or abruptly as an observation point moves through the sample. Fifth, we have assumed that only the external field is significant except during actual collisions, and, further, we have assumed that collisions are well-defined events and that we can tell (in principle at least) when a particle is undergoing a collision and when it is not. One of the primary forces of interaction between charged particles, however, is the Coulomb force, which is a very long-range force. Even when particles are far apart, their Coulomb interaction may be significant. Certainly, this interaction interferes with any clear-cut separation of moments of collision from moments of free motion in an external field. Finally, we have assumed that the microscopic carriers experience an electric field defined by a "macroscopic" test charge. Although that assumption is known to be valid for microscopic particles moving in free space (e.g., electrons in a cathode ray tube), it may in fact not be correct for carriers in the interior of some conductor. Having enumerated these sources of possible difficulty, one might even be surprised that Ohm's law works as well as it does over such a broad range of materials and applied fields.

This discussion of conduction is far from complete. We have outlined the essential descriptive framework, and we have seen some of how this general framework can be molded to describe specific types of matter more explicitly. What remains is more detailed quantum mechanical calculations of mobilities and collision times, a full examination (both experimentally and theoretically) of departures from Ohm's law, and a cataloging of measured properties of specific materials. Further pursuit of these topics, however, would distract us from our main intent of exposing the basic theoretical framework. The interested reader is directed to the references listed below.

PROBLEMS

P9-1. Find the dimensions of conductivity in cgs-esu and ponder the implications of your result.

P9-2. It is desired to use a nichrome wire having a cross-sectional area of 0.2 mm^2 to make a heating coil that will produce 600 W when operated at 120 V. How long must the wire in the coil be?

P9-3. A conducting wire of radius a and length L is surrounded by a coaxial conducting sheath of radius b. The space between the wire and the sheath is filled with a weakly ionized gas that has a (small) conductivity g. Let the wire be maintained at a potential V with respect to the sheath. Ignoring fringing, find the current flowing between the wire and the sheath and show that the resistance of the gas is given by $[\ell n\,(b/a)/2\pi Lg]$.

P9-4. Steady currents in a homogeneous, ohmic medium placed in a static electric field are described by the equations $\nabla \cdot \mathbf{J} = 0$, $\nabla \times \mathbf{E} = 0$, and $\mathbf{J} = g\mathbf{E}$. (Why?) Show that the electric field can be derived from a scalar potential that satisfies Laplace's equation, thus justifying the use of field trays to study electric field patterns.

P9-5. Suppose that at time zero the charge density inside a conductor happens to have the nonzero (nonequilibrium) value $\rho_0(\mathbf{r})$, where $\mathbf{r}$ is a point in the conductor. (a) Assume $\rho(\mathbf{r}, t) = \rho_0(\mathbf{r})f(t)$ and combine the equation of continuity and Ohm's law with Gauss's law—the presently available form will do, although strictly you should use the modified form to be developed in Chapter 10—to obtain a differential equation for $f(t)$, show that $f(t) = e^{-t/\tau_r}$, and find an explicit expression for the *relaxation time* τ_r. (b) Calculate τ_r numerically for aluminum. In essence, this result demonstrates that aluminum and other good conductors dissipate nonequilibrium charge distributions very rapidly.

P9-6. An aluminum wire at room temperature ($\approx$ 300°K) has a cross-sectional area of 0.1 mm^2 and carries a current of 5×10^{-4} A. (a) If each aluminum atom contributes three conduction electrons and all the current is carried by these electrons, calculate the electronic drift velocity in the wire.

Hints: Avogadro's number is 6.02×10^{23} atoms/mole; the atomic weight of aluminum is 27; the density of aluminum is 2.70 g/cm³. (b) Estimate the thermal speed $\sqrt{3kT/m_e}$ of the electrons and compare the result with the drift velocity. Here k is Boltzmann's constant and m_e is the electron mass. (c) Use Eq. (9-16) to estimate the collision time from the measured conductivity. (d) Approximately how far does an electron move between collisions? (e) Use Eq. (9-16) to estimate the electron mobility from the measured conductivity and then estimate the electric field necessary to bring about the given current. Compare this field with the field necessary to produce a drift velocity comparable to the thermal velocity and compare each with typical laboratory fields.

P9-7. A current I is passed through a sample of conducting material in a uniform magnetic induction $\mathbf{B} = B\hat{\mathbf{k}}$ that is perpendicular to the direction of current flow (Fig. P9-7). Assume the current is carried by a single type of

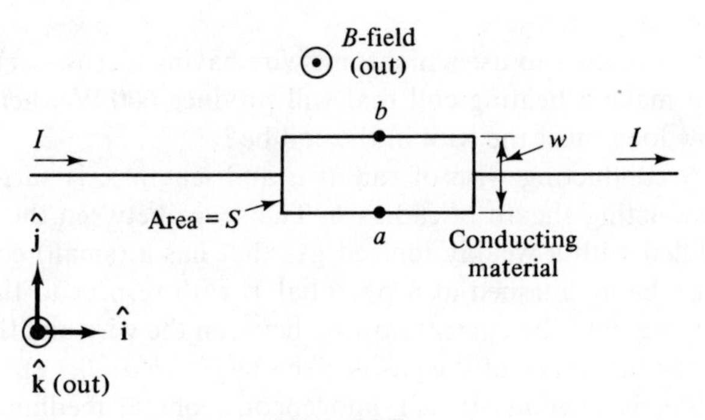

Figure P9-7

carrier with charge q, density n, and drift velocity $v\hat{\mathbf{i}}$, and let the points a and b be separated by a distance w. (a) Argue qualitatively that an electric field $\mathbf{E} = E\hat{\mathbf{j}}$ is induced, show that $E = vB$, and calculate the potential difference ΔV between points a and b. Which point is at the higher potential? (b) The Hall coefficient for this sample of material is defined by $R_H = E/JB$, where J is the $\hat{\mathbf{i}}$-component of the current density. Show that $R_H = 1/qn$. (c) Show also that $R_H = \mu/g$, where μ is the carrier mobility and g is the conductivity of the sample. Since R_H and g are both measurable quantities, this so-called *Hall effect* can be used to determine the sign of charge carriers, carrier densities, and carrier mobilities, particularly in semiconductors. Further, once R_H is known for a particular sample, the effect can be turned around and measurements of ΔV can be used to determine magnetic fields; used in this way, the device is called a *Hall probe*.

P9-8. In the classical model of conduction introduced by Drude (1900), the resistive forces on the carriers in a conducting material are represented by a viscous damping term in the equation of motion, so that the individual carriers of mass m and charge q move in accordance with

$$m\frac{d\mathbf{v}}{dt} = q\mathbf{E} - b\mathbf{v} \tag{1}$$

when an external **E**-field is applied. (a) Justify this view by showing that Eq. (1) predicts a terminal (drift) velocity, relate b to the carrier mobility μ, and, assuming that the carriers are electrons and that $\mu = -1.2 \times 10^{-3}\ \text{m}^2/\text{V}\cdot\text{sec}$, estimate how quickly the terminal velocity will be reached. *Hint*: Consider motion only in the x direction. (b) Assume that this same approach applies when a uniform external **B**-field is added, so that individual carriers now move in accordance with

$$m\frac{d\mathbf{v}}{dt} = q[\mathbf{E} + \mathbf{v} \times \mathbf{B}] - b\mathbf{v} \tag{2}$$

Take $\mathbf{B} = B\hat{\mathbf{k}}$ and let $\mathbf{E}$ have all three components, $\mathbf{E} = E_1\hat{\mathbf{i}} + E_2\hat{\mathbf{j}} + E_3\hat{\mathbf{k}}$. Determine the terminal drift velocity $\mathbf{v}_d$ in terms of the components of **E**, and show that

$$\begin{pmatrix} v_{d1} \\ v_{d2} \\ v_{d3} \end{pmatrix} = \begin{pmatrix} \dfrac{\alpha b}{q} & \alpha B & 0 \\ -\alpha B & \dfrac{\alpha b}{q} & 0 \\ 0 & 0 & \dfrac{q}{b} \end{pmatrix} \begin{pmatrix} E_1 \\ E_2 \\ E_3 \end{pmatrix}$$

where $\alpha = 1/[B^2 + (b^2/q^2)]$. Finally, infer that for this case the mobility μ and the conductivity $g = qn\mu$ are not scalars, and show that **J** is *not* parallel to **E**.

REFERENCES

C. Kittel, *Introduction to Solid State Physics* (John Wiley & Sons, Inc., New York, 1966), Third Edition, Chapters 7–11.

E. M. Purcell, *Electricity and Magnetism* (McGraw-Hill Book Company, New York, 1965), Chapter 4.

V. F. Weisskopf, "On the Theory of the Electric Resistance of Metals," *Am. J. Phys.* **11**, 1 (1943).

10

Properties of Matter II: Dielectric Polarization

In considering the response of matter to applied electric fields, it is useful to distinguish two ideal types of matter. We have already introduced the perfect conductor. Microscopically, it contains large numbers of free charges that move and produce a macroscopic current whenever an external electric field is applied. Real materials, however, conduct only imperfectly and many materials, called insulators or *dielectrics*, conduct very poorly indeed. A perfect dielectric, again an idealization, contains no free (microscopic) charges, and an external electric field induces not a current but rather a redistribution of the charge in each molecule of the dielectric and perhaps a reorientation of these molecules. This dielectric response, called *dielectric polarization*, is the subject of this chapter. Although additional charge beyond that intrinsic to its atoms may be placed in or on a dielectric, we shall consider primarily the electrically neutral dielectric. Further, we shall confine our discussion here to static external fields, postponing the treatment of time-dependent polarization to Chapter 12.

In contrast to conduction, which is brought about by one field (the electric field) but itself produces a *different* field (the magnetic induction field), dielectric polarization is brought about by the *same* field that it also produces. Thus, when a dielectric is placed in an external electric field and is thereby polarized, the dielectric produces additional electric fields that modify the original fields both inside and outside the dielectric. This modification of the fields in turn alters the response of the dielectric, which again modifies the fields, and so on. Physically, the system ultimately reaches a steady state in which the dielectric is polarized to the extent required by the *final* field. From a calculational point of view, however, it would seem that we need to know the final field in order to determine the final polarization but we cannot determine the final polarization until we know the final field. The treatment of dielectric polarization is thus more complicated than that of conduction because we cannot ignore the fields produced by the dielectric as we determine the response of the dielectric to an external field.

Because of the difficulty pointed out in the previous paragraph, it is convenient to divide the overall treatment of dielectric polarization into two parts. In the first part we shall develop a model for the structure of a dielectric and introduce descriptive concepts so that we can

(1) describe quantitatively what might be called a *field-causing* state of the dielectric, and
(2) determine from this field-causing state the field that it causes, both inside and outside the dielectric.

In this part, we do not ask how a particular field-causing state of the dielectric is brought about or maintained; we simply describe this state by giving values to the appropriate descriptive concepts and then utilize this description to calculate the resulting field. In the second part of the treatment of dielectric polarization, however, the establishment of the field-causing state is of primary concern; we seek methods by which to

(3) determine the response of a dielectric when the *final* field is assumed known.

Although steps (1) and (2) in this program can be treated quite generally, step (3) involves consideration of specific materials and will of necessity be more complicated, less general, and more empirical than the treatment of steps (1) and (2). As we shall see, for example, in Section 10-5, the results of these three steps taken together will permit us to calculate final fields and polarizations even though neither may be known before the calculation is begun.

As with conduction, there are two different approaches to the treatment of dielectric polarization. The more involved of these is the *microscopic approach*, which begins by adopting a detailed microscopic model for the composition of matter[1]—matter consists of extremely small molecules, each of which has electric and/or magnetic properties that are presumably understood. These molecules are taken to exist in free space. In this view, there is consequently no intrinsic difference between free space occupied here and there by free charge and/or free currents and free space occupied by matter

[1]We use the word matter rather than dielectric and we speak of electric and magnetic properties so that this discussion can be applied also to the phenomenon of magnetization.

since (insofar as its electromagnetic properties are concerned) matter is composed ultimately of (microscopic) charges and currents. One describes the field-causing state of matter by specifying the location of every molecule and the (microscopic) charge and current distributions associated with each molecule. With this information as input, the formalism of Chapters 4–6 permits a detailed calculation of the corresponding electromagnetic field, which is called the *microscopic electromagnetic field.* It is the field experienced by test charges whose dimensions are comparable to molecular dimensions, and it fluctuates rapidly and erratically both in space (because on a molecular scale the charge and current distributions can hardly be regarded as smoothly varying) and in time (because thermal agitation of the molecules introduces an erratic time variation of charge and current distributions on the microscopic scale).

Although in principle the microscopic field can be fully determined by methods already available, in practice the very large number ($\approx 10^{23}$) of molecules typically involved precludes its actual evaluation by these means. Fortunately, for *macroscopic* problems, it is usually not necessary to know the *microscopic* state of affairs in detail. Only rarely are we interested in the variations of the field over distances comparable to molecular dimensions or on a time scale determined by the thermal fluctuations. In most instances, the presence of fields is detected by test charges that are large compared to molecules (though perhaps small macroscopically) and that move slowly compared to the velocities of thermal vibration (i.e., that stay in one place for a long time if times are measured in an atomic time scale). Such test charges are sensitive not to the detailed microscopic field but to the time and space average of the microscopic field, where the averages are taken over temporal and spatial regions large microscopically but small macroscopically. The next step in the development of the full theory in the microscopic approach is therefore to introduce a *macroscopic field* as an average of the microscopic field. The detailed evaluation of this average turns out rather naturally to involve making a distinction between free charge and currents placed on matter and the so-called bound charge and current distributions that result from the charges that are an intrinsic part of the matter itself. In evaluating the average, one ultimately (and quite naturally) identifies a pair of auxiliary fields whose sources are essentially the free charges and currents alone. These auxiliary fields differ from the electric field and the magnetic induction field, whose sources are the free and bound distributions combined, by terms that involve directly and explicitly the time and space averages of the microscopic properties of the matter. Finding the macroscopic fields then involves finding expressions for these averages in terms of the fields.

In the formal treatment of the electromagnetic field in matter the microscopic approach is extremely valuable. It provides a deep insight into the structure of the macroscopic fields in the presence of matter, it begins with the vacuum form of Maxwell's equations, and the auxiliary fields emerge naturally as the averages of the microscopic fields are evaluated. Although we shall use microscopic arguments occasionally and we shall begin with an examination of the microscopic response of molecules to external fields, we shall not adopt the microscopic approach in our basic development.[2] Instead, we adopt the *macroscopic approach* at the very outset. The price we pay is the inapplicability (at least directly) of the vacuum form of Maxwell's equations. We must introduce in a rather ad hoc fashion the descriptive terminology that would have arisen naturally had we adopted the microscopic approach. Once we have done so, however, the subject develops easily and with rather less mathematical complexity than from the microscopic view, although the final results are identical.

PROBLEM

P10-1. Explain qualitatively how the macroscopic fields defined as averages of the microscopic fields may be static even though the microscopic fields certainly are time-dependent.

10-1 The Microscopic Description: Electric Polarizability

As a prelude to a more detailed macroscopic development of dielectric theory, we shall consider briefly the response of individual molecules to static electric fields. We assume that the field varies slowly enough in space to be regarded as a constant over the region occupied by a single molecule. Then the molecules, each of which is overall electrically neutral, experience no net force from the field. Even an electrically neutral molecule, however, may have an electric dipole moment and hence may experience torques in an external field. We are thus led to explore the electric dipole moments of molecules as a possible microscopic basis for understanding the dielectric response of macroscopic matter.

Molecules can be divided into two categories by the character of their dipole moments. (For the rest of this chapter we understand that the *electric* dipole moment is meant.) Some molecules, such as the water molecule, have a naturally asymmetric even though still neutral internal charge distribution and hence possess a *permanent* dipole moment. When placed in an electric field (whatever its origin), these molecules will tend to align themselves with their dipole moments parallel to the field (P3-10). This tendency of permanent dipoles to align themselves, however, is opposed by the thermal tendency to randomness. Thus, a given permanent dipole assumes an *average* dipole

[2]The interested reader is referred, for example, to J. D. Jackson, *Classical Electrodynamics* (John Wiley & Sons, Inc., New York, 1962), p. 103ff.

moment that would be zero except that in the external field the dipole favors a preferred orientation as it undergoes thermal agitation. To express this discussion more quantitatively, consider the following situation: A dielectric at absolute temperature T composed of permanent dipoles having dipole moments of magnitude p_0 is placed in an external electric field. The dipoles adjust to some sort of average equilibrium position, in which state a particular dipole experiences a total field consisting of the applied external field *plus* the time-averaged *microscopic* field established by all *other* dipoles when the external field is present.[3] Let the *time* average (over the microscopic fluctuations) of this total field be $\mathbf{E}_m$. Combination of statistical mechanics with considerations of electrostatic energy (P10-2) leads ultimately to the result that the *average* dipole moment of a single dipole is given by

$$\langle \mathbf{p}_0 \rangle_{\text{perm}} = \alpha \mathbf{E}_m; \qquad \alpha = \frac{p_0^2}{3kT} \tag{10-1}$$

where k is Boltzmann's constant and the condition $p_0 E_m \ll kT$ has been imposed. (This apparent restriction to "high" temperature is not as much of a restriction as it might seem. Typically p_0 is quite small and fields approaching the very large value of 10^9 V/m must be applied before the condition is violated even at room temperature.) The constant α in Eq. (10-1) might be called an *orientational* molecular polarizability, since it measures the extent to which a molecule possessing a permanent dipole moment orients itself in a polarizing field. In calculating this result, we assumed the dipoles to interact dominantly with the external field and only weakly with one another. The interaction of dipoles with dipoles has been taken into account only through the contribution of the internally generated microscopic field and through recognition that these interactions are responsible for the thermal agitation that each molecule undergoes.

A second category of molecule contains those molecules that do not possess a permanent dipole moment. When placed in an electric field, these molecules acquire an induced dipole moment that depends on the strength of the applied field. In effect, the external field causes a distortion of the molecule because the positive nucleus experiences a force in one direction while the negative electron cloud experiences a force in the opposite direction. Provided that the distortion is not too great, the induced dipole moment is in many cases proportional to the total field experienced by the molecule, i.e.,

$$\mathbf{p}_{\text{ind}} = \alpha_0 \mathbf{E}_m \tag{10-2}$$

Here α_0 might be called a *deformational* molecular polarizability. A very simple classical model that *predicts* Eq. (10-2) is explored in P10-3.

[3] Actually, the situation is even a bit more complicated. We really want the microscopic field established at a particular molecular site by all *other* molecules when the other molecules, however, *have the orientations that they have when the molecule in question is present.*

In a still more general situation, a molecule may exhibit both a permanent dipole moment and a capacity to acquire an additional induced moment when placed in an external field. For such molecules, the total dipole moment is related to the field $\mathbf{E}_m$ by

$$\mathbf{p} = \langle \mathbf{p} \rangle_{\text{perm}} + \mathbf{p}_{\text{ind}} = \left(\frac{p_0^2}{3kT} + \alpha_0 \right) \mathbf{E}_m = \alpha_t \mathbf{E}_m \qquad \text{(10-3)}$$

where the *total* molecular polarizability α_t is given by

$$\alpha_t = \frac{p_0^2}{3kT} + \alpha_0 \qquad \text{(10-4)}$$

For some dielectrics, this molecular polarizability can be determined from macroscopically measurable quantities by using the Clausius-Mossotti equation (Section 10-6). Sometimes the orientational polarizability and the deformational polarizability can be separated if measurements of the total polarizability over a range of temperatures are made.

In this section, we have been concerned mostly with the qualitative responses of molecules to external electric fields. Our quantitative statements have therefore pertained to the simplest dielectrics. In more general circumstances, we would have to allow for a possible field dependence in both α and α_0 and for possible anisotropies that could make α and/or α_0 depend on the direction of the external field (P10-25). A detailed treatment of these more complicated cases or of the more satisfactory quantum mechanical approach to determining molecular response would, however, lead us away from our primary examination of the *macroscopic* dielectric response.

Fortunately, we need not anticipate a need to look at multipole moments of higher order than the dipole moment. A molecule with a *permanent* moment no lower than the quadrupole moment experiences neither force nor torque in a field whose spatial variation is slow enough that the field can be regarded as constant over the molecule (P10-5). In this field, such a molecule exhibits no response arising from its permanent moment. Instead, the primary response of this molecule and also of molecules with no permanent moments whatever is to acquire an induced *dipole* moment, and we have returned to a case already considered. In essence, we can ignore moments higher than the dipole moment in treating the response of molecules (and hence of dielectrics) to external fields having a slow spatial variation. Further, once the dipole moment is present, its contribution to the resulting total field dominates the contribution of any possible higher moments as soon as the observation point is more than a few molecular dimensions away from the molecule itself. Thus, at observation points that are at least a few molecular dimensions away from a molecule (which is almost always the case), we can ignore moments higher than the dipole moment in calculating the contribution of the molecule to the total field.

PROBLEMS

P10-2. Accepting (1) that the energy of an electric dipole of moment $\mathbf{p}$ in an electric field $\mathbf{E}$ is $-\mathbf{p}\cdot\mathbf{E}$ (P4-17) and (2) that the distribution of dipoles over these energy states is given in statistical thermodynamics by exp (−energy/kT), where k is Boltzmann's constant and T is the absolute temperature, show that the average value $\langle\mathbf{p}_0\rangle$ of the dipole moment of a *permanent* dipole in an electric field $\mathbf{E}_m = E_m\hat{\mathbf{k}}$ is given by $p_0\hat{\mathbf{k}}\mathcal{L}(p_0E_m/kT)$, where the *Langevin function* $\mathcal{L}(\lambda)$ is defined by

$$\mathcal{L}(\lambda) = \coth\lambda - \frac{1}{\lambda}$$

Here p_0 is the magnitude of the permanent dipole moment of the dipole. Show also that Eq. (10-1) emerges from this expression when $p_0E_m \ll kT$. How large can the field E_m be before this condition is violated at room temperature? *Hints:* (1) Let $\mathbf{p} = p_0(\sin\theta\cos\phi\hat{\mathbf{i}} + \sin\theta\sin\phi\hat{\mathbf{j}} + \cos\theta\hat{\mathbf{k}})$. (2) Note from statistical mechanics that

$$\langle\mathbf{p}_0\rangle = \frac{\int \mathbf{p}e^{-\mathbf{p}\cdot\mathbf{E}_m/kT}\sin\theta\,d\theta\,d\phi}{\int e^{-\mathbf{p}\cdot\mathbf{E}_m/kT}\sin\theta\,d\theta\,d\phi}$$

where θ and ϕ are the usual spherical angles with θ being the angle between $\mathbf{p}$ and $\mathbf{E}_m$. (3) Typical values for permanent molecular dipole moments are on the order of 10^{-30} C·m.

P10-3. An extremely crude calculation of the deformational molecular polarizability can be made for a single atom by the following model. Let the atom consist of a positive nucleus of charge q surrounded by an electron charge cloud of radius r containing a total charge $-q$ that is distributed *uniformly* throughout a sphere of radius r. In the absence of a polarizing field, the nucleus is located at the center of the spherical charge cloud. When a polarizing field $\mathbf{E}_m$ is present, suppose that the electron cloud is displaced without distortion. Thus, in the presence of a polarizing field, the atom (in

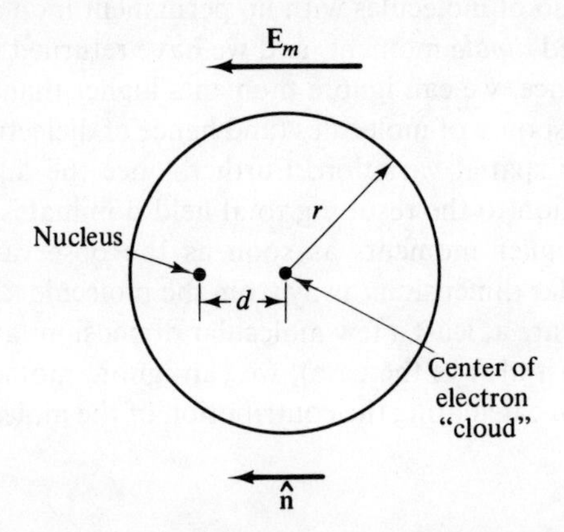

Figure P10-3

some crude sense) will have the appearance shown in Fig. P10-3. The induced dipole moment of the atom is therefore $qd\hat{\mathbf{n}}$. Since the atom is at rest, the nucleus in particular must experience zero force. (a) Find the electric field established by the electron charge cloud at the position of the nucleus. (b) Determine the total force on the nucleus and then show that $\alpha_0 = 4\pi\epsilon_0 r^3$. (c) Insert a typical value of r into the answer to part (b) and estimate the value of α_0. (d) Use this value of α_0 to estimate the induced dipole moment in a fairly large field, say 10^6 V/m. (e) Taking q to be the charge on the proton, estimate the displacement of the electron cloud from the nucleus and compare the result with typical atomic dimensions.

P10-4. Estimate the electric field established by the nucleus of a hydrogen atom at the position of the electron and compare the value with typical laboratory fields. What does this comparison suggest about the possibility of effecting severe distortion of a molecule through the application of external fields?

P10-5. Describe the behavior of a linear quadrupole of arbitrary initial orientation in a uniform electric field. Do your conclusions apply also to more general quadrupoles? Defend your answer.

10-2 The Macroscopic Description: Dielectric Polarization

As suggested in Section 10-1, we shall now picture a dielectric as a material composed microscopically of electrically neutral molecules that have either a permanent dipole moment or a capability for acquiring an induced dipole moment (or perhaps both). Without asking how to relate a particular field-causing or *polarized* state of this material to the causes of that state (see Section 10-5), let us consider a dielectric sample that is already polarized. Microscopically the molecular dipoles in this dielectric then possess a nonzero average dipole moment. Now, if $\mathbf{p}_i$ is the *average* dipole moment of the ith molecule, the total dipole moment $\Delta\mathbf{p}$ of a selected small volume Δv within the dielectric is given by

$$\Delta\mathbf{p} = \sum_i \mathbf{p}_i \tag{10-5}$$

where the sum extends over all the molecules in Δv. In our subsequent theory, however, it is more convenient to work instead with a quantity determined by the average properties of the dielectric over a small volume. We therefore introduce the dipole moment per unit volume or *dielectric polarization* $\mathbf{P}(\mathbf{r})$, defined by

$$\mathbf{P}(\mathbf{r}) = \lim_{\Delta v \to 0 \text{ about } \mathbf{r}} \frac{\Delta\mathbf{p}}{\Delta v} \tag{10-6}$$

where the meaning of the limit is the same as always: Δv becomes macroscopically small but remains large enough to contain a large number of microscopic dipoles. The polarization—usually the adjective *dielectric* is suppressed—is a *macroscopic* concept that plays the same role in specifying the *macroscopic* state of a polarized dielectric as the charge density plays in specifying the state of a general charge distribution. Of course, the polarization may vary throughout the dielectric and indeed may be time-dependent as well. It does not, however, depend on the origin of the coordinate system chosen for its expression, for the dipole moment of a charge distribution that is overall electrically neutral is invariant to translation of the coordinate system (P3-18).[4]

Among other things (Sections 10-3 and 10-4), knowledge of the polarization permits calculation of the macroscopic dipole moment of the dielectric. From Eq. (10-6) we infer that the dipole moment of a small volume element Δv centered at $\mathbf{r}$ is given by $\Delta\mathbf{p} \approx \mathbf{P}(\mathbf{r})\,\Delta v$. Hence, the dipole moment of a macroscopic region of the dielectric is given by

$$\mathbf{p} = \int \mathbf{P}(\mathbf{r})\,dv \tag{10-7}$$

where the integral extends over the volume whose dipole moment is sought and in particular may extend over the entire dielectric.

PROBLEM

P10-6. The dipole moment of the water molecule is about 6.2×10^{-30} C·m. Find the *maximum* polarization of water vapor at a temperature of 100°C and a pressure of 1 atmosphere.

10-3 The Macroscopic Scalar Potential and Electric Field at a Point Exterior to a Polarized Dielectric

In Sections 10-3 and 10-4, we shall calculate the contribution made to the macroscopic electric field by a polarized dielectric whose polarization $\mathbf{P}(\mathbf{r}')$ at every point $\mathbf{r}'$ within the dielectric is known. Consider first an observation point $\mathbf{r}$ *exterior* to the dielectric. In microscopic terms, such an observation point is far away from *all* of the dipoles composing the dielectric. Thus, a small element of the dielectric having volume $\Delta v'$ and located at $\mathbf{r}'$ (Fig. 10-1) can be treated as a dipole of moment $\mathbf{P}(\mathbf{r}')\,\Delta v'$, regardless of where within the dielectric the point $\mathbf{r}'$ lies. The contribution of this element of the

[4]The definition of polarization given in this paragraph is adopted in all systems of units with which the author is familiar.

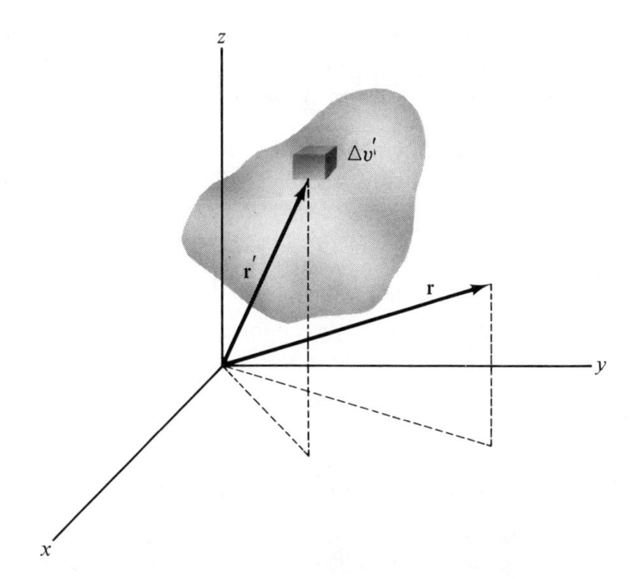

Fig. 10-1. Coordinates and vectors for calculating the electric field at a point exterior to a polarized dielectric.

dielectric to the electrostatic potential at **r** is then given by

$$\Delta V = \frac{\mathbf{P}(\mathbf{r}')\Delta v' \cdot (\mathbf{r} - \mathbf{r}')}{4\pi\epsilon_0 |\mathbf{r} - \mathbf{r}'|^3} \tag{10-8}$$

[see Eq. (4-44)] and integration of Eq. (10-8) over the volume of the dielectric yields

$$V(\mathbf{r}) = \frac{1}{4\pi\epsilon_0} \int \frac{\mathbf{P}(\mathbf{r}') \cdot (\mathbf{r} - \mathbf{r}')}{|\mathbf{r} - \mathbf{r}'|^3} \, dv' \tag{10-9}$$

for the contribution of the entire dielectric to the total electrostatic potential at **r**.

Equation (10-9) can be given a more useful physical interpretation if it is rewritten. First, we substitute the identity

$$\boldsymbol{\nabla}' \frac{1}{|\mathbf{r} - \mathbf{r}'|} = \frac{\mathbf{r} - \mathbf{r}'}{|\mathbf{r} - \mathbf{r}'|^3} \tag{10-10}$$

[compare Eq. (4-45)] to obtain

$$V(\mathbf{r}) = \frac{1}{4\pi\epsilon_0} \int \left[\mathbf{P}(\mathbf{r}') \cdot \boldsymbol{\nabla}' \frac{1}{|\mathbf{r} - \mathbf{r}'|} \right] dv' \tag{10-11}$$

Then we use the vector identity in Eq. (C-11) in the form

$$\mathbf{Q} \cdot \boldsymbol{\nabla}' \boldsymbol{\Phi} = \boldsymbol{\nabla}' \cdot (\boldsymbol{\Phi} \mathbf{Q}) - \boldsymbol{\Phi} \, \boldsymbol{\nabla}' \cdot \mathbf{Q} \tag{10-12}$$

with $\boldsymbol{\Phi} = 1/|\mathbf{r} - \mathbf{r}'|$ and $\mathbf{Q} = \mathbf{P}$, to obtain

$$V(\mathbf{r}) = \frac{1}{4\pi\epsilon_0} \int \left(\boldsymbol{\nabla}' \cdot \frac{\mathbf{P}(\mathbf{r}')}{|\mathbf{r} - \mathbf{r}'|} - \frac{\boldsymbol{\nabla}' \cdot \mathbf{P}(\mathbf{r}')}{|\mathbf{r} - \mathbf{r}'|} \right) dv' \tag{10-13}$$

Continuing, we use the divergence theorem in the first term to find that

$$V(\mathbf{r}) = \frac{1}{4\pi\epsilon_0} \oint \frac{\mathbf{P}(\mathbf{r}') \cdot d\mathbf{S}'}{|\mathbf{r} - \mathbf{r}'|} + \frac{1}{4\pi\epsilon_0} \int \frac{[-\nabla' \cdot \mathbf{P}(\mathbf{r}')]}{|\mathbf{r} - \mathbf{r}'|} dv' \qquad (10\text{-}14)$$

where the surface integral extends over the surface bounding the dielectric. Finally, we introduce the unit *outward* normal $\hat{\mathbf{n}}(\mathbf{r}')$ to the surface at the point $\mathbf{r}'$. Then, with $dS' = |d\mathbf{S}'|$, we have that $d\mathbf{S}' = \hat{\mathbf{n}}(\mathbf{r}')\, dS'$ and further that

$$V(\mathbf{r}) = \frac{1}{4\pi\epsilon_0} \oint \frac{\mathbf{P}(\mathbf{r}') \cdot \hat{\mathbf{n}}(\mathbf{r}')}{|\mathbf{r} - \mathbf{r}'|} dS' + \frac{1}{4\pi\epsilon_0} \int \frac{[-\nabla' \cdot \mathbf{P}(\mathbf{r}')]}{|\mathbf{r} - \mathbf{r}'|} dv' \qquad (10\text{-}15)$$

This result, however, is immediately recognized as the potential established by a charge distribution described by a volume charge density

$$\rho_b(\mathbf{r}) = -\nabla \cdot \mathbf{P}(\mathbf{r}) \qquad (10\text{-}16)$$

and a surface charge density

$$\sigma_b(\mathbf{r}) = \mathbf{P}(\mathbf{r}) \cdot \hat{\mathbf{n}}(\mathbf{r}) \quad (\mathbf{r} \text{ on the surface}) \qquad (10\text{-}17)$$

[See Eq. (4-49).] Thus, at exterior points, the polarized dielectric produces the same potential as these so-called *bound* charges, and we can replace our description in terms of the macroscopic polarization with this equivalent description in terms of bound charges if we find it convenient to do so.[5]

We can even see qualitatively how these bound charges arise. Consider the surface distribution first. When a dielectric is polarized, all of the dipoles have a preferred orientation. Thus, at some point on the surface of the polarized dielectric, all of the dipoles tend to point the same way, say with the positive charges out. A net positive charge thus appears on the surface at that point [Fig. 10-2(b)]. In the unpolarized state, some of these dipoles would point the other way, leaving the surface apparently neutral to a macroscopic observer [Fig. 10-2(a)]. Turning now to a qualitative interpretation of the

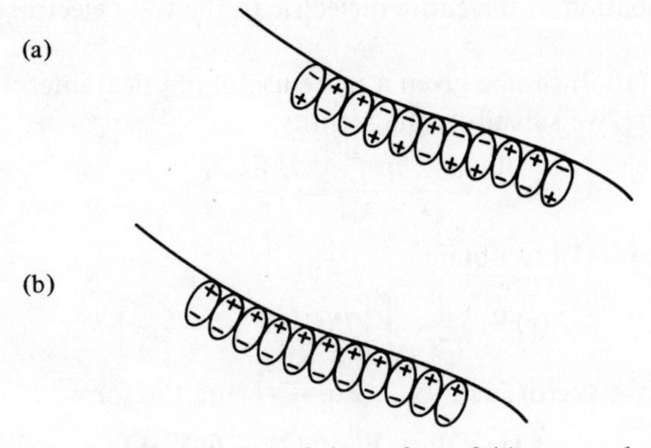

Fig. 10-2. An enlarged view of the surface of (a) an unpolarized and (b) a polarized dielectric.

[5]The expressions in Eqs. (10-16) and (10-17) for the bound charge densities apply in all systems of units with which the author is familiar.

volume charge density, we note that $\rho_b = 0$ unless $\mathbf{P}(\mathbf{r})$ is nonuniform. Assume a dielectric that is initially unpolarized. As the material is polarized, the interior charges are redistributed. Some of the charge in a selected but fixed volume element moves out one side while charge moves in from the surrounding volumes at the other side. If the polarization is uniform, the influx balances the outgo and the volume element remains electrically neutral despite the polarized state of the matter. If the polarization is not uniform, however, the two charge motions do not balance and the volume element acquires a net charge of one sign or the other. It is shown in P10-7 that the qualitative mechanisms described here yield Eqs. (10-16) and (10-17) when made quantitative. Further, it is shown in P10-8 that polarization of an initially neutral dielectric does not change that overall neutrality, i.e., that the sum of all bound charges is zero.

Given the potential, we can readily calculate the electric field established at an exterior point by a polarized dielectric; we simply take the negative gradient of the potential, finding that

$$\mathbf{E}(\mathbf{r}) = \frac{1}{4\pi\epsilon_0} \oint \frac{\sigma_b(\mathbf{r}')[\mathbf{r} - \mathbf{r}']}{|\mathbf{r} - \mathbf{r}'|^3} \, dS' + \frac{1}{4\pi\epsilon_0} \int \frac{\rho_b(\mathbf{r}')[\mathbf{r} - \mathbf{r}']}{|\mathbf{r} - \mathbf{r}'|^3} \, dv' \qquad (10\text{-}18)$$

Equation (10-18) for the field and Eq. (10-15) for the potential of course give only the contributions of the bound charges on the polarized dielectric; contributions from other sources must be added to those given by these equations.

PROBLEMS

P10-7. Convert the qualitative discussion of the physical origin of bound charges as presented in Section 10-3 into a quantitative derivation of Eqs. (10-16) and (10-17).

P10-8. Show that the total *bound* charge on a dielectric with arbitrary polarization is zero.

P10-9. A dielectric cube of side $2a$ has its center at the origin and its side parallel to the coordinate planes, and it has a radial polarization $\mathbf{P} = b\mathbf{r}$, where b is a constant. Find the charge distribution equivalent to this dielectric and show that the total bound charge is zero.

P10-10. A dielectric cylinder of length L and radius R is uniformly polarized with the polarization parallel to the axis of the cylinder. For convenience let the rod lie along the z axis with its center at the origin and take the polarization to be $\mathbf{P} = P\hat{\mathbf{k}}$. (a) Find the charge distribution equivalent to this dielectric. (b) Find the electrostatic potential at the point $(0, 0, z)$ and sketch a graph of this potential as a function of z/R for various values of L/R over the range $-\infty < z/R < \infty$. *Hint:* See P4-25. (c) Find the limiting form of the electrostatic potential at the point $(0, 0, z)$ in the region $z \gg L$, $z \gg R$ and comment on the multipole character of the distribution. (d) What is the dipole moment of the distribution as inferred from the polarization? Is your result in agreement with the dipole moment inferred from the limiting form of the potential?

P10-11. A spherical dielectric with its center at the origin has radius a and uniform polarization $\mathbf{P} = P\hat{\mathbf{k}}$. (a) Determine the equivalent charge distribution. (b) Calculate the dipole moment of this dielectric both from the polarization and from the equivalent charge distribution.

10-4 The Macroscopic Electric Field at a Point Interior to a Polarized Dielectric

The calculation of the *macroscopic* electric field at a point *interior* to a polarized dielectric is more complicated than the calculation of the field at an exterior point because the interior point is microscopically close to at least some of the molecules in the dielectric. Since these "near" molecules must be treated differently than the more numerous molecules that are microscopically remote from the interior point, we therefore divide the calculation of the interior field into two parts by identifying a sphere Σ_{II} of radius R centered at the interior point $\mathbf{r}$, choosing R large enough in microscopic terms so that all molecules outside of Σ_{II} are microscopically far away from $\mathbf{r}$. (See Fig. 10-3.) We then imagine the molecules within Σ_{II} to be removed temporarily, without, however, changing the polarization of the remaining dielectric. Once the near molecules have been removed, the point $\mathbf{r}$ is an exterior point; therefore, the contribution $\mathbf{E}_{\text{I}}(\mathbf{r})$ made to the macroscopic field at $\mathbf{r}$ by the molecules outside of Σ_{II} can be evaluated by the methods of Section 10-3.

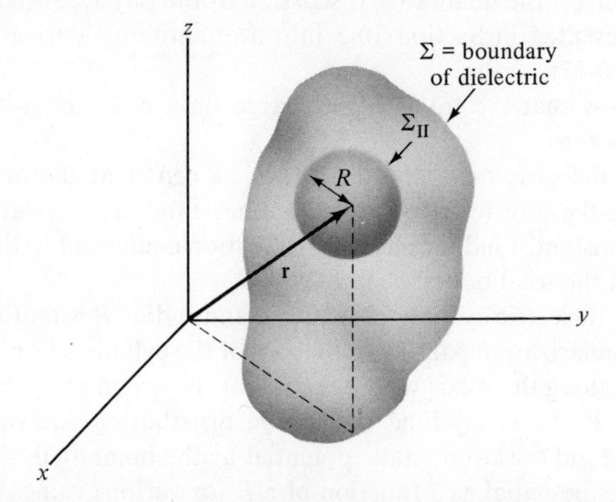

Fig. 10-3. Geometry for calculating the electric field at a point interior to a polarized dielectric. The spherical surface Σ_{II} centered at $\mathbf{r}$ divides the dielectric into "near" and "far" regions as described in the text.

We now replace the molecules that were temporarily removed and calculate their contribution $\mathbf{E}_{II}(\mathbf{r})$ to the macroscopic field at $\mathbf{r}$ by an averaging process to be described later in this section. In effect, we express the interior field $\mathbf{E}_{\text{interior}}(\mathbf{r})$ as the sum of two parts,

$$\mathbf{E}_{\text{interior}}(\mathbf{r}) = \mathbf{E}_{I}(\mathbf{r}) + \mathbf{E}_{II}(\mathbf{r}) \tag{10-19}$$

where $\mathbf{E}_I(\mathbf{r})$ is the contribution of the "distant" molecules and $\mathbf{E}_{II}(\mathbf{r})$ is the contribution of the near molecules. We then adopt different methods for evaluating the two contributions. In some respects we are *defining* what is meant by the electric field at an interior point—macroscopic test charges cannot easily be placed inside a dielectric—and the procedure as so far outlined will certainly not be satisfactory unless the final result for $\mathbf{E}_{\text{interior}}(\mathbf{r})$ is independent of the radius R of the (microscopically large but macroscopically small) sphere Σ_{II}.

Because the calculations are involved, we elect to present the comparatively simple final results before justifying them. We shall find that the two contributions to the total interior field are given, respectively, by

$$\mathbf{E}_I(\mathbf{r}) = \mathbf{E}(\mathbf{r}) + \frac{\mathbf{P}(\mathbf{r})}{3\epsilon_0}, \qquad \mathbf{E}_{II}(\mathbf{r}) = -\frac{\mathbf{P}(\mathbf{r})}{3\epsilon_0} \tag{10-20}$$

where $\mathbf{E}$ is the field contributed by the charge distribution equivalent to the entire dielectric as given by Eq. (10-18) (evaluated, however, at the *interior* point) and $\mathbf{P}$ is the polarization of the dielectric. Substituting Eq. (10-20) into Eq. (10-19), we find that

$$\mathbf{E}_{\text{interior}}(\mathbf{r}) = \mathbf{E}(\mathbf{r}) \tag{10-21}$$

and the validity of Eq. (10-18) for *interior* points is therefore established.

We shall now turn to the detailed derivation of Eq. (10-20). To facilitate the discussion, we introduce the notation

$$\begin{aligned}
v &= \text{volume of the original dielectric} \\
v_{II} &= \text{volume of sphere of radius } R \\
v_I &= v - v_{II} \\
\Sigma &= \text{surface of original dielectric} \\
\Sigma_{II} &= \text{surface of sphere of radius } R
\end{aligned}$$

as shown in Fig. 10-3.

Consider first the field $\mathbf{E}_I(\mathbf{r})$. Since $\mathbf{r}$ is exterior to the volume v_I and further is microscopically remote from every point of that volume, the results of Section 10-3 apply and we find from Eq. (10-18) that

$$\begin{aligned}
\mathbf{E}_I = \frac{1}{4\pi\epsilon_0}\int_{v_I} \frac{[-\nabla'\cdot\mathbf{P}(\mathbf{r}')]}{|\mathbf{r}-\mathbf{r}'|^3}(\mathbf{r}-\mathbf{r}')\,dv' \\
+ \frac{1}{4\pi\epsilon_0}\int_{\Sigma} \frac{\mathbf{P}(\mathbf{r}')\cdot\hat{\mathbf{n}}(\mathbf{r}')}{|\mathbf{r}-\mathbf{r}'|^3}(\mathbf{r}-\mathbf{r}')\,dS' \\
+ \frac{1}{4\pi\epsilon_0}\int_{\Sigma_{II}} \frac{\mathbf{P}(\mathbf{r}')\cdot\hat{\mathbf{n}}(\mathbf{r}')}{|\mathbf{r}-\mathbf{r}'|^3}(\mathbf{r}-\mathbf{r}')\,dS'
\end{aligned} \tag{10-22}$$

where on Σ and Σ_{II} the direction of the unit outward normal is reckoned relative to a point inside the volume v_I. The third integral in Eq. (10-22) now has a simple evaluation. Since the volume v_{II} is by hypothesis macroscopically small, the polarization $\mathbf{P}(\mathbf{r}')$ does not vary macroscopically over its surface Σ_{II}. Thus, in this third integral, $\mathbf{P}(\mathbf{r}')$ may be replaced by its value $\mathbf{P}(\mathbf{r})$ at the center of the sphere. Furthermore, the vector $\hat{\mathbf{n}}(\mathbf{r}')$ at a point $\mathbf{r}'$ on Σ_{II} is given by

$$\hat{\mathbf{n}}(\mathbf{r}') = \frac{\mathbf{r} - \mathbf{r}'}{|\mathbf{r} - \mathbf{r}'|} \tag{10-23}$$

where the outward direction at a point on Σ_{II} is a direction *toward* the point $\mathbf{r}$. Thus, the third integral $\mathbf{I}_3$ in Eq. (10-22) can be written in the form

$$\mathbf{I}_3 = \frac{1}{4\pi\epsilon_0}\int_{\Sigma_{II}} \frac{\mathbf{P}(\mathbf{r})\cdot(\mathbf{r} - \mathbf{r}')}{|\mathbf{r} - \mathbf{r}'|^4}(\mathbf{r} - \mathbf{r}')\, dS' \tag{10-24}$$

Now, introduce a spherical polar coordinate system with its center at the point $\mathbf{r}$ and its polar axis in the direction of $\mathbf{P}(\mathbf{r})$ (Fig. 10-4). Let $\hat{\mathbf{s}}$ be a unit vector in the radial direction in this coordinate system. Then, we find that

$$\mathbf{r} - \mathbf{r}' = -R\hat{\mathbf{s}}; \qquad |\mathbf{r} - \mathbf{r}'| = R; \qquad dS' = R^2 \sin\theta\, d\theta\, d\phi \tag{10-25}$$

and Eq. (10-24) becomes

$$\begin{aligned}\mathbf{I}_3 &= \frac{1}{4\pi\epsilon_0}\int [\mathbf{P}(\mathbf{r})\cdot\hat{\mathbf{s}}]\hat{\mathbf{s}} \sin\theta\, d\theta\, d\phi \\ &= \frac{P(\mathbf{r})}{4\pi\epsilon_0}\int \hat{\mathbf{s}} \cos\theta \sin\theta\, d\theta\, d\phi\end{aligned} \tag{10-26}$$

where we have recognized that the angle between $\mathbf{P}(\mathbf{r})$ and $\hat{\mathbf{s}}$ is θ. Now, in terms of Cartesian unit vectors $\hat{\boldsymbol{\xi}}, \hat{\boldsymbol{\eta}}, \hat{\boldsymbol{\zeta}}$ in the three coordinate directions shown in Fig. 10-4, the unit radial vector $\hat{\mathbf{s}}$ has the expression

$$\hat{\mathbf{s}} = \sin\theta\cos\phi\hat{\boldsymbol{\xi}} + \sin\theta\sin\phi\hat{\boldsymbol{\eta}} + \cos\theta\hat{\boldsymbol{\zeta}} \tag{10-27}$$

which on substitution into Eq. (10-26) leads to the evaluation

$$\mathbf{I}_3 = \frac{\mathbf{P}(\mathbf{r})}{3\epsilon_0} \tag{10-28}$$

We therefore find from Eq. (10-22) that

$$\begin{aligned}\mathbf{E}_I(\mathbf{r}) &= \frac{1}{4\pi\epsilon_0}\int_{v_I} \frac{[-\nabla'\cdot\mathbf{P}(\mathbf{r}')]}{|\mathbf{r} - \mathbf{r}'|^3}(\mathbf{r} - \mathbf{r}')\, dv' \\ &\quad + \frac{1}{4\pi\epsilon_0}\int_{\Sigma} \frac{\mathbf{P}(\mathbf{r}')\cdot\hat{\mathbf{n}}(\mathbf{r}')}{|\mathbf{r} - \mathbf{r}'|^3}(\mathbf{r} - \mathbf{r}')\, dS' + \frac{\mathbf{P}(\mathbf{r})}{3\epsilon_0}\end{aligned} \tag{10-29}$$

Equation (10-29) would be further simplified if the integral over v_I could be extended to include the volume v_{II} of the small sphere. To do so would

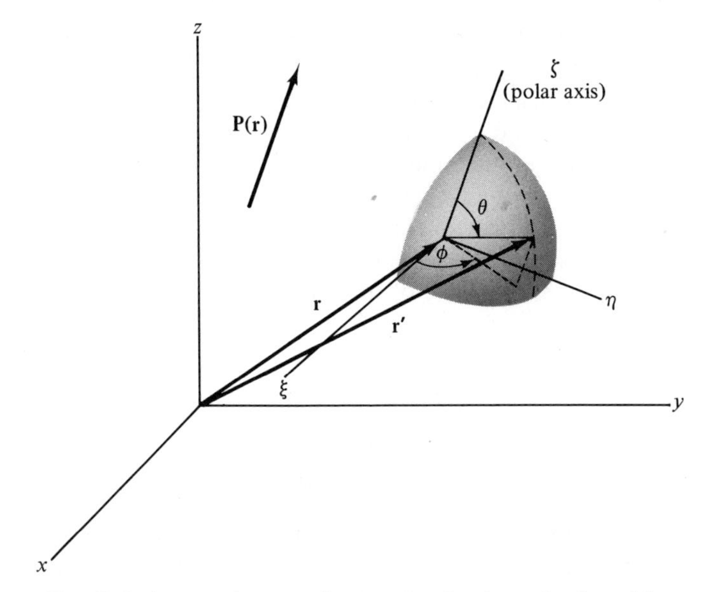

Fig. 10-4. A convenient coordinate system for the evaluation of the integral in Eq. (10-24).

require adding the integral

$$\mathbf{I}_4 = \frac{1}{4\pi\epsilon_0}\int_{v_{\text{II}}} \frac{[-\boldsymbol{\nabla}'\cdot\mathbf{P}(\mathbf{r}')]}{|\mathbf{r}-\mathbf{r}'|^3}(\mathbf{r}-\mathbf{r}')\,dv' \tag{10-30}$$

to Eq. (10-29). Over the macroscopically small volume v_{II}, however, we can treat $-\boldsymbol{\nabla}'\cdot\mathbf{P}(\mathbf{r}')$ as a constant. The integral $\mathbf{I}_4$ then represents the electric field at the center of a uniformly charged sphere of radius R (Why?), and consequently $\mathbf{I}_4 = 0$. (See P10-12.) Thus, $\mathbf{I}_4$ can be added to Eq. (10-29) with no change and we have finally that

$$\begin{aligned}\mathbf{E}_{\text{I}}(\mathbf{r}) &= \frac{1}{4\pi\epsilon_0}\int_{v} \frac{-\boldsymbol{\nabla}'\cdot\mathbf{P}(\mathbf{r}')}{|\mathbf{r}-\mathbf{r}'|^3}(\mathbf{r}-\mathbf{r}')\,dv' \\ &\quad + \frac{1}{4\pi\epsilon_0}\int_{\Sigma} \frac{\mathbf{P}(\mathbf{r}')\cdot\hat{\mathbf{n}}(\mathbf{r}')}{|\mathbf{r}-\mathbf{r}'|^3}(\mathbf{r}-\mathbf{r}')\,dS' + \frac{\mathbf{P}(\mathbf{r})}{3\epsilon_0} \\ &= \mathbf{E}(\mathbf{r}) + \frac{\mathbf{P}(\mathbf{r})}{3\epsilon_0}\end{aligned} \tag{10-31}$$

where $\mathbf{E}(\mathbf{r})$ is the field given at the *interior* point $\mathbf{r}$ by evaluating the exterior expression, Eq. (10-18), at the interior point; i.e., $\mathbf{E}(\mathbf{r})$ is the field established at the interior point by the bound charge distribution equivalent to the entire dielectric. Equation (10-31) confirms the first member of Eq. (10-20) for the contribution made by the molecules outside Σ_{II}.

To evaluate the contribution made by the molecules *within* Σ_{II} to the macroscopic field at the point **r**, we note first that this contribution must be insensitive to the precise microscopic location of Σ_{II} within the dielectric. That is, *microscopic* displacements of the point **r** within the dielectric cannot alter the *macroscopic* field $\mathbf{E}_{II}(\mathbf{r})$. We therefore *define* $\mathbf{E}_{II}(\mathbf{r})$ by the following *average*. Let **r** be located randomly—perhaps within a molecule, perhaps in the space between molecules—and assume that all molecules within Σ_{II} are characterized by the same time-averaged microscopic charge distribution (which is equivalent to assuming that the volume v_{II} is macroscopically small and that all molecules in v_{II} are of the same chemical species[6]). Finally, let the volume v_{II} contain N molecules. Then the probability of finding a molecule centered within the infinitesimal volume element dv' centered at the point $\mathbf{r}'$ somewhere in v_{II} is $N\,dv'/v_{II}$, and the *average* contribution made by the element dv' at $\mathbf{r}'$ to the microscopic field at **r** is given by

$$d\mathbf{E}_{II} = \frac{N\,dv'}{v_{II}}\mathbf{E}_{mol}(\mathbf{r}, \mathbf{r}') \tag{10-32}$$

where $\mathbf{E}_{mol}(\mathbf{r}, \mathbf{r}')$ is the time average of the microscopic field produced at **r** by a molecule whose center is at $\mathbf{r}'$. All of the molecules in v_{II} thus produce an *average* microscopic field at **r** given by

$$\mathbf{E}_{II}(\mathbf{r}) = \frac{N}{v_{II}}\int_{v_{II}} \mathbf{E}_{mol}(\mathbf{r}, \mathbf{r}')\,dv' \tag{10-33}$$

and we take this *average microscopic* field as the contribution of the molecules in v_{II} to the *macroscopic* field at **r**.

To obtain a more explicit evaluation of Eq. (10-33), we use the above assumption that all molecules in v_{II} can be characterized by the *same* time-averaged molecular charge density, which we take to be given at the point $\mathbf{r}_1$ for a molecule whose center is at $\mathbf{r}_c$ by $\rho_{mol}(\mathbf{r}_1 - \mathbf{r}_c)$. Then, the field $\mathbf{E}_{mol}(\mathbf{r}, \mathbf{r}')$ is given by Eq. (4-7), viz.,

$$\mathbf{E}_{mol}(\mathbf{r}, \mathbf{r}') = \frac{1}{4\pi\epsilon_0}\int \rho_{mol}(\mathbf{r}''' - \mathbf{r}')\frac{\mathbf{r} - \mathbf{r}'''}{|\mathbf{r} - \mathbf{r}'''|^3}\,dv''' \tag{10-34}$$

where $\mathbf{r}'''$ locates a point in the molecule and the proper limits are effectively imposed by the function ρ_{mol}, which is zero outside the molecule. More conveniently, we can write Eq. (10-34) as an integral on the variable $\mathbf{r}'' = \mathbf{r}''' - \mathbf{r}'$, finding

$$\mathbf{E}_{mol}(\mathbf{r}, \mathbf{r}') = \frac{1}{4\pi\epsilon_0}\int \rho_{mol}(\mathbf{r}'')\frac{\mathbf{r} - \mathbf{r}' - \mathbf{r}''}{|\mathbf{r} - \mathbf{r}' - \mathbf{r}''|^3}\,dv'' \tag{10-35}$$

The relationships among $\mathbf{r}$, $\mathbf{r}'$, $\mathbf{r}''$, and $\mathbf{r}'''$ are illustrated in Fig. 10-5. In terms of the (unknown) molecular charge density, Eq. (10-33) now becomes

$$\mathbf{E}_{II}(\mathbf{r}) = \frac{N}{4\pi\epsilon_0 v_{II}}\int \rho_{mol}(\mathbf{r}'')\left[\int_{v_{II}} \frac{\mathbf{r} - \mathbf{r}' - \mathbf{r}''}{|\mathbf{r} - \mathbf{r}' - \mathbf{r}''|^3}\,dv'\right]dv'' \tag{10-36}$$

[6]If several different types of molecule are present, the argument must be extended to include a sum over these several types.

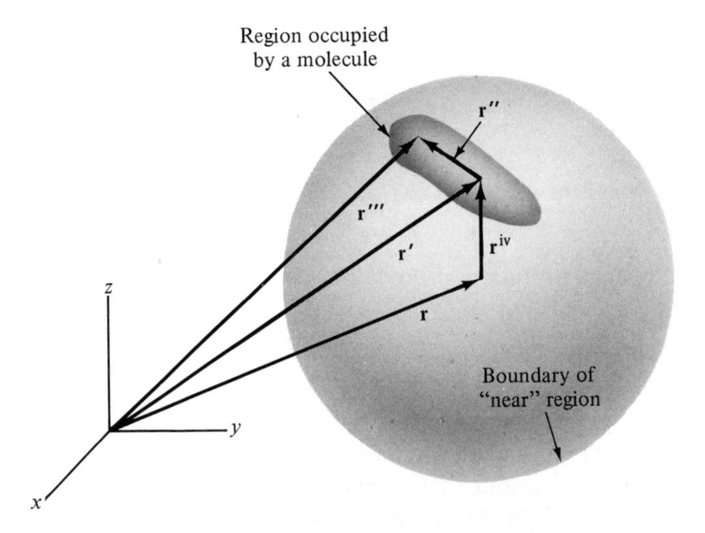

Fig. 10-5. An exaggerated view of the "near" region in a dielectric. The region occupied by a single molecule is shown.

where we have chosen to do the integral on the single primed variables first, since the full dependence of the integrand on that variable is known. The integral $\mathbf{I}_1$ in square brackets, however, is more conveniently evaluated if we write it as an integral on the variable $\mathbf{r}^{iv} = \mathbf{r}' - \mathbf{r}$ which locates the point $\mathbf{r}'$ relative to the center of v_{II} (Fig. 10-5); we thus have that

$$\mathbf{I}_1 = \int_{v_{\text{II}}} \frac{(-\mathbf{r}'') - \mathbf{r}^{iv}}{|(-\mathbf{r}'') - \mathbf{r}^{iv}|^3}\, dv^{iv} \tag{10-37}$$

which is written to facilitate the recognition that $\mathbf{I}_1$ is *numerically* the same as the electric field at the point $(-\mathbf{r}'')$ inside a spherical charge distribution having the uniform charge density $\rho(\mathbf{r}^{iv}) = 4\pi\epsilon_0$. [Compare Eq. (4-7) with $dq' \longrightarrow \rho(\mathbf{r}^{iv})\, dv^{iv}$, $\mathbf{r}' \longrightarrow \mathbf{r}^{iv}$, and $\mathbf{r} \longrightarrow -\mathbf{r}''$.] We can therefore evaluate $\mathbf{I}_1$ by exploiting Gauss's law to calculate the corresponding electric field. Because this source distribution is invariant to rotation about *any* diameter, the field itself can be a function only of the distance R from the center to the point of observation and must be radially directed, i.e.,

$$\mathbf{E}(\mathbf{R}) = E(R)\hat{\mathbf{R}} \tag{10-38}$$

where $\mathbf{R}$ is the position vector of the observation point relative to the center of the sphere. Since the flux out of a spherical surface Σ' having radius R and centered at the center of the distribution is therefore given by

$$\oint_{\Sigma'} \mathbf{E} \cdot d\mathbf{S} = 4\pi R^2 E(R) \tag{10-39}$$

and the total charge inside the volume V bounded by Σ' is given by

$$\int_V \rho\, dv = (4\pi\epsilon_0)(\tfrac{4}{3}\pi R^3) \tag{10-40}$$

Gauss's law, which requires the flux to be $(1/\epsilon_0)$ times the total charge, yields that

$$4\pi R^2 E(R) = 4\pi(\tfrac{4}{3}\pi R^3)$$

$$\Longrightarrow E(R) = \tfrac{4}{3}\pi R \Longrightarrow \mathbf{E}(\mathbf{R}) = \tfrac{4}{3}\pi R\hat{\mathbf{R}} = \tfrac{4}{3}\pi\mathbf{R} \qquad (10\text{-}41)$$

Thus, we obtain the value of $\mathbf{I}_1$ as

$$\mathbf{I}_1 = \mathbf{E}(-\mathbf{r}'') = -\tfrac{4}{3}\pi\mathbf{r}'' \qquad (10\text{-}42)$$

and, finally, on substitution of Eq. (10-42) for the term in square brackets in Eq. (10-36), we find that

$$\mathbf{E}_{\text{II}}(\mathbf{r}) = \frac{N}{4\pi\epsilon_0 v_{\text{II}}}\left(-\frac{4\pi}{3}\right)\int \mathbf{r}''\rho_{\text{mol}}(\mathbf{r}'')\,dv'' \qquad (10\text{-}43)$$

The remaining integral, however, is the dipole moment of a single molecule located near the point $\mathbf{r}$ in the dielectric. Thus, the integral times N is the total dipole moment in v_{II} and that product divided by v_{II} is the *macroscopic* polarization $\mathbf{P}(\mathbf{r})$ of the dielectric at $\mathbf{r}$. Equation (10-43) therefore has the macroscopic evaluation

$$\mathbf{E}_{\text{II}}(\mathbf{r}) = -\frac{\mathbf{P}(\mathbf{r})}{3\epsilon_0} \qquad (10\text{-}44)$$

thus verifying the second member of Eq. (10-20) and also confirming the conclusion that the *macroscopic interior* field is given by the same expression that gives the *exterior* field, i.e., by Eq. (10-18).

PROBLEM

P10-12. Assuming that v_{II} in Eq. (10-30) is small enough so that $\nabla'\cdot\mathbf{P}(\mathbf{r}')$ can be regarded as constant over the volume, show that the integral in Eq. (10-30) is zero (a) by relating the integral to the field at the center of a uniformly charged sphere and invoking symmetry, (b) by direct evaluation of the integral (*Hint:* Pick a spherical polar coordinate system with its origin at the center of v_{II}), and (c) by noting Eqs. (10-10) and (C-21) and evaluating the resulting surface integral.

10-5 The Basic Equations of Electrostatics When Dielectrics Are Present

In this section, we shall translate the basic equations of electrostatics into a convenient form for treating problems involving dielectrics. We have already established that the static electric field produced at any point in space by a polarized dielectric can be viewed as originating in a suitable distribution of static charges in free space. Thus, the static electric field $\mathbf{E}$

established jointly by a polarized dielectric and by any simultaneously present distributions of free charge still satisfies both the restricted Faraday law

$$\oint \mathbf{E}\cdot d\boldsymbol{\ell} = 0 \qquad \nabla \times \mathbf{E} = 0 \tag{10-45}$$

and Gauss's law

$$\oint \mathbf{E}\cdot d\mathbf{S} = \frac{1}{\epsilon_0}\int \rho_t\, dv \qquad \nabla\cdot\mathbf{E} = \frac{\rho_t}{\epsilon_0} \tag{10-46}$$

provided we now interpret the charge density ρ_t in Gauss's law as the *total* charge density, which includes any "free" charges placed in space *and* any bound charges present on polarized dielectrics. Since we know how to relate the bound charge to the polarization **P**, however, we can reexpress Eq. (10-46) in a better form. Let ρ now denote only the free charge density. Then

$$\rho_t = \rho + \rho_b = \rho - \nabla\cdot\mathbf{P} \tag{10-47}$$

Hence, with some rearrangement, Eq. (10-46) yields

$$\oint (\epsilon_0\mathbf{E} + \mathbf{P})\cdot d\mathbf{S} = \int \rho\, dv \qquad \nabla\cdot(\epsilon_0\mathbf{E} + \mathbf{P}) = \rho \tag{10-48}$$

(The integral form is obtained after using the divergence theorem.) We stress again that the charge density appearing on the right-hand side is the *free* charge density; the bound charges have been explicitly introduced and appear now on the left-hand side in somewhat disguised form. We can now push the polarization, which we usually do not know initially, into the background altogether by introducing the *displacement field* **D** defined in mks units by[7]

$$\mathbf{D} = \epsilon_0\mathbf{E} + \mathbf{P} \tag{10-49}$$

At least the flux of **D** over closed surfaces or equivalently the divergence of **D** is determined solely by the free charge, as evidenced by the new form of Gauss's law,

$$\oint \mathbf{D}\cdot d\mathbf{S} = \int \rho\, dv \qquad \nabla\cdot\mathbf{D} = \rho \tag{10-50}$$

obtained by substituting Eq. (10-49) into Eq. (10-48). The displacement field **D** is the first of the two auxiliary fields mentioned in the introductory paragraphs of Chapter 10. Although we shall continue to regard **E** as the basic field, arguments supporting the opposite view can be presented. Whichever field is viewed as basic, the field **E** remains the field that determines the force on a charged particle.

We have now determined that the static electric field in the presence of

[7]The vector **D** is defined differently in different systems of units but is always some linear combination of **E** and **P**. In cgs-esu, cgs-emu, Gaussian units, and Heaviside-Lorentz units **D** is defined so that $\nabla\cdot\mathbf{D} = 4\pi\rho$, $4\pi\rho$, $4\pi\rho$, and ρ, respectively. (See P10-24.)

dielectrics satisfies the basic equations

$$\oint \mathbf{E}\cdot d\boldsymbol{\ell} = 0 \qquad \nabla \times \mathbf{E} = 0 \tag{10-51}$$

$$\oint \mathbf{D}\cdot d\mathbf{S} = \int \rho\, dv \qquad \nabla\cdot\mathbf{D} = \rho \tag{10-52}$$

$$\mathbf{D} = \epsilon_0\mathbf{E} + \mathbf{P} \tag{10-53}$$

This system, of course, reduces to the vacuum form when there are no dielectrics and **P** is therefore zero. In that case, **D** and **E** are trivially different and Eqs. (10-51) and (10-52) give sufficient information about the field in vacuum to determine that field. When polarizable dielectrics are present, however, **D** may differ nontrivially from **E**, the two differential equations involve two essentially different vectors, and we can make little progress toward a solution of these equations until we know a relationship between **D** and **E**. Equation (10-53) is a step toward specifying that relationship, but it is not a complete specification because the vector **P** is still not known. The specific properties of the dielectric involved enter at this point because the extent to which a dielectric acquires a macroscopic polarization depends not only on the polarizing field but on the intrinsic nature of the material as well. Since the polarization depends on the forces that the charges experience when they are immersed in an electric field, it is reasonable to expect some *constitutive relation* of the general form

$$\mathbf{P} = \mathbf{P}(\mathbf{E}) \tag{10-54}$$

to describe the response of the dielectric, even though the individual molecules in fact polarize in response to the microscopic rather than the macroscopic field. A more explicit form for Eq. (10-54) can be determined only by an *empirical* study of specific dielectrics or, in some cases, by a detailed quantum mechanical calculation. Once the form of Eq. (10-54) is known for a particular material, Eqs. (10-51)–(10-53) provide sufficient information to determine the two fields **E** and **D** if the *free* charge distribution and/or suitable boundary conditions—see Section 12-7—are known.

The constitutive relation appropriate to a particular material may in fact be quite complicated. A material composed of molecules with a permanent dipole moment may be characterized by a different constitutive relation than a material composed of nonpolar molecules; some materials develop polarizations that are not in the same direction as the polarizing field; especially when the applied fields are strong, the polarization may not depend in any simple way on **E**; some materials do not respond in the same way to fields applied in different directions; some few materials even exhibit a polarization in the absence of a polarizing field (P10-22). Even if materials displaying these and other complications are excluded, however, there remain many materials that develop a polarization in the same direction as the polarizing field. For these dielectrics, the constitutive relation has the form

$$\mathbf{P} = \chi_e(E)\mathbf{E} \tag{10-55}$$

where the *static dielectric susceptibility* $\chi_e(E)$ may still depend on the field strength. For many materials, however, χ_e happens to be constant, at least if E is not too large. These simplest of all dielectrics are called *linear* dielectrics and the displacement vector in such a dielectric is proportional to the electric field,

$$\mathbf{D} = \epsilon_0\mathbf{E} + \mathbf{P} = (\epsilon_0 + \chi_e)\mathbf{E} = \epsilon\mathbf{E} \tag{10-56}$$

where the *static permittivity of the dielectric* ϵ is defined by

$$\epsilon = \epsilon_0 + \chi_e \tag{10-57}$$

Since χ_e is necessarily positive (Why?), the permittivity ϵ always exceeds ϵ_0 and the *static dielectric constant* (or the relative permittivity) K_e, defined by

$$K_e = \frac{\epsilon}{\epsilon_0} = 1 + \frac{\chi_e}{\epsilon_0} \tag{10-58}$$

always exceeds 1. Values of K_e for a selection of common dielectric materials are shown in Table 10-1.

We shall conclude this section with an example that not only illustrates the use of the new form of Gauss's law but also provides the basis of one method for measuring dielectric constants. Consider a parallel plate capacitor (P4-16) having plates of area A and separation d, let the space between the plates be filled with a linear dielectric of permittivity ϵ (Fig. 10-6), and ignore fringing so that the translational and rotational invariances of the whole arrangement require the fields to be perpendicular to the plates. We seek the polarization **P**, the electric field **E**, and the displacement **D** in the region between the plates. If we knew **P**, we could calculate the bound charge densities from **P** and then calculate **E** from the total charge. But we do not know **P**. Let us therefore begin by calculating **D**, the flux of which is determined solely by the (known) free charge distribution. Applying Gauss's law, Eq.

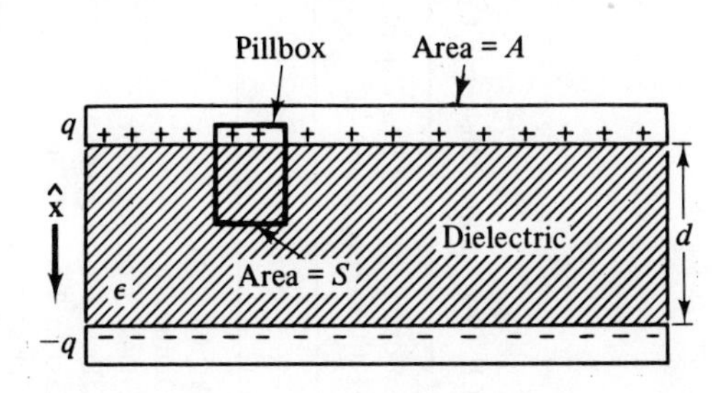

Fig. 10-6. A parallel plate capacitor with a dielectric filler.

TABLE 10-1 Static Dielectric Constants for Selected Materials

The values in this table are quoted from the *Handbook of Chemistry and Physics* (Chemical Rubber Publishing Company, Cleveland, 1965), Forty-sixth Edition, pp. (E-49)–(E-55), and are used by permission of the Chemical Rubber Company.

Material	*Temperature*	K_e	*Material*	*Temperature*	K_e
Air (1 atm)	Room	1.000590	Oxygen (liquid)	80.0°K	1.507
Diamond	Room	5.5	Oxygen (1 atm)	100°C	1.000523
Glass: Corning 8870	Room	9.5	Polyethylene	Room	2.3
Glass: Corning 0080	Room	6.75	Porcelain	Room	6–10[a]
Glass: Pyrex 7070	Room	4.00	Paraffin	Room	2.0–2.5[a]
Glass: Fused quartz	Room	3.75–4.1[a]	Rubber (hard)	Room	2.8
Helium (liquid)	4.19°K	1.048	Water (liquid)	0°C	88.00
Helium (1 atm)	140°C	1.0000684	Water (liquid)	25°C	78.54
Hydrogen (liquid)	20.4°K	1.228	Water (liquid)	50°C	69.94
Hydrogen (1 atm)	100°C	1.000264	Water (liquid)	100°C	55.33
Nylon (F.M. 3001)	Room	3.5	Water (steam; 1 atm)	110°C	1.0126

[a]Depending on composition.

(10-52), in integral form to the pillbox shown in Fig. 10-6, we find that

$$\oint \mathbf{D}\cdot d\mathbf{S} = DS = \int \rho\, dv = \sigma S = \frac{qS}{A}$$

$$\Longrightarrow D = \sigma = \frac{q}{A} \Longrightarrow \mathbf{D} = \frac{q}{A}\hat{\mathbf{x}} \tag{10-59}$$

where S is the cross-sectional area of the pillbox and σ is the surface density of free charge and q the total free charge on the upper plate of the capacitor. [The surface integral in Eq. (10-59) has a contribution only from the lower face of the pillbox because $\mathbf{D}$ is parallel to the vertical face and the top face is in a conductor where $\mathbf{D} = 0$.] Thus,

$$\mathbf{E} = \frac{\mathbf{D}}{\epsilon} = \frac{q}{\epsilon A}\hat{\mathbf{x}} \tag{10-60}$$

and

$$\mathbf{P} = \chi_e \mathbf{E} = \frac{\epsilon - \epsilon_0}{\epsilon}\frac{q}{A}\hat{\mathbf{x}} \tag{10-61}$$

Equations (10-16) and (10-17) then give

$$\begin{aligned} \rho_b &= 0 \\ \sigma_b(\text{upper}) &= \mathbf{P}\cdot(-\hat{\mathbf{x}}) = -\frac{\epsilon - \epsilon_0}{\epsilon}\frac{q}{A} \\ \sigma_b(\text{lower}) &= \mathbf{P}\cdot(\hat{\mathbf{x}}) = \frac{\epsilon - \epsilon_0}{\epsilon}\frac{q}{A} \end{aligned} \tag{10-62}$$

for the bound charge distribution induced on the dielectric. Knowing $\mathbf{E}$ as in Eq. (10-60), we find that the potential difference ΔV between the plates is given by

$$\Delta V = -\int \mathbf{E}\cdot d\boldsymbol{\ell} = \frac{qd}{\epsilon A} \tag{10-63}$$

and further the capacitance of the arrangement as defined by Eq. (4-53) is given by

$$C = \frac{q}{\Delta V} = \epsilon\frac{A}{d} \tag{10-64}$$

In particular the capacitance of the arrangement when $\epsilon = \epsilon_0$ (no dielectric) is $C_0 = \epsilon_0 A/d$ and

$$\frac{C}{C_0} = \frac{\epsilon}{\epsilon_0} = K_e \tag{10-65}$$

Since $K_e > 1$, placing a dielectric between the plates of a capacitor *increases* its capacitance. Further, Eq. (10-65) shows that careful measurements of capacitance with and without a dielectric filler can yield a numerical value for the dielectric constant of the filler.

PROBLEMS

P10-13. Apply Gauss's law, Eq. (10-52), to an infinite linear dielectric and derive Coulomb's law for the force between two point charges in such a medium. Is this force larger or smaller than the force between the same two charges at the same separation in vacuum? By what factor is the force changed?

P10-14. Suppose a dielectric is linear but inhomogeneous (ϵ is independent of field strength but depends on position in the dielectric). Starting with Eqs. (10-51)–(10-53), argue the existence of an electrostatic potential V and show that it satisfies the equation $\nabla \cdot (\epsilon \nabla V) = -\rho$, which reduces to Poisson's equation $\nabla^2 V = -\rho/\epsilon$ if the dielectric is homogeneous (ϵ independent of position). *Hint:* See item (3) in Section 2-5.

P10-15. Replace the dielectric in Fig. 10-6 with the equivalent charge distribution in Eq. (10-62) and determine the electric field between the plates by applying Gauss's law in free space. Show that the result agrees with Eq. (10-60).

P10-16. The dielectric constant of the material filling the space between the two plates of a parallel plate capacitor varies linearly from the value K_1 at one plate to the value K_2 at the other. If the plates have separation d and area A, determine the capacitance and verify that your result gives the correct value in the limiting case $K_1 = K_2$. *Hint:* Ignore fringing.

P10-17. Two parallel conducting plates of area A are separated by a gap of thickness d in which is placed a dielectric slab of uniform thickness $t\,(< d)$ and of dielectric constant K. Let the potential difference between the plates be ΔV. (a) Determine **E** and **D** at all points between the plates. *Hint:* Ignore fringing. (b) What is the free charge density on the conducting plates? What is the bound charge density on the surfaces of the dielectric? (c) Determine the capacitance of the arrangement and compare it with the capacitance that the arrangement would have if the dielectric were removed.

P10-18. A conducting sphere of radius a carries a total charge q and is embedded in a concentric spherical shell of inner radius a and outer radius b made of a linear dielectric of permittivity ϵ. (a) Determine **D**, **E**, and **P** at points within the dielectric. *Hint:* Note the symmetry and apply Gauss's law for the displacement vector. (b) What is the equivalent charge distribution for this dielectric? (c) Determine **D** and **E** at points in the free space outside the dielectric. (d) Calculate **D** and **E** at a point infinitesimally beyond the outer surface of the dielectric and compare with **D** and **E** at a point infinitesimally inside that surface. Show that **E** has a discontinuity and relate that discontinuity to the bound surface charge density. (e) A conducting spherical shell of radius b is now placed about the arrangement and carries a free charge $-q$. Determine the capacitance of this new arrangement and compare the result with the capacitance that the device would have if the dielectric were removed.

10-6 Connecting the Microscopic Polarizability and the Macroscopic Dielectric Constant: The Clausius-Mossotti Relation

We shall now derive the Clausius-Mossotti relation, which permits a determination of the microscopic molecular polarizabilities of *some* (but not all) materials from measured values of the macroscopic dielectric constants. We shall begin by finding the field $\mathbf{E}_m$ that a single molecule in a dielectric experiences, i.e., by finding the microscopic field established *at* the position of a molecule by all molecules except the one located at that site. Unfortunately, the field $\mathbf{E}_m$ is not the same field as the interior macroscopic field calculated in Section 10-4, because the macroscopic field involves an average of the microscopic field over points within some volume and hence includes contributions from the microscopic field not only at molecular sites but also at other points in the dielectric. The procedure used in Section 10-4, however, can be modified to apply to the present situation. Again, we remove the molecules from a macroscopically small but microscopically large spherical region centered this time on a selected molecular site but we do not permit the part of the dielectric remaining to readjust to the absence of the molecules removed. Then we use macroscopic methods to calculate the contributions $\mathbf{E}_{\mathrm{I}}$ made by the "far" molecules *and* by any free charges to the field $\mathbf{E}_m$. Finally, we replace the molecules removed (except for the one at the selected molecular site) and add the field $\mathbf{E}'$ contributed by these near molecules. In brief, we express $\mathbf{E}_m$ as a sum of two parts,

$$\mathbf{E}_m = \mathbf{E}_{\mathrm{I}} + \mathbf{E}' \tag{10-66}$$

and evaluate each part separately. In fact, we have already calculated $\mathbf{E}_{\mathrm{I}}$; its definition here is essentially identical with its definition in Section 10-4 and it is therefore given by Eq. (10-31) except that $\mathbf{E}$ must now be interpreted to include contributions not only from the bound charge on the dielectric but also from any free charges present; i.e., $\mathbf{E}$ is the *total* macroscopic field at the molecular site. Thus,

$$\mathbf{E}_m = \mathbf{E} + \frac{\mathbf{P}}{3\epsilon_0} + \mathbf{E}' \tag{10-67}$$

The difference between the present calculation and the calculation in Section 10-4 arises because the field $\mathbf{E}'$ here is very different from the field $\mathbf{E}_{\mathrm{II}}$ there, for we are now interested in the field at a more specific point, namely at the site of a molecule. In fact, $\mathbf{E}'$ is very difficult to evaluate for any particular case. Its value depends on the properties of the specific molecule, on the relative placement of the molecules in the structure of the dielectric, and on the interrelations among these molecules. For gases and liquids (in which the near molecules are randomly positioned) and for some crystals (in which very strong translational symmetries exist), however, $\mathbf{E}'$ turns out to be zero (P10-23). *For those materials*, the so-called *Lorentz form* of the field $\mathbf{E}_m$, where

$$\mathbf{E}_m = \mathbf{E} + \frac{\mathbf{P}}{3\epsilon_0} \quad (\text{if } \mathbf{E}' = 0) \tag{10-68}$$

applies and $\mathbf{E}_m$ differs from $\mathbf{E}$ by a term proportional to the polarization.

With an expression for $\mathbf{E}_m$ now available, we can develop an equation relating the molecular polarizability α_t of Eq. (10-4) to the dielectric constant K_e. We simply note from Eq. (10-3) that the dipole moment $\mathbf{p}$ of a single molecule is given by

$$\mathbf{p} = \alpha_t \mathbf{E}_m = \alpha_t\left(\mathbf{E} + \frac{\mathbf{P}}{3\epsilon_0}\right) \tag{10-69}$$

Then, the polarization of the dielectric is given by

$$\mathbf{P} = N\mathbf{p} = N\alpha_t\left(\mathbf{E} + \frac{\mathbf{P}}{3\epsilon_0}\right) \tag{10-70}$$

where N is the number of molecules per unit volume in the dielectric. Equation (10-70), however, has the solution

$$\mathbf{P} = \frac{N\alpha_t}{1 - (N\alpha_t/3\epsilon_0)}\mathbf{E} \tag{10-71}$$

for $\mathbf{P}$. Finally on comparing Eq. (10-71) with Eq. (10-55), we find that

$$\chi_e = \frac{N\alpha_t}{1 - (N\alpha_t/3\epsilon_0)} = \epsilon_0(K_e - 1) \tag{10-72}$$

which can be solved either for α_t in terms of K_e or vice versa, giving that

$$\alpha_t = \frac{3\epsilon_0}{N}\frac{K_e - 1}{K_e + 2}, \qquad K_e - 1 = \frac{N\alpha_t/\epsilon_0}{1 - (N\alpha_t/3\epsilon_0)} \tag{10-73}$$

Equation (10-72) is known as the *Clausius-Mossotti relation* when applied to nonpolar molecules and the *Debye equation* when applied to polar molecules. For materials conforming to the Lorentz form of $\mathbf{E}_m$ in Eq. (10-68), the (microscopic) molecular polarizability can therefore be determined from measurements of the (macroscopic) dielectric constant. More complicated relations than Eq. (10-73) can be derived for materials that do not conform to Eq. (10-68), but both a fuller theoretical study of the connection between macroscopic and microscopic properties for all dielectrics and a complete cataloging of known experimental properties of dielectrics lie outside the scope of this book.

PROBLEMS

P10-19. Derive Eq. (10-73) from Eq. (10-72).

P10-20. Combine the Clausius-Mossotti relation with the result in P10-3(b) to obtain an expression for the radius of a nonpolar atom. Given that the dielectric constant of air at 1 atmosphere pressure is 1.00059, estimate the radius of an atom in an air molecule.

P10-21. Nitrogen gas at room temperature ($\approx 27°C$) and 1 atmosphere pressure has a dielectric constant $K_e = 1.000580$. The density of liquid nitrogen at its boiling point ($-195.8°C$) is 0.808 g/cm³. (a) Show from Eq. (10-73) that $K_e - 1 \approx N\alpha_t/\epsilon_0$ when $K_e \approx 1$ and find $N\alpha_t/\epsilon_0$ for nitrogen gas. (b) Use the ideal gas laws to determine the density of molecules in a gas at room temperature and then use part (a) to find α_t for the nitrogen molecule. (c) Assuming that α_t as a property of individual *molecules* is approximately the same for molecules in the liquid as for those in the gas, find $N\alpha_t/\epsilon_0$ for liquid nitrogen and then use Eq. (10-73) without approximation to predict the dielectric constant of liquid nitrogen at its boiling point. The measured value is $K_e = 1.474$. *Optional:* Write a computer program to predict the dielectric constants of liquids by the method of this problem, look up data on other substances (e.g., A, C_6H_6, CS_2, O_2, He, Br_2, CCl_4) in the *Handbook of Chemistry and Physics*, and compare your predictions with measured values.

P10-22. Consider a dielectric crystal of the type to which Eq. (10-71) applies and let the individual molecules have a polarizability dominated by the orientational part, $\alpha_t = p_0^2/3kT$. Sketch a graph of χ_e versus T. What happens when $kT \leq Np_0^2/9\epsilon_0$? Do negative values of χ_e make physical sense? *Hint:* Some few materials, called *ferroelectrics*, exhibit a spontaneous permanent polarization at temperatures below a critical temperature whose value depends on the material.

P10-23. Because the distortion of a molecule brought about even by fairly strong fields is very small compared to molecular dimensions [P10-3(e) and P10-4], the contribution made by a near molecule to the field $\mathbf{E}'$ at a molecular site can be computed by considering the near molecule to be a dipole whose field is given by Eq. (4-10). Introduce a coordinate system with its origin at the molecular site and its z axis parallel to the macroscopic polarization at the molecular site. Then all molecules contributing to $\mathbf{E}'$ have the same dipole moment $\mathbf{p} = p\hat{\mathbf{k}}$. Finally, let the ith contributing molecule be at $\mathbf{r}_i$. (a) Show that the field $\mathbf{E}'$ produced at the origin by all contributing molecules is given by

$$\mathbf{E}' = \frac{p}{4\pi\epsilon_0} \sum_i \frac{3z_i x_i \hat{\mathbf{i}} + 3z_i y_i \hat{\mathbf{j}} + (3z_i^2 - r_i^2)\hat{\mathbf{k}}}{r_i^5}$$

(b) Show that $\mathbf{E}' = 0$ if the near molecules are located randomly, as, for example, in a liquid or gas. *Hint:* With random location $\sum (z_i x_i/r_i^5) = \sum (z_i y_i/r_i^5) = 0$ and $\sum (z_i^2/r_i^5) = \frac{1}{3} \sum (r_i^2/r_i^5)$. Why? (c) Show that $\mathbf{E}' = 0$ if the molecules are located at the sites of a cubic lattice of side a and the z axis is defined by $\mathbf{p}$, which coincides with a crystal axis; i.e., show that $\mathbf{E}' = 0$ if all coordinates x_i, y_i, and z_i are (positive or negative) integer multiples of a. *Hint:* Consider the contributions from nearest neighbors, next nearest neighbors, etc., separately. *Optional:* Generalize part (c) to a proof when the dipole moment is not parallel to one of the crystal axes.

Supplementary Problems

P10-24. Given that $\rho_b = -\nabla \cdot \mathbf{P}$ in all systems of units and that $\nabla \cdot \mathbf{E} = 4\pi\rho_t$, $4\pi c^2\rho_t$, $4\pi\rho_t$, and ρ_t in cgs-esu, cgs-emu, Gaussian units, and Heaviside-Lorentz units, respectively, find the relationship among **D**, **E**, and **P** in these systems of units. *Hint:* See Footnote 7.

P10-25. Consider an anisotropic crystal made up of molecules consisting of a *fixed* nucleus carrying charge q and an electron cloud (charge $-q$) that is attracted to its nucleus by a *harmonic* force $\mathbf{F} = -k_x x\hat{\mathbf{i}} - k_y y\hat{\mathbf{j}} - k_z z\hat{\mathbf{k}}$, where k_x, k_y, and k_z may be different. Introduce a 3×3 matrix K whose diagonal elements are k_x, k_y, and k_z and whose off-diagonal elements are zero. (a) Show that $\mathbf{F} = -K\mathbf{r}$. (b) Show that a polarizing field $\mathbf{E}_m$ induces a dipole moment given by $\mathbf{p} = q^2K^{-1}\mathbf{E}_m$. (c) Assuming the crystal is one for which $\mathbf{E}' = 0$ in Eq. (10-67), show that the dielectric constant of this material is given by the matrix

$$K_e = \frac{\epsilon}{\epsilon_0} = I + \frac{Nq^2}{\epsilon_0}\left(K - \frac{Nq^2}{3\epsilon_0}I\right)^{-1}$$

where I is the 3×3 unit matrix and N is the number of molecules per unit volume in the crystal. (d) For the sake of a more concrete example, consider a probably unrealistic material for which $Nq^2/\epsilon_0 = 1$, $k_x = k_y = 1$, $k_z = 2$, and **E** lies in the y-z plane making an angle β with the y axis, i.e., $\mathbf{E} = E(\cos\beta\hat{\mathbf{j}} + \sin\beta\hat{\mathbf{k}})$. Obtain a careful graph of the angle between **D** and **E** as a function of β. *Hint:* Use a desk calculator or computer. *Optional:* Assuming that $Nq^2/\epsilon_0 \ll k_x$, k_y, and k_z, find an expression for the angle between **D** and **E**. *Hint:* Write $\mathbf{D} = \epsilon_0(I + K')\mathbf{E}$, where K' is a matrix with small elements, and expand in powers of K'.

REFERENCES

R. P. FEYNMAN, R. B. LEIGHTON, and M. SANDS, *The Feynman Lectures on Physics* (Addison-Wesley Publishing Company, Inc., Reading Mass., 1964), Volume II, Lectures 10 and 11.

C. KITTEL, *Introduction to Solid State Physics* (John Wiley & Sons, Inc., New York, 1966), Third Edition, Chapter 12.

E. M. PURCELL, *Electricity and Magnetism* (McGraw-Hill Book Company, New York, 1965), Chapter 9.

C. P. SMYTH, *Dielectric Behavior and Structure* (McGraw-Hill Book Company, New York, 1955).

11

Properties of Matter III: Magnetization

In this chapter, we shall consider the response of matter to externally applied static magnetic induction fields. Because magnetic monopoles are not found in the physical world, there is no magnetic response corresponding to electric conduction and we might therefore conclude that magnetic materials could be considered in a single category. Such a hasty conclusion, however, is not supported by real magnetic materials. Just as we found two different microscopic mechanisms—molecular orientation and molecular deformation—that contributed to the response of a dielectric in an electric field, we must here distinguish two mechanisms contributing to the response of matter in a magnetic induction field. In sharp contrast to the dielectric mechanisms (both of which gave rise to essentially the same macroscopic effect), the two magnetic mechanisms produce two qualitatively different effects. Thus, even in the absence of magnetic "conduction", two basic types of magnetic material (paramagnetic and diamagnetic) can be differentiated.

As with conduction and dielectric polarization, the magnetic response of matter can also be examined from either a microscopic or a macroscopic viewpoint. The microscopic approach is essentially identical with that described in the introduction to Chapter 10, and that discussion will not be repeated. We shall again adopt the macroscopic approach, but not before exploring briefly the microscopic response of individual molecules to externally applied magnetic induction fields. The development will once again be divided into a *descriptive* part, concerned solely with describing field-causing states of matter and with determining the fields caused by matter in a given state, and a *causal* part, concerned with the ways in which field-causing states are brought into being and maintained. Together, the descriptive and causal parts constitute the theoretical framework for finding magnetic induction fields in the presence of magnetizable matter, even though the contribution of the matter to the total field is not in general known until after the total field has been found.

11-1 The Microscopic Description: Magnetic Polarizability

From a microscopic point of view, a sample of magnetically responsive material is composed of molecules residing in free space. A discussion similar to that in the opening and closing paragraphs of Section 10-1 supports the adequacy of characterizing each molecule solely by a magnetic dipole moment. Magnetically responsive matter is then viewed microscopically as a dense assemblage of molecular or atomic magnetic dipoles located in free space. In this section, the response of individual molecules to their presence in an external magnetic induction field will be explored.

The magnetic moment of a single atom or molecule arises from two contributions. On the one hand, each electron circulating the nucleus constitutes a current loop and hence contributes to the magnetic moment of the atom or molecule. On the other hand, the electrons and the nuclei in the molecule possess intrinsic angular momenta or spins. With each intrinsic spin is associated a further magnetic moment that adds to the moment produced by the orbiting electrons. For some molecules, these contributions may add up to produce a net permanent magnetic moment; for others, the contributions may cancel, leaving an atom or molecule with no permanent magnetic moment but with the potential for developing an induced magnetic moment when an external field is applied.

Consider first the response of a molecule possessing a permanent dipole moment. (For the rest of this chapter, we understand that the *magnetic* dipole moment is meant.) We shall confine our discussion to a classical treatment, recognizing, however, that the only fully correct approach is a quantum mechanical approach (P11-29). When the molecule of concern is placed in a magnetic induction field, it experiences a torque tending to align its magnetic moment with the direction of the field (P3-13). In an assembly of permanent dipoles, however, this tendency to alignment is opposed by the thermal tendency to randomness. Thus, a given permanent dipole assumes an *average* dipole moment that would be zero except that in the external field each dipole favors a preferred orientation as it undergoes thermal agitation. To express this discussion more quantitatively, consider the following situation: A magnetic material at absolute temperature T composed of permanent dipoles having dipole moments of magnitude m_0 is placed in an external magnetic induction field. The dipoles adjust to some sort of average equilibrium position, in which state a particular dipole experiences a total field consisting of the applied external field *plus* the time-averaged *microscopic* field established by all *other* dipoles when the external field is present.[1] Let the time average (over the microscopic fluctuations) of this total field be $\mathbf{B}_m$. Combination of statistical mechanics with considerations of magnetostatic energy (P11-1) leads ultimately to the result that the *average* dipole moment of a single dipole is given by

$$\langle \mathbf{m}_0 \rangle_{\text{perm}} = \beta \mathbf{B}_m, \qquad \beta = \frac{m_0^2}{3kT} \tag{11-1}$$

where k is Boltzmann's constant and the condition $m_0 B_m \ll kT$ has been imposed. [Compare Eq. (10-1).] The constant β might be called an *orientational* (magnetic) molecular polarizability, since it measures the extent to which a molecule possessing a permanent dipole moment is oriented in a polarizing field. The orientational molecular polarizability is positive, and materials composed of permanent dipoles are said to be *paramagnetic*. The paramagnetic response of molecules to an external field results in an average dipole moment having the same direction as the field, and the strength of this response decreases with increasing temperature.

Molecules that possess no permanent dipole moment exhibit a different response. When placed in a magnetic induction field, these molecules develop an induced dipole moment that is *opposite* in direction to the polarizing field. Materials exhibiting this property are said to be *diamagnetic*. We shall see that *all* materials should exhibit a diamagnetic response. In materials composed of polar molecules, however, the stronger paramagnetic response usually masks the weaker diamagnetic response of the material. Some separation of the two responses can be made because the diamagnetic response is not temperature-dependent, but experiments to distinguish the two responses on this basis are difficult because of the weakness of the diamagnetic compared to the paramagnetic response.

Diamagnetism arises from the response of individual electronic orbits as these orbits adjust to new equilibria in the presence of an external field. Although this response is extremely complicated and can be fully treated only by using quantum mechanics, a classical model that leads to a crude microscopic understanding of the phenomenon can be constructed. Consider first the qualitative features involved when an electron in a circular orbit is placed in an external magnetic induction field at right angles to the plane of the orbit. Some interaction—Coulomb or otherwise—provides the centripetal force that keeps the electron in its orbit when no external fields are present. When the field is turned on, the circulating electron experiences an additional radially directed force that may tend to expand or contract the orbit, depend-

[1] See footnote 3, Chapter 10.

TABLE 11-1 Effects on the Orbit of an Electron when a Polarizing Field is Turned On

Direction of Final **B**	*Direction of Electron Circulation*	*Direction of Resulting Magnetic Force*	*Direction of Induced Electric Field*
$+z$		Radially out (smaller centripetal force)	Slows electron down
$+z$		Radially in (greater centripetal force)	Speeds electron up
$-z$		Radially in (greater centripetal force)	Speeds electron up
$-z$		Radially out (smaller centripetal force)	Slows electron down

In this table, the viewer is assumed to be located at a point on the positive z axis in Fig. 11-1. Since the electron has a negative charge, the force that it experiences in an electric field is opposite in direction to that of the field.

ing on the direction in which the electron is circulating. Simultaneously, as the field is turned on, the electron experiences tangential forces from the induced electric field (Faraday law), and these forces may speed up or slow down the electron, again depending on the direction in which the electron is circulating. Table 11-1 summarizes qualitatively the direction of these two effects for various directions of the field and of the circulation of the electron.

Now consider the adjustments that the orbit might be expected to undergo as a result of these two effects. From Table 11-1, we conclude that a final magnetic induction that reduces the centripetal force is accompanied by an induced electric effect that slows down the electron and vice versa. But a slower electron can stay at the same radius if the centripetal force is suitably reduced and similarly a faster electron can stay at the same radius if the centripetal force is appropriately increased. Classically, the change in radius as the **B**-field is turned on is therefore likely to be less significant than the change in electron speed, since the radius is affected by two phenomena that tend to compensate. Let us then assume that the radius does not change at all. (It is shown in P11-2 that in fact this assumption is correct provided that the change in angular velocity is small compared to the initial angular velocity.)

To make our discussion more quantitative, consider an electron with charge $-q_e$ and mass m_e circulating in the x-y plane with angular velocity ω in an orbit of radius r (Fig. 11-1). The circulating electron constitutes a

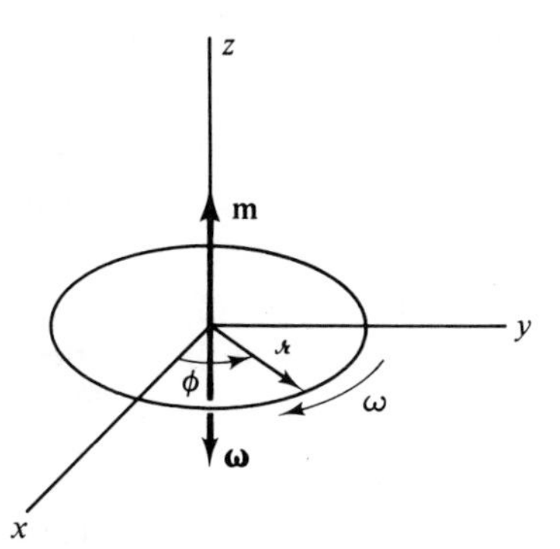

Fig. 11-1. Path of an electron circulating in the x-y plane with angular velocity $\boldsymbol{\omega}$.

current of strength $I = q_e\omega/2\pi$ in the *counter*clockwise direction as seen from a point on the positive z axis. Hence, the magnetic moment of the resulting current loop is given by

$$\mathbf{m} = (\text{current})(\textbf{area}) = \frac{q_e\omega}{2\pi}(\pi r^2\hat{\mathbf{k}}) = q_e\frac{r^2}{2}\omega\hat{\mathbf{k}} = -q_e\frac{r^2}{2}\boldsymbol{\omega} \tag{11-2}$$

where $\boldsymbol{\omega} = -\omega\hat{\mathbf{k}}$ is the angular velocity of the electron and ω is positive for an electron circulating clockwise and negative for an electron circulating counterclockwise.

We shall now calculate the *change* induced in this magnetic moment when an external magnetic induction field perpendicular to the plane of the orbit is turned on. Let the field change from zero to a final value B_m during the time interval $0 < t < t_f$ in accordance with

$$\mathbf{B} = B(t)\hat{\mathbf{k}} \tag{11-3}$$

where $B(0) = 0$ and $B(t_f) = B_m$. As this field is turned on, an induced electric field tangential to the orbit of the electron is produced. If we assume that the electron is circulating very rapidly compared to the time scale associated with the changing field, the electron makes many revolutions before the field has changed appreciably. It is then appropriate to calculate the electric field induced at an *instant* of time and use it to determine the work done on the electron over short *intervals* of time. Applied to the orbit of the electron, Faraday's law,

$$\oint \mathbf{E}\cdot d\boldsymbol{\ell} = -\frac{d}{dt}\int \mathbf{B}\cdot d\mathbf{S} \tag{11-4}$$

gives

$$2\pi r E_\phi = -\frac{d}{dt}(\pi r^2 B) \tag{11-5}$$

where we have assumed that $\mathbf{E} = E_\phi(r)\hat{\boldsymbol{\phi}}$, $\hat{\boldsymbol{\phi}}$ being a unit vector in the direction of increasing ϕ. Hence, we find first that

$$\mathbf{E} = -\left[\frac{1}{2\pi r}\frac{d}{dt}(\pi r^2 B)\right]\hat{\boldsymbol{\phi}} \tag{11-6}$$

and then that the rate P at which this field does work on the electron is given by

$$P = \mathbf{F}\cdot\mathbf{v} = (-q_e\mathbf{E})\cdot(-\omega r\hat{\boldsymbol{\phi}})$$
$$= -\frac{q_e\omega}{2\pi}\frac{d}{dt}(\pi r^2 B) \tag{11-7}$$

The rate at which the energy of the electron increases, however, is also given by the time rate of change of the kinetic energy of the electron, i.e.,

$$P = \frac{d}{dt}\left(\frac{1}{2}m_e r^2\omega^2\right) \tag{11-8}$$

Thus, on combining Eqs. (11-7) and (11-8), we find that the motion of the electron in this (slowly) changing field must satisfy the equation

$$\frac{d}{dt}(r^2\omega^2) = -\frac{q_e\omega}{m_e}\frac{d}{dt}(r^2 B) \tag{11-9}$$

Assuming now that r is constant, we find from Eq. (11-9) that

$$\frac{d\omega}{dt} = -\frac{q_e}{2m_e}\frac{dB}{dt} \tag{11-10}$$

Direct integration then yields

$$\Delta\omega = \omega_f - \omega_0 = -\frac{q_e B_m}{2m_e} = \pm\omega_L \tag{11-11}$$

where ω_f is the final angular velocity of the electron, ω_0 is its initial angular velocity, and $\omega_L = q_e|B_m|/2m_e$ is the so-called *Larmor frequency*; in Eq. (11-11), the upper (lower) sign applies when $B_m < 0$ ($B_m > 0$) and the condition of slowly changing fields is more explicitly stated by the requirement $|\Delta\omega| \ll \omega_0$. (Compare P11-2.) Finally, we find from Eq. (11-2) that the change in magnetic moment induced by the turning on of this external field is given by

$$\Delta\mathbf{m} = q_e\frac{r^2}{2}\Delta\omega\hat{\mathbf{k}} = q_e\frac{r^2}{2}(\omega_f - \omega_0)\hat{\mathbf{k}}$$
$$= q_e\frac{r^2}{2}\left(-\frac{q_e B_m}{2m_e}\right)\hat{\mathbf{k}} = -\frac{q_e^2 r^2}{4m_e}\mathbf{B}_m \tag{11-12}$$

Thus, when an external **B**-field perpendicular to the orbit is turned on, the induced change in the magnetic moment of a circulating charge is independent of ω (and hence does not depend on the direction of circulation), is independent of the sign of the charge, and finally is opposite in direction to the inducing field [as evidenced by the minus sign in Eq. (11-12)]. This minus sign should be contrasted with the plus sign in Eq. (11-1); it is this difference in sign that distinguishes diamagnetic from paramagnetic response.

Despite its classical origin, Eq. (11-12) can be converted into an expression that happens to be quantum mechanically correct, at least for molecules with a single nucleus (i.e., atoms) in which the electron distribution about the nucleus is spherically symmetric. Quantum mechanically, an electron orbit does not have a well-defined radius. We would expect then that $\imath^2$ in Eq. (11-12) should be replaced by $\langle \imath^2 \rangle = \langle x^2 \rangle + \langle y^2 \rangle$, where $\langle \cdots \rangle$ denotes an average over the electron wave function ψ, e.g., $\langle x^2 \rangle = \int \psi^* x^2 \psi \, dv$. But for a spherically symmetric wave function, $\langle x^2 \rangle = \langle y^2 \rangle = \langle z^2 \rangle = \frac{1}{3}\langle r^2 \rangle$, where r is the distance of the point (x, y, z) from the center of the atom. Thus $\langle \imath^2 \rangle = \frac{2}{3}\langle r^2 \rangle$ and Eq. (11-12) is replaced with

$$\Delta \mathbf{m} = -\frac{q_e^2 \langle r^2 \rangle}{6m_e} \mathbf{B}_m \tag{11-13}$$

The task of quantum mechanics is then to calculate $\langle r^2 \rangle$ for each electron in the atom.

Whether Eq. (11-13) or a more complicated equation applies, the total *change* in the magnetic moment of the molecule when the field is turned on is obtained by summing the contributions of each electron. Since we have been discussing nonpolar molecules, which have no initial magnetic moment, this change is equal to the final induced moment. Thus, if Eq. (11-13) applies, we have that

$$\mathbf{m}_{\text{ind}} = \left(-\frac{q_e^2}{6m_e} \sum_i \langle r^2 \rangle_i \right) \mathbf{B}_m = \beta_0 \mathbf{B}_m \tag{11-14}$$

where $\langle r^2 \rangle_i$ is the average value of r^2 for the ith electron. Since all electrons contribute to changing $\mathbf{m}$ in the *same* direction, a net change results. The coefficient β_0 in Eq. (11-14) multiplying $\mathbf{B}_m$ might be called a deformational (magnetic) molecular polarizability because it relates to a magnetic moment produced through deformation of the internal structure of the molecule.

If both paramagnetic and diamagnetic effects are present, the total molecular polarizability is obtained by combining Eqs. (11-1) and (11-14) to find that

$$\mathbf{m} = \langle \mathbf{m}_0 \rangle_{\text{perm}} + \mathbf{m}_{\text{ind}} = \beta_t \mathbf{B}_m \tag{11-15}$$

where the total (magnetic) molecular polarizability β_t is given by

$$\beta_t = \frac{m_0^2}{3kT} - \frac{q_e^2}{6m_e} \sum_i \langle r^2 \rangle_i \tag{11-16}$$

As we have already remarked, more precise and more general expressions for these polarizabilities can be calculated only by using quantum mechanics; we leave those calculations to other authors.

PROBLEMS

P11-1. Accepting (1) that the energy of a magnetic dipole of moment $\mathbf{m}$ in a magnetic induction field $\mathbf{B}$ is $-\mathbf{m} \cdot \mathbf{B}$ (P3-20) and (2) that the distribution of dipoles over these energy states is given in statistical thermodynamics by $\exp(-\text{energy}/kT)$, where k is Boltzmann's constant and T is the absolute temperature, determine the average value of the dipole moment $\langle \mathbf{m}_0 \rangle$ of a dipole in a magnetic induction field $\mathbf{B}_m$. Show also that Eq. (11-1) emerges when $m_0 B_m \ll kT$. How large can the field B_m be before this condition is violated at room temperature? *Hints*: (1) See P10-2. (2) Typical values for permanent molecular dipole moments are on the order of 10^{-23} J/T (mks units).

P11-2. The appropriateness of our assumption of a constant orbital radius at Eq. (11-10) can be shown as follows. A particle of mass m_e moving in a circle of radius $\imath$ with angular velocity ω_0 must have acting on it a centripetal force $\mathbf{F}_c = -m_e\omega_0^2 \imath\hat{\boldsymbol{\imath}}$. Suppose ω_0 is changed to $\omega_0 + \Delta\Omega$. (a) What must be the change $\Delta\mathbf{F}_c$ in $\mathbf{F}_c$ if the particle is to continue to move in a circular orbit of the same radius? Assume $\Delta\Omega \ll \omega_0$. (b) Show that turning on the magnetic induction $B_m\hat{\mathbf{k}}$ in Fig. 11-1 changes $\mathbf{F}_c$ by $\Delta\mathbf{F}_c = q_e\omega_0 \imath B_m\hat{\boldsymbol{\imath}}$. (c) Combine parts (a) and (b) to show that, if $\Delta\Omega = -q_e B_m/2m_e$, then adding the magnetic induction will not change the radius. (d) Compare this particular $\Delta\Omega$ with $\Delta\omega$ as given in Eq. (11-11) to conclude that $\imath$ is constant provided $|\Delta\omega| \ll \omega_0$. (e) Estimate ω_0 numerically for the electron in the hydrogen atom and then find how large B_m can become before $|\Delta\omega|$ exceeds $0.01\omega_0$. Compare this value with typical laboratory fields, say 10,000 G.

P11-3. (a) Show from Eq. (11-2) that $\mathbf{m} = -q_e\mathbf{L}/2m_e$, where $\mathbf{L}$ is the orbital angular momentum of the electron. (b) Quantum mechanically, the component of $\mathbf{L}$ parallel to a magnetic induction field can assume only the values $m_l\hbar$, where m_l is an integer and $\hbar$ is Planck's constant divided by 2π. The Bohr magneton M_B, which is a convenient unit for expressing atomic magnetic moments, is defined so that the component of the magnetic moment parallel to the field is given by $-m_l M_B$. Show that $M_B = q_e\hbar/2m_e$. (c) Suppose an electron has $m_l = 1$. What (numerically) is its magnetic moment? Use this result in Eq. (11-2) to estimate its angular velocity. (d) How large can B_m become in Eq. (11-11) before $\Delta\omega$ ceases to be small compared to ω_0, thereby invalidating the assumption that the field changes slowly?

P11-4. Given that $\mathbf{m} = -q_e\mathbf{L}/2m_e$, where $\mathbf{L}$ is the orbital angular momentum of the electron (see P11-3 and P3-21), evaluate the torque $\mathbf{N}$ exerted on the electron by the field in Eq. (11-6) and integrate Newton's second law $d\mathbf{L}/dt = \mathbf{N}$ to show ultimately that the magnetic dipole moment of the electron is changed by an amount given by Eq. (11-12) when the magnetic induction field is turned on. Assume $\imath$ is constant.

11-2 The Macroscopic Description: Magnetization

The macroscopic approach to magnetic properties is founded on a concept called magnetization, which is analogous to the concept of polarization introduced in the discussion of dielectrics. Each elementary magnetic dipole in a given volume element Δv makes an additive contribution to the total dipole moment of that volume element. If the ith molecular dipole has *average* dipole moment $\mathbf{m}_i$, then the *magnetization* (magnetic dipole moment per unit volume) is defined by

$$\mathbf{M}(\mathbf{r}) = \lim_{\Delta v \to 0 \text{ about } \mathbf{r}} \left(\frac{1}{\Delta v} \sum_i \mathbf{m}_i \right) \tag{11-17}$$

where the sum extends over all the molecules in the volume element and the meaning of the limit is the same as always—Δv becomes macroscopically small but remains large enough to contain a large number of microscopic dipoles.[2] In terms of the magnetization, each small volume element Δv therefore has associated with it a magnetic dipole moment $\Delta\mathbf{m} \approx \mathbf{M}(\mathbf{r})\,\Delta v$ and the total magnetic dipole moment of the sample will be

$$\mathbf{m} = \int \mathbf{M}(\mathbf{r})\,dv \tag{11-18}$$

11-3 The Macroscopic Vector Potential and Magnetic Induction Field at a Point Exterior to a Magnetized Object; Bound Currents

In Sections 11-3, 11-4, and 11-5, we shall calculate the contribution made to the macroscopic magnetic induction field by a magnetized object whose magnetization $\mathbf{M}(\mathbf{r}')$ at every point $\mathbf{r}'$ within the object is known. Consider first an observation point $\mathbf{r}$ exterior to the object. As in the analogous dielectric case, such an observation point is microscopically far from *all* of the dipoles in the object. Thus, a small element of the object having volume $\Delta v'$ and located at $\mathbf{r}'$ (Fig. 10-1) can be treated as a dipole of moment $\mathbf{M}(\mathbf{r}')\,\Delta v'$, regardless of where within the object the point $\mathbf{r}'$ lies. The contribution of this element of the object to the magnetic vector potential at $\mathbf{r}$ is then given by

$$\Delta\mathbf{A} = \frac{\mu_0}{4\pi} \frac{\mathbf{M}(\mathbf{r}')\Delta v' \times (\mathbf{r} - \mathbf{r}')}{|\mathbf{r} - \mathbf{r}'|^3} \tag{11-19}$$

[see Eq. (5-49)] and integration of Eq. (11-19) over the volume of the object yields

$$\mathbf{A}(\mathbf{r}) = \frac{\mu_0}{4\pi} \int \frac{\mathbf{M}(\mathbf{r}') \times (\mathbf{r} - \mathbf{r}')}{|\mathbf{r} - \mathbf{r}'|^3}\,dv' \tag{11-20}$$

for the contribution of the entire object to the total vector potential at $\mathbf{r}$.

Equation (11-20) can be rewritten in two different ways, each of which leads to a useful physical interpretation. We shall postpone treating the reformulation that leads to an identification of equivalent magnetic poles to the next section and shall treat here the reformulation leading to the identification of bound currents. First, we substitute the identity in Eq. (10-10) into

[2]The definition of magnetization given in this paragraph is adopted in all systems of units with which the author is familiar.

Eq. (11-20) to obtain

$$\mathbf{A}(\mathbf{r}) = \frac{\mu_0}{4\pi}\int\left[\mathbf{M}(\mathbf{r}') \times \nabla'\left(\frac{1}{|\mathbf{r}-\mathbf{r}'|}\right)\right]dv' \tag{11-21}$$

Then we use the vector identity in Eq. (C-8) in the form

$$\mathbf{Q} \times \nabla'\Phi = \Phi\nabla' \times \mathbf{Q} - \nabla' \times (\Phi\mathbf{Q}) \tag{11-22}$$

with $\Phi = 1/|\mathbf{r}-\mathbf{r}'|$ and $\mathbf{Q} = \mathbf{M}$, to obtain

$$\mathbf{A}(\mathbf{r}) = \frac{\mu_0}{4\pi}\int\frac{\nabla' \times \mathbf{M}(\mathbf{r}')}{|\mathbf{r}-\mathbf{r}'|}dv' - \frac{\mu_0}{4\pi}\int\nabla' \times \left(\frac{\mathbf{M}(\mathbf{r}')}{|\mathbf{r}-\mathbf{r}'|}\right)dv' \tag{11-23}$$

Finally, we use the identity in Eq. (C-25) in the second term to find that

$$\mathbf{A}(\mathbf{r}) = \frac{\mu_0}{4\pi}\int\frac{\nabla' \times \mathbf{M}(\mathbf{r}')}{|\mathbf{r}-\mathbf{r}'|}dv' + \frac{\mu_0}{4\pi}\oint\frac{\mathbf{M}(\mathbf{r}') \times \hat{\mathbf{n}}(\mathbf{r}')}{|\mathbf{r}-\mathbf{r}'|}dS' \tag{11-24}$$

where the surface integral extends over the surface bounding the magnetized object, $\hat{\mathbf{n}}(\mathbf{r}')$ is a unit outward normal to that surface at $\mathbf{r}'$, and $dS' = |d\mathbf{S}'|$. This result, however, is the vector potential established by a current distribution described by a volume current density

$$\mathbf{J}_m(\mathbf{r}) = \nabla \times \mathbf{M}(\mathbf{r}) \tag{11-25}$$

and a surface current density

$$\mathbf{j}_m(\mathbf{r}) = \mathbf{M}(\mathbf{r}) \times \hat{\mathbf{n}}(\mathbf{r}) \quad (\mathbf{r} \text{ on the surface}) \tag{11-26}$$

[See Eq. (5-41) and P5-32.[3]] Thus, at exterior points, the magnetized object produces the same vector potential as these so-called *bound* (or *Amperian*) currents and we can replace our description in terms of the macroscopic magnetization with this equivalent description in terms of bound currents if we find it convenient to do so.

These bound currents can also be obtained by a longer but less formal derivation that makes their physical origin more clear. A suitable model for a magnetized object is shown in Fig. 11-2. Each volume element Δv into which the object is divided has a magnetic dipole moment $\Delta\mathbf{m} \approx \mathbf{M}\,\Delta v$ associated with it, and we can view this moment as arising from a current circulating about the boundary of the volume element. The proper current to associate with a volume element at which the magnetization is $\mathbf{M}$ is found by computing the magnetic dipole moment of the element in two ways. On the one hand, for example, the y-component of the magnetic moment is given by $M_y\,\Delta v = M_y\,\Delta x\,\Delta y\,\Delta z$. On the other hand, if we let i_{xz} be the current circulating about the volume element in orbits whose plane is parallel to the x-z plane and adopt the usual sign convention that $i_{xz} > 0$ when the current circulates in the counterclockwise direction as viewed from a point far away on the positive y axis, then the y-component of the magnetic moment of the volume element is also

[3]Equations (11-25) and (11-26) retain the same form in cgs-esu and in cgs-emu. In Gaussian and Heaviside-Lorentz units, we would find that $\mathbf{J}_m = c\nabla \times \mathbf{M}$ and $\mathbf{j}_m = c\mathbf{M} \times \hat{\mathbf{n}}$, where c is the speed of light in cm/sec.

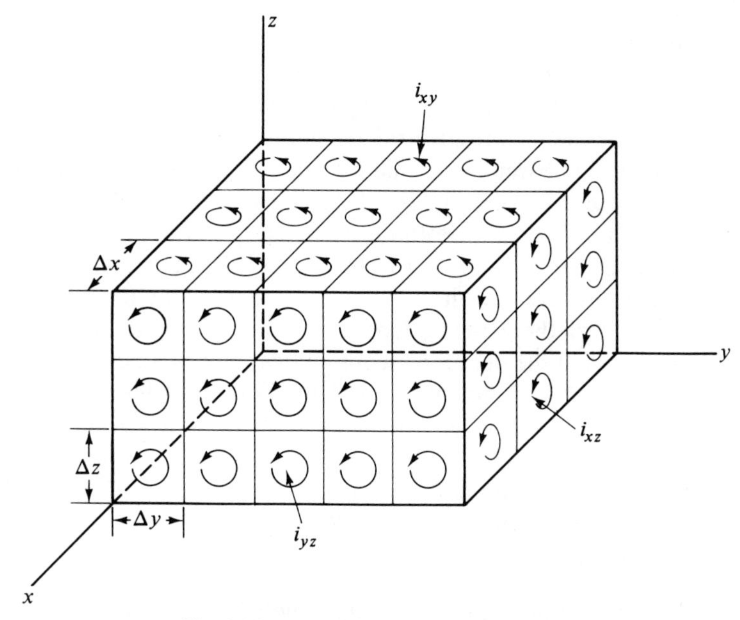

Fig. 11-2. A model of magnetized matter.

given by the current i_{xz} times the area $\Delta x\ \Delta z$ of the face of the element in the y-z plane. Equating these two evaluations of the y-component of the magnetic moment gives

$$M_y\,\Delta x\,\Delta y\,\Delta z = i_{xz}\,\Delta x\,\Delta z \Longrightarrow i_{xz} = M_y\,\Delta y \tag{11-27}$$

By applying similar arguments to the other two coordinate directions, we find that

$$i_{xy} = M_z\,\Delta z, \qquad i_{yz} = M_x\,\Delta x \tag{11-28}$$

An enlarged view of these currents at a single volume element is shown in Fig. 11-3.

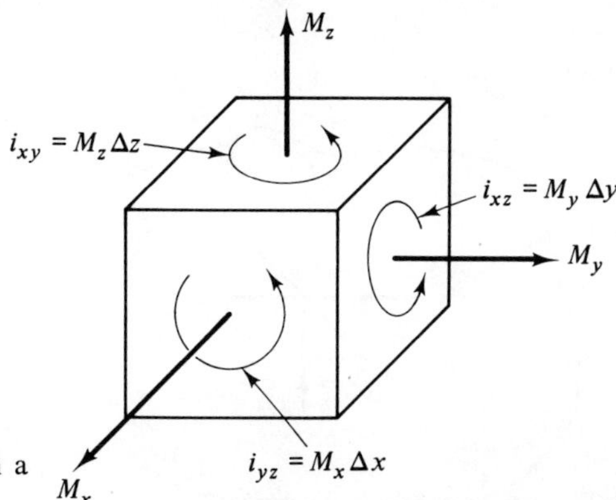

Fig. 11-3. The bound currents in a single volume element.

In this model, bound currents arise because it is possible for these circulating currents to combine in such a way that there appears to be a macroscopic current even though there is no net macroscopic transport of charge. Consider first the bound surface current. A volume element at the surface has at least one face—the face toward the outside—on which the supposed circulating current is apparent from the outside. If, for example, the outside face parallel to the *y-z* plane in Fig. 11-2 is viewed, one might see currents flowing as shown in Fig. 11-4. Both the currents i_{xy} and the currents i_{xz} have the appearance of currents flowing in the surface. Even though no charges are transported physically from one "cell" to the next, the net macroscopic appearance of all of the contributions from i_{xy} is a current flowing to the right in the figure and the net macroscopic appearance of all of the contributions from i_{xz} is a current flowing toward the bottom of the picture. Now, surface currents have a direction and are described by a density similar to volume current densities (P2-25). With reference to Fig. 11-5, let us define a surface current density **j** so that the current i flowing across a (short) line in the surface can be represented by the equation

$$i = \mathbf{j} \cdot \boldsymbol{\ell} \tag{11-29}$$

where $\boldsymbol{\ell}$ is a vector whose magnitude is the length of the line and whose direction is *perpendicular* to the line in the direction of the current desired. In

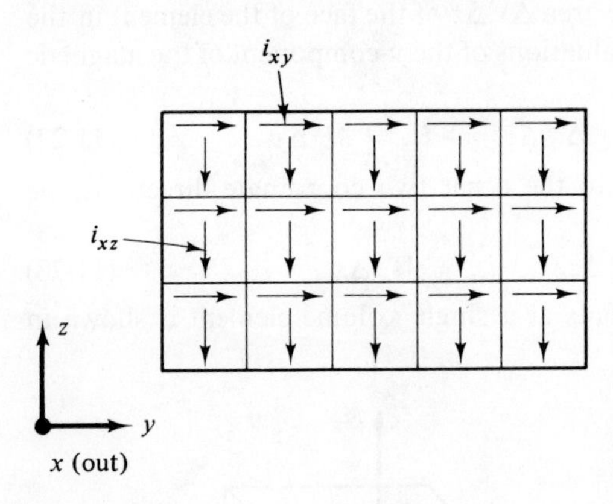

Fig. 11-4. Currents in a surface of a magnetized material. The surface shown is in a plane parallel to the *y-z* plane. The current i_{yz}, which circulates in this plane, is not shown because it makes no contribution to the bound surface currents in this plane.

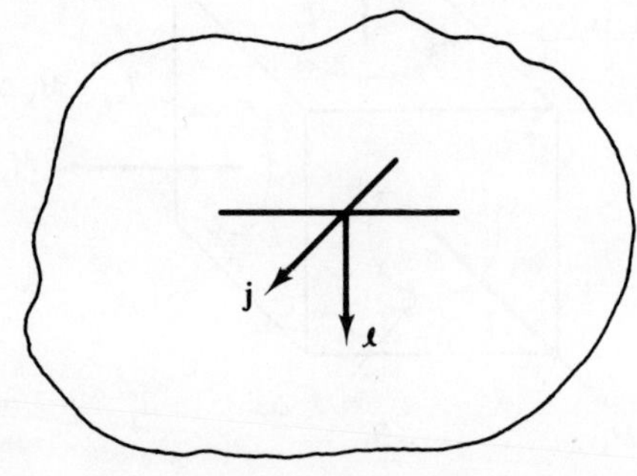

Fig. 11-5. Vectors for defining the surface current density. Both vectors shown lie *in* the plane of the paper.

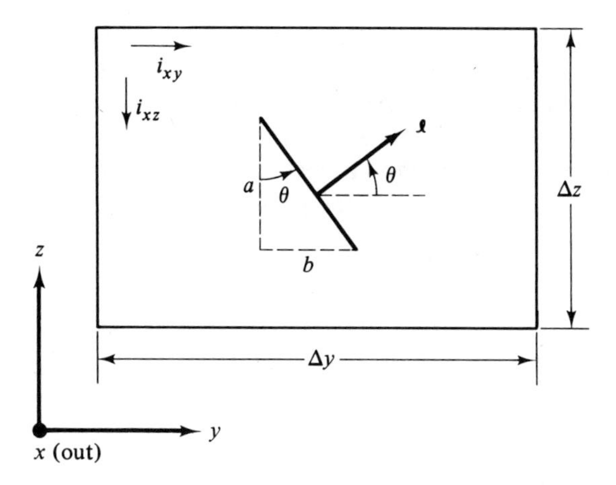

Fig. 11-6. Currents in one of the surface elements shown in Fig. 11-4.

effect, **j** measures the rate at which charge is transported across a line of unit length perpendicular to **j**. We now look at the surface of one of the blocks illustrated in Fig. 11-4. A more detailed diagram is shown in Fig. 11-6. Only the currents that contribute to an apparent macroscopic charge transport in the surface are shown. The current i flowing across the line $\boldsymbol{\ell}$ in Fig. 11-6 is composed of two parts, and we find that

$$\begin{aligned} i &= i_{xy}\frac{a}{\Delta z} - i_{xz}\frac{b}{\Delta y} \\ &= M_z a - M_y b \quad \text{[Eqs. (11-27) and (11-28)]} \\ &= (M_z\hat{\mathbf{j}} - M_y\hat{\mathbf{k}})\cdot(a\hat{\mathbf{j}} + b\hat{\mathbf{k}}) \\ &= (\mathbf{M} \times \hat{\mathbf{i}})\cdot\boldsymbol{\ell} \end{aligned} \tag{11-30}$$

where the final form follows because $\boldsymbol{\ell} = a\hat{\mathbf{j}} + b\hat{\mathbf{k}}$. (Why?) Direct comparison of Eq. (11-30) with Eq. (11-29) now leads us to identify a surface current density

$$\mathbf{j}_m = \mathbf{M} \times \hat{\mathbf{i}} \tag{11-31}$$

in the surface parallel to the y-z plane. In this result, of course, $\hat{\mathbf{i}}$ is the unit vector normal to the surface used in the example. More generally, we would expect that the unit normal $\hat{\mathbf{n}}$ would appear instead of $\hat{\mathbf{i}}$ and we have obtained Eq. (11-26) by arguments based on the model in Fig. 11-2.

Bound volume currents arise when the magnetization is not uniform. Under such conditions, the current associated with a particular volume element is different from that associated with an adjacent volume element and the currents flowing in opposite directions along the boundary between the two elements do not cancel. The result is an apparent macroscopic transport of charge in the interior of the magnetized object. To obtain a quantitative expression for this volume current density, consider an area having sides Δx

and Δy and lying in a plane parallel to the x-y plane. Let this area be positioned with its center at the point (x, y, z). Now, divide the magnetized object into elements having sides Δx, Δy, and Δz in such a way that the selected area lies in two of these elements as shown in Fig. 11-7. The circulating currents that contribute to charge flow across the selected surface are also shown. In balance, the net current flow in the positive z direction is

$$\begin{aligned} i_1 &= i_{xz}(x + \tfrac{1}{2}\Delta x) - i_{xz}(x - \tfrac{1}{2}\Delta x) \\ &= \frac{\partial i_{xz}}{\partial x}\Delta x = \frac{\partial M_y}{\partial x}\Delta x\,\Delta y \end{aligned} \tag{11-32}$$

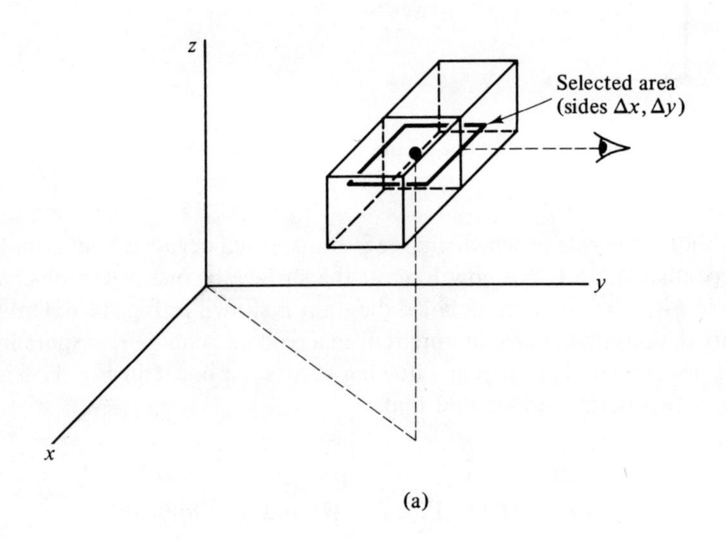

(a)

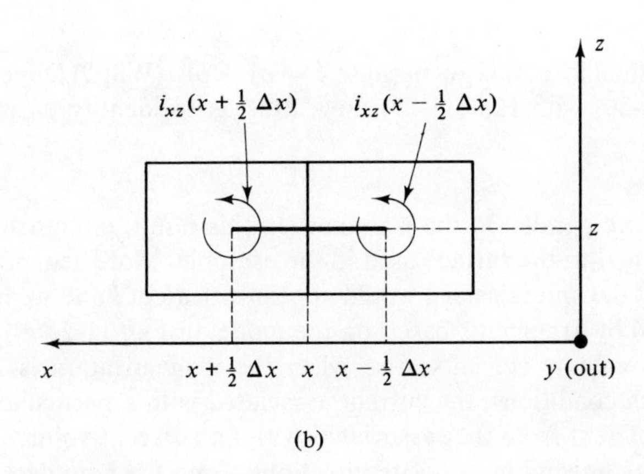

(b)

Fig. 11-7. A selected area and two volume elements for determining part of the bound volume current across that area. Part (a) shows the two elements and part (b) shows the view as seen from the point marked ➤ in part (a).

where Eq. (11-27) has been used to relate i_{xz} to M_y. An additional contribution to the net current flow across this surface in the positive z direction is made by currents circulating in the y-z plane (Fig. 11-8) and is given by

$$i_2 = i_{yz}(y - \tfrac{1}{2}\Delta y) - i_{yz}(y + \tfrac{1}{2}\Delta y)$$
$$= -\frac{\partial i_{yz}}{\partial y}\Delta y = -\frac{\partial M_x}{\partial y}\Delta x\,\Delta y \qquad (11\text{-}33)$$

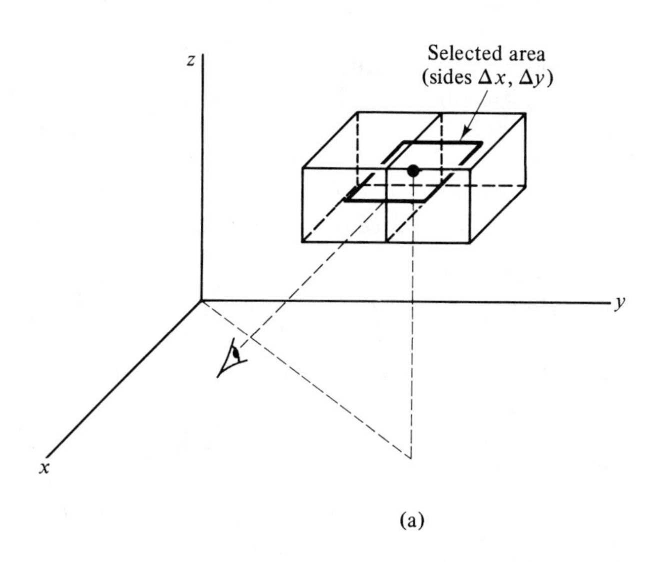

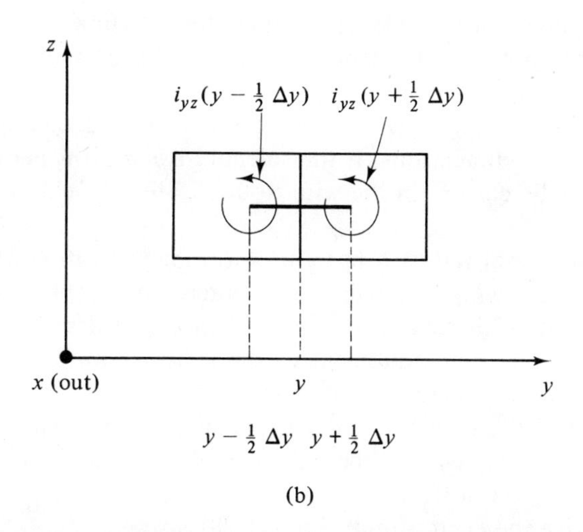

Fig. 11-8. Two volume elements for determining a second contribution to the bound volume current across the same area shown in Fig. 11-7. Part (a) shows the two elements and part (b) shows the view as seen from the point marked ➤ in part (a).

where Eq. (11-28) has been used to relate i_{yz} to M_x. Combining Eqs. (11-32) and (11-33), we find that the net current across the surface in the positive z direction is given by

$$i_1 + i_2 = \left(\frac{\partial M_y}{\partial x} - \frac{\partial M_x}{\partial y}\right)\Delta x\,\Delta y = (\nabla \times \mathbf{M})\cdot\hat{\mathbf{k}}\,\Delta x\,\Delta y \tag{11-34}$$

Since the current across a small surface $\Delta\mathbf{S}$ placed in a current flow described by the current density $\mathbf{J}$ is given by $\mathbf{J}\cdot\Delta\mathbf{S}$ and since in the present example $\Delta\mathbf{S} = \hat{\mathbf{k}}\,\Delta x\,\Delta y$, we deduce from Eq. (11-34) that the circulating currents give rise to a bound volume current density given by $\mathbf{J}_m = \nabla \times \mathbf{M}$. Strictly, of course, our development supports this conclusion only for the z-component of $\mathbf{J}_m$, but the other components can be similarly derived (P11-9). We have therefore obtained Eq. (11-25) by arguments based on the model in Fig. 11-2.

Given the vector potential, we can readily calculate the field established at an exterior point by a magnetized object; we take the curl of the potential, finding that

$$\mathbf{B} = \frac{\mu_0}{4\pi}\int \frac{\mathbf{J}_m(\mathbf{r}') \times (\mathbf{r} - \mathbf{r}')}{|\mathbf{r} - \mathbf{r}'|^3}\,dv' + \frac{\mu_0}{4\pi}\oint \frac{\mathbf{j}_m(\mathbf{r}') \times (\mathbf{r} - \mathbf{r}')}{|\mathbf{r} - \mathbf{r}'|^3}\,dS' \tag{11-35}$$

(P11-10). Equation (11-35) for the field and Eq. (11-24) for the vector potential of course give only the contributions of the bound currents on the magnetized object; contributions from other sources (if any) must be added to those given by these equations.

PROBLEMS

P11-5. A cylindrical rod of length L and radius R is uniformly magnetized with magnetization $\mathbf{M}$ directed along its axis. Let the axis coincide with the z axis. Find the bound current densities and then argue that the exterior field produced by this bar magnet is identical to the field produced by a solenoid (P5-11) of the same dimensions. If the solenoid has n turns per unit length, what should be the current in the wire to duplicate the field of the magnet exactly?

P11-6. A sphere of radius R has a uniform magnetization $\mathbf{M}$. Choosing a coordinate system with its origin at the center of the sphere and its polar axis parallel to $\mathbf{M}$ so that $\mathbf{M} = M\hat{\mathbf{k}}$, (a) find the equivalent bound currents and (b) calculate the dipole moment of the sphere both from the magnetization and from the bound currents.

P11-7. Show that the total bound current crossing an open surface within a magnetized object is given by the line integral of the magnetization about the path bounding the surface.

P11-8. Derive the expression in Eq. (11-28) for i_{xy}.

P11-9. Following the pattern illustrated in the text, start from the model in Fig. 11-2 and derive the expression $J_{mx} = (\nabla \times \mathbf{M})_x$ for the x-component of the bound current density.

P11-10. Evaluate the curl of Eq. (11-24) to derive Eq. (11-35).

P11-11. A spherical cavity of radius R is cut in the interior of an object with uniform magnetization $\mathbf{M}$. Find the equivalent currents on the surface of that cavity and show that these currents contribute an amount $-\frac{2}{3}\mu_0\mathbf{M}$ to the field at the center of the cavity.

11-4 An Alternative Approach to the Exterior Field; Equivalent Poles

Although the expression for the exterior field in terms of bound currents is at least in one way more general than the expression we now derive (see Section 11-5), the integrals contain cross products and their evaluation is often very complicated. A formalism that expresses $\mathbf{B}$ as the gradient of a scalar is simpler and more useful in some cases than the expression in Eq. (11-35). We therefore describe an alternative rewriting of Eq. (11-20). First we replace $(\mathbf{r} - \mathbf{r}')/|\mathbf{r} - \mathbf{r}'|^3$ by its equivalent $-\nabla(1/|\mathbf{r} - \mathbf{r}'|)$ as in Eq. (4-45) and then we calculate $\mathbf{B}$ directly by evaluating $\nabla \times \mathbf{A}$. We find that

$$\mathbf{B}(\mathbf{r}) = \frac{\mu_0}{4\pi}\int \nabla \times \left[\mathbf{M}(\mathbf{r}') \times (-\nabla)\frac{1}{|\mathbf{r} - \mathbf{r}'|}\right] dv'$$
$$= -\frac{\mu_0}{4\pi}\int \left[\mathbf{M}(\mathbf{r}')\, \nabla^2\left(\frac{1}{|\mathbf{r} - \mathbf{r}'|}\right) - (\mathbf{M}(\mathbf{r}')\cdot\nabla)\,\nabla\left(\frac{1}{|\mathbf{r} - \mathbf{r}'|}\right)\right] dv' \quad (11\text{-}36)$$

where to obtain the second form we used the vector identity in Eq. (C-16) with $\mathbf{Q} = \mathbf{M}(\mathbf{r}')$ and $\mathbf{R} = \nabla(1/|\mathbf{r} - \mathbf{r}'|)$ and we noted that such combinations as $\nabla\cdot\mathbf{M}(\mathbf{r}')$ are zero because ∇ differentiates with respect to the components of $\mathbf{r}$. Now, for points *exterior* to the magnetized object, $\mathbf{r}'$ is never equal to $\mathbf{r}$. Hence, $\nabla^2(1/|\mathbf{r} - \mathbf{r}'|) = 0$ throughout the domain of integration and Eq. (11-36) therefore reduces to

$$\mathbf{B}(\mathbf{r}) = \frac{\mu_0}{4\pi}\int (\mathbf{M}(\mathbf{r}')\cdot\nabla)\,\nabla\left(\frac{1}{|\mathbf{r} - \mathbf{r}'|}\right) dv' \quad (11\text{-}37)$$

Continuing, we use the vector identity in Eq. (C-15) with $\mathbf{Q}$ and $\mathbf{R}$ as before to rewrite Eq. (11-37) in the form

$$\mathbf{B}(\mathbf{r}) = \mu_0\,\nabla\left[\frac{1}{4\pi}\int \mathbf{M}(\mathbf{r}')\cdot\nabla\left(\frac{1}{|\mathbf{r} - \mathbf{r}'|}\right) dv'\right] \quad (11\text{-}38)$$

[All but one term in the right-hand side of Eq. (C-15) is zero either because ∇ acts on a function only of $\mathbf{r}'$ or because $\nabla \times \nabla(\cdots) = 0$.] Equation (11-38), however, expresses the field at an exterior point as the gradient of a scalar. Thus, if we introduce the *magnetic* scalar potential $V^{(m)}$ in such a way that

$$\mathbf{B}(\mathbf{r}) = -\mu_0\,\nabla V^{(m)}(\mathbf{r}) \quad (11\text{-}39)$$

we find from Eq. (11-38) that, apart from an arbitrary additive constant,

$$V^{(m)}(\mathbf{r}) = \frac{1}{4\pi}\int \mathbf{M}(\mathbf{r}')\cdot\nabla'\left(\frac{1}{|\mathbf{r} - \mathbf{r}'|}\right) dv' \quad (11\text{-}40)$$

where the operator ∇ acting on the function $1/|\mathbf{r} - \mathbf{r}'|$ has been replaced by its equivalent $-\nabla'$. Since Eq. (11-40) is now identical in form with Eq. (10-11), further rewriting of Eq. (11-40) follows the same path as was described for Eq. (10-11). Analogous to Eq. (10-15) we ultimately find that

$$V^{(m)}(\mathbf{r}) = \frac{1}{4\pi}\oint \frac{\mathbf{M}(\mathbf{r}')\cdot\hat{\mathbf{n}}(\mathbf{r}')}{|\mathbf{r} - \mathbf{r}'|}\,dS' + \frac{1}{4\pi}\int \frac{[-\nabla'\cdot\mathbf{M}(\mathbf{r}')]}{|\mathbf{r} - \mathbf{r}'|}\,dv' \tag{11-41}$$

where $\hat{\mathbf{n}}(\mathbf{r}')$ is a unit vector normal to the surface bounding the magnetized object (P11-12). Equation (11-41) suggests that, at least for purposes of calculation if not as a physically real description, we introduce a volume density of magnetic *mono*poles given by

$$\rho_m(\mathbf{r}) = -\nabla\cdot\mathbf{M}(\mathbf{r}) \tag{11-42}$$

and a surface density of magnetic monopoles given by

$$\sigma_m(\mathbf{r}) = \mathbf{M}(\mathbf{r})\cdot\hat{\mathbf{n}}(\mathbf{r}) \tag{11-43}$$

Thus, insofar as the *exterior* field is concerned, the magnetized object can be replaced either by an equivalent distribution of currents [Eqs. (11-25) and (11-26)] or by an equivalent distribution of (fictitious) magnetic (mono)poles [Eqs. (11-42) and (11-43)]. If the poles are adopted, the field, calculated by taking the gradient of the magnetic scalar potential, is given by

$$\mathbf{B}(\mathbf{r}) = \frac{\mu_0}{4\pi}\oint \frac{\sigma_m(\mathbf{r}')(\mathbf{r} - \mathbf{r}')}{|\mathbf{r} - \mathbf{r}'|^3}\,dS' + \frac{\mu_0}{4\pi}\int \frac{\rho_m(\mathbf{r}')(\mathbf{r} - \mathbf{r}')}{|\mathbf{r} - \mathbf{r}'|^3}\,dv \tag{11-44}$$

[Compare Eq. (10-18).] Because of the similarity of Eqs. (11-41) and (11-44) to expressions for the electrostatic potential and field, many problems in magnetostatics correspond to analogous problems in electrostatics and solutions in one area can be directly translated to the other area. (See P11-13.) Of course, Eqs. (11-41) and (11-44) give only the contribution of the magnetized object to the **B**-field; the contributions of any other sources must be added to these equations to obtain the total field.

PROBLEMS

P11-12. Patterning your argument after the development associated with Eq. (10-11), derive Eq. (11-41) from Eq. (11-40).

P11-13. Find the distribution of poles equivalent to the magnetized cylinder in P11-5 and then find the magnetic scalar potential at the point $(0, 0, z)$ by direct transcription of the solution to the analogous electrostatic problem in P10-10.

P11-14. Find the distribution of poles equivalent to the uniformly magnetized sphere of P11-6 and calculate the dipole moment of the sphere from these poles. *Hint*: By analogy with electrostatics, $\mathbf{m} = \int \mathbf{r}\rho_m(\mathbf{r})\,dv$ for a volume distribution of poles.

P11-15. Show that the magnetic moment **m** of an object with magnetization **M** can be calculated from the equivalent magnetic poles, i.e.,

$$\mathbf{m} = \int \mathbf{r}(-\nabla \cdot \mathbf{M})\, dv + \oint \mathbf{r}(\mathbf{M} \cdot \hat{\mathbf{n}})\, dS$$

Hint: Integrate Eq. (C-11) over the volume of the object, setting $\mathbf{Q} = \mathbf{M}$ and $\Phi = x, y$, and z successively. Then, use the divergence theorem and note Eq. (11-18).

P11-16. Derive Eq. (11-44) for the **B**-field by evaluating $-\mu_0 \nabla V^{(m)}$, with $V^{(m)}$ given by Eq. (11-41).

P11-17. A spherical cavity of radius R is cut in the interior of an object with uniform magnetization **M**. Find the equivalent poles on the surface of that cavity and show that these poles contribute an amount $\mathbf{B} = \frac{1}{3}\mu_0 \mathbf{M}$ to the field at the center of the cavity.

11-5 The Macroscopic Magnetic Induction Field at a Point Interior to a Magnetized Object

With two alternative expressions—Eqs. (11-35) and (11-44)—for the magnetic induction field at a point exterior to a magnetized object, we have two starting points for the calculation of the field at an interior point. In both cases, we proceed as we did for the dielectric by separating the material into near and far regions, treating the far regions macroscopically by the methods described in one or the other of the preceding two sections and considering the near region by methods similar to those used in Section 10-4. In this section we shall outline the starting points for the two calculations of the interior field and then state the results. Detailed derivation of these results is left to the problems.

If we elect to express the field in terms of equivalent currents, then we find from Eq. (11-35) that

$$\mathbf{B}_{\text{interior}}(\mathbf{r}) = \frac{\mu_0}{4\pi} \int_{v_{\mathrm{I}}} \frac{\mathbf{J}_m(\mathbf{r}') \times (\mathbf{r} - \mathbf{r}')}{|\mathbf{r} - \mathbf{r}'|^3}\, dv' + \frac{\mu_0}{4\pi} \int_{\Sigma} \frac{\mathbf{j}_m(\mathbf{r}') \times (\mathbf{r} - \mathbf{r}')}{|\mathbf{r} - \mathbf{r}'|^3}\, dS' + \frac{\mu_0}{4\pi} \int_{\Sigma_{\mathrm{II}}} \frac{\mathbf{j}_m(\mathbf{r}') \times (\mathbf{r} - \mathbf{r}')}{|\mathbf{r} - \mathbf{r}'|^3}\, dS' + \mathbf{B}_{\mathrm{II}}(\mathbf{r}) \qquad (11\text{-}45)$$

where v_{I} is the volume of the magnetized material *excluding* a spherical volume of radius R that contains the near molecules, Σ is the surface of the original object, Σ_{II} is the surface of the cavity that remains in the magnetized object when the near molecules are removed, and $\mathbf{B}_{\mathrm{II}}$ is the contribution to the macroscopic **B**-field made by the near molecules. (The notation is the same as that used in Section 10-4; see Fig. 10-3.) Now, since $\mathbf{j}_m(\mathbf{r}') = \mathbf{M}(\mathbf{r}') \times \hat{\mathbf{n}}(\mathbf{r}')$ and $\mathbf{M}(\mathbf{r}')$ can be considered constant over the macroscopically small surface Σ_{II}, the third integral in Eq. (11-45) gives the contribution made by the equivalent currents on the surface of a spherical cavity in a uniformly magnetized object to the **B**-field at the center of the cavity, and it has the value $-\frac{2}{3}\mu_0\mathbf{M}$ (P11-11). Further, as in the dielectric case, extending the integral over v_{I} to include the volume v_{II} enclosed by the sphere Σ_{II} adds zero to the overall expression (P11-18). Thus, Eq. (11-45) reduces to the simpler expression

$$\mathbf{B}_{\text{interior}}(\mathbf{r}) = \mathbf{B}(\mathbf{r}) - \tfrac{2}{3}\mu_0\mathbf{M}(\mathbf{r}) + \mathbf{B}_{\text{II}}(\mathbf{r}) \qquad (11\text{-}46)$$

where $\mathbf{B}(\mathbf{r})$ is the field given at the *interior* point $\mathbf{r}$ by evaluating the exterior expression, Eq. (11-35), at the interior point; i.e., $\mathbf{B}(\mathbf{r})$ is the field established at the interior point by the bound current distribution equivalent to the entire magnetized object.

If, on the other hand, we elect to express the field in terms of equivalent poles, then we find from Eq. (11-44) that

$$\begin{aligned}\mathbf{B}_{\text{interior}}(\mathbf{r}) = \frac{\mu_0}{4\pi}\int_{v_{\text{I}}} \rho_m(\mathbf{r}')\frac{\mathbf{r}-\mathbf{r}'}{|\mathbf{r}-\mathbf{r}'|^3}\,dv' &\\ + \frac{\mu_0}{4\pi}\int_{\Sigma} \sigma_m(\mathbf{r}')\frac{\mathbf{r}-\mathbf{r}'}{|\mathbf{r}-\mathbf{r}'|^3}\,dS' &\\ + \frac{\mu_0}{4\pi}\int_{\Sigma_{\text{II}}} \sigma_m(\mathbf{r}')\frac{\mathbf{r}-\mathbf{r}'}{|\mathbf{r}-\mathbf{r}'|^3}\,dS' + \mathbf{B}_{\text{II}}(\mathbf{r}) & \qquad (11\text{-}47)\end{aligned}$$

where the volumes and surfaces are the same as those described in connection with Eq. (11-45). Now, since $\sigma_m(\mathbf{r}') = \mathbf{M}(\mathbf{r}')\cdot\hat{\mathbf{n}}(\mathbf{r}')$ and $\mathbf{M}(\mathbf{r}')$ can be considered constant over the macroscopically small surface Σ_{II}, the third integral in Eq. (11-47) gives the contribution made by the equivalent poles on the surface of a spherical cavity in a uniformly magnetized object to the **B**-field at the center of the cavity, and it has the value $\frac{1}{3}\mu_0\mathbf{M}$ (P11-17). Further, the integral over v_{I} can again be extended to include the volume v_{II} (P11-19). Thus, Eq. (11-47) reduces to the simpler expression

$$\mathbf{B}_{\text{interior}}(\mathbf{r}) = -\mu_0\,\nabla V^{(m)}(\mathbf{r}) + \tfrac{1}{3}\mu_0\mathbf{M}(\mathbf{r}) + \mathbf{B}_{\text{II}}(\mathbf{r}) \qquad (11\text{-}48)$$

where $V^{(m)}(\mathbf{r})$ is the magnetic scalar potential given at the *interior* point $\mathbf{r}$ by evaluating the exterior expression, Eq. (11-41), at the interior point; i.e., $V^{(m)}(\mathbf{r})$ is the potential established at the interior point by the distribution of poles equivalent to the entire magnetized object.

The contribution $\mathbf{B}_{\text{II}}(\mathbf{r})$ made by the near molecules is the same, whichever approach to calculating the far field is adopted. Following the pattern in Section 10-4, we define

$$\mathbf{B}_{\text{II}}(\mathbf{r}) = \frac{N}{v_{\text{II}}}\int_{v_{\text{II}}} \mathbf{B}_{\text{mol}}(\mathbf{r}, \mathbf{r}')\,dv' \qquad (11\text{-}49)$$

where N is the number of molecules in v_{II} and $\mathbf{B}_{\text{mol}}(\mathbf{r}, \mathbf{r}')$ is the time average of the microscopic field produced at $\mathbf{r}$ by a molecule at $\mathbf{r}'$. [Compare Eq. (10-33).] Now, we introduce the time-averaged molecular current density which we take to be given at the point $\mathbf{r}_1$ for a molecule whose center is at $\mathbf{r}_c$ by $\mathbf{J}_{\text{mol}}(\mathbf{r}_1 - \mathbf{r}_c)$. Then the field $\mathbf{B}_{\text{mol}}(\mathbf{r}, \mathbf{r}')$ is given by Eq. (5-10), viz.,

$$\mathbf{B}_{\text{mol}}(\mathbf{r}, \mathbf{r}') = \frac{\mu_0}{4\pi} \int \mathbf{J}_{\text{mol}}(\mathbf{r}''' - \mathbf{r}') \times \frac{\mathbf{r} - \mathbf{r}'''}{|\mathbf{r} - \mathbf{r}'''|^3} \, dv''' \tag{11-50}$$

[Compare Eq. (10-34).] Now, we rewrite Eq. (11-50) as an integral over the variable $\mathbf{r}'' = \mathbf{r}''' - \mathbf{r}'$ [compare Eq. (10-35)] and then substitute the result into Eq. (11-49) to find that

$$\mathbf{B}_{\text{II}}(\mathbf{r}) = \frac{N\mu_0}{4\pi v_{\text{II}}} \int \mathbf{J}_{\text{mol}}(\mathbf{r}'') \times \left[\int_{v_{\text{II}}} \frac{\mathbf{r} - \mathbf{r}' - \mathbf{r}''}{|\mathbf{r} - \mathbf{r}' - \mathbf{r}''|^3} \, dv' \right] dv'' \tag{11-51}$$

[Compare Eq. (10-36).] The integral in square brackets, however, is the integral $\mathbf{I}_1$, whose value is $\mathbf{I}_1 = -\frac{4}{3}\pi\mathbf{r}''$ [Eq. (10-42)]. Thus, Eq. (11-51) reduces to

$$\mathbf{B}_{\text{II}}(\mathbf{r}) = \frac{N\mu_0}{4\pi v_{\text{II}}} \left(-\frac{4\pi}{3}\right) \int \mathbf{J}_{\text{mol}}(\mathbf{r}'') \times \mathbf{r}'' \, dv'' \tag{11-52}$$

We note finally that the remaining integral is nothing but the negative of *twice* the dipole moment of a single molecule located near the point $\mathbf{r}$ in the magnetized object. [Replace $I\,d\mathbf{r}$ by $\mathbf{J}\,dv$ in Eq. (3-36).] Thus, N times the integral is the negative of *twice* the total dipole moment in v_{II} and that product divided by v_{II} is the negative of *twice* the macroscopic magnetization $\mathbf{M}(\mathbf{r})$ of the object at $\mathbf{r}$. Equation (11-52) therefore becomes

$$\mathbf{B}_{\text{II}}(\mathbf{r}) = \tfrac{2}{3}\mu_0\mathbf{M} \tag{11-53}$$

The results for the macroscopic interior field now have different expressions depending on whether we adopt bound currents or equivalent poles. Substituting Eq. (11-53) into Eq. (11-46), we find that

$$\mathbf{B}_{\text{interior}}(\mathbf{r}) = \mathbf{B}(\mathbf{r}) \tag{11-54}$$

and the interior field and the exterior field are both given correctly by the bound currents alone. In contrast, if we substitute Eq. (11-53) into Eq. (11-48), we find that

$$\mathbf{B}_{\text{interior}}(\mathbf{r}) = -\mu_0 \nabla V^{(m)}(\mathbf{r}) + \mu_0\mathbf{M}(\mathbf{r}) \tag{11-55}$$

Although this result is valid outside the object as well as inside (the magnetization is zero outside the material), we nevertheless find that the equivalent poles by themselves are not sufficient to account for the entire field of a magnetized object. The object can for *all* purposes be replaced by a set of bound currents, but it cannot be replaced by a set of equivalent poles without recognizing the additional term in Eq. (11-55).

PROBLEMS

P11-18. Show that the integral over v_{I} in Eq. (11-45) can be extended to include the volume v_{II} with no change in value.

P11-19. Show that the integral over v_{I} in Eq. (11-47) can be extended to include the volume v_{II} with no change in value.

P11-20. In the text, we represented the molecule by a current distribution in order to calculate $\mathbf{B}_{\mathrm{II}}(\mathbf{r})$. Suppose instead we represented the molecule by a volume distribution of magnetic poles, say $\rho_{\mathrm{mol}}(\mathbf{r}_1 - \mathbf{r}_c)$. Then, in accordance with Eq. (11-44), we would write

$$\mathbf{B}_{\mathrm{mol}}(\mathbf{r}, \mathbf{r}') = \frac{\mu_0}{4\pi} \int \rho_{\mathrm{mol}}(\mathbf{r}''' - \mathbf{r}') \frac{\mathbf{r} - \mathbf{r}'''}{|\mathbf{r} - \mathbf{r}'''|^3} \, dv'''$$

instead of Eq. (11-50). The resulting equation for $\mathbf{B}_{\mathrm{II}}(\mathbf{r})$ is identical in form with Eq. (10-36) and hence has the value $\mathbf{B}_{\mathrm{II}}(\mathbf{r}) = -\frac{1}{3}\mu_0 \mathbf{M}(\mathbf{r})$, as inferred from Eq. (10-44). This result differs significantly from the result in Eq. (11-53). Explain why Eq. (11-53) is correct and the result obtained by representing the molecule in terms of equivalent poles is incorrect.

11-6 The Basic Equations of Magnetostatics When Magnetically Responsive Matter Is Present

In this section, we shall translate the basic equations of magnetostatics into a convenient form for treating problems involving magnetically responsive matter. We have already established that the static magnetic induction field produced at any point in space by a magnetized object can be viewed as originating in a suitable distribution of (steady) currents in free space. Thus, the static magnetic induction field **B** established jointly by a magnetized object and by any simultaneously present free currents still satisfies both the magnetic flux law,

$$\oint \mathbf{B} \cdot d\mathbf{S} = 0 \qquad \nabla \cdot \mathbf{B} = 0 \tag{11-56}$$

and Ampere's circuital law,

$$\oint \mathbf{B} \cdot d\boldsymbol{\ell} = \mu_0 \int \mathbf{J}_t \cdot d\mathbf{S} \qquad \nabla \times \mathbf{B} = \mu_0 \mathbf{J}_t \tag{11-57}$$

provided we now interpret the current density $\mathbf{J}_t$ as the *total* current density, which includes any free currents placed in space *and* any bound currents present on magnetized objects. Since we know how to relate the bound currents to the magnetization **M**, however, we can reexpress Eq. (11-57) in a better form. Let **J** now denote only the free current density. Then

$$\mathbf{J}_t = \mathbf{J} + \mathbf{J}_m = \mathbf{J} + \nabla \times \mathbf{M} \tag{11-58}$$

Hence, with some rearrangement, Eq. (11-57) yields

$$\oint \left(\frac{\mathbf{B}}{\mu_0} - \mathbf{M}\right) \cdot d\boldsymbol{\ell} = \int \mathbf{J} \cdot d\mathbf{S} \qquad \nabla \times \left(\frac{\mathbf{B}}{\mu_0} - \mathbf{M}\right) = \mathbf{J} \tag{11-59}$$

(The integral form is obtained after using Stokes' theorem.) We stress again that the current density appearing on the right-hand side is the *free* current density; the bound currents have been explicitly introduced and appear now on the left-hand side in somewhat disguised form. We can now push the magnetization, which we usually do not know initially, into the background altogether by introducing the *magnetic field intensity* **H** defined in mks units by[4]

$$\mathbf{H} = \frac{\mathbf{B}}{\mu_0} - \mathbf{M} \tag{11-60}$$

At least the line integral of **H** about a closed path or equivalently its curl is determined solely by the free currents, as evidenced by the new form of Ampere's circuital law,

$$\oint \mathbf{H}\cdot d\boldsymbol{\ell} = \int \mathbf{J}\cdot d\mathbf{S} \qquad \nabla \times \mathbf{H} = \mathbf{J} \tag{11-61}$$

obtained by substituting Eq. (11-60) into Eq. (11-59). The magnetic intensity **H** is the second of the two auxiliary fields mentioned in the introductory paragraphs of Chapter 10. Although we shall continue to regard **B** as the basic field, arguments supporting the opposite view can be presented. Whichever field is viewed as basic, the field **B** remains the field that determines the force on a charged particle.

We have now determined that the static magnetic induction field in the presence of magnetically responsive matter satisfies the basic equations

$$\oint \mathbf{H}\cdot d\boldsymbol{\ell} = \int \mathbf{J}\cdot d\mathbf{S} \qquad \nabla \times \mathbf{H} = \mathbf{J} \tag{11-62}$$

$$\oint \mathbf{B}\cdot d\mathbf{S} = 0 \qquad \nabla\cdot\mathbf{B} = 0 \tag{11-63}$$

$$\mathbf{H} = \frac{\mathbf{B}}{\mu_0} - \mathbf{M} \qquad \mathbf{B} = \mu_0(\mathbf{H} + \mathbf{M}) \tag{11-64}$$

As with the analogous electrostatic equations, this system reduces to the vacuum form when no magnetically responsive matter is present and **M** is therefore zero. In that case **B** and **H** are trivially different and Eqs. (11-62) and (11-63) are sufficient to determine the field. When magnetizable matter is present, however, **B** may differ nontrivially from **H**, the two differential equations involve two essentially different fields, and we can make little progress toward a solution of these equations until we know a relationship between **B** and **H**. Equation (11-64) is a step toward specifying that relationship, but it is not a complete specification because the vector **M** is still not

[4]The vector **H** is defined differently in different systems of units but is always some linear combination of **B** and **M**. In cgs-esu, cgs-emu, Gaussian units, and Heaviside-Lorentz units, **H** is defined so that $\nabla \times \mathbf{H} = 4\pi\mathbf{J}$, $4\pi\mathbf{J}$, $4\pi\mathbf{J}/c$, and $\mathbf{J}/c$, respectively, where c is the speed of light in cm/sec. In particular, the Gaussian unit of **H**, which is both dimensionally and numerically the same thing as the gauss, is nevertheless commonly called the oersted; an **H**-field of 1 ampere-turn/m in mks units is the same field as an **H**-field of $4\pi \times 10^{-3}$ oersted in Gaussian units.

known. As with dielectrics, the specific properties of matter enter the theory at this point. Following established custom we write the necessary constitutive relation in the general form

$$\mathbf{M} = \mathbf{M}(\mathbf{H}) \tag{11-65}$$

thinking of $\mathbf{M}$ more directly as a function of $\mathbf{H}$ rather than of $\mathbf{B}$. A more specific form for Eq. (11-65) can be determined only by an *empirical* study of specific magnetizable objects or, in some cases, by a detailed quantum mechanical calculation. Once the form of Eq. (11-65) is known for a specific material, Eqs. (11-62)–(11-64) provide sufficient information to determine the two fields $\mathbf{B}$ and $\mathbf{H}$ if the *free* currents and/or suitable boundary conditions—see Section 12-7—are known.

As in the analogous dielectric case, the specific form of the constitutive relation appropriate to a given material may be quite complicated. Magnetically responsive materials may be anisotropic (different in different directions), they may exhibit a permanent or spontaneous magnetization even in the absence of a magnetizing field, they may show properties determined in part by their past history, etc. Even though magnetic materials displaying these complications are more common and more important than materials displaying the analogous dielectric complications, there are nonetheless many magnetic materials that develop a magnetization that is parallel to and also simply related to the magnetizing field. For these materials, the constitutive relation has the form

$$\mathbf{M} = \chi_m(H)\mathbf{H} \tag{11-66}$$

where the *static magnetic susceptibility* $\chi_m(H)$ may still depend on the field strength. For many materials, however, χ_m is constant, at least if H is not too large. These simplest of all magnetic materials are called *linear* materials. They are diamagnetic when $\chi_m < 0$ and paramagnetic when $\chi_m > 0$. (Why?) Further, the magnetic induction field in such a material is proportional to the magnetic field intensity,

$$\mathbf{B} = \mu_0(\mathbf{H} + \mathbf{M}) = \mu_0(1 + \chi_m)\mathbf{H} = \mu\mathbf{H} \tag{11-67}$$

where the *static magnetic permeability* μ of the material is defined by

$$\mu = \mu_0(1 + \chi_m) \tag{11-68}$$

Since χ_m may be positive or negative, μ may be larger or smaller than μ_0 and the *relative permeability* defined by

$$K_m = \frac{\mu}{\mu_0} = 1 + \chi_m \tag{11-69}$$

may be either larger or smaller than unity. Values of χ_m for a selection of common materials are shown in Table 11-2.

We shall conclude this section with an example that not only illustrates the new form of the circuital law but also provides the basis of one method for measuring magnetic permeabilities. Consider the so-called *Rowland ring*,

TABLE 11-2 Static Magnetic Susceptibilities for Selected Materials

Material	*Temperature*	$\chi_m = K_m - 1$	*Material*	*Temperature*	$\chi_m = K_m - 1$
Air (1 atm)	20°C	$+30.36 \times 10^{-5}$	Hydrogen (1 atm)	20°C	-2.48×10^{-5}
Aluminum	18°C	$+0.82 \times 10^{-5}$	Magnesium	18°C	$+0.69 \times 10^{-5}$
Argon (1 atm)	20°C	-0.56×10^{-5}	Neon (1 atm)	20°C	-0.41×10^{-5}
Bismuth	18°C	-1.70×10^{-5}	Nitrogen (1 atm)	20°C	-0.430×10^{-5}
Cobaltous chloride	25°C	$+114 \times 10^{-5}$	Oxygen (liquid)	−219°C	$+390 \times 10^{-5}$
Copper	18°C	-0.11×10^{-5}	Oxygen (1 atm)	20°C	$+133 \times 10^{-5}$
Diamond	20°C	-0.62×10^{-5}	Paraffin	Room	-0.75×10^{-5}
Dysprosium oxide	16°C	$+288 \times 10^{-5}$	Silicon	18°C	-0.16×10^{-5}
Ferric chloride	20°C	$+108 \times 10^{-5}$	Silver	18°C	-0.25×10^{-5}
Germanium	18°C	-0.15×10^{-5}	Sodium	18°C	$+0.64 \times 10^{-5}$
Glass (crown)	Room	-1.1×10^{-5}	Titanium	20°C	$+1.57 \times 10^{-5}$
Helium (1 atm)	20°C	-0.59×10^{-5}	Tungsten	18°C	$+0.35 \times 10^{-5}$

The values in this table are quoted from the *Handbook of Chemistry and Physics* (Chemical Rubber Publishing Company, Cleveland, 1955), Thirty-seventh Edition, pp. 2390–2400, and are used by permission of the Chemical Rubber Company. Values from the handbook (in cgs units) must be multiplied by 4π to obtain the values tabulated here (in mks units). Note that paramagnetic materials ($\chi_m > 0$) typically exhibit stronger response than diamagnetic materials ($\chi_m < 0$) and that both paramagnetic and diamagnetic metals may be found.

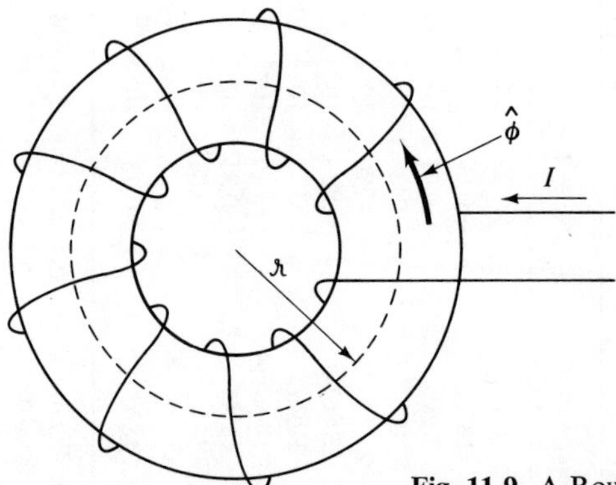

Fig. 11-9. A Rowland ring.

which consists of a coil wound around a toroidal core of some magnetic material (Fig. 11-9). Using symmetries by now familiar for these geometries, we apply the new form of the circuital law, Eq. (11-62), to a circular path of radius $\imath$ in the material of the core, shown dashed in Fig. 11-9, and find that

$$\oint \mathbf{H}\cdot d\boldsymbol{\ell} = 2\pi\imath H_\phi = NI \Longrightarrow \mathbf{H} = \frac{NI}{2\pi\imath}\hat{\boldsymbol{\phi}} \tag{11-70}$$

where I is the current in the coil, N is the total number of turns, and $\hat{\boldsymbol{\phi}}$ is a unit vector tangent to the path. From Eqs. (11-67) and (11-69), we then have that

$$\mathbf{B} = \mu\mathbf{H} = \frac{\mu NI}{2\pi\imath}\hat{\boldsymbol{\phi}} = K_m\frac{\mu_0 NI}{2\pi\imath}\hat{\boldsymbol{\phi}} = K_m\mathbf{B}_0 \tag{11-71}$$

where μ is the permeability of the core and $\mathbf{B}_0$ is the magnetic induction field that would be present if the core were removed. The self-inductance L of this coil, which is defined in Eq. (6-28), is now given by

$$L = \frac{d}{dI}\left[N\int_\Sigma \mathbf{B}\cdot d\mathbf{S}\right] = K_m\frac{d}{dI}\left[N\int_\Sigma \mathbf{B}_0\cdot d\mathbf{S}\right]$$
$$= K_m L_0 \tag{11-72}$$

where Σ is a surface bounded by a single turn of the coil and L_0 is the self-inductance that the coil would have if the core were removed. Thus, measurement of two self-inductances, one with the core present and the second with the core removed, leads to a determination of the relative permeability of the material composing the core. Unlike the corresponding dielectric case, it is here possible for L to be less than L_0 (diamagnetic core) or for L to be greater than L_0 (paramagnetic core). We should realize, however, that for many materials K_m does not differ from unity by very much, and the change in self-inductance may be small enough to require sophisticated measuring techniques.

PROBLEMS

P11-21. (a) Add to Eq. (11-55) a term giving the contribution of free currents with current density $\mathbf{J}(\mathbf{r})$ and then show that

$$\begin{aligned}\mathbf{H} &= -\nabla V^{(m)} + \frac{1}{4\pi}\int \frac{\mathbf{J}(\mathbf{r}') \times (\mathbf{r} - \mathbf{r}')}{|\mathbf{r} - \mathbf{r}'|^3}\, dv' \\ &= -\nabla V^{(m)} + \frac{1}{\mu_0}\nabla \times \mathbf{A}_{\text{free}}\end{aligned}$$

where $\mathbf{A}_{\text{free}}$ is the vector potential established by the free currents. (Note that these relationships are valid both inside and outside magnetized matter.) (b) Show that $\nabla \times \mathbf{H} = \mathbf{J}$ from this result.

P11-22. Show from Eqs. (11-62) and (11-63) that the magnetic field intensity in a region containing no free currents ($\mathbf{J} = 0$) but filled with a linear material having permeability μ can be derived from a scalar potential $\mathcal{V}^{(m)}$ by $\mathbf{H} = -\nabla\mathcal{V}^{(m)}$ and that $\mathcal{V}^{(m)}$ satisfies $\nabla\cdot[\mu\, \nabla\mathcal{V}^{(m)}] = 0$, which reduces to Laplace's equation if μ is independent of position. *Hint:* See item (3) in Section 2-5. *Optional:* Reconcile $\mathbf{H} = -\nabla\mathcal{V}^{(m)}$ with the result in P11-21, in which the integral in general makes a contribution to **H** even at points where **J** is zero. *Hint:* Imagine the free currents to be the currents equivalent to a suitably magnetized object and then replace the object instead with a set of equivalent poles. At what points in space is this replacement valid?

P11-23. An infinite solenoid (P5-11) of radius R has N turns per unit length, carries current I, and is wound on a core of linear material that has permeability μ. Let the axis of the solenoid coincide with the z axis. (a) Calculate **H**, **B**, and **M** in the core. (b) Show that **H** has the same value that it would have if the core were vacuum but that **B** is K_m times the vacuum value. (c) Calculate the bound surface current on the core, add this current to the free current, and then calculate **B** using Eq. (11-57).

11-7 Connecting the Microscopic Polarizability with the Macroscopic Relative Permeability

A relation similar to the Clausius-Mossotti relation in Section 10-6 can be derived to determine the microscopic (magnetic) molecular polarizability of *some* (but not all) materials from measured values of the macroscopic relative permeability. Following the pattern of Section 10-6, we begin by finding the field $\mathbf{B}_m$ that a single molecule in a magnetic material experiences, i.e., by finding the (time-averaged) microscopic field established at the position of a molecule by all molecules except the one located at that site. The same separation of the material into near and far molecules that led to Eq.

(10-66) here leads to the expression

$$\mathbf{B}_m = \mathbf{B}_\mathrm{I} + \mathbf{B}' \tag{11-73}$$

where $\mathbf{B}_\mathrm{I}$ is the contribution of the far molecules and $\mathbf{B}'$ the contribution of the near molecules. In fact, we have already calculated $\mathbf{B}_\mathrm{I}$ by two different methods; its definition here is essentially identical with the definition of $\mathbf{B}_{\text{interior}} - \mathbf{B}_{\mathrm{II}}$ in Section 11-5 except that we must include any contributions $\mathbf{B}_{\text{free}}$ from free currents. Thus, we have from Eq. (11-46) that

$$\mathbf{B}_m = \mathbf{B} - \tfrac{2}{3}\mu_0\mathbf{M} + \mathbf{B}' \tag{11-74}$$

where $\mathbf{B}$ now *includes* $\mathbf{B}_{\text{free}}$, or we have from Eq. (11-48) that

$$\mathbf{B}_m = -\mu_0 \nabla V^{(m)} + \tfrac{1}{3}\mu_0\mathbf{M} + \mathbf{B}_{\text{free}} + \mathbf{B}' \tag{11-75}$$

Continuing as in Section 10-6, we confine our attention to those materials for which $\mathbf{B}' = 0$ (see P10-23). Further, since the fields $\mathbf{B}_m$ and $\mathbf{B}_{\text{free}}$ are regarded to exist in free space, $\mathbf{B}_m = \mu_0\mathbf{H}_m$ and $\mathbf{B}_{\text{free}} = \mu_0\mathbf{H}_{\text{free}}$. Thus, Eqs. (11-74) and (11-75) both give the expressions

$$\mathbf{H}_m = \mathbf{H} + \tfrac{1}{3}\mathbf{M} \qquad \mathbf{B}_m = \mu_0(\mathbf{H} + \tfrac{1}{3}\mathbf{M}) \tag{11-76}$$

for the microscopic field at a molecular site; here, $\mathbf{H}$ is the *total* $\mathbf{H}$-field, including contributions of the free currents and of the magnetized matter. Finally, again as in Section 10-6, we combine Eq. (11-76) with Eq. (11-15) to find that

$$\mathbf{M} = N\mathbf{m} = N\beta_t\mathbf{B}_m = N\beta_t\mu_0(\mathbf{H} + \tfrac{1}{3}\mathbf{M})$$

$$\Longrightarrow \mathbf{M} = \frac{N\beta_t\mu_0}{1 - \tfrac{1}{3}N\beta_t\mu_0}\mathbf{H} \tag{11-77}$$

where N is the number of molecules per unit volume. Thus, we have by comparison with Eq. (11-66) that

$$\chi_m = \frac{N\beta_t\mu_0}{1 - \tfrac{1}{3}N\beta_t\mu_0} \tag{11-78}$$

and, further, by combining Eq. (11-78) with Eq. (11-69), we have that

$$K_m = \frac{1 + \tfrac{2}{3}N\beta_t\mu_0}{1 - \tfrac{1}{3}N\beta_t\mu_0}, \qquad \beta_t = \frac{3}{N\mu_0}\frac{K_m - 1}{K_m + 2} \tag{11-79}$$

In contrast to the dielectric case, however, the materials for which Eqs. (11-78) and (11-79) work well are the diamagnetic and paramagnetic materials for which $|\chi_m| \ll 1$ or, equivalently, $K_m \approx 1$. For these materials, $N\beta_t\mu_0$ is small compared to one (Why?), and Eqs. (11-78) and (11-79) can be approximated to give

$$\chi_m \approx N\beta_t\mu_0, \qquad \beta_t \approx \frac{1}{N\mu_0}(K_m - 1) \tag{11-80}$$

which we would have obtained directly by ignoring the difference between $\mathbf{H}$ and $\mathbf{H}_m$ altogether. When they apply, Eqs. (11-79) and (11-80) express the (microscopic) molecular polarizability in terms of the (macroscopic) relative permeability and are therefore analogous to the Clausius-Mossotti relation of Section 10-6. Unfortunately, magnetic materials that fail to conform with the assumptions of this section are more common than dielectrics that fail to conform with the analogous assumptions of Section 10-6, and the above equations are therefore less generally useful than the Clausius-Mossotti relation.

PROBLEMS

P11-24. Derive Eq. (11-76) for $\mathbf{H}_m$ and $\mathbf{B}_m$ (a) from Eq. (11-74) and (b) from Eq. (11-75).

P11-25. For a paramagnetic material, with $\beta_t = m_0^2/3kT$ [Eq. (11-1)], Eq. (11-77) suggests that at the right temperature T_c, $\mathbf{M}$ can be nonzero even if $\mathbf{H}$ is vanishingly small, and at $T \leq T_c$ we might therefore expect the material to exhibit a spontaneous magnetization. (a) Show that $T_c = N\mu_0 m_0^2/9k$. (b) Very small volumes of some materials, called *ferromagnetic materials*, in fact do exhibit spontaneous magnetization at temperatures below some critical temperature. Calculate T_c for pure iron ($m_0 = 2.05 \times 10^{-23}$ J/T, which is the mks unit of magnetic moment; specific gravity at 20°C = 7.874) and compare your result with the measured value 1043°K. (c) In the Weiss theory of ferromagnetism, the factor of $\frac{1}{3}$ in Eq. (11-76) is replaced by a different factor γ so that $\mathbf{H}_m = \mathbf{H} + \gamma\mathbf{M}$. What value must γ have in order to predict the critical temperature of iron correctly?

11-8 Ferromagnetism

Some materials, known as *ferromagnetic materials*, exhibit a very extreme nonlinear paramagnetic response that arises from a very strong inclination of the constituent permanent dipoles to orient themselves parallel to an externally applied magnetizing field. The extreme strength of this ferromagnetic response can be understood only by quantum mechanical arguments and the phenomenon itself is therefore a macroscopic manifestation of a purely quantum mechanical behavior; ferromagnetism cannot even be approximately understood by classical arguments. We cannot digress here to present the full quantum mechanical treatment. In essence, however, it turns out that, under the conditions existing in ferromagnetic materials, parallel orientation of all of the molecular dipoles within macroscopically small but still microscopically large volumes of the material is energetically preferred to a more random orientation of these dipoles. Thus, a macroscopic sample of ferromagnetic material breaks up *spontaneously* into macroscopically small but microscopically large *domains*, each of which contains a very large number of molecular dipoles all oriented in the same way. Now, in a

(a)

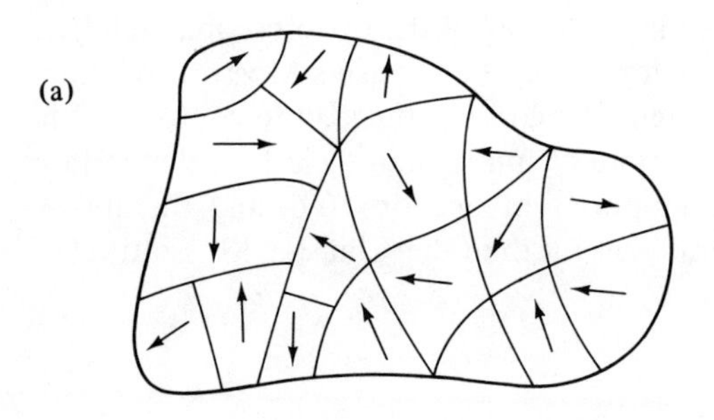

(b)

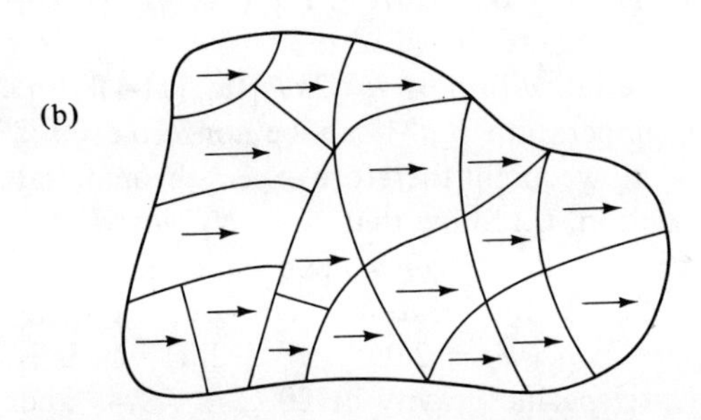

Fig. 11-10. Domains in (a) an unmagnetized and (b) a (completely) magnetized ferromagnetic sample.

macroscopically *unmagnetized* sample, these domains are randomly oriented within the constraints imposed by the crystal structure of the sample [Fig. 11-10(a)]. When placed in an external field, however, the sample magnetizes by preferential orientation of domains or by the motion of domain walls so as to enlarge some domains at the expense of others [Fig. 11-10(b)]. The resulting *collective* response of huge numbers of molecules can produce an extremely large magnetization; the value of K_m—to the extent that it is meaningful at all—may be as large as 10^5 or 10^6.

Although the relationship $\mathbf{B} = \mu_0(\mathbf{H} + \mathbf{M})$ is always correct, for ferromagnetic materials the relationship between **M** and **H** and consequently also the relationship between **B** and **H** is very complicated. Even the qualitative properties of ferromagnetic materials are therefore somewhat involved. Ferromagnetic response in a particular ferromagnetic material, for example, is not observed at all unless the temperature T is below the critical *Curie temperature* for that material; at temperatures above the Curie temperature ferromagnetic materials exhibit much weaker paramagnetic response. Even when $T \leq T_c$, **M** and hence **B** may both be nonzero even if **H** is zero, and we have a permanent magnet. Further, whether **H** is zero or not (but T is still below T_c), the values of **M** and **B** in general are not unique, for the ferromagnetic response is usually not even approximately a single-valued function of **H**; it depends both on how **H** is built up from zero and on the initial state of magnetization of the sample. Finally, ferromagnetic materials may be anisotropic, so that **B**, **H**, and **M** need not be parallel except possibly when **H** is applied in one of a few very special directions in the material. Clearly, we need to know quite

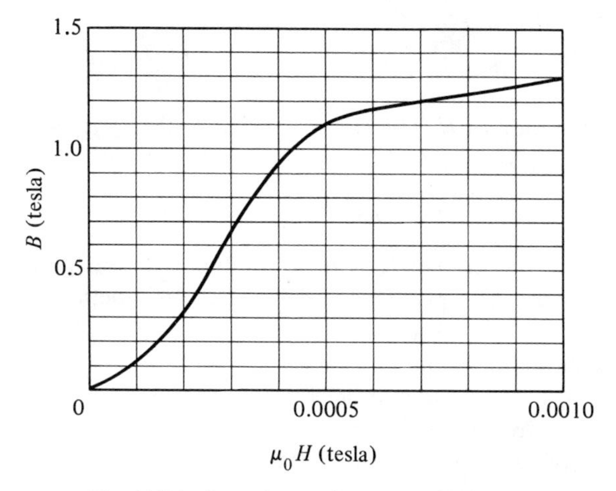

Fig. 11-11. A representative magnetization curve.

a bit about the properties of a particular ferromagnetic sample, about its present environment, and about its past history before we can determine **B** and **M** from knowledge of **H**.

A common method for presenting information about the relationship between **B** and **H** in a particular ferromagnetic material, at least when **B** and **H** are parallel, is to display a graph of B versus H, where B and H here mean *not* the *magnitudes* of **B** and **H** but the *components* of **B** and **H** along a line parallel to **B** and **H**. (In mks units, it is perhaps better to plot B versus $\mu_0 H$ so that the axes are labeled with quantities having the same dimensions.) Even a graph of B versus H, however, depends on the past history of the sample to which it applies. For the sake of standardization, let us therefore start with a macroscopically *unmagnetized* sample, which we place, say, as the core of a Rowland ring (Fig. 11-9). Now, let H be gradually (and monotonically) increased by gradually increasing the current. If we measure B for various values of H,[5] we obtain a relationship having the general form of Fig. 11-11, which is called the *magnetization curve* for the particular material involved. The quantity B increases *nonlinearly* with H, and ultimately a point is reached above which further increases in H do not alter B very significantly. Above this point, the material has magnetized to saturation, all of the domains are fully aligned, M cannot be further increased, and the effect of increasing H is no longer amplified by the response of the material; $B = \mu_0(H + M)$ increases only slowly with further increases in H. Two permeabilities are sometimes used to describe the material to which Fig. 11-11 applies. The permeability analogous to the permeability used for diamagnetic and paramagnetic materials is geometrically the slope of a line from the origin to some

[5]One way to determine B in the core is to use, for example, a Hall probe (P9-7) to measure B in a *narrow* gap in the core. B in the gap is the same as B in the core (P12-19).

point on the graph; it is defined by

$$B = \mu(H)H \quad \text{or} \quad \mu(H) = \frac{B(H)}{H} \tag{11-81}$$

and is a strong function of H. The *differential* permeability μ_d is geometrically the slope of the graph

$$\mu_d(H) = \frac{dB}{dH} \tag{11-82}$$

and is also a strong function of H. Which (if either) of these permeabilities is best suited to a particular problem involving ferromagnetic materials must be determined by examining the problem.

We can calculate the amount of energy required to move from one point on the magnetization curve to another by using Faraday's law, which in fact is valid even in the presence of matter (see Section 12-1). Thus, the emf $\mathscr{E}$ induced in the Rowland ring of Fig. 11-9 is given by $\mathscr{E} = -Nd\Phi_m/dt$, where N is the number of turns and Φ_m is the flux of $\mathbf{B}$ across a single turn of the coil. The power input dW/dt as given by Eq. (6-23) then is

$$\frac{dW}{dt} = NI\frac{d\Phi_m}{dt} \tag{11-83}$$

To simplify the argument, suppose the ring is large enough so that the fields are essentially uniform within its core. Then $\Phi_m = BS$, where S is the area of the cross section of the core and B is the tangential component of $\mathbf{B}$. Further, from Eq. (11-70), the tangential component of $\mathbf{H}$, which we now denote merely H, is given by $H = NI/2\pi r = NI/\ell$, where ℓ is the (median) length of the core. Thus, Eq. (11-83) can be written in the form

$$\frac{dW}{dt} = (\ell H)\frac{d}{dt}(BS) = VH\frac{dB}{dt} \tag{11-84}$$

where $V = \ell S$ is the volume of the core. Consequently, the energy input *per unit volume* required to increase the $\mathbf{B}$-field from B to $B + dB$ when the $\mathbf{H}$-field is H is given by

$$d\left(\frac{W}{V}\right) = H\,dB \tag{11-85}$$

and the total energy required *per unit volume* to change the $\mathbf{B}$-field from $B_{\text{initial}} = B_i$ to $B_{\text{final}} = B_f$ is given by

$$\frac{W}{V} = \int_{B_i}^{B_f} H(B)\,dB \tag{11-86}$$

which can be interpreted geometrically as the area bounded by the magnetization curve, the vertical axis, and the lines $B = B_i$ and $B = B_f$. If $B_f > B_i$ in Fig. 11-11, work must be done on the fields; if $B_f < B_i$, energy will be taken out of the fields.

The magnetization curve, of course, does not convey all properties of a ferromagnetic material. Suppose now that we gradually reduce H from the

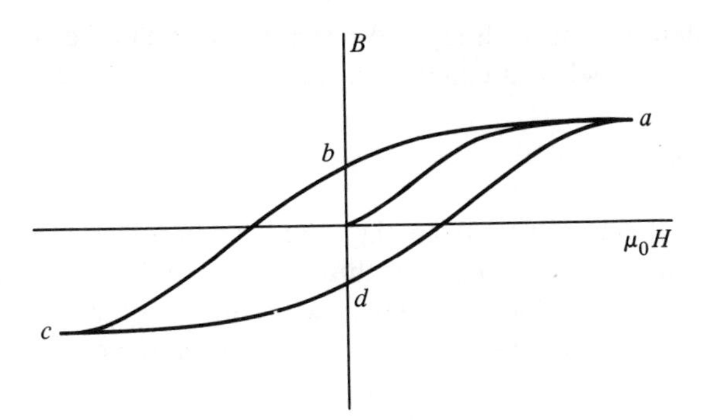

Fig. 11-12. A representative hysteresis curve.

maximum value reached in Fig. 11-11. We find that B does not trace back to zero along the curve already obtained. Instead, B follows a curve such as that shown from a to b in Fig. 11-12. When H has been reduced to *zero*, B is still nonzero, and a permanent magnetization has resulted. Application of an **H**-field in the opposite direction brings about further reduction of B, and finally M saturates in the other direction at some point c. If H is now increased from this point, B again rises but still does not go to zero when H has again become zero. The curve passes through some point d as H passes through zero. Further increase in H causes B ultimately to rise again to the point a where M saturates. If this cycle is now repeated many times in succession, after awhile the curve traced out in the B-H plane stabilizes with a shape and an extent determined by the material and by the range through which H varies. The resulting stable curve is known as a *hysteresis loop*, the word *hysteresis* itself referring to the lag between the magnetization M and the field H that produces M. The loop is a (actually *the*) major hysteresis loop if H becomes large enough to produce saturation and a minor hysteresis loop otherwise. Two parameters of the major hysteresis loop are often given to specify at least part of the properties of a ferromagnetic material: The *remanence* is the (positive) value of B when H is zero, and the *coercive force* is the magnitude of H when B is zero. Together, the remanence and the coercive force of a particular material outline the approximate region occupied by the major hysteresis loop of the material.

We now note from Eq. (11-86) that the amount of energy input to the fields as a ferromagnetic sample is moved along *cda* in Fig. 11-12 is greater than the amount recovered as the sample is returned to its initial point along *abc*. Thus, some energy is converted to heat in the sample as it is carried around its hysteresis loop. It is shown in P11-27 that in cycling a ferromagnetic material once around its hysteresis loop, an amount of energy given by the integral

$$\text{energy dissipated per unit volume per cycle} = \oint H\,dB \qquad (11\text{-}87)$$

about the hysteresis loop is dissipated per unit volume in the material. Geometrically, this energy loss is equal to the area of the hysteresis loop.

PROBLEMS

P11-26. (a) Reading data from the magnetization curve in Fig. 11-11, obtain a graph of μ/μ_0 versus $\mu_0 H$ for this material. (b) Suppose this material now composes the core of an infinitely long solenoid with 200 turns per unit length. Find the magnetic induction in the core when the current in the coil is .5, 2, and 4 A

P11-27. Show that the energy dissipated per unit volume in a ferromagnetic material cycled once around its hysteresis loop is given by $\oint H\,dB$ and verify that this integral expresses the area of the hysteresis loop.

P11-28. Describe the general characteristics of the hysteresis loop of ferromagnetic materials that are particularly good for making (a) permanent magnets and (b) transformer cores. Defend your answers.

Supplementary Problems

P11-29. Quantum mechanically, the magnetic moment $\mathbf{m}$ of an atomic system can be written in terms of the total angular momentum $\mathbf{J}$ of the system in the form $\mathbf{m} = -g(q_e/2m_e)\mathbf{J}$, where q_e is the magnitude of the electronic charge, m_e is the mass of the electron, and g is a dimensionless factor—called the *Landé g-factor*—that characterizes the specific system. (In particular, g has the value 1 if the angular momentum is purely orbital—compare P11-3—the value 2 if the angular momentum originates entirely in spin, and some intermediate value if the angular momentum has both orbital and spin components.) Further, when an external magnetic induction $\mathbf{B}_m = B_m\hat{\mathbf{k}}$ is applied to the system, its energy is changed by an amount

$$E = -\mathbf{m}\cdot\mathbf{B}_m = -m_z B_m = g\frac{q_e\hbar}{2m_e}B_m\frac{J_z}{\hbar} \tag{1}$$

where $\hbar$ is Planck's constant divided by 2π. Now, for a system with total angular momentum quantum number j (which must have one of the values $0, \frac{1}{2}, 1, \frac{3}{2}, \ldots$), $J_z/\hbar$ can assume only the $2j+1$ values $s = -j, -j+1, \ldots, j-1, j$. Thus, only the energies $E_s = gM_BsB_m$, where $M_B = q_e\hbar/2m_e$ is the Bohr magneton, are allowed. Further, the z-component of the magnetic moment can assume only the values $m_z = -gM_Bs$. Finally, statistically,

$$\langle m_z\rangle = \frac{\sum_s(-gM_Bs)e^{-E_s/kT}}{\sum_s e^{-E_s/kT}} \tag{2}$$

where k is Boltzmann's constant and T is the absolute temperature. (a) Show that

$$\langle m_z\rangle = m_0\tanh\left(\frac{m_0B_m}{kT}\right) \tag{3}$$

when $j = \frac{1}{2}$. Here, $m_0 = \frac{1}{2}gM_B$, $\tanh x = \sinh x/\cosh x$ and $\sinh x$ and $\cosh x$ are defined in footnote 1, Chapter 8. (b) If there are n atoms per unit volume, then the magnetization M is given by $n\langle m_z \rangle$. Sketch a graph of M versus m_0B_m/kT, and show that the saturation value of the magnetization is $M_s = nm_0$. (c) Writing $H_m = H + \gamma M$ (as in P11-25), show that the magnetization of this sample is determined by the solution to the transcendental system

$$\frac{M}{M_s} = \tanh x \qquad \frac{M}{M_s} = \frac{kT}{\gamma\mu_0 n m_0^2}x - \frac{H}{\gamma n m_0} \tag{4}$$

(d) On the same axes, sketch graphs of M/M_s versus x as given by both of these equations and then, by allowing the straight line to move, obtain qualitative data and sketch graphs of M/M_s versus H for various fixed temperatures. (e) Now let $H = 0$ and obtain a sketch of M/M_s versus T. Show in particular that $M/M_s = 0$ unless $T < T_c = \gamma\mu_0 n m_0^2/k$. *Optional:* (1) Use a computer to find M/M_s versus T/T_c for $H = 0$ to three significant figures and plot a careful graph. (2) Find and explore an expression analogous to Eq. (3) for some other value for j.

P11-30. Write Eq. (11-41) in the form

$$V^{(m)}(\mathbf{r}) = \frac{1}{4\pi}\int \frac{dq'_m}{|\mathbf{r} - \mathbf{r}'|}$$

where $dq'_m = \rho_m\, dv'$ or $\sigma_m\, dS'$ as appropriate to the distribution, follow the development in Section 4-6, and derive a multipole expansion for the magnetic scalar potential. Make sure to present a convincing *mathematical* argument for the absence of the monopole term.

P11-31. Consider a cylindrical bar magnet of length ℓ and cross-sectional area S uniformly magnetized to magnetization $\mathbf{M}$. Let this magnet be placed in a uniform external magnetic induction $\mathbf{B}$, with $\mathbf{B}$ perpendicular to $\mathbf{M}$. (a) Calculate the torque on the magnet from the expression $\mathbf{m} \times \mathbf{B}$, where $\mathbf{m} = \mathbf{M}S\ell$ is the dipole moment of the magnet. (b) Calculate the torque also by replacing the magnet with equivalent poles and assuming a magnetic force $\mathbf{F}_m$ parallel to $\mathbf{B}$ on the positive pole and an equal but opposite force on the negative pole. (c) Equate these two evaluations of the torque to show that a magnetic charge q_m in the field $\mathbf{B}$ experiences a magnetic force given by $q_m\mathbf{B}$. (d) Show from Eq. (11-44) that the $\mathbf{B}$-field established at $\mathbf{r}$ by a magnetic pole of strength q_m at the origin is given by $\mathbf{B} = \mu_0 q'_m \mathbf{r}/4\pi r^3$ and infer "Coulomb's" law $\mathbf{F}(q_m, q'_m) = \mu_0 q_m q'_m \mathbf{r}/4\pi r^3$ for the force between two magnetic poles. (e) Under what conditions would you expect to observe such an inverse square force experimentally?

P11-32. If Eq. (11-36) is applied to a point interior to the magnetized object, $\mathbf{r}'$ assumes the value $\mathbf{r}$ somewhere in the domain of integration and the first term no longer contributes zero. However, $\nabla^2(1/|\mathbf{r} - \mathbf{r}'|)$ is still zero except at $\mathbf{r}' = \mathbf{r}$, so the first term can be reduced to an integral over a small volume Δv centered on the point $\mathbf{r}$. If Δv is small enough, $\mathbf{M}(\mathbf{r}')$ is essentially constant throughout Δv and can be removed from the integral with the value

M(**r**). Evaluate the remaining integral formally and combine your result with Eq. (11-39), which expresses the second term in Eq. (11-36), to derive Eq. (11-55) for the *interior* field. Criticize this method of deriving the interior field. *Hints:* (1) Write $\nabla^2 (1/|\mathbf{r} - \mathbf{r}'|)$ as $\nabla' \cdot [(\mathbf{r} - \mathbf{r}')/|\mathbf{r} - \mathbf{r}'|^3]$. (2) Use the divergence theorem. (3) Use Eq. (2-37).

P11-33. Given that $\mathbf{J}_m = \nabla \times \mathbf{M}, \nabla \times \mathbf{M}, c\,\nabla \times \mathbf{M}$, and $c\,\nabla \times \mathbf{M}$ and that $\nabla \times \mathbf{B} = 4\pi \mathbf{J}_t/c^2, 4\pi \mathbf{J}_t, 4\pi \mathbf{J}_t/c$, and $\mathbf{J}_t/c$ in cgs-esu, cgs-emu, Gaussian units, and Heaviside-Lorentz units, respectively, find the equation relating **H**, **B**, and **M** in each of these systems of units. *Hint:* See Footnote 4.

REFERENCES

R. P. FEYNMAN, R. B. LEIGHTON, and M. SANDS, *The Feynman Lectures on Physics* (Addison-Wesley Publishing Company, Inc., Reading, Mass., 1964), Volume II, Lectures 34, 35, 36, and 37.

C. KITTEL, *Introduction to Solid State Physics* (John Wiley & Sons, Inc., New York, 1966), Third Edition, Chapters 14 and 15.

E. M. PURCELL, *Electricity and Magnetism* (McGraw-Hill Book Company, New York, 1965), Chapter 10.

J. H. VAN VLECK, *The Theory of Electric and Magnetic Susceptibilities* (Oxford University Press, London, 1932).

12

Time-Dependent Fields When Matter Is Present: Maxwell's Equations Revised

In Chapters 10 and 11, we introduced the displacement field **D** and the magnetic intensity **H**, which supplement the electric intensity **E** and the magnetic induction **B** when matter is present. When the fields are static (which requires also that the polarization **P** and magnetization **M** be static), the four fields satisfy the equations

$$\oint \mathbf{D} \cdot d\mathbf{S} = \int \rho \, dv \qquad \nabla \cdot \mathbf{D} = \rho \tag{12-1}$$

$$\oint \mathbf{E} \cdot d\boldsymbol{\ell} = 0 \qquad \nabla \times \mathbf{E} = 0 \tag{12-2}$$

$$\oint \mathbf{H} \cdot d\boldsymbol{\ell} = \int \mathbf{J} \cdot d\mathbf{S} \qquad \nabla \times \mathbf{H} = \mathbf{J} \tag{12-3}$$

$$\oint \mathbf{B} \cdot d\mathbf{S} = 0 \qquad \nabla \cdot \mathbf{B} = 0 \tag{12-4}$$

where ρ is the *free* charge density and **J** is the *free* current density. Supplemented with definitions of **D** and **H** and with relationships determining **P** and **M**, these equations constitute the formal framework of electricity and magnetism when matter is present, but this framework will not be fully general until we have extended it to include time-dependent fields. In some ways, this generalization is simple: All of the quantities in Eqs. (12-1)–(12-4)—including **P** and **M**, whose explicit appearance has been suppressed by introducing the vectors **D** and **H**—merely acquire a time dependence. In other ways, this generalization is far less simple, for additional terms appear in the equations. We have already seen the form of these additional terms in vacuum. We shall now determine their form when matter is present. In addition, we shall examine some of the general consequences of the revised equations.

12-1 Maxwell's Equations in Matter

Equations (12-2) and (12-3) acquire additional terms when the full set of equations is generalized to include time-dependent fields. Consider first the generalization of Eq. (12-2). As with the analogous generalization in Section 6-1 for the fields in vacuum, we begin with the concept of emf, which we still define as the work done on a unit charge moved once around the path described in the first subsection of Section 6-1. Throughout our development of properties of matter, however, the fields **E** and **B** have retained their connection with forces experienced by charged particles. Thus, even in the presence of matter, the electromagnetic part of the emf is still given by

$$\begin{aligned}\mathscr{E}_{\text{em}}(t) &= \oint_{\Gamma} \frac{\mathbf{F}_{\text{em}}}{q} \cdot d\boldsymbol{\ell} \\ &= \oint_{\Gamma} [\mathbf{E}(\mathbf{r}, t) + \mathbf{v}(\mathbf{r}, t) \times \mathbf{B}(\mathbf{r}, t)] \cdot d\boldsymbol{\ell} \end{aligned} \tag{12-5}$$

where $\mathbf{F}_{\text{em}}$ is the force on a particle of charge q in the electromagnetic field **E**, **B** and $\mathbf{v}(\mathbf{r}, t)$ is the velocity of the path $\boldsymbol{\Gamma}$ at the point **r** at the time t. In particular, this definition includes a motional emf that we can still write in the form

$$\mathscr{E}_{\text{em}}^{\text{mot}} = \oint_{\Gamma} (\mathbf{v} \times \mathbf{B}) \cdot d\boldsymbol{\ell} = -\frac{d}{dt} \int_{\Sigma} \mathbf{B} \cdot d\mathbf{S} = -\frac{d\Phi_m}{dt} \tag{12-6}$$

where $\boldsymbol{\Sigma}$ is a surface bounded by $\boldsymbol{\Gamma}$. [See Eqs. (6-13)–(6-17).] Thus, at least the motional emf remains unchanged by the presence of matter. The more general validity of the analogous expression in Eq. (6-12) suggests that Eq. (12-6) may also apply to emf's originating in time-dependent fields as well as to motional emf's. Assuming that generality, we would then have that

$$\mathscr{E}_{\text{em}} = -\frac{d\Phi_m}{dt} \tag{12-7}$$

or, equivalently, that

$$\oint \mathbf{E} \cdot d\boldsymbol{\ell} = -\int \frac{\partial \mathbf{B}}{\partial t} \cdot d\mathbf{S} \qquad \nabla \times \mathbf{E} = -\frac{\partial \mathbf{B}}{\partial t} \tag{12-8}$$

[Compare Eqs. (6-18)–(6-21) and Eq. (6-47).] Equations (12-7) and (12-8), of course, contain new elements that are not present in Eq. (12-6), and the support for Eqs. (12-7) and (12-8) lies not in the argument that led to them—that argument is a plausibility argument, not a derivation—but rather in an experimental confirmation of the predictions based on these equations. To date no experimental contradictions have been found, and we therefore conclude that Faraday's law assumes the same form whether matter is present or not.

Consider now the generalization of Eq. (12-3) to time-dependent fields. The necessary modification is determined most directly if we return temporarily to the equation relating the curl of **B** to the *total* current density, whatever its origin, viz., to

$$\nabla \times \mathbf{B} = \mu_0(\mathbf{J} + \mathbf{J}_m) \tag{12-9}$$

This equation is applicable only to static fields. When the fields (and the macroscopic polarization and magnetization vectors) become time-dependent, additional currents appear. We already know (Section 6-2) that a time-dependent electric field has associated with it a current density $\mathbf{J}_e$ given by $\epsilon_0\, \partial \mathbf{E}/\partial t$. But a time-dependent polarization, which involves a time-dependent readjustment of charge within the dielectric, also gives rise to an apparent macroscopic current. To obtain a quantitative expression for this polarization current density, we recognize first that the total bound charge Q_b contained within an arbitrary (but fixed) volume of a polarized dielectric is given by

$$Q_b = \int \rho_b\, dv = -\int \nabla \cdot \mathbf{P}\, dv = -\oint \mathbf{P} \cdot d\mathbf{S} \tag{12-10}$$

where the final form is obtained by using the divergence theorem. Consequently, the net rate at which bound charge flows *out* of this volume is given by

$$-\frac{\partial Q_b}{\partial t} = \oint \frac{\partial \mathbf{P}}{\partial t} \cdot d\mathbf{S} \tag{12-11}$$

and we infer that a time-dependent polarization gives rise to a current density $\mathbf{J}_p$ given by $\partial \mathbf{P}/\partial t$. In extending Eq. (12-9) to time-dependent fields, both $\mathbf{J}_e$ and $\mathbf{J}_p$ must be included, and we find that

$$\begin{aligned}\nabla \times \mathbf{B} &= \mu_0(\mathbf{J} + \mathbf{J}_m + \mathbf{J}_e + \mathbf{J}_p)\\ &= \mu_0\left(\mathbf{J} + \nabla \times \mathbf{M} + \epsilon_0 \frac{\partial \mathbf{E}}{\partial t} + \frac{\partial \mathbf{P}}{\partial t}\right)\\ &\Longrightarrow \nabla \times \left(\frac{\mathbf{B}}{\mu_0} - \mathbf{M}\right) = \mathbf{J} + \frac{\partial}{\partial t}(\epsilon_0 \mathbf{E} + \mathbf{P})\end{aligned} \tag{12-12}$$

Finally, returning to the fields $\mathbf{H} = \mu_0^{-1}\mathbf{B} - \mathbf{M}$ and $\mathbf{D} = \epsilon_0 \mathbf{E} + \mathbf{P}$, we have that

$$\oint \mathbf{H} \cdot d\boldsymbol{\ell} = \int \left(\mathbf{J} + \frac{\partial \mathbf{D}}{\partial t}\right) \cdot d\mathbf{S} \qquad \nabla \times \mathbf{H} = \mathbf{J} + \frac{\partial \mathbf{D}}{\partial t} \tag{12-13}$$

As with Faraday's law, the real test of this equation lies much more in the experimental confirmation of its predictions than in the above "derivation".

On the basis of our present knowledge, the only other terms that we might expect in Maxwell's equations would be a *free* magnetic charge density in Eq. (12-4) and a *free* magnetic current density in Eq. (12-2). Both of these terms are ruled out by the experimental absence of free magnetic charge.

Thus, Maxwell's equations for time-dependent fields in matter are

$$\oint \mathbf{D}\cdot d\mathbf{S} = \int \rho \, dv \qquad \nabla\cdot\mathbf{D} = \rho \tag{12-14}$$

$$\oint \mathbf{E}\cdot d\boldsymbol{\ell} = -\int \frac{\partial \mathbf{B}}{\partial t}\cdot d\mathbf{S} \qquad \nabla\times\mathbf{E} = -\frac{\partial \mathbf{B}}{\partial t} \tag{12-15}$$

$$\oint \mathbf{H}\cdot d\boldsymbol{\ell} = \int \left(\mathbf{J} + \frac{\partial \mathbf{D}}{\partial t}\right)\cdot d\mathbf{S} \qquad \nabla\times\mathbf{H} = \mathbf{J} + \frac{\partial \mathbf{D}}{\partial t} \tag{12-16}$$

$$\oint \mathbf{B}\cdot d\mathbf{S} = 0 \qquad \nabla\cdot\mathbf{B} = 0 \tag{12-17}$$

These equations are to be supplemented with the definitions

$$\mathbf{D} = \epsilon_0 \mathbf{E} + \mathbf{P} \tag{12-18}$$

$$\mathbf{H} = \frac{\mathbf{B}}{\mu_0} - \mathbf{M} \tag{12-19}$$

and with the Lorentz force

$$\mathbf{F} = q[\mathbf{E} + \mathbf{v}\times\mathbf{B}] \tag{12-20}$$

on a particle with charge q moving with velocity $\mathbf{v}$ in the fields. Equations (12-14)–(12-20) are the fundamental equations of electricity and magnetism when matter is present. They are valid *regardless* of the properties of whatever matter may be present, although they are restricted to use in a coordinate system in which all matter is at rest. Because these equations, as the mathematical expression of particular experimental results, have a very broad range of applicability, they have come to be regarded as a very basic expression of the properties of the electromagnetic field. From expressions of experimental results, these equations have been elevated to the status of basic principles and are regarded essentially as the starting point for all theoretical considerations involving the electromagnetic field. Indeed, in some developments of the subject, Maxwell's equations are taken as postulates and the entire subject is developed from these equations as an assumed starting point.

Of course, it is not possible to obtain solutions to these equations in the absence of more specific information about ρ, $\mathbf{J}$, $\mathbf{P}$, and $\mathbf{M}$. For some problems these quantities will be part of the given information. In other circumstances, they may be determined by the fields and by the response of whatever matter is present to those fields. If the latter is the case, empirical information about the response of that matter is usually needed. The *simplest* materials are *linear*. For these materials, the *constitutive relations* have the form

$$\mathbf{P} = \chi_e \mathbf{E}, \qquad \mathbf{M} = \chi_m \mathbf{H}, \qquad \mathbf{J} = g\mathbf{E} \tag{12-21}$$

and Eqs. (12-18) and (12-19) become

$$\mathbf{D} = \epsilon \mathbf{E}, \qquad \mathbf{B} = \mu \mathbf{H} \tag{12-22}$$

In contrast to Eqs. (12-14)–(12-20), Eqs. (12-21) and (12-22) are valid *only* for specific materials. More complicated constitutive relations may be needed to describe the empirical properties of other materials adequately.

PROBLEMS

P12-1. Show that Maxwell's equations admit a principle of superposition; that is, show that if $\mathbf{E}_i$, $\mathbf{H}_i$, $\mathbf{D}_i$, and $\mathbf{B}_i$ satisfy the equations with sources $\mathbf{J}_i$ and ρ_i, then $\sum \mathbf{E}_i$, $\sum \mathbf{H}_i$, $\sum \mathbf{D}_i$, and $\sum \mathbf{B}_i$ satisfy the equations with sources $\sum \mathbf{J}_i$ and $\sum \rho_i$.

P12-2. Show that the fields $\mathbf{E}' = \beta\mu\mathbf{H}$, $\mathbf{H}' = -\beta\epsilon\mathbf{E}$ satisfy Maxwell's equations in a linear medium with $\mathbf{J} = 0$ and $\rho = 0$ if $\mathbf{E}$ and $\mathbf{H}$ satisfy the equations. Here, β is a constant. Apart from a few constants and a minus sign, the fields $\mathbf{E}$ and $\mathbf{H}$ can therefore be exchanged.

P12-3. Reexpress Eqs. (12-14)–(12-17) by eliminating $\mathbf{D}$ and $\mathbf{H}$ in favor of $\mathbf{E}$ and $\mathbf{B}$, thereby displaying explicitly where $\mathbf{P}$ and $\mathbf{M}$ appear in the equations.

P12-4. Starting with Maxwell's equations, derive a law for the force between two point charges in a linear dielectric. *Hint*: Use Gauss's law in integral form.

P12-5. A conductor of arbitrary shape is surrounded by a linear dielectric. Show that, provided the fields are static, the *free* surface charge density $\sigma(P)$ at a point P on the surface of the conductor is given by $\sigma(P) = \mathbf{D}(P)\cdot\hat{\mathbf{n}}(P)$, where $\hat{\mathbf{n}}(P)$ is a unit vector normal to the conductor at P.

P12-6. (a) Show that in linear matter with no free charge or currents all Cartesian components of the electric field and all Cartesian components of the magnetic induction satisfy the homogeneous wave equation. *Hint*: Take the curl of Eqs. (12-15) and (12-16) and use Eq. (C-19). (b) What is the speed of propagation of an electromagnetic wave in linear matter characterized by a permittivity ϵ and a permeability μ? Predict the index of refraction n for this matter, where n is defined as the ratio of the speed of light in vacuum to the speed of light in the matter.

12-2 The Equation of Continuity

In the next several sections, we shall consider how the present form of Maxwell's equations alters various expressions with which we are already familiar in the form applicable when matter is not present. Consider first the equation of continuity. Just as in the absence of matter, this equation expressing conservation of charge can be derived from Maxwell's equations. Since the divergence of a curl is automatically zero, the divergence of Eq. (12-16) yields the result

$$\nabla\cdot\mathbf{J} + \frac{\partial}{\partial t}(\nabla\cdot\mathbf{D}) = 0 \tag{12-23}$$

Substitution from Eq. (12-14) then gives

$$\oint \mathbf{J}\cdot d\mathbf{S} = -\int \frac{\partial\rho}{\partial t}\,dv \qquad \nabla\cdot\mathbf{J} + \frac{\partial\rho}{\partial t} = 0 \tag{12-24}$$

which has the same form as the equation of continuity in vacuum except that ρ and $\mathbf{J}$ now refer to the *free* charge density and *free* current density. Because of their origin, the bound charge and current distributions automatically satisfy a conservation law and consequently Eq. (12-24) does not include that component of the charge distribution (P12-7).

PROBLEM

P12-7. Explain why the equation of continuity involves only *free* charges and currents. *Hint*: Write equations of continuity for the polarization and magnetization currents, and show that these equations are automatically satisfied.

12-3 The Energy Theorem

An equation that can be interpreted as an energy theorem can be derived from Maxwell's equations even when matter is present. Following the pattern of Section 6-4, we subtract the dot product of Eq. (12-16) with $\mathbf{E}$ from the dot product of Eq. (12-15) with $\mathbf{H}$ and use Eq. (C-12) to find that

$$\mathbf{H}\cdot\nabla\times\mathbf{E} - \mathbf{E}\cdot\nabla\times\mathbf{H} = \nabla\cdot(\mathbf{E}\times\mathbf{H}) = -\mathbf{J}\cdot\mathbf{E} - \mathbf{E}\cdot\frac{\partial\mathbf{D}}{\partial t} - \mathbf{H}\cdot\frac{\partial\mathbf{B}}{\partial t} \tag{12-25}$$

We then integrate Eq. (12-25) over some volume and use the divergence theorem to obtain

$$-\oint_{\Sigma}(\mathbf{E}\times\mathbf{H})\cdot d\mathbf{S} = \int_V \mathbf{J}\cdot\mathbf{E}\,dv + \int_V\left(\mathbf{E}\cdot\frac{\partial\mathbf{D}}{\partial t} + \mathbf{H}\cdot\frac{\partial\mathbf{B}}{\partial t}\right)dv \tag{12-26}$$

Each of the terms in this result can be given an interpretation similar to that of the corresponding term in Eq. (6-51). Since, for example, presence of matter does not alter the relationship between the fields $\mathbf{E}$ and $\mathbf{B}$ and the forces experienced by charged particles, the arguments associated with Eq. (3-39) apply also to the present case and $\int \mathbf{J}\cdot\mathbf{E}\,dv$ still represents the rate at which work is done by the fields on the particles in the volume over which the integral extends.

The second integral on the right in Eq. (12-26) is more difficult to interpret. *If* the medium is *linear*, so that $\mathbf{D} = \epsilon\mathbf{E}$ and $\mathbf{B} = \mu\mathbf{H}$, with ϵ and μ constants, then

$$\mathbf{E}\cdot\frac{\partial\mathbf{D}}{\partial t} = \epsilon\mathbf{E}\cdot\frac{\partial\mathbf{E}}{\partial t} = \frac{1}{2}\frac{\partial}{\partial t}(\mathbf{E}\cdot\epsilon\mathbf{E}) = \frac{\partial}{\partial t}\left(\frac{1}{2}\mathbf{E}\cdot\mathbf{D}\right) \tag{12-27}$$

and similarly

$$\mathbf{H}\cdot\frac{\partial\mathbf{B}}{\partial t} = \frac{\partial}{\partial t}\left(\frac{1}{2}\mathbf{H}\cdot\mathbf{B}\right) \tag{12-28}$$

For these media, the final term in Eq. (12-26) assumes the form

$$\frac{\partial}{\partial t}\int_V\left(\frac{1}{2}\mathbf{E}\cdot\mathbf{D}+\frac{1}{2}\mathbf{H}\cdot\mathbf{B}\right)dv \tag{12-29}$$

from which by analogy with the similar term in Eq. (6-51) we are led to associate the energy density

$$u_{\mathrm{EM}}=\tfrac{1}{2}\mathbf{E}\cdot\mathbf{D}+\tfrac{1}{2}\mathbf{H}\cdot\mathbf{B} \tag{12-30}$$

with the electromagnetic field. Further, we are led to interpret the final term in Eq. (12-26) as the rate at which energy stored in the fields is increasing. Equations (12-29) and (12-30) remain valid even for those anisotropic media in which the components of **D** are related to those of **E** by a symmetric permittivity *tensor* ϵ_{ij} through the equation $D_i=\sum_j \epsilon_{ij}E_j$, where ϵ_{ij} is time-independent and field-independent but may depend on the spatial coordinates (P12-8). Even in more general situations for which an energy density cannot be identified, the final term in Eq. (12-26) is still interpreted as the rate at which energy stored in the fields in a particular volume is increasing.

Accepting these interpretations, we conclude that the right-hand side of Eq. (12-26) is the total energy per unit time that must be coming into the volume over which the integrals extend. Since we have (tacitly) assumed that there are no sources or sinks of energy within any volume, the power input represented by the right-hand side of Eq. (12-26) must arise from a flow of energy across the boundary of the volume. Conveniently, the left-hand side of Eq. (12-26) involves a surface integral, and we therefore identify that integral with the rate at which energy is transported into the volume by virtue of energy flow across the bounding surface. Replacing the minus sign with a plus sign, we conclude that

$$\begin{pmatrix}\text{rate at which energy flows out of } V\\ \text{across the bounding surface } \Sigma\end{pmatrix}=\oint_\Sigma(\mathbf{E}\times\mathbf{H})\cdot d\mathbf{S} \tag{12-31}$$

which in turn leads to the identification of the vector $\mathbf{S}$

$$\mathbf{S}=\mathbf{E}\times\mathbf{H} \tag{12-32}$$

—still called the *Poynting vector*—as a vector representing energy transport by the electromagnetic fields. It is common to interpret the vector $\mathbf{S}$ as representing a density of energy flow across surfaces, but there has been considerable discussion about the appropriateness and correctness of this interpretation. [See the remarks following Eq. (6-55) and the references given there.] Note, however, that this interpretation of the Poynting vector has not been specific to any particular medium; it applies whatever form the constitutive relations may happen to assume. Note also that energy transport in an electromagnetic field is most directly discussed by using the vectors **E** and **H**.

PROBLEMS

P12-8. Show that Eq. (12-27) applies if the relationship between the components E_j of **E** and D_i of **D** is given by $D_i=\sum_j \epsilon_{ij}E_j$, where ϵ_{ij} is a time-independent and field-independent but spatially dependent *symmetric* second-rank tensor. In fact, an argument based on conservation of energy can be constructed to show that ϵ_{ij} is *necessarily* symmetric.

P12-9. Suppose that $\mathbf{B} = \mathbf{B}(\mathbf{H})$ as in ferromagnetic materials. Show from the last term in Eq. (12-26) that the energy required to change the fields in a ferromagnetic material is given by $\int \mathbf{H} \cdot d\mathbf{B}$ and compare this result with Eq. (11-86). Carefully enumerate any assumptions made.

P12-10. A dielectric slab of permittivity ϵ fits snugly (but moves without friction) between the (rectangular) plates of a parallel plate capacitor (P4-16). Let the plates have separation d, length a, and width b, and let them be maintained at a constant potential difference ΔV by a battery. Finally, suppose the slab is partially removed by sliding it a distance x parallel to the edge of length a. Calculate the force on the dielectric slab and determine whether the force pulls the slab into the plates or repels it still further. *Hints*: (1) Neglect fringing. (2) Imagine a small, virtual displacement and calculate the change in energy. (3) Equate that change to the work done by the force on the slab, after making suitable correction for any work done by the battery to maintain the specified potential.

12-4 The Momentum Theorem

In Section 6-5, the momentum to be associated with an electromagnetic field in free space was inferred by considering the forces experienced by a charge and current distribution by virtue of the interaction of that distribution with the fields that the distribution itself establishes. Although in the presence of matter, the force $\mathbf{F}$ experienced by a charge and current distribution in an electromagnetic field is still given by

$$\mathbf{F} = \int_V [\rho \mathbf{E} + \mathbf{J} \times \mathbf{B}] \, dv \tag{12-33}$$

and this force is still the rate of change of the linear momentum $\mathbf{P}$ of the matter within the volume over which the integral extends, Maxwell's equations *with no restrictions on the constitutive properties of whatever matter is present* lead now to the result

$$\frac{d}{dt}\left(\mathbf{P} + \int_V (\mathbf{D} \times \mathbf{B}) \, dv\right) = \int_V [\mathbf{E}(\nabla \cdot \mathbf{D}) - \mathbf{D} \times (\nabla \times \mathbf{E}) - \mathbf{B} \times (\nabla \times \mathbf{H})] \, dv \tag{12-34}$$

instead of to Eq. (6-59). Because the integral under the time derivative appears on the same footing as the momentum $\mathbf{P}$, Eq. (12-34) suggests that the fields in the volume V should be assigned a momentum density $\mathcal{G}$ given by

$$\mathcal{G} = \mathbf{D} \times \mathbf{B} \tag{12-35}$$

Since no restricting assumptions have been made about the properties of whatever matter is present, this identification is independent of how **D** and **B** are related to **E** and **H**. Thus, the momentum density $\mathcal{G}$ (determined from **D** and **B**) and the Poynting vector $\mathbf{S}$ (determined from **E** and **H**) are in general *not* simply related. If, however, the medium involved is *linear*, then

$$\mathcal{G} = \epsilon \mathbf{E} \times \mu \mathbf{H} = \epsilon\mu \mathbf{E} \times \mathbf{H} = \epsilon\mu \mathbf{S} \tag{12-36}$$

which reduces to Eq. (6-62) in vacuum where $\epsilon = \epsilon_0$ and $\mu = \mu_0$.

An alternative expression for the quantity appearing on the right-hand side in Eq. (12-34) is explored in P12-28.

12-5 On Which Fields Are Basic

Four fields—**E**, **D**, **B**, **H**—have been introduced but only two of these four are independent. The other two are determined by Eqs. (12-18) and (12-19). We have so far regarded **E** and **B** as basic and **D** and **H** as auxiliary. Now that we have seen how these fields enter into various aspects of the general formalism, we point out that the choice of which fields to regard as basic is in part arbitrary, this choice being primarily determined by the circumstances of the problem at hand. If, for example, we are interested in

———forces on charged particles, we take **E** and **B** as basic.
———response of matter, we take **E** and **H** as basic.
———energy transport, we take **E** and **H** as basic.
———momentum in the fields, we take **D** and **B** as basic.
———field problems when bound charge and current distributions are initially unknown, we take **D** and **H** as basic.
———symmetric discussion of Maxwell's equations, we take either **D** and **B** or **E** and **H** as basic.

Perhaps other situations could be added to this list. These few observations should, however, be sufficient to point out that the choice of basic fields is as much a matter of taste and convenience as of the intrinsic theory.

12-6 The Potentials

Even when matter (of whatever type) is present, it is possible to express the fields **E** and **B** in terms of scalar and vector potentials V and **A**. The arguments are identical with those presented to support Eqs. (6-63)–(6-65). In essence,

$$\nabla \cdot \mathbf{B} = 0 \Longrightarrow \mathbf{B} = \nabla \times \mathbf{A} \tag{12-37}$$

and then

$$\nabla \times \mathbf{E} = -\frac{\partial \mathbf{B}}{\partial t} \Longrightarrow \nabla \times \left(\mathbf{E} + \frac{\partial \mathbf{A}}{\partial t}\right) = 0$$

$$\Longrightarrow \mathbf{E} + \frac{\partial \mathbf{A}}{\partial t} = -\nabla V$$

$$\Longrightarrow \mathbf{E} = -\nabla V - \frac{\partial \mathbf{A}}{\partial t} \tag{12-38}$$

[See items (3) and (4) in Section 2-5.] Further, the fields continue to be invariant to the gauge transformation

$$\mathbf{A}_2 = \mathbf{A}_1 + \nabla \Lambda, \qquad V_2 = V_1 - \frac{\partial \Lambda}{\partial t} \tag{12-39}$$

[compare Eqs. (6-66) and (6-68)] and the divergence of **A** remains arbitrary.

As in Section 6-6, two of Maxwell's equations—Eqs. (12-15) and (12-17)—are automatically satisfied when the fields are expressed in terms of the potentials **A** and V; the remaining two generate differential equations whose solutions determine the potentials. When matter is present, however, this second pair of equations—Eqs. (12-14) and (12-16)—involves the fields **D** and **H**, which are *not* given directly by the potentials. Thus, the form of the equations for **A** and V depends on the form of the relationships between **D** and **E** and between **H** and **B**. If **D** and **B** are linearly related to **E** and **H**, as in Eq. (12-22), then combining the potentials with Eqs. (12-14) and (12-16) ultimately yields that

$$\left(\nabla^2 - \mu\epsilon\frac{\partial^2}{\partial t^2}\right)\mathbf{A} = -\mu\mathbf{J} \tag{12-40}$$

and that

$$\left(\nabla^2 - \mu\epsilon\frac{\partial^2}{\partial t^2}\right)V = -\frac{\rho}{\epsilon} \tag{12-41}$$

provided the potentials satisfy the Lorentz condition

$$\nabla \cdot \mathbf{A} + \mu\epsilon\frac{\partial V}{\partial t} = 0 \tag{12-42}$$

[Compare Eqs. (6-72), (6-73), and (6-71).] For more general relationships between **D** and **E** and between **H** and **B**, the resulting equations for **A** and V may be more complicated, but they can always be obtained by substituting Eqs. (12-37) and (12-38) into Eqs. (12-18) and (12-19) and then substituting those results into Eqs. (12-14) and (12-16).

Although the vector and scalar potentials **A** and V are the only potentials that can always be identified without imposing special conditions on the fields, the sources, the polarization, or the magnetization, other potentials with more limited applicability can be identified. We have, for example, already introduced the magnetic scalar potential, which is useful for static fields when $\mathbf{J} = 0$ (for then $\nabla \times \mathbf{H} = 0$, which implies that $\mathbf{H} = -\nabla V^{(m)}$). One other potential that is occasionally useful is explored in P12-12.

PROBLEMS

P12-11. Verify that Eqs. (12-40) and (12-41) for the potentials in homogeneous, linear matter follow from Eqs. (12-14) and (12-16) provided that the condition in Eq. (12-42) applies.

P12-12. Consider a situation in which there are no free charges or currents but the polarization throughout all space is a specified function of position and time. Let the permeability be that of free space. Show that all of Maxwell's equations are satisfied if we introduce a single potential $\mathbf{\Pi}$, called the *Hertz potential* or the *polarization potential*, in terms of which

$$\mathbf{D} = \epsilon_0 \nabla \times (\nabla \times \mathbf{\Pi}) \qquad \mathbf{B} = \mu_0\epsilon_0 \nabla \times \frac{\partial \mathbf{\Pi}}{\partial t}$$

provided $\mathbf{\Pi}$ satisfies

$$\left(\nabla^2 - \mu_0\epsilon_0\frac{\partial^2}{\partial t^2}\right)\mathbf{\Pi} = -\frac{\mathbf{P}}{\epsilon_0}$$

Show also that the fields can be derived from the vector and scalar potentials

$$\mathbf{A} = \mu_0\epsilon_0\frac{\partial \mathbf{\Pi}}{\partial t}, \qquad V = -\nabla\cdot\mathbf{\Pi}$$

and that these potentials satisfy the condition in Eq. (12-42) with $\mu\epsilon = \mu_0\epsilon_0$.

12-7 Boundary Conditions at Discontinuities in the Medium

Because Maxwell's equations are differential equations, the complete statement of a problem in electromagnetism requires the specification of conditions that acceptable solutions must satisfy at the boundaries of the domain of the problem. In one-dimensional problems, these conditions are adequately stated by giving the value of the solution and/or sometimes the value of its derivative at *both* ends of the region of interest. In n dimensions, values and maybe derivatives must be specified over the $(n - 1)$-dimensional *surface* that bounds the n-dimensional volume in which the solution is sought. If, for example, the region of interest is the interior of a sphere of radius a, the boundary condition might be that the electric field vanish on the boundary, expressed analytically in spherical coordinates by the equation

$$\mathbf{E}(a, \theta, \phi) = 0 \tag{12-43}$$

If the volume of interest happens to be infinite, the bounding surface is a surface at infinity. In such cases the boundary condition may be stated not by giving a value but by requiring that the fields approach a particular asymptotic limit as the boundary at infinity is approached; for example, we might require that

$$\mathbf{E}(r, \theta, \phi) \longrightarrow \frac{1}{r^2}\hat{\mathbf{r}} \tag{12-44}$$

as $r \longrightarrow \infty$. The particular character of boundary conditions of this type is very specific to the problem being solved and we can say little more without specifying the problem more explicitly.

In principle, the differential form of Maxwell's equations and the appropriate conditions at the boundaries of some region are sufficient to determine the fields in this region. In fact, such a formulation of an electromagnetic problem is suitable only if the properties of space in the region of interest vary smoothly from point to point. In many problems of interest, however, a boundary in the form of an interface between two media having different properties occurs *within* the volume of the problem. If we can regard this boundary to be spread out over some region and if we can regard the properties of the medium to change smoothly in some *known* way from one side of this region to the other, then Maxwell's equations can be integrated across the boundary with no difficulty. Despite its physical impossibility, however, it is much more convenient to regard an interface between two different media (one of which may be vacuum) to be abrupt. For such an idealistic interface, the differential form of Maxwell's equations cannot be easily integrated across the interface and we must find another way to express the connection between the fields on one side of the interface and the fields on the other side of the interface. As we shall show in this section, the *integral* form of Maxwell's equations contains the necessary information.

Consider, then, an interface between two different media (Fig. 12-1). In a sufficiently small region about a selected point P on this interface, the boundary between the two media can be regarded as plane. Introduce a unit vector $\hat{\mathbf{n}}$ normal to the interface at the point P and, for definiteness, direct the normal vector *from* medium 1 *to* medium 2. Finally, imagine a small pillbox with its two plane faces perpendicular to the unit vector $\hat{\mathbf{n}}$ and its remaining side perpendicular to the interface. Let the plane faces of this pillbox have area ΔS and let the pillbox enclose a volume Δv. Now, apply Gauss's law, Eq. (12-14), to this pillbox. The left-hand side of this equation has the more explicit evaluation

$$\oint \mathbf{D}\cdot d\mathbf{S} \approx \mathbf{D}_2(P)\cdot\hat{\mathbf{n}}\,\Delta S + \mathbf{D}_1(P)\cdot(-\hat{\mathbf{n}})\,\Delta S + \begin{pmatrix}\text{contribution from}\\ \text{the cylinder}\end{pmatrix} \tag{12-45}$$

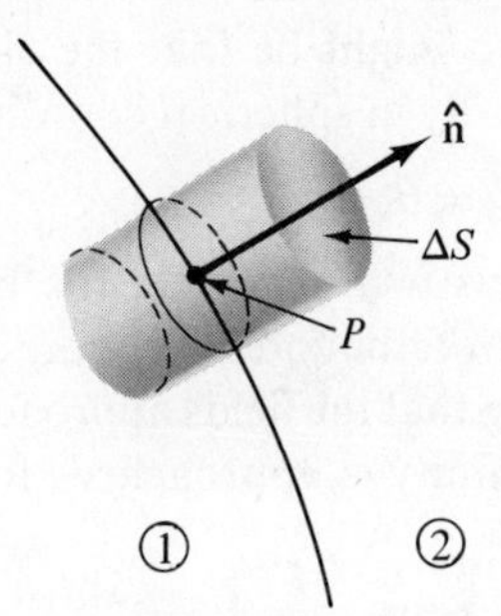

Fig. 12-1. Interface between two media. The pillbox used to obtain boundary conditions on the normal components of **D** and **B** is shown.

where $\mathbf{D}_1(P)$ is the displacement vector at P in medium 1 and $\mathbf{D}_2(P)$ is the displacement vector at P in medium 2; the right-hand side has the evaluation

$$\begin{aligned}\int \rho\, dv &= \text{total free charge in pillbox} \\ &\approx \rho_1(P)\,\Delta v_1 + \rho_2(P)\,\Delta v_2 + \sigma(P)\,\Delta S \end{aligned} \qquad (12\text{-}46)$$

where $\rho_1(P)$ and $\rho_2(P)$ are the *free* volume charge densities at P in medium 1 and medium 2, $\sigma(P)$ is the *free* surface charge density at P, and Δv_1 and Δv_2 are the portions of Δv lying in medium 1 and in medium 2, respectively. All volumes and surfaces have been assumed small enough to justify treating the fields and charge densities as constants throughout the region of integration. Substituting Eqs. (12-45) and (12-46) into Eq. (12-14), we find that

$$\begin{aligned}[\mathbf{D}_2(P)\cdot\hat{\mathbf{n}} - \mathbf{D}_1(P)\cdot\hat{\mathbf{n}}]\,\Delta S &+ (\text{contribution from the cylinder}) \\ &\approx \rho_1(P)\,\Delta v_1 + \rho_2(P)\,\Delta v_2 + \sigma(P)\,\Delta S \end{aligned} \qquad (12\text{-}47)$$

Now let the two faces of the pillbox move arbitrarily close to the interface, each remaining on its own side of the interface. In that limit, Δv_1 and Δv_2 approach zero and the contribution to the flux from the cylindrical wall of the pillbox also disappears. In this limit, Eq. (12-47) becomes

$$\mathbf{D}_2(P)\cdot\hat{\mathbf{n}} - \mathbf{D}_1(P)\cdot\hat{\mathbf{n}} = \sigma(P) \qquad (12\text{-}48)$$

Thus, the *normal component* of the *displacement vector* is discontinuous across an interface by an amount equal to the *free* charge density on the interface. If this charge density is zero, then the normal component of the displacement vector is continuous across the boundary.

An essentially identical argument that begins with the application of the integral form of the magnetic flux law, Eq. (12-17), to the pillbox of Fig. 12-1 leads finally to the conclusion that

$$\mathbf{B}_2(P)\cdot\hat{\mathbf{n}} - \mathbf{B}_1(P)\cdot\hat{\mathbf{n}} = 0 \qquad (12\text{-}49)$$

Thus, the *normal component* of the *magnetic induction field* is *continuous* across the interface. Because no free magnetic charge exists, the normal component of the magnetic induction field is never discontinuous.

To obtain two additional equations relating the fields on opposite sides of an interface, we apply the remaining two Maxwell equations (in integral form) to a rectangular path whose plane contains the unit vector $\hat{\mathbf{n}}$ normal to the interface at point P (Fig. 12-2). Let the path enclose a (plane) area that is small enough so that two of its sides can be regarded as perpendicular to the interface and two as parallel to the interface, and let $\hat{\mathbf{t}}$ be a unit vector tangent to the surface and in the plane of the rectangle (which may be any plane containing the vector $\hat{\mathbf{n}}$). Now, apply Faraday's law, Eq. (12-15), to this path. The left-hand side of this equation has the evaluation

$$\oint \mathbf{E}\cdot d\boldsymbol{\ell} \approx \mathbf{E}_2(P)\cdot\hat{\mathbf{t}}\,\Delta\ell + \mathbf{E}_1(P)\cdot(-\hat{\mathbf{t}})\,\Delta\ell + \begin{pmatrix}\text{contribution} \\ \text{from sides 2 and 4}\end{pmatrix} \qquad (12\text{-}50)$$

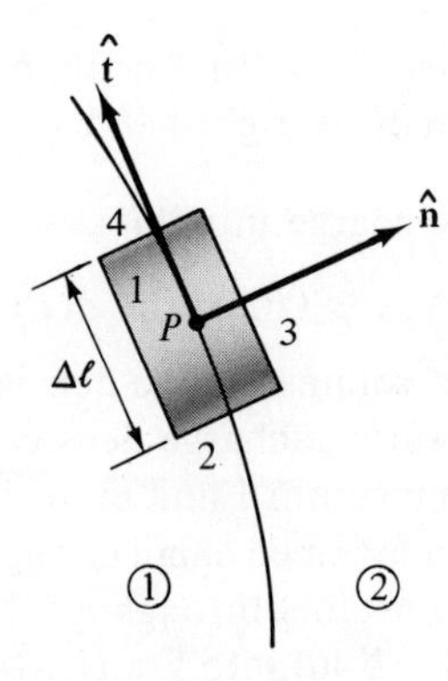

Fig. 12-2. Interface between two media. The path used to obtain boundary conditions on the tangential components of **E** and **H** is shown.

and the right-hand side has the evaluation

$$-\int \frac{\partial \mathbf{B}}{\partial t} \cdot d\mathbf{S} \approx (\text{some finite factor})\Delta S \tag{12-51}$$

Substituting Eqs. (12-50) and (12-51) into Faraday's law, we find that

$$[\mathbf{E}_2(P)\cdot\hat{\mathbf{t}} - \mathbf{E}_1(P)\cdot\hat{\mathbf{t}}]\,\Delta\ell + \begin{pmatrix}\text{contribution}\\ \text{from sides 2 and 4}\end{pmatrix} = \begin{pmatrix}\text{some finite}\\ \text{factor}\end{pmatrix}\Delta S \tag{12-52}$$

Here, the line integral has been evaluated by traversing the path in a direction that makes $\hat{\mathbf{n}} \times \hat{\mathbf{t}}$ the direction to assign to the surface bounded by the path. Now let the two sides parallel to the interface move arbitrarily close to the interface, each remaining on its own side of the interface. In this limit, the surface area ΔS goes to zero and the contribution to the line integral from sides 2 and 4 of the path also disappears; Eq. (12-52) yields

$$\mathbf{E}_2(P)\cdot\hat{\mathbf{t}} - \mathbf{E}_1(P)\cdot\hat{\mathbf{t}} = 0 \tag{12-53}$$

where $\hat{\mathbf{t}}$ is perpendicular to $\hat{\mathbf{n}}$ but is otherwise arbitrary. Thus, *all tangential* components of the electric field (and particularly the two tangential components in mutually orthogonal directions) are continuous across an interface between two different media. The alternative expression

$$\hat{\mathbf{n}} \times [\mathbf{E}_2(P) - \mathbf{E}_1(P)] = 0 \tag{12-54}$$

of this condition combines Eq. (12-53) for all $\hat{\mathbf{t}}$ into a single *vector* expression and is derived in P12-14.

Finally, application of the circuital law, Eq. (12-16), to the path shown in Fig. 12-2 gives

$$\mathbf{H}_2(P)\cdot\hat{\mathbf{t}}\,\Delta\ell - \mathbf{H}_1(P)\cdot\hat{\mathbf{t}}\,\Delta\ell + \begin{pmatrix}\text{contribution}\\ \text{from sides 2 and 4}\end{pmatrix}$$
$$= \begin{pmatrix}\text{some finite}\\ \text{factor}\end{pmatrix}\Delta S + \mathbf{j}\cdot(\hat{\mathbf{n}} \times \hat{\mathbf{t}})\,\Delta\ell \tag{12-55}$$

where the final term represents a contribution to the current across the surface bounded by the path from a possible *free surface* current on the interface between the two media. Again moving the two sides that are parallel to the interface arbitrarily close to the interface, we can ignore the contributions from sides 2 and 4 and also the contribution multiplied by ΔS in Eq. (12-55); Eq. (12-55) then yields that

$$\mathbf{H}_2(P)\cdot\hat{\mathbf{t}} - \mathbf{H}_1(P)\cdot\hat{\mathbf{t}} = \mathbf{j}\cdot(\hat{\mathbf{n}} \times \hat{\mathbf{t}}) \tag{12-56}$$

which applies for any unit vector $\hat{\mathbf{t}}$ tangent to the interface at the point P. Thus, the two *tangential* components of the magnetic field intensity may be discontinuous across an interface, but will be discontinuous only if there is a free surface current in the interface. The alternative expression

$$\hat{\mathbf{n}} \times [\mathbf{H}_2(P) - \mathbf{H}_1(P)] = \mathbf{j} \tag{12-57}$$

of Eq. (12-56) is derived in P12-14. We remind the reader that, throughout this section, $\hat{\mathbf{n}}$ is a normal vector directed *from* medium 1 *to* medium 2.

Although not related directly to the fields, one additional condition, obtained by applying the equation of continuity, Eq. (12-24), to the pillbox in Fig. 12-1, should be included in this section; we find that

$$\mathbf{J}_2(P)\cdot\hat{\mathbf{n}}\,\Delta S + \mathbf{J}_1(P)\cdot(-\hat{\mathbf{n}}\,\Delta S) + \begin{pmatrix}\text{contribution}\\ \text{from the cylinder}\end{pmatrix}$$
$$\approx -\frac{\partial}{\partial t}[\rho_1(P)\,\Delta v_1 + \rho_2(P)\,\Delta v_2 + \sigma(P)\,\Delta S] \tag{12-58}$$

which in turn results in the condition

$$\mathbf{J}_2(P)\cdot\hat{\mathbf{n}} - \mathbf{J}_1(P)\cdot\hat{\mathbf{n}} = -\frac{\partial\sigma(P)}{\partial t} \tag{12-59}$$

when the plane surfaces of the pillbox are allowed to approach arbitrarily close to the interface.

A summary of the boundary conditions that must be imposed on the fields at an abrupt interface between two different media follows. In this summary, we suppress explicit indication of the point P on the boundary at which the fields are evaluated. Further, the vector $\hat{\mathbf{n}}$ is a unit vector directed *from* medium 1 *to* medium 2. The relations expressed in this summary have been derived without reference to any special or restrictive conditions and hence are perfectly general (although if the fields vary at very high frequencies, complications will probably arise). In general terms, the necessary boundary conditions are

$$(\mathbf{D}_2 - \mathbf{D}_1)\cdot\hat{\mathbf{n}} = \sigma \qquad \begin{pmatrix}\text{normal component of }\mathbf{D}\\ \text{discontinuous by the free}\\ \text{surface charge density}\end{pmatrix} \tag{12-60}$$

$$\hat{\mathbf{n}} \times (\mathbf{E}_2 - \mathbf{E}_1) = 0 \qquad \begin{pmatrix}\text{tangential components of}\\ \mathbf{E}\text{ continuous}\end{pmatrix} \tag{12-61}$$

$$(\mathbf{B}_2 - \mathbf{B}_1)\cdot\hat{\mathbf{n}} = 0 \qquad \begin{pmatrix}\text{normal component of}\\ \mathbf{B}\text{ continuous}\end{pmatrix} \tag{12-62}$$

$$\hat{\mathbf{n}} \times (\mathbf{H}_2 - \mathbf{H}_1) = \mathbf{j} \qquad \begin{pmatrix}\text{tangential components of}\\ \mathbf{H}\text{ discontinuous by free}\\ \text{surface current density}\end{pmatrix} \tag{12-63}$$

$$(\mathbf{J}_2 - \mathbf{J}_1)\cdot\hat{\mathbf{n}} = -\frac{\partial\sigma}{\partial t} \quad \begin{pmatrix}\text{normal component of } \mathbf{J} \\ \text{discontinuous by rate at} \\ \text{which free charge builds} \\ \text{up on surface}\end{pmatrix} \tag{12-64}$$

In terms of a specific Cartesian coordinate system whose z axis coincides with the vector $\hat{\mathbf{n}}$ and whose x and y axes are tangent to the interface at P, the boundary conditions in Eqs. (12-60)–(12-64) have the more explicit expressions

$$\begin{aligned} E_{2x} - E_{1x} &= 0 \qquad & E_{2y} - E_{1y} &= 0 \qquad & D_{2z} - D_{1z} &= \sigma \\ H_{2x} - H_{1x} &= j_y \qquad & H_{2y} - H_{1y} &= -j_x \qquad & B_{2z} - B_{1z} &= 0 \\ & & & & J_{2z} - J_{1z} &= -\frac{\partial\sigma}{\partial t} \end{aligned} \tag{12-65}$$

Although the conditions on the normal and tangential components do not involve the same fields, either in the electric or the magnetic domain, **E** and **D** and also **B** and **H** are related on each side of the interface by the properties of the corresponding medium. Taking those relationships into account, we therefore have conditions on all three components of all four field vectors and Eqs. (12-60)–(12-63) fully determine the fields on one side of an interface from those on the other side of the interface.

We shall mention two remaining boundary conditions. When the fields are *static*, the difference in the *scalar* potential V between two points gives the amount of work required to move a unit charge from one point to the other. Since physically the electric field cannot become infinite at an interface, the amount of work required to move a unit charge an infinitesimal distance from one side of the interface to the other must be infinitesimal, and consequently, with P a point on the interface,

$$V_2(P) - V_1(P) = 0 \tag{12-66}$$

i.e., the electro*static* potential must be continuous across the interface between two different media. In a way, this condition is not independent of Eq. (12-61), for Eq. (12-61) implies the validity of Eq. (12-66) at *all* points P if the potential is continuous at but a *single* point. (Why?) It almost goes without saying that, when the fields are *static* and the media are *linear*, the condition on the normal component of **D** is equivalent to the condition

$$[\epsilon_2 \nabla V_2(P) - \epsilon_1 \nabla V_1(P)]\cdot\hat{\mathbf{n}} = -\sigma \tag{12-67}$$

on the normal derivative of the potential. (Why?)

To illustrate the use of these boundary conditions, suppose two linear dielectric media with permittivities ϵ_1 and ϵ_2 meet in a plane interface (Fig. 12-3). Let the electric fields $\mathbf{E}_1$ and $\mathbf{E}_2$ lie in the plane of the paper and make angles θ_1 and θ_2 with the normal to the interface. We seek a relationship between θ_1 and θ_2 when there is no free charge on the interface. Continuity of

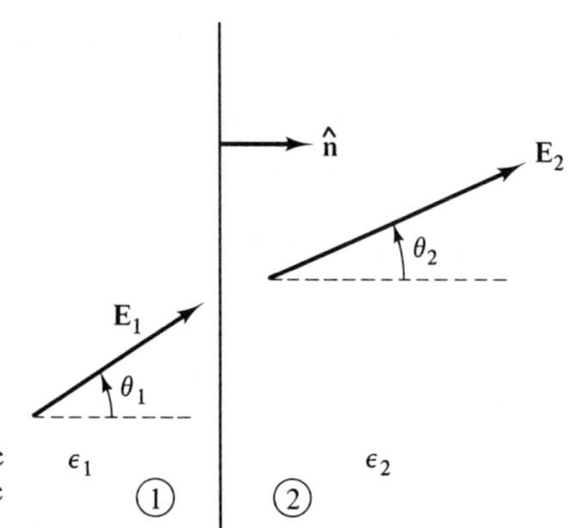

Fig. 12-3. Refraction of a static electric field at a plane dielectric interface.

the normal component of $\mathbf{D} = \epsilon\mathbf{E}$ then requires that

$$\epsilon_1 E_1 \cos\theta_1 = \epsilon_2 E_2 \cos\theta_2 \tag{12-68}$$

and continuity of the tangential component of $\mathbf{E}$ requires that

$$E_1 \sin\theta_1 = E_2 \sin\theta_2 \tag{12-69}$$

Here, $E_i = |\mathbf{E}_i|$. Thus, dividing Eq. (12-68) by Eq. (12-69), we find that θ_1 and θ_2 are related by

$$\epsilon_1 \cot\theta_1 = \epsilon_2 \cot\theta_2 \tag{12-70}$$

regardless of the strength of the fields. In effect, an electric field is "refracted" at a dielectric interface.

PROBLEMS

P12-13. The way Eq. (12-48) is written, exchange of what we call medium 1 with what we call medium 2, which interchanges $\mathbf{D}_1$ and $\mathbf{D}_2$, seems to change the sign on the left side but will certainly not affect the right side. Explain where this argument is incomplete.

P12-14. Derive (a) Eq. (12-54) from Eq. (12-53) and (b) Eq. (12-57) from Eq. (12-56). *Hint*: See P0-5 and P0-7.

P12-15. Derive Eq. (12-65) from Eqs. (12-60)–(12-64).

P12-16. Solve Eq. (12-70) for θ_2 as a function of θ_1 and, using a computer to obtain the necessary data, plot graphs of θ_2 versus θ_1 over the range $0 \leq \theta_1 \leq 90°$ for various values of ϵ_1/ϵ_2. Include $\epsilon_1/\epsilon_2 < 1$, $= 1$, and > 1 in your study.

P12-17. Verify that all applicable boundary conditions are satisfied by your solution to P10-18.

P12-18. Suppose that the media in Fig. 12-3 are linear magnetic media with permeabilities μ_1 and μ_2. Replace $\mathbf{E}_1$ and $\mathbf{E}_2$ by $\mathbf{B}_1$ and $\mathbf{B}_2$ and let there be no free surface currents on the interface. Find a relationship analogous to Eq. (12-70). *Optional*: Explore this relationship after the pattern suggested in P12-16.

P12-19. Suppose that the toroidal core in Fig. 11-9 has a narrow gap of length d as measured along a center line of the toroid. Determine **H** and **B** for points along this center line, both inside and outside the toroid. State clearly any assumptions made. *Hint*: Ignore fringing in the gap.

12-8 Static Potentials

The electro*static* field in a linear dielectric statisfies $\nabla \times \mathbf{E} = 0$ and $\nabla \cdot (\epsilon \mathbf{E}) = \rho$, the first of which implies the existence of a scalar potential in terms of which $\mathbf{E} = -\nabla V$ and the second of which then requires that this potential satisfy Poisson's equation,

$$\nabla^2 V = -\frac{\rho}{\epsilon} \tag{12-71}$$

Further, the electro*static* field in a region containing a *uniformly* polarized dielectric ($\mathbf{P}$ = constant) satisfies $\nabla \times \mathbf{E} = 0$ and $\nabla \cdot \mathbf{D} = \nabla \cdot (\epsilon_0 \mathbf{E} + \mathbf{P}) = \nabla \cdot (\epsilon_0 \mathbf{E}) = \rho$ (since $\nabla \cdot \mathbf{P} = 0$), the first of which again implies that $\mathbf{E} = -\nabla V$ and the second of which then requires that V satisfy Eq. (12-71) with ϵ replaced by ϵ_0. Yet again, the magneto*static* field intensity in a region containing linear matter but no free currents satisfies $\nabla \times \mathbf{H} = 0$ and $\nabla \cdot (\mu \mathbf{H}) = 0$, the first of which implies the existence of a (magnetic) scalar potential $V^{(m)}$ in terms of which $\mathbf{H} = -\nabla V^{(m)}$ and the second of which then requires that this potential satisfy Laplace's equation

$$\nabla^2 V^{(m)} = 0 \tag{12-72}$$

Finally, the magneto*static* field in a region containing *uniformly* magnetized matter ($\mathbf{M}$ = constant) but no free currents satisfies $\nabla \times \mathbf{H} = 0$ and $\nabla \cdot \mathbf{B} = \nabla \cdot [\mu_0(\mathbf{H} + \mathbf{M})] = \nabla \cdot \mu_0 \mathbf{H} = 0$ (since $\nabla \cdot \mathbf{M} = 0$), the first of which again implies that $\mathbf{H} = -\nabla V^{(m)}$ and the second of which then requires $V^{(m)}$ to satisfy Eq. (12-72). (See also P9-4.) Thus, many electro*static* and magneto*static* problems *in matter* reduce to solving Poisson's or Laplace's equation in various regions of space and using the boundary conditions developed in Section 12-7 to match the solutions together at any interfaces. Many of the methods discussed in Chapter 8 can therefore be used even when matter is present.

We shall illustrate this type of problem with a single example. Consider an *uncharged* sphere made of a *linear* dielectric and placed in a previously uniform electric field. We seek the resulting field both inside and outside the dielectric. If we take the direction of the uniform field $\mathbf{E}_0$ to define the polar axis of a spherical coordinate system so that

$$\mathbf{E}_0 = E_0 \hat{\mathbf{k}} \tag{12-73}$$

and further if we place the origin at the center of the sphere, the external field and the sphere are invariant to rotation about the z axis, nothing can depend on the coordinate ϕ, and the general solution to Laplace's equation has the form of Eq. (8-41). We shall find that all boundary conditions can be satisfied by taking only the terms for which $n = 0$ and 1, and we shall not prove explicitly that the other coefficients must be zero. Thus, our solution for the electrostatic potential has two pieces, viz.,

$$\begin{aligned} V_{\text{in}}(r, \theta) &= \left(a_0 + \frac{b_0}{r}\right) + \left(a_1 r + \frac{b_1}{r^2}\right) \cos\theta \\ V_{\text{out}}(r, \theta) &= \left(A_0 + \frac{B_0}{r}\right) + \left(A_1 r + \frac{B_1}{r^2}\right) \cos\theta \end{aligned} \tag{12-74}$$

where V_{in} refers to the region $r < a$, with a the radius of the sphere, and V_{out} refers to the region $r > a$. Several boundary conditions must be satisfied by Eq. (12-74), and each imposes constraints on the undetermined constants. First, V_{out} must approach the potential $-\mathbf{E}\cdot\mathbf{r} = -E_0 r \cos\theta$ of a uniform field as $r \longrightarrow \infty$; thus $A_0 = 0$ and $A_1 = -E_0$. Next, the sphere was required to be uncharged, and hence $B_0 = 0$. Thus, V_{out} reduces to

$$V_{\text{out}}(r, \theta) = \left(-E_0 r + \frac{B_1}{r^2}\right) \cos\theta \tag{12-75}$$

A third boundary condition requires that $V_{\text{in}}(r, \theta)$ be finite everywhere, particularly at $r = 0$; therefore, $b_0 = b_1 = 0$, and $V_{\text{in}}(r, \theta)$ reduces to

$$V_{\text{in}}(r, \theta) = a_0 + a_1 r \cos\theta \tag{12-76}$$

Finally, we must require that V and the normal component of $\mathbf{D}$ be continuous at $r = a$ for *all* θ. Mathematically, the first of these conditions requires that

$$\begin{aligned} &V_{\text{in}}(a, \theta) = V_{\text{out}}(a, \theta) \\ &\qquad \Longrightarrow a_0 + a_1 a \cos\theta = \left(-E_0 a + \frac{B_1}{a^2}\right) \cos\theta \\ &\qquad \Longrightarrow a_0 = 0, \qquad a a_1 = -E_0 a + \frac{B_1}{a^2} \end{aligned} \tag{12-77}$$

where the final conclusion follows because a condition having this form cannot be satisfied for all θ unless the coefficients of $\cos n\theta$, $n = 0, 1, 2, \ldots,$ are separately equal on both sides of the equation. The condition on the normal component of $\mathbf{D}$ provides a second equation involving a_1 and B_1, viz.,

$$\begin{aligned} &\epsilon \frac{\partial V_{\text{in}}(a, \theta)}{\partial r} = \epsilon_0 \frac{\partial V_{\text{out}}(a, \theta)}{\partial r} \\ &\qquad \Longrightarrow \epsilon a_1 \cos\theta = \epsilon_0\left(-E_0 - 2\frac{B_1}{a^3}\right) \cos\theta \\ &\qquad \Longrightarrow \epsilon a_1 = -\epsilon_0\left(E_0 + 2\frac{B_1}{a^3}\right) \end{aligned} \tag{12-78}$$

Together, Eqs. (12-77) and (12-78) give values for a_1 and B_1, and we find on substituting these values and the value $a_0 = 0$ into Eqs. (12-75) and (12-76) that

$$\begin{aligned} V_{\text{in}}(r, \theta) &= -\frac{3\epsilon_0}{\epsilon + 2\epsilon_0} E_0 r \cos\theta \\ V_{\text{out}}(r, \theta) &= -\left(1 - \frac{\epsilon - \epsilon_0}{\epsilon + 2\epsilon_0}\frac{a^3}{r^3}\right) E_0 r \cos\theta \end{aligned} \tag{12-79}$$

In particular, the electric field *inside* the sphere is *uniform* and parallel to the original exterior field, as is shown by

$$\mathbf{E}_{\text{in}} = -\nabla V_{\text{in}} = \frac{3\epsilon_0}{\epsilon + 2\epsilon_0}\mathbf{E}_0 \tag{12-80}$$

(Write $r \cos\theta = z$ and calculate the gradient directly in Cartesian coordinates.) Other aspects of this field are explored in P12-20.

PROBLEMS

P12-20. (a) Solve Eqs. (12-77) and (12-78) for a_1 and B_1 and then obtain Eq. (12-79) from Eqs. (12-75) and (12-76). (b) Find the electric field corresponding to Eq. (12-79) at all points in space and, in particular, show that the boundary condition on the tangential components of $\mathbf{E}$ at $r = a$ is automatically satisfied. (c) Find the equivalent charge distribution with which the dielectric sphere of Section 12-8 might be replaced. (d) Show that the exterior field computed in part (b) consists of the field of a dipole superimposed on the original uniform field, and find the dipole moment describing the dipole contribution. *Optional*: Use a program such as that described in P4-56 to obtain graphs of field lines and equipotentials for this field.

P12-21. An infinitely long, uncharged cylinder of radius a made of a linear dielectric having dielectric constant K_e is placed in a previously uniform electric field $\mathbf{E} = E_0\hat{\mathbf{i}}$ with its axis coincident with the z axis. Determine the electrostatic potential and the electrostatic field at points both inside and outside the cylinder.

P12-22. An uncharged hollow sphere with inner radius a and outer radius b is made of a linear dielectric having dielectric constant K_e. Let this sphere be placed in a uniform external electric field $\mathbf{E}_0$. (a) Show that the field in the hollow interior is constant and is given by

$$\mathbf{E}_1 = \frac{9K_e}{(2K_e + 1)(K_e + 2) - 2(a/b)^3(K_e - 1)^2}\mathbf{E}_0$$

Note: You do not need to calculate *all* of the unknown constants in order to determine the required field. (b) Show that $\mathbf{E}_1$ reduces to $\mathbf{E}_0$ when $K_e = 1$. *Optional*: Use a computer to obtain graphs of the coefficient in the above equation as a function of K_e for various values of a/b. (Note that physically $a/b < 1$.) Suggest a possible use for the arrangement of this problem.

P12-23. A uniform field $\mathbf{E}_m$ is established in a linear dielectric having dielectric constant K_e. A spherical cavity is now cut in the dielectric. Show that the field in the cavity is given by $3K_e\mathbf{E}_m/(2K_e + 1)$. *Hint*: Exchange of ϵ and ϵ_0 converts the dielectric sphere treated in the text into this problem.

P12-24. A sphere of radius a made of a linear magnetic material having relative permeability K_m is placed with its center at the origin in a previously uniform magnetic induction $\mathbf{B} = B_0\hat{\mathbf{k}}$. Determine the magnetic scalar potential and the magnetic induction field at points inside and outside the sphere. Show that the added field is the field of a dipole and determine the magnetic dipole moment of that dipole.

P12-25. Calculate the fields $\mathbf{B}$ and $\mathbf{H}$ both inside and outside a sphere of radius a having uniform magnetization $\mathbf{M} = M\hat{\mathbf{k}}$.

P12-26. By solving Laplace's equation in spherical coordinates, find a series expansion for the magnetic scalar potential at points off the axis of a circular current loop of radius a carrying current I. Consider particularly the region $r > a$, where r is the distance from the center of the loop to the observation point. *Hints*: (1) The **B**-field on the axis of the loop is given in P5-4. What should $V^{(m)}$ reduce to on the axis? (2) See P8-17. (3) Let the loop lie in the x-y plane with its center at the origin. *Optional*: Consider the region $r < a$.

Supplementary Problems

P12-27. Describe an experimental setup in which the B-H curve for a ferromagnetic material could be displayed directly on the screen of a cathode ray oscilloscope.

P12-28. Show that the ith component of the right-hand side of Eq. (12-34) can be written in the form $\sum_j \oint T_{ij}\, dS_j$, where

$$T_{ij} = E_i D_j + H_i B_j - \tfrac{1}{2}\delta_{ij}(\mathbf{D}\cdot\mathbf{E} + \mathbf{H}\cdot\mathbf{B})$$

provided the material present is characterized by constant, scalar permittivities and permeabilities. The quantity T_{ij} is called the Maxwell stress tensor. (Compare P6-37.)

P12-29. The *depolarization factor* L of a dielectric object in a uniform external field $\mathbf{E}_0$ is defined by $\epsilon_0\mathbf{E}_{\text{in}} = \epsilon_0\mathbf{E}_0 - L\mathbf{P}_{\text{in}}$, where $\mathbf{E}_{\text{in}}$ and $\mathbf{P}_{\text{in}}$ are the electric field and the polarization in the object. Find the depolarization factor of a sphere made of a linear dielectric. *Note*: A similar concept, called the *demagnetization factor*, is often used in discussing objects made of magnetic material.

P12-30. The method of images can sometimes be extended to solve problems involving dielectrics. Suppose, for example, that two dielectrics of permittivity ϵ_1 and ϵ_2 meet in a plane interface, say the x-z plane. Let medium 1 occupy the region $y > 0$ and medium 2 the region $y < 0$. Finally, let a point charge q be placed at $(0, d, 0)$ with $d > 0$. (*Suggestion*: Sketch a figure.) We seek separate expressions V_1 and V_2 for the potential in the two media. In seeking V_1, we can place image charges in the region $y < 0$ and, further, we can regard that region to be filled with a dielectric of permittivity ϵ_1. Similarly, in seeking V_2, we can place image charges in the region $y > 0$ and regard that region to be filled with a dielectric of permittivity ϵ_2. In both cases, the region modified lies outside the domain in which the resulting potential applies. Thus, in seeking V_1 we might place an image charge q_1 at $(0, -d_1, 0)$, with $d_1 > 0$, and have

$$V_1(\mathbf{r}) = \frac{q}{4\pi\epsilon_1} \frac{1}{|\mathbf{r} - d\hat{\mathbf{j}}|} + \frac{q_1}{4\pi\epsilon_1} \frac{1}{|\mathbf{r} + d_1\hat{\mathbf{j}}|}$$

Further, in seeking V_2, we might place an image charge q_2 at $(0, d_2, 0)$, with $d_2 > 0$, and have

$$V_2(\mathbf{r}) = \frac{q_2}{4\pi\epsilon_2} \frac{1}{|\mathbf{r} - d_2\hat{\mathbf{j}}|}$$

Show that these potentials will satisfy the necessary boundary conditions at the interface ($y = 0$) for *all* x and z provided that d_1, d_2, q_1, and q_2 are properly chosen, and find these quantities in terms of d, q, ϵ_1, and ϵ_2. *Hint*: Write $|\mathbf{r} - d\hat{\mathbf{j}}|$, for example, as $[x^2 + (y - d)^2 + z^2]^{1/2}$. *Optional*: Obtain graphs of the field lines and equipotentials in the y-z plane. (See P4-56.)

13

Plane Electromagnetic Waves in Linear Matter

In Chapter 13, we shall transfer our attention from general properties shared by all time-dependent electromagnetic fields to the specific description of fields that vary sinusoidally with time. These *monochromatic* fields are of particular interest for several reasons: They are analytically simpler than fields having a more general time dependence; they can be superposed to build up more complicated fields; the response of matter to monochromatic fields is often much simpler than its response to more general fields; and monochromatic fields (especially monochromatic plane waves) play an important role in applications of the theoretical framework developed in the previous chapters to many physical phenomena, including optical phenomena. We shall begin by reducing Maxwell's equations and the associated boundary conditions to forms that are convenient for treating monochromatic fields. Next we shall obtain analytic expressions for *plane* monochromatic waves and shall illustrate the application of these expressions to problems in macroscopic optics. Then we shall examine briefly the propagation of monochromatic fields in regions bounded by good conductors. Finally, we shall discuss plane waves produced by the superposition of two or more monochromatic plane waves.

13-1 Maxwell's Equations for Monochromatic Fields in Linear Matter

A field whose time dependence is sinusoidal or, more specifically, a field whose time dependence is characterized by a single (constant) angular frequency ω is called a *monochromatic field* and is conveniently represented as the real part of a complex field. If we introduce the notation

$$\begin{aligned} \mathbf{E}(\mathbf{r}, t) &= \operatorname{Re} \boldsymbol{\mathcal{E}}(\mathbf{r}, t) \qquad & \mathbf{D}(\mathbf{r}, t) &= \operatorname{Re} \boldsymbol{\mathcal{D}}(\mathbf{r}, t) \\ \mathbf{B}(\mathbf{r}, t) &= \operatorname{Re} \boldsymbol{\mathcal{B}}(\mathbf{r}, t) \qquad & \mathbf{H}(\mathbf{r}, t) &= \operatorname{Re} \boldsymbol{\mathcal{H}}(\mathbf{r}, t) \end{aligned} \tag{13-1}$$

then our assumption that the *physical* fields **E**, **D**, **B**, and **H** are monochromatic is expressed analytically by letting the *complex* fields $\boldsymbol{\mathcal{E}}$, $\boldsymbol{\mathcal{D}}$, $\boldsymbol{\mathcal{B}}$, and $\boldsymbol{\mathcal{H}}$ have the form

$$\begin{aligned} \boldsymbol{\mathcal{E}}(\mathbf{r}, t) &= \boldsymbol{\mathcal{E}}_0(\mathbf{r})e^{-i\omega t} \qquad & \boldsymbol{\mathcal{D}}(\mathbf{r}, t) &= \boldsymbol{\mathcal{D}}_0(\mathbf{r})e^{-i\omega t} \\ \boldsymbol{\mathcal{B}}(\mathbf{r}, t) &= \boldsymbol{\mathcal{B}}_0(\mathbf{r})e^{-i\omega t} \qquad & \boldsymbol{\mathcal{H}}(\mathbf{r}, t) &= \boldsymbol{\mathcal{H}}_0(\mathbf{r})e^{-i\omega t} \end{aligned} \tag{13-2}$$

In general, the coefficient multiplying the exponential factor in each equation will be complex. Further, if the fields vary sinusoidally with time, Maxwell's equations cannot be satisfied unless the free charge and current densities also vary sinusoidally with time (P13-1). We therefore represent the physical free current density by the real part of a complex current density

$$\mathbf{J}(\mathbf{r}, t) = \operatorname{Re} \boldsymbol{\mathcal{J}}(\mathbf{r}, t) \qquad \boldsymbol{\mathcal{J}}(\mathbf{r}, t) = \boldsymbol{\mathcal{J}}_0(\mathbf{r})e^{-i\omega t} \tag{13-3}$$

and we would do the same for the charge density were it not more appropriate to the circumstances of interest to set the *volume* density of *free* charge equal to zero,

$$\rho(\mathbf{r}, t) = 0 \tag{13-4}$$

(which will be the case if all dielectrics are initially uncharged and the fields vary slowly enough so that all conductors can be assumed to respond instantly and completely to changes in the external field[1]). Now, since Maxwell's equations are linear, we can require the *complex* fields to satisfy these equations with the assurance that the *physical* fields (i.e., the real part of the complex fields) will therefore automatically satisfy the equations. Thus, substituting Eqs. (13-2)–(13-4) into Maxwell's equations, Eqs. (12-14)–(12-17), and canceling a common factor $e^{-i\omega t}$ in each equation, we find that the coefficients in Eq. (13-2) are determined by the equations

$$\boldsymbol{\nabla}\cdot\boldsymbol{\mathcal{D}}_0(\mathbf{r}) = 0 \qquad \boldsymbol{\nabla}\times\boldsymbol{\mathcal{E}}_0(\mathbf{r}) = i\omega\boldsymbol{\mathcal{B}}_0(\mathbf{r}) \tag{13-5), (13-6}$$

$$\boldsymbol{\nabla}\cdot\boldsymbol{\mathcal{B}}_0(\mathbf{r}) = 0 \qquad \boldsymbol{\nabla}\times\boldsymbol{\mathcal{H}}_0(\mathbf{r}) = \boldsymbol{\mathcal{J}}_0(\mathbf{r}) - i\omega\boldsymbol{\mathcal{D}}_0(\mathbf{r}) \tag{13-7), (13-8}$$

[1] The latter assumption is not as restrictive as it might seem. It was found in P9-5 that a free charge distribution *in* any conductor decays *spontaneously* to zero and that in good conductors, the time scale of that decay is on the order of 10^{-18} sec. Thus, as long as $\omega \ll 2\pi \times 10^{18}\ \text{sec}^{-1}$, which is easily satisfied even for frequencies in the visible spectrum

One of the ways in which considering monochromatic solutions simplifies the discussion of time-dependent fields in matter arises because in many media the relationships between **E** and **D** and between **B** and **H** when the fields vary *sinusoidally* in time can be written by assuming frequency-dependent permittivities and permeabilities. Further, in many media, the current density can be related to a *sinusoidal* electric field that produces the current by assuming a frequency-dependent conductivity. These media are adequately described by the relationships

$$\mathfrak{D}_0(\mathbf{r}) = \epsilon(\omega)\mathcal{E}_0(\mathbf{r}) \qquad \mathcal{B}_0(\mathbf{r}) = \mu(\omega)\mathcal{H}_0(\mathbf{r}) \qquad (13\text{-}9), (13\text{-}10)$$

$$\mathcal{J}_0(\mathbf{r}) = g(\omega)\mathcal{E}_0(\mathbf{r}) \qquad (13\text{-}11)$$

even though the forms $\mathbf{D}(\mathbf{r}, t) = \epsilon\mathbf{E}(\mathbf{r}, t)$, etc., are *not* applicable when the fields have a more general time dependence (P13-45).

By way of illustration, we shall consider a simple classical model that predicts a frequency-dependent conductivity, recognizing, however, that the model has now been replaced with better quantum mechanical treatments. As in P9-8, suppose that the resistive forces on an electron (mass m_e, charge $-q_e$) in a conducting medium can be represented by a viscous damping term in the equation of motion. When an external electric field $\mathcal{E}$ is applied to the conducting medium, the individual electrons in the medium then move in accordance with the equation

$$m_e \frac{d\mathbf{v}}{dt} + b\mathbf{v} = -q_e\mathcal{E} \qquad (13\text{-}12)$$

where $\mathbf{v}$ is the (complex) electron velocity and b is the viscous damping constant. Suppose $\mathcal{E} = E_0\hat{\mathbf{i}}e^{-i\omega t}$ and, ignoring the (short-lived) transient effects associated with turning the field on, assume $\mathbf{v} = v_0\hat{\mathbf{i}}e^{-i\omega t}$. Then, Eq. (13-12) gives

$$(-i\omega m_e + b)v_0 = -q_eE_0 \Longrightarrow v_0 = -\left(\frac{q_e}{b - i\omega m_e}\right)E_0 \qquad (13\text{-}13)$$

where the factor $e^{-i\omega t}$ common to all terms has been canceled. Thus, the (complex) mobility μ_e of the electron is given by

$$\mu_e = \frac{v_0}{E_0} = -\frac{q_e}{b - i\omega m_e} \qquad (13\text{-}14)$$

and the conductivity of the material involved is given by

$$g(\omega) = -q_e n\mu_e = \frac{q_e^2 n}{b - i\omega m_e} \qquad (13\text{-}15)$$

where n is the density of electrons in the material [Eq. (9-16)]. For this model, the predicted conductivity is not only frequency-dependent but also complex; it is expressed in polar form by the equation

$$g(\omega) = |g(\omega)|e^{i\theta(\omega)} \qquad (13\text{-}16)$$

($\approx 10^{15}$ sec^{-1}), the assumption of zero *volume* charge density is realistic. *Surface* charge densities, of course, may appear, but these will be treated through the boundary conditions.

where

$$|g(\omega)| = \frac{q_e^2 n/b}{\sqrt{1 + (\omega m_e/b)^2}} \qquad \theta(\omega) = \tan^{-1} \frac{\omega m_e}{b} \tag{13-17}$$

To obtain a physical interpretation, note that in this example

$$\begin{aligned} \mathbf{E} &= \text{Re}\{E_0 \hat{\mathbf{i}} e^{-i\omega t}\} = E_0 \hat{\mathbf{i}} \cos \omega t \\ \mathbf{J} &= \text{Re}\{g(\omega) E_0 \hat{\mathbf{i}} e^{-i\omega t}\} = |g(\omega)| E_0 \hat{\mathbf{i}} \cos (\omega t - \theta) \end{aligned} \tag{13-18}$$

Thus, we see that $|g(\omega)|$ relates the amplitude of the conduction current density to that of the applied field and $\theta(\omega)$ expresses a phase difference between the current density and the applied field. For a given applied field, the amplitude of the current density falls off and the phase difference between the current density and the applied field increases in magnitude with increasing ω. As shown in Fig. 13-1, however, $|g(\omega)| \approx q_e^2 n/b$ and $\theta(\omega) \approx 0$ if $\omega \ll b/m_e$; at "low" frequencies, the conductivity is real and independent of frequency, and the current density and the applied field are in phase. The numerical value of $\omega_c = b/m_e$ (which divides "low" from "high" frequencies) is determined by the properties of the material. For good conductors, $b \approx 10^{-16}$ kg/sec, and (with $m_e = 9.1 \times 10^{-31}$ kg) $\omega_c \approx 10^{14}$ sec^{-1}; thus in the microwave region ($\approx 10^{10}$ sec^{-1}) and at lower frequencies, we expect the conductivity of good conductors to be real and independent of frequency, but in the visible region ($\approx 10^{15}$ sec^{-1}) we expect the conductivity of good conductors to be complex and dependent on frequency. For dielectrics, on the other hand, b is effectively infinite and $g(\omega) = 0$ at all frequencies.

A classical model that yields a frequency-dependent permittivity is explored in P13-2. In part, this model predicts (1) that the static dielectric constant should be given by

$$K_e(0) = \frac{\epsilon(0)}{\epsilon_0} = 1 + \frac{nq_e^2}{m_e \epsilon_0 \omega_0^2} \tag{13-19}$$

where n, q_e, and m_e are as in the previous paragraph and ω_0 is the natural frequency associated with the individual atoms composing the dielectric; i.e., ω_0 is the empirical frequency at which the atoms absorb electromagnetic radiation; (2) that at very high frequencies

$$K_e(\infty) = 1 \tag{13-20}$$

and (3) that for $\omega \ll \omega_0$, $K_e(\omega) \approx K_e(0)$ and for $\omega \gg \omega_0$, $K_e(\omega) \approx 1$. Thus, according to this model, the dielectric constant and also the permittivity are real and independent of frequency except when $\omega \approx \omega_0$. Different substances, of course, are characterized by different values of ω_0. Indeed, quantum mechanically, substances exhibit more than one absorption line. Although the classical model then breaks down in detail, we would still expect the dielectric constant to exhibit "funny" behavior only at frequencies near to each of the absorption frequencies.

For the remainder of this chapter, we shall confine our attention to matter for which Eqs. (13-9)–(13-11) provide an adequate description. Further, we

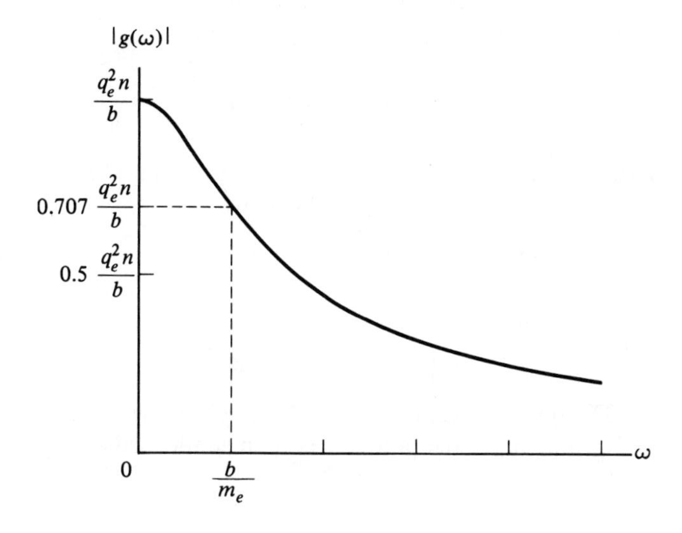

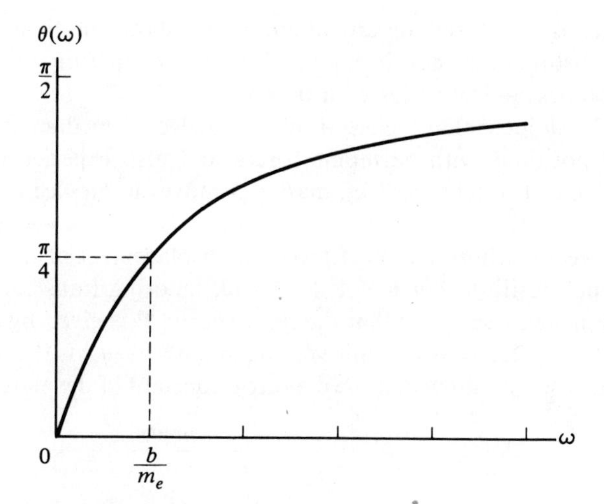

Fig. 13-1. Amplitude and phase of the complex conductivity as functions of frequency.

shall assume that no free currents other than those induced on conductors by the field $\boldsymbol{\mathcal{E}}_0(\mathbf{r})$ are present. Thus, Eqs. (13-5)–(13-8) reduce to

$$\nabla \cdot \boldsymbol{\mathcal{E}}_0(\mathbf{r}) = 0 \qquad \nabla \times \boldsymbol{\mathcal{E}}_0(\mathbf{r}) = i\omega\mu\boldsymbol{\mathcal{H}}_0(\mathbf{r}) \tag{13-21), (13-22}$$

$$\nabla \cdot \boldsymbol{\mathcal{H}}_0(\mathbf{r}) = 0 \qquad \nabla \times \boldsymbol{\mathcal{H}}_0(\mathbf{r}) = -i\omega\left(\epsilon + i\frac{g}{\omega}\right)\boldsymbol{\mathcal{E}}_0(\mathbf{r}) \tag{13-23), (13-24}$$

Explicit indication of the possible dependence of ϵ, μ, and g on ω has been suppressed. We can also show that the fields satisfy

$$\left[\nabla^2 + \mu\epsilon\omega^2\left(1 + i\frac{g}{\epsilon\omega}\right)\right]\mathcal{E}_0(\mathbf{r}) = 0$$
$$\left[\nabla^2 + \mu\epsilon\omega^2\left(1 + i\frac{g}{\epsilon\omega}\right)\right]\mathcal{H}_0(\mathbf{r}) = 0 \tag{13-25}$$

(P13-4) and that the time-averaged Poynting vector $\langle\mathbf{S}\rangle$ and the time-averaged energy density $\langle u_{\text{EM}}\rangle$ are given for these fields by

$$\langle\mathbf{S}\rangle = \langle\mathbf{E}\times\mathbf{H}\rangle = \tfrac{1}{2}\text{Re}\,(\mathcal{E}_0^* \times \mathcal{H}_0) \tag{13-26}$$

and

$$\langle u_{\text{EM}}\rangle = \tfrac{1}{2}\langle\mathbf{E}\cdot\mathbf{D}\rangle + \tfrac{1}{2}\langle\mathbf{H}\cdot\mathbf{B}\rangle$$
$$= \tfrac{1}{4}\,\text{Re}\,(\epsilon\mathcal{E}_0^*\cdot\mathcal{E}_0 + \mu\mathcal{H}_0^*\cdot\mathcal{H}_0) \tag{13-27}$$

[See Eqs. (12-32) and (12-30) and compare P7-26.] In these equations, explicit evaluation of $\mathcal{E}_0$ and $\mathcal{H}_0$ at the point $\mathbf{r}$ has been suppressed.

PROBLEMS

P13-1. Without making any assumptions about the time dependence of $\mathcal{J}(\mathbf{r}, t)$, substitute the fields in Eq. (13-2) and show that Eq. (12-16) cannot be satisfied unless $\mathcal{J}(\mathbf{r}, t)$ varies with time as $e^{-i\omega t}$.

P13-2. Suppose that the electrons in a dielectric medium are tied to their nominal positions with harmonic forces and also experience damping, so that each electron (charge $-q_e$, mass m_e) moves in an electric field $\mathcal{E}$ in accordance with the equation of motion $m_e\ddot{\mathbf{r}} + b\dot{\mathbf{r}} + k\mathbf{r} = -q_e\mathcal{E}$, where $\mathbf{r}$ is a complex vector whose real part gives the displacement of the electron from its nominal position. Let $\mathcal{E} = \mathbf{E}_0 e^{-i\omega t}$ and, ignoring transients, assume $\mathbf{r} = \mathbf{r}_0 e^{-i\omega t}$. Further, remember that the polarization $\mathbf{P}$ is given by $n\mathbf{p}$, where n is the number of electrons per unit volume and $\mathbf{p} = -q_e\mathbf{r}$ is the dipole moment of the electron. (a) Show that the dielectric constant of the material is given by

$$K_e(\omega) = \frac{\epsilon(\omega)}{\epsilon_0} = 1 + \frac{nq_e^2/m_e\epsilon_0}{(\omega_0^2 - \omega^2) - i(b\omega/m_e)}$$

where $\omega_0 = \sqrt{k/m_e}$ is the natural frequency of the oscillating system. (b) Verify the two limits given in Eqs. (13-19) and (13-20). (c) Explain what it means physically for the permittivity $\epsilon(\omega)$ to be complex and support your explanation with an analytic argument. (d) *Sketch* graphs of Re $K_e(\omega)$ and of Im $K_e(\omega)$ for various values of b, starting with $b = 0$. Note in particular that Im $K_e(\omega)$ is related to the absorption spectrum of the oscillating system and that values of b can be inferred from measurements of the width of the absorption curve.

P13-3. Estimate ω_0 for hydrogen (a) by setting the ionization potential, 13.6 eV, equal to $\hbar\omega_0$, where $\hbar$ is Planck's constant divided by 2π, and (b) by determining the force constant k using the model discussed in P10-3. (c) Use

Eq. (13-19) to predict the static dielectric constant of hydrogen gas at 100°C and 1 atmosphere pressure, comparing the results with the measured value 1.000264. Is the agreement good or bad? Suggest possible sources for any discrepancy. *Hints:* (1) Assume that the two atoms in each hydrogen molecule respond independently to the applied field. (2) The Bohr radius for the ground state of the hydrogen atom is .528 Å.

P13-4. Verify that Eq. (13-25) follows from Eqs. (13-21)–(13-24). *Hint:* Take the curl of Eqs. (13-22) and (13-24) and then use Eq. (C-19).

P13-5. Show that the average power dissipated per unit volume by a monochromatic field in a conducting medium is given by $\langle P \rangle = \frac{1}{2}$ Re $\mathcal{J}_0 \cdot \mathcal{E}_0^*$, and then show that $\langle P \rangle = 0$ when $g(\omega)$ is purely imaginary (i.e., when the electric field and the current are 90° out of phase, as in the high-frequency domain shown in Fig. 13-1).

13-2 Boundary Conditions on Monochromatic Fields

As with all differential equations, those in Section 13-1 for the amplitudes of the complex fields become specific to a particular problem only when supplemented with whatever boundary conditions the problem imposes. Since in this chapter we are interested ultimately in the optical problem of reflection and transmission of monochromatic fields at the interface between two media and in the confinement of monochromatic fields to regions bounded by conductors, we need in particular to rewrite the general boundary conditions developed in Section 12-7 in a form applicable to monochromatic fields. Essentially, we must write the boundary conditions at an interface so that they apply to the complex fields (which is easy, because the boundary conditions are linear in the fields and hence apply directly to the complex fields) and we must examine what happens if the medium on one side (or perhaps the media on both sides) of the interface has nonzero or even infinite conductivity. For our present purposes, it is sufficient to assume a plane interface that we take to define the x-y plane (Fig. 13-2). Then, assuming that all fields, current densities, and charge densities have the time dependence $e^{-i\omega t}$ and recognizing that the frequency must be the same on both sides of the interface,[2] we can reexpress the boundary conditions in Eq. (12-65) in the form

$$\mathcal{E}_{0x}^{(1)} = \mathcal{E}_{0x}^{(2)} \qquad \mathcal{H}_{0x}^{(2)} - \mathcal{H}_{0x}^{(1)} = \mathscr{j}_{0y} \qquad \text{(13-28), (13-29)}$$

$$\mathcal{E}_{0y}^{(1)} = \mathcal{E}_{0y}^{(2)} \qquad \mathcal{H}_{0y}^{(2)} - \mathcal{H}_{0y}^{(1)} = -\mathscr{j}_{0x} \qquad \text{(13-30), (13-31)}$$

[2]The boundary conditions cannot be satisfied for all times if the frequencies are different. Continuity of a tangential component of $\mathcal{E}(\mathbf{r}, t)$, for example, requires that $\mathcal{E}_{0x}^{(1)}(\mathbf{r})$ exp $[-i\omega_1 t] = \mathcal{E}_{0x}^{(2)}(\mathbf{r})$ exp $[-i\omega_2 t]$ or equivalently that $\mathcal{E}_{0x}^{(1)}(\mathbf{r}) = \mathcal{E}_{0x}^{(2)}(\mathbf{r})$ exp $[-i(\omega_2 - \omega_1)t]$. Since the left-hand side of this latter form is independent of t, the right-hand side must also be independent of t and will be so only if $\omega_1 = \omega_2$. Q.E.D.

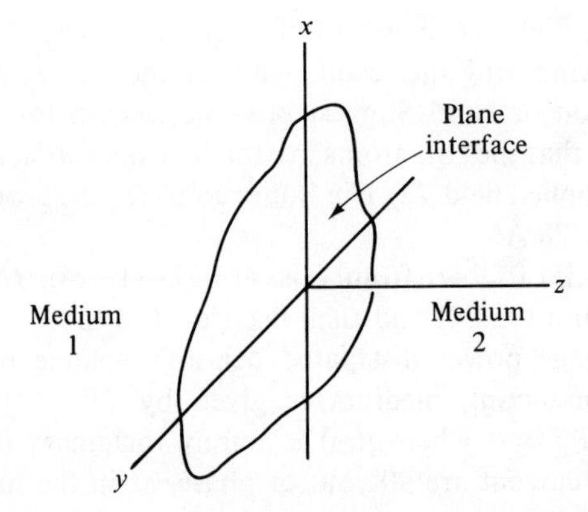

Fig. 13-2. Plane interface between two different media.

$$\epsilon_2 \mathcal{E}_{0z}^{(2)} - \epsilon_1 \mathcal{E}_{0z}^{(1)} = \sigma_0 \qquad \mu_1 \mathcal{H}_{0z}^{(1)} = \mu_2 \mathcal{H}_{0z}^{(2)} \quad \text{(13-32), (13-33)}$$

$$\mathcal{J}_{0z}^{(2)} - \mathcal{J}_{0z}^{(1)} = g_2 \mathcal{E}_{0z}^{(2)} - g_1 \mathcal{E}_{0z}^{(1)} = i\omega\sigma_0 \quad \text{(13-34)}$$

where we have (1) suppressed explicit indication that all quantities are to be evaluated at a point on the interface, (2) moved the specification of the medium to a superscript to avoid a triple subscript, (3) canceled the factor $e^{-i\omega t}$ from every term, (4) written all fields in terms of $\boldsymbol{\mathcal{E}}_0$ and $\boldsymbol{\mathcal{H}}_0$, and (5) written $\sigma(\mathbf{r}, t) = \text{Re}\, \sigma_0(\mathbf{r})e^{-i\omega t}$, $j_x(\mathbf{r}, t) = \text{Re}\, \mathcal{j}_{0x}(\mathbf{r})e^{-i\omega t}$, etc.

It is useful to comment specifically on further constraints that more specific stipulation of the two media imposes on Eqs. (13-28)–(13-34).[3] If, for example, both media are nonconductors so that $g_1 = g_2 = 0$, then there can be neither induced surface charge nor induced surface currents on the interface. Consequently, $\sigma_0 = \mathcal{j}_{0x} = \mathcal{j}_{0y} = 0$ and we conclude that the tangential components of $\boldsymbol{\mathcal{E}}_0$ and $\boldsymbol{\mathcal{H}}_0$ and the normal components of $\epsilon\boldsymbol{\mathcal{E}}_0$ and $\mu\boldsymbol{\mathcal{H}}_0$ must be continuous across an interface between two nonconductors; i.e., when $g_1 = g_2 = 0$,

$$\begin{array}{lll} \mathcal{E}_{0x}^{(1)} = \mathcal{E}_{0x}^{(2)} & \mathcal{E}_{0y}^{(1)} = \mathcal{E}_{0y}^{(2)} & \epsilon_1 \mathcal{E}_{0z}^{(1)} = \epsilon_2 \mathcal{E}_{0z}^{(2)} \\ \mathcal{H}_{0x}^{(1)} = \mathcal{H}_{0x}^{(2)} & \mathcal{H}_{0y}^{(1)} = \mathcal{H}_{0y}^{(2)} & \mu_1 \mathcal{H}_{0z}^{(1)} = \mu_2 \mathcal{H}_{0z}^{(2)} \end{array} \quad \text{(13-35)}$$

In a second useful special case, g_1 and g_2 may be nonzero but *neither* is infinite. Since a medium with *finite* conductivity cannot support *surface* currents (existence of a *surface* current requires the *volume* current density to be infinite, but Ohm's law $\mathbf{J} = g\mathbf{E}$ precludes infinite $\mathbf{J}$ unless $g = \infty$, since physically $\mathbf{E}$ must be finite; *surface* currents are an idealization that can occur only in conjunction with the idealization of a *perfect* conductor), $\mathcal{j}_{0x} = \mathcal{j}_{0y} = 0$ for this case. The surface charge density σ_0, however, may or may not be zero. If $\sigma_0 = 0$, then Eqs. (13-32) and (13-34) are contradictory unless

[3]The classification of boundary conditions presented in this section is patterned after a similar classification in Chapter 16 of *Foundations of Electromagnetic Theory* by J. R. Reitz and F. J. Milford (Addison-Wesley Publishing Company, Inc., Reading, Mass., 1967), Second Edition, and is used here by permission of Addison-Wesley Publishing Company, Inc.

$\epsilon_1 g_2 = \epsilon_2 g_1$. (Why?) When this particular relationship among ϵ_1, ϵ_2, g_1, and g_2 obtains, the boundary conditions to be imposed are the same as those in Eq. (13-35). More often, however, this special relationship will not be satisfied. Then $\sigma_0 \neq 0$ and Eqs. (13-32) and (13-34) can be combined to give a single condition that does not involve σ_0; except for the special case $\epsilon_1 g_2 = \epsilon_2 g_1$, we thus find that, when g_1 and g_2 are finite and at least one differs from zero, the fields must satisfy

$$\begin{aligned} &\mathcal{E}_{0x}^{(1)} = \mathcal{E}_{0x}^{(2)} \qquad \mathcal{E}_{0y}^{(1)} = \mathcal{E}_{0y}^{(2)} \qquad \left(\epsilon_1 + i\frac{g_1}{\omega}\right)\mathcal{E}_{0z}^{(1)} = \left(\epsilon_2 + i\frac{g_2}{\omega}\right)\mathcal{E}_{0z}^{(2)} \\ &\mathcal{H}_{0x}^{(1)} = \mathcal{H}_{0x}^{(2)} \qquad \mathcal{H}_{0y}^{(1)} = \mathcal{H}_{0y}^{(2)} \qquad \mu_1\mathcal{H}_{0z}^{(1)} = \mu_2\mathcal{H}_{0z}^{(2)} \end{aligned} \tag{13-36}$$

In words, the tangential components of $\boldsymbol{\mathcal{E}}_0$ and $\boldsymbol{\mathcal{H}}_0$ and the normal components of $[\epsilon + i(g/\omega)]\boldsymbol{\mathcal{E}}_0$ and $\mu\boldsymbol{\mathcal{H}}_0$ must be continuous across an interface between two imperfect conductors. Once the fields have been found, Eq. (13-32) can be used to calculate the charge density on the interface.

A third special case occurs when one of the media, say medium 2, can be approximated as a *perfect* conductor ($g_2 = \infty$) but the other medium is either an imperfect conductor or a nonconductor. In this case both surface currents and surface charges may appear on the interface, and the simplifications lie in the requirement that all fields be zero in the perfect conductor. To prove that $\boldsymbol{\mathcal{E}}_0^{(2)} = 0$, for example, suppose first that g_2 is finite. Then Eq. (13-24) gives

$$\boldsymbol{\mathcal{E}}_0^{(2)} = \frac{\nabla \times \boldsymbol{\mathcal{H}}_0^{(2)}}{g_2 - i\omega\epsilon} \tag{13-37}$$

Thus, if $\nabla \times \boldsymbol{\mathcal{H}}_0^{(2)}$ is finite (which must be required on physical grounds), $\boldsymbol{\mathcal{E}}_0^{(2)} \longrightarrow 0$ as $g_2 \longrightarrow \infty$. Coupled with the vanishing of $\boldsymbol{\mathcal{E}}_0^{(2)}$ in the (perfect) conductor, Eq. (13-22) yields the conclusion that $\boldsymbol{\mathcal{H}}_0^{(2)}$ also approaches zero as $g_2 \longrightarrow \infty$. Thus, on the interface between a general medium ($g_1 \neq \infty$) and a perfect conductor ($g_2 = \infty$), the fields must satisfy

$$\begin{aligned} &\mathcal{E}_{0x}^{(1)} = \mathcal{E}_{0x}^{(2)} = 0 \qquad \mathcal{E}_{0y}^{(1)} = \mathcal{E}_{0y}^{(2)} = 0 \qquad \epsilon_1\mathcal{E}_{0z}^{(1)} = -\sigma_0 \\ &\qquad\qquad\qquad\qquad\qquad\qquad\qquad\qquad\quad \epsilon_2\mathcal{E}_{0z}^{(2)} = 0 \\ &\mathcal{H}_{0x}^{(1)} = -\mathcal{J}_{0y} \qquad \mathcal{H}_{0y}^{(1)} = \mathcal{J}_{0x} \\ &\qquad\qquad\qquad\qquad\qquad\qquad\qquad\qquad\quad \mu_1\mathcal{H}_{0z}^{(1)} = \mu_2\mathcal{H}_{0z}^{(2)} = 0 \\ &\mathcal{H}_{0x}^{(2)} = 0 \qquad \mathcal{H}_{0y}^{(2)} = 0 \end{aligned} \tag{13-38}$$

the tangential components of $\boldsymbol{\mathcal{E}}_0$ and the normal component of $\mu\boldsymbol{\mathcal{H}}_0$ must be zero on both sides of an interface between a general medium and a perfect conductor; the normal component of $\epsilon\boldsymbol{\mathcal{E}}_0$ and the tangential components of $\boldsymbol{\mathcal{H}}_0$ must be zero in the perfect conductor and will be nonzero in the general medium by amounts determined by the charge and current distributions that the fields themselves induce on the interface. These charge and current densities induced on the interface are, of course, *not* part of the information given at the beginning of a problem and the above boundary conditions often are used to determine these densities *after* the fields have been found by other means.

13-3 Plane Monochromatic Waves in Unbounded, Isotropic, Homogeneous, Linear Media

Before we can apply the boundary conditions developed in Section 13-2 to the optical problems of transmission and reflection, we need to obtain analytic expressions for more explicit solutions to Eqs. (13-21)–(13-25). In this section we shall examine the properties of *plane* wave solutions in *unbounded, isotropic, homogeneous, linear* media. In Chapter 7, we expressed the fields in a sinusoidal plane wave propagating in the z direction as trigonometric functions of the argument $(\kappa z - \omega t)$, where κ represents the wave number. More generally, for a plane wave propagating in the direction of the unit vector $\hat{\mathbf{n}}$, we expect that the coordinate $\hat{\mathbf{n}}\cdot\mathbf{r}$, which measures the distance of the point $\mathbf{r}$ from a plane through the origin whose normal is $\hat{\mathbf{n}}$, will replace the coordinate $z = \hat{\mathbf{k}}\cdot\mathbf{r}$, which measures the distance of $\mathbf{r}$ from the x-y plane (whose normal is $\hat{\mathbf{k}}$). Thus, we expect the more general plane wave to be described analytically by trigonometric functions of the argument $K\hat{\mathbf{n}}\cdot\mathbf{r} - \omega t = \mathbf{K}\cdot\mathbf{r} - \omega t$, where K is a constant analogous to κ—we use a different symbol because we shall find that K is sometimes complex—and the *propagation vector* $\mathbf{K}$ is defined by

$$\mathbf{K} = K\hat{\mathbf{n}} \tag{13-39}$$

Thus, in terms of this vector, whose meaning is yet to be explored, a sinusoidal plane wave propagating in the direction $\hat{\mathbf{n}}$ is expressed as the real part of the quantity

$$(\text{constant})e^{i(\mathbf{K}\cdot\mathbf{r}-\omega t)} \tag{13-40}$$

where the constant may be complex and contains information about both the phase and the amplitude of the corresponding wave. The general monochromatic field expressed in Eqs. (13-2) and (13-3) is therefore reduced to a sinusoidal plane wave by requiring the spatially dependent coefficients to have the form

$$\begin{aligned} \boldsymbol{\mathcal{E}}_0(\mathbf{r}) &= \boldsymbol{\mathcal{E}}_0 e^{i\mathbf{K}\cdot\mathbf{r}} \qquad \boldsymbol{\mathcal{D}}_0(\mathbf{r}) = \boldsymbol{\mathcal{D}}_0 e^{i\mathbf{K}\cdot\mathbf{r}} \\ \boldsymbol{\mathcal{B}}_0(\mathbf{r}) &= \boldsymbol{\mathcal{B}}_0 e^{i\mathbf{K}\cdot\mathbf{r}} \qquad \boldsymbol{\mathcal{H}}_0(\mathbf{r}) = \boldsymbol{\mathcal{H}}_0 e^{i\mathbf{K}\cdot\mathbf{r}} \\ \boldsymbol{\mathcal{J}}_0(\mathbf{r}) &= \boldsymbol{\mathcal{J}}_0 e^{i\mathbf{K}\cdot\mathbf{r}} \end{aligned} \tag{13-41}$$

where the *amplitudes* $\boldsymbol{\mathcal{E}}_0$, $\boldsymbol{\mathcal{D}}_0$, $\boldsymbol{\mathcal{B}}_0$, $\boldsymbol{\mathcal{H}}_0$, and $\boldsymbol{\mathcal{J}}_0$ are now constant vectors, although they may have complex components. Further, substitution of Eq. (13-41) into Eqs. (13-9)–(13-11) yields the relationships

$$\boldsymbol{\mathcal{D}}_0 = \epsilon\boldsymbol{\mathcal{E}}_0 \qquad \boldsymbol{\mathcal{B}}_0 = \mu\boldsymbol{\mathcal{H}}_0 \qquad \boldsymbol{\mathcal{J}}_0 = g\boldsymbol{\mathcal{E}}_0 \tag{13-42}$$

among these constant amplitudes and the parameters ϵ, μ, and g that characterize the medium.

Instead of solving Maxwell's equations directly and systematically, we have in a sense guessed the spatial dependence of the complex fields representing a monochromatic plane wave. We must now determine whether these fields actually satisfy Maxwell's equations. Note first that

$$\nabla[e^{i\mathbf{K}\cdot\mathbf{r}}] = \nabla[e^{i(K_x x + K_y y + K_z z)}]$$
$$= i(K_x\hat{\mathbf{i}} + K_y\hat{\mathbf{j}} + K_z\hat{\mathbf{k}})e^{i(\cdots)} = i\mathbf{K}e^{i\mathbf{K}\cdot\mathbf{r}} \qquad (13\text{-}43)$$

Thus, substitution of Eq. (13-41) into Eqs. (13-21)–(13-24), use of the vector identities in Eqs. (C-8) and (C-11), and cancellation of the factor $e^{i\mathbf{K}\cdot\mathbf{r}}$ yields the relationships

$$\mathbf{K}\cdot\boldsymbol{\mathcal{E}}_0 = 0 \qquad \mathbf{K}\times\boldsymbol{\mathcal{E}}_0 = \omega\mu\boldsymbol{\mathcal{H}}_0 \qquad (13\text{-}44), (13\text{-}45)$$

$$\mathbf{K}\cdot\boldsymbol{\mathcal{H}}_0 = 0 \qquad \mathbf{K}\times\boldsymbol{\mathcal{H}}_0 = -\omega\left(\epsilon + i\frac{g}{\omega}\right)\boldsymbol{\mathcal{E}}_0 \qquad (13\text{-}46), (13\text{-}47)$$

that must be satisfied by the constant vectors $\boldsymbol{\mathcal{E}}_0$ and $\boldsymbol{\mathcal{H}}_0$ if the assumed fields are to satisfy Maxwell's equations. The so-called *dispersion relation*

$$K^2 = \mu\omega^2\left(\epsilon + i\frac{g}{\omega}\right) \qquad (13\text{-}48)$$

connecting K and ω is implied by Eqs. (13-44)–(13-47). [Evaluate the cross product of Eq. (13-45) with $\mathbf{K}$ and then use Eq. (C-1); see P13-6.] In words, Eqs. (13-44) and (13-46) require that both $\boldsymbol{\mathcal{E}}_0$ and $\boldsymbol{\mathcal{H}}_0$ be perpendicular to $\hat{\mathbf{n}}$, which is the direction of propagation [see Eq. (13-39)]; the plane monochromatic electromagnetic wave is thus a *transverse wave*. Further, Eqs. (13-45) and (13-47) require $\boldsymbol{\mathcal{H}}_0$ and $\boldsymbol{\mathcal{E}}_0$ to be mutually perpendicular. Finally, Eq. (13-48) determines K if ω is given. Paralleling the discussion in Section 7-5, we can think of this plane wave as being defined as follows:

(1) Specify the direction of propagation $\hat{\mathbf{n}}$ and the frequency ω *arbitrarily*.
(2) Specify $\boldsymbol{\mathcal{E}}_0$ *arbitrarily*, subject only to the constraint that $\boldsymbol{\mathcal{E}}_0$ be perpendicular to $\hat{\mathbf{n}}$.
(3) Determine $\boldsymbol{\mathcal{H}}_0$ from Eq. (13-45) and K from Eq. (13-48).

Substituting Eqs. (13-42) and (13-44)–(13-47) into Eq. (13-41) and reinstating the exponential time factor, we find finally that the general plane monochromatic solution to Maxwell's equations in unbounded, isotropic, homogeneous, linear media is represented analytically by the real part of the complex fields

$$\boldsymbol{\mathcal{E}}(\mathbf{r}, t) = \frac{\boldsymbol{\mathcal{D}}(\mathbf{r}, t)}{\epsilon(\omega)} = \boldsymbol{\mathcal{E}}_0 e^{i(\mathbf{K}\cdot\mathbf{r} - \omega t)} \qquad (13\text{-}49)$$

$$\boldsymbol{\mathcal{H}}(\mathbf{r}, t) = \frac{\boldsymbol{\mathcal{B}}(\mathbf{r}, t)}{\mu(\omega)} = \frac{\mathbf{K}\times\boldsymbol{\mathcal{E}}_0}{\omega\mu(\omega)} e^{i(\mathbf{K}\cdot\mathbf{r} - \omega t)} \qquad (13\text{-}50)$$

where K and ω must satisfy Eq. (13-48) and $\boldsymbol{\mathcal{E}}_0$ must be perpendicular to $\mathbf{K}$. It is shown in P13-8 that the Poynting vector and the energy density for these fields are given by

$$\langle\boldsymbol{\mathcal{S}}\rangle = \frac{|\boldsymbol{\mathcal{E}}_0|^2 \operatorname{Re} K}{2\omega\mu}\hat{\mathbf{n}}e^{-2(\hat{\mathbf{n}}\cdot\mathbf{r})\operatorname{Im} K} \qquad (13\text{-}51)$$

and

$$\langle u_{\text{EM}} \rangle = \frac{1}{4} |\mathbf{\mathcal{E}}_0|^2 \left(\epsilon + \frac{|K|^2}{\mu \omega^2} \right) e^{-2\hat{\mathbf{n}}\cdot\mathbf{r}\,\text{Im}\,K} \tag{13-52}$$

where $|\mathbf{\mathcal{E}}_0|^2 = \mathbf{\mathcal{E}}_0 \cdot \mathbf{\mathcal{E}}_0^*$, $\mathbf{K}$ has been written as in Eq. (13-39), and we have assumed that ϵ and μ are real. Note that whenever K has a nonzero imaginary part, both $\langle \mathbf{S} \rangle$ and $\langle u_{\text{EM}} \rangle$ (and in fact also the fields—see the next paragraph) have a *real* exponential factor. Since physically the fields cannot increase in amplitude as the wave propagates into the medium, we anticipate that Im $K > 0$.[4]

As we have already recognized, plane waves propagating in a general medium are characterized by a propagation vector $\mathbf{K}$ that is complex. The physical significance of this vector is therefore to be sought in the separate significances of its real and imaginary parts, which we introduce explicitly by writing

$$\mathbf{K} = K\hat{\mathbf{n}} = (\kappa + i\alpha)\hat{\mathbf{n}} = \boldsymbol{\kappa} + i\boldsymbol{\alpha} \tag{13-53}$$

where $\boldsymbol{\kappa} = \kappa\hat{\mathbf{n}}$ and $\boldsymbol{\alpha} = \alpha\hat{\mathbf{n}}$, and κ and α are both real. The exponential factor in Eqs. (13-49) and (13-50) then has the more interpretable expression

$$e^{i(\mathbf{K}\cdot\mathbf{r} - \omega t)} = e^{i(\kappa\xi - \omega t)} e^{-\alpha\xi} \tag{13-54}$$

where $\xi = \hat{\mathbf{n}} \cdot \mathbf{r}$ is in essence the single spatial coordinate on which the plane wave depends (analogous to z when the wave propagates in the z direction). Further, the exponential factor $e^{-2\hat{\mathbf{n}}\cdot\mathbf{r}\,\text{Im}\,K}$ in $\langle \mathbf{S} \rangle$ and $\langle u_{\text{EM}} \rangle$ as given by Eqs. (13-51) and (13-52) reduces to $e^{-2\alpha\xi}$. Clearly, $\kappa = \text{Re}\,K$ relates to the oscillations of the wave in space and $\alpha = \text{Im}\,K$ describes an attenuation or absorption of the wave. (We assume $\alpha > 0$.) More specifically, κ determines the wavelength λ and the velocity of propagation[5] v_p of the monochromatic wave by

$$\lambda = \frac{2\pi}{\kappa} \qquad v_p = \frac{\omega}{\kappa} \tag{13-55}$$

[compare Eqs. (7-21) and (7-20)] and α determines the *penetration length* or *skin depth* δ by

$$\delta = \frac{1}{\alpha} \tag{13-56}$$

where δ is the distance within which the *amplitude* of the wave decays to $1/e$ of its value at $\xi = 0$. With c representing the speed of light in vacuum, we also define the (ordinary) *index of refraction* of the medium by

$$n = \frac{c}{v_p} = \frac{c\kappa}{\omega} \tag{13-57}$$

and the *complex index of refraction* by

[4]In some treatments of this subject, the exponential factor in plane waves is written in the form $\exp[i(\omega t - \mathbf{K}\cdot\mathbf{r})]$. In that form, decaying waves occur when Im $K < 0$.

[5]In Section 13-6, we shall refer to this velocity as the *phase* velocity to distinguish it from the *group* velocity to be defined there. Hence, we use the subscript p at the outset.

$$\eta = \frac{cK}{\omega} = \frac{c}{\omega}(\kappa + i\alpha) = n + i\frac{c}{\omega\delta} \tag{13-58}$$

both of which are in general dependent on frequency. The complex index of refraction is particularly useful in theoretical considerations, because it combines both the (ordinary) index of refraction and the penetration depth in a single quantity that reduces to the ordinary index of refraction in the absence of absorption (i.e., when $\alpha = 0$). However we choose to express these properties analytically, the essential *physical* features are a wavelength, velocity of propagation, and an (ordinary) index of refraction determined by $\kappa = \text{Re } K$ and an attenuation determined $\alpha = \text{Im } K$. Note also that the factor $\mathbf{K} \times \boldsymbol{\mathcal{E}}_0 = (\kappa + i\alpha)\hat{\mathbf{n}} \times \boldsymbol{\mathcal{E}}_0$ in Eq. (13-50) is complex when $\alpha \neq 0$, and attenuation of the wave is thus accompanied by a phase difference between $\boldsymbol{\mathcal{E}}$ and $\boldsymbol{\mathcal{H}}$.

To connect α and κ more directly with the physical characteristics of the medium, we begin by substituting $K = \kappa + i\alpha$ into the dispersion relation, Eq. (13-48), obtaining

$$\kappa^2 - \alpha^2 + 2i\alpha\kappa = \mu\omega^2\left(\epsilon + i\frac{g}{\omega}\right) \tag{13-59}$$

We now explicitly assume that ϵ, μ, and g are real at the particular frequency involved. (An interesting case in which this assumption is violated is treated in P13-57.) The real and imaginary parts of Eq. (13-59) are then readily separated to give the two equations

$$\kappa^2 - \alpha^2 = \mu\epsilon\omega^2 \qquad 2\alpha\kappa = \mu g\omega \tag{13-60}$$

which can be solved for κ and α. In terms of the quantity

$$\beta(\omega) = 1 + \sqrt{1 + \left(\frac{g}{\epsilon\omega}\right)^2} \tag{13-61}$$

the results are

$$\kappa^2 = \frac{1}{2}\mu\epsilon\omega^2\beta \qquad \alpha^2 = \frac{\mu\epsilon}{2\beta}\left(\frac{g}{\epsilon}\right)^2 \qquad \text{(13-62), (13-63)}$$

(P13-9). Thus, nonzero α and the consequent attenuation of the wave occur physically in media with nonzero conductivity. Further, in terms of the properties of the medium, the wave length λ, the velocity of propagation v_p, the penetration length δ, and the (ordinary) index of refraction n are given by

$$\lambda = \frac{2\pi}{\kappa} = \left(\frac{2\pi}{g}\sqrt{\frac{\epsilon}{\mu}}\right)\frac{g}{\epsilon\omega}\sqrt{\frac{2}{\beta}} \tag{13-64}$$

$$v_p = \frac{\omega}{\kappa} = \frac{1}{\sqrt{\mu\epsilon}}\sqrt{\frac{2}{\beta}} \tag{13-65}$$

$$\delta = \frac{1}{\alpha} = \left(\frac{2}{g}\sqrt{\frac{\epsilon}{\mu}}\right)\sqrt{\frac{\beta}{2}} \tag{13-66}$$

$$n = \frac{c}{v_p} = (\sqrt{\mu\epsilon}\,c)\sqrt{\frac{\beta}{2}} \tag{13-67}$$

Equations (13-64)–(13-67) are of particular interest in two special cases. Suppose first that the medium is a good conductor; i.e., suppose that g is large, specifically $g \gg \epsilon\omega$. (Equivalently, we can think of this case as a low-frequency limit, $\omega \ll g/\epsilon$; see P13-16.) In this limit β as given by Eq. (13-61) approaches $g/\epsilon\omega$, and Eqs. (13-64)–(13-67) have the limits

$$\lambda \longrightarrow 2\pi\sqrt{\frac{2}{\mu g\omega}} \qquad \frac{v_p}{c} \longrightarrow \sqrt{\frac{2\omega}{\mu g c^2}} \qquad \text{(13-68), (13-69)}$$

$$\delta \longrightarrow \sqrt{\frac{2}{\mu g\omega}} \qquad n \longrightarrow \sqrt{\frac{\mu g c^2}{2\omega}} \qquad \text{(13-70), (13-71)}$$

(P13-10). In particular, we expect these limits to apply for good conductors ($\mu \approx \mu_0 = 4\pi \times 10^{-7}$ N/A^2; $g \approx 10^7$ mho/m) in the microwave region ($\omega \approx 10^{10}$ sec^{-1}; $\lambda_0 =$ wavelength in vacuum ≈ 20 cm) where g is real (Section 13-1). With these values for the parameters, $\lambda \approx 10^{-4}\lambda_0$, $v_p \approx 10^{-4}c$, $\delta \approx 4 \times 10^{-4}$ cm, and $n \approx 10^4$. Microwaves in good conductors therefore have much smaller wavelengths than in vacuum, travel slowly (for electromagnetic waves), penetrate only very little, and are characterized by a very large index of refraction. Another aspect of wave propagation in good conductors is explored in P13-12.

The general results in Eqs. (13-64)–(13-67) can also be simplified when the medium is nonconducting ($g = 0$) or approximately so ($g \ll \epsilon\omega$). (Equivalently, we can think of this case as a high-frequency limit, $\omega \gg g/\epsilon$; see P13-16.) In this limit, β as given by Eq. (13-61) approaches the value 2, and Eqs. (13-64)–(13-67) reduce to

$$\lambda = \frac{2\pi}{\sqrt{\mu\epsilon}\,\omega} \qquad v_p = \frac{1}{\sqrt{\mu\epsilon}} \qquad \text{(13-72), (13-73)}$$

$$\delta = \sqrt{\frac{4\epsilon}{\mu g^2}} \longrightarrow \infty \qquad \text{(13-74)}$$

$$n = \sqrt{\mu\epsilon c^2} = \sqrt{\frac{\mu\epsilon}{\mu_0\epsilon_0}} = \sqrt{K_m K_e} \approx \sqrt{K_e} \qquad \text{(13-75)}$$

[See Eq. (7-20).] The final form in Eq. (13-75) follows because most optically transparent nonconductors are nonmagnetic and K_m is therefore approximately unity for the cases of interest. Although ω appears explicitly only in the expression for λ, all of these quantities may nonetheless depend on ω because ϵ, μ, and g may depend on ω. Tabulated indices of refraction, for example, can therefore be expected to agree with the square roots of tabulated dielectric constants only if the compared values are measured at the same frequency. Numerically, for a typical nonmagnetic transparent dielectric (a glass), for which $K_e \approx 2$, $K_m = 1$, and $g \approx 10^{-12}$ mho/m in the visible spectrum ($\omega \approx 4 \times 10^{15}$ sec^{-1}, $\lambda_0 =$ wavelength in vacuum $= 5000$ Å), we find that

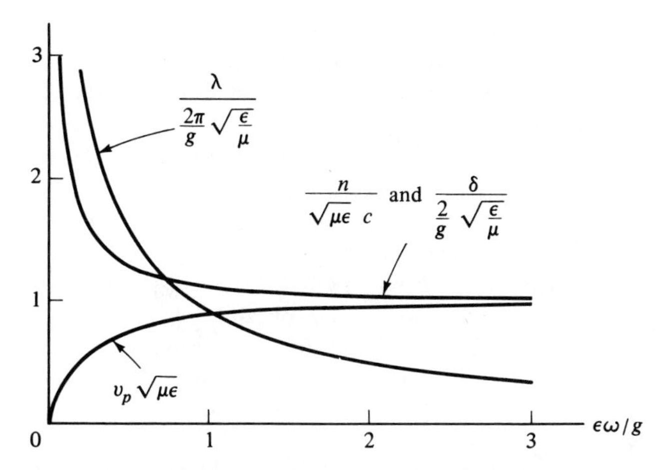

Fig. 13-3. Wavelength, velocity of propagation, penetration length, and index of refraction versus frequency for a plane wave in a conducting medium. The graphs show the indicated quantities under the assumption that ϵ, μ, and g are *independent* of frequency.

$\lambda \approx .7\lambda_0$, $v_p \approx .7c$, $\delta \approx 7 \times 10^9$ m(!), and $n \approx 1.4$. Graphs of λ, v_p, δ, and n reflecting the limiting forms in Eqs. (13-68)–(13-75) are shown in Fig. 13-3.

A further aspect of the limit $g \approx 0$ is that α is small enough to be set equal to zero [Eq. (13-63)]. Thus, the dispersion relation in Eq. (13-48) becomes

$$\kappa = \sqrt{\mu\epsilon}\,\omega \tag{13-76}$$

and the vector **K**, given by Eq. (13-53) as

$$\mathbf{K} = K\hat{\mathbf{n}} = \kappa\hat{\mathbf{n}} = \boldsymbol{\kappa} \tag{13-77}$$

becomes *real.* The results in Eqs. (13-49)–(13-52) therefore reduce to

$$\boldsymbol{\mathcal{E}}(\mathbf{r}, t) = \boldsymbol{\mathcal{E}}_0 e^{i(\boldsymbol{\kappa}\cdot\mathbf{r} - \omega t)} \tag{13-78}$$

$$\boldsymbol{\mathcal{H}}(\mathbf{r}, t) = \sqrt{\frac{\epsilon}{\mu}}\,\hat{\mathbf{n}} \times \boldsymbol{\mathcal{E}}_0 e^{i(\boldsymbol{\kappa}\cdot\mathbf{r} - \omega t)} \tag{13-79}$$

$$\langle \boldsymbol{\mathcal{S}} \rangle = \frac{|\boldsymbol{\mathcal{E}}_0|^2}{2\omega\mu}\boldsymbol{\kappa} = \frac{1}{2}\sqrt{\frac{\epsilon}{\mu}}\,|\boldsymbol{\mathcal{E}}_0|^2\hat{\mathbf{n}} \tag{13-80}$$

$$\langle u_{\mathrm{EM}} \rangle = \frac{1}{2}\epsilon|\boldsymbol{\mathcal{E}}_0|^2 \tag{13-81}$$

and there is no attenuation.

PROBLEMS

P13-6. (a) Substitute Eq. (13-41) into Eqs. (13-21)–(13-24) to derive the conditions of Eqs. (13-44)–(13-47) on the constant amplitudes of a plane monochromatic field. (b) Derive the dispersion relation, Eq. (13-48), by the method outlined in the text and also by evaluating the cross product of Eq. (13-47) with **K**. (c) Show that Eq. (13-48) also follows if the fields in Eq. (13-41) are required to satisfy Eq. (13-25).

P13-7. Given the dispersion relation, Eq. (13-48), rewrite the boundary condition on $\mathcal{E}_{0z}$ in Eq. (13-36) in terms of K^2.

P13-8. Derive Eqs. (13-51) and (13-52) for $\langle \mathbf{S} \rangle$ and $\langle u_{\mathrm{EM}} \rangle$ from Eqs. (13-26) and (13-27). *Hints:* (1) Allow for the possibility of complex **K**, so that, for example, $[\exp(i\mathbf{K}\cdot\mathbf{r})]^* = \exp(-i\mathbf{K}^*\cdot\mathbf{r})$. (2) See Eq. (D-10).

P13-9. Solve Eq. (13-60) for κ and α, obtaining Eqs. (13-62) and (13-63). *Hints:* (1) Solve $2\alpha\kappa = \mu g\omega$ for α and substitute into $\kappa^2 - \alpha^2 = \mu\epsilon\omega^2$. The result is an equation quadratic in κ^2. (2) Physically, κ is real and so $\kappa^2 > 0$.

P13-10. Verify that Eqs. (13-68)–(13-71) follow from Eqs. (13-64)–(13-67) when $g \gg \epsilon\omega$.

P13-11. The conductivity of sea water is about 4.3 mho/m. Assume $\mu = \mu_0$ and take $\epsilon \approx 75\epsilon_0$, which is about the value of ϵ for pure water at low frequency. (a) For what frequencies can Eq. (13-70) be applied? (b) Calculate the penetration depth in sea water at the typical radio frequency of 10^5 Hz ($\omega = 2\pi \times 10^5\ \mathrm{sec}^{-1}$) and comment on the suitability of ordinary radio signals as a means of underwater communication, say, between submarines.

P13-12. Show that for a good conductor ($g \gg \epsilon\omega$) $K = \kappa + i\alpha \rightarrow \sqrt{\mu g\omega}\, \exp(i\pi/4)$ and hence infer from Eqs. (13-49) and (13-50) that the $\mathcal{H}$- and $\mathcal{E}$-fields are out of phase by 45° in a good conductor.

P13-13. Show that when ϵ, μ, and g are real, the complex index of refraction can be written in the form $\eta = n[1 + (2i/\mu g\omega\delta^2)]$. What does η become in the two limits $g \gg \epsilon\omega$ and $g \ll \epsilon\omega$?

P13-14. Use a computer to determine representative points on the graphs in Fig. 13-3 and verify the quantitative correctness of those graphs.

P13-15. Examine κ and α as given by Eqs. (13-62) and (13-63) in the limits $g \ll \epsilon\omega$ and $g \gg \epsilon\omega$ and sketch graphs of κ and α versus $\epsilon\omega/g$.

P13-16. In the context of this section, high and low frequencies are divided by the frequency $\omega_c = g/\epsilon$. Numerically, what is this frequency for good conductors ($g \approx 10^7$ mho/m; $\epsilon \approx \epsilon_0$) and for poor conductors ($g \approx 10^{-12}$ mho/m; $\epsilon \approx 2\epsilon_0$)? In what region of the spectrum does ω_c lie in each case?

P13-17. Show that a plane monochromatic wave having frequency ω and wave number κ, propagating in a nonconductor in a direction $\hat{\mathbf{n}}$ lying in the x-z plane at an angle θ to the z axis, and linearly polarized in the y direction is described by the fields

$$\mathcal{E}(\mathbf{r}, t) = \mathcal{E}_0 \hat{\mathbf{j}} e^{i[\kappa(z\cos\theta + x\sin\theta) - \omega t]}$$

$$\mathcal{H}(\mathbf{r}, t) = \sqrt{\frac{\epsilon}{\mu}}\, \mathcal{E}_0(-\cos\theta\,\hat{\mathbf{i}} + \sin\theta\,\hat{\mathbf{k}}) e^{i[\kappa(z\cos\theta + x\sin\theta) - \omega t]}$$

The angle θ is positive when measured from the positive z axis toward the positive x axis.

13-4
Transmission and Reflection at Plane Interfaces

When a beam of light is incident on the interface between two different media, in general some of the incident energy is reflected back into the medium of incidence and some is transmitted across the interface. An important problem in classical optics therefore is to relate the intensity of the transmitted and reflected beams to the characteristics of the incident beam and to the properties of the two media. Representing beams of light (which we assume to be monochromatic) by electromagnetic waves, we can now bring classical electromagnetic theory to bear on this optical problem. To be sure, we should represent a physical beam of light by a wave of finite (temporal) duration rather than by a monochromatic wave of infinite duration. Even when we recognize that macroscopic beams of light are usually made up of individual wave trains whose duration is comparable to the lifetime of excited atomic states ($\approx 10^{-8}$ sec), however, the individual wave trains contain very many ($\approx 6 \times 10^7$ at 5000 Å) complete cycles and each wave train is difficult indeed to distinguish from one that contains an infinite number of cycles. If the "light bulb" used as a source is a laser, in which wave trains from individual atoms are locked together into much longer wave trains, the approximation of infinite wave trains is all the more applicable. We do, however, confine our attention to interfaces whose area is large compared to the square of the wavelength of the light (so we can ignore diffraction) and to wavelengths that are large compared to atomic dimensions (so we can treat matter as a macroscopic continuum). Although we shall consider reflection and transmission only at plane interfaces, in fact (properly interpreted) our results will be valid also for curved interfaces provided the curvature is not significant over distances on the order of several wavelengths. With these restrictions on applicability, we therefore conclude that a suitable qualitative electromagnetic model for treating an optical transmission and reflection problem theoretically involves picturing three plane monochromatic electromagnetic waves: an incident wave, a reflected wave, and a transmitted wave (Fig. 13-4). In a general problem, the amplitude, polarization, direction, and frequency of the incident wave will be given and we seek these characteristics of the reflected and transmitted waves. A suitable general strategy involves the following steps:

(1) Using whichever of the results in Section 13-3 is appropriate, write analytic expressions for the incident, reflected, and transmitted waves. Several unknown amplitudes, directions, etc., will be introduced.
(2) Obtain a set of equations for the unknowns by imposing the appropriate boundary conditions at the interface.
(3) Solve the resulting equations for the unknowns in terms of the (known) characteristics of the incident wave.

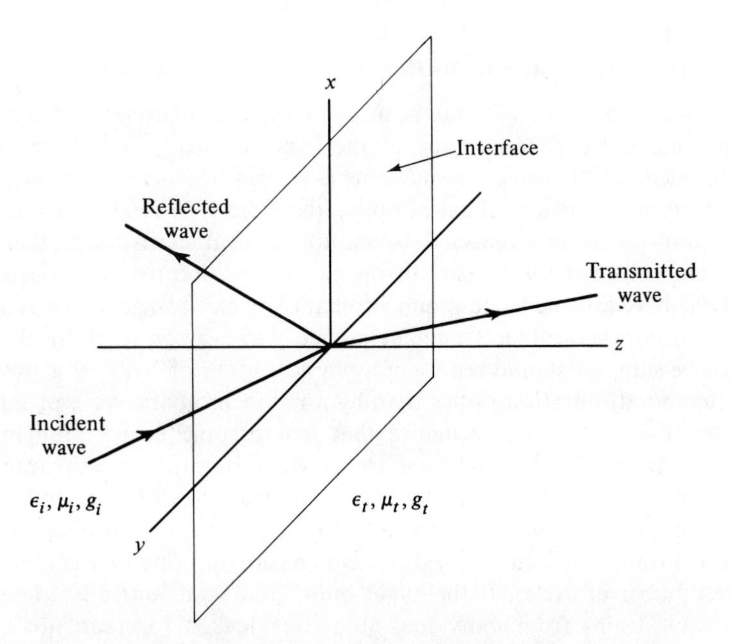

Fig. 13-4. The three waves present in a reflection/transmission problem. The medium of incidence lies in the region $z < 0$ and the medium of transmission lies in the region $z > 0$. The plane interface dividing these two media coincides with the x-y plane.

(4) Calculate the rates $I_i = |\langle \mathbf{S}_i \rangle \cdot \hat{\mathbf{k}}|$, $I_r = |\langle \mathbf{S}_r \rangle \cdot \hat{\mathbf{k}}|$, and $I_t = |\langle \mathbf{S}_t \rangle \cdot \hat{\mathbf{k}}|$ at which the incident, reflected, and transmitted waves transport energy in a direction *along the normal* to the interface. Here, $\hat{\mathbf{k}}$ is the unit vector normal to the interface in Fig. 13-4.

(5) Calculate the reflection and transmission coefficients R and T by applying the definitions

$$R = \begin{pmatrix} \text{fraction of incident} \\ \text{energy that is reflected} \end{pmatrix} = \frac{I_r}{I_i} = \frac{|\langle \mathbf{S}_r \rangle \cdot \hat{\mathbf{k}}|}{|\langle \mathbf{S}_i \rangle \cdot \hat{\mathbf{k}}|} \tag{13-82}$$

$$T = \begin{pmatrix} \text{fraction of incident} \\ \text{energy that is transmitted} \end{pmatrix} = \frac{I_t}{I_i} = \frac{|\langle \mathbf{S}_t \rangle \cdot \hat{\mathbf{k}}|}{|\langle \mathbf{S}_i \rangle \cdot \hat{\mathbf{k}}|} \tag{13-83}$$

Let us illustrate the approach with a simple example before discussing the more general case of Fig. 13-4. Consider a linearly polarized plane wave propagating in a nonconductor and incident normally on the plane surface of an imperfect conductor. Let the incident direction define the z axis and let the interface be at $z = 0$. The three waves involved are shown in Fig. 13-5. We first write analytic expressions for these three waves. If the incident wave is polarized in the x direction, we take $\boldsymbol{\mathcal{E}}_{i0} = \mathcal{E}_{i0}\hat{\mathbf{i}}$, $\boldsymbol{\kappa} = \kappa_i\hat{\mathbf{k}}$, and $\hat{\mathbf{n}} = \hat{\mathbf{k}}$ in Eqs. (13-78) and (13-79) to find that the incident wave is represented analyti-

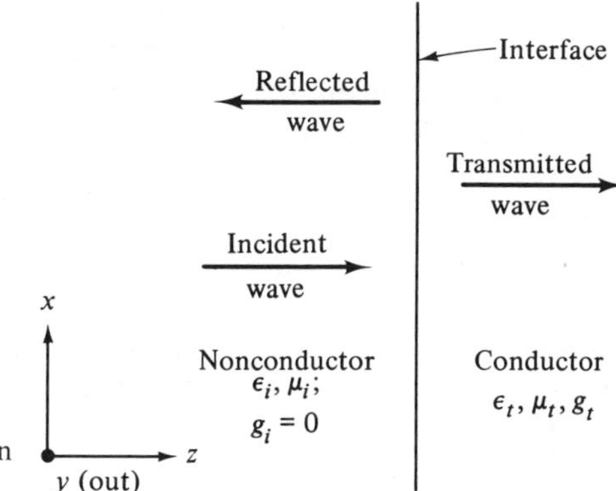

Fig. 13-5. A reflection/transmission problem at normal incidence.

cally by

$$\mathcal{E}_i(z, t) = \mathcal{E}_{i0}\hat{\mathbf{i}}e^{i(\kappa_i z - \omega t)} \qquad \text{(incident)} \qquad (13\text{-}84)$$
$$\mathcal{H}_i(z, t) = \sqrt{\frac{\epsilon_i}{\mu_i}}\,\mathcal{E}_{i0}\hat{\mathbf{j}}e^{i(\kappa_i z - \omega t)}$$

Now, the frequencies of all three waves must be the same (Section 13-2) and hence by Eq. (13-76) the wave numbers of the incident and reflected waves must be the same. Thus, with $\boldsymbol{\mathcal{E}}_{r0} = \mathcal{E}_{r0}\hat{\mathbf{i}}$, $\hat{\mathbf{n}} = -\hat{\mathbf{k}}$, and $\boldsymbol{\kappa} = -\kappa_i\hat{\mathbf{k}}$, we find from Eqs. (13-78) and (13-79) that the reflected wave is represented analytically by the fields

$$\mathcal{E}_r(z, t) = \mathcal{E}_{r0}\hat{\mathbf{i}}e^{-i(\kappa_i z + \omega t)} \qquad \text{(reflected)} \qquad (13\text{-}85)$$
$$\mathcal{H}_r(z, t) = -\sqrt{\frac{\epsilon_i}{\mu_i}}\,\mathcal{E}_{r0}\hat{\mathbf{j}}e^{-i(\kappa_i z + \omega t)}$$

Finally, the analytic representation

$$\mathcal{E}_t(z, t) = \mathcal{E}_{t0}\hat{\mathbf{i}}e^{i(K_t z - \omega t)} \qquad \text{(transmitted)} \qquad (13\text{-}86)$$
$$\mathcal{H}_t(z, t) = \frac{K_t}{\omega\mu_t}\,\mathcal{E}_{t0}\hat{\mathbf{j}}e^{i(K_t z - \omega t)}$$

for the transmitted wave follows from Eqs. (13-49) and (13-50) on setting $\boldsymbol{\mathcal{E}}_{t0} = \mathcal{E}_{t0}\hat{\mathbf{i}}$ and $\mathbf{K} = K_t\hat{\mathbf{k}}$. In Eqs. (13-85) and (13-86), we have made the reasonable (but tacit) "assumption" that the reflected and transmitted waves are polarized in the same direction as the incident wave. (It is shown in P13-18 that, in fact, this requirement follows from the boundary conditions.) Throughout Eqs. (13-84)–(13-86), the dispersion relations

$$\kappa_i = \omega\sqrt{\mu_i\epsilon_i} \qquad K_t^2 = \mu_t\omega^2\left(\epsilon_t + i\frac{g_t}{\omega}\right) \qquad (13\text{-}87)$$

are assumed.

We now find the unknown amplitudes in the reflected and transmitted waves by imposing the proper boundary conditions on the fields at $z = 0$.

We must, of course, add the incident and reflected waves to find the total field in the medium of incidence. Then, identifying medium 1 with the medium of incidence and medium 2 with the medium of transmission, we find by applying the conditions on $\mathcal{E}_{0x}$ and $\mathcal{H}_{0y}$ in Eq. (13-36) that

$$\begin{gathered}\mathcal{E}_{i0} + \mathcal{E}_{r0} = \mathcal{E}_{t0} \\ \sqrt{\frac{\epsilon_i}{\mu_i}}\,(\mathcal{E}_{i0} - \mathcal{E}_{r0}) = \frac{K_t}{\omega\mu_t}\,\mathcal{E}_{t0}\end{gathered} \tag{13-88}$$

(*remember:* $z = 0$); all of the remaining conditions in Eq. (13-36) are automatically satisfied by the above fields. The solution of Eq. (13-88) for the unknowns $\mathcal{E}_{r0}$ and $\mathcal{E}_{t0}$ now is

$$\mathcal{E}_{r0} = \frac{1-\zeta}{1+\zeta}\,\mathcal{E}_{i0} \qquad \mathcal{E}_{t0} = \frac{2}{1+\zeta}\,\mathcal{E}_{i0}; \qquad \zeta = \frac{K_t}{\omega\mu_t}\sqrt{\frac{\mu_i}{\epsilon_i}} \tag{13-89}$$

and we have found the amplitudes of the reflected and transmitted waves.

Finally, we find the reflection and transmission coefficients. From Eqs. (13-80) and (13-51) with $\hat{\mathbf{n}} = \pm\hat{\mathbf{k}}$, we find first that

$$\begin{aligned} I_i &= |\langle \mathbf{S}_i \rangle \cdot \hat{\mathbf{k}}| = \frac{\kappa_i |\mathcal{E}_{i0}|^2}{2\omega\mu_i} \\ I_r &= |\langle \mathbf{S}_r \rangle \cdot \hat{\mathbf{k}}| = \frac{\kappa_i |\mathcal{E}_{r0}|^2}{2\omega\mu_i} \\ I_t &= |\langle \mathbf{S}_t \rangle \cdot \hat{\mathbf{k}}| = \frac{|\mathcal{E}_{t0}|^2 \operatorname{Re} K_t}{2\omega\mu_t} \end{aligned} \tag{13-90}$$

(*Remember again:* $z = \hat{\mathbf{n}}\cdot\mathbf{r} = 0$.) Substituting Eq. (13-90) into the definitions in Eqs. (13-82) and (13-83), we then find the expressions

$$\begin{aligned} R &= \frac{I_r}{I_i} = \frac{|\mathcal{E}_{r0}|^2}{|\mathcal{E}_{i0}|^2} = \left|\frac{1-\zeta}{1+\zeta}\right|^2 \\ T &= \frac{I_t}{I_i} = \frac{\mu_i \operatorname{Re} K_t}{\mu_t \kappa_i}\,\frac{|\mathcal{E}_{t0}|^2}{|\mathcal{E}_{i0}|^2} = \frac{4\mu_i \operatorname{Re} K_t}{\mu_t \kappa_i |1+\zeta|^2} \end{aligned} \tag{13-91}$$

for the reflection and transmission coefficients R and T.

Two limiting cases of these general results are of interest. In the first case, the medium of transmission is a perfect conductor, for which $g_t \longrightarrow \infty$ and $K_t \longrightarrow \sqrt{\mu_t g_t \omega}\, e^{i\pi/4}$ (P13-12). Thus, ζ as given by Eq. (13-89) becomes infinite in proportion to $\sqrt{g_t}$ and, in accordance with Eqs. (13-89) and (13-91), we find that, as $g_t \longrightarrow \infty$,

$$\mathcal{E}_{r0} \longrightarrow -\mathcal{E}_{i0} \qquad \mathcal{E}_{t0} \longrightarrow 0 \tag{13-92}$$

$$R \longrightarrow 1 \qquad T \longrightarrow 0 \tag{13-93}$$

Thus, if the medium of transmission is a good conductor, the incident wave is (almost) completely reflected and there is a 180° phase shift between the incident and reflected waves.

The second interesting limiting case occurs when the medium of transmission is a nonconductor, for which $g_t = 0$. Then $K_t = \kappa_t = \omega\sqrt{\mu_t\epsilon_t}$ [Eq. (13-87)], and

$$\zeta = \sqrt{\frac{\epsilon_t\mu_i}{\epsilon_i\mu_t}} \approx \sqrt{\frac{\epsilon_t/\epsilon_0}{\epsilon_i/\epsilon_0}} = \frac{n_t}{n_i} \tag{13-94}$$

where n_i and n_t are the indices of refraction of the two media. [See Eq. (13-89), recall that μ_i and $\mu_t \approx \mu_0$ for nonmagnetic, optically transparent dielectrics, and then note Eq. (13-75).] For reflection and transmission at normal incidence on an interface between two dielectrics, we therefore find from Eqs. (13-89) and (13-91) that

$$\mathcal{E}_{r0} = \frac{n_i - n_t}{n_i + n_t}\,\mathcal{E}_{i0} \qquad \mathcal{E}_{t0} = \frac{2\,n_i}{n_i + n_t}\,\mathcal{E}_{i0} \tag{13-95}$$

$$R = \left(\frac{n_i - n_t}{n_i + n_t}\right)^2 \qquad T = \frac{4n_i\,n_t}{(n_i + n_t)^2} \tag{13-96}$$

Thus, at normal incidence on a real interface ($n_t \neq n_i$) between two dielectrics, both the reflected and transmitted waves have nonzero amplitude and there is a phase shift of 0° or 180° between the incident and reflected waves depending on whether $n_i > n_t$ or $n_i < n_t$, respectively; the transmitted wave is always in phase with the incident wave. Graphs of R and T are shown in Fig. 13-6, and it is shown in P13-19 that $R + T = 1$.

We now turn to the more general problem illustrated in Fig. 13-4. Let the interface define the x-y plane; let the incident, reflected, and transmitted waves be characterized, respectively, by propagation vectors $\mathbf{K}_i$, $\mathbf{K}_r$, and $\mathbf{K}_t$, making angles θ_i, θ_r, and θ_t with the normal $\hat{\mathbf{n}}$ (or here $\hat{\mathbf{k}}$) to the interface;

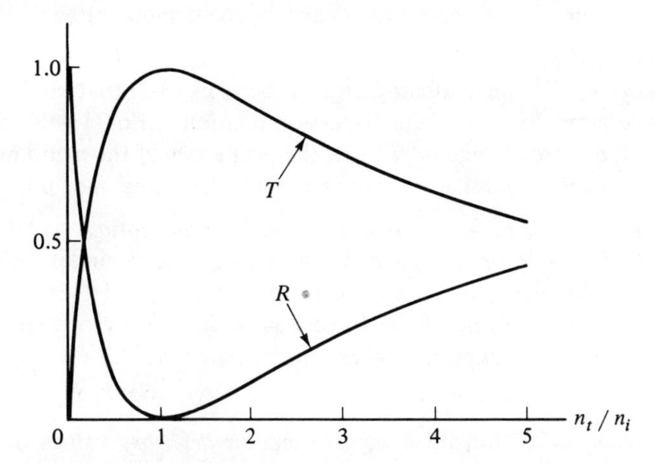

Fig. 13-6. Reflection and transmission coefficients for a wave incident normally on a dielectric interface.

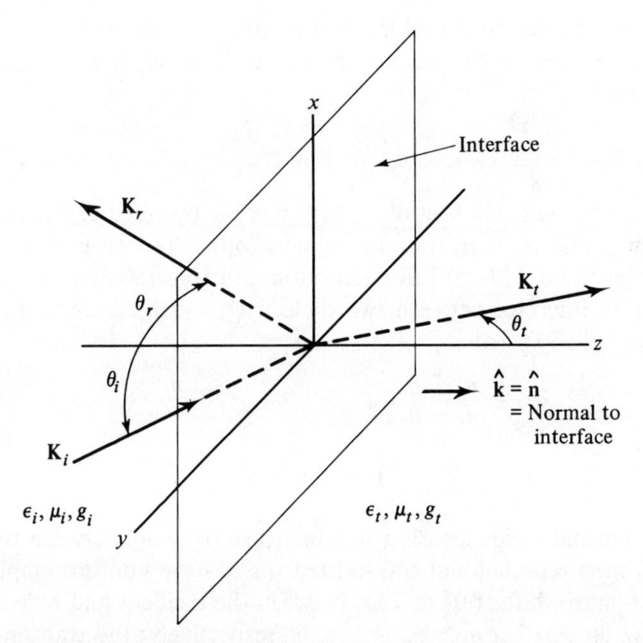

Fig. 13-7. Coordinates and vectors for treating a general reflection/transmission problem.

and let the *plane of incidence* determined by the vectors $\mathbf{K}_i$ and $\hat{\mathbf{n}}$ define the y-z plane. The angles θ_i, θ_r, and θ_t are measured positively as shown by the arrows in Fig. 13-7 and are called the *angle of incidence*, the *angle of reflection*, and the *angle of transmission* (or the *angle of refraction*), respectively. Several general requirements on these waves can now be demonstrated:

(1) *The incident and reflected waves are characterized by the* same *value of* K, *i.e.*, $K_i = K_r$. We have already argued (Section 13-2) that all frequencies must be the same. In view of the dispersion relation in Eq. (13-48), the quantity K can therefore change only when the properties of the medium change. But the incident and reflected waves are in the *same* medium. Q.E.D.

(2) *The incident, reflected, and transmitted propagation vectors are all parallel to the plane of incidence.* Each term in the boundary conditions at $z = 0$ will involve a factor of the form $e^{i\mathbf{K}\cdot\mathbf{r}}|_{z=0} = e^{i(K_x x + K_y y)}$, where the vector $\mathbf{K}$ will refer to one of the three waves. Continuity of the tangential component of $\boldsymbol{\mathcal{E}}$, for example, will require in part that

$$\mathcal{E}_{0x}^{(i)} e^{i(K_{ix}x + K_{iy}y)} + \mathcal{E}_{0x}^{(r)} e^{i(K_{rx}x + K_{ry}y)} = \mathcal{E}_{0x}^{(t)} e^{i(K_{tx}x + K_{ty}y)} \tag{13-97}$$

This boundary condition must be satisfied for *arbitrary* values of x and y (Why?), which is equivalent to requiring that the exponential factors cancel from Eq. (13-97). Cancellation will occur for arbitrary x and y, however, only if

$$K_{ix} = K_{rx} = K_{tx} \tag{13-98}$$

and

$$K_{iy} = K_{ry} = K_{ty} \tag{13-99}$$

Now, the incident wave certainly lies in the plane of incidence, so $K_{ix} = 0$. Equation (13-98) then requires that $K_{rx} = K_{tx} = 0$. Q.E.D.

(3) *The angle of incidence is equal to the angle of reflection.* From the geometry of Fig. 13-7,

$$K_{iy} = -K_i \sin \theta_i \qquad K_{ry} = -K_r \sin \theta_r \tag{13-100}$$

Since $K_i = K_r$ [item (1) above], Eqs. (13-99) and (13-100) together yield the proof:

$$K_{iy} = K_{ry} \Longrightarrow \sin \theta_i = \sin \theta_r \Longrightarrow \theta_i = \theta_r \tag{13-101}$$

This property is, of course, the classical law of specular reflection.

(4) *The angle of incidence and the angle of transmission satisfy Snell's law,* $\eta_i \sin \theta_i = \eta_t \sin \theta_t$, where η_i and η_t are the *complex* indices of refraction of the two media. [See Eq. (13-58).] Again, combining the geometry of Fig. 13-7 with Eq. (13-99), we find that

$$\begin{aligned} K_{iy} = K_{ty} &\Longrightarrow K_i \sin \theta_i = K_t \sin \theta_t \\ &\Longrightarrow \eta_i \sin \theta_i = \eta_t \sin \theta_t \end{aligned} \tag{13-102}$$

where the final form follows after multiplying the original form by c/ω and using Eq. (13-58). Q.E.D. In particular, when *both* media are nonconductors, η_i and η_t reduce to the ordinary (real) indices of refraction n_i and n_t, and Eq. (13-102) becomes

$$n_i \sin \theta_i = n_t \sin \theta_t \tag{13-103}$$

Depending on the nature of the two media, Eq. (13-102) or Eq. (13-103) determines θ_t when θ_i is given. Now, θ_i can range from 0 to $\frac{1}{2}\pi$ and is a real angle. Since η_i and η_t are in general complex, however, Eq. (13-102) cannot be satisfied unless we sometimes interpret θ_t as a complex angle. (See Appendix D.) Even when Eq. (13-103) applies and n_i and n_t are *real*, θ_t will be complex when $(n_i/n_t) \sin \theta_i > 1$, which occurs when $n_i > n_t$ and $\sin \theta_i > n_t/n_i$. In the case of two nonconductors, complex θ_t corresponds physically to what is called *total reflection;* there is a "transmitted" wave, but it has a number of curious properties and in particular all of the incident energy appears in the reflected wave (P13-51).

(5) *The incident wave may be polarized at any direction in a plane perpendicular to* $\mathbf{K}_i$, *and in particular an incident wave with an arbitrary polarization can be regarded as a superposition of a wave polarized* perpendicular *to the plane of incidence* [Fig. 13-8(a)] *and a wave polarized* parallel *to the plane of incidence* [Fig. 13-8(b)]. It is therefore sufficient to treat only these two special cases.

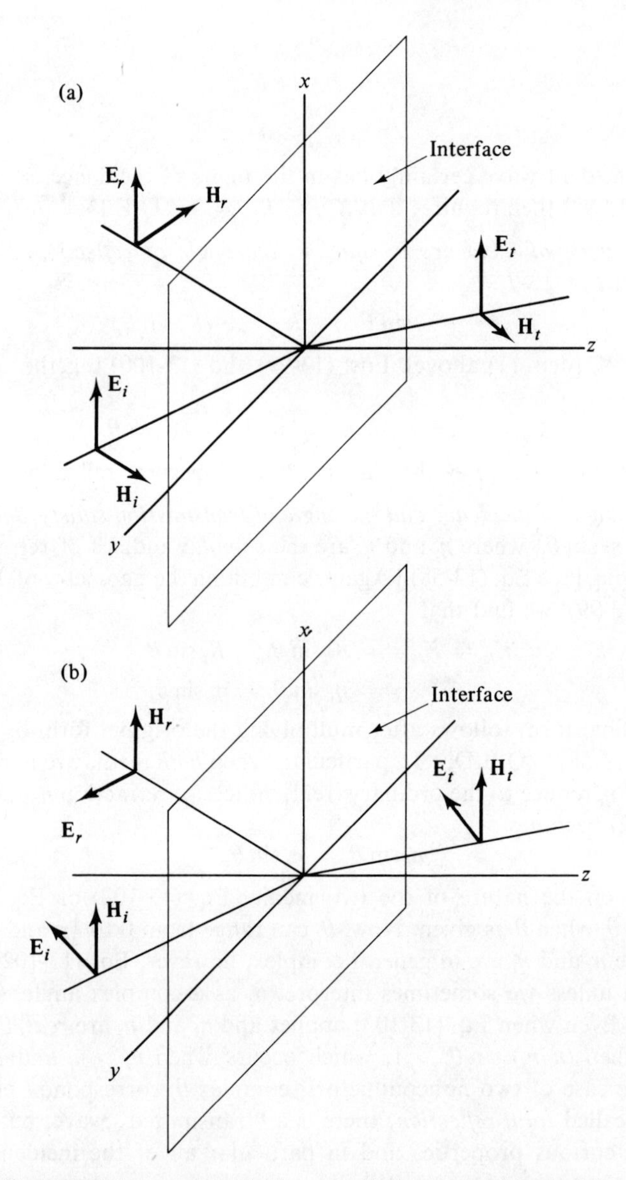

Fig. 13-8. Waves polarized (a) perpendicular to and (b) parallel to the plane of incidence.

(6) *If the incident wave is polarized parallel (perpendicular) to the plane of incidence, the reflected and transmitted waves are also polarized parallel (perpendicular) to the plane of incidence*, as shown in Fig. 13-8. To prove this property, assume an incident wave containing *only one* polarization, assume reflected and transmitted waves containing *both* polarizations, and then show that the boundary conditions at the interface cannot be satisfied unless the

"wrong" polarizations in the reflected and transmitted waves have zero amplitude. The details are left to P13-24.

To illustrate the more general calculation, let us determine the amplitudes of the reflected and transmitted waves when the incident wave is polarized perpendicular to the plane of incidence. To find analytic expressions for the three waves, we set

$$\boldsymbol{\mathcal{E}}_{i0} = \mathcal{E}_{i0}\hat{\mathbf{i}} \qquad \boldsymbol{\mathcal{E}}_{r0} = \mathcal{E}_{r0}\hat{\mathbf{i}} \qquad \boldsymbol{\mathcal{E}}_{t0} = \mathcal{E}_{t0}\hat{\mathbf{i}} \tag{13-104}$$

and

$$\begin{aligned} \mathbf{K}_i &= K_i[\cos\theta_i\hat{\mathbf{k}} - \sin\theta_i\hat{\mathbf{j}}] \\ \mathbf{K}_r &= K_i[-\cos\theta_i\hat{\mathbf{k}} - \sin\theta_i\hat{\mathbf{j}}] \\ \mathbf{K}_t &= K_t[\cos\theta_t\hat{\mathbf{k}} - \sin\theta_t\hat{\mathbf{j}}] \end{aligned} \tag{13-105}$$

(*Remember:* $\theta_r = \theta_i$ and $K_r = K_i$.) The vectors in square brackets are unit vectors $\hat{\mathbf{n}}_i$, $\hat{\mathbf{n}}_r$, and $\hat{\mathbf{n}}_t$ in the direction of propagation of the three waves, and all of these vectors were found by examining the geometry of Fig. 13-7. We now find analytic expressions for the waves by substituting Eqs. (13-104) and (13-105) into Eqs. (13-49) and (13-50); the results are

$$\begin{aligned} \boldsymbol{\mathcal{E}}_i(\mathbf{r}, t) &= \mathcal{E}_{i0}\hat{\mathbf{i}}e^{i(\mathbf{K}_i\cdot\mathbf{r}-\omega t)} \\ \boldsymbol{\mathcal{H}}_i(\mathbf{r}, t) &= \frac{K_i}{\omega\mu_i}\mathcal{E}_{i0}[\cos\theta_i\hat{\mathbf{j}} + \sin\theta_i\hat{\mathbf{k}}]e^{i(\mathbf{K}_i\cdot\mathbf{r}-\omega t)} \end{aligned} \tag{13-106}$$

$$\begin{aligned} \boldsymbol{\mathcal{E}}_r(\mathbf{r}, t) &= \mathcal{E}_{r0}\hat{\mathbf{i}}e^{i(\mathbf{K}_r\cdot\mathbf{r}-\omega t)} \\ \boldsymbol{\mathcal{H}}_r(\mathbf{r}, t) &= \frac{K_i}{\omega\mu_i}\mathcal{E}_{r0}[-\cos\theta_i\hat{\mathbf{j}} + \sin\theta_i\hat{\mathbf{k}}]e^{i(\mathbf{K}_r\cdot\mathbf{r}-\omega t)} \end{aligned} \tag{13-107}$$

$$\begin{aligned} \boldsymbol{\mathcal{E}}_t(\mathbf{r}, t) &= \mathcal{E}_{t0}\hat{\mathbf{i}}e^{i(\mathbf{K}_t\cdot\mathbf{r}-\omega t)} \\ \boldsymbol{\mathcal{H}}_t(\mathbf{r}, t) &= \frac{K_t}{\omega\mu_t}\mathcal{E}_{t0}[\cos\theta_t\hat{\mathbf{j}} + \sin\theta_t\hat{\mathbf{k}}]e^{i(\mathbf{K}_t\cdot\mathbf{r}-\omega t)} \end{aligned} \tag{13-108}$$

The boundary conditions in Eq. (13-36) are applicable to the present case, with medium 1 identified with the medium of incidence and medium 2 identified with the medium of transmission. Many of these conditions are automatically satisfied; only the conditions on $\mathcal{E}_{0x}$, $\mathcal{H}_{0y}$, and $\mathcal{H}_{0z}$ are nontrivial, and they yield

$$\mathcal{E}_{i0} + \mathcal{E}_{r0} = \mathcal{E}_{t0} \tag{13-109}$$

$$\frac{K_i}{\mu_i}(\mathcal{E}_{i0} - \mathcal{E}_{r0})\cos\theta_i = \frac{K_t}{\mu_t}\mathcal{E}_{t0}\cos\theta_t \tag{13-110}$$

$$K_i(\mathcal{E}_{i0} + \mathcal{E}_{r0})\sin\theta_i = K_t\mathcal{E}_{t0}\sin\theta_t \tag{13-111}$$

where the exponential factors must cancel and will if (as we have already assumed) the conditions discussed in items (2)–(4) above are satisfied. Now, since $K_i \sin\theta_i = K_t \sin\theta_t$ [Eq. (13-102)], Eq. (13-111) is identical with Eq. (13-109). Thus, Eqs. (13-109) and (13-110) are two equations for the two unknowns $\mathcal{E}_{r0}$ and $\mathcal{E}_{t0}$. In terms of the complex index of refraction [Eq.

(13-58)], Eq. (13-110) assumes the more convenient form

$$\mu_t\eta_i(\mathcal{E}_{i0} - \mathcal{E}_{r0})\cos\theta_i = \mu_i\eta_t\mathcal{E}_{t0}\cos\theta_t \tag{13-112}$$

Finally, simultaneous solution of Eqs. (13-109) and (13-112) yields the results

$$\mathcal{E}_{r0}^{\perp} = \frac{\mu_t\eta_i\cos\theta_i - \mu_i\eta_t\cos\theta_t}{\mu_t\eta_i\cos\theta_i + \mu_i\eta_t\cos\theta_t}\,\mathcal{E}_{i0}^{\perp} \tag{13-113}$$

$$\mathcal{E}_{t0}^{\perp} = \frac{2\mu_t\eta_i\cos\theta_i}{\mu_t\eta_i\cos\theta_i + \mu_i\eta_t\cos\theta_t}\,\mathcal{E}_{i0}^{\perp} \tag{13-114}$$

where the superscript $\perp$ has been added as a reminder that these results apply when the incident wave is polarized perpendicular to the plane of incidence. Equations (13-113) and (13-114) are two of *Fresnel's equations*, the other two being

$$\mathcal{E}_{r0}^{\parallel} = \frac{-\mu_i\eta_t\cos\theta_i + \mu_t\eta_i\cos\theta_t}{\mu_i\eta_t\cos\theta_i + \mu_t\eta_i\cos\theta_t}\,\mathcal{E}_{i0}^{\parallel} \tag{13-115}$$

$$\mathcal{E}_{t0}^{\parallel} = \frac{2\mu_t\eta_i\cos\theta_i}{\mu_i\eta_t\cos\theta_i + \mu_t\eta_i\cos\theta_t}\,\mathcal{E}_{i0}^{\parallel} \tag{13-116}$$

which apply to the case of an incident wave polarized parallel to the plane of incidence provided unit vectors in the direction of the electric field are chosen to point in the directions indicated in Fig. 13-8(b) (P13-25). These equations, which give the amplitudes of the reflected and transmitted waves for the two polarizations, are quite general; they apply not only when η_i and η_t are complex but also when θ_t must be interpreted as a complex angle.

We shall examine the predictions of Fresnel's equations only for the special case in which both media are nonmagnetic ($\mu_i = \mu_t = \mu_0$) and are nonconductors ($\eta_i = n_i$; $\eta_t = n_t$). Further, we shall restrict our considerations to circumstances for which θ_t is real (i.e., we do not consider total reflection). Under these circumstances, Fresnel's equations reduce to

$$\mathcal{E}_{r0}^{\perp} = \frac{n_i\cos\theta_i - n_t\cos\theta_t}{n_i\cos\theta_i + n_t\cos\theta_t}\,\mathcal{E}_{i0}^{\perp} \tag{13-117}$$

$$\mathcal{E}_{t0}^{\perp} = \frac{2n_i\cos\theta_i}{n_i\cos\theta_i + n_t\cos\theta_t}\,\mathcal{E}_{i0}^{\perp} \tag{13-118}$$

$$\mathcal{E}_{r0}^{\parallel} = \frac{-n_t\cos\theta_i + n_i\cos\theta_t}{n_t\cos\theta_i + n_i\cos\theta_t}\,\mathcal{E}_{i0}^{\parallel} \tag{13-119}$$

$$\mathcal{E}_{t0}^{\parallel} = \frac{2n_i\cos\theta_i}{n_t\cos\theta_i + n_i\cos\theta_t}\,\mathcal{E}_{i0}^{\parallel} \tag{13-120}$$

Further, we find from the definitions in Eqs. (13-82) and (13-83) that the reflection and transmission coefficients for the two polarizations are given by

$$R_{\perp} = \frac{I_r}{I_i} = \frac{|\langle\mathbf{S}_r\rangle\cdot\hat{\mathbf{k}}|}{|\langle\mathbf{S}_i\rangle\cdot\hat{\mathbf{k}}|} = \frac{|\mathcal{E}_{r0}^{\perp}|^2\cos\theta_i}{|\mathcal{E}_{i0}^{\perp}|^2\cos\theta_i}$$

$$= \left(\frac{n_i\cos\theta_i - n_t\cos\theta_t}{n_i\cos\theta_i + n_t\cos\theta_t}\right)^2 \tag{13-121}$$

$$T_\perp = \frac{I_t}{I_i} = \frac{|\langle \mathbf{S}_t \rangle \cdot \hat{\mathbf{k}}|}{|\langle \mathbf{S}_i \rangle \cdot \hat{\mathbf{k}}|} = \sqrt{\frac{\epsilon_t \mu_i}{\mu_t \epsilon_i}} \frac{|\mathcal{E}_{t0}^\perp|^2 \cos\theta_t}{|\mathcal{E}_{i0}^\perp|^2 \cos\theta_i}$$

$$= \frac{4 n_i n_t \cos\theta_i \cos\theta_t}{(n_i \cos\theta_i + n_t \cos\theta_t)^2} \tag{13-122}$$

$$R_\parallel = \left(\frac{-n_t \cos\theta_i + n_i \cos\theta_t}{n_t \cos\theta_i + n_i \cos\theta_t} \right)^2 \tag{13-123}$$

$$T_\parallel = \frac{4 n_i n_t \cos\theta_i \cos\theta_t}{(n_t \cos\theta_i + n_i \cos\theta_t)^2} \tag{13-124}$$

where the Poynting vectors have been evaluated by substituting Eqs. (13-104) and (13-105) (and the analogous expressions for the parallel polarization) into Eq. (13-80) and the multiplying square root in Eq. (13-122) has been written using the approximations $\mu = \mu_0$ and $\sqrt{\epsilon} = n\sqrt{\epsilon_0}$ for both media. We note now the following properties of these results:

(7) *All of these results for* both *polarizations reduce* to Eqs. (13-95) and (13-96) at normal incidence ($\theta_i = \theta_t = 0$), for which special case the plane of incidence is not defined.

(8) For *both* polarizations, *the transmitted wave and the incident wave are in phase*, because the coefficients multiplying $\mathcal{E}_{i0}^\perp$ and $\mathcal{E}_{i0}^\parallel$ in Eqs. (13-118) and (13-120) are real and positive.

(9) For the *perpendicular* polarization, *the reflected wave* and *the incident wave are in phase when* $n_t < n_i$ *and* 180° *out of phase when* $n_t > n_i$. At normal incidence ($\theta_i = \theta_t = 0$), the coefficient in Eq. (13-117) is positive when $n_t < n_i$ and negative when $n_t > n_i$. At oblique incidence, the coefficient will be positive when

$$n_i \cos\theta_i - n_t \cos\theta_t > 0 \tag{13-125}$$

or (in view of Snell's law and some trigonometric identities) when

$$n_i \cos\theta_i - \frac{n_i \sin\theta_i}{\sin\theta_t} \cos\theta_t = \frac{n_i \sin(\theta_t - \theta_i)}{\sin\theta_t} > 0$$

$$\Longrightarrow \theta_t > \theta_i \tag{13-126}$$

Again from Snell's law, however, $\theta_t > \theta_i$ requires that $n_t < n_i$. In reverse, the coefficient will be negative when $\theta_t < \theta_i$ or $n_t > n_i$, and all aspects of property (9) are established.

(10) For the *parallel* polarization, *the reflected wave and the incident wave are in phase* (a) *if* $n_t < n_i$ *and* $\theta_t + \theta_i < \frac{1}{2}\pi$ *or* (b) *if* $n_t > n_i$ *and* $\theta_t + \theta_i > \frac{1}{2}\pi$ *and out of phase otherwise*. For normal incidence, $\theta_t + \theta_i = 0$ and case (a) applies. At normal incidence, the coefficient in Eq. (13-119) is positive for $n_t < n_i$ and negative for $n_t > n_i$. At oblique incidence, we use Snell's law (and some trigonometric identities) to rewrite the numerator of that coefficient in the form

$$-n_t \cos\theta_i + n_i \cos\theta_t = \frac{n_i[\sin 2\theta_t - \sin 2\theta_i]}{2\sin\theta_t}$$

$$= \frac{n_i \sin(\theta_t - \theta_i)\cos(\theta_t + \theta_i)}{2\sin\theta_t} \qquad (13\text{-}127)$$

which can be positive only if $\theta_t > \theta_i$ ($n_t < n_i$) *and* $\theta_t + \theta_i < \frac{1}{2}\pi$ or if $\theta_t < \theta_i$ ($n_t > n_i$) *and* $\theta_t + \theta_i > \frac{1}{2}\pi$. Q.E.D.

(11) For the *perpendicular* polarization, *the transmitted wave has zero amplitude only at grazing incidence* ($\theta_i = \frac{1}{2}\pi$) *and the reflected wave has zero amplitude only in the trivial case of no interface,* $n_t = n_i$. From Eq. (13-118), $\mathcal{E}_{t0}^{\perp} = 0$ only if $\theta_i = \frac{1}{2}\pi$. For the reflected wave at normal incidence ($\theta_t = \theta_i = 0$), $\mathcal{E}_{r0}^{\perp}$ as given by Eq. (13-117) will be zero only if $n_t = n_i$. For the reflected wave at oblique incidence, $\mathcal{E}_{r0}^{\perp}$ can be zero only if $\theta_t = \theta_i$ [see Eq. (13-126)], which implies $n_t = n_i$. Q.E.D.

(12) For the *parallel* polarization, *the transmitted wave has zero amplitude only at grazing incidence* ($\theta_i = \frac{1}{2}\pi$) *and the reflected wave has zero amplitude not only for the trivial case of no interface,* $n_t = n_i$, *but also when* $n_t \neq n_i$ *and* $\theta_t + \theta_i = \frac{1}{2}\pi$, equivalent to $\theta_i = \tan^{-1}(n_t/n_i)$. The proofs for the transmitted wave and for the reflected wave when $n_t = n_i$ are similar to those in item (11) above and are left to the reader. The remainder of this property follows from Eq. (13-127), which gives the numerator in Eq. (13-119). When $n_t \neq n_i$ ($\theta_t \neq \theta_i$), Eq. (13-127) yields zero for that numerator if $\theta_t + \theta_i = \frac{1}{2}\pi$. Q.E.D. Snell's law in turn then gives

$$n_i \sin\theta_i = n_t \sin(\tfrac{1}{2}\pi - \theta_i) = n_t \cos\theta_i \Longrightarrow \theta_i = \tan^{-1}(n_t/n_i) \qquad (13\text{-}128)$$

The critical angle $\theta_B = \tan^{-1}(n_t/n_i)$ is called *Brewster's angle. Unpolarized* light incident at this angle results in a reflected beam *polarized* perpendicular to the plane of incidence, and this phenomenon can be used to produce polarized light.

Several graphs showing various aspects of the results in Eqs. (13-117)–(13-124) appear in Fig. 13-9.

We shall conclude this section with a reminder that all of the results contained herein have been obtained by formal application of the principles of electromagnetic theory. We have made *no* reference to direct results of optical experiments per se. Thus, *as we have obtained them*, the results of this section are *predictions* of optical behavior, not summaries of experimental observations, even though in most cases the optical properties were known long before electromagnetic theory was developed. The agreement of these predictions with observations is therefore substantial indirect evidence supporting the correctness of Maxwell's equations and represents one of the major achievements of late nineteenth-century physics.

PROBLEMS

P13-18. Add waves polarized in the y direction to Eqs. (13-85) and (13-86) and show that the boundary conditions cannot be satisfied unless the amplitudes of these added waves are zero.

P13-19. Show that $R + T = 1$, with R and T given by Eq. (13-96). *Optional:* Show that $R + T = 1$, with R and T given by Eq. (13-91).

P13-20. Find an expression for the transmission coefficient given by Eq. (13-91) in the limit of large (but not infinite) conductivity. Take the medium of incidence to be vacuum and assume $\mu_t = \mu_0$. Substitute a typical conductivity for a good conductor ($g_t \approx 10^7$ mho/m) and determine numerical values of T for radio waves, microwaves, and visible light.

P13-21. Substitute Eq. (13-92) into Eqs. (13-84)–(13-86) to obtain expressions for the incident and reflected waves when the plane wave in Eq. (13-84) is incident normally on a *perfect* conductor and then use the boundary conditions in Eq. (13-38) to find the surface charge density and the surface current density induced on the surface of the conductor.

P13-22. Obtain graphs of θ_t versus θ_i as given by Eq. (13-103) for several different values of n_i/n_t. Include values both larger and smaller than unity. *Suggestion:* Use a computer.

P13-23. Assume θ_i and θ_t are real in Eq. (13-102). Then the real part of the equation is identical with Eq. (13-103). Show that the imaginary part of the equation can be written in the form $(\mu_i g_i/n_i) \sin \theta_i = (\mu_t g_t/n_t) \sin \theta_t$ and that this form is identical with Eq. (13-103) *when both media are good conductors.*

P13-24. Following the method outlined in the text, prove property (6) for one or the other of the polarizations.

P13-25. Write down equations analogous to Eqs. (13-104) and (13-105) but describing waves polarized parallel to the plane of incidence, find the fields in the three waves, and apply the boundary conditions to derive Eqs. (13-115) and (13-116).

P13-26. Derive the law of reflection from a perfect conductor by considering a wave that is linearly polarized perpendicular to the plane of incidence and incident obliquely on the plane surface of a perfect conductor. Find also the surface charge density σ and the surface current density $\mathbf{j}$ induced on the surface of the conductor. *Optional:* Do the same for an incident wave polarized in the plane of incidence.

P13-27. A linearly polarized, monochromatic plane wave having wavelength λ (in vacuum) is incident normally on a thin, transparent dielectric film in vacuum. Let the film have thickness d and index of refraction n. Determine the transmission and reflection coefficients and sketch graphs of each as functions of λ, assuming n to be independent of frequency. *Hint:* Take the plane of the film to be vertical and introduce five waves: incident and reflected waves in the medium of incidence, left and right traveling waves in the film,

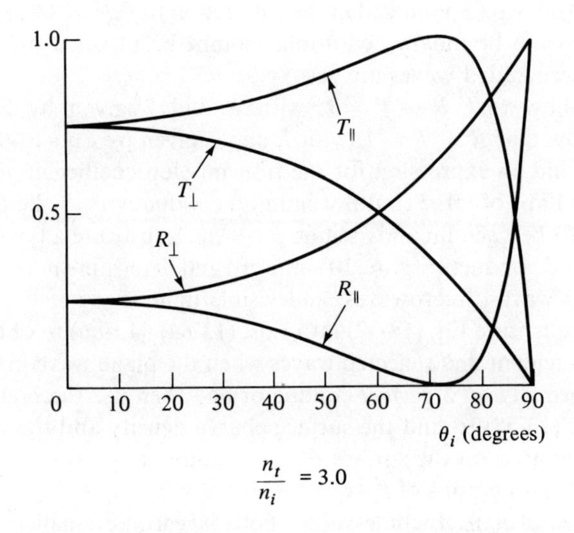

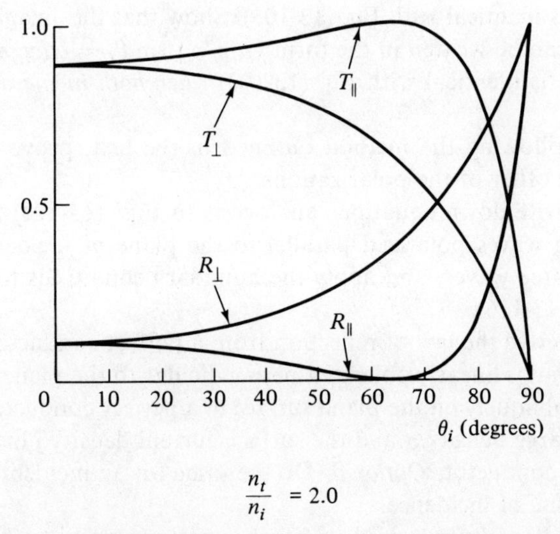

Fig. 13-9. Reflection and transmission coefficients as a function of angle of incidence. Graphs are shown for both incident polarizations and for several values of n_t/n_i.

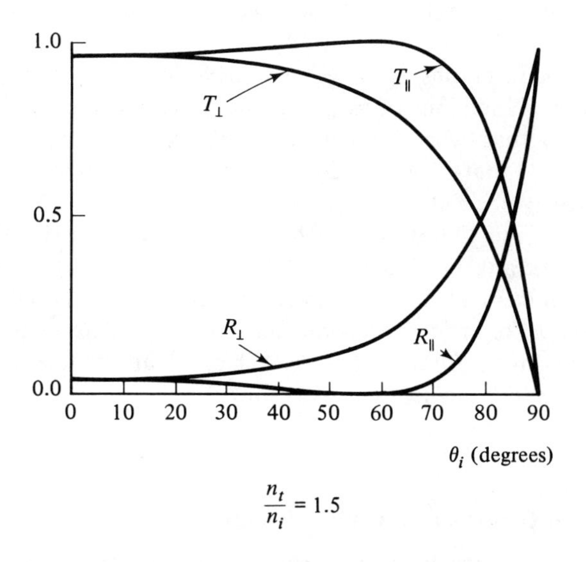

$$\frac{n_t}{n_i} = 1.5$$

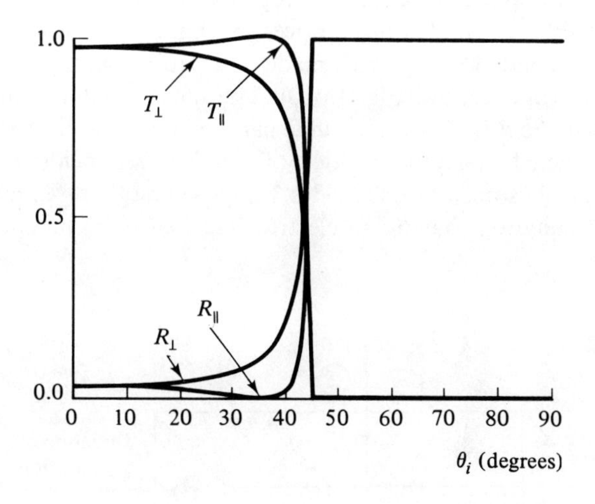

$$\frac{n_t}{n_i} = 0.7$$

Fig. 13-9. (Continued)

and a transmitted wave in the medium of transmission. This problem involves techniques similar to those used in the theory of nonreflecting coatings.

P13-28. Obtain graphs of Brewster's angle and of the critical angle for total reflection as functions of n_t/n_i over the range $0.1 < n_t/n_i < 10$ and, in particular, show that (when both angles exist) Brewster's angle is always the smaller one. Note that only real angles in the range 0° to 90° are physically meaningful. *Suggestion:* Use a computer.

P13-29. (a) Find the expression to which $R_\perp$ as given by Eq. (13-121) reduces when the angle of incidence is Brewster's angle. (b) Find Brewster's angle and $R_\perp$ numerically if $n_i = 1$ and $n_t = 1.5$. *Optional:* (1) Obtain a graph of $R_\perp$ versus n_t/n_i for incidence at Brewster's angle. (2) Three glass plates with $n = 1.45$, 1.55, and 1.70 are available. Which would you select to build a polarizer to use with incident beams in air? Why?

13-5
Wave Guides and Cavity Resonators

We shall consider now some characteristics of monochromatic solutions to Maxwell's equations in regions that are partially or totally bounded by perfectly conducting surfaces, beginning with a *wave guide* consisting of an evacuated hollow pipe made of a perfect conductor and having a rectangular cross section (Fig. 13-10). The most general monochromatic electromagnetic field in this wave guide can be very complicated, but it can also be constructed as a superposition of simpler basic fields or *modes*. We shall examine only some of these modes. Suppose first that we seek a solution to Maxwell's equations having an electric field that is (1) independent of x,

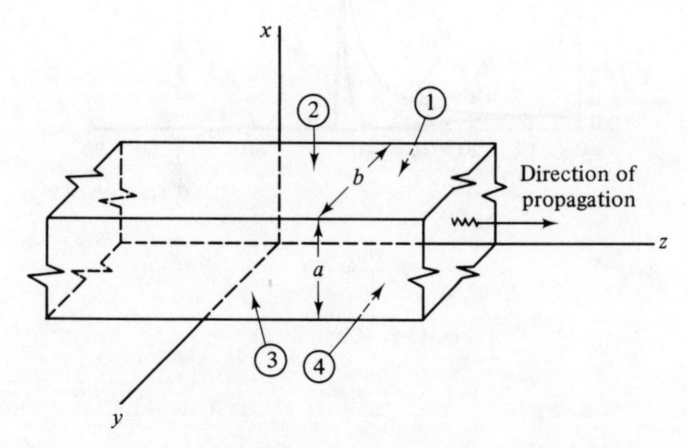

Fig. 13-10. A wave guide with a rectangular cross section. All four walls of the wave guide are constructed of perfect conductors.

(2) polarized parallel to the x axis, and (3) propagating in the positive z direction along the guide. The (complex) electric field in such a solution has the analytic representation

$$\boldsymbol{\mathcal{E}}(\mathbf{r}, t) = \mathcal{E}_0(y)\hat{\mathbf{i}}e^{i(\kappa_z z - \omega t)} \tag{13-129}$$

and, in accordance with Eq. (13-22), the associated (complex) magnetic intensity is

$$\begin{aligned}\boldsymbol{\mathcal{H}}(\mathbf{r}, t) &= \frac{\nabla \times \boldsymbol{\mathcal{E}}(\mathbf{r}, t)}{i\omega\mu_0} \\ &= \frac{1}{i\omega\mu_0}\left(i\kappa_z\mathcal{E}_0(y)\hat{\mathbf{j}} - \frac{\partial\mathcal{E}_0(y)}{\partial y}\hat{\mathbf{k}}\right)e^{i(\kappa_z z - \omega t)}\end{aligned} \tag{13-130}$$

Now, the field $\boldsymbol{\mathcal{E}}_0(\mathbf{r})$ obtained from Eq. (13-129) by deleting the factor $e^{-i\omega t}$ must satisfy Eq. (13-25) with $\mu\epsilon = \mu_0\epsilon_0 = 1/c^2$ and $g = 0$. Thus, if we set

$$\kappa_y^2 = -\kappa_z^2 + \frac{\omega^2}{c^2} \tag{13-131}$$

then $\mathcal{E}_0(y)$ must satisfy

$$\frac{d^2\mathcal{E}_0}{dy^2} + \kappa_y^2\mathcal{E}_0 = 0 \tag{13-132}$$

and we find that

$$\mathcal{E}_0(y) = A \sin \kappa_y y + B \cos \kappa_y y \tag{13-133}$$

As in Eq. (13-38), the tangential components of $\boldsymbol{\mathcal{E}}_0(\mathbf{r})$ must now be made zero at all (perfectly) conducting surfaces. In particular, the x-component of $\boldsymbol{\mathcal{E}}_0(\mathbf{r})$ must be zero on sides 1 and 3 in Fig. 13-10 and will be so only if $\mathcal{E}_0(0) = \mathcal{E}_0(b) = 0$ or if $B = 0$ and $\kappa_y b = n\pi$, where $n = 1, 2, 3, \cdots$. With these restrictions on $\mathcal{E}_0(y)$, we find finally that the fields in the wave guide are given by

$$\boldsymbol{\mathcal{E}}_{0n}^{\mathrm{TE}}(\mathbf{r}, t) = A \sin\left(\frac{n\pi y}{b}\right)\hat{\mathbf{i}}e^{i(\kappa_z z - \omega t)} \tag{13-134}$$

$$\boldsymbol{\mathcal{H}}_{0n}^{\mathrm{TE}}(\mathbf{r}, t) = \frac{A}{i\omega\mu_0}\left[i\kappa_z \sin\left(\frac{n\pi y}{b}\right)\hat{\mathbf{j}} - \frac{n\pi}{b}\cos\left(\frac{n\pi y}{b}\right)\hat{\mathbf{k}}\right]e^{i(\kappa_z z - \omega t)} \tag{13-135}$$

Here the subscripts and superscripts have been added to specify the mode to which these fields apply: The superscript TE indicates that the mode is a *transverse electric* mode, in which the electric field has no component in the direction of propagation along the guide (i.e., the electric field is transverse to this direction of propagation); the subscript $0n$ is a conventional two-index subscript that relates to the dependence of the fields on x and y, the value 0 here for the first index indicating that the fields are those in a mode that does not depend on x. Finally, the equation

$$\kappa_z^2 = \frac{\omega^2}{c^2} - \left(\frac{n\pi}{b}\right)^2 \tag{13-136}$$

determining κ_z from ω for the TE_{0n} mode follows from Eq. (13-131) when κ_y is set equal to $n\pi/b$. The reader may now verify that all of the boundary conditions in Eq. (13-38) either are satisfied by the fields in Eqs. (13-134) and (13-135) or yield values for surface currents and surface charges at each conducting boundary of the wave guide. Thus, Eqs. (13-134)–(13-136) certainly express *a* solution for the fields in the guide, even though they do not express the *only* solution.

Several properties of this guided wave are of interest. Note first that κ_z as given by Eq. (13-136) is purely imaginary if

$$\omega < \omega_c = \frac{n\pi c}{b} \tag{13-137}$$

where ω_c is a (lower) *cutoff frequency* below which the exponential factor in Eqs. (13-134) and (13-135) expresses a decay rather than a sinusoidal oscillation. Thus, waves for which $\omega < \omega_c$ or, equivalently, waves having a vacuum wavelength $\lambda > \lambda_c = 2b/n$ are attenuated rather than propagated in the guide. Practically, guides whose dimensions are on the order of centimeters are most common. Thus, the typical laboratory guide has a cutoff wavelength on the order of a few centimeters and propagates waves in the microwave region (and at shorter wavelengths) but attenuates waves at longer wavelengths.

A second property of guided waves can be inferred if we write the sine in Eq. (13-134) in its complex exponential form, finding that

$$\boldsymbol{\mathcal{E}}_{0n}^{\mathrm{TE}}(\mathbf{r}, t) = \frac{A}{2i}\hat{\mathbf{i}}(e^{i[(n\pi y/b)+\kappa_z z-\omega t]} - e^{i[-(n\pi y/b)+\kappa_z z-\omega t]}) \tag{13-138}$$

We can thus interpret the solution as the superposition of two plane waves, the first having a propagation vector

$$\boldsymbol{\kappa}_1 = \frac{n\pi}{b}\hat{\mathbf{j}} + \kappa_z\hat{\mathbf{k}} \tag{13-139}$$

and the second, whose electric field is 180° out of phase with that of the first, having a propagation vector

$$\boldsymbol{\kappa}_2 = -\frac{n\pi}{b}\hat{\mathbf{j}} + \kappa_z\hat{\mathbf{k}} \tag{13-140}$$

The relative orientation of these two vectors is shown in Fig. 13-11. Since $\boldsymbol{\kappa}_1$ and $\boldsymbol{\kappa}_2$ have the same z-component and have y-components differing only in sign, they both make the same angle with the walls, namely

$$\theta_1 = \theta_2 = \tan^{-1}\left(\frac{\kappa_z b}{n\pi}\right) \tag{13-141}$$

and we can therefore think of either wave as the reflection of the other in one of the walls. Even the phase difference between the two waves is consistent with that view. Propagation of this wave along the guide therefore involves successive reflections of the wave first from wall 1, then from wall 3, then from wall 1 again, and so on. The angle at which the wave is incident on the walls decreases as the mode number n increases, so higher modes involve more

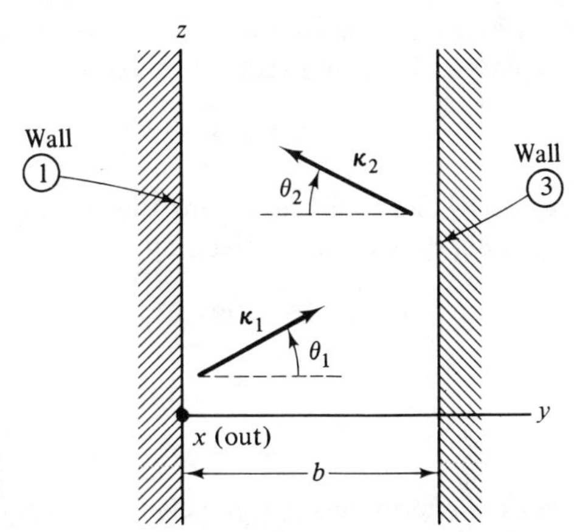

Fig. 13-11. The two propagation vectors for waves in a rectangular wave guide.

reflections per unit length along the guide than lower modes. At the cutoff frequency for the nth mode ($\kappa_z = 0$) Eq. (13-141) gives $\theta_1 = \theta_2 = 0$, and the two waves described in this paragraph are incident normally on the walls; $\boldsymbol{\kappa}_1$ and $\boldsymbol{\kappa}_2$ have no component along the guide and no wave is propagated.

Several different velocities and several different wavelengths are used to describe the guided wave in Eqs. (13-134) and (13-135). As measured in the direction of the vectors $\boldsymbol{\kappa}_1$ and $\boldsymbol{\kappa}_2$, for example, the wave is characterized by a wavelength λ_0 given by

$$\lambda_0 = \frac{2\pi}{\sqrt{\kappa_y^2 + \kappa_z^2}} = \frac{2\pi c}{\omega} \tag{13-142}$$

and propagates with speed c; the wavelength λ_0 is the vacuum wavelength corresponding to the frequency ω. As measured across the guide, the wave is characterized by a wavelength $\lambda_\perp$ given by

$$\lambda_\perp = \frac{2\pi}{\kappa_y} = \frac{2b}{n} \Longrightarrow b = \frac{1}{2} n\lambda_\perp \tag{13-143}$$

which supports the conclusion that the allowed wavelengths for a given guide are those that establish standing wave patterns between two of the walls. Finally, as measured along the guide, the wave is characterized by a wavelength $\lambda_\parallel$ given by

$$\lambda_\parallel = \frac{2\pi}{\kappa_z} = \frac{2\pi}{\sqrt{(\omega^2/c^2) - (n\pi/b)^2}} = \frac{\lambda_0}{\sqrt{1 - (\lambda_0/\lambda_\perp)^2}} \tag{13-144}$$

and propagates with a speed $v_\parallel$ given by

$$v_\parallel = \frac{\omega}{\kappa_z} = \frac{\omega}{\sqrt{(\omega^2/c^2) - (n\pi/b)^2}} = \frac{c}{\sqrt{1 - (n\pi c/b\omega)^2}} \tag{13-145}$$

Note that $\lambda_\parallel > \lambda_0$ and $v_\parallel > c$. This apparent contradiction of the limitations imposed by special relativity will be resolved in the next paragraph when we find that information (i.e., energy) is transmitted along the guide not with speed $v_\parallel$ but with a different speed that in fact is less than c.

Consider now the energy in the fields in this guide. We find first that the time-averaged Poynting vector is given by

$$\langle \mathbf{S} \rangle = \frac{1}{2} \operatorname{Re} (\mathbf{E} \times \mathbf{H}^*) = \frac{\kappa_z |A|^2}{2\omega\mu_0} \sin^2 \left(\frac{n\pi y}{b}\right) \hat{\mathbf{k}} \tag{13-146}$$

As expected, $\langle \mathbf{S} \rangle$ is directed along the guide. Similarly, we find that the time-averaged energy density is given by

$$\begin{aligned} \langle u_{\mathrm{EM}} \rangle &= \frac{1}{4} \operatorname{Re} (\mathbf{E} \cdot \mathbf{D}^* + \mathbf{B} \cdot \mathbf{H}^*) \\ &= \frac{1}{4} |A|^2 \left[\left(\epsilon_0 + \frac{\kappa_z^2}{\mu_0 \omega^2} \right) \sin^2 \left(\frac{n\pi y}{b}\right) + \frac{1}{\mu_0 \omega^2} \left(\frac{n\pi}{b}\right)^2 \cos^2 \left(\frac{n\pi y}{b}\right) \right] \end{aligned} \tag{13-147}$$

Neither the energy flux nor the energy density is uniform over the cross section of the guide. We can, however, calculate the average rate at which energy is transported along the guide by integrating $\langle \mathbf{S} \rangle$ over that cross section; we find that

$$\begin{aligned} \begin{pmatrix} \text{average rate of} \\ \text{energy transport} \end{pmatrix} &= \int \langle \mathbf{S} \rangle \cdot d\mathbf{S} = \int_0^b \frac{\kappa_z |A|^2}{2\omega\mu_0} \sin^2 \left(\frac{n\pi y}{b}\right) (a\, dy) \\ &= \frac{ab\kappa_z |A|^2}{4\omega\mu_0} \end{aligned} \tag{13-148}$$

Similarly, we find that the average energy stored in the fields in unit length of the guide is given by

$$\begin{pmatrix} \text{average energy} \\ \text{in unit length} \end{pmatrix} = \int_0^b \langle u_{\mathrm{EM}} \rangle a\, dy = \frac{1}{4} ab\epsilon_0 |A|^2 \tag{13-149}$$

If we now think of the average energy given by Eq. (13-149) as propagating down the guide with speed v_g so as to produce the average energy transport given by Eq. (13-148), it must be that

$$\begin{pmatrix} \text{average rate of} \\ \text{energy transport} \end{pmatrix} = v_g \begin{pmatrix} \text{average energy} \\ \text{in unit length} \end{pmatrix} \tag{13-150}$$

and thus that

$$v_g = \frac{\kappa_z}{\omega\mu_0\epsilon_0} = \frac{c^2}{v_\parallel} \Longrightarrow v_\parallel v_g = c^2 \tag{13-151}$$

Since $v_\parallel > c$, the speed v_g at which energy is propagated along the guide is smaller than c, in agreement with the limits imposed by special relativity. Another aspect of energy in guided waves relates to losses arising from the finite (even though large) conductivities of any real material used for the walls; the essential idea of a method for treating this complication is explored in P13-53.

As our notation indicates, we have considered only a very few of the possible waves in a rectangular wave guide. There are additional TE modes

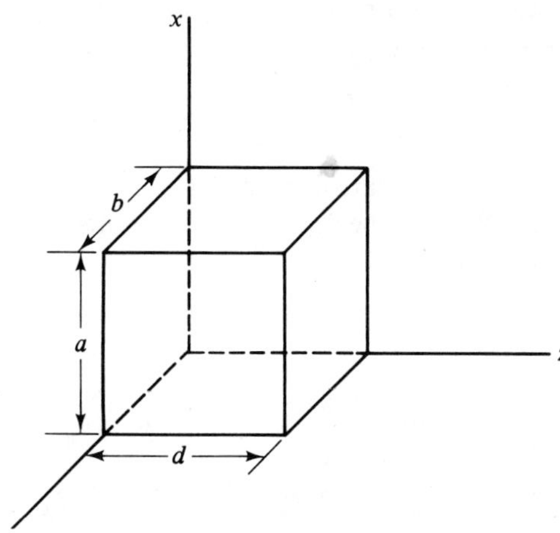

Fig. 13-12. A simple cavity resonator.

depending only on x and not on y (the TE_{m0} modes), there are more involved TE modes depending on both x and y (the TE_{mn} modes), and there is an analogous family of transverse magnetic modes (TM_{mn}) in which the magnetic intensity $\mathcal{H}$ is normal to the direction of propagation along the guide. Further, for each mode representing propagation toward $z = +\infty$ there is a corresponding mode representing propagation toward $z = -\infty$. The main characteristics of the propagation of guided waves, however, have emerged in our simple example and we therefore leave a treatment of these other modes (and also of wave guides having nonrectangular cross sections) to the problems and to other authors. We leave a discussion of wave guides that turn corners, change dimension, or contain dielectric fillers entirely to other authors.

As a second example of fields in a region bounded by conductors, consider a *cavity resonator*, which consists of an evacuated volume *completely* surrounded by conducting walls. The simplest such devices are shaped like rectangular parallelopipeds, right circular cylinders, and spheres. We shall consider here only the first shape (Fig. 13-12). Again we seek simple solutions in the interior of this region, realizing that more general solutions can be constructed by superposition. Thus, let us assume that

$$\boldsymbol{\mathcal{E}}(\mathbf{r}, t) = \boldsymbol{\mathcal{E}}_0(\mathbf{r})e^{-i\omega t} \tag{13-152}$$

with

$$\boldsymbol{\mathcal{E}}_0(\mathbf{r}) = \mathcal{E}_{0x}(\mathbf{r})\hat{\mathbf{i}} + \mathcal{E}_{0y}(\mathbf{r})\hat{\mathbf{j}} + \mathcal{E}_{0z}(\mathbf{r})\hat{\mathbf{k}} \tag{13-153}$$

Now, $\boldsymbol{\mathcal{E}}_0(\mathbf{r})$ and hence each of its (Cartesian) components must satisfy Eq. (13-25) with $\mu\epsilon = \mu_0\epsilon_0 = 1/c^2$ and $g = 0$; e.g.,

$$\left(\frac{\partial^2}{\partial x^2} + \frac{\partial^2}{\partial y^2} + \frac{\partial^2}{\partial z^2} + \frac{\omega^2}{c^2}\right)\mathcal{E}_{0x}(x, y, z) = 0 \tag{13-154}$$

Assuming that $\mathcal{E}_{0x}(x, y, z)$ can be factored into the form $X(x)Y(y)Z(z)$ (as in Section 8-4), we find that Eq. (13-154) is equivalent to

$$\frac{1}{X}\frac{d^2X}{dx^2} + \frac{1}{Y}\frac{d^2Y}{dy^2} + \frac{1}{Z}\frac{d^2Z}{dz^2} + \frac{\omega^2}{c^2} = 0 \tag{13-155}$$

which can be satisfied only if each of the first three terms is a constant by itself. Writing the separation constants as $-\kappa_x^2$, $-\kappa_y^2$, and $-\kappa_z^2$, we thus have that

$$\frac{d^2X}{dx^2} + \kappa_x^2 X = 0 \Longrightarrow X(x) = A \sin \kappa_x x + B \cos \kappa_x x \tag{13-156}$$

$$\frac{d^2Y}{dy^2} + \kappa_y^2 Y = 0 \Longrightarrow Y(y) = A' \sin \kappa_y y + B' \cos \kappa_y y \tag{13-157}$$

$$\frac{d^2Z}{dz^2} + \kappa_z^2 Z = 0 \Longrightarrow Z(z) = A'' \sin \kappa_z z + B'' \cos \kappa_z z \tag{13-158}$$

where the separation constants must satisfy

$$\kappa_x^2 + \kappa_y^2 + \kappa_z^2 = \frac{\omega^2}{c^2} \tag{13-159}$$

The relevant boundary conditions, Eq. (13-38), require that the tangential components of $\boldsymbol{\mathcal{E}}$ be zero on all conducting surfaces. Since $\mathcal{E}_{0x}(x, y, z)$ is a tangential component on four of the six surfaces, we must then have that

$$\mathcal{E}_{0x}(x, 0, z) = 0 \Longrightarrow Y(0) = 0 \Longrightarrow B' = 0 \tag{13-160}$$

$$\mathcal{E}_{0x}(x, b, z) = 0 \Longrightarrow Y(b) = 0 \Longrightarrow \kappa_y b = n\pi \tag{13-161}$$

$$\mathcal{E}_{0x}(x, y, 0) = 0 \Longrightarrow Z(0) = 0 \Longrightarrow B'' = 0 \tag{13-162}$$

$$\mathcal{E}_{0x}(x, y, d) = 0 \Longrightarrow Z(d) = 0 \Longrightarrow \kappa_z d = p\pi \tag{13-163}$$

where n and p are nonnegative integers. Further, we must require that

$$\nabla \cdot \boldsymbol{\mathcal{E}}_0 = 0 \Longrightarrow \frac{\partial \mathcal{E}_{0y}}{\partial y} + \frac{\partial \mathcal{E}_{0z}}{\partial z} = -\frac{\partial \mathcal{E}_{0x}}{\partial x} \tag{13-164}$$

In particular, this equation must be satisfied when $x = 0^+$ and $x = a^-$, for which coordinates $\mathcal{E}_{0y}$ and $\mathcal{E}_{0z}$ must both be zero *for all y and z* because they both are tangential components to the planes $x = 0$ and $x = a$. Consequently, neither $\mathcal{E}_{0y}$ nor $\mathcal{E}_{0z}$ change with y and z if $x = 0^+$ or $x = a^-$ and Eq. (13-164) reduces to $\partial \mathcal{E}_{0x}/\partial x = 0$ at $x = 0$ and $x = a$. This condition in turn requires that

$$\left.\frac{dX}{dx}\right|_{x=0} = 0 \Longrightarrow A = 0 \tag{13-165}$$

$$\left.\frac{dX}{dx}\right|_{x=a} = 0 \Longrightarrow \kappa_x a = m\pi \tag{13-166}$$

where m is a nonnegative integer. Thus, the boundary conditions reduce X, Y, and Z substantially and their product, which gives $\mathcal{E}_{0x}$, now is

$$\mathcal{E}_{0x}(x, y, z) = E_x \cos \kappa_x x \sin \kappa_y y \sin \kappa_z z \tag{13-167}$$

Similar calculations give

$$\mathcal{E}_{0y}(x, y, z) = E_y \sin \kappa_x x \cos \kappa_y y \sin \kappa_z z \tag{13-168}$$

$$\mathcal{E}_{0z}(x, y, z) = E_z \sin \kappa_x x \sin \kappa_y y \cos \kappa_z z \tag{13-169}$$

for the remaining two components of the $\boldsymbol{\mathcal{E}}$-field. Here E_x, E_y, and E_z are constants that are arbitrary except for the condition

$$\kappa_x E_x + \kappa_y E_y + \kappa_z E_z = 0 \tag{13-170}$$

imposed by Maxwell's equation $\nabla \cdot \boldsymbol{\mathcal{E}}_0 = 0$. Throughout these expressions, κ_x, κ_y, and κ_z can assume only the values

$$\kappa_x = \frac{m\pi}{a} \qquad \kappa_y = \frac{n\pi}{b} \qquad \kappa_z = \frac{p\pi}{d} \tag{13-171}$$

where m, n, and p are nonnegative integers, at most one of which can be zero. (If two or more are zero, the field is zero.) Evaluation of the $\boldsymbol{\mathcal{H}}$-field is left to P13-37. Note finally that when $m = 0$, $\mathcal{E}_{0y} = \mathcal{E}_{0z} = 0$ and the $\boldsymbol{\mathcal{E}}$-field in the cavity reduces to

$$\begin{aligned} \boldsymbol{\mathcal{E}}(\mathbf{r}, t) &= E_x \sin \kappa_y y \sin \kappa_z z \hat{\mathbf{i}} e^{-i\omega t} \\ &= \frac{E_x}{2i} \sin \kappa_y y \hat{\mathbf{i}} [e^{i(\kappa_z z - \omega t)} - e^{i(-\kappa_z z - \omega t)}] \end{aligned} \tag{13-172}$$

which can be viewed as a superposition of two waves of the TE_{0n} mode of the rectangular wave guide [Eq. (13-134)], one propagating toward $z = +\infty$ and the other propagating toward $z = -\infty$. The coefficients $(+1, -1)$ in the superposition and the values of κ_z, however, are restricted so that (1) each TE_{0n} wave can be regarded as a reflection of the other in the planes $z = 0$ and $z = d$ and (2) the tangential component of the resulting $\boldsymbol{\mathcal{E}}$-field is zero on the planes $z = 0$ and $z = d$. Those more general modes that follow from the above fields when $m \neq 0$ but $E_z = 0$ can also be viewed as a similar superposition of the TE_{mn} modes of the rectangular wave guide and are referred to as the TE_{mnp} modes of the resonator. The remaining modes of the resonator (in which E_z is not zero but H_z is zero) arise from a superposition of TM_{mn} modes of the guide and are called the TM_{mnp} modes of the resonator. The fields in all of these modes can be viewed as forming standing wave patterns in all three coordinate directions.

In contrast to the wave guide, in which *all* frequencies exceeding the cut-off frequency are allowed, the cavity resonator has a discrete frequency spectrum. Substitution of Eq. (13-171) into Eq. (13-159) gives

$$\omega_{mnp}^2 = \pi^2 c^2 \left(\frac{m^2}{a^2} + \frac{n^2}{b^2} + \frac{p^2}{d^2} \right) \tag{13-173}$$

for the frequency ω_{mnp} characterizing the TE_{mnp} mode. Since this frequency also characterizes the TM_{mnp} mode, there are at least two different modes corresponding to the frequency ω_{mnp} and all frequencies in this cavity are at least doubly degenerate. Additional degeneracies will occur when a, b, and d

are so related that ω_{mnp} has the same value for two or more different values of mnp. (For example, $\omega_{623} = \omega_{433}$ when $a = 2b$.) Finally, since only one of m, n, and p can be zero, the lowest resonant frequency of the cavity is that one of ω_{110}, ω_{101}, and ω_{011} for which the zero corresponds to the direction of the smallest dimension of the cavity. Thus, for example, a cavity with dimensions 1 cm $\times$ 2 cm $\times$ 4 cm will have a lowest resonant frequency ω_f given by

$$\omega_f^2 = \pi^2 c^2 \left(\frac{1}{2^2} + \frac{1}{4^2}\right) \text{cm}^{-2} \Longrightarrow \omega_f = 5.3 \times 10^{10} \text{sec}^{-1}$$

which corresponds to a vacuum wavelength of 3.5 cm.

As we have seen, wave guides and cavity resonators of reasonable dimensions are characterized by frequencies in the microwave region of the electromagnetic spectrum. This property contributes significantly to the importance of these devices in experimental work involving microwaves, for wave guides and cavity resonators are distinctly superior to conventional circuitry at these frequencies.

PROBLEMS

P13-30. Derive the final expressions in Eqs. (13-146) and (13-147) for $\langle \mathbf{S} \rangle$ and $\langle u_{\text{EM}} \rangle$ in a rectangular guide carrying a TE_{0n} wave.

P13-31. Find the (surface) charge and current densities induced on all walls in Fig. 13-10 when the guide carries the TE_{0n} wave.

P13-32. (a) Draw three figures like Fig. 13-11 (but without the propagation vectors) and then draw lines along which the physical **E**-field is zero at time $t = 0$ for the modes TE_{01}, TE_{02}, and TE_{03}, marking each region with a + or − sign to indicate the direction of the field in that region. (b) Find the physical **H**-field for these modes and sketch the lines of **H** in each figure.

P13-33. (a) For the rectangular wave guide of Fig. 13-10, find the complex fields for the transverse electric modes TE_{m0} in which the $\boldsymbol{\mathcal{E}}$-field is in the $\hat{\mathbf{j}}$ direction and all fields are independent of y. (b) Sketch figures showing the fields in the TE_{10}, TE_{20}, and TE_{30} modes.

P13-34. Find the complex fields for the transverse electric modes TE_{mn} in the rectangular wave guide of Fig. 13-10. *Hints:* (1) Let $\boldsymbol{\mathcal{E}}(\mathbf{r}, t) = [\mathcal{E}_{0x}(x, y)\hat{\mathbf{i}} + \mathcal{E}_{0y}(x, y)\hat{\mathbf{j}}] \exp[i(\kappa_z z - \omega t)]$ and use separation of variables on the equation analogous to Eq. (13-132). (2) The tangential components of the electric field must be zero at all conducting walls. (3) $\nabla \cdot \boldsymbol{\mathcal{E}} = 0$ for all x and y. (4) Find $\boldsymbol{\mathcal{E}}$ first; then find $\boldsymbol{\mathcal{H}}$ from Eq. (13-22).

P13-35. Let a cylindrical wave guide of radius a have its axis along the z axis. Assuming an electric field of the general form

$$\boldsymbol{\mathcal{E}}(\mathbf{r}, t) = \mathcal{E}_0(\imath)\hat{\boldsymbol{\phi}} e^{i(\kappa_z z - \omega t)}$$

find the fields and then find the longest three cutoff wavelengths if $a = 5$ cm. *Hints:* (1) Be careful with the Laplacian of a *vector* in cylindrical coordinates.

(2) The equation for $\mathcal{E}_0(r)$ is Bessel's equation of order *one*, and the only solution that is finite at the origin is the first-order Bessel function usually denoted by $J_1(x)$, which is a standard, tabulated function. In particular, $J_1(x) = 0$ for $x = 3.8317, 7.0156, 10.1735, \cdots$.

P13-36. The systematic development of Eqs. (13-168) and (13-169) by the method used to obtain Eq. (13-167) does not by itself require κ_x, κ_y, and κ_z to be the same in all three components of the field. Present a supplementary argument proving that κ_x, κ_y, and κ_z in fact *must* have the same value in all three components.

P13-37. Find the $\mathcal{H}$-field corresponding to the $\boldsymbol{\mathcal{E}}$-field in Eqs. (13-167)–(13-169).

P13-38. Let $a = 1$ cm, $b = 2$ cm, and $d = 3$ cm; use a computer to calculate ω_{mnp} for $0 \leq m, n, p \leq 4$, and show the allowed frequencies as horizontal lines in a diagram with a vertical frequency axis, labeling each line with the number(s) of the mode(s) to which it corresponds. This diagram is analogous to a quantum mechanical energy level diagram. *Suggestion:* Use the computer also to arrange the frequencies in increasing order.

13-6 Superposition of Waves of Different Frequency: Dispersion

Plane monochromatic waves in matter can be superposed in all of the ways that were discussed in Sections 7-3 and 7-4 for plane monochromatic waves in vacuum. Waves in matter therefore can be circularly and elliptically polarized and they can interfere with one another. They can also be superposed by summing or integrating over a spectrum of wave numbers to produce plane waves that are no longer sinusoidal. In this section, we shall consider briefly a phenomenon that occurs in matter but not in vacuum, restricting our consideration to unattenuated waves and therefore to nonconducting matter. The phenomenon is called *dispersion*, and it arises because *in matter* the velocity of propagation of a *purely mono*chromatic wave *depends on frequency*. Thus, the different frequencies composing a *poly*chromatic wave travel through matter at *different* speeds and, as a result, the overall wave form (think of a pulse of some initial shape) changes shape as the wave propagates through the matter. Further, as we shall see, the speed of propagation of the envelope of a polychromatic pulse will be different from—perhaps even quite different from—the speed of propagation of any one of its component frequencies. We must therefore distinguish two velocities: (1) the *phase velocity* v_p, which is the speed at which a purely monochromatic wave propagates and is given by

$$v_p = \frac{\omega}{\kappa} \tag{13-174}$$

and (2) the *group velocity* v_g, which is the speed at which the envelope of a polychromatic pulse propagates.

An expression for the group velocity is quickly obtained by considering a simple superposition of two linearly polarized waves having the same amplitude but slightly different frequencies and propagating toward $z = +\infty$. If the wave number κ depends on frequency, then two different wave numbers

$$\kappa = \kappa(\omega) \tag{13-175}$$

and

$$\kappa' = \kappa(\omega') = \kappa(\omega + \Delta\omega) = \kappa + \Delta\kappa \tag{13-176}$$

must be introduced. The electric field in this superposed wave is then given by

$$\begin{aligned}\boldsymbol{\mathcal{E}}(z, t) &= \mathbf{E}_0[e^{i(\kappa z - \omega t)} + e^{i(\kappa' z - \omega' t)}] \\ &= \mathbf{E}_0 e^{i(\bar{\kappa} z - \bar{\omega} t)}[e^{(1/2)i(\Delta\kappa\, z - \Delta\omega\, t)} + e^{-(1/2)i(\Delta\kappa\, z - \Delta\omega\, t)}] \\ &= 2\mathbf{E}_0 e^{i(\bar{\kappa} z - \bar{\omega} t)} \cos[\tfrac{1}{2}(\Delta\kappa\, z - \Delta\omega\, t)]\end{aligned} \tag{13-177}$$

where

$$\bar{\kappa} = \tfrac{1}{2}(\kappa + \kappa') \tag{13-178}$$

and

$$\bar{\omega} = \tfrac{1}{2}(\omega + \omega') \tag{13-179}$$

Physically, Eq. (13-177) represents a sinusoidal wave of frequency $\bar{\omega}$ propagating with a *phase velocity*

$$v_p = \frac{\bar{\omega}}{\bar{\kappa}} = \frac{\omega + \omega'}{\kappa + \kappa'} \xrightarrow[\Delta\omega \to 0]{} \frac{\omega}{\kappa} \tag{13-180}$$

on which has been superimposed an envelope (the cosine factor) that modulates the wave (Fig. 13-13). The envelope, however, propagates with the *group velocity*

$$v_g = \frac{\Delta\omega}{\Delta\kappa} \xrightarrow[\Delta\omega \to 0]{} \frac{d\omega}{d\kappa} \tag{13-181}$$

which is equal to the phase velocity when ω is proportional to κ (P13-40) but more generally may be either larger or smaller than the phase velocity. Since a sinusoidal wave can be considered to transmit information (i.e., energy) only to the extent that it is modulated, we conclude that any information resides in the modulation and hence must be propagated through matter at the *group* velocity.

We shall conclude this section by examining briefly the propagation of a linearly polarized pulse in a medium in which v_p depends on frequency, i.e., in a *dispersive medium*. In general, we can represent the electric field in a pulse propagating toward $z = +\infty$ and linearly polarized in the x direction by the superposition

$$\boldsymbol{\mathcal{E}}(z, t) = \hat{\mathbf{i}} \int_0^\infty \mathcal{E}_0(\kappa) e^{i(\kappa z - \omega t)}\, d\kappa \tag{13-182}$$

$$= \hat{\mathbf{i}} \int_0^\infty \mathcal{E}_0(\kappa) e^{i\kappa(z - v_p t)}\, d\kappa \tag{13-183}$$

where we assume that each frequency present propagates at the appropriate phase velocity (which now, however, we must think of as a function of κ). For our present purposes, we shall not need the accompanying magnetic intensity. As in Chapter 7, it is here convenient to express the field as an integral over the range $-\infty < \kappa < \infty$, which we accomplish by extracting the physical field, finding

$$\begin{aligned}
\mathbf{E}(z, t) &= \tfrac{1}{2}[\boldsymbol{\mathcal{E}}(z, t) + \boldsymbol{\mathcal{E}}^*(z, t)] \\
&= \tfrac{1}{2}\hat{\mathbf{i}} \int_0^{\infty} \mathcal{E}_0(\kappa) e^{i\kappa[z - v_p(\kappa)t]}\, d\kappa \\
&\quad + \tfrac{1}{2}\hat{\mathbf{i}} \int_0^{\infty} \mathcal{E}_0^*(\kappa) e^{-i\kappa[z - v_p(\kappa)t]}\, d\kappa \\
&= \tfrac{1}{2}\hat{\mathbf{i}} \int_0^{\infty} \mathcal{E}_0(\kappa) e^{i\kappa[z - v_p(\kappa)t]}\, d\kappa \\
&\quad + \tfrac{1}{2}\hat{\mathbf{i}} \int_{-\infty}^{0} \mathcal{E}_0^*(-\kappa) e^{i\kappa[z - v_p(-\kappa)t]}\, d\kappa \\
&= \hat{\mathbf{i}} \int_{-\infty}^{\infty} A(\kappa) e^{i\kappa[z - V_p(\kappa)t]}\, \frac{d\kappa}{2\pi}
\end{aligned} \tag{13-184}$$

where we have defined the *spectral function* $A(\kappa)$ as in Eq. (7-63),

$$A(\kappa) = \begin{cases} \pi\mathcal{E}_0(\kappa), & \kappa > 0 \\ \pi\mathcal{E}_0^*(-\kappa), & \kappa < 0 \end{cases} \tag{13-185}$$

and further we have defined a phase velocity $V_p(\kappa)$ that applies over the entire range of κ by

$$V_p(\kappa) = \begin{cases} v_p(\kappa), & \kappa > 0 \\ v_p(-\kappa), & \kappa < 0 \end{cases} \tag{13-186}$$

This "capital" phase velocity is thus the *even* extension of $v_p(\kappa)$ into the region $\kappa < 0$. Given Eq. (13-184), we conclude in particular that the field at $t = 0$ is given by

$$\mathbf{E}(z, 0) = \hat{\mathbf{i}} \int_{-\infty}^{\infty} A(\kappa) e^{i\kappa z}\, \frac{d\kappa}{2\pi} \tag{13-187}$$

which by Fourier inversion (Appendix D) yields the expression

$$A(\kappa) = \int_{-\infty}^{\infty} E_x(z, 0) e^{-i\kappa z}\, dz \tag{13-188}$$

where $\mathbf{E}(z, 0) = E_x(z, 0)\hat{\mathbf{i}}$, for $A(\kappa)$. Thus, if we know the field $E_x(z, 0)$ at time zero and we know the phase velocity $V_p(\kappa)$ as a function of κ, we can find the field at other times by calculating $A(\kappa)$ from Eq. (13-188) and then calculating $E_x(z, t)$ from Eq. (13-184). In general, numerical methods will be needed for at least part of this calculation.[6]

[6]See, for example, J. R. Merrill, *Am. J. Phys.* **39**, 539 (1971).

To illustrate even more clearly the nature of wave propagation in a dispersive medium, suppose that at $t = 0$ the pulse in the previous paragraph is the triangular pulse for which

$$E_x(z, 0) = \begin{cases} 10(1 - |z|), & |z| < 1 \\ 0, & |z| > 1 \end{cases} \tag{13-189}$$

as shown in Fig. 13-14(a) and is propagating toward $z = +\infty$. By using Eq.

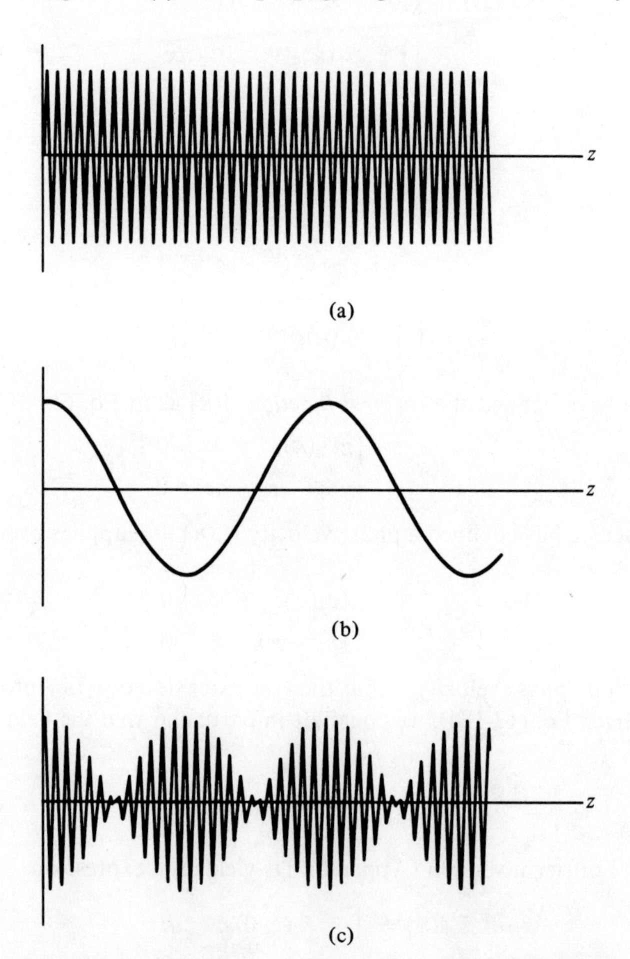

Fig. 13-13. Superposition of two sinusoidal waves of slightly different frequency. Part (a) shows the real part of the rapidly oscillating factor in Eq. (13-177) as a function of z at $t = 0$; part (b) shows the modulating envelope at $t = 0$; and part (c) shows the superposition [i.e., the product of (a) and (b)]. These graphs are drawn for $\Delta\kappa/\bar{\kappa} = 0.1$. The rapid oscillations in part (a) propagate with speed v_p and the envelope in part (b) propagates with speed v_g.

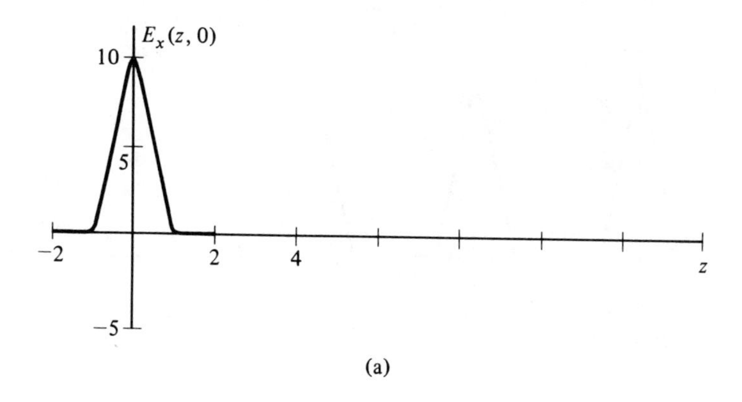

(a)

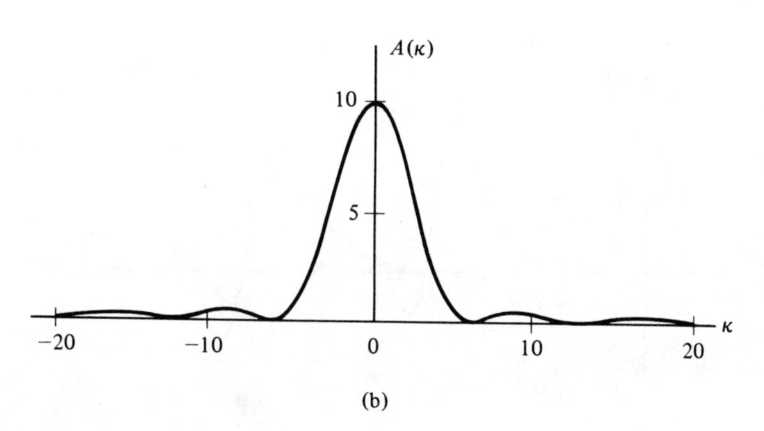

(b)

Fig. 13-14. Propagation of a pulse in various dispersive media.

(13-188), we find that the spectral function is

$$A(\kappa) = 10 \int_{-1}^{1} (1 - |z|)e^{-i\kappa z}\, dz = 20\, \frac{1 - \cos \kappa}{\kappa^2} \qquad (13\text{-}190)$$

which happens for this pulse to be a real, even function of κ; it is graphed in Fig. 13-14(b). Continuing, we then find from Eq. (13-184) that

$$E_x(z, t) = 20 \int_{-\infty}^{\infty} \frac{1 - \cos \kappa}{\kappa^2}\, e^{i\kappa[z - V_p(\kappa)t]}\, \frac{d\kappa}{2\pi}$$

$$= \frac{20}{\pi} \int_{0}^{\infty} \frac{1 - \cos \kappa}{\kappa^2} \cos \{\kappa[z - v_p(\kappa)t]\}\, d\kappa \qquad (13\text{-}191)$$

where the second form follows by writing the exponent as $\cos(\cdots) + i \sin(\cdots)$ and then noting that the real part of the resulting integrand is even

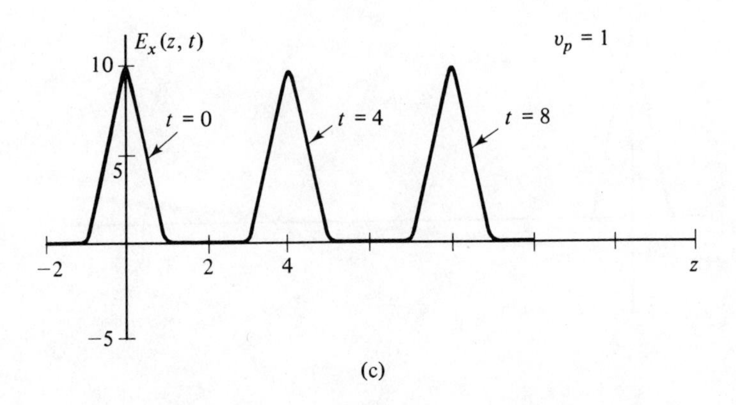

(c)

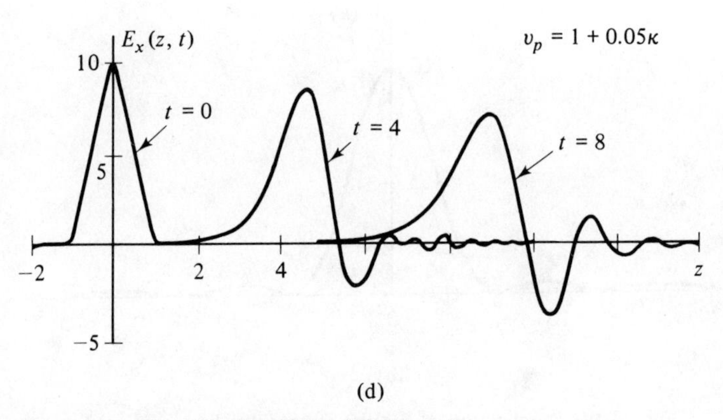

(d)

Fig. 13-14. (Continued)

in κ and the imaginary part is odd in κ. Although analytically Eq. (13-191) is essentially intractable for all but the simplest functions $v_p(\kappa)$, numerical integration yields $E_x(z, t)$ quite easily for *any* function $v_p(\kappa)$, provided the infinite interval of integration is truncated to some finite range, say $0 < \kappa < \kappa_{\max}$, outside of which $A(\kappa)$ is approximately zero. From the many possible formulas for numerical integration, we select Simpson's rule, which gives[7]

$$E_x(z, t) \approx \tfrac{1}{3}\,\Delta\kappa \sum_{i=0}^{N} f_i g(\kappa_i, z, t) \tag{13-192}$$

where N (which must be *even*) is the number of subintervals into which $0 < \kappa < \kappa_{\max}$ is divided, $\Delta\kappa = \kappa_{\max}/N$, $\kappa_i = i\,\Delta\kappa, f_i = 1, 4, 2, 4, 2, \cdots,$

[7]See, for example, H. D. Peckham, *Computers, BASIC, and Physics* (Addison-Wesley Publishing Company, Inc., Reading, Mass., 1971), Chapter 5.

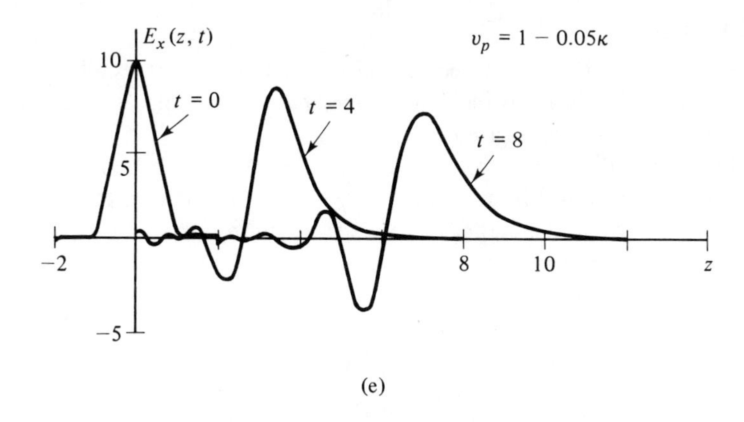

(e)

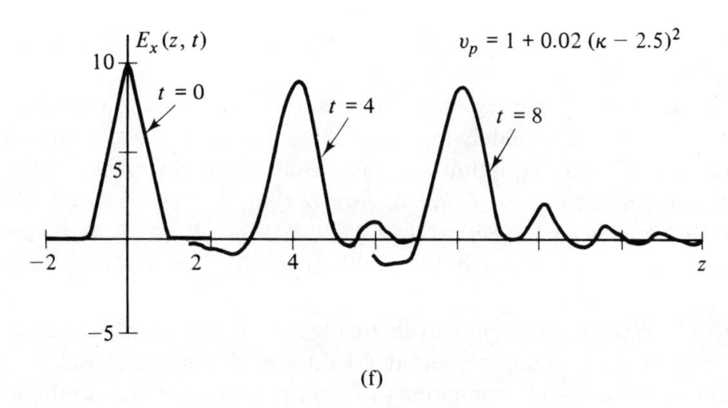

(f)

Fig. 13-14. (Continued)

2, 4, 1 as i increases from $i = 0$ to $i = N$, and

$$g(\kappa, z, t) = \frac{20}{\pi}\frac{1 - \cos \kappa}{\kappa^2} \cos \{\kappa[z - v_p(\kappa)t]\} \tag{13-193}$$

The writing of a computer program to evaluate this sum as a function of z for selected t and v_p is left to P13-43. Figure 13-14(c)–(f) shows the output of such a program when $v_p = 1$ (no dispersion), when $v_p = 1 + .05\kappa$ (phase velocity increases steadily over the range $0 < \kappa < \kappa_{\max}$), when $v_p = 1 - .05\kappa$ (phase velocity decreases steadily over the range $0 < \kappa < \kappa_{\max}$), and when $v_p = 1 + .02(\kappa - 2.5)^2$ (phase velocity has a minimum at $\kappa = 2.5$). In dispersive media, the pulse becomes less sharp and oscillatory leading and/or trailing "edges" become more pronounced as the pulse propagates further into the medium.

PROBLEMS

P13-39. Show that the reflected and transmitted waves resulting when a *circularly* polarized wave is incident nonnormally on a dielectric interface are in general *elliptically* polarized.

P13-40. Let the dispersion relation for a particular medium be $\omega = a\kappa$, where a is a constant. Show that $v_g = v_p = a$. In particular, this dispersion relation describes electromagnetic waves in vacuum if $a = c$.

P13-41. The group velocity of electromagnetic waves can be expressed in many different ways. Show that

$$\begin{aligned} v_g &= -\frac{\lambda^2}{2\pi}\frac{d\omega}{d\lambda} = c\left(\frac{1}{n} + \frac{\lambda}{n^2}\frac{dn}{d\lambda}\right) \\ &= \frac{c}{n}\left(1 + \frac{d(\ell n\, n)}{d(\ell n\, \lambda)}\right) = v_p + \kappa\frac{dv_p}{d\kappa} \\ &= v_p - \lambda\frac{dv_p}{d\lambda} \end{aligned}$$

where $\lambda = 2\pi/\kappa$ is the wavelength *in the medium* and $n = c/v_p = \kappa c/\omega$ is the index of refraction at the frequency ω.

P13-42. In the visible region of the spectrum, the index of refraction n of some media can be adequately represented as a function of wavelength by the two-constant Cauchy equation $n = A + B/\lambda^2$. Find the phase and group velocities as functions of λ. *Optional:* For barium flint glass, $n = 1.58848$ at $\lambda = 6563$ Å and $n = 1.60870$ at $\lambda = 3988$ Å. Find A and B and then plot graphs of v_g and v_p versus λ in the visible spectrum. *Suggestion:* Use a computer.

P13-43. Write a program to evaluate the sum in Eq. (13-192) as a function of z for specified t, determine suitable values of N and κ_{max}, and run your program on an available computer to reproduce some of the results in Fig. 13-14. *Hint:* The function $(1 - \cos\kappa)/\kappa^2$ is indeterminate at $\kappa = 0$. Show that

$$\frac{20}{\pi}\frac{1 - \cos\kappa}{\kappa^2} = \frac{10}{\pi}\left[1 - \frac{\kappa^2}{12}\left(1 - \frac{\kappa^2}{30}\right)\right]$$

to within about $\pm 2 \times 10^{-10}$ when $|\kappa| < .1$ and use this series for that range of κ. *Optional:* (1) Try other values of N and κ_{max} and other functions $v_p(\kappa)$. (2) Try to increase the running speed of your program by thinking of ways to minimize the number of calls to cosine and sine routines. Note, for example, that $\cos[(\kappa + \Delta\kappa)z] = \cos(\kappa z)\cos(\Delta\kappa\, z) - \sin(\kappa z)\sin(\Delta\kappa\, z)$. How can this and similar identities be used to determine $\cos(m\,\Delta\kappa\, z)$ for all m using only *two* calls to the sine and cosine routines? Note also that cosines and sines that are used repeatedly can be calculated once and stored.

P13-44. Without evaluating any integrals, sketch a labeled graph of the likely spectral function for the pulse shown in Fig. P13-44. How would increasing the number of cycles to 10 and to 100 affect the spectral function?

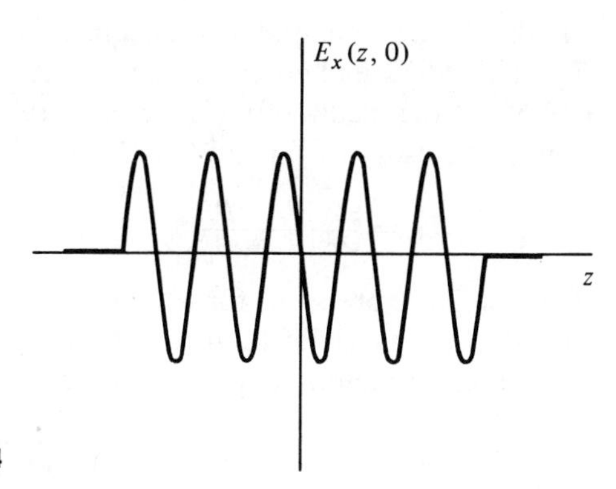

Figure P13-44

Supplementary Problems

P13-45. At a particular point in space, the electric field is given as a function of time by $\boldsymbol{\mathcal{E}}(t) = \mathcal{E}(t)\hat{\mathbf{i}}$, where $\mathcal{E}(t)$ can be expressed in terms of a Fourier transform

$$\tilde{\mathcal{E}}(\omega) = \int_{-\infty}^{\infty} \mathcal{E}(t)e^{i\omega t}\, dt$$

Suppose the medium is such that **D** and **E** are related by a frequency-dependent permittivity $\tilde{\epsilon}(\omega)$ so that $\tilde{\boldsymbol{\mathfrak{D}}}(\omega) = \tilde{\epsilon}(\omega)\tilde{\boldsymbol{\mathcal{E}}}(\omega)$. Show that

$$\mathfrak{D}(t) = \int_{-\infty}^{\infty} \epsilon(t')\mathcal{E}(t - t')\, dt'$$

where

$$\epsilon(t) = \frac{1}{2\pi}\int_{-\infty}^{\infty} \tilde{\epsilon}(\omega)e^{-i\omega t}\, d\omega$$

etc. The relationship between $\mathfrak{D}(t)$ and $\mathcal{E}(t)$ in this case is far less simple than the relationship between $\tilde{\mathfrak{D}}(\omega)$ and $\tilde{\mathcal{E}}(\omega)$.

P13-46. The Q of a medium is defined as the absolute value of the ratio of the displacement current density to the conduction current density. Consider a linear medium carrying a monochromatic plane wave. (a) Show that $Q = \epsilon\omega/g$. (b) Express κ and α as given in Eqs. (13-62) and (13-63) in terms of Q by eliminating ω and then express $K = \kappa + i\alpha$ in the form $|K| \exp(i\phi)$, finding $|K|$ and ϕ in terms of Q. (c) Sketch graphs of κ, α, $|K|$, and ϕ as functions of Q. (d) Determine the ratio of the electric to the magnetic energy densities in the wave of Eqs. (13-49) and (13-50), expressing the result in terms of Q, and sketch a graph of this ratio as a function of Q.

P13-47. Letting $Q = \epsilon_t\omega/g_t$ (compare P13-46) and $\gamma = \sqrt{\mu_i\epsilon_t/\epsilon_i\mu_t}$, and taking $K_t = \kappa_t + i\alpha_t$ with κ_t and α_t given by Eqs. (13-62) and (13-63), find ζ [Eq. (13-89)] and then obtain graphs of R and T [Eq. (13-91)] as functions of Q for typical values of γ, including $\gamma = 1$.

P13-48. Show that for a good nonmagnetic conductor Eq. (13-91) for the reflection coefficient at normal incidence reduces to $R = 1 - \sqrt{8\epsilon_i\omega/g_t}$. *Hint:* Evaluate T first; then find R from R = 1 − T.

P13-49. Using a computer to calculate the points, obtain a few graphs of $R_\perp$, $T_\perp$, $R_\parallel$, and $T_\parallel$ and compare your results with Fig. 13-9.

P13-50. If the media of incidence and transmission are both nonmagnetic, show that Eqs. (13-113) and (13-114) can be written in the form

$$\mathcal{E}_{r0}^{\perp} = \frac{\sin(\theta_t - \theta_i)}{\sin(\theta_t + \theta_i)} \mathcal{E}_{i0}^{\perp} \qquad \mathcal{E}_{t0}^{\perp} = \frac{2\cos\theta_i \sin\theta_t}{\sin(\theta_t + \theta_i)} \mathcal{E}_{i0}^{\perp}$$

and find similar expressions for $\mathcal{E}_{r0}^{\parallel}$ and $\mathcal{E}_{t0}^{\parallel}$.

P13-51. In the case of total reflection at an interface between two nonmagnetic nonconductors, Snell's law requires θ_t to be compex and

$$\cos\theta_t = \sqrt{1 - \sin^2\theta_t} = i\sqrt{\left(\frac{n_i}{n_t}\right)^2 \sin^2\theta_i - 1}$$

Show from Eqs. (13-113) and (13-115) that the reflection coefficients for *both* polarizations are unity.

P13-52. A plane wave in air is incident on a nonmagnetic nonconducting slab as shown in Fig. P13-52, producing a reflected wave and also a transmitted wave. The transmitted wave is subsequently partially reflected and partially transmitted at the second interface. Let the incident wave be polarized perpendicular to the plane of incidence and have amplitude a. Finally, introduce reflection and transmission coefficients r, t, r', and t' at the two interfaces so that the *amplitudes* of the several waves are as shown in the figure. Using Eqs. (13-113) and (13-114), find r, t, r', and t' in terms of θ_i, θ_t, and n and then show that $r = -r'$ and that $tt' - rr' = 1$. Do these final two results also apply for parallel polarization? Defend your answer.

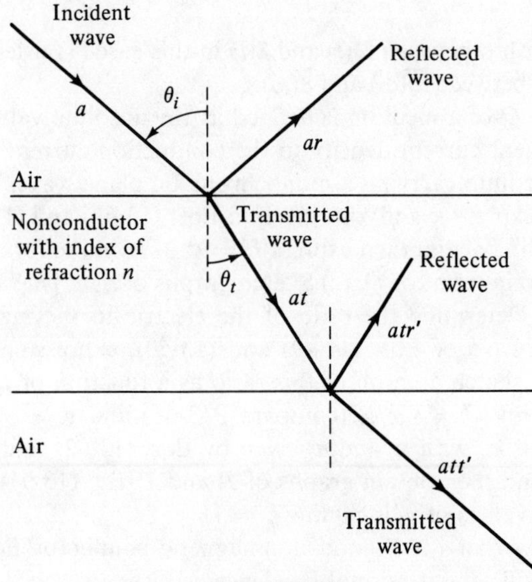

Figure P13-52

P13-53. The following procedure is sometimes used to calculate energy losses when plane waves are incident on *nearly* perfect conductors: (1) Solve the problem assuming the conductor to be perfect; (2) determine the surface current density **j** in the perfect conductor and then, assuming that the same total current appears for the real conductor and that that current is distributed in a layer of thickness δ at the surface, take the current density **J** in that layer to satisfy $\mathbf{j} = \delta \mathbf{J}$; (3) estimate the electric field in that layer from Ohm's law, $\mathbf{J} = g\mathbf{E}$; and (4) estimate the magnetic intensity in that layer by requiring continuity of the tangential component of **H** across the interface. (a) Apply this procedure to estimate $\boldsymbol{\mathcal{J}}$, $\boldsymbol{\mathcal{E}}$, and $\boldsymbol{\mathcal{H}}$ in the surface layer of a nearly perfect conductor when the incident wave strikes the surface normally. (b) Find the average power dissipated in an area A of the conducting surface. *Optional:* Consider an area A of the conducting surface bounded by a rectangle having sides L and W, with the sides of length L directed perpendicular to the direction of **J**. (1) Find the root mean square (rms) value of the total current flowing between the two sides of length L. (2) Express the average power dissipated in the area bounded by this rectangle in terms of the rms current and identify a resistance to associate with the surface. (3) What is the surface resistance of a square sheet between two parallel edges? Can you understand why this resistance does not depend on the size of the sheet? (*Note:* The quantity $1/g\delta$ is sometimes called the surface resistivity.)

P13-54. Determine the fields in the TM_{mnp} modes of the cavity resonator in Fig. 13-12. *Hint:* See P13-37 and require $\mathcal{H}_{0z}$ to be zero.

P13-55. Suppose a portion of the cylindrical wave guide in P13-35 is closed off by conducting planes at $z = 0$ and $z = b$ to make a cylindrical cavity resonator. What are the frequencies of the resonant modes corresponding to the traveling modes found in P13-35? *Optional:* Choose reasonable dimensions for the cavity, calculate several of the lowest resonant frequencies, and plot a "frequency level" diagram as described in P13-38.

P13-56. Let a general monochromatic wave in an evacuated wave guide be given by

$$\boldsymbol{\mathcal{E}}(\mathbf{r}, t) = \boldsymbol{\mathcal{E}}_0(x, y)e^{i(\kappa_z z - \omega t)}$$

$$\boldsymbol{\mathcal{H}}(r, t) = \boldsymbol{\mathcal{H}}_0(x, y)e^{i(\kappa_z z - \omega t)}$$

where $\boldsymbol{\mathcal{E}}_0$ and $\boldsymbol{\mathcal{H}}_0$ may have components in all three coordinate directions. Show from Maxwell's equations [Eqs. (13-5)–(13-8)] that

$$\mathcal{E}_{0x} = \frac{i}{\omega\epsilon_0}\left(1 - \frac{\kappa_z^2 c^2}{\omega^2}\right)^{-1}\left(\frac{\partial \mathcal{H}_{0z}}{\partial y} + \frac{\kappa_z}{\omega\mu_0}\frac{\partial \mathcal{E}_{0z}}{\partial x}\right)$$

find similar expressions determining $\mathcal{E}_{0y}$, $\mathcal{H}_{0x}$, and $\mathcal{H}_{0y}$ from $\mathcal{E}_{0z}$ and $\mathcal{H}_{0z}$, and hence conclude that all components of a guided wave are known if the z-components of the two fields are known. In particular, TE modes have $\mathcal{E}_{0z} = 0$ and TM modes have $\mathcal{H}_{0z} = 0$.

P13-57. In a tenuous plasma, collisions between particles are infrequent enough that resistive forces can be neglected and $\epsilon \approx \epsilon_0$, $\mu \approx \mu_0$. In such a case $b = 0$ in Eq. (13-12) and the conductivity given by Eq. (13-15) becomes $g = iq_e^2N/\omega m_e$ and is *purely imaginary*. (We switch from n to N for particle density in order to avoid confusion with the index of refraction n.) (a) What is the dispersion relation [Eq. (13-48)] for this plasma? Write your result in terms of the so-called *plasma frequency* ω_p defined by $\omega_p^2 = q_e^2N/m_e\epsilon_0$. (b) Sketch a graph of κ versus ω. (c) Describe qualitatively the nature of the waves given by Eqs. (13-49) and (13-50) for this plasma. Consider both $\omega > \omega_p$ and $\omega < \omega_p$. (d) In what region of the electromagnetic spectrum does the frequency ω_p lie when $N \approx 10^{11}$ particles/m^3, a value characteristic of the ionosphere? What limitations does the ionosphere impose on the frequencies used for communication with moon-bound astronauts? (e) Recognizing that $\kappa = n\omega/c$, where n is the index of refraction, find n for this plasma.

REFERENCES

J. D. Jackson, *Classical Electrodynamics* (John Wiley & Sons, Inc., New York, 1962), Chapters 7 and 8.

P. Lorrain and D. Corson, *Electromagnetic Fields and Waves* (W. H. Freeman and Company, San Francisco, 1970), Second Edition, Chapters 11, 12, and 13.

S. Ramo and J. R. Whinnery, *Fields and Waves in Modern Radio* (John Wiley & Sons, Inc., New York, 1953), Second Edition.

14

Radiation from Prescribed Sources in Vacuum

Maxwell's equations not only predict that electromagnetic waves propagate through space but also provide the theoretical framework for relating those waves to their ultimate source in some charge and current distribution. This relationship is the subject of this chapter. The discussion is confined to radiation produced by free charges and currents in space free of matter, and Maxwell's equations therefore assume the form

$$\nabla\cdot\mathbf{E} = \frac{\rho}{\epsilon_0} \qquad \nabla\times\mathbf{E} = -\frac{\partial\mathbf{B}}{\partial t} \qquad \text{(14-1), (14-2)}$$

$$\nabla\cdot\mathbf{B} = 0 \qquad \nabla\times\mathbf{B} = \mu_0\mathbf{J} + \frac{1}{c^2}\frac{\partial\mathbf{E}}{\partial t} \qquad \text{(14-3), (14-4)}$$

where $1/c^2 = \mu_0\epsilon_0$. In treating radiation problems, however, it is usually convenient to seek first the scalar and vector potentials V and $\mathbf{A}$ and then derive the fields by applying the relationships

$$\mathbf{E} = -\nabla V - \frac{\partial\mathbf{A}}{\partial t} \qquad \mathbf{B} = \nabla\times\mathbf{A} \qquad \text{(14-5), (14-6)}$$

obtained in Section 6-6. We elect to work in a Lorentz gauge. The potentials therefore satisfy the inhomogeneous wave equations

$$\left(\nabla^2 - \frac{1}{c^2}\frac{\partial^2}{\partial t^2}\right)\mathbf{A} = -\mu_0\mathbf{J} \qquad \text{(14-7)}$$

$$\left(\nabla^2 - \frac{1}{c^2}\frac{\partial^2}{\partial t^2}\right)V = -\frac{1}{\epsilon_0}\rho \qquad \text{(14-8)}$$

and, in addition, are related by the Lorentz condition,

$$\nabla \cdot \mathbf{A} + \frac{1}{c^2}\frac{\partial V}{\partial t} = 0 \tag{14-9}$$

[Compare Eqs. (6-71)–(6-73).] In brief, the objectives of this chapter are (1) to solve Eqs. (14-7) and (14-8) for the potentials and (2) to examine these solutions and the corresponding fields for a few representative source distributions.

14-1 The General Solution of the Inhomogeneous Wave Equation; Retardation

In this section, we shall seek general solutions to Eqs. (14-7) and (14-8). The results, of course, will be general solutions to Maxwell's equations in vacuum and, once we have obtained these solutions, *all* problems of electromagnetism in vacuum are solved, at least in principle. We have already found this general solution *when the potentials are static*, namely

$$\mathbf{A}(\mathbf{r}) = \frac{\mu_0}{4\pi}\int \frac{\mathbf{J}(\mathbf{r}')}{|\mathbf{r}-\mathbf{r}'|}\,dv' \tag{14-10}$$

$$V(\mathbf{r}) = \frac{1}{4\pi\epsilon_0}\int \frac{\rho(\mathbf{r}')}{|\mathbf{r}-\mathbf{r}'|}\,dv' \tag{14-11}$$

[Compare Eqs. (5-41) and (4-50).] The solution for the time-dependent potentials, however, is more complicated. Because of the time *derivatives* in the basic equations, we *cannot* obtain the solutions in the time-dependent case simply by allowing the sources in Eqs. (14-10) and (14-11) to become time-dependent. The effect of the time derivatives is more subtle, but it is not at all surprising once it has been obtained. Any one of several methods of solution might be pursued. Mathematically elegant methods exist but cannot be used here without an extended digression to develop the necessary mathematical techniques (but see P14-26). We adopt instead an approach that draws on no more mathematics than has already been introduced.

Consider first a small, *spherically symmetric* charge distribution, which we shall ultimately allow to approach a point. If we can find the potentials established by this distribution, we can obtain the solution for a more realistic distribution by superposition. Let the charge be located at the origin (we can translate coordinates any time we wish) and, disregarding temporarily the requirements of charge conservation, suppose that the total charge in the distribution varies with time in accordance with

$$q = q(t) \tag{14-12}$$

If this distribution is confined to a region of radius a centered at the origin, then $\rho(\mathbf{r}, t)$ is zero outside this region and the scalar potential V satisfies

$$\left(\nabla^2 - \frac{1}{c^2}\frac{\partial^2}{\partial t^2}\right)V(\mathbf{r}, t) = 0 \tag{14-13}$$

when $r > a$. Since the supposed spherical symmetry of the charge distribution implies that V depends only on r and t, we can write this equation more simply in spherical coordinates, obtaining

$$\frac{1}{r^2}\frac{\partial}{\partial r}\left(r^2\frac{\partial V}{\partial r}\right) - \frac{1}{c^2}\frac{\partial^2 V}{\partial t^2} = 0 \tag{14-14}$$

The substitution $V = h(r, t)/r$ reduces Eq. (14-14) to

$$\frac{\partial^2 h}{\partial r^2} = \frac{1}{c^2}\frac{\partial^2 h}{\partial t^2} \tag{14-15}$$

whose solution

$$h(r, t) = f(r - ct) + g(r + ct) \tag{14-16}$$

where f and g are arbitrary functions, is immediate (P7-1). Thus, the general solution to Eq. (14-14) is

$$V(\mathbf{r}, t) = \frac{1}{r}f(r - ct) + \frac{1}{r}g(r + ct) \tag{14-17}$$

(Compare P7-27.)

Two physical conditions must yet be imposed on this solution. First we must set $g = 0$, for that term represents a wave propagating *inward* and such a wave is incompatible with our expectation on physical grounds that the electromagnetic wave produced by a changing charge cannot be present at any point in space *before* the charge has changed.[1] Thus, the most general, *physically acceptable* solution to Eq. (14-14) is

$$V(\mathbf{r}, t) = \frac{1}{r}f(r - ct) \tag{14-18}$$

The second condition to be imposed on $V(\mathbf{r}, t)$ determines the function f. Essentially, we must require that the solution, Eq. (14-18), reflect the presence of the charge $q(t)$ within the (small) sphere $r < a$. If a is small enough, however, the field established by the *changing* charge in the region $r < a$ propagates to the surface $r = a$ in negligible time. Thus, if a is small enough, the potential at and just outside of $r = a$ follows changes in the charge with negligible time lag. At each instant, the potential at and just outside of $r = a$ is therefore equal to the static potential appropriate to the value of the charge at that instant. More specifically, *near* $r = a$, the *time-dependent* potential should be given by

$$V(\mathbf{r}, t) = \frac{1}{4\pi\epsilon_0}\frac{q(t)}{r} \qquad (r \approx a \longrightarrow 0) \tag{14-19}$$

As $r \longrightarrow 0$, however, Eq. (14-18) becomes

$$V(\mathbf{r}, t) = \frac{f(-ct)}{r} \qquad (r \approx a \longrightarrow 0) \tag{14-20}$$

[1]This general conviction of the physicist that event A at $(\mathbf{r}_1, t_1)$ is a possible cause of event B at point $(\mathbf{r}_2, t_2)$ only if $t_1 < t_2$ is usually called the *principle of causality*, and it plays a particularly significant role in some aspects of quantum field theory.

Equations (14-19) and (14-20) are the same only if

$$f(-ct) = \frac{1}{4\pi\epsilon_0} q(t) \Longrightarrow f(\xi) = \frac{1}{4\pi\epsilon_0} q\left(-\frac{\xi}{c}\right) \tag{14-21}$$

and the function f is determined for any argument ξ. The potential expressed in Eq. (14-18) now assumes the more explicit form

$$V(\mathbf{r}, t) = \frac{1}{4\pi\epsilon_0 r} q\left(t - \frac{r}{c}\right) \tag{14-22}$$

More generally, if the localized (point) charge is at $\mathbf{r}'$, we would interpret r in Eq. (14-22) as $|\mathbf{r} - \mathbf{r}'|$ and conclude that

$$V(\mathbf{r}, t) = \frac{1}{4\pi\epsilon_0 |\mathbf{r} - \mathbf{r}'|} q\left(t - \frac{1}{c}|\mathbf{r} - \mathbf{r}'|\right) \tag{14-23}$$

A potential of this form is referred to as a *retarded potential* because the potential at time t is determined by the state of the charge at the so-called *retarded time* t' given by

$$t' = t - \frac{1}{c}|\mathbf{r} - \mathbf{r}'| \tag{14-24}$$

which, not surprisingly, is earlier than time t by an interval precisely sufficient for an electromagnetic wave to have propagated from the source point to the observation point by the time of observation. In effect a change in a charge at $\mathbf{r}'$ is not detected at some other point $\mathbf{r}$ until sufficient time has elapsed after the change for an electromagnetic signal to propagate from $\mathbf{r}'$ to $\mathbf{r}$.

Since Eq. (14-8) is linear, we find its general solution by adding up elemental contributions having the form of Eq. (14-23). Because different points in space have associated with them different retarded times, however, it is here more suitable to imagine space (rather than the charge distribution itself) to be divided into infinitesimal elements, with the element of volume dv' being centered at $\mathbf{r}'$ and having retarded time t' associated with it. If the charge density at $(\mathbf{r}, t)$ is $\rho(\mathbf{r}, t)$, then the charge dq in dv' at the retarded time t' is given by $\rho(\mathbf{r}', t')\, dv'$ and the contribution made by the charge in this volume element to the potential at $(\mathbf{r}, t)$ is given by

$$dV(\mathbf{r}, t) = \frac{\rho(\mathbf{r}', t')}{4\pi\epsilon_0 |\mathbf{r} - \mathbf{r}'|}\, dv' \tag{14-25}$$

Finally, the total potential at $(\mathbf{r}, t)$ is obtained by integrating Eq. (14-25) over all volume elements; we find that

$$V(\mathbf{r}, t) = \frac{1}{4\pi\epsilon_0} \int \frac{\rho(\mathbf{r}', t')}{|\mathbf{r} - \mathbf{r}'|}\, dv' \tag{14-26}$$

By similar arguments, we find that the general solution to Eq. (14-7) is given by

$$\mathbf{A}(\mathbf{r}, t) = \frac{\mu_0}{4\pi} \int \frac{\mathbf{J}(\mathbf{r}', t')}{|\mathbf{r} - \mathbf{r}'|}\, dv' \tag{14-27}$$

Extension of the solutions in Eqs. (14-10) and (14-11) to time-dependent potentials thus not only involves allowing the sources to become time-dependent but also requires evaluating these sources at the *retarded* time. Unfortunately for the more explicit evaluation of Eqs. (14-26) and (14-27), evaluation of ρ and $\mathbf{J}$ at the retarded time complicates the expressions considerably for at least two reasons. First, since we seek potentials at a fixed (even though general) observation time t, the integrals must be evaluated at fixed t. For fixed t, however, t' as given by Eq. (14-24) varies from point to point in the distribution. Hence, $\rho(\mathbf{r}', t')$ and $\mathbf{J}(\mathbf{r}', t')$ depend not only *explicitly* on $\mathbf{r}'$ but also *implicitly* on $\mathbf{r}'$ through a dependence of t' on $\mathbf{r}'$. Even for simple charge distributions, the integrands in Eqs. (14-26) and (14-27) may be complicated functions of $\mathbf{r}'$.

The appearance of the retarded time also complicates finding the region of space over which the *nominally* infinite integrals in Eqs. (14-26) and (14-27) *actually* extend. Generally charge distributions are specified by giving $\rho(\mathbf{r}', t')$ and $\mathbf{J}(\mathbf{r}', t')$ as functions of $\mathbf{r}'$ for fixed t', which in part involves specifying the regions, say R' and S', within which $\rho(\mathbf{r}', t')$ and $\mathbf{J}(\mathbf{r}', t')$ differ from zero at fixed t'. The integrals of interest, however, extend over the regions, say R and S, in which $\rho(\mathbf{r}', t')$ and $\mathbf{J}(\mathbf{r}', t')$ differ from zero at fixed t. Since the relationship between t and t' is generally complicated, the relationship between the (probably simple) regions R' and S' and the regions R and S is also complicated.

In summary, we have in this section found general solutions for the potentials established by arbitrary charge and current distributions in vacuum. In principle we therefore have found solutions to all electromagnetic problems in vacuum. In practice, however, the phenomenon of retardation makes the integrands and regions of integration so complicated that explicit evaluation of these general solutions for particular sources is typically very difficult. Some specific situations that *can* be treated fairly simply are discussed in the remainder of this chapter.

PROBLEM

P14-1. Show that the potentials given by Eqs. (14-26) and (14-27) satisfy the Lorentz condition, Eq. (14-9).

14-2
Radiation from Monochromatic Sources: The Oscillating Electric Dipole

If in particular the source distribution varies sinusoidally (monochromatically) with time, it is convenient to express the charge and current densities as the real parts of the complex densities

$$\rho(\mathbf{r}, t) = \rho_0(\mathbf{r})e^{-i\omega t} \qquad \mathcal{J}(\mathbf{r}, t) = \mathcal{J}_0(\mathbf{r})e^{-i\omega t} \tag{14-28}$$

the potentials as the real parts of the complex potentials

$$\mathcal{V}(\mathbf{r}, t) = \mathcal{V}_0(\mathbf{r})e^{-i\omega t} \qquad \mathcal{A}(\mathbf{r}, t) = \mathcal{A}_0(\mathbf{r})e^{-i\omega t} \tag{14-29}$$

and the fields as the real parts of the complex fields

$$\mathcal{E}(\mathbf{r}, t) = \mathcal{E}_0(\mathbf{r})e^{-i\omega t} \qquad \mathcal{B}(\mathbf{r}, t) = \mathcal{B}_0(\mathbf{r})e^{-i\omega t} \tag{14-30}$$

Equations (14-24), (14-26), and (14-27) then yield that the potentials are given by

$$\mathcal{V}_0(\mathbf{r}) = \frac{1}{4\pi\epsilon_0}\int \frac{\rho_0(\mathbf{r}')e^{i\kappa|\mathbf{r}-\mathbf{r}'|}}{|\mathbf{r}-\mathbf{r}'|}\,dv' \tag{14-31}$$

and

$$\mathcal{A}_0(\mathbf{r}) = \frac{\mu_0}{4\pi}\int \frac{\mathcal{J}_0(\mathbf{r}')e^{i\kappa|\mathbf{r}-\mathbf{r}'|}}{|\mathbf{r}-\mathbf{r}'|}\,dv' \tag{14-32}$$

where $\kappa = \omega/c$. Further, from Eqs. (14-5) and (14-6) we find that the fields are given by

$$\mathcal{E}_0(\mathbf{r}) = -\nabla\mathcal{V}_0(\mathbf{r}) + i\omega\mathcal{A}_0(\mathbf{r}) \tag{14-33}$$

$$\mathcal{B}_0(\mathbf{r}) = \nabla \times \mathcal{A}_0(\mathbf{r}) \tag{14-34}$$

although *outside the source* (where $\mathbf{J} = 0$) the electric field can often be more readily obtained from the expression

$$\mathcal{E}_0(\mathbf{r}) = \frac{ic}{\kappa}\nabla \times \mathcal{B}_0(\mathbf{r}) \tag{14-35}$$

derived from Eq. (14-4).

Consider now the special case of the oscillating electric dipole shown in Fig. 14-1. Let the charge at $z = \frac{1}{2}a$ vary sinusoidally in accordance with the

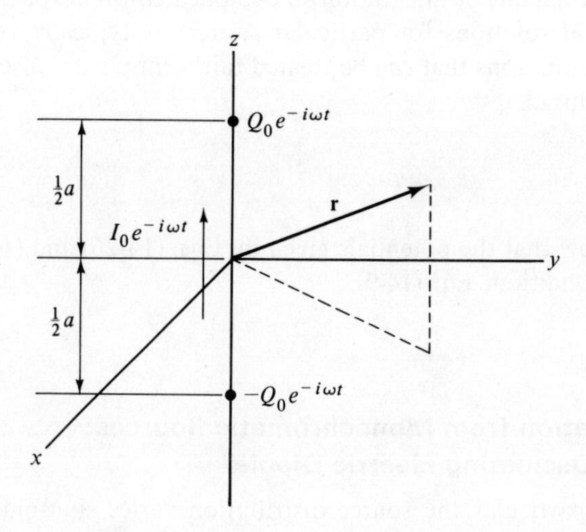

Fig. 14-1. Coordinates and vectors for evaluating the fields of an oscillating dipole.

expression $\mathcal{Q}_1 = Q_0 e^{-i\omega t}$, let the charge at $z = -\frac{1}{2}a$ vary in accordance with $\mathcal{Q}_2 = -Q_0 e^{-i\omega t}$, and let the wire connecting the two charges run along the z axis and carry the current $\mathcal{I} = \mathcal{I}_0 e^{-i\omega t}$. The equation of continuity applied to a spherical surface surrounding the charge at $z = \frac{1}{2}a$ requires that $\mathcal{I}_0 = -i\omega Q_0$. Replacing $\mathcal{J}_0(\mathbf{r}')\,dv'$ by $\mathcal{I}_0(\mathbf{r}')\,d\boldsymbol{\ell}$ in Eq. (14-32), we find that

$$\mathcal{A}_0(\mathbf{r}) = \frac{\mu_0}{4\pi}\int_{-a/2}^{a/2} \frac{\mathcal{I}_0 e^{i\kappa|\mathbf{r}-z'\hat{\mathbf{k}}|}}{|\mathbf{r}-z'\hat{\mathbf{k}}|}\,dz'\hat{\mathbf{k}} \tag{14-36}$$

We now confine our attention to distant observation points, i.e., to the region $r \gg a$, in which case $z'/r \ll 1$ for all values of z' within the range of integration. Thus, ignoring terms of order $(a/r)^2$, we find the expansion

$$\mathcal{A}_0(\mathbf{r}) = \frac{\mu_0 \mathcal{I}_0}{4\pi r}\hat{\mathbf{k}}\int_{-a/2}^{a/2} \frac{e^{i\kappa r[1-(z'/r)\cos\theta+\cdots]}}{1-(z'/r)\cos\theta+\cdots}\,dz' \tag{14-37}$$

$$= \frac{\mu_0 \mathcal{I}_0}{4\pi r}e^{i\kappa r}\hat{\mathbf{k}}\int_{-a/2}^{a/2}\left(1+\frac{z'}{r}\cos\theta+\cdots\right)e^{-i\kappa z'\cos\theta+\cdots}\,dz' \tag{14-38}$$

where $\cos\theta = \hat{\mathbf{k}}\cdot\hat{\mathbf{r}}$. Further, we confine our attention to radiated wavelengths λ that are long compared to the dimensions of the dipole, i.e., to wavelengths for which $\kappa a = 2\pi a/\lambda \ll 1$, in which case $\kappa z' \ll 1$ within the range of integration and the Taylor expansion

$$e^{-i\kappa z'\cos\theta} = 1 - i\kappa z'\cos\theta + \cdots \tag{14-39}$$

can be employed. Under the conditions $r \gg a$ and $\kappa a \ll 1$, the vector potential is therefore given approximately by

$$\begin{aligned}\mathcal{A}_0(\mathbf{r}) &= \frac{\mu_0 \mathcal{I}_0}{4\pi r}e^{i\kappa r}\hat{\mathbf{k}}\int_{-a/2}^{a/2}\left[1+\frac{z'}{r}\cos\theta + O\left(\frac{a^2}{r^2}\right)\right] \\ &\quad \times [1 - i\kappa z'\cos\theta + O(\kappa^2 a^2)]\,dz' \\ &= -i\omega\frac{\mu_0 \mathbf{p}_0}{4\pi r}e^{i\kappa r} + O\left(\frac{a^2}{r^2}, \kappa^2 a^2, \frac{\kappa a^2}{r}\right)\end{aligned} \tag{14-40}$$

where in the second expression $\mathcal{I}_0$ has been replaced by its evaluation in terms of Q_0, namely, by $-i\omega Q_0$ and $\mathbf{p}_0 = Q_0 a\hat{\mathbf{k}}$ is the maximum dipole moment of the oscillating distribution. In both expressions the notation $O(\cdots)$ indicates the lowest order of the terms that have been neglected. Finally, using Eqs. (14-34) and (14-35), we find that the fields are given by

$$\mathcal{B}_0 = \frac{\mu_0}{4\pi}\omega\kappa\frac{e^{i\kappa r}}{r}\left(1-\frac{1}{i\kappa r}\right)\hat{\mathbf{r}}\times\mathbf{p}_0 \tag{14-41}$$

$$\begin{aligned}\mathcal{E}_0 &= \frac{\mu_0}{4\pi}\omega\kappa c\frac{e^{i\kappa r}}{r}(\hat{\mathbf{r}}\times\mathbf{p}_0)\times\hat{\mathbf{r}} \\ &\quad + \frac{1}{4\pi\epsilon_0}(1-i\kappa r)\frac{e^{i\kappa r}}{r^3}[3(\hat{\mathbf{r}}\cdot\mathbf{p}_0)\hat{\mathbf{r}} - \mathbf{p}_0]\end{aligned} \tag{14-42}$$

Note that $\mathcal{B}_0$ (and hence $\mathbf{B}$) is perpendicular to $\hat{\mathbf{r}}$ at all points but that $\mathcal{E}_0$ in general has a component parallel to $\hat{\mathbf{r}}$.

To simplify the discussion of these fields, we have assumed (1) that the dimensions of the source are small compared to the observation distance, $a \ll r$, and (2) that the dimensions of the source are small compared to the wavelength of the emitted radiation, $a \ll \lambda = 2\pi/\kappa$. These conditions do not impose any constraint on the relationship between the wavelength and the distance of observation. It is useful to distinguish three zones as follows:

$\kappa r \ll 1$: the near (static) zone, where $r \ll \lambda$
$\kappa r \approx 1$: the intermediate (induction) zone, where $r \approx \lambda$
$\kappa r \gg 1$: the far (radiation) zone, where $r \gg \lambda$

In the near zone the fields have the limiting expressions

$$\mathcal{B}_0 = i\omega\frac{\mu_0}{4\pi}\frac{\hat{\mathbf{r}} \times \mathbf{p}_0}{r^2} \tag{14-43}$$

$$\mathcal{E}_0 = \frac{1}{4\pi\epsilon_0 r^3}[3(\hat{\mathbf{r}}\cdot\mathbf{p}_0)\hat{\mathbf{r}} - \mathbf{p}_0] \tag{14-44}$$

Since $\kappa r \ll 1$ and r is therefore small in the near zone, the additional power of r in the denominator of Eq. (14-44) means that the electric field dominates the magnetic induction in the near zone. Further, apart from the sinusoidal time dependence whose explicit appearance in Eqs. (14-43) and (14-44) has been suppressed [Compare Eq. (14-30)], the field given by Eq. (14-44) is the field established by a static electric dipole. Thus at any instant in the near zone the fields are essentially those of a static charge distribution in which the charge density is the charge density in the distribution at that instant. In the particular limit $\kappa \longrightarrow 0$ (in which case $\omega \longrightarrow 0$ also) the fields are truly static and the near zone extends to $r = \infty$.

In contrast to Eqs. (14-43) and (14-44), the fields in the far zone have the limiting expressions

$$\mathcal{B}_0 = \frac{\mu_0}{4\pi}\omega\kappa\frac{e^{i\kappa r}}{r}\hat{\mathbf{r}} \times \mathbf{p}_0 \tag{14-45}$$

$$\mathcal{E}_0 = \frac{\mu_0}{4\pi}\omega\kappa c\frac{e^{i\kappa r}}{r}(\hat{\mathbf{r}} \times \mathbf{p}_0) \times \hat{\mathbf{r}} = c\mathcal{B}_0 \times \hat{\mathbf{r}} \tag{14-46}$$

Both are perpendicular to $\hat{\mathbf{r}}$, and each is perpendicular to the other; both vary as $1/r$; and the electric field at a particular observation point is polarized in the direction given by $(\hat{\mathbf{r}} \times \mathbf{p}_0) \times \hat{\mathbf{r}}$, which is the direction of $-\hat{\boldsymbol{\theta}}$ if $\mathbf{p}_0$ defines the polar axis. Further, since the time average of the Poynting vector in the far zone is given by

$$\langle \mathcal{S} \rangle = \frac{1}{2\mu_0}\,\mathrm{Re}\,\mathcal{E}_0 \times \mathcal{B}_0^* = \frac{\mu_0\omega^4\,|\hat{\mathbf{r}} \times \mathbf{p}_0|^2}{32\pi^2 c r^2}\hat{\mathbf{r}} \tag{14-47}$$

(see P7-26), the fields in the far zone describe a transport of energy radially away from the dipole; the time average of the total radiated power, which is given by the integral of Eq. (14-47) over a large sphere centered at the dipole, then has the explicit evaluation

$$\langle\text{radiated power}\rangle = \lim_{r\to\infty} \int_0^{2\pi}\int_0^{\pi} (\langle \mathbf{S}\rangle\cdot\hat{\mathbf{r}})r^2 \sin\theta\, d\theta\, d\phi \tag{14-48}$$

$$= \frac{4}{3}\left(\frac{\mu_0\omega^4|p_0|^2}{16\pi c}\right) \tag{14-49}$$

where a coordinate system in which $\mathbf{p}_0$ defines the polar axis ($\mathbf{p}_0 = p_0\hat{\mathbf{k}}$) has been chosen for evaluating the integral. Note particularly the dependence of the radiated power on the *fourth* power of the frequency; broadcasting at low frequencies is difficult by comparison with broadcasting at higher frequencies. Further, we conclude from Eq. (14-48) that the time average of the power radiated *per unit solid angle* by this dipole is given as a function of θ and ϕ by

$$\left\langle\frac{dP}{d\Omega}\right\rangle = \lim_{r\to\infty} r^2\langle\mathbf{S}\rangle\cdot\hat{\mathbf{r}} \tag{14-50}$$

$$= \frac{\mu_0\omega^4}{32\pi^2 c}|\hat{\mathbf{r}}\times\mathbf{p}_0|^2 = \frac{\mu_0\omega^4|p_0|^2}{32\pi^2 c}\sin^2\theta \tag{14-51}$$

where $d\Omega = \sin\theta\, d\theta\, d\phi$ is an element of solid angle.[2] Because of their connection with emitted radiation, the fields given by Eqs. (14-45) and (14-46) are referred to as the *radiation fields* of an oscillating dipole.

In the intermediate zone, the fields undergo transition from the quasi-static fields of the near zone to the radiation fields of the far zone, but the details of that transition are very involved and will not be treated here.

PROBLEMS

P14-2. Verify that Eqs. (14-31) and (14-32) follow from Eqs. (14-26) and (14-27).

P14-3. Letting $\mathbf{p}_0 = p_0\hat{\mathbf{k}}$ with p_0 real, work out the spherical components of the *physical* fields $\mathbf{E}(\mathbf{r}, t)$ and $\mathbf{B}(\mathbf{r}, t)$ represented by the complex fields in Eqs. (14-45) and (14-46). Describe the polarization of the radiation fields. A sketch, as, for example, Fig. 5-10 in J. M. Stone, *Radiation and Optics* (McGraw-Hill Book Company, New York, 1963), might be helpful.

P14-4. Starting with Eq. (14-51), draw a careful polar graph of the angular distribution of the radiation emitted by an oscillating dipole. Are the lobes in the graph circles?

P14-5. Several aspects of the derivation of the fields of an oscillating dipole were not worked out in detail in the text. To complete this derivation, (a) derive Eqs. (14-33)–(14-35) from the starting points indicated in the text, (b) derive the fields given in Eqs. (14-41) and (14-42) from the potential in

[2]Although Eq. (14-50) can be applied to any monochromatic field, it is not correct for fields exhibiting a more general time dependence. For these more general fields, the phenomenon of retardation must be more carefully handled; see Section 14-5.

Eq. (14-40) by using Eqs. (14-34) and (14-35), (c) derive the final term in Eq. (14-47), and (d) carry out the integration to obtain Eq. (14-49).

P14-6. From the point of view of an oscillating dipole, radiation to distant space looks power dissipation. Express the total power given by Eq. (14-49) in terms of the current in the dipole and, recalling that the power dissipated in a resistive circuit can be expressed as the product of a resistance and the square of the rms value of the current, infer that an oscillating dipole can be thought of as radiating into a resistance R_{eff} given by

$$R_{\text{eff}} = \frac{2\pi}{3}\sqrt{\frac{\mu_0}{\epsilon_0}}\left(\frac{a}{\lambda}\right)^2 = 790\left(\frac{a}{\lambda}\right)^2 \text{ ohms}$$

P14-7. Let the source of an electromagnetic field be a sinusoidally varying current of amplitude $\mathscr{I}_0$ in a circular loop of radius a, and let the loop lie in the x-y plane with its center at the origin. Following the pattern illustrated in Section 14-2, find the magnetic vector potential $\mathbf{\mathfrak{A}}_0$ established by this source in the region where $r \gg a$ and $\kappa a \ll 1$, find the corresponding fields $\mathbf{\mathfrak{B}}_0$ and $\mathbf{\mathcal{E}}_0$, and compare the results with Eqs. (14-41) and (14-42). Finally, determine the fields in the far zone and compare the angular distribution of energy and the polarization of the radiation fields for this oscillating *magnetic* dipole with those of an oscillating *electric* dipole.

14-3 The Liénard-Wiechert Potentials

One of the few situations for which the scalar and vector potentials given by Eqs. (14-26) and (14-27) can be evaluated in closed form occurs when the source distribution is confined to a *small* region of space and is observed from far away. (Ultimately, we shall reduce the distribution to a point charge and *all* observation points will be far away.) For such an arrangement, the integral in Eq. (14-26), for example, extends over a small region of space centered about the nominal *retarded* location $\mathbf{R}(t')$ of the charge distribution and, since $\mathbf{r}' \approx \mathbf{R}(t')$ throughout the region of integration, we can approximate $|\mathbf{r} - \mathbf{r}'|$ by $|\mathbf{r} - \mathbf{R}(t')|$ to find that

$$V(\mathbf{r}, t) \approx \frac{1}{4\pi\epsilon_0|\mathbf{r} - \mathbf{R}(t')|}\int \rho(\mathbf{r}', t')\, dv' \tag{14-52}$$

The approximation, of course, becomes better as the region occupied by the charge distribution becomes smaller. It is now very tempting to replace the integral in Eq. (14-52) with the total charge in the distribution. This evaluation would be correct, however, *only* if the integral extended over the region of space in which $\rho(\mathbf{r}', t')$ is nonzero at some fixed *retarded* time t'. Unfortunately (as we remarked in Section 14-1), the integral extends over the volume in which $\rho(\mathbf{r}', t')$ differs from zero at some fixed *observation* time t. This integral is therefore more complicated than one might at first sight suppose,

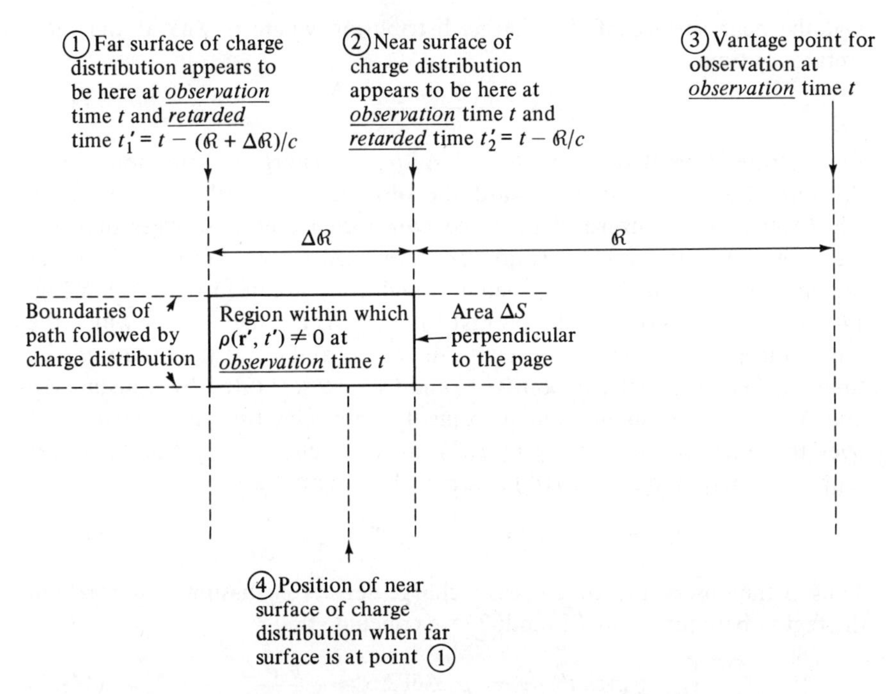

Fig. 14-2. Positions of a moving, charged rectangular parallelopiped at a fixed observation time.

and (as we shall see) the complications do *not* become negligible as the charge distribution shrinks to a point.

To obtain a correct evaluation of the integral under debate, suppose first that the charge distribution (at fixed t') has the shape of a rectangular parallelopiped and that (at the retarded time) it is moving directly *toward* a remote observation point along the path shown in Fig. 14-2. Further, let this charge distribution be observed from the observation point at *observation* time t. In Fig. 14-2, the points labeled ① and ② at distances $\mathcal{R} + \Delta\mathcal{R}$ and $\mathcal{R}$ from point ③ are the positions at which an observer at point ③ sees the far and near surfaces of the charge distributions to be located at the fixed observation time t. These two points therefore bound the region of space within which $\rho(\mathbf{r}', t')$ differs from zero at the fixed observation time t and hence also bound the volume over which the integral in Eq. (14-52) extends. These two points do *not*, however, bound the region within which the charge distribution is confined at any fixed instant of *retarded* time, for the observer at a fixed *observation* time t sees the far surface of the charge distribution where it was at the (retarded) time

$$t'_1 = t - \frac{\mathcal{R} + \Delta\mathcal{R}}{c} \tag{14-53}$$

and the near surface of the charge distribution where it was at the *later* (retarded) time

$$t'_2 = t - \frac{\mathcal{R}}{c} = t'_1 + \frac{\Delta\mathcal{R}}{c} \tag{14-54}$$

In the time interval $\Delta t' = t'_2 - t'_1 = \Delta\mathcal{R}/c$, the charge distribution moves (by our above hypothesis) toward the observer. Thus, the region within which $\rho(\mathbf{r}', t')$ is nonzero at the fixed *observation* time t is larger than the region within which $\rho(\mathbf{r}', t')$ is nonzero at a fixed *retarded* time t', the latter region at the retarded time t'_1 being bounded by points ① and ④ in Fig. 14-2. [The region bounded by points ① and ④ is the region over which the integral in Eq. (14-52) would extend *if* it were evaluated at the fixed *retarded* time t'_1.] Quantitatively, if points ① and ④ are separated by a distance a and $\Delta t' = \Delta\mathcal{R}/c$ is small enough to justify approximating the velocity $v(t')$ over the interval $t'_1 < t' < t'_2$ by $v(t'_1)$, then the charge distribution moves (approximately) a distance $v(t'_1)\,\Delta\mathcal{R}/c$ in the time $\Delta t'$ and

$$\Delta\mathcal{R} \approx a + v(t'_1)\frac{\Delta\mathcal{R}}{c} \Longrightarrow \Delta\mathcal{R} \approx \frac{a}{1 - v(t'_1)/c} \tag{14-55}$$

Thus, if the observer at time t sees a charge density ρ_0 (assumed uniform) in the region between points ① and ②, we conclude that

$$\int_{\text{fixed } t} \rho(\mathbf{r}', t')\, dv' = \rho_0\, \Delta\mathcal{R}\, \Delta S \approx \frac{\rho_0 a\, \Delta S}{1 - v(t'_1)/c} \tag{14-56}$$

where ΔS is the cross-sectional area of the charge distribution in a plane at right angles to the direction of view. Finally, however, the charge density ρ_0 seen by the observer is exactly the same as the charge density existing within the charge itself.[3] Thus, $\rho_0 a\, \Delta S$ is the actual total charge q in the distribution (as reckoned by an observation made at a fixed *retarded* time) and we find that

$$\int_{\text{fixed } t} \rho(\mathbf{r}', t')\, dv' \approx \frac{q}{1 - v(t'_1)/c} \tag{14-57}$$

Substituting Eq. (14-57) into Eq. (14-52), we find then that the scalar potential established by a small charge distribution is given approximately by

$$V(\mathbf{r}, t) \approx \frac{1}{4\pi\epsilon_0} \frac{q}{|\mathbf{r} - \mathbf{R}(t')|\,[1 - v(t'_1)/c]} \tag{14-58}$$

[3] In looking at a fixed observation time t, the observer is *not* seeing the same total charge merely distributed over a larger volume. He is seeing the combination of several portions of this charge at *different retarded* times. Since this charge is moving, the portions that the observer sees at different retarded times will contain some of the same fundamental charges. Thus, when he looks at a fixed observation time, the observer will in general see some of the same charges in more than one place (if the charges are moving toward him). Hence, the apparent total charge at a fixed *observation* time will be larger than the actual total charge at a fixed *retarded* time. Anyone who has tried to count monkeys at the zoo has encountered a similar phenomenon.

where t' is a retarded time somewhere in the interval $t'_1 \leq t' \leq t'_2$. This approximation, of course, becomes progressively more accurate as the dimensions of the charge distribution shrink to zero; it becomes exact for an idealized point charge, for which $t'_1 \longrightarrow t'_2$ and

$$V(\mathbf{r}, t) = \frac{1}{4\pi\epsilon_0} \frac{q}{|\mathbf{r} - \mathbf{R}(t')|\,[1 - v(t')/c]} \tag{14-59}$$

where $\mathbf{R}(t')$ is the position and $v(t')$ the speed of the charge q at the retarded time t' determined (as a function of t) by the equation

$$t' = t - \frac{1}{c}|\mathbf{r} - \mathbf{R}(t')| \tag{14-60}$$

[Compare Eq. (14-24).]

The generalization of Eq. (14-59) to point charges not moving directly toward the observation point is immediate. We replace the speed $v(t')$ by the component of the velocity in the direction of the observation point; i.e., we make the replacement

$$v(t') \longrightarrow \mathbf{v}(t') \cdot \frac{\mathbf{r} - \mathbf{R}(t')}{|\mathbf{r} - \mathbf{R}(t')|} = \mathbf{v}(t') \cdot \hat{\mathbf{n}}(t') \tag{14-61}$$

where

$$\hat{\mathbf{n}}(t') = \frac{\mathbf{r} - \mathbf{R}(t')}{|\mathbf{r} - \mathbf{R}(t')|} \tag{14-62}$$

is a unit vector directed toward the observation point from the (retarded) position of the charge. Thus, defining

$$\mathcal{R} = |\mathbf{r} - \mathbf{R}(t')| \qquad \xi = 1 - \frac{\mathbf{v}(t')}{c} \cdot \hat{\mathbf{n}}(t') \qquad \text{(14-63), (14-64)}$$

we find finally from Eq. (14-59) that the scalar potential established at $(\mathbf{r}, t)$ by a moving point charge is given by

$$\begin{aligned} V(\mathbf{r}, t) &= \frac{1}{4\pi\epsilon_0} \frac{q}{|\mathbf{r} - \mathbf{R}(t')|\left(1 - \dfrac{\mathbf{v}(t')}{c} \cdot \hat{\mathbf{n}}(t')\right)} \\ &= \frac{q}{4\pi\epsilon_0 \mathcal{R}\xi} \end{aligned} \tag{14-65}$$

Explicit indication that $\mathcal{R}$ and ξ are to be evaluated at the *retarded* time is suppressed.

By arguments similar to those presented above, we find that a moving point charge establishes a vector potential at $(\mathbf{r}, t)$ given by

$$\mathbf{A}(\mathbf{r}, t) = \frac{\mu_0 q \mathbf{v}(t')}{4\pi\mathcal{R}\xi} \tag{14-66}$$

Together, the potentials given by Eqs. (14-65) and (14-66) are called the *Liénard-Wiechert potentials* and they apply specifically to a moving *point* charge.

PROBLEMS

P14-8. Derive Eq. (14-66) from Eq. (14-27).

P14-9. In deriving the Liénard-Wiechert potentials, we assumed (1) that the charge was shaped like a rectangular parallelopiped and (2) that the charge density within the charge was uniform. Present an argument showing that both of these assumptions can be relaxed without changing the final results.

P14-10. Consider a particle carrying charge q and moving with constant speed along the z axis toward $z = +\infty$. Imagine further several observers stationed at points on the circumference of a circle of radius a lying in the y-z plane and centered at the origin. Each observer measures the scalar potential at that instant of observation time t when the particle appears to him to be at the origin. (a) Show that all observers make their measurements simultaneously. (b) Calculate the value obtained by an observer located at polar angle θ from the z axis and sketch graphs of the scalar potential versus θ for several different values of v/c. Draw a careful graph for at least one value of v/c. *Hint*: Present your graphs in dimensionless form by plotting $V/(q/4\pi\epsilon_0 a)$ versus θ. (c) Describe the differences between these graphs in the nonrelativistic ($v/c \ll 1$) and extreme relativistic ($v/c \sim 1$) regions.

P14-11. A charged particle moves with constant speed v along the z axis toward $z = +\infty$ and at the retarded time t' is located at the point $\mathbf{R}(t') = vt'\hat{\mathbf{k}}$. Show that the resulting scalar and vector potentials are given at the point $\mathbf{r}$ and observation time t by

$$V(\mathbf{r}, t) = \frac{q}{4\pi\epsilon_0} \frac{1}{\sqrt{[1 - (v^2/c^2)](x^2 + y^2) + (z - vt)^2}}$$

$$\mathbf{A}(\mathbf{r}, t) = \frac{\mu_0 q}{4\pi} \frac{v\hat{\mathbf{k}}}{\sqrt{[1 - (v^2/c^2)](x^2 + y^2) + (z - vt)^2}}$$

where x, y, and z are the Cartesian coordinates of the point $\mathbf{r}$. *Optional*: Examine the equipotential surfaces of $V(\mathbf{r}, 0)$ for selected values of v/c.

14-4 The Fields of a Moving Point Charge

Once the potentials established by a particular source distribution have been determined, finding the fields "merely" involves substituting the potentials into Eqs. (14-5) and (14-6). In applying this method to determine the fields established by a moving point charge, however, we must exercise considerable mathematical care, first, because Eqs. (14-5) and (14-6) require spatial derivatives at constant *observation* time t (*not* at constant *retarded* time t') and, second, because the derivative $\partial/\partial t$ (not $\partial/\partial t'$) is needed. In evaluating the necessary derivatives of the potential, we must therefore not overlook the dependence of the retarded time (in terms of which the potentials have been expressed) on both $\mathbf{r}$ and t. [See Eq. (14-60).] In this section, we shall summarize the mathematical manipulations that yield the fields when the potentials in Eqs. (14-65) and (14-66) are substituted into Eqs. (14-5) and (14-6); several of the detailed mathematical evaluations, however, are relegated to P14-12 and P14-13.

Let us begin by evaluating the negative gradient of $V(\mathbf{r}, t)$. From Eq. (14-65) and then Eqs. (14-62)–(14-64), we find that

$$-\nabla V = -\frac{q}{4\pi\epsilon_0}\nabla\left(\frac{1}{\mathcal{R}\xi}\right) = \frac{q}{4\pi\epsilon_0\mathcal{R}^2\xi^2}\nabla(\mathcal{R}\xi)$$
$$= \frac{q}{4\pi\epsilon_0(\mathcal{R}\xi)^2}\nabla\left(|\mathbf{r} - \mathbf{R}(t')| - \frac{\mathbf{v}(t')}{c}\cdot[\mathbf{r} - \mathbf{R}(t')]\right) \tag{14-67}$$

which breaks into two terms. Writing the first of these terms in the form

$$\nabla|\mathbf{r} - \mathbf{R}| = \frac{1}{2|\mathbf{r} - \mathbf{R}|}\nabla[|\mathbf{r} - \mathbf{R}|^2]$$
$$= \frac{1}{2|\mathbf{r} - \mathbf{R}|}\nabla[(\mathbf{r} - \mathbf{R})\cdot(\mathbf{r} - \mathbf{R})] \tag{14-68}$$

[see Eq. (C-6)] and then using the vector identity in Eq. (C-15) and the two identities $\nabla \times \mathbf{r} = 0$ and $(\hat{\mathbf{n}}\cdot\nabla)\mathbf{r} = \hat{\mathbf{n}}$ (P14-12), we find that

$$\nabla|\mathbf{r} - \mathbf{R}| = (\hat{\mathbf{n}}\cdot\nabla)(\mathbf{r} - \mathbf{R}) + \hat{\mathbf{n}} \times [\nabla \times (\mathbf{r} - \mathbf{R})]$$
$$= \hat{\mathbf{n}} - (\hat{\mathbf{n}}\cdot\nabla)\mathbf{R} - \hat{\mathbf{n}} \times (\nabla \times \mathbf{R}) \tag{14-69}$$

where $\hat{\mathbf{n}}$ is given by Eq. (14-62). Further, remembering that $\mathbf{R}$ is a function of t' and is therefore implicitly a function of $\mathbf{r}$, we find that

$$(\hat{\mathbf{n}}\cdot\nabla)\mathbf{R} = \sum_{i,j=1}^{3} n_i\frac{\partial}{\partial x_i}R_j\hat{\mathbf{e}}_j$$
$$= \sum_{i,j=1}^{3} n_i\frac{dR_j}{dt'}\frac{\partial t'}{\partial x_i}\hat{\mathbf{e}}_j = (\hat{\mathbf{n}}\cdot\nabla t')\mathbf{v} \tag{14-70}$$

where $\mathbf{v}$ is the velocity of the particle *evaluated at the retarded time*. Similarly, we find that

$$\nabla \times \mathbf{R} = \nabla t' \times \mathbf{v} \tag{14-71}$$

(P14-12). Finally, substituting Eqs. (14-70) and (14-71) into Eq. (14-69) and using the vector identity in Eq. (C-1), we find that the first part of the gradient in Eq. (14-67) has the alternative expression

$$\nabla|\mathbf{r} - \mathbf{R}| = \hat{\mathbf{n}} - (\hat{\mathbf{n}}\cdot\mathbf{v})\,\nabla t' \tag{14-72}$$

Turning now to the second part of the gradient in Eq. (14-67) and using vector identities already referred to above, we find that

$$\nabla\left[\frac{\mathbf{v}}{c}\cdot(\mathbf{r}-\mathbf{R})\right]=\left(\frac{\mathbf{v}}{c}\cdot\nabla\right)(\mathbf{r}-\mathbf{R})+[(\mathbf{r}-\mathbf{R})\cdot\nabla]\frac{\mathbf{v}}{c}$$
$$+\frac{\mathbf{v}}{c}\times[\nabla\times(\mathbf{r}-\mathbf{R})]+(\mathbf{r}-\mathbf{R})\times\left[\nabla\times\frac{\mathbf{v}}{c}\right]$$
$$=\frac{\mathbf{v}}{c}-\left(\frac{\mathbf{v}}{c}\cdot\nabla\right)\mathbf{R}+[(\mathbf{r}-\mathbf{R})\cdot\nabla]\frac{\mathbf{v}}{c}$$
$$-\frac{\mathbf{v}}{c}\times[\nabla\times\mathbf{R}]+(\mathbf{r}-\mathbf{R})\times\left[\nabla\times\frac{\mathbf{v}}{c}\right] \qquad (14\text{-}73)$$

Then, using results similar to Eqs. (14-70) and (14-71), we find that

$$\nabla\left[\frac{\mathbf{v}}{c}\cdot(\mathbf{r}-\mathbf{R})\right]=\frac{\mathbf{v}}{c}-\left(\frac{\mathbf{v}}{c}\cdot\nabla t'\right)\mathbf{v}+\left(\mathcal{R}\hat{\mathbf{n}}\cdot\nabla t'\right)\frac{\dot{\mathbf{v}}}{c}$$
$$-\frac{\mathbf{v}}{c}\times\left(\nabla t'\times\mathbf{v}\right)+\mathcal{R}\hat{\mathbf{n}}\times\left(\nabla t'\times\frac{\dot{\mathbf{v}}}{c}\right)$$
$$=\frac{\mathbf{v}}{c}-\frac{v^2}{c}\nabla t'+\left(\mathcal{R}\hat{\mathbf{n}}\cdot\nabla t'\right)\frac{\dot{\mathbf{v}}}{c}+\mathcal{R}\hat{\mathbf{n}}\times\left(\nabla t'\times\frac{\dot{\mathbf{v}}}{c}\right)$$
$$(14\text{-}74)$$

where $\mathcal{R}$ is given by Eq. (14-63) and

$$\dot{\mathbf{v}}=\frac{d\mathbf{v}(t')}{dt'} \qquad (14\text{-}75)$$

is the acceleration of the particle at the *retarded* time.

Finally, substituting Eqs. (14-72) and (14-74) into Eq. (14-67), we find that

$$-\nabla V=\frac{q}{4\pi\epsilon_0\mathcal{R}^2\xi^2}\left[\hat{\mathbf{n}}-\frac{\mathbf{v}}{c}-\left(\hat{\mathbf{n}}\cdot\mathbf{v}-\frac{v^2}{c}\right)\nabla t'\right]$$
$$-\frac{q}{4\pi\epsilon_0\mathcal{R}\xi^2}\left[(\hat{\mathbf{n}}\cdot\nabla t')\frac{\dot{\mathbf{v}}}{c}+\hat{\mathbf{n}}\times\left(\nabla t'\times\frac{\dot{\mathbf{v}}}{c}\right)\right] \qquad (14\text{-}76)$$

By a similar calculation, we find from Eq. (14-66) that

$$\frac{\partial A}{\partial t}=\frac{\partial A}{\partial t'}\frac{\partial t'}{\partial t}$$
$$=\frac{q}{4\pi\epsilon_0\mathcal{R}^2\xi^2c^2}\left[(\hat{\mathbf{n}}\cdot\mathbf{v})\mathbf{v}-\frac{v^2}{c}\mathbf{v}\right]\frac{\partial t'}{\partial t}$$
$$+\frac{q}{4\pi\epsilon_0\mathcal{R}\xi^2c^2}\left[\dot{\mathbf{v}}\left(1-\frac{\mathbf{v}}{c}\cdot\hat{\mathbf{n}}\right)+\mathbf{v}\left(\frac{\dot{\mathbf{v}}}{c}\cdot\hat{\mathbf{n}}\right)\right]\frac{\partial t'}{\partial t} \qquad (14\text{-}77)$$

(See P14-12.) Further, we find from Eq. (14-60) not only that

$$\frac{\partial t'}{\partial t}=1-\frac{1}{c}\frac{\partial}{\partial t'}|\mathbf{r}-\mathbf{R}(t')|\frac{\partial t'}{\partial t}$$
$$=1+\frac{\hat{\mathbf{n}}\cdot\mathbf{v}}{c}\frac{\partial t'}{\partial t}\Longrightarrow\frac{\partial t'}{\partial t}=\frac{1}{1-\hat{\mathbf{n}}\cdot\mathbf{v}/c}=\frac{1}{\xi} \qquad (14\text{-}78)$$

but also that

$$\nabla t'=-\frac{1}{c}\nabla|\mathbf{r}-\mathbf{R}(t')|$$

$$= -\frac{1}{c}[\hat{\mathbf{n}} - (\hat{\mathbf{n}}\cdot\mathbf{v})\,\nabla t']$$

$$\Longrightarrow \nabla t' = \frac{-\hat{\mathbf{n}}/c}{1 - \hat{\mathbf{n}}\cdot\mathbf{v}/c} = -\frac{\hat{\mathbf{n}}}{\xi c} \tag{14-79}$$

Thus, substituting Eqs. (14-78) and (14-79) into Eqs. (14-76) and (14-77), substituting those results in turn into Eq. (14-5), and finally using a number of by now familiar vector identities, we find that the electric field established by a point charge is given by

$$\mathbf{E}(\mathbf{r}, t) = -\nabla V(\mathbf{r}, t) - \frac{\partial \mathbf{A}(\mathbf{r}, t)}{\partial t}$$

$$= \frac{q\mathbf{N}}{4\pi\epsilon_0 \mathcal{R}^2 \xi^3} + \frac{q\mathbf{M}}{4\pi\epsilon_0 c^2 \mathcal{R} \xi^3} \tag{14-80}$$

where, for brevity, we have introduced the quantities

$$\mathbf{N} = \left(\hat{\mathbf{n}} - \frac{\mathbf{v}}{c}\right)\left(1 - \frac{v^2}{c^2}\right) \tag{14-81}$$

and

$$\mathbf{M} = \hat{\mathbf{n}} \times \left[\left(\hat{\mathbf{n}} - \frac{\mathbf{v}}{c}\right) \times \dot{\mathbf{v}}\right] \tag{14-82}$$

(See P14-12.) An equally lengthy but similar argument leads ultimately to the expression

$$\mathbf{B}(\mathbf{r}, t) = \nabla \times \mathbf{A}(\mathbf{r}, t) = \frac{1}{c}\hat{\mathbf{n}} \times \mathbf{E}(\mathbf{r}, t) \tag{14-83}$$

for the magnetic induction field accompanying the electric field in Eq. (14-80). (See P14-13.) We remind the reader that all quantities with unspecified time arguments in Eqs. (14-80)–(14-83) are evaluated at the *retarded* time.

The fields produced by a point charge in uniform motion along a straight line are calculated explicitly in P14-15. Pictures of the field lines produced not only by charges in uniform motion but also by accelerated charges may be found, for example, in E. M. Purcell, *Electricity and Magnetism* (McGraw-Hill Book Company, New York, 1965), pp. 163–165; in J. C. Hamilton and J. L. Schwartz, *Am. J. Phys.* **39**, 1540 (1971); and in R. Y. Tsien, *Am. J. Phys.* **40**, 46 (1972).

PROBLEMS

P14-12. To complete the derivation of the electric field established by a moving point charge, (a) show that $(\hat{\mathbf{n}}\cdot\nabla)\mathbf{r} = \hat{\mathbf{n}}$, (b) derive Eq. (14-71), (c) derive Eq. (14-77), and (d) combine the expressions indicated in the text to derive Eq. (14-80).

P14-13. Derive Eq. (14-83). Shorten the labor as much as possible by using results already obtained.

P14-14. Show that the fields established by a particle moving nonrelativistically are given by

$$\mathbf{E}(\mathbf{r}, t) = \frac{q}{4\pi\epsilon_0}\left[\frac{\hat{\mathbf{n}}}{\mathcal{R}^2} + \frac{3(\mathbf{v}\cdot\hat{\mathbf{n}})\hat{\mathbf{n}} - \mathbf{v}}{c\mathcal{R}^2} + \frac{\hat{\mathbf{n}} \times (\hat{\mathbf{n}} \times \dot{\mathbf{v}})}{c^2\mathcal{R}}\right]$$

$$\mathbf{B}(\mathbf{r}, t) = \frac{q}{4\pi\epsilon_0}\left[\frac{\mathbf{v} \times \hat{\mathbf{n}}}{c^2\mathcal{R}^2} + \frac{\dot{\mathbf{v}} \times \hat{\mathbf{n}}}{c^3\mathcal{R}}\right]$$

where all quantities on the *right*-hand side of these equations are evaluated at the retarded time. A discussion of the properties of these fields is presented in C. W. Sherwin, *Basic Concepts of Physics* (Holt, Rinehart and Winston, Inc., New York, 1961), Chapter 5, and also in J. B. Brackenridge and R. M. Rosenberg, *Principles of Physics and Chemistry* (McGraw-Hill Book Company, New York, 1970), Chapter 16.

P14-15. The position of a charge q as a function of time t is given by $\mathbf{R} = vt\hat{\mathbf{k}}$, where v is a positive constant and $\hat{\mathbf{k}}$ is a unit vector in the z direction. Let the observation point have coordinates $(\imath, \phi, z, t)$ in cylindrical coordinates. (a) Show that the fields established by this charge and current distribution are given by

$$\begin{Bmatrix} \mathbf{E}(\imath, \phi, z, t) \\ \mathbf{B}(\imath, \phi, z, t) \end{Bmatrix} = \frac{\dfrac{q}{4\pi\epsilon_0}\left(1 - \dfrac{v^2}{c^2}\right)}{\left[\imath^2\left(1 - \dfrac{v^2}{c^2}\right) + (z - vt)^2\right]^{3/2}} \begin{Bmatrix} \imath\hat{\boldsymbol{\imath}} + (z - vt)\hat{\mathbf{k}} \\ \dfrac{\imath v}{c^2}\hat{\boldsymbol{\phi}} \end{Bmatrix}$$

(b) Calculate $|\mathbf{E}(\imath, \phi, z, t = 0)|$, express the result in terms of the *spherical* polar coordinates of the observation point, and draw a polar graph of this quantity as a function of the polar angle for fixed radial coordinate. Describe how this graph is altered as v moves from nonrelativistic values to relativistic values. It might be wise to make a careful plot of this quantity for one value of v/c, say $v/c = 0.707$. *Optional*: Derive the above fields by applying Eqs. (14-5) and (14-6) to the results of P14-11.

14-5
Radiation from Accelerated Point Charges

The energy transported by any electromagnetic field is determined by the Poynting vector $\mathbf{S} = (\mathbf{E} \times \mathbf{B})/\mu_0$. For the fields of a point charge [Eqs. (14-80) and (14-83)], the Poynting vector has the more explicit evaluation

$$\begin{aligned}\mathbf{S} &= \left[\frac{q}{4\pi\epsilon_0\xi^3}\right]^2 \frac{1}{\mu_0 c}\left[\frac{\mathbf{N}}{\mathcal{R}^2} + \frac{\mathbf{M}}{c^2\mathcal{R}}\right] \times \left[\frac{\hat{\mathbf{n}} \times \mathbf{N}}{\mathcal{R}^2} + \frac{\hat{\mathbf{n}} \times \mathbf{M}}{c^2\mathcal{R}}\right] \\ &= \frac{q^2[\mathbf{M} \times (\hat{\mathbf{n}} \times \mathbf{M})]}{16\pi^2\epsilon_0 c^3\mathcal{R}^2\xi^6} + O\left(\frac{1}{\mathcal{R}^3}\right) \\ &= \frac{q^2[M^2\hat{\mathbf{n}} - (\mathbf{M}\cdot\hat{\mathbf{n}})\mathbf{M}]}{16\pi^2\epsilon_0 c^3\mathcal{R}^2\xi^6} + O\left(\frac{1}{\mathcal{R}^3}\right) \\ &= \frac{q^2M^2}{16\pi^2\epsilon_0 c^3\mathcal{R}^2\xi^6}\hat{\mathbf{n}} + O\left(\frac{1}{\mathcal{R}^3}\right) \qquad (14\text{-}84)\end{aligned}$$

where (1) the final form follows because $\mathbf{M}$ and $\hat{\mathbf{n}}$ are perpendicular [see Eq. (14-82)] and (2) terms in $1/\mathcal{R}^3$ and $1/\mathcal{R}^4$ have been indicated only symbolically because they will prove to play no role in determining the radiation emitted by the charge. As we expected, the dominant term in Eq. (14-84) is directed away from the (retarded) position of the charge.

To connect the Poynting vector more explicitly with the distribution of radiated energy, note first that an observer who measures the total energy ΔW transported across a small surface of area ΔS during the (observation) time interval $t_1 \leq t \leq t_2$ will obtain a result given by

$$\Delta W = \int_{t_1}^{t_2} (\mathbf{S} \cdot \Delta\mathbf{S})\, dt \tag{14-85}$$

It is, however, more appropriate for our present application to rewrite this integral as an integral over the retarded time interval $t'_1 \leq t' \leq t'_2$, where

$$t'_1 = t_1 - \frac{1}{c}|\mathbf{r} - \mathbf{R}(t'_1)| \qquad t'_2 = t_2 - \frac{1}{c}|\mathbf{r} - \mathbf{R}(t'_2)| \tag{14-86}$$

and $\mathbf{r}$ is the position of the area element $\Delta\mathbf{S}$. Since

$$dt = \frac{dt}{dt'}\, dt' = \xi\, dt' \tag{14-87}$$

[see Eq. (14-78)], we find that

$$\Delta W = \int_{t_1'}^{t_2'} \xi(\mathbf{S} \cdot \Delta\mathbf{S})\, dt' \tag{14-88}$$

from which we conclude that the rate at which energy is radiated toward the surface $\Delta\mathbf{S}$ [or equivalently the power $\Delta P(t')$ radiated toward $\Delta\mathbf{S}$ at the retarded time t'] is given by

$$\Delta P(t') = \xi \mathbf{S} \cdot \Delta\mathbf{S} \tag{14-89}$$

Now let $\Delta\mathbf{S}$ specifically represent a surface of area $|\Delta\mathbf{S}|$ oriented with its plane perpendicular to the vector from the retarded position of the charge to the surface; i.e., let $\Delta\mathbf{S} = \hat{\mathbf{n}}\,|\Delta\mathbf{S}|$. Then, $|\Delta\mathbf{S}| = \mathcal{R}^2\, \Delta\Omega$, where $\Delta\Omega$ is the solid angle subtended by the surface as viewed from the retarded position of the charge (Section 2-3), and we find from Eq. (14-89) that

$$\begin{aligned} \Delta P(t') &= \xi \mathbf{S} \cdot \hat{\mathbf{n}} \mathcal{R}^2\, \Delta\Omega \\ \Longrightarrow \frac{dP(t')}{d\Omega} &= \xi \mathcal{R}^2 \mathbf{S} \cdot \hat{\mathbf{n}} \end{aligned} \tag{14-90}$$

where $dP(t')/d\Omega$ gives the power emitted per unit solid angle at the retarded time t'. More usefully, if $\mathcal{R} \longrightarrow \infty$, Eq. (14-90) then gives that portion of the power emitted per unit solid angle at the retarded time t' that is destined to escape from the charge altogether; i.e.,

$$\frac{dP(t')}{d\Omega} = \lim_{\mathcal{R}\to\infty} \xi \mathcal{R}^2 \mathbf{S} \cdot \hat{\mathbf{n}} \tag{14-91}$$

gives the angular distribution of the power actually radiated away from the charge. Specifically, from the Poynting vector in Eq. (14-84), we find that

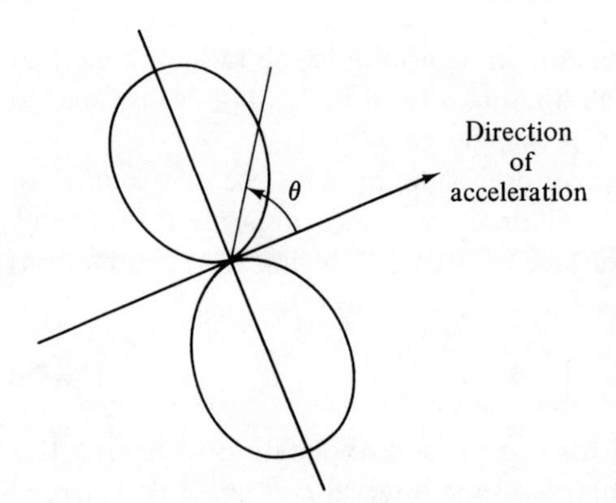

Fig. 14-3. Angular distribution of the radiation from an accelerated point charge when the motion is non-relativistic. The distance from the origin to the curve in the direction at an angle θ to the direction of the acceleration is proportional to $dP(t')/d\Omega$ at that angle.

$$\frac{dP(t')}{d\Omega} = \frac{q^2 M^2}{16\pi^2 \epsilon_0 c^3 \xi^5} = \frac{q^2 \left| \hat{\mathbf{n}} \times \left[\left(\hat{\mathbf{n}} - \frac{\mathbf{v}}{c} \right) \times \dot{\mathbf{v}} \right] \right|^2}{16\pi^2 \epsilon_0 c^3 \xi^5} \tag{14-92}$$

By way of reminder, all time-dependent quantities in this equation are evaluated at the *retarded* time. Further, the vector whose square magnitude appears has the same direction as the radiative part of the electric field given in Eq. (14-80) and hence conveys the polarization of the radiation at the observation point. Finally, note that there is *no* radiated power at all ($dP/d\Omega = 0$) unless the charged particle is *accelerated*.

Equation (14-92) assumes its simplest form in the nonrelativistic limit ($v/c \ll 1$), in which case $\xi \longrightarrow 1$ [Eq. (14-64)]. Further, if we assume that the particle is near the origin, then $\hat{\mathbf{n}} = \hat{\mathbf{r}}$ and Eq. (14-92) reduces to

$$\frac{dP(t')}{d\Omega} = \frac{q^2 |\hat{\mathbf{r}} \times \dot{\mathbf{v}}|^2}{16\pi^2 \epsilon_0 c^3} = \frac{q^2 \dot{v}^2 \sin^2 \theta}{16\pi^2 \epsilon_0 c^3} \tag{14-93}$$

where (in the final form) θ is the angle between the direction of the *acceleration* and the direction to the observation point. The radiated power is therefore distributed in space as shown in Fig. 14-3. Further, the total power emitted into all directions is given by the integral of Eq. (14-93) over all solid angles, viz., by

$$P(t') = \int_0^{2\pi} \int_0^{\pi} \frac{dP(t')}{d\Omega} \sin \theta \, d\theta \, d\phi = \frac{q^2 \dot{v}^2}{6\pi \epsilon_0 c^3} \tag{14-94}$$

which is known as *Larmor's formula*; its generalization to include relativistic motion of the charge is

$$P(t') = \frac{q^2}{6\pi \epsilon_0 c^3} \frac{1}{[1 - (v/c)^2]^3} \left[\dot{v}^2 - \frac{|\mathbf{v} \times \dot{\mathbf{v}}|^2}{c^2} \right] \tag{14-95}$$

and is known as *Liénard's formula*.[4]

[4]See J. D. Jackson, *Classical Electrodynamics* (John Wiley & Sons, Inc., New York, 1962), Section 14-2; see also P15-40.

Equation (14-85) also provides the starting point for determining the frequency distribution of the emitted radiation. Again, we take $\Delta\mathbf{S} = \hat{\mathbf{n}}\mathcal{R}^2\Delta\Omega$ [see the sentences following Eq. (14-89)] but this time we retain the *observer's* time and extend the integral over all time. We find that the *total* energy emitted per unit solid angle is given by

$$\frac{dW}{d\Omega} = \int_{-\infty}^{\infty} \lim_{\mathcal{R}\to\infty} (\mathbf{S}\cdot\hat{\mathbf{n}})\mathcal{R}^2\, dt \tag{14-96}$$

Substituting from Eq. (14-84), however, we find more explicitly that

$$\frac{dW}{d\Omega} = \int_{-\infty}^{\infty} |\mathbf{Q}(t)|^2\, dt \tag{14-97}$$

where

$$\mathbf{Q}(t) = \left(\frac{q^2}{16\pi^2\epsilon_0 c^3}\right)^{1/2}\frac{\mathbf{M}}{\xi^3} \tag{14-98}$$

and $\mathbf{M}$ and ξ continue to be evaluated at the retarded time. To obtain the frequency distribution of the radiated energy, we first introduce the Fourier transform

$$\tilde{\mathbf{Q}}(\omega) = \int_{-\infty}^{\infty} e^{i\omega t}\mathbf{Q}(t)\, dt \tag{14-99}$$

(See Appendix D, but note that, in keeping with the established convention for this topic, we use the exponential $e^{i\omega t}$ rather than $e^{-i\omega t}$ in defining the transform.) Then, we use Parseval's theorem [Eq. (D-34), which holds no matter which sign is taken in the exponential in Eq. (14-99)] to write Eq. (14-97) as the integral

$$\frac{dW}{d\Omega} = \frac{1}{2\pi}\int_{-\infty}^{\infty} |\tilde{\mathbf{Q}}(\omega)|^2\, d\omega \tag{14-100}$$

where $|\tilde{\mathbf{Q}}(\omega)|^2 = \tilde{\mathbf{Q}}^*(\omega)\cdot\tilde{\mathbf{Q}}(\omega)$. Next we write Eq. (14-100) as an integral over positive frequencies only, i.e., as

$$\frac{dW}{d\Omega} = \frac{1}{2\pi}\int_{0}^{\infty} [|\tilde{\mathbf{Q}}(\omega)|^2 + |\tilde{\mathbf{Q}}(-\omega)|^2]\, d\omega \tag{14-101}$$

which simplifies to

$$\frac{dW}{d\Omega} = \frac{1}{\pi}\int_{0}^{\infty} |\tilde{\mathbf{Q}}(\omega)|^2\, d\omega \tag{14-102}$$

because $\mathbf{Q}(t)$ is real and Eq. (14-99) then yields that

$$\tilde{\mathbf{Q}}^*(\omega) = \tilde{\mathbf{Q}}(-\omega) \Longrightarrow |\tilde{\mathbf{Q}}(-\omega)|^2 = |\tilde{\mathbf{Q}}(\omega)|^2 \tag{14-103}$$

Finally, we note that the integrand in Eq. (14-102) is just the distribution of $dW/d\Omega$ in frequency. Thus, the frequency distribution of the total energy emitted per unit solid angle is given by

$$\frac{dW}{d\Omega\, d\omega} = \frac{1}{\pi} |\tilde{\mathbf{Q}}(\omega)|^2 \tag{14-104}$$

Evaluation of $\tilde{\mathbf{Q}}(\omega)$ can be simplified in two ways. First, we substitute from Eq. (14-98) and then rewrite Eq. (14-99) as an integral over retarded times, obtaining

$$\tilde{\mathbf{Q}}(\omega) = \left(\frac{q^2}{16\pi^2\epsilon_0 c^3}\right)^{1/2} \int_{-\infty}^{\infty} \frac{\mathbf{M}(t')}{\xi^2} e^{i\omega[t' + |\mathbf{r} - \mathbf{R}(t')|/c]}\, dt' \tag{14-105}$$

[Compare Eq. (14-87).] Second, we assume that the observation point is remote from the charge and that motion of the charge is confined to a region close to the origin of coordinates. Under these conditions, $\hat{\mathbf{n}}$ is approximately constant and $\mathbf{r} \approx r\hat{\mathbf{n}}$ so that

$$|\mathbf{r} - \mathbf{R}(t')| \approx r \left| \hat{\mathbf{n}} - \frac{\mathbf{R}(t')}{r} \right| \approx r - \hat{\mathbf{n}} \cdot \mathbf{R}(t') \tag{14-106}$$

and, *apart from a phase factor* $e^{i\omega r/c}$ [which we can ignore because only the square magnitude of $\tilde{\mathbf{Q}}(\omega)$ is given a physical interpretation], we find from Eqs. (14-105), (14-82), and (14-64) that

$$\tilde{\mathbf{Q}}(\omega) = \left(\frac{q^2}{16\pi^2\epsilon_0 c^3}\right)^{1/2} \int_{-\infty}^{\infty} \frac{\hat{\mathbf{n}} \times \left[\left(\hat{\mathbf{n}} - \dfrac{\mathbf{v}}{c}\right) \times \dot{\mathbf{v}}\right]}{[1 - \mathbf{v}\cdot\hat{\mathbf{n}}/c]^2} e^{i\omega(t' - \hat{\mathbf{n}}\cdot\mathbf{R}/c)}\, dt' \tag{14-107}$$

in which $\mathbf{R}$, $\mathbf{v}$, $\dot{\mathbf{v}}$, and $\hat{\mathbf{n}}$ are all evaluated at the retarded time t'. Thus, in principle, the problem of finding the distribution in frequency and in solid angle of the energy emitted by a charged particle following a prescribed trajectory is solved: We evaluate $\tilde{\mathbf{Q}}(\omega)$ from Eq. (14-107) and substitute the result into Eq. (14-104). In practice the task is rarely trivial; two examples may be found in P14-21 and P14-22.

PROBLEMS

P14-16. Do the necessary integration and obtain Eq. (14-94).

P14-17. (a) Show that an electron moving nonrelativistically in a constant magnetic induction field of magnitude B no larger than about 10,000 G radiates a very small fraction of its total energy per cycle of its orbit. (b) Obtain a differential equation for the energy E of the particle as a function of time and integrate the equation. *Hint*: The total power radiated is equal to the rate of change of kinetic energy. Why?

P14-18. A particle experiences a *constant* acceleration in the direction of its velocity. Determine the angular distribution of the emitted power as given by Eq. (14-92). Do *not* assume the motion to be nonrelativistic. Show the results in a polar graph and indicate how this graph changes as the particle velocity approaches the speed of light. *Optional*: Determine the angle at which the maximum power appears and draw a graph of that angle as a function of v/c.

P14-19. Consider an electron moving (perhaps relativistically) along a straight line, as, for example, in a linear accelerator, and let the line define the x axis. (a) Taking the momentum of the electron to be given by $p = mv/\sqrt{1-(v/c)^2}$, show from Eq. (14-95) that the total power radiated is given by

$$P = \frac{q^2}{6\pi\epsilon_0 m^2 c^3}\left(\frac{dp}{dt}\right)^2$$

(b) Show for this particle that the rate P at which energy is radiated bears the ratio

$$\frac{\text{power radiated}}{\text{power input}} = \frac{P}{dE/dt} = \frac{q^2}{6\pi\epsilon_0 c^3 m^2 v}\frac{dE}{dx}$$

to the rate dE/dt at which external forces do work on the particle. *Hint*: $dp/dt = dE/dx$. Why? (c) In a typical linear accelerator, $dE/dx \lesssim 10$ MeV/m. Estimate $P/(dE/dt)$ numerically for such an accelerator, taking $v \approx c$, and comment on the importance of radiative losses in linear accelerators.

P14-20. Once the speed v of a particle moving in a circle of radius r approaches c, the acceleration of the particle is approximately perpendicular to the velocity and has magnitude v^2/r. (a) Show from Eq. (14-95) that the energy radiated per revolution δW is given by

$$\delta W = \frac{q^2}{3\epsilon_0 r}\left(\frac{v}{c}\right)^3\left(\frac{E}{mc^2}\right)^4$$

where $E = mc^2/\sqrt{1-(v/c)^2}$ and m is the (rest) mass of the particle. (b) To what maximum energy (in MeV) can an electron be accelerated in a synchrotron having a radius of 10 m and capable of supplying about 5 MeV/revolution to the particle? Comment on the importance of radiative losses in circular particle accelerators. *Hint*: Assume that the maximum energy is high enough so that $v \approx c$.

P14-21. Consider a charged particle constrained to move along the z axis so that the acceleration and the velocity necessarily are either parallel or antiparallel. Suppose further that the velocity of the particle is as follows:

$$\begin{aligned}\mathbf{v}(t') &= v_0\hat{\mathbf{k}}, & t' < 0\\ &= \frac{1}{2}v_0\left[1+\cos\left(\frac{\pi t'}{T}\right)\right]\hat{\mathbf{k}}, & 0 < t' < T\\ &= 0\hat{\mathbf{k}}, & t' > T\end{aligned}$$

which represents a *smooth* deceleration from some initial (constant) velocity to a condition of rest at time T. Calculate the distribution of the emitted energy in frequency and in solid angle; i.e., calculate $dW/d\Omega\, d\omega$. Then calculate the integral of this quantity over all solid angles, sketch a graph of the result as a function of ω, and determine how this graph is affected by changes in the time T required for the particle to be brought to rest. Assume that the motion is nonrelativistic. *Hint*: For nonrelativistic motion, $v/c \ll 1$, and, in particular $\mathbf{R}(t')$ is on the order of $\langle v \rangle T$, where $\langle v \rangle$ is a characteristic velocity of the particle. Thus, $\hat{\mathbf{n}} \cdot \mathbf{R}(t')/c \approx \langle v \rangle t'/c$ and can be ignored relative to t' in the exponent in Eq. (14-107). This neglect of $\hat{\mathbf{n}} \cdot \mathbf{R}$ is called the *dipole approximation.*

P14-22. A charged particle moves in a circular path described by the trajectory

$$\mathbf{R}(t') = R_0(\cos \omega_0 t' \hat{\mathbf{i}} + \sin \omega_0 t' \hat{\mathbf{j}})$$

during the time interval $-T < t < T$ and is at rest for $|t| > T$. Assume that the motion is nonrelativistic. (a) Show that the term $\hat{\mathbf{n}} \cdot \mathbf{R}(t')/c$ in the exponent in Eq. (14-107) can be neglected. *Hint*: See the hint in P14-21. (b) Find $dW/d\Omega\, d\omega$ and the integral of this quantity over all solid angles. Sketch the latter distribution as a function of ω for several different values of T and describe the emitted spectrum in words. Radiation by this particular particle is called (nonrelativistic) cyclotron radiation.

P14-23. Assume that the electron in the hydrogen atom moves nonrelativistically in a circular orbit of radius r and that the amount of energy radiated *per cycle* is a small fraction of the total energy. (a) Show that the total energy E of the electron and the total power radiated P are given by

$$E = -\frac{q^2}{8\pi\epsilon_0 r} \qquad P = \frac{q^6}{96\pi^3\epsilon_0^3 c^3 m^2}\frac{1}{r^4}$$

where q and m are the charge and mass of the electron. (b) Argue that $dE/dt = -P$, obtain a differential equation for r, and solve that equation for r as a function of t. (c) How long does it take for the initial radius—say r_0—to be reduced to zero? Insert appropriate numerical values for all quantities and estimate the (classical) lifetime of the hydrogen atom.

14-6 The Radiation Reaction

In the previous sections, we have confined our attention to calculating the radiation emitted by a charged particle following a *prescribed* trajectory. In those cases (e.g., P14-17 and P14-23) where the trajectory was determined by forces on the particle from external electric and magnetic fields, we first calculated the trajectory *by assuming that there was no radiation* and then we calculated the radiation by assuming that the trajectory in the presence of radiation differs only slightly from the trajectory in the absence of radiation. Frequently, this two-step solution of a complicated radiation problem yields a valid approximation, because the energy radiated in some characteristic time is often a small fraction of the total energy of the particle and changes in the trajectory arising because of radiative losses can therefore be treated quasi-statically. Strictly, however, this two-step procedure is not correct, for the radiated field carries off energy and momentum and hence must result in a back force or a radiative force (usually called the *radiation reaction*) on the radiating particle. This radiation reaction, of course, affects the trajectory of the particle and should be taken into account when that trajectory is calculated. The complete problem, however, is extremely difficult and—to the author's knowledge—no fully satisfactory solution has yet been developed. In this book, we shall be content with a very simple, approximate treatment that is restricted to nonrelativistic motion. Essentially, we add a radiation reaction $\mathbf{F}_{\text{rad}}$ to the external forces $\mathbf{F}$ on the particle and write

$$m\dot{\mathbf{v}} = \mathbf{F} + \mathbf{F}_{\text{rad}} \tag{14-108}$$

for the equation of motion. We then determine $\mathbf{F}_{\text{rad}}$ so that it accounts properly for the total energy radiated; i.e., we choose $\mathbf{F}_{\text{rad}}$ so that

$$\int_{t_1}^{t_2} \mathbf{F}_{\text{rad}} \cdot \mathbf{v}\, dt = -\int_{t_1}^{t_2} \frac{q^2}{6\pi\epsilon_0 c^3} \dot{\mathbf{v}} \cdot \dot{\mathbf{v}}\, dt \tag{14-109}$$

where the left-hand side is the work done on the particle by $\mathbf{F}_{\text{rad}}$ and the right-hand side is the negative of the energy radiated away [Compare Eq. (14-94)] in the time interval $t_1 \leq t \leq t_2$. Integration of the right-hand side of Eq. (14-109) by parts gives

$$\int_{t_1}^{t_2} \mathbf{F}_{\text{rad}} \cdot \mathbf{v}\, dt = -\frac{q^2}{6\pi\epsilon_0 c^3} \dot{\mathbf{v}} \cdot \mathbf{v} \Big|_{t_1}^{t_2} + \int_{t_1}^{t_2} \frac{q^2 \ddot{\mathbf{v}}}{6\pi\epsilon_0 c^3} \cdot \mathbf{v}\, dt \tag{14-110}$$

We now assume that the interval $t_1 \leq t \leq t_2$ covers the entire time during which the particle is accelerated so that $\dot{\mathbf{v}}(t_1) = \dot{\mathbf{v}}(t_2) = 0$. Then, the first term on the right in Eq. (14-110) is zero and we find that the total energy radiated throughout the motion is given by either side of the equation

$$\int_{t_1}^{t_2} \mathbf{F}_{\text{rad}} \cdot \mathbf{v}\, dt = \int_{t_1}^{t_2} \frac{q^2 \ddot{\mathbf{v}}}{6\pi\epsilon_0 c^3} \cdot \mathbf{v}\, dt \tag{14-111}$$

This equation in turn *suggests* that we take

$$\mathbf{F}_{\text{rad}} = \frac{q^2 \ddot{\mathbf{v}}}{6\pi\epsilon_0 c^3} \tag{14-112}$$

and thus the equation of motion for a *radiating* charged particle [Eq. (14-108)] becomes

$$m\left(\dot{\mathbf{v}} - \frac{q^2}{6\pi\epsilon_0 m c^3} \ddot{\mathbf{v}}\right) = \mathbf{F} \tag{14-113}$$

which is called the *Abraham-Lorentz equation of motion*. By our derivation, of course, we cannot expect Eq. (14-112) to be correct except in the average sense expressed by Eq. (14-111), but a more detailed treatment must be left to other authors.[5] A few properties of the solutions to Eq. (14-113) are explored in P14-24 and P14-25.

[5]See, for example, J. D. Jackson, *Classical Electrodynamics* (John Wiley & Sons, Inc., New York, 1962), Chapter 17.

PROBLEMS

P14-24. (a) Dimensionally, the quantity $q^2/6\pi\epsilon_0 mc^3 = \tau$ in Eq. (14-113) must be a time. What is the numerical value of τ for an electron? (b) Assuming that the force **F** in Eq. (14-113) is zero, show that the *runaway solution* $\dot{\mathbf{v}} = (\mathbf{constant}) \exp(t/\tau)$ exists and discuss whether this solution is physically admissible.

P14-25. Assume that the force **F** in Eq. (14-113) is known as a function of time and let $\tau = q^2/6\pi\epsilon_0 mc^3$. Show that

$$m\dot{\mathbf{v}} = \frac{e^{t/\tau}}{\tau} \int_t^\infty e^{-t'/\tau}\mathbf{F}(t')\, dt'$$

where the upper limit has been chosen so that as $\tau \longrightarrow 0$ (no radiation) the equation approaches $m\dot{\mathbf{v}} = \mathbf{F}$. Discuss what it means for the acceleration at a given time t to be determined by the forces acting at times into the indefinite *future.*

Supplementary Problems

P14-26. The four-dimensional Fourier transform of a function $\phi(\mathbf{r}, t)$ is defined by

$$\tilde{\phi}(\boldsymbol{\kappa}, \omega) = \int \phi(\mathbf{r}, t) e^{-i(\boldsymbol{\kappa}\cdot\mathbf{r} - \omega t)}\, dv\, dt$$

where the integral extends over all space and time. (a) Apply the Fourier inversion theorem to all four variables of integration independently to show that

$$\phi(\mathbf{r}, t) = (2\pi)^{-4} \int \tilde{\phi}(\boldsymbol{\kappa}, \omega) e^{i(\boldsymbol{\kappa}\cdot\mathbf{r} - \omega t)}\, d\kappa_x\, d\kappa_y\, d\kappa_z\, d\omega$$

(b) Multiply the wave equation for the scalar potential, Eq. (14-8), by exp $[-i(\boldsymbol{\kappa}\cdot\mathbf{r} - \omega t)]$ and integrate the result over all space and time to obtain an equation from which an explicit solution for the Fourier transform of the potential can be found. (c) Finally, express the solution for the potential itself as an integral in $(\boldsymbol{\kappa}, \omega)$-space. *Note*: Strictly, the result obtained here is only a start toward a solution, because the integral obtained in part (c) is an improper integral. The careful treatment of this integral is most easily accomplished using the theory of contour integration in the complex plane and involves also an examination of the requirements of causality (that cause precede effect). Correctly manipulated, the result in part (c) leads directly to Eq. (14-26).

P14-27. A charged particle interacts with an incident linearly polarized plane electromagnetic wave whose electric field is

$$\mathbf{E} = E_0(\cos\psi\hat{\mathbf{i}} + \sin\psi\hat{\mathbf{j}})e^{i(\kappa z - \omega t')}$$

and hence experiences an acceleration given by

$$\dot{\mathbf{v}} = \frac{q}{m} E_0(\cos\psi\hat{\mathbf{i}} + \sin\psi\hat{\mathbf{j}})e^{i(\kappa z - \omega t')}$$

where q and m are the charge and mass of the particle. Assume that the motion is nonrelativistic, note that the particle moves in the plane $z = 0$ (why?), and let θ and ϕ be the polar and azimuthal angles of a distant observation point in a coordinate system with the particle near the origin. (a) Calculate $\langle dP(t')/d\Omega \rangle$ for this particle, where the average is evaluated over a single cycle of the motion. (b) The differential cross section for scattering of incident light by a free particle is defined by

$$\frac{d\sigma}{d\Omega} = \frac{\text{power radiated/unit solid angle}}{\text{incident energy/unit area/unit time}} = \frac{\langle dP/d\Omega \rangle}{|\langle \mathbf{S} \rangle|}$$

where $\mathbf{S}$ is the Poynting vector of the incident wave. Evaluate $d\sigma/d\Omega$ for the present case. (c) The cross section for scattering of *unpolarized* incident light is obtained by averaging the result of part (b) over ψ. Determine this cross section and sketch a graph showing this cross section as a function of θ. The resulting expression is referred to as the *Thomson formula.* (d) The total cross section σ for scattering is obtained by integrating $d\sigma/d\Omega$ over all solid angles. Obtain the total cross section corresponding to the result of part (c). This result is referred to as the *Thomson cross section.* Evaluate the Thomson cross section numerically if the particle is an electron. *Note*: The discussion of this problem is valid only at low frequencies. At higher frequencies, when the photons in the electromagnetic wave have energies comparable to or exceeding the rest energy of the particle, quantum and relativistic effects become important and classical scattering goes over into *Compton scattering,* the cross section for which is given by the so-called *Klein-Nishina formula.* [See, for example, J. M. Jauch and F. Rohrlich, *The Theory of Photons and Electrons* (Addison-Wesley Publishing Company, Inc., Reading, Mass., 1955), Chapter 11.]

REFERENCES

R. P. Feynman, R. B. Leighton, and M. Sands, *The Feynman Lectures on Physics* (Addison-Wesley Publishing Company, Inc., Reading, Mass., 1964), Volume II, Lecture 21.

J. D. Jackson, *Classical Electrodynamics* (John Wiley & Sons, Inc., New York, 1963), Chapters 9, 14, and 17.

P. Lorrain and D. R. Corson, *Electromagnetic Fields and Waves* (W. H. Freeman and Company, San Francisco, 1970), Second Edition, Chapter 14.

J. Schwinger, "On the Classical Radiation of Accelerated Charges," *Phys. Rev.* **75**, 1912 (1949).

J. M. Stone, *Radiation and Optics* (McGraw-Hill Book Company, New York, 1963), Chapter 5.

15

Relativistic Formulation of Maxwell's Equations

When Maxwell's equations were first presented in the late nineteenth century, physicists were convinced that *all* physical phenomena would ultimately be explained by applying Newton's laws and known (or to be discovered) forces to predict the behavior of suitable particles. Within this *mechanical* view, any wave phenomenon required a medium to support and transmit the wave. The late nineteenth-century physicist therefore introduced a medium to transmit electromagnetic waves. This *ether* was imagined to pervade all of space, the electromagnetic field was interpreted as a stress or distortion of the ether from its normal equilibrium "state" in the absence of a field, and electromagnetic waves were thought of as propagating "oscillations" in the ether. Collectively, physicists invested considerable effort in attempts to convert this qualitative view of the electromagnetic field into a satisfactory, quantitative model. But there were many difficulties with this model, and, by the early twentieth century, the ether (and with it the hope of a completely mechanical physical world) had generally been abandoned. The electromagnetic field had been accepted as a physical manifestation that could exist in otherwise *empty* space—a view that we have, in fact, adopted without serious question throughout this book.

To set the stage for this chapter, we shall describe briefly just one of the fundamental difficulties that plagued the concept of a mechanical ether. A classical, mechanical wave propagating in a medium moves through that medium with a fixed velocity, say $\mathbf{v}_W$, *relative to the medium* regardless of the

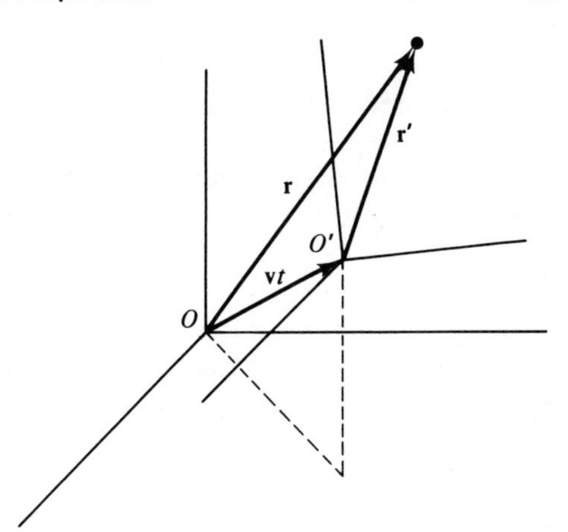

Fig. 15-1. Two general frames of reference in uniform relative motion.

state of motion of its source, and an observer at rest *relative to the medium* sees the wave propagate past him with this *same* velocity. If, however, this (classical) observer is moving *relative to the medium* with velocity $\mathbf{v}_0$, he will measure the velocity of the wave *relative to him* to be $\mathbf{v}_W - \mathbf{v}_0$. This result is, of course, in complete agreement (P15-1) with the predictions of the (classical) *Galilean transformation*

$$\begin{aligned} \mathbf{r}' &= \mathbf{r} - \mathbf{v}t \\ t' &= t \end{aligned} \tag{15-1}$$

relating the coordinates of an event in space-time as measured in two different frames of reference, where the origins of the two frames are assumed to coincide at the common time $t = t' = 0$ but (as shown in Fig. 15-1) the primed frame is in uniform motion relative to the unprimed frame with velocity $\mathbf{v}$. If, now, the electromagnetic wave is a classical wave in a mechanical ether, it should propagate *relative to the ether* with a fixed speed and an observer who is moving relative to the ether should measure some *different* value for the speed of electromagnetic waves relative to him. Indeed, from such measurements, an observer should be able to determine his speed relative to the ether. Equivalently, it should be possible to locate a frame of reference fixed with respect to the ether (and hence to locate a viable candidate for the absolute frame of reference that Newton assumed was fixed with respect to distant "fixed" stars). Furthermore, if the speed of electromagnetic waves is different for observers in relative motion, Maxwell's equations as we have developed them cannot be correct for all observers, because the correctness of these equations for all observers would imply that the speed of electromagnetic waves should be the same for all observers and this implication contradicts the above-outlined properties of classical waves. Thus, the classical physicist inferred that there must be a preferred frame of reference (presumably a frame fixed with respect to the ether) in which Maxwell's equations are valid, and a search for Newton's absolute frame of reference is then also a search for that preferred frame of reference in which Maxwell's equations are valid.

One of the early crucial experiments bearing on the search for this preferred frame of reference was the Michelson-Morley experiment,[1] which used the interference of light in attempts to detect any difference between the speeds of light (relative to the earth) in two directions at right angles to one another. Measurements were taken for various "absolute" orientations of the apparatus and at all times of the year in order to make sure that all possible motions of the earth relative to the ether were examined. In repetitions of the experiment by others, both stellar and terrestrial sources of light have been used. Overall, the experiment was very carefully designed and was more than adequate to detect differences in the speed of light comparable to the orbital speed of the earth (about 1 part in 10,000), and the results of numerous measurements were unequivocal: The speed of light *relative to the earth* is the same in all directions and at all times of the year. This (at the time) startling observation means either (1) that the earth is *at all times* at rest relative to the ether and the experiment provides no information about the form of Maxwell's equations in a frame of reference moving relative to the ether or (2) that Maxwell's equations are, in fact, valid in all frames of reference in uniform relative motion and something is wrong with the Galilean transformation (and hence also with Newtonian mechanics). The first of these conclusions can be accepted only if the earth is imagined to drag the ether along with it; accepting the second conclusion requires drastic conceptual changes in a long-standing and very successful world view. Neither alternative was particularly palatable when the results of the Michelson-Morley experiment were first presented. There is, however, at least one further pertinent experimental observation: The apparent direction from the earth to any star is slightly different at different times of the year, even after correction for the different positions of the earth with each observation. This phenomenon, called *aberration* (P15-7 and P15-39), is readily and quantitatively explained if the earth moves through the ether but cannot be understood at all easily if the earth drags the ether along. Apparently, we are forced to conclusion (2) by direct experimental observations. The full story is, of course, not quite that simple. It is more accurate to say that, although numerous ingenious attempts were made to preserve the ether, Newtonian mechanics, and the Galilean transformation, the complexity and artificiality of these attempts gradually convinced the physical community that conclusion (2) was the simpler and the more attractive alternative. After a few decades of hope that a preferred frame of reference in absolute rest might at last be located, attempts to locate this frame were abandoned and physicists returned to an earlier conception that *no* experiment performed in a given frame of reference could determine the absolute motion of that frame of reference. Absolute

[1] *American Journal of Science* **34**, 333 (1887); *Phil. Mag.* **24**, 449 (1887).

motion once again became undetectable and the Galilean principle of relativity, which asserts that the laws of physics are the same for all observers in uniform relative motion, was returned to the roster of basic beliefs about the nature of the physical world. This principle, however, did not emerge unchanged from the several decades of turmoil over the ether. Prior to these decades, the principle applied to the laws of mechanics; following this period, the principle was taken to apply to *all* laws of physics, including in particular the laws of mechanics and the laws of electromagnetism, and in this extended form the principle is more properly called the Einsteinian principle of relativity[2].

In summary, we have decided in this introduction to adopt the point of view that *all* laws of physics must conform to the Einsteinian principle of relativity. Further, from the observation that the speed of light is the same for all observers regardless of their relative motion, we are led to assume that Maxwell's equations as we have developed them are already consistent with this broad principle. If that is the case, however, the Galilean transformation (which contradicts the universal validity of Maxwell's equations) cannot be the correct transformation relating coordinates in two frames of reference that are in uniform relative motion. Still further, if the Galilean transformation is replaced by a new transformation that is consistent with the universal validity of Maxwell's equations, this new transformation is almost certain to be inconsistent with the assumption that Newton's laws have universal validity. Thus, replacing the Galilean transformation with another transformation to "save" Maxwell's equations will almost certainly "lose" Newton's laws in their classical form, and, in addition to seeking a replacement for the Galilean transformation, we should also seek at least a modified form of Newton's laws to preserve their invariance under whatever transformation replaces the Galilean transformation. We assume here that the reader is already acquainted with the concepts of Einstein's theory of special relativity and that he knows that the correct transformation relating coordinates in two frames of reference in uniform motion is the Lorentz transformation. In this chapter, we shall summarize the Lorentz transformation, describe some of the mathematical machinery that is useful in working with the Lorentz transformation, and then develop a reexpression of Maxwell's equations that makes their invariance to the Lorentz transformation (and hence their conformity with the principle of relativity) manifestly obvious. We shall also find how several electromagnetic quantities, such as charge density, current density, vector and scalar potentials, and the electromagnetic field itself, transform when one shifts his point of view from one frame of reference to a

[2]In fairness to the earlier form of the principle, it should be noted, however, that in the mechanical view of its time, the laws of mechanics were the *only* laws of physics; *everything* was ultimately mechanical. Thus, the Galilean principle of relativity by implication really did include the laws of electromagnetism. The real difference resulting from the years of turmoil was the discovery that electromagnetism could not be interpreted (easily) as a mechanical phenomenon.

second in uniform motion relative to the first. We shall leave to other authors a description of how Newton's laws must be modified to make them invariant to the Lorentz transformation instead of to the Galilean transformation.

PROBLEMS

P15-1. A particle follows a trajectory given by $\mathbf{R}(t)$ in the unprimed frame of reference and $\mathbf{R}'(t')$ in the primed frame of reference, where coordinates in the two frames of reference are related by the Galilean transformation, Eq. (15-1). Show that the velocities $\mathbf{V}$ and $\mathbf{V}'$ and accelerations $\mathbf{A}$ and $\mathbf{A}'$ as determined in the two frames are related by

$$\mathbf{V}' = \mathbf{V} - \mathbf{v} \qquad \mathbf{A}' = \mathbf{A}$$

and hence argue that force defined by Newton's second law is a Galilean invariant. What assumptions did you make about the behavior of mass under a Galilean transformation?

P15-2. Explain why it should be Maxwell's equations and not Newton's laws that turn out to be consistent automatically with the Einsteinian principle of relativity.

P15-3. Suppose a scalar function $u(\mathbf{r}, t)$ satisfies the wave equation

$$\left(\nabla^2 - \frac{1}{c^2}\frac{\partial^2}{\partial t^2}\right)u = 0$$

in the unprimed frame of reference in Eq. (15-1). What equation does this function satisfy in the primed frame of reference? *Hint*: Assume that the x, y, and z axes are parallel to the x', y', and z' axes, respectively, and let the x axis be chosen so that $\mathbf{v} = v\hat{\mathbf{i}}$.

15-1
A Review of Special Relativity

Although Lorentz is credited with discovering a coordinate transformation and an associated transformation of the electromagnetic field that together leave Maxwell's equations unchanged in form, the theory of special relativity advanced by Einstein[3] in 1905 placed this transformation on a firm theoretical and philosophical footing by deriving it directly from the fundamental postulates described in the above introductory paragraphs. The usual statement of these postulates involves the notion of inertial frames of reference, which in totality consist of *any* single frame of reference in which the laws of physics have their simplest form and *all* other frames of reference in translation at constant velocity relative to the first frame of reference. The postulates of special relativity themselves are

[3] *Annalen der Physik* **17** (1905). This original paper is extremely readable and is available in English translation in a collection of reprints titled *The Principle of Relativity* (Dover Publications, Inc., New York).

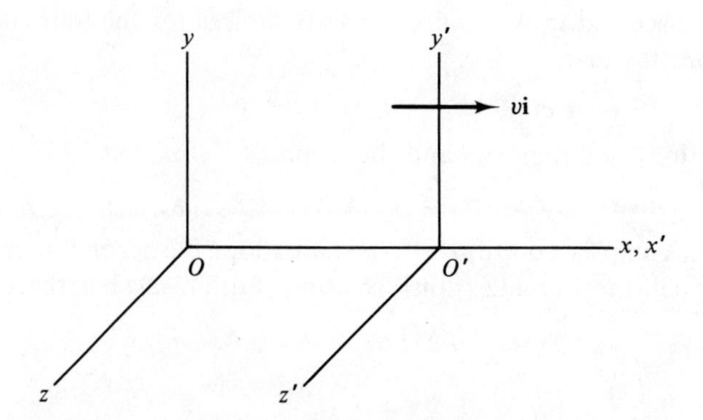

Fig. 15-2. Two frames of reference in uniform relative motion along a common axis. O' moves with velocity $v\hat{\mathbf{i}}$ relative to O.

(1) that the laws of physics are the same in *all* inertial frames of reference, and
(2) that the speed of light is a universal constant, the same in all inertial frames of reference regardless of the motion of these frames relative to the source of the light.

Together these postulates dictate uniquely the transformation that must relate the spatial and temporal coordinates of an event observed from two different inertial frames of reference. If (for simplicity) the two frames of reference are related as in Fig. 15-2, with the x, y, and z axes parallel to the x', y' and z' axes, respectively, and with their origins coinciding when $t = t' = 0$, then the resulting *Lorentz transformation* between the coordinates (x, y, z, t) and the coordinates (x', y', z', t') is

$$\begin{aligned} x' &= \gamma(x - \beta ct) \qquad & z' &= z \\ y' &= y \qquad & ct' &= \gamma(ct - \beta x) \end{aligned} \tag{15-2}$$

where

$$\gamma = \frac{1}{\sqrt{1 - \beta^2}} \qquad \beta = \frac{v}{c} \tag{15-3}$$

and v is the velocity of the primed frame relative to the unprimed frame, this velocity being in the (common) x and x' directions.[4]

For our purposes, it is useful to stress that the Lorentz transformation is a *linear* transformation between the *four* primed and *four* unprimed coordinates and to express that transformation in a matrix form. (The algebra of

[4]A full deduction of this transformation from the basic postulates may be found in many sources, including, for example, the original paper of Einstein (see footnote 3); R. Resnick, *Introduction to Special Relativity* (John Wiley & Sons, Inc., New York, 1968), Section 2.2; and J. D. Jackson, *Classical Electrodynamics* (John Wiley & Sons, Inc., New York, 1962), Section 11.2.

matrices is discussed in Appendix A.) Two choices for the four coordinates are common; the first,

$$x_0 = ct \qquad x_1 = x \qquad x_2 = y \qquad x_3 = z \tag{15-4}$$

involves entirely real numbers and the second,

$$x_1 = x \qquad x_2 = y \qquad x_3 = z \qquad x_4 = ict \tag{15-5}$$

introduces a complex coordinate. We shall adopt the second alternative, in terms of which the Lorentz transformation, Eq. (15-2), has the expression

$$\begin{aligned} x_1' &= \gamma(x_1 + i\beta x_4) \qquad & x_3' &= x_3 \\ x_2' &= x_2 & x_4' &= \gamma(x_4 - i\beta x_1) \end{aligned} \tag{15-6}$$

or, equivalently,

$$\begin{pmatrix} x_1' \\ x_2' \\ x_3' \\ x_4' \end{pmatrix} = \begin{pmatrix} \gamma & 0 & 0 & i\gamma\beta \\ 0 & 1 & 0 & 0 \\ 0 & 0 & 1 & 0 \\ -i\gamma\beta & 0 & 0 & \gamma \end{pmatrix} \begin{pmatrix} x_1 \\ x_2 \\ x_3 \\ x_4 \end{pmatrix} \tag{15-7}$$

Thus, if we introduce the 4×4 transformation matrix

$$\mathfrak{L}(\beta) = \begin{pmatrix} \gamma & 0 & 0 & i\gamma\beta \\ 0 & 1 & 0 & 0 \\ 0 & 0 & 1 & 0 \\ -i\gamma\beta & 0 & 0 & \gamma \end{pmatrix} \tag{15-8}$$

whose elements we denote by $\mathfrak{L}_{\mu\nu}(\beta)$ or by $\mathfrak{L}_{\mu\nu}$, we can express the Lorentz transformation in the form

$$x_\mu' = \mathfrak{L}_{\mu\nu}(\beta) x_\nu \qquad \mathbf{x}' = \mathfrak{L}(\beta)\mathbf{x} \tag{15-9}$$

where we have adopted the summation convention in which a sum from 1 to 4 is understood on every *repeated* Greek index. Further, we shall use boldface *sans serif* characters (e.g., $\mathbf{x}$) for *four*-component vectors. Finally, we understand without explicit mention that any index appearing only *once* in a given term assumes all four possible values in succession; the first member of Eq. (15-9), for example, expresses four separate equations.

Physically, the *inverse* transformation (from the primed to the unprimed coordinates) must be obtainable by replacing v by $-v$ (or β by $-\beta$) in the matrix $\mathfrak{L}$. Thus, we must have that

$$\mathbf{x} = [\mathfrak{L}(\beta)]^{-1}\mathbf{x}' = \mathfrak{L}(-\beta)\mathbf{x}' \Longrightarrow [\mathfrak{L}(\beta)]^{-1} = \mathfrak{L}(-\beta) \tag{15-10}$$

Direct observation of Eq. (15-8), however, reveals that $\mathfrak{L}(-\beta) = [\mathfrak{L}(\beta)]^T$, the superscript T denoting the transposed matrix. We conclude that

$$[\mathfrak{L}(\beta)]^{-1} = [\mathfrak{L}(\beta)]^T \tag{15-11}$$

and the matrix $\mathfrak{L}$ is therefore a (complex) orthogonal matrix (Appendix A). Thus, the Lorentz transformation in Eq. (15-7) can be interpreted as a rigid

rotation of the *four* coordinate axes in a *four*-dimensional space. (Compare PA-12 in Appendix A.) In fact, the Lorentz transformation between frames of reference whose axes are *not* parallel and whose relative velocity is in some arbitrary direction is also represented by a (complex) orthogonal matrix and is therefore also equivalent to a rigid rotation in a four-dimensional space; a portion of this more general conclusion is explored in P15-35.

Typically, physical quantities of interest in a given frame of reference are functions of the spatial and temporal coordinates in that frame of reference, and the way in which these physical quantities transform under a Lorentz transformation provides a useful means to classify them. The simplest such quantities are *scalars*, which we can define by considering a physical quantity that we believe to have but a single component. Let $\phi(\mathbf{x}) = \phi(x_1, x_2, x_3, x_4)$ express the spatial and temporal dependence of this quantity as measured in the unprimed frame of reference and $\phi'(\mathbf{x}')$ express the spatial and temporal dependence of this same quantity as measured in the primed frame of reference. The quantity ϕ is a scalar if

$$\phi'(\mathbf{x}') = \phi(\mathbf{x}) \tag{15-12}$$

where $\mathbf{x}'$ and $\mathbf{x}$ are related by the Lorentz transformation. More concretely, a physical quantity is a scalar if it has the same numerical value at corresponding points in space-time in any two frames of reference related by a Lorentz transformation. For this reason, a scalar is often said to be invariant to the Lorentz transformation. Usually, Eq. (15-12) is written $\phi' = \phi$ and explicit indication that ϕ' and ϕ are most directly functions of the primed and unprimed coordinates, respectively, is suppressed.

The next most complicated type of physical quantity has four components that become entangled with one another when the frame of reference is subjected to a Lorentz transformation. The four-component vector $\mathbf{x} = (x_1, x_2, x_3, x_4)$ is the prototype of these quantities; they are called *four-vectors* and their components transform under the Lorentz transformation just as the components of $\mathbf{x}$ transform. That is, the four-component quantity $\mathbf{A} = (A_1, A_2, A_3, A_4)$ is a four-vector if its components $\mathbf{A}' = (A'_1, A'_2, A'_3, A'_4)$ in the primed frame of reference are determined from those in the unprimed frame by

$$\mathbf{A}' = \mathfrak{L}\mathbf{A} \quad \text{or} \quad A'_\mu = \mathfrak{L}_{\mu\nu} A_\nu \tag{15-13}$$

where here (hereafter) the above-mentioned summation convention is (will be) employed. The first three components of a four-vector are often called the *spatial* components; the fourth component is the *temporal* component. Occasionally we shall use the symbol **A** for the *three*-dimensional vector formed from the spatial components of the *four*-vector $\mathbf{A}$.

Among other quantities, the four-dimensional gradient of a scalar ϕ, whose four components are $\partial\phi/\partial x_\mu$, is a four-vector. To prove this contention, we note first that differentiation of Eq. (15-12) with respect to x'_μ gives

$$\frac{\partial \phi'}{\partial x'_\mu} = \frac{\partial \phi}{\partial x'_\mu} = \frac{\partial \phi}{\partial x_\nu}\frac{\partial x_\nu}{\partial x'_\mu} \tag{15-14}$$

From Eqs. (15-10) and (15-11), however, we find that

$$\begin{aligned} x_\nu &= (\mathfrak{L}^{-1})_{\nu\mu} x'_\mu = \mathfrak{L}_{\mu\nu} x'_\mu \\ \Longrightarrow \frac{\partial x_\nu}{\partial x'_\mu} &= \mathfrak{L}_{\mu\nu} \end{aligned} \tag{15-15}$$

and Eq. (15-14) becomes

$$\frac{\partial \phi'}{\partial x'_\mu} = \mathfrak{L}_{\mu\nu}\frac{\partial \phi}{\partial x_\nu} \tag{15-16}$$

which states that the components of the (four-dimensional) gradient in two different frames of reference are related by Eq. (15-13) and hence that this gradient is in fact a four-vector. Q.E.D. Since $\phi = \phi'$, we can write Eq. (15-16) as a relationship among differential operators

$$\frac{\partial}{\partial x'_\mu} = \mathfrak{L}_{\mu\nu}\frac{\partial}{\partial x_\nu} \tag{15-17}$$

and think of the operator whose four components are $\partial/\partial x_\mu$ as a four-vector. We shall frequently use the notation ∂'_μ for $\partial/\partial x'_\mu$ and ∂_ν for $\partial/\partial x_\nu$, so that Eq. (15-17) may also be written in the more compact form $\partial'_\mu = \mathfrak{L}_{\mu\nu}\,\partial_\nu$.

A third useful type of physical quantity that is identified by the way it behaves under a Lorentz transformation is a 16-component entity called a *second-rank tensor*. It is usual to arrange these 16 components into a 4×4 matrix whose elements are, say, $F_{\mu\nu}$ in the unprimed frame of reference and $F'_{\mu\nu}$ in the primed frame of reference. A prototype second-rank tensor is a quantity whose 16 components are the 16 possible products of a component of a four-vector **A** with a component of another four-vector **B** so that $F_{\mu\nu} = A_\mu B_\nu$, but there are many second-rank tensors not so easily factored. Whatever its origin, a 16-component quantity F is a second-rank tensor by definition if

$$F'_{\mu\nu} = \mathfrak{L}_{\mu\sigma}\mathfrak{L}_{\nu\tau}F_{\sigma\tau} \tag{15-18}$$

where *two* sums (on σ and τ) are implied. Equivalently, this transformation rule can be written as a matrix product by noting that

$$\begin{aligned} F'_{\mu\nu} &= \mathfrak{L}_{\mu\sigma}F_{\sigma\tau}(\mathfrak{L}^T)_{\tau\nu} = (\mathfrak{L}F\mathfrak{L}^T)_{\mu\nu} \\ \Longrightarrow F' &= \mathfrak{L}F\mathfrak{L}^T \end{aligned} \tag{15-19}$$

The pattern illustrated in Eq. (15-18) can be extended to third- and higher-rank tensors, but we shall have no need for these more complicated entities.

We shall conclude this summary of the mathematical machinery of special relativity by identifying several important scalars (i.e., Lorentz invariants). Many invariants can be constructed directly from known vectors and tensors. For example, let **A** and **B** be four-vectors and consider the quantity $A_\mu B_\mu$ (implicit sum on μ). We find that

$$\begin{aligned} A'_\mu B'_\mu &= (\mathfrak{L}_{\mu\sigma}A_\sigma)(\mathfrak{L}_{\mu\tau}B_\tau) \\ &= A_\sigma(\mathfrak{L}^T)_{\sigma\mu}\mathfrak{L}_{\mu\tau}B_\tau \\ &= A_\sigma(\mathfrak{L}^{-1}\mathfrak{L})_{\sigma\tau}B_\tau \\ &= A_\sigma B_\sigma \end{aligned} \tag{15-20}$$

since $\mathfrak{L}^{-1}\mathfrak{L}$ is the identity matrix. Equation (15-20), however, states that the quantity $A_\mu B_\mu$ (which is a four-dimensional "dot product") not only has the *same value* in two different frames but is computed in *each* frame by applying the *same prescription* to the components of the vector in *that* frame. This dot product is both invariant in *value* (i.e., is a scalar) and invariant in *form* when the constituent vectors are subjected to the same Lorentz transformation. By similar arguments, we can demonstrate that the *trace* $F_{\mu\mu}$ of a second-rank tensor, the four-divergence

$$\partial_\mu A_\mu = \nabla\cdot\mathbf{A} + \frac{\partial}{\partial t}\left(\frac{A_4}{ic}\right) \tag{15-21}$$

of a vector **A**, and the four-dimensional "Laplacian" (which is called the *d'Alembertian*)[5]

$$\Box^2 = \partial_\mu\partial_\mu = \nabla^2 - \frac{1}{c^2}\frac{\partial^2}{\partial t^2} \tag{15-22}$$

are all Lorentz invariants (P15-10).

Finally, if we assume familiarity with the theorem that multidimensional volume elements transform under a change of variables from $u_1, u_2, \ldots, u_n$ to $v_1, v_2, \ldots, v_n$ by[6]

$$dv_1\,dv_2\cdots dv_n = \left|J\left(\frac{v_1, v_2, \ldots, v_n}{u_1, u_2, \ldots, u_n}\right)\right| du_1\,du_2\cdots du_n \tag{15-23}$$

where the Jacobian $J(\cdots)$ is the determinant of a matrix whose ij element is $\partial v_i/\partial u_j$, we can demonstrate that the four-dimensional volume element

$$d^4\mathbf{x} = dx_1\,dx_2\,dx_3\,dx_4 \tag{15-24}$$

is invariant to the Lorentz transformation. From Eq. (15-23), we have that

$$d^4\mathbf{x}' = \left|J\left(\frac{x'_1, x'_2, x'_3, x'_4}{x_1, x_2, x_3, x_4}\right)\right| d^4\mathbf{x} \tag{15-25}$$

The Jacobian appearing here, however, is the determinant of a matrix whose $\mu\nu$ element $\partial x'_\mu/\partial x_\nu$ is $\mathfrak{L}_{\mu\nu}$. [See Eq. (15-9).] Thus, the Jacobian in Eq. (15-25) is the determinant of the matrix $\mathfrak{L}$. Since $\mathfrak{L}$ is orthogonal, however, its determinant can only be ± 1 (PA-10). Equation (15-25) therefore reduces to

[5] We use the notation $\Box^2$ for the d'Alembertian because of its similarity to ∇^2; the symbol $\Box$ (without the exponent) is also used for this operator.

[6] L. Brand, *Advanced Calculus* (John Wiley & Sons, Inc., New York, 1955), Section 169; W. Kaplan, *Advanced Calculus* (Addison-Wesley Publishing Company, Inc., Reading, Mass., 1952), Section 5-14.

$$dx'_1\, dx'_2\, dx'_3\, dx'_4 = dx_1\, dx_2\, dx_3\, dx_4 \tag{15-26}$$

and we have established the Lorentz invariance of the (four-dimensional) volume element.

PROBLEMS

P15-4. Show that the Lorentz transformation reduces to the Galilean transformation in the nonrelativistic limit $v/c \ll 1$.

P15-5. Suppose a scalar function $u(\mathbf{r}, t)$ satisfies the wave equation

$$\left(\nabla^2 - \frac{1}{c^2}\frac{\partial^2}{\partial t^2}\right)u = 0$$

in the unprimed frame of reference in Eq. (15-2). Show by direct manipulation with the equation that this function also satisfies the wave equation in the primed frame of reference and hence conclude that the scalar wave equation is invariant to the Lorentz transformation. Compare this problem with P15-3.

P15-6. (a) Show that two events occurring at the same *point* ($\Delta x' = 0$) in the primed frame of reference but separated by a temporal interval $\Delta t'$ are separated in the unprimed frame of reference by a temporal interval $\Delta t = \gamma\, \Delta t'$. Note that $\Delta t > \Delta t'$; this phenomenon is referred to as the relativistic *time dilation.* (b) Show that two events occurring at the same *time* ($\Delta t' = 0$) in the primed frame of reference but separated by a spatial interval $\Delta x'$ are separated in the unprimed frame by a temporal interval $\Delta t = \gamma\beta\, \Delta x'/c$. Thus, simultaneous events in the primed frame are simultaneous in the unprimed frame only if $\Delta x' = 0$. (c) A rod of length L' lies along the x' axis and is at rest in the primed frame. An observer in the unprimed frame measures the length of this rod by measuring the time Δt required for the rod to move past him, calculating the length as $L = v\, \Delta t$. Show that $L = L'/\gamma$. Note that $L < L'$; this phenomenon is referred to as the *Lorentz-Fitzgerald contraction. Hint*: Consider the two events defined by those two instants when the one or the other end of the rod is opposite the unprimed observer.

P15-7. The position of a moving object is given by $\mathbf{R}(t)$ and $\mathbf{R}'(t')$ in the unprimed and primed frames of reference, respectively. Let the velocity of the object be $\mathbf{V}(t) = d\mathbf{R}/dt$ and $\mathbf{V}'(t') = d\mathbf{R}'/dt'$ in the two frames. (a) Show from the Lorentz transformation in Eq. (15-2) that

$$\begin{pmatrix} V'_x \\ V'_y \\ V'_z \end{pmatrix} = \frac{1}{1 - V_x v/c^2} \begin{pmatrix} V_x - v \\ \dfrac{V_y}{\gamma} \\ \dfrac{V_z}{\gamma} \end{pmatrix}$$

(b) Now consider a light beam propagating in the x-y plane at speed c in a direction making an angle θ with the x axis in Fig. 15-2. Find the speed and the direction θ' of propagation of this beam in the primed frame of reference. The phenomenon of *aberration* of starlight is predicted by the difference between θ and θ'. *Optional*: Use a computer to determine θ' as a function of β for selected values of θ and plot graphs of these relationships.

P15-8. In the Fizeau experiment (1859), light passes through a moving liquid that has index of refraction n (Fig. P15-8). Using an interferometer,

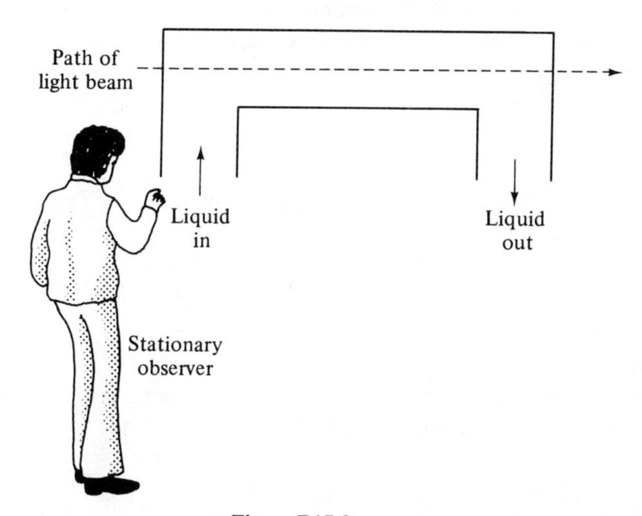

Figure P15-8

one can measure the velocity of the light relative to a stationary observer. For realizable (i.e., nonrelativistic) speeds v of the liquid, Fizeau found experimentally that

$$\text{speed of light relative to laboratory} = \frac{c}{n} + v\left(1 - \frac{1}{n^2}\right)$$

Predict this result from the Lorentz transformation. *Hints*: (1) What is the speed of light relative to the liquid? (2) See P15-7.

P15-9. Prove the following theorem: If $A_\mu B_\mu$ is a scalar for *arbitrary* four-vectors **B**, then **A** is also a four-vector.

P15-10. Verify that the four-divergence and the d'Alembertian are correctly given by the right-hand sides of Eqs. (15-21) and (15-22) and present complete arguments showing that these quantities are invariant to the Lorentz transformation.

P15-11. Let *two* events be separated by the infinitesimal *intervals* dx_μ in the unprimed frame and dx'_μ in the primed frame. (a) Recognizing that the coordinates of *each* event are related in the two frames by the Lorentz transformation, show that $dx'_\mu = \mathfrak{L}_{\mu\nu}\, dx_\nu$. (b) Show that the so-called *proper time* interval $d\tau$ defined by

$$c^2d\tau^2 = c^2dt^2 - dx^2 - dy^2 - dz^2$$

is a Lorentz invariant. (c) Argue that the quantity whose four components are $u_\mu = dx_\mu/d\tau$ is a four-vector and show that these components are $(\Gamma V_x, \Gamma V_y, \Gamma V_z, ic\Gamma)$, where $\Gamma = (1 - V^2/c^2)^{-1/2}$, $V_x = dx/dt$, $V_y = dy/dt$, and $V_z = dz/dt$. The four-vector $\mathbf{u}$ is called the *world velocity*. (d) Show that $u_\mu u_\mu = -c^2$.

P15-12. In relativistic mechanics, the momentum $\mathbf{p}$ and *total* energy E of a particle of *rest* mass m moving with velocity $\mathbf{V}$ are defined by $\mathbf{p} = \Gamma m\mathbf{V}$ and $E = \Gamma mc^2$, where $\Gamma = [1 - V^2/c^2]^{-1/2}$. Further, in a second frame of reference, the momentum $\mathbf{p}'$ and energy E' are given by $\mathbf{p}' = \Gamma' m\mathbf{V}'$ and $E' = \Gamma' mc^2$, where $\Gamma' = [1 - (V')^2/c^2]^{-1/2}$. Let the primed and unprimed frames be related by Eq. (15-2). (a) Show that $\Gamma' = \gamma\Gamma(1 - V_x v/c^2)$. *Hint*: See P15-7. (b) Show that the four-component quantity $(p_x, p_y, p_z, iE/c)$ is a four-vector; i.e., show that

$$p'_x = \gamma\left(p_x - \frac{v}{c^2}E\right) \qquad p'_z = p_z$$

$$p'_y = p_y \qquad E' = \gamma(E - vp_x)$$

(c) Recognizing that $p_\mu p_\mu$ is a scalar, show that $p^2c^2 - E^2$ is a Lorentz invariant and that it has the value $-m^2c^4$.

15-2 Maxwell's Equations in Covariant Form; The Electromagnetic Field Tensor

As we developed them in Chapter 6, Maxwell's equations for the electromagnetic field in the absence of matter are

$$\nabla\cdot\mathbf{B} = 0 \qquad \nabla\times\mathbf{E} = -\frac{\partial\mathbf{B}}{\partial t} \qquad (15\text{-}27), (15\text{-}28)$$

$$\nabla\cdot\mathbf{E} = \frac{\rho}{\epsilon_0} \qquad \nabla\times\mathbf{B} = \mu_0\mathbf{J} + \frac{1}{c^2}\frac{\partial\mathbf{E}}{\partial t} \qquad (15\text{-}29), (15\text{-}30)$$

Further, the fields can be derived from the potentials using the equations.

$$\mathbf{B} = \nabla\times\mathbf{A} \qquad \mathbf{E} = -\nabla V - \frac{\partial\mathbf{A}}{\partial t} \qquad (15\text{-}31), (15\text{-}32)$$

and, in a Lorentz gauge defined by the requirement that

$$\nabla\cdot\mathbf{A} + \frac{1}{c^2}\frac{\partial V}{\partial t} = 0 \qquad (15\text{-}33)$$

the potentials satisfy the equations

$$\Box^2\mathbf{A} = -\mu_0\mathbf{J} \qquad \Box^2 V = -\frac{\rho}{\epsilon_0} \qquad (15\text{-}34), (15\text{-}35)$$

Finally, the current and charge densities are related by the equation of continuity

$$\nabla \cdot \mathbf{J} + \frac{\partial \rho}{\partial t} = 0 \tag{15-36}$$

The object of this section now is to identify suitable four-vectors and tensors and to reexpress the above equations in terms of these quantities, thereby making the behavior of each equation under a Lorentz transformation directly determinable from the (known) behavior of four-vectors and tensors. As we identify each four-vector or tensor, we must be certain that it has the requisite transformation properties. (Mere possession of the proper number of components offers no assurances that those components behave correctly under a Lorentz transformation.)

We introduce first a four-vector that combines the current density and the charge density. *Experimentally*, observers in different frames of reference measure the *same* total charge in a uniquely defined region of space, even though each measures a different charge density in the region and assigns a different volume to the region. More mathematically, this empirical observation states that the charge dq measured in an infinitesimal volume in the unprimed frame of reference and the charge dq' measured in the same region of space in the primed frame of reference are equal, $dq = dq'$, or that

$$\rho\, dx_1\, dx_2\, dx_3 = \rho'\, dx_1'\, dx_2'\, dx_3' \tag{15-37}$$

Equivalently, the quantity $\rho\ dx_1\ dx_2\ dx_3$ is a Lorentz invariant. Thus, the four-component entity

$$ic(\rho\, dx_1\, dx_2\, dx_3)\, dx_\mu = ic\rho(dx_1\, dx_2\, dx_3\, dx_4)\frac{dx_\mu}{dx_4} \tag{15-38}$$

which is the product of a scalar and a four-vector, is itself a four-vector. We established in the last section, however, that the four-dimensional volume element in parentheses on the right in Eq. (15-38) is a scalar. Consequently, the quantity whose components are

$$J_\mu = ic\rho\frac{dx_\mu}{dx_4} = \rho\frac{dx_\mu}{dt} \tag{15-39}$$

must be a four-vector. Now, if dx_μ in Eq. (15-39) is interpreted as the infinitesimal separation between two successive space-time "positions" of one of the particles in the charge and current distribution,[7] dx_μ/dt for $\mu = 1, 2$, and 3 yields the components of the (drift) velocity of the particle and $dx_4/dt = ic$. We have therefore that

$$\begin{aligned} \mathsf{J} &= (\rho v_x, \rho v_y, \rho v_z, ic\rho) \\ &= (J_x, J_y, J_z, ic\rho) = (\mathbf{J}, ic\rho) \end{aligned} \tag{15-40}$$

[7]Strictly, we should consider different types of particles separately, obtaining an evaluation of J_μ for each type and then summing the separate results. We elect, however, to simplify the discussion by assuming that only a single type of particle is present.

[See Eqs. (2-10) and (2-18).] In short, experimental evidence on the invariance of electric charge leads to the conclusion that the quantity **J** constructed out of the ordinary current and charge densities via Eq. (15-40) is a four-vector; we shall call it the *four-current*. Finally, in terms of the four-current, the equation of continuity assumes the form

$$\nabla \cdot \mathbf{J} + \frac{\partial(ic\rho)}{\partial(ict)} = \nabla \cdot \mathbf{J} + \frac{\partial J_4}{\partial x_4} = 0$$

$$\Longrightarrow \partial_\mu J_\mu = 0 \tag{15-41}$$

In this form, the equation of continuity clearly transforms to the *same* equation $\partial'_\mu J'_\mu = 0$ under a Lorentz transformation and hence is properly invariant under a change in the frame of reference.

We next introduce a four-vector that combines the potentials **A** and V. Essentially, we note that Eq. (15-35) for V can be written in the form

$$\Box^2 V = -\mu_0\left(\frac{1}{\mu_0\epsilon_0}\right)\rho = -\mu_0 c^2 \rho = -\mu_0(-ic)(ic\rho)$$

$$\Longrightarrow \Box^2\left(i\frac{V}{c}\right) = -\mu_0 J_4 \tag{15-42}$$

Thus, if we introduce a four-component quantity

$$\mathsf{A} = \left(A_x, A_y, A_z, i\frac{V}{c}\right) = \left(\mathbf{A}, i\frac{V}{c}\right) \tag{15-43}$$

we can combine Eqs. (15-34) and (15-42) into the single four-component equation

$$\Box^2 A_\mu = -\mu_0 J_\mu \tag{15-44}$$

Accepting the (experimentally supported) correctness of Maxwell's equations in all frames of reference, we finally conclude from Eq. (15-44) that **A** must be a four-vector, because $\Box^2$ is a Lorentz invariant and $\Box^2 A_\mu$ can therefore be equal to J_μ in *all* frames of reference only if **A** and **J**—the latter a known four-vector—have the same transformation properties. We shall call **A** the *four-potential*; its four-divergence

$$\partial_\mu A_\mu = \nabla \cdot \mathbf{A} + \frac{\partial A_4}{\partial x_4} = \nabla \cdot \mathbf{A} + \frac{\partial(iV/c)}{\partial(ict)}$$

$$= \nabla \cdot \mathbf{A} + \frac{1}{c^2}\frac{\partial V}{\partial t} \tag{15-45}$$

is exactly the quantity appearing in the Lorentz condition, Eq. (15-33). Thus, the Lorentz condition may be written in the form

$$\partial_\mu A_\mu = 0 \tag{15-46}$$

and the invariance of this equation under a Lorentz transformation assures that Lorentz gauge potentials in one frame of reference will transform into Lorentz gauge potentials in any other frame of reference.

A third important quantity, which looks *something* like a four-dimensional curl, is defined by

$$F_{\mu\nu} = \partial_\mu A_\nu - \partial_\nu A_\mu \tag{15-47}$$

and is manifestly a second-rank tensor (being constructed of the sums of products of two quantities known to transform as vectors). Clearly, the tensor F is antisymmetric ($F_{\mu\nu} = -F_{\nu\mu}$) and in particular its diagonal elements, F_{11}, F_{22}, etc., are zero. Of potentially 16 elements, only six (say $F_{12}, F_{13}, F_{14}, F_{23}, F_{24}$, and F_{34}) can have arbitrary values. We shall now show that these 6 elements are related to the 6 components of the electromagnetic field. If neither index in Eq. (15-47) is a 4, we find that $F_{\mu\nu}$ is a component of the magnetic induction field, for example,

$$F_{12} = -F_{21} = \partial_1 A_2 - \partial_2 A_1 = \frac{\partial A_y}{\partial x} - \frac{\partial A_x}{\partial y} = B_z \tag{15-48}$$

[Eq. (15-31)]; similarly, $F_{13} = -F_{31} = -B_y$ and $F_{23} = -F_{32} = B_x$. If, on the other hand, one of the indices in Eq. (15-47) is a 4, we find that $F_{\mu\nu}$ is related to a component of the electric field, for example,

$$\begin{aligned} F_{14} = -F_{41} = \partial_1 A_4 - \partial_4 A_1 &= \frac{i}{c}\frac{\partial V}{\partial x} - \frac{1}{ic}\frac{\partial A_x}{\partial t} \\ &= \frac{i}{c}\left(\frac{\partial V}{\partial x} + \frac{\partial A_x}{\partial t}\right) = -\frac{i}{c}E_x \end{aligned} \tag{15-49}$$

[Eq. (15-32)]; similarly, $F_{24} = -F_{42} = -iE_y/c$ and $F_{34} = -F_{43} = -iE_z/c$. In total, we find that the components of the electric and magnetic induction fields are related to those of the *electromagnetic field tensor* F as indicated in the matrix

$$F = \begin{pmatrix} 0 & B_z & -B_y & -\dfrac{iE_x}{c} \\ -B_z & 0 & B_x & -\dfrac{iE_y}{c} \\ B_y & -B_x & 0 & -\dfrac{iE_z}{c} \\ \dfrac{iE_x}{c} & \dfrac{iE_y}{c} & \dfrac{iE_z}{c} & 0 \end{pmatrix} \tag{15-50}$$

Furthermore, because we accept the correctness of Maxwell's equations in all frames of reference, the correspondence between field components and elements of F indicated in Eq. (15-50) applies equally in all frames of reference.

It remains to express the field equations in terms of the electromagnetic field tensor. Note that

$$\begin{aligned} \partial_\mu F_{\mu 1} &= \partial_1 F_{11} + \partial_2 F_{21} + \partial_3 F_{31} + \partial_4 F_{41} \\ &= 0 - \frac{\partial B_z}{\partial y} + \frac{\partial B_y}{\partial z} + \frac{1}{ic}\frac{\partial (iE_x/c)}{\partial t} \\ &= -(\nabla \times B)_x + \frac{1}{c^2}\frac{\partial E_x}{\partial t} = -\mu_0 J_x \\ \Longrightarrow \partial_\mu F_{\mu 1} &= -\mu_0 J_1 \end{aligned} \tag{15-51}$$

which suggests that the equation

$$\partial_\mu F_{\mu\nu} = -\mu_0 J_\nu \tag{15-52}$$

combines Eqs. (15-29) and (15-30) into a single equation whose behavior under the Lorentz transformation is obvious from the notation; both sides of the equation are four-vectors. The correctness of Eq. (15-52) for all ν is demonstrated in P15-14. Note also that

$$\nabla \cdot \mathbf{B} = \frac{\partial F_{23}}{\partial x} + \frac{\partial F_{31}}{\partial y} + \frac{\partial F_{12}}{\partial z} = \partial_1 F_{23} + \partial_2 F_{31} + \partial_3 F_{12} = 0 \tag{15-53}$$

which suggests that the four equations

$$\partial_\mu F_{\nu\sigma} + \partial_\nu F_{\sigma\mu} + \partial_\sigma F_{\mu\nu} = 0 \tag{15-54}$$

where (μ, ν, σ) assumes the values (1, 2, 3), (1, 2, 4), (1, 3, 4), and (2, 3, 4), might combine the two homogeneous Maxwell equations, Eqs. (15-27) and (15-28); the correctness of this suggestion is demonstrated in P15-17. Note also that Eq. (15-54) is correct when two or more indices are the same, but that in those cases it is identically correct and provides no further constraints on the field tensor.

In summary, we have shown that we can introduce two four-vectors and a tensor, namely

$$\mathbf{J} = (\mathbf{J}, ic\rho) \qquad \mathbf{A} = \left(\mathbf{A}, i\frac{V}{c}\right)$$
$$F_{\mu\nu} = \partial_\mu A_\nu - \partial_\nu A_\mu \tag{15-55}$$

and that the equations of electromagnetism can be written in terms of these quantities as in Table 15-1. Although the physical content of these equations

TABLE 15-1 Some Equations of Electromagnetism in Relativistic Notation

Maxwell's Equations

$$\left.\begin{aligned} \nabla \cdot \mathbf{E} &= \frac{\rho}{\epsilon_0} \\ \nabla \times \mathbf{B} &= \mu_0 \mathbf{J} + \frac{1}{c^2}\frac{\partial \mathbf{E}}{\partial t} \end{aligned}\right\} \Rightarrow \partial_\mu F_{\mu\nu} = -\mu_0 J_\nu$$

$$\left.\begin{aligned} \nabla \cdot \mathbf{B} &= 0 \\ \nabla \times \mathbf{E} &= -\frac{\partial \mathbf{B}}{\partial t} \end{aligned}\right\} \Rightarrow \partial_\mu F_{\nu\sigma} + \partial_\nu F_{\sigma\mu} + \partial_\sigma F_{\mu\nu} = 0$$

Equation of Continuity

$$\nabla \cdot \mathbf{J} + \frac{\partial \rho}{\partial t} = 0 \Rightarrow \partial_\mu J_\mu = 0$$

Lorentz Condition

$$\nabla \cdot \mathbf{A} + \frac{1}{c^2}\frac{\partial V}{\partial t} = 0 \Rightarrow \partial_\mu A_\mu = 0$$

Equation for the (Lorentz Gauge) Potentials

$$\left.\begin{aligned} \Box^2 \mathbf{A} &= -\mu_0 \mathbf{J} \\ \Box^2 V &= -\frac{\rho}{\epsilon_0} \end{aligned}\right\} \Rightarrow \Box^2 A_\mu = -\mu_0 J_\mu$$

is no different from that of Eqs. (15-27)–(15-36), their form is now such that their correctness in *all* frames of reference is apparent from the way in which vectors and tensors transform. Further, by examining these transformation properties, we have gained considerable insight into the relationship between the electric and magnetic induction fields.

PROBLEMS

P15-13. Confirm the correctness of all entries in Eq. (15-50) by working out each element of $F_{\mu\nu}$ from Eq. (15-47).

P15-14. Verify that $\partial_\mu F_{\mu\nu} = -\mu_0 J_\nu$ reduces to the three components of Eq. (15-30) and to Eq. (15-29) for $\nu = 1, 2, 3$, and 4, respectively.

P15-15. Derive the equation $\partial_\mu F_{\mu\nu} = -\mu_0 J_\nu$ by formal differentiation of Eq. (15-47) defining $F_{\mu\nu}$. *Hint:* See Eqs. (15-44) and (15-46).

P15-16. Derive the equation of continuity $\partial_\nu J_\nu = 0$ directly from Eq. (15-52). *Hint: F* is an antisymmetric tensor.

P15-17. Verify that Eq. (15-54) correctly reduces to Eq. (15-27) and to the three components of Eq. (15-28) when (μ, ν, σ) assume the values given in the text.

P15-18. (a) Show that $E^2 - c^2B^2$ is a Lorentz invariant. *Hint:* $F_{\mu\nu}F_{\mu\nu}$ is an invariant. Why? (b) Show that an electromagnetic field that is purely magnetic in one frame of reference cannot be transformed into a purely electric field by changing to a different frame of reference.

15-3
Transformation of the Electromagnetic Field

Now that we have established that the components of the electromagnetic field combine to form a second-rank tensor F as in Eq. (15-50), determination of the way in which the electromagnetic field transforms under a Lorentz transformation from an unprimed to a primed frame of reference involves merely evaluating the matrix product $\mathfrak{L}F\mathfrak{L}^T$ as in Eq. (15-19). In particular, for $\mathfrak{L}$ given by Eq. (15-8) and F by Eq. (15-50), we find that

$$F' = \mathfrak{L}F\mathfrak{L}^T$$

$$= \begin{pmatrix} 0 & \gamma\left(B_z - \frac{\beta}{c}E_y\right) & -\gamma\left(B_y + \frac{\beta}{c}E_z\right) & -\frac{i}{c}E_x \\ -\gamma\left(B_z - \frac{\beta}{c}E_y\right) & 0 & B_x & -\frac{i\gamma}{c}(E_y - \beta c B_z) \\ \gamma\left(B_y + \frac{\beta}{c}E_z\right) & -B_x & 0 & -\frac{i\gamma}{c}(E_z + \beta c B_y) \\ \frac{i}{c}E_x & \frac{i\gamma}{c}(E_y - \beta c B_z) & \frac{i\gamma}{c}(E_z + \beta c B_y) & 0 \end{pmatrix} \tag{15-56}$$

Imagining primes on all entries in Eq. (15-50) and comparing the resulting matrix with Eq. (15-56), we conclude that the field in the primed frame of reference is determined from the field in the unprimed frame of reference by

$$\begin{aligned} E'_x &= E_x & B'_x &= B_x \\ E'_y &= \gamma(E_y - \beta c B_z) & B'_y &= \gamma\left(B_y + \frac{\beta}{c}E_z\right) \\ E'_z &= \gamma(E_z + \beta c B_y) & B'_z &= \gamma\left(B_z - \frac{\beta}{c}E_y\right) \end{aligned} \tag{15-57}$$

More conveniently for some purposes (and more generally), we introduce vector components parallel ($\|$) and perpendicular ($\perp$) to the relative velocity $\mathbf{v}$ (compare P0-5) and find that Eq. (15-57) is equivalent to the expressions

$$\begin{aligned} \mathbf{E}'_{\|} &= \mathbf{E}_{\|} & \mathbf{B}'_{\|} &= \mathbf{B}_{\|} \\ \mathbf{E}'_{\perp} &= \gamma(\mathbf{E}_{\perp} + \mathbf{v} \times \mathbf{B}) & \mathbf{B}'_{\perp} &= \gamma\left(\mathbf{B}_{\perp} - \frac{1}{c^2}\mathbf{v} \times \mathbf{E}\right) \end{aligned} \tag{15-58}$$

(See P15-20.) Several applications of these transformations are explored in the problems.

PROBLEMS

P15-19. Evaluate the matrix product $\mathfrak{L}F\mathfrak{L}^T$ and verify Eq. (15-56).

P15-20. Show that Eq. (15-58) reduces to Eq. (15-57) when $\mathbf{v} = v\hat{\mathbf{i}}$.

P15-21. Show that if the electric and magnetic induction fields are perpendicular for one observer, they are perpendicular for *all* observers. *Hint:* By manipulating with Eq. (15-58) show first that $\mathbf{E}\cdot\mathbf{B}$ is a Lorentz invariant.

P15-22. Suppose constant fields $\mathbf{E}$ and $\mathbf{B}$ exist in a particular frame of reference. Argue that a second frame of reference in which $\mathbf{B}' = 0$ can be found only if $E > cB$ *and* $\mathbf{E}$ is perpendicular to $\mathbf{B}$, and find a possible velocity of this second frame relative to the first. Is the second frame unique? *Hint:* Use the invariance of $E^2 - c^2B^2$ (P15-18), the invariance of $\mathbf{E}\cdot\mathbf{B}$ (P15-21), and Eq. (15-58).

P15-23. A point charge q moves with a uniform speed v in the positive x direction. (a) Use the transformation rules for the electromagnetic field to show that the fields established by this charge are given by

$$\mathbf{E} = \frac{q(1-\beta^2)}{4\pi\epsilon_0}\frac{(x-vt)\hat{\mathbf{i}} + y\hat{\mathbf{j}} + z\hat{\mathbf{k}}}{[(x-vt)^2 + (1-\beta^2)(y^2+z^2)]^{3/2}}$$

$$\mathbf{B} = \frac{1}{c^2}\mathbf{v} \times \mathbf{E}$$

Compare these results with those obtained in P14-15. *Hint:* Write down first the fields produced in a frame of reference in which the charge is at rest and then transform the fields to the original frame of reference. (b) Show that the expression for $\mathbf{B}$ reduces to the Biot-Savart law, $\mathbf{B} = (\mu_0 q\mathbf{v} \times \mathbf{r})/(4\pi r^3)$ in the nonrelativistic limit. Here, **r** is the position of the observation point relative to the position of the particle.

P15-24. (a) Assuming the four-vector character of the four-potential **A**, show that, under the Lorentz transformation in Eq. (15-6), the potentials (**A**, V) become (**A**′, V'), where

$$A'_x = \gamma\left(A_x - \frac{v}{c^2}V\right) \qquad A'_z = A_z$$

$$A'_y = A_y \qquad V' = \gamma(V - vA_x)$$

(b) Use these results to find the potentials **A** and V established by the moving charge in P15-23. Compare your results with those given in P14-11. *Hint:* First write down the potentials **A**′ and V' in a frame of reference in which the charge is at rest.

P15-25. Assuming the four-vector character of the four-current **J**, show that, under the Lorentz transformation in Eq. (15-6), the current density and charge density (**J**, ρ) become (**J**′, ρ'), where

$$J'_x = \gamma(J_x - v\rho) \qquad J'_z = J_z$$

$$J'_y = J_y \qquad \rho' = \gamma\left(\rho - \frac{v}{c^2}J_x\right)$$

15-4 The Stress-Energy-Momentum Tensor

In Chapter 6, we concluded our discussion of the electromagnetic field by finding expressions for energy and momentum in the fields. Relativistically, the energy and momentum in an electromagnetic field turn out to be elements of a (four-dimensional) second-rank tensor that also contains the elements of the (three-dimensional) stress tensor explored in P6-37. We identify this new tensor by noting first that the components of the force density, $\rho\mathbf{E} + \mathbf{J} \times \mathbf{B}$, are the spatial components of the four-vector whose components are $F_{\mu\nu}J_\nu$. For example,

$$\begin{aligned}(\rho\mathbf{E} + \mathbf{J} \times \mathbf{B})_x &= \rho E_x + J_yB_z - J_zB_y \\ &= (icF_{14})\left(\frac{J_4}{ic}\right) + F_{12}J_2 + F_{13}J_3 \\ &= F_{1\nu}J_\nu \end{aligned} \tag{15-59}$$

The fourth component of this vector is related to the power input per unit volume to the particles composing the charge and current distribution that is the source of the fields, i.e.,

$$F_{4\nu}J_\nu = F_{41}J_1 + F_{42}J_2 + F_{43}J_3 = \frac{i}{c}\mathbf{E}\cdot\mathbf{J} \tag{15-60}$$

[See Eq. (3-39).]

We next show that the quantities $F_{\mu\nu}J_\nu$ can be written as the (four-) divergence of a (symmetric) second-rank tensor, i.e., that

$$F_{\mu\nu}J_\nu = \partial_\alpha T_{\mu\alpha} \tag{15-61}$$

with $T_{\mu\alpha} = T_{\alpha\mu}$. A direct substitution from Eq. (15-52) for J_ν yields that

$$\begin{aligned} F_{\mu\nu}J_\nu &= -\frac{1}{\mu_0}F_{\mu\nu}\partial_\alpha F_{\alpha\nu} \\ \Longrightarrow \mu_0 F_{\mu\nu}J_\nu &= F_{\mu\nu}\partial_\alpha F_{\nu\alpha} \\ &= \partial_\alpha(F_{\mu\nu}F_{\nu\alpha}) - (\partial_\alpha F_{\mu\nu})F_{\nu\alpha} \end{aligned} \tag{15-62}$$

(Remember that F is *anti*symmetric.) Now, since it involves merely a renaming of summation indices, we find that

$$(\partial_\alpha F_{\mu\nu})F_{\nu\alpha} = (\partial_\nu F_{\mu\alpha})F_{\alpha\nu} \tag{15-63}$$

whence

$$\begin{aligned} (\partial_\alpha F_{\mu\nu})F_{\nu\alpha} &= \tfrac{1}{2}[(\partial_\alpha F_{\mu\nu})F_{\nu\alpha} + (\partial_\nu F_{\mu\alpha})F_{\alpha\nu}] \\ &= \tfrac{1}{2}[\partial_\alpha F_{\mu\nu} + \partial_\nu F_{\alpha\mu}]F_{\nu\alpha} \\ &= -\tfrac{1}{2}(\partial_\mu F_{\nu\alpha})F_{\nu\alpha} \\ &= -\tfrac{1}{4}\partial_\mu(F_{\sigma\tau}F_{\sigma\tau}) \end{aligned} \tag{15-64}$$

where the next to last form follows after using Eq. (15-54) and in the final form the summation indices ν and α have been changed to σ and τ. Finally, we substitute Eq. (15-64) into Eq. (15-62) to find that

$$\mu_0 F_{\mu\nu}J_\nu = \partial_\alpha[F_{\mu\nu}F_{\nu\alpha} + \tfrac{1}{4}\delta_{\mu\alpha}F_{\sigma\tau}F_{\sigma\tau}] \tag{15-65}$$

where $\delta_{\mu\alpha} = 1$ if $\mu = \alpha$ and zero otherwise. Thus, by comparison with Eq. (15-61), we conclude that

$$T_{\mu\alpha} = \frac{1}{\mu_0}\left[F_{\mu\nu}F_{\nu\alpha} + \frac{1}{4}\delta_{\mu\alpha}F_{\sigma\tau}F_{\sigma\tau}\right] \tag{15-66}$$

which by its construction is demonstrably a tensor of the second rank (P15-26) and is also symmetric. Its elements can be worked out from knowledge of F (P15-27); one finds that

$$T = \begin{pmatrix} T_{11} & T_{12} & T_{13} & -ic\mathcal{G}_x \\ T_{21} & T_{22} & T_{23} & -ic\mathcal{G}_y \\ T_{31} & T_{32} & T_{33} & -ic\mathcal{G}_z \\ -\frac{i}{c}\mathcal{S}_x & -\frac{i}{c}\mathcal{S}_y & -\frac{i}{c}\mathcal{S}_z & u_{\text{EM}} \end{pmatrix} \tag{15-67}$$

where

$$T_{ij} = \epsilon_0 E_i E_j + \frac{1}{\mu_0}B_i B_j - \frac{1}{2}\delta_{ij}\left(\epsilon_0 E^2 + \frac{1}{\mu_0}B^2\right) \tag{15-68}$$

for $i, j \leq 3$ are the elements of the (three-dimensional) stress tensor explored in P6-37,

$$\mathbf{S} = \frac{1}{\mu_0}\mathbf{E} \times \mathbf{B} \tag{15-69}$$

is the Poynting vector [Eq. (6-55)],

$$\mathbf{G} = \epsilon_0 \mathbf{E} \times \mathbf{B} = \frac{1}{c^2}\mathbf{S} \tag{15-70}$$

is the momentum density (Eq. (6-61)), and

$$u_{\text{EM}} = \frac{1}{2}\epsilon_0 E^2 + \frac{1}{2\mu_0}B^2 \tag{15-71}$$

is the electromagnetic energy density [Eq. (6-52)]. The tensor $T_{\mu\alpha}$ is called the *electromagnetic stress-energy-momentum tensor*, and it combines these several attributes of the electromagnetic field into a single 16-component entity whose components become entangled with one another under a Lorentz transformation of the frame of reference.

The conservation laws of energy and momentum, which we have already treated in Sections 6-4 and 6-5, are contained within Eq. (15-61). For example, when $\mu = 4$, Eq. (15-61) becomes

$$\begin{aligned} F_{4\nu}J_\nu &= \partial_\alpha T_{4\alpha} \\ \Longrightarrow \frac{i}{c}\mathbf{E}\cdot\mathbf{J} &= \partial_1 T_{41} + \partial_2 T_{42} + \partial_3 T_{43} + \partial_4 T_{44} \\ &= \nabla\cdot\left(-\frac{i}{c}\mathbf{S}\right) + \frac{1}{ic}\frac{\partial u_{EM}}{\partial t} \\ \Longrightarrow \mathbf{E}\cdot\mathbf{J} &= -\nabla\cdot\mathbf{S} - \frac{\partial u_{EM}}{\partial t} \end{aligned} \tag{15-72}$$

which is Eq. (6-50)—an expression we have already interpreted as an equation of energy balance. In like fashion (P15-29), Eq. (15-61) for $\mu = 1, 2,$ and 3 yields statements equivalent to Eq. (6-59), which we have interpreted as an equation of momentum balance.

PROBLEMS

P15-26. Assuming that $F_{\mu\nu}$ is a second-rank tensor, show that $T_{\mu\alpha}$ as defined by Eq. (15-66) is a second-rank tensor. *Hint:* Show first that $\delta_{\mu\alpha}$ is a second-rank tensor.

P15-27. Demonstrate that the elements of $T_{\mu\alpha}$ given in Eq. (15-67) follow from F as given in Eq. (15-50). *Hints:* (1) Show that $F_{\sigma\tau}F_{\sigma\tau} = 2(B^2 - E^2/c^2)$. (2) Show that $F_{\mu\nu}F_{\nu\alpha}$ is the $\mu\alpha$ element of the *matrix* product F^2.

P15-28. Show that the trace $T_{\mu\mu}$ of the stress-energy-momentum tensor is zero.

P15-29. Show that Eq. (15-61) with $\mu = 1, 2,$ and 3 is equivalent to Eq. (6-59).

15-5
A New Viewpoint: The Law of Biot-Savart Revisited

Until Chapter 15, we considered electric and magnetic phenomena as observed in a *single* frame of reference, and we started the study of *each* topic by quoting the results of a few pertinent experiments. In Chapter 15, we have discovered that the division of a particular electromagnetic field into separate electric and magnetic induction fields is not intrinsic to the fields themselves but depends as well on the frame of reference within which that division is accomplished. Purely electric fields in one frame of reference, for example, will give rise to both magnetic and electric effects when viewed from some other frame of reference. This observation suggests that Coulomb's law and the theory of relativity should be sufficient to permit *prediction* of at least some magnetic phenomena *without* introducing the experimental evidence used in our previous development. In this section, we shall illustrate the accuracy of this suggestion by showing how one might infer the law of Biot-Savart from application of special relativity to Coulomb's law.[8]

As a preliminary to the main argument, we must develop the rule for transforming forces from one frame of reference to another. The starting point is contained in the results of P15-11 and P15-12, where it was shown (1) that the momentum $\mathbf{p}$ and energy E defined for a particle of (rest) mass m and velocity $\mathbf{V}$ by

$$\mathbf{p} = \frac{m\mathbf{V}}{\sqrt{1-(V^2/c^2)}} = \Gamma m\mathbf{V} \qquad E = \Gamma mc^2 \tag{15-73}$$

combine to form a four-vector

$$\mathsf{p} = \left(p_x, p_y, p_z, \frac{i}{c}E\right) \tag{15-74}$$

(2) that Γ and V in the unprimed frame of reference in Fig. 15-2 are related to Γ' and V' in the primed frame of reference by

$$\Gamma' = \gamma\Gamma\left(1 - \frac{V_x v}{c^2}\right) \tag{15-75}$$

or, equivalently (exchanging primes and "unprimes" and replacing v with $-v$), by

$$\Gamma = \gamma\Gamma'\left(1 + \frac{V'_x v}{c^2}\right) \tag{15-76}$$

and (3) that $d\tau$ defined by

[8] We shall present only a very limited development from the point of view of special relativity. More detailed treatments may be found in E. M. Purcell, *Electricity and Magnetism* (McGraw-Hill Book Company, New York, 1965), Chapter 5; in P. Lorrain and D. R. Corson, *Electromagnetic Fields and Waves* (W. H. Freeman and Company, San Francisco, 1970), Chapter 6; and in numerous other sources.

$$c^2\,d\tau^2 = c^2\,dt^2 - dx^2 - dy^2 - dz^2 = c^2\,dt^2\left(1 - \frac{V^2}{c^2}\right)$$

$$\Longrightarrow d\tau = \frac{1}{\Gamma}\,dt \tag{15-77}$$

is a scalar. Finally, we note that the relativistic force remains by definition the time derivative of momentum,

$$\mathbf{F} = \frac{d\mathbf{p}}{dt} \qquad \mathbf{F}' = \frac{d\mathbf{p}'}{dt'} \tag{15-78}$$

and that the quantity $dp_\mu/d\tau$—a vector divided by a scalar—is a four-vector. Thus, we find that

$$\frac{dp_\mu}{d\tau} = (\mathfrak{L}^{-1})_{\mu\nu}\frac{dp'_\nu}{d\tau'} \tag{15-79}$$

or, by virtue of Eqs. (15-77) and (15-76), that

$$\Gamma\frac{dp_\mu}{dt} = \Gamma'(\mathfrak{L}^{-1})_{\mu\nu}\frac{dp'_\nu}{dt'}$$

$$\Longrightarrow \frac{dp_\mu}{dt} = \frac{1}{\gamma(1 + V'_x v/c^2)}(\mathfrak{L}^{-1})_{\mu\nu}\frac{dp'_\nu}{dt'} \tag{15-80}$$

Since $\mathfrak{L}^{-1}$ is the transpose of the matrix in Eq. (15-8) [see Eq. (15-11)], we find from Eq. (15-80) that

$$F_x = \frac{F'_x - i\beta\,dp'_4/dt'}{(1 + V'_x v/c^2)} \tag{15-81}$$

$$F_y = \frac{F'_y}{\gamma(1 + V'_x v/c^2)} \qquad F_z = \frac{F'_z}{\gamma(1 + V'_x v/c^2)} \tag{15-82), (15-83}$$

Equation (15-81), however, has a more useful form obtained by noting that

$$\frac{dp'_4}{dt'} = \frac{i}{c}\frac{dE'}{dt'} = \frac{i}{c}\mathbf{F}'\cdot\mathbf{V}' \tag{15-84}$$

since dE'/dt' is the rate at which the force $\mathbf{F}'$ does work on the particle. Substituting Eq. (15-84) into Eq. (15-81), we find finally that

$$F_x = \frac{F'_x + (v/c^2)\mathbf{F}'\cdot\mathbf{V}'}{(1 + V'_x v/c^2)} = F'_x + \frac{v(V'_y F'_y + V'_z F'_z)}{c^2 + V'_x v} \tag{15-85}$$

A derivation of the inverse transformation is the subject of P15-30.

We shall now derive the law of Biot-Savart by considering the force between two charges q_1 and q_2 *at rest* at the origin and at (x', y', z') in the primed frame of reference (Fig. 15-3). Using Coulomb's law in the primed frame, we find that the force on q_2 as measured in that frame is given by

$$\{F'_x, F'_y, F'_z\} = \frac{q_1 q_2}{4\pi\epsilon_0}\frac{\{x', y', z'\}}{[(x')^2 + (y')^2 + (z')^2]^{3/2}} \tag{15-86}$$

Written in terms of the *un*primed coordinates (assuming the measurement is

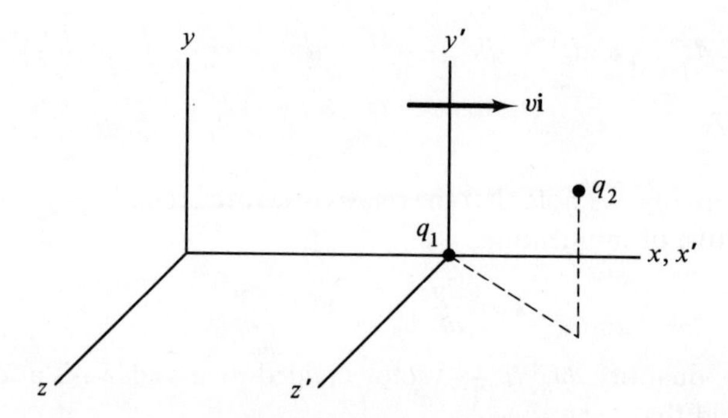

Fig. 15-3. Two point charges observed from two frames of reference. The charges are at rest in the primed frame.

made at time zero for simplicity), this force is also given by

$$\{F'_x, F'_y, F'_z\} = \frac{q_1 q_2}{4\pi\epsilon_0 R^3}\{\gamma x, y, z\} \tag{15-87}$$

where

$$R = [\gamma^2 x^2 + y^2 + z^2]^{1/2} \tag{15-88}$$

Equations (15-82), (15-83), and (15-85) with $\mathbf{V}' = 0$ then give the expression

$$\{F_x, F_y, F_z\} = \frac{q_1 q_2 \gamma}{4\pi\epsilon_0 R^3}\left\{x, \frac{y}{\gamma^2}, \frac{z}{\gamma^2}\right\}$$

$$\Longrightarrow \mathbf{F} = \frac{q_1 q_2 \gamma}{4\pi\epsilon_0 R^3}\left[\mathbf{r} - \frac{v^2}{c^2}(y\hat{\mathbf{j}} + z\hat{\mathbf{k}})\right] \tag{15-89}$$

where we have used again the invariance of the charges q_1 and q_2. Now, since the velocity of both charges in the unprimed frame is $\mathbf{v} = v\hat{\mathbf{i}}$, we find that

$$\begin{aligned}\mathbf{v} \times (\mathbf{v} \times \mathbf{r}) &= (\mathbf{v}\cdot\mathbf{r})\mathbf{v} - v^2\mathbf{r} \\ &= v^2 x\hat{\mathbf{i}} - v^2\mathbf{r} = -v^2(y\hat{\mathbf{j}} + z\hat{\mathbf{k}})\end{aligned} \tag{15-90}$$

and Eq. (15-89) can be written in the form

$$\mathbf{F} = q_2\left[\frac{q_1\gamma\mathbf{r}}{4\pi\epsilon_0 R^3} + \mathbf{v} \times \left\{\frac{q_1\gamma(\mathbf{v} \times \mathbf{r})}{4\pi\epsilon_0 c^2 R^3}\right\}\right] \tag{15-91}$$

Thus, *solely on the basis of a relativistic transformation of Coulomb's law*, we are led to introduce *two* fields

$$\mathbf{E} = \frac{q_1\gamma\mathbf{r}}{4\pi\epsilon_0 R^3} \qquad \mathbf{B} = \frac{q_1\gamma(\mathbf{v} \times \mathbf{r})}{4\pi\epsilon_0 c^2 R^3} \tag{15-92), (15-93}$$

established in the unprimed frame by the moving charge q_1. Further, we determine the force on the *moving* charge q_2 in this frame from the expression

$$\mathbf{F} = q_2(\mathbf{E} + \mathbf{v} \times \mathbf{B}) \tag{15-94}$$

Starting from Coulomb's law, the invariance of charge, and special relativity, we have thus *proved* that magnetic induction fields must exist. More specifically, writing μ_0 for $1/\epsilon_0 c^2$ and assuming $v/c \ll 1$ so that $\gamma \longrightarrow 1$, we find from Eq. (15-93) that the **B**-field established by a single charge moving nonrelativistically through the origin with velocity **v** is given by

$$\mathbf{B} = \frac{\mu_0}{4\pi} \frac{q_1 \mathbf{v} \times \mathbf{r}}{r^3} \tag{15-95}$$

where $r^2 = x^2 + y^2 + z^2$. More generally, if a source charge q' is moving through the point $\mathbf{r}'$ rather than through the origin, the resulting **B**-field at point **r** is obtained from Eq. (15-95) by interpreting **r** as the vector from the source point to the observation point; we find that

$$\mathbf{B}(\mathbf{r}) = \frac{\mu_0}{4\pi} \frac{q' \mathbf{v} \times (\mathbf{r} - \mathbf{r}')}{|\mathbf{r} - \mathbf{r}'|^3} \tag{15-96}$$

Finally, we determine the **B**-field produced by a general current distribution by using Eq. (15-96) and the principle of superposition. We divide the general distribution of interest into volume elements, with the element dv' centered at $\mathbf{r}'$. Within dv', there may be particles of several different types, those of type a carrying charge q_a and having density $n^{(a)}(\mathbf{r}')$ and (drift) velocity $\mathbf{v}^{(a)}(\mathbf{r}')$. Summing Eq. (15-96) over all particles yields

$$\mathbf{B}(\mathbf{r}) = \frac{\mu_0}{4\pi} \int \frac{\mathbf{J}(\mathbf{r}') \times (\mathbf{r} - \mathbf{r}')}{|\mathbf{r} - \mathbf{r}'|^3} \, dv' \tag{15-97}$$

where, as in Eq. (2-18),

$$\mathbf{J}(\mathbf{r}') = \sum_a q_a n^{(a)}(\mathbf{r}') \mathbf{v}^{(a)}(\mathbf{r}') \tag{15-98}$$

is the current density describing the distribution. Equation (15-97) is the law of Biot-Savart, and we have achieved the objective of this paragraph by *deriving* this law from Coulomb's law, special relativity, the invariance of charge, and the principle of superposition.

This demonstration that the law of Biot-Savart can in a sense be viewed as an aspect of Coulomb's law provides a deep insight into the relationship between electric and magnetic induction fields. These fields are not only tied together *within* a given frame of reference by their simultaneous occurrence in some of Maxwell's equations; they are also connected *across* frames of reference by the requirements of the theory of special relativity. Some magnetic induction fields can even be thought of as arising from a relativistic transformation of a purely electric field, and one might argue that such magnetic induction fields exist only because the observer has (unwisely) picked the "wrong" frame of reference. Not every magnetic induction field, however, can be transformed to a purely electric field (see P15-18), so we cannot eliminate magnetic induction fields altogether by thinking of them solely as relativistic consequences of electric fields. Nonetheless, starting with Coulomb's law and the theory of special relativity, it is possible to demonstrate the existence of the magnetic induction field and to obtain the Lorentz force without drawing on the results of any experiments specifically involving magnetic forces. It is also possible to deduce the transformation equations for the fields (P15-32) and to deduce Maxwell's equations from the beginnings discussed in this section. (See footnote 8.) Thus, beginning with a pair of fields originally conceived to be independent, we have been led first to Maxwell's equations in a single frame of reference and then—partly by apparent contradictions between the predictions of Maxwell's equations and the behavior of the physical world—to the theory of special relativity. Finally, from the vantage point provided by special relativity, we have in this section looked back on our starting points and discovered very significant but originally unsuspected connections between electric and magnetic induction fields.

PROBLEMS

P15-30. Derive a transformation rule giving the *primed* components of a force in terms of the *unprimed* components of the force and the *unprimed* components of the velocity of the particle; i.e., derive the inverse of Eqs. (15-82), (15-83), and (15-85).

P15-31. Suppose the charge q_2 in Fig. 15-3 is moving with some velocity $\mathbf{V}'$ in the primed reference frame. Show that the force on this charge in the unprimed frame is given by $q_2\mathbf{E} + q_2\mathbf{V} \times \mathbf{B}$ with $\mathbf{E}$ and $\mathbf{B}$ given by Eqs. (15-92) and (15-93). *Hints:* (1) Since q_1 is still at rest in the primed frame, the force on q_2 in that frame remains solely an electrostatic force. (2) See P15-7.

P15-32. Given the transformation rules for force and for velocity (P15-7) and knowing that $\mathbf{F}' = q(\mathbf{E}' + \mathbf{V}' \times \mathbf{B}')$ while $\mathbf{F} = q(\mathbf{E} + \mathbf{V} \times \mathbf{B})$, find the transformation rules for the fields.

P15-33. (a) Show from Eqs. (15-92) and (15-93) that the fields of a moving point charge are given by

$$\mathbf{E} = \frac{q_1}{4\pi\epsilon_0}\frac{\hat{\mathbf{r}}}{r^2}\frac{1-\beta^2}{(1-\beta^2\sin^2\alpha)^{3/2}}$$

$$\mathbf{B} = \frac{q_1}{4\pi\epsilon_0 c}\frac{\hat{\mathbf{n}}}{r^2}\frac{\beta(1-\beta^2)\sin\alpha}{(1-\beta^2\sin^2\alpha)^{3/2}}$$

where $r^2 = x^2 + y^2 + z^2$, α is the angle between $\mathbf{r}$ and the (positive) x axis ($=$ the direction of $\mathbf{v}$), and $\hat{\mathbf{n}}$ is a unit vector in the direction of $\hat{\mathbf{i}} \times \hat{\mathbf{r}}$. (b) Sketch graphs of $|\mathbf{E}|$ and $|\mathbf{B}|$ as functions of α for fixed r and various values of β.

Supplementary Problems

P15-34. Add to Fig. 15-2 a double-primed frame of reference moving with speed v_1 relative to the primed frame along the common positive $x - x' - x''$ direction. Find the single transformation relating double-primed coordinates directly to unprimed coordinates and show that the velocity of the double-primed origin relative to the unprimed origin is given by $(v_1 + v)/(1 + vv_1/c^2)$.

Hints: (1) $x''_\mu = \mathfrak{L}_{\mu\nu}(v_1/c)\mathfrak{L}_{\nu\sigma}(v/c)x_\sigma$. Why? (2) Note that $\mathfrak{L}_{\mu\nu}(v_1/c)\mathfrak{L}_{\nu\sigma}(v/c)$ is the $\mu\sigma$ element of a matrix product.

P15-35. Let **A** be the three-dimensional vector whose components are the spatial components of the four-vector **A**. (a) Show that the Lorentz transformation of the four-vector can be expressed in the form

$$\mathbf{A}'\cdot\boldsymbol{\beta} = \gamma(\mathbf{A}\cdot\boldsymbol{\beta} + i\beta^2 A_4)$$

$$\boldsymbol{\beta}\times(\mathbf{A}'\times\boldsymbol{\beta}) = \boldsymbol{\beta}\times(\mathbf{A}\times\boldsymbol{\beta})$$

$$A'_4 = \gamma(A_4 - i\mathbf{A}\cdot\boldsymbol{\beta})$$

where $\boldsymbol{\beta}$ is a vector of magnitude $\beta = v/c$ in the direction of the velocity of the primed frame relative to the unprimed frame. (b) Find a 4×4 matrix expressing the Lorentz transformation between two frames of reference whose axes are parallel but whose relative motion is described by $\boldsymbol{\beta} = \beta_x\hat{\mathbf{i}} + \beta_y\hat{\mathbf{j}}$. (c) Show that your result reduces to Eq. (15-8) when $\beta_y = 0$ and $\beta_x = \beta$. (d) Verify that your result is a (complex) orthogonal matrix.

P15-36. Show that the gauge transformation [Eqs. (6-66) and (6-68)] between two equivalent sets of potentials ($\mathbf{A}_1$, V_1) and ($\mathbf{A}_2$, V_2) in a *single* frame of reference can be written in terms of the four-potential in the form $A_{2\mu} = A_{1\mu} + \partial_\mu\Lambda$, where Λ is an arbitrary function of the spatial and temporal coordinates.

P15-37. The homogeneous Maxwell equations, Eq. (15-54), can be written more compactly if we introduce the notion of a dual tensor. Let the quantity $\epsilon_{\alpha\beta\gamma\delta}$ be defined to be 1 when (α, β, γ, δ) is an even permutation of (1, 2, 3, 4), -1 when (α, β, γ, δ) is an odd permutation of (1, 2, 3, 4), and 0 when (α, β, γ, δ) has any other combination of allowed values. (a) Show that $\epsilon_{\alpha\beta\gamma\delta}$ is a fourth-rank tensor under the Lorentz transformation in Eq. (15-8). (b) The tensor $\bar{F}$ defined by $\bar{F}_{\alpha\beta} = \frac{1}{2}\epsilon_{\alpha\beta\mu\nu}F_{\mu\nu}$ is called the tensor dual to F. Use the results of part (a) to argue that $\bar{F}_{\alpha\beta}$ is a second-rank tensor under the Lorentz transformation in Eq. (15-8), find the matrix similar to Eq. (15-50) expressing the relationship between $\bar{F}$ and the components of the electromagnetic field, and show that Eq. (15-54) has the alternative expression $\partial_\alpha\bar{F}_{\alpha\beta} = 0$. (c) Show that $\mathbf{E}\cdot\mathbf{B}$ is a Lorentz invariant. *Hint:* $F_{\mu\nu}\bar{F}_{\mu\nu}$ is a scalar. Why? *Note:* In part tacitly, the conclusions of this problem are confined to behavior under the so-called *proper* Lorentz transformations, which do not involve reflections of the coordinates. If the *improper* transformations, which involve reflection of one or of three coordinates, are allowed, then it is necessary to distinguish two kinds of tensors, the first of which transforms, for example, as in Eq. (15-18) under *both* types of transformation and the second of which transforms as in Eq. (15-18) only under the proper transformations. Members of the second group are called *pseudo*tensors. Strictly, the tensor $\epsilon_{\alpha\beta\gamma\delta}$ is a fourth-rank pseudotensor and consequently $\bar{F}$ is a second-rank pseudotensor and $F_{\mu\nu}\bar{F}_{\mu\nu}$ is a pseudoscalar.

P15-38. (a) Transform the electrostatic field of a long uniformly charged line (P4-6) to a frame of reference moving with speed v parallel to the wire.

Suggestion: Let the wire lie along the x axis in Fig. 15-2 and be at rest in the unprimed frame. (b) Compare the resulting fields with those produced by an infinitely long current [see Eq. (5-11)] and discuss the difference between a current and a moving charged rod.

P15-39. Consider a region of space devoid of charges and currents. (a) Show that Eq. (15-44) for the potentials is satisfied by the four-component plane wave

$$A_\mu(\mathbf{r}, t) = \alpha_\mu e^{i(\boldsymbol{\kappa}\cdot\mathbf{r}-\omega t)} \tag{1}$$

where α_μ is a component of a *constant* four-vector, provided $\kappa^2 c^2 = \omega^2$. (b) Argue that a plane wave in one frame of reference must transform to a plane wave in another frame of reference. Thus,

$$A'_\mu(\mathbf{r}', t') = \alpha'_\mu e^{i(\boldsymbol{\kappa}'\cdot\mathbf{r}'-\omega' t')} = \mathfrak{L}_{\mu\nu}\alpha_\nu e^{i(\boldsymbol{\kappa}\cdot\mathbf{r}-\omega t)} \tag{2}$$

(c) Show that Eq. (2) cannot be true for arbitrary $\mathbf{r}$ and t unless $\boldsymbol{\kappa}\cdot\mathbf{r} - \omega t = \boldsymbol{\kappa}'\cdot\mathbf{r}' - \omega' t'$ and, hence, show that $\kappa_\mu = (\kappa_x, \kappa_y, \kappa_z, i\omega/c)$ is a four-vector. (See P15-9.) (d) Write

$$F_{\mu\nu}(\mathbf{r}, t) = F^0_{\mu\nu} e^{i\kappa_\sigma x_\sigma} \tag{3}$$

and find the matrix $F^0_{\mu\nu}$. (e) Suppose a wave with frequency ω is propagating in the unprimed frame of Fig. 15-2 in a direction making an angle α with the x axis. Use the transformation rules for a four-vector to show that

$$\tan\alpha' = \frac{\sqrt{1-\beta^2}\sin\alpha}{(\cos\alpha) - \beta} \qquad \omega' = \omega\frac{1-\beta\cos\alpha}{\sqrt{1-\beta^2}}$$

where, in the primed frame, ω' is the frequency of the wave and α' is the angle between the direction of propagation and the x' axis. The first of these expressions predicts the phenomenon of aberration, as explored also in P15-7; the second predicts the relativistic Doppler effect. (f) The *longitudinal* Doppler effect is analogous to the classical Doppler effect and occurs when the relative motion and the direction of propagation are the same ($\alpha = 0$). This effect is responsible for the *red shift* in the spectrum of light from distant galaxies. Astronomers define the *percent red shift* of a galaxy fixed in the *un*primed frame and observed in the primed frame in terms of the wavelength λ by $100(\lambda' - \lambda)/\lambda$. Show that the percent red shift is given by

$$P = 100\left[\sqrt{\frac{1+\beta}{1-\beta}} - 1\right]$$

Sketch a graph of P versus β over the range $-1 < \beta < 1$, and, in particular, calculate the velocity of a galaxy for which $P = 50\%$. (g) The *transverse* Doppler effect occurs when $\alpha = 90°$ and has no classical analog. Sketch a graph of λ'/λ versus β for the transverse effect. Note that the longitudinal effect is first order in β while the transverse effect is second order in β.

P15-40. Let $\mathbf{p} = (\mathbf{p}, iE/c)$ be the momentum four-vector of a particle (P15-12). (a) Show that the expression

$$P = \frac{q^2}{6\pi\epsilon_0 m^2 c^3}\frac{dp_\mu}{d\tau}\cdot\frac{dp_\mu}{d\tau}$$

reduces to Eq. (14-94) in the limit of $v/c \ll 1$. Thus, this invariant expression is a natural relativistic generalization of Eq. (14-94) for the total power radiated by an accelerated particle. (b) Show that this expression can be written in the alternative form given in Eq. (14-95).

P15-41. A particle starts from rest at the origin and moves relativistically in a uniform electric field $\mathbf{E} = E\hat{\mathbf{i}}$. Solve the equation of motion

$$\frac{d}{dt}\left(\frac{mv}{\sqrt{1-(v/c)^2}}\right) = qE, \qquad v = \frac{dx}{dt}$$

to find the position of the particle as a function of time. Sketch graphs of x and of v versus t and compare these graphs with the corresponding nonrelativistic results.

P15-42. A charged particle moves relativistically in a constant magnetic induction field $\mathbf{B} = B\hat{\mathbf{k}}$. There is no electric field. The relativistic equation of motion therefore is $d\mathbf{p}/dt = q\mathbf{v} \times \mathbf{B}$, where $\mathbf{p} = m\mathbf{v}/\sqrt{1-(v/c)^2}$ is the relativistic momentum of the particle (P15-12). (a) Show that the energy E of this particle, given by $E^2 = p^2c^2 + m^2c^4$, is constant and hence that $\gamma = [1-(v/c)^2]^{-1/2}$ is constant. (b) Show that, even relativistically, a charged particle in a uniform magnetic induction field moves in a helical path. (c) Show that the frequency of circulation about the helical path is given by $\omega = qB/m\gamma$. Show also that, when the velocity is perpendicular to the **B**-field and the path is a circle, the radius of that circle is given by $R = mv\gamma/qB$. (d) Sketch graphs of ω and of R versus v/c and compare these graphs with the corresponding nonrelativistic results.

REFERENCES

D. Bohm, *The Special Theory of Relativity* (W. A. Benjamin, Inc., Reading, Mass., 1965).

C. V. Durell, *Readable Relativity* (Harper & Row, Publishers, New York, 1960). (Reprinting of a book published in 1926 by G. Bell & Sons Ltd., London.)

R. P. Feynman, R. B. Leighton, and M. Sands, *The Feynman Lectures on Physics* (Addison-Wesley Publishing Company, Inc., Reading, Mass., 1963), Volume I, Lectures 15–17; Volume II, Lectures 25 and 26.

J. D. Jackson, *Classical Electrodynamics* (John Wiley & Sons, Inc., New York, 1962), Chapter 11.

P. Lorrain and D. R. Corson, *Electromagnetic Fields and Waves* (W. H. Freeman and Company, San Francisco, 1970), Chapters 5 and 6.

W. K. H. Panofsky and M. Phillips, *Classical Electricity and Magnetism* (Addison-Wesley Publishing Company, Inc., Reading, Mass., 1955), Chapters 14–17.

R. Resnick, *Introduction to Special Relativity* (John Wiley & Sons, Inc., New York, 1968).

C. W. Sherwin, *Basic Concepts of Physics* (Holt, Rinehart and Winston, Inc., New York, 1961), Chapters 4 and 5 and Appendix I.

E. F. Taylor and J. A. Wheeler, *Spacetime Physics* (W. H. Freeman and Company, San Francisco, 1966).

Appendices

A

Linear Equations, Determinants, and Matrices

This appendix contains a summary of some properties of determinants and matrices. More detailed discussions may be found in the references at the end of the appendix.

A-1 Simultaneous Linear Equations and Determinants

Consider the general system of two linear equations

$$\begin{aligned} a_{11}x_1 + a_{12}x_2 &= b_1 \\ a_{21}x_1 + a_{22}x_2 &= b_2 \end{aligned} \tag{A-1}$$

with *coefficients* $\begin{pmatrix} a_{11} & a_{12} \\ a_{21} & a_{22} \end{pmatrix}$, *inhomogeneities* $\begin{pmatrix} b_1 \\ b_2 \end{pmatrix}$ and *unknowns* $\begin{pmatrix} x_1 \\ x_2 \end{pmatrix}$. Systematic elimination first of x_2 and then of x_1 yields the solution

$$x_1 = \frac{b_1 a_{22} - b_2 a_{12}}{a_{11}a_{22} - a_{21}a_{12}}, \quad x_2 = \frac{b_2 a_{11} - b_1 a_{21}}{a_{11}a_{22} - a_{21}a_{12}} \tag{A-2}$$

If we now introduce what is called a *second-order* (or 2×2) *determinant*, defined for an array with elements q_{ij} by

$$\begin{vmatrix} q_{11} & q_{12} \\ q_{21} & q_{22} \end{vmatrix} = q_{11}q_{22} - q_{21}q_{12} \tag{A-3}$$

we can write Eq. (A-2) in the form

$$x_1 = \frac{\begin{vmatrix} b_1 & a_{12} \\ b_2 & a_{22} \end{vmatrix}}{\begin{vmatrix} a_{11} & a_{12} \\ a_{21} & a_{22} \end{vmatrix}}, \quad x_2 = \frac{\begin{vmatrix} a_{11} & b_1 \\ a_{21} & b_2 \end{vmatrix}}{\begin{vmatrix} a_{11} & a_{12} \\ a_{21} & a_{22} \end{vmatrix}} \tag{A-4}$$

Each member of the solution to two simultaneous linear equations can therefore be immediately obtained as the ratio of two determinants. The denominator is the determinant of the coefficients; the numerator is obtained from the denominator by replacing one column with the column of inhomogeneities—the first column for the first variable and the second column for the second variable.

Although the algebra is more tedious for the general system of three linear equations,

$$\begin{aligned} a_{11}x_1 + a_{12}x_2 + a_{13}x_3 &= b_1 \\ a_{21}x_1 + a_{22}x_2 + a_{23}x_3 &= b_2 \\ a_{31}x_1 + a_{32}x_2 + a_{33}x_3 &= b_3 \end{aligned} \tag{A-5}$$

systematic elimination of any two unknowns nevertheless leads to a solution for the remaining unknown that can be expressed as a ratio of two *third-order* (or 3 × 3) *determinants*, provided that the 3 × 3 determinant of an array with elements q_{ij} is defined by

$$\begin{vmatrix} q_{11} & q_{12} & q_{13} \\ q_{21} & q_{22} & q_{23} \\ q_{31} & q_{32} & q_{33} \end{vmatrix} = q_{11}\begin{vmatrix} q_{22} & q_{23} \\ q_{32} & q_{33} \end{vmatrix} - q_{12}\begin{vmatrix} q_{21} & q_{23} \\ q_{31} & q_{33} \end{vmatrix} + q_{13}\begin{vmatrix} q_{21} & q_{22} \\ q_{31} & q_{32} \end{vmatrix} \tag{A-6}$$

Here each 2 × 2 determinant is obtained from the original 3 × 3 array by deleting the row and column containing its premultiplier. (*Warning*: The minus sign before the second term is easily forgotten.) With this definition, the solution to Eq. (A-5) can be immediately obtained by the same rule as was described for the system of two equations. For example, the solution for x_2 is

$$x_2 = \frac{\begin{vmatrix} a_{11} & b_1 & a_{13} \\ a_{21} & b_2 & a_{23} \\ a_{31} & b_3 & a_{33} \end{vmatrix}}{\begin{vmatrix} a_{11} & a_{12} & a_{13} \\ a_{21} & a_{22} & a_{23} \\ a_{31} & a_{32} & a_{33} \end{vmatrix}} \tag{A-7}$$

the denominator being the determinant of the coefficients and the numerator being obtained from the denominator by replacing the second column with the column of inhomogeneities. The rules for reducing the solution of a set of simultaneous linear equations to ratios of determinants are known as *Cramer's rules.* Equations (A-4) and (A-7) are examples of these rules, but Nth order (or $N \times N$) determinants can be defined so that these rules apply to systems of equations of any size.

Determinants of all orders share many interesting properties that can be exploited to facilitate their evaluation. Among these properties are the following:

(1) The value of a determinant is unaltered by the exchange of rows for columns.
(2) The value of a determinant is changed in sign if any two rows (or columns) are exchanged.
(3) The value of a determinant is zero if every element in any row (or column) is zero.
(4) The value of a determinant is unaltered if every element in one row (or column) is multiplied by a constant and the result is added element by element to another row (or column).
(5) The value of a determinant is multiplied by the constant c if *every* element in a *single* row (or column) is replaced by c times that element.
(6) A determinant D of order N may be reduced to a sum of determinants of order $N - 1$ by an expansion similar to Eq. (A-6) in which the premultipliers are the elements of *any* row (or column). Let q_{ij} be the element in the ith row and jth column of the determinant of order N and let M_{ij}—called the *minor* of q_{ij}—be the determinant of order $N - 1$ obtained by deleting the ith row and jth column from the original $N \times N$ array. Then, the *Laplace development* of D states that

$$D = \sum_{j=1}^{N} (-1)^{i+j} q_{ij} M_{ij} \quad \text{(expansion on the } i\text{th row)} \tag{A-8}$$

$$= \sum_{i=1}^{N} (-1)^{i+j} q_{ij} M_{ij} \quad \text{(expansion on the } j\text{th column)} \tag{A-9}$$

Equation (A-6), of course, is the special case of Eq. (A-8) obtained with $N = 3$ and $i = 1$.

A system of linear equations is said to be *homogeneous* if all inhomogeneities are zero [$b_1 = b_2 = b_3 = 0$ in the 3×3 case of Eq. (A-5)]. If a system is homogeneous, the numerator in the solution given by Cramer's rules for every unknown contains a column of zeros. By property (3) above, all of the unknowns in the most general homogeneous system will then be zero and the solution is said to be *trivial.* For some special systems, however, it may happen that the denominator of the solutions given by Cramer's rules also vanishes. In such cases, these solutions are indeterminate and nontrivial solutions can in fact be found. Since the critical denominator is the determinant of the coefficients, we have made plausible the theorem that *nontrivial solutions of a homogeneous system of n linear (algebraic) equations in n unknowns exist if and only if the determinant of the coefficients vanishes.* When this condition is satisfied, at least one of the equations in the system can be obtained as a linear combination of the other equations. Any such equation contains superfluous information and can be discarded.

PROBLEMS

PA-1. (a) What would you say about a pair of equations for which the denominator in Eq. (A-4) is zero but the numerators are different from zero? (b) What would you say if the numerators and the denominator in Eq. (A-4) were all zero? (c) Invent an example of each type of system.

PA-2. Using property (2) in Section A-1, show that the value of a determinant is zero if each element of some row (or column) is equal to the corresponding element of some other row (or column).

PA-3. Use Cramer's rules to solve the equations

$$
\begin{aligned}
3x_1 + x_2 - x_3 &= -6 \\
-x_1 + 2x_2 + x_3 &= 10 \\
5x_1 - x_2 + x_3 &= -2
\end{aligned}
$$

Use a different method (expansion on row 1, expansion on column 2, adding rows, etc.) to evaluate each of the four determinants arising.

PA-4. Show that the system

$$
\begin{aligned}
2x_1 - 3x_2 + x_3 &= 0 \\
x_1 + x_2 - x_3 &= 0 \\
9x_1 - x_2 - 3x_3 &= 0
\end{aligned}
$$

possesses nontrivial solutions and find *three* different sets of numbers (x_1, x_2, x_3) not including (0, 0, 0) satisfying these equations.

PA-5. Write a general computer program that uses Cramer's rules to solve Eq. (A-5). In brief, your program might accept specific values for a_{ij} and b_i as input from the user, and then calculate and print out the values of x_1, x_2, and x_3. Test your program with the system in problem PA-3 and on other systems of your choosing. *Optional*: Modify your program so that it tests the determinant of the coefficients and prints a suitable message if this determinant is zero. *Important note*: Cramer's rules become computationally very inefficient as the size of the system increases. Further, computer evaluation of large determinants directly from primary definitions such as Eq. (A-6) may be subject to disastrous roundoff errors. Better methods for solving linear equations involve systematic elimination and are described in books on numerical analysis, e.g., S.D. Conte, *Elementary Numerical Analysis* (McGraw-Hill Book Company, New York, 1965).

A-2 Matrix Algebra

Let Q symbolize a rectangular array of (possibly complex) numbers having m rows and n columns, and let q_{ij}, where $1 \leq i \leq m$ and $1 \leq j \leq n$, be the entry in the ith row and jth column. Such an array is normally presented in the form

$$\begin{pmatrix} q_{11} & q_{12} & \cdots & q_{1j} & \cdots & q_{1n} \\ q_{21} & q_{22} & \cdots & q_{2j} & \cdots & q_{2n} \\ \vdots & \vdots & & \vdots & & \vdots \\ q_{i1} & q_{i2} & \cdots & q_{ij} & \cdots & q_{in} \\ \vdots & \vdots & & \vdots & & \vdots \\ q_{m1} & q_{m2} & \cdots & q_{mj} & \cdots & q_{mn} \end{pmatrix}$$

and is called a *matrix*. The individual numbers q_{ij} are called the *elements* (or in some cases the *components*) of the matrix, and the number of rows and columns m and n are called its *dimensions.*

Since we are in effect inventing matrices, we may assign to them whatever algebraic properties we wish. Without attempting to motivate our particular choice here, we shall adopt the following properties:

(1) *Equality.* Two matrices Q and R *having the same dimensions* are defined to be equal if and only if every element of Q is equal to the corresponding element of R; i.e., $Q = R$ if and only if $q_{ij} = r_{ij}$ for *all* i and j. Equality is not defined for matrices having different dimensions.

(2) *Addition.* A matrix S is defined to be the sum of two other matrices Q and R *having the same dimensions* if and only if every element of S is the sum of the corresponding elements of Q and R; i.e., $S = Q + R$ if and only if $s_{ij} = q_{ij} + r_{ij}$ for *all* i and j. It then follows from the corresponding properties of ordinary addition that matrix addition is commutative, $Q + R = R + Q$, and associative, $(Q + R) + T = Q + (R + T)$. The sum of two matrices having different dimensions is not defined.

(3) *Multiplication of a matrix by a scalar.* A matrix R is defined to be the product of a scalar s and another matrix Q if and only if *every* element of R is equal to s times the corresponding element of Q, i.e., $R = sQ$ if and only if $r_{ij} = sq_{ij}$ for *all* i and j; the product Qs is defined to be equal to sQ. It then follows from the properties of ordinary multiplication that multiplication of a matrix by a scalar is distributive, $(s + t)Q = sQ + tQ$ and $s(Q + R) = sQ + sR$. The difference between two matrices having the same dimensions may now be defined by $Q - R = Q + (-R)$.

(4) *Multiplication of a matrix by a matrix.* Let m_q and n_q be the number of rows and columns in the matrix Q and let m_r and n_r be the number of rows and columns in the matrix R. If $n_q = m_r$ (number of *columns* in Q = number of *rows* in R), the matrix S whose elements are given by

$$s_{ij} = \sum_{k=1}^{n_q} q_{ik} r_{kj}, \quad 1 \leq i \leq m_q,\ 1 \leq j \leq n_r \tag{A-10}$$

is defined to be the matrix product QR of the matrices Q and R; i.e., $S = QR$ if Eq. (A-10) holds. Less formally, Eq. (A-10) simply states that the element appearing in the ith row and jth column of S is obtained by multiplying each element in the ith *row* of Q by the corresponding element in the jth *column* of R (first by first, second by second, etc.) and adding these products. Even more specifically, we have the product

$$\begin{pmatrix} 1 & 4 & 3 \\ \mathbf{-2} & \mathbf{1} & \mathbf{6} \end{pmatrix} \begin{pmatrix} -1 & 1 & \mathbf{0} & -3 \\ 4 & 3 & \mathbf{1} & -1 \\ -2 & -2 & \mathbf{2} & 1 \end{pmatrix} = \begin{pmatrix} 9 & 7 & 10 & -4 \\ -6 & -11 & \mathbf{13} & 11 \end{pmatrix} \quad \text{(A-11)}$$

in which, for example, the 2, 3 element (shown in boldface) is obtained by multiplying corresponding elements of the boldface row and column and adding the products, $13 = (-2)(0) + (1)(1) + (6)(2)$. Provided rows and columns match appropriately so that the necessary products are defined, it then follows from this definition that matrix multiplication is distributive, $Q(R + T) = QR + QT$, and associative, $(QR)T = Q(RT)$. Even when both products QR and RQ are defined (i.e., when $n_q = m_r$ *and* $n_r = m_q$), matrix multiplication is *not* in general commutative, $QR \neq RQ$. The product of two matrices is not defined if the number of columns in the first factor differs from the number of rows in the second factor.

Several additional terms are used in discussing matrices. Let Q again be a matrix with m rows and n columns. Particularly important special forms for Q include the *square matrix* ($m = n$), the *column matrix* or *column vector* ($n = 1$), and the *row matrix* or *row vector* ($m = 1$). The *complex conjugate* Q^* is that matrix whose elements are the complex conjugates of the elements of Q, the *transpose* Q^T is that matrix whose *columns* are the *rows* of Q, and the *adjoint* $Q^\dagger$ is the transpose of the complex conjugate, $Q^\dagger = (Q^*)^T$, which is equal to the complex conjugate of the transpose $(Q^T)^*$. A *square* matrix is said to be *symmetric if* $Q^T = Q$, *antisymmetric* (or skew symmetric) if $Q^T = -Q$, *Hermitian* if $Q^\dagger = Q$, and *antiHermitian* (or skew Hermitian) if $Q^\dagger = -Q$. A *diagonal* matrix is a *square* matrix in which only the *diagonal elements*—those elements q_{ij} for which $i = j$—are different from zero, and a *unit matrix*, often symbolized by I, is a diagonal matrix all of whose diagonal elements are unity. It is readily verified that $IQ = QI = Q$ for any square matrix Q. Finally, the *inverse* Q^{-1} of a *square* matrix Q is that matrix having the property $Q^{-1}Q = QQ^{-1} = I$, and a matrix is said to be *orthogonal* if $Q^{-1} = Q^T$ and *unitary* if $Q^{-1} = Q^\dagger$.

PROBLEMS

PA-6. Verify all the elements in the right-hand side of Eq. (A-11).

PA-7. Let

$$A = \begin{pmatrix} 0 & 1 & 0 \\ 1 & 0 & 1 \\ 0 & 1 & 0 \end{pmatrix}, \quad B = \begin{pmatrix} 3 & 1 & -1 \\ -1 & 2 & 1 \\ 5 & -1 & 1 \end{pmatrix}, \quad \mathbf{x} = \begin{pmatrix} -1 \\ 2 \\ 5 \end{pmatrix}, \quad \mathbf{y} = \begin{pmatrix} 1 \\ -1 \\ 2 \end{pmatrix}$$

and evaluate $\mathbf{x}^T\mathbf{y}$, $\mathbf{y}\mathbf{x}^T$, $B\mathbf{x}$, and $A(A + B)$.

PA-8. The trace $\text{tr}(Q)$ of a square matrix Q is defined to be the sum of the diagonal elements of Q. Let A and B be square matrices of the same dimension and show that $\text{tr}(AB) = \text{tr}(BA)$, even if $AB \neq BA$.

PA-9. Show that the inverse of a diagonal matrix D is itself a diagonal matrix whose diagonal elements are the reciprocals of the diagonal elements of D.

PA-10. Given that $\det(QR) = [\det(Q)]\,[\det(R)]$, where Q and R are $n \times n$ matrices and $\det(\cdots)$ denotes the determinant, show that (a) $\det(sQ) = s^n \det(Q)$, (*b*) $\det(Q^T) = \det(Q)$, (c) $\det(Q^{-1}) = 1/\det(Q)$, and (d) $\det(Q) = \pm 1$ *if Q is orthogonal.*

PA-11. (a) Show that Eq. (A-1) can be written in the form $A\mathbf{x} = \mathbf{b}$, where

$$A = \begin{pmatrix} a_{11} & a_{12} \\ a_{21} & a_{22} \end{pmatrix}, \quad \mathbf{x} = \begin{pmatrix} x_1 \\ x_2 \end{pmatrix}, \quad \mathbf{b} = \begin{pmatrix} b_1 \\ b_2 \end{pmatrix}$$

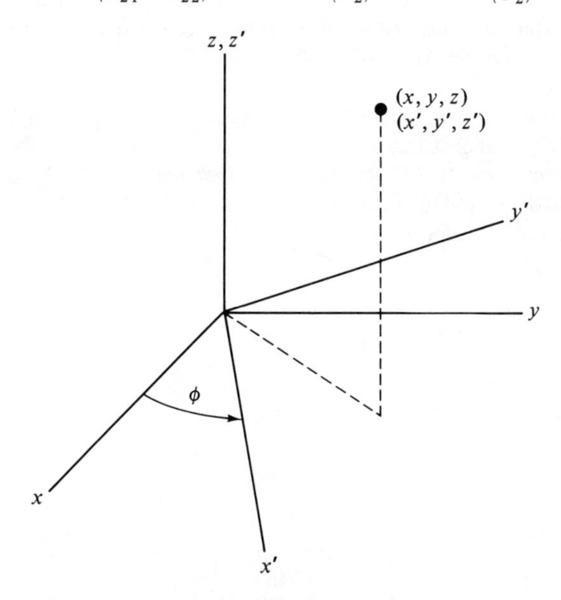

Figure PA-12

(b) Multiplication of this form by A^{-1} from the left gives the solution $\mathbf{x} = A^{-1}\mathbf{b}$. Infer the 2×2 matrix A^{-1} from Eq. (A-2). (c) Can you infer a general rule for finding the inverse of a larger (square) matrix?

PA-12. Let a primed Cartesian coordinate system be obtained from an unprimed system by rigid rotation through an angle ϕ about the z axis (Fig. PA-12). Show that the coordinates of a (fixed) point in space in these two systems are related by

$$\begin{pmatrix} x' \\ y' \\ z' \end{pmatrix} = \begin{pmatrix} \cos\phi & \sin\phi & 0 \\ -\sin\phi & \cos\phi & 0 \\ 0 & 0 & 1 \end{pmatrix} \begin{pmatrix} x \\ y \\ z \end{pmatrix}$$

and verify that the matrix expressing this rigid rotation is orthogonal.

PA-13. (a) Convince yourself that the result of multiplying an n-component column vector $\mathbf{x}$ by an $n \times n$ matrix Q is another n-component column vector $\mathbf{y}$, $Q\mathbf{x} = \mathbf{y}$. (b) For many matrices Q there exist characteristic vectors, called *eigenvectors* of Q. For the ith eigenvector $\mathbf{x}_i$, $Q\mathbf{x}_i$ is merely a (particular) multiple λ_i of $\mathbf{x}_i$, i.e., $Q\mathbf{x}_i = \lambda_i\mathbf{x}_i$; λ_i is called an *eigenvalue* of Q. Show that an eigenvector can be found only if λ_i is one of the values of λ satisfying $\det(Q - \lambda I) = 0$, where $\det(\cdots)$ denotes a determinant. (c) Find the three eigenvalues and the corresponding eigenvectors of the matrix A in PA-7.

REFERENCES

A. C. AITKEN, *Determinants and Matrices* (Oliver & Boyd Ltd., Edinburgh, 1956).

G. ARFKEN, *Mathematical Methods for Physicists* (Academic Press, Inc., New York, 1966), Chapter 4.

M. L. BOAS, *Mathematical Methods in the Physical Sciences* (John Wiley & Sons, Inc., New York, 1966), Chapter 3.

I. S. SOKOLNIKOFF and R. M. REDHEFFER, *Mathematics of Physics and Modern Engineering* (McGraw-Hill Book Company, New York, 1966), Second Edition, Appendix A and Chapter 4.

B

Binomial and Taylor Expansions

Very often in physics we encounter a complicated function of some variable, say x, but are really interested only in the behavior of this function for small values of x or perhaps for values of x deviating by some small amount from a specific value a. In such cases, an expansion of the function in powers of x or of $x - a$ coupled with neglect of all but the earliest few terms in this expansion may simplify the complicated function without sacrificing any of its essential characteristics. This appendix contains a brief statement of two techniques for obtaining such expansions. More detailed discussions may be found in the references at the end of the appendix.

The *binomial theorem* states that the binomial $(1 + x)^n$ has the expansion

$$(1 + x)^n = 1 + nx + \frac{n(n-1)}{2!}x^2 + \cdots + \frac{n(n-1)(n-2)\cdots(n-m+1)}{m!}x^m + \cdots \qquad \text{(B-1)}$$

where $m!$, read m factorial, stands for the product of all positive integers up to and including m. Thus, we have that

$$\frac{1}{1+f} = (1+f)^{-1} = 1 - f + \frac{(-1)(-2)}{2!}f^2 + O(f^3) = 1 - f + f^2 + O(f^3) \qquad \text{(B-2)}$$

$$(1 - y^2)^{-3/2} = 1 + \left(-\frac{3}{2}\right)(-y^2) + \frac{1}{2!}\left(-\frac{3}{2}\right)\left(-\frac{5}{2}\right)(-y^2)^2 + O(y^6)$$
$$= 1 + \frac{3}{2}y^2 + \frac{15}{8}y^4 + O(y^6) \tag{B-3}$$

and so on. Here, the notation $O(f^3)$, for example, indicates that the omitted terms contain powers of f with exponents no smaller than 3. In these examples, infinite series have resulted and the convergence of the binomial series should therefore be examined. If n is a positive integer, the expansion in Eq. (B-1) ends with the term x^n (why?), and the series certainly converges for all x. More generally, when this expansion does not terminate, it converges absolutely if $|x| < 1$, diverges if $|x| > 1$, and may either converge or diverge when $x = \pm 1$ depending on the value of n. In the present context, the variable x is restricted by the condition $|x| \ll 1$ and the series in Eq. (B-1) can usually be truncated after the first few terms without significant loss of accuracy. The number of terms one must retain to provide a given accuracy must be determined individually for each case that arises.

The *Taylor expansion* of a function $f(x)$ of a single variable x about the point a is given by

$$f(x) = \sum_{n=0}^{\infty} \frac{1}{n!} \left.\frac{d^n f}{dx^n}\right|_{x=a} (x - a)^n$$
$$= f(a) + f'(a)(x - a) + \tfrac{1}{2}f''(a)(x - a)^2 + \cdots \tag{B-4}$$

(with the understanding that $0! = 1$); the Taylor expansion of a function $f(x, y)$ of two variables x and y about the point (a, b) begins with the terms

$$f(x, y) = f(a, b) + f_x(a, b)(x - a) + f_y(a, b)(y - b)$$
$$+ \tfrac{1}{2}f_{xx}(a, b)(x - a)^2 + f_{xy}(a, b)(x - a)(y - b)$$
$$+ \tfrac{1}{2}f_{yy}(a, b)(y - b)^2 + \cdots \tag{B-5}$$

In these equations, $f'(a)$ and $f''(a)$ stand for the first and second derivatives of $f(x)$ evaluated at $x = a$, $f_x(a, b)$ stands for $\partial f(x, y)/\partial x$ evaluated at $(x, y) = (a, b)$, etc. When $|x - a|$ and $|y - b|$ are sufficiently small, these series can often be truncated after a very few terms without significant loss of accuracy. As with the binomial series, however, the number of terms to be retained in a given Taylor expansion must be separately determined for each case that arises.

PROBLEMS

PB-1. Generalize Eq. (B-1) to obtain an expansion for $(a + b)^n$.

PB-2. Use mathematical induction to *prove* the binomial theorem for n a positive integer. That is, (1) show that Eq. (B-1) is true for $n = 1$ and (2) assume that Eq. (B-1) is correct for $n = N$ and prove that it is correct for $n = N + 1$.

PB-3. Use the binomial theorem to obtain the first three nonvanishing terms in the expansion of the quantity

$$mc^2\left(\frac{1}{\sqrt{1-(v/c)^2}}-1\right)$$

in powers of $(v/c)^2$. This quantity, in which c is the speed of light, is the relativistic expression for the kinetic energy of a particle of rest mass m moving with speed v.

PB-4. Within what range must f be confined if the approximation $(1+f)^{-1} \approx (1-f)$ is to have an error no larger than 1%?

PB-5. The coefficients $P_n(t)$ in the expansion

$$\frac{1}{\sqrt{1-2ut+u^2}} = \sum_{n=0}^{\infty} P_n(t)u^n$$

are called the *Legendre polynomials.* Identify x in Eq. (B-1) with $(-2ut+u^2)$ and expand the quantity on the left to find the first four Legendre polynomials.

PB-6. Obtain the first three nonvanishing terms in the Taylor expansion about the origin ($a=0$) of $\sin x$, e^x, and $(1+x)^q$. The last result is, of course, the binomial expansion. When done rigorously, this approach *proves* the binomial theorem even if q is not a positive integer.

REFERENCES

L. Brand, *Advanced Calculus* (John Wiley & Sons, Inc., New York, 1955), Sections 39, 65, and 91.

W. Kaplan, *Advanced Calculus* (Addison-Wesley Publishing Company, Inc., Reading, Mass., 1952), Sections 6-16, 6-17, and 6-21.

J. M. H. Olmsted, *Advanced Calculus* (Appleton-Century-Crofts, New York, 1961), Sections 1203 and 1215.

A. E. Taylor and W. R. Mann, *Advanced Calculus* (Xerox College Publishing, Lexington, Mass., 1972), Second Edition, especially Section 7.5.

C

Vector Identities and Relationships

In this appendix, **A**, **B**, **C**, **D**, **Q**, and **R** represent vector fields, Φ and Ψ represent scalar fields, **r** represents the position vector, and $\hat{\mathbf{e}}_i$ is a *Cartesian* unit vector.

$$\mathbf{A} \times (\mathbf{B} \times \mathbf{C}) = (\mathbf{A}\cdot\mathbf{C})\mathbf{B} - (\mathbf{A}\cdot\mathbf{B})\mathbf{C} \tag{C-1}$$

$$(\mathbf{A} \times \mathbf{B}) \times \mathbf{C} = (\mathbf{A}\cdot\mathbf{C})\mathbf{B} - (\mathbf{B}\cdot\mathbf{C})\mathbf{A} \tag{C-2}$$

$$\mathbf{A}\cdot(\mathbf{B} \times \mathbf{C}) = (\mathbf{A} \times \mathbf{B})\cdot\mathbf{C} \tag{C-3}$$

$$(\mathbf{A} \times \mathbf{B})\cdot(\mathbf{C} \times \mathbf{D}) = \begin{vmatrix} \mathbf{A}\cdot\mathbf{C} & \mathbf{A}\cdot\mathbf{D} \\ \mathbf{B}\cdot\mathbf{C} & \mathbf{B}\cdot\mathbf{D} \end{vmatrix} \tag{C-4}$$

$$\nabla(\Psi + \Phi) = \nabla\Psi + \nabla\Phi \tag{C-5}$$

$$\nabla(\Psi\Phi) = \Phi\,\nabla\Psi + \Psi\,\nabla\Phi \tag{C-6}$$

$$\nabla \times (\mathbf{Q} + \mathbf{R}) = \nabla \times \mathbf{Q} + \nabla \times \mathbf{R} \tag{C-7}$$

$$\nabla \times (\Phi\mathbf{Q}) = \nabla\Phi \times \mathbf{Q} + \Phi\,\nabla \times \mathbf{Q} \tag{C-8}$$

$$\nabla \times \mathbf{r} = 0 \tag{C-9}$$

$$\nabla\cdot(\mathbf{Q} + \mathbf{R}) = \nabla\cdot\mathbf{Q} + \nabla\cdot\mathbf{R} \tag{C-10}$$

$$\nabla\cdot(\Phi\mathbf{Q}) = \nabla\Phi\cdot\mathbf{Q} + \Phi\,\nabla\cdot\mathbf{Q} \tag{C-11}$$

$$\nabla\cdot(\mathbf{Q} \times \mathbf{R}) = \mathbf{R}\cdot(\nabla \times \mathbf{Q}) - \mathbf{Q}\cdot(\nabla \times \mathbf{R}) \tag{C-12}$$

$$\nabla\cdot\mathbf{r} = 3 \tag{C-13}$$

$$\nabla\cdot(\mathbf{r}/r^3) = 0, \qquad r \neq 0 \tag{C-14}$$

$$\nabla(\mathbf{Q}\cdot\mathbf{R}) = (\mathbf{Q}\cdot\nabla)\mathbf{R} + (\mathbf{R}\cdot\nabla)\mathbf{Q} + \mathbf{Q} \times (\nabla \times \mathbf{R}) + \mathbf{R} \times (\nabla \times \mathbf{Q}) \tag{C-15}$$

$$\nabla \times (\mathbf{Q} \times \mathbf{R}) = \mathbf{Q}(\nabla\cdot\mathbf{R}) - \mathbf{R}(\nabla\cdot\mathbf{Q}) + (\mathbf{R}\cdot\nabla)\mathbf{Q} - (\mathbf{Q}\cdot\nabla)\mathbf{R} \tag{C-16}$$

$$\nabla \times (\nabla\Phi) = 0 \tag{C-17}$$

$$\nabla \cdot (\nabla \times \mathbf{Q}) = 0 \tag{C-18}$$

$$\nabla \times (\nabla \times \mathbf{Q}) = \nabla(\nabla \cdot \mathbf{Q}) - \nabla^2 \mathbf{Q} \tag{C-19}$$

$$\oint \Phi \, d\boldsymbol{\ell} = \int d\mathbf{S} \times \nabla \Phi \tag*{(C-20)*}$$

$$\oint \Phi \, d\mathbf{S} = \int \nabla \Phi \, dv \tag*{(C-21)†}$$

$$\oint \mathbf{Q} \cdot d\boldsymbol{\ell} = \int \nabla \times \mathbf{Q} \cdot d\mathbf{S} \qquad \text{(Stokes' theorem)} \tag*{(C-22)*}$$

$$\oint \mathbf{Q} \cdot d\mathbf{S} = \int \nabla \cdot \mathbf{Q} \, dv \qquad \text{(divergence theorem)} \tag*{(C-23)†}$$

$$\oint \mathbf{Q} \times d\boldsymbol{\ell} = \int (\nabla \cdot \mathbf{Q}) \, d\mathbf{S} - \sum_i \hat{\mathbf{e}}_i \int [(\hat{\mathbf{e}}_i \cdot \nabla)\mathbf{Q}] \cdot d\mathbf{S} \qquad \text{(Cartesian coordinates only)} \tag*{(C-24)*}$$

$$\oint d\mathbf{S} \times \mathbf{Q} = \int \nabla \times \mathbf{Q} \, dv \tag*{(C-25)†}$$

$$\oint \mathbf{Q}(\mathbf{R} \cdot d\mathbf{S}) = \int [\mathbf{Q}(\nabla \cdot \mathbf{R}) + (\mathbf{R} \cdot \nabla)\mathbf{Q}] \, dv \tag*{(C-26)†}$$

$$\oint (\Phi \, \nabla \Psi - \Psi \, \nabla \Phi) \cdot d\mathbf{S} = \int (\Phi \, \nabla^2 \Psi - \Psi \, \nabla^2 \Phi) \, dv \qquad \text{(Green's theorem)} \tag*{(C-27)†}$$

*In all integral relationships involving a line integral about a *closed* path, $d\mathbf{S}$ by convention has the direction determined by the thumb of the right hand when the fingers point in the direction of $d\boldsymbol{\ell}$ and the palm faces the area enclosed by the path.

†In all integral relationships involving a surface integral over a *closed* surface, $d\mathbf{S}$ by convention has the direction of the outward normal.

D

Complex Numbers and Fourier Analysis

A *complex number* z—not to be confused with the z coordinate of a point in three-dimensional space—is an ordered pair of real numbers (x, y), x being called the *real* part of z, denoted Re z, and y being called the *imaginary* part of z, denoted Im z, i.e.,

$$z = (x, y) \longleftrightarrow \text{Re } z = x, \qquad \text{Im } z = y \tag{D-1}$$

In this appendix, we shall summarize the algebraic properties that define complex numbers more fully and then very briefly develop those few applications that we need in the main part of this book. More detailed discussions may be found in the references at the end of the appendix.

D-1 The Algebra of Complex Numbers

We adopt the point of view that we are in effect *inventing* complex numbers. Thus, we may assign to them whatever algebraic properties we wish. Without attempting here to motivate our particular choice, we shall assign the following properties:

(1) *Equality*. Two complex numbers $z_1 = (x_1, y_1)$ and $z_2 = (x_2, y_2)$ are defined to be equal if and only if the real and imaginary parts are separately equal; i.e., $z_1 = z_2$ if and only if $x_1 = x_2$ *and* $y_1 = y_2$.

(2) *Addition.* A complex number z is defined to be the sum of two other complex numbers z_1 and z_2 if and only if the real and imaginary parts of z are separately the sums of the real and imaginary parts of z_1 and z_2; i.e., $z = z_1 + z_2$ if and only if $x = x_1 + x_2$ *and* $y = y_1 + y_2$. It then follows from the corresponding properties of real numbers that addition of complex numbers is commutative, $z_1 + z_2 = z_2 + z_1$, and associative, $z_1 + (z_2 + z_3) = (z_1 + z_2) + z_3$. By this definition, the complex number $(0, 0)$ has the property that $(0, 0) + z_1 = z_1$ and is therefore the identity element for addition. Finally, if we define $-z$ by requiring that $z + (-z) = (0, 0)$, we find that $-z = (-x, -y)$ and the definition $z_1 - z_2 = z_1 + (-z_2)$ for the *difference* between two complex numbers is unambiguous.

(3) *Multiplication.* A complex number z is defined to be the product of two other complex numbers z_1 and z_2, $z = z_1 z_2$, if and only if $x = (x_1 x_2 - y_1 y_2)$ *and* $y = y_1 x_2 + x_1 y_2$. It then follows (PD-1) that multiplication of complex numbers is commutative, $z_1 z_2 = z_2 z_1$; associative, $z_1(z_2 z_3) = (z_1 z_2)z_3$; and distributive with respect to addition, $z_1(z_2 + z_3) = z_1 z_2 + z_1 z_3$. By this definition, the complex number $(1, 0)$ has the property that $(1, 0)\cdot z = z\cdot(1, 0) = z$ and is therefore the identity element for multiplication. Finally, if we define the *reciprocal* $z_r = 1/z = (x_r, y_r)$ of the complex number z by requiring that $zz_r = (x, y)\cdot(x_r, y_r) = (1, 0)$, we find by applying the definition of multiplication that $xx_r - yy_r = 1$ and $xy_r + yx_r = 0$ or, on solution for x_r and y_r, that $x_r = x/(x^2 + y^2)$ and $y_r = -y/(x^2 + y^2)$. Thus, the definition $z_1/z_2 = z_1\cdot(1/z_2)$ for the *quotient* of two complex numbers is now unambiguous.

It follows from the definitions of the previous paragraphs that complex numbers of the form $(x, 0)$ with zero imaginary part have the same algebraic properties as the real numbers x. We are therefore justified in putting $(x, 0)$ into correspondence with x and indeed in denoting $(x, 0)$ more simply by x. The complex numbers thus include the real numbers, but they also include more general numbers. For example, the above rule for multiplication gives $(0, 1)^2 = (-1, 0) = -1$, a property possessed by no real number. The number $(0, 1)$ is usually denoted by i and is called the imaginary unit; it has the curious property that $i^2 = -1$. Further, by the above rules, we can write

$$
\begin{aligned}
z = (x, y) &= (x, 0)(1, 0) + (0, 1)(y, 0) \\
&= x + iy \qquad \text{(D-2)}
\end{aligned}
$$

and the complex number can be viewed as a (complex) sum of two parts, the second containing the imaginary unit i as a factor. Further, all of the above rules for algebraic manipulation of complex numbers can be summarized in the statement that complex numbers written in the form of Eq. (D-2) can be treated as ordinary binomials except that i^2 can be replaced by -1 and equality between two such binomials implies separate equality of the real and imaginary parts. Thus, for example,

$$(x_1 + iy_1) + (x_2 + iy_2) = (x_1 + x_2) + i(y_1 + y_2) \tag{D-3}$$

$$(x_1 + iy_1)(x_2 + iy_2) = (x_1x_2 - y_1y_2) + i(y_1x_2 + x_1y_2) \tag{D-4}$$

$$x_1 + iy_1 = x_2 + iy_2 \Longleftrightarrow x_1 = x_2 \quad and \quad y_1 = y_2 \tag{D-5}$$

Division of two numbers written in the form of Eq. (D-2) is facilitated by introducing the complex conjugate z^* of z, obtained by making the replacement $i \longrightarrow -i$, i.e.,

$$z = x + iy \Longrightarrow z^* = x - iy \tag{D-6}$$

and then noting that

$$zz^* = x^2 + y^2 \tag{D-7}$$

is a *real* number. Thus, the sequence of operations

$$\frac{z_1}{z_2} = \frac{z_1}{z_2} \cdot \frac{z_2^*}{z_2^*} = \frac{(x_1 + iy_1)(x_2 - iy_2)}{x_2^2 + y_2^2} \tag{D-8}$$

$$= \frac{x_1x_2 + y_1y_2}{x_2^2 + y_2^2} + i\frac{y_1x_2 - x_1y_2}{x_2^2 + y_2^2} \tag{D-9}$$

reduces the quotient z_1/z_2 to the form $a + ib$ with a and b real and thus verifies that the quotient of two complex numbers is itself a complex number. Finally, in terms of z and z^*, we can now derive the relationships

$$x = \text{Re}\, z = \frac{1}{2}(z + z^*), \qquad y = \text{Im}\, z = \frac{1}{2i}(z - z^*) \tag{D-10}$$

for the real and imaginary parts (PD-7).[1]

A graphical meaning for the complex number $z = x + iy$ emerges if we represent z as the point (x, y) in the *complex z plane* defined by a horizontal *real axis*, along which x is plotted, and a vertical *imaginary axis*, along which y is plotted (Fig. D-1). Equivalently, we can think of z as a *vector* in this

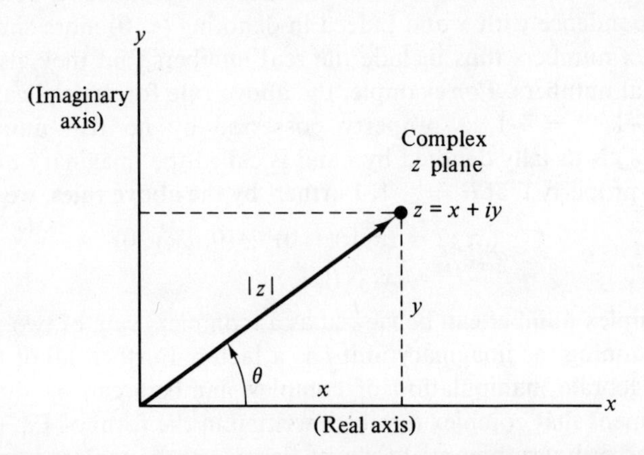

Fig. D-1. A graphical representation of the complex number z.

[1]Note incidentally that the imaginary part of z is the *coefficient* of i; *it does not include the i.*

plane, this vector having horizontal and vertical components x and y, having length $|z|$, usually called the *magnitude* or the *absolute value* of z, and making an angle θ, usually called the *argument* or *phase* of z, with the positive real axis. Analytically,

$$|z| = \sqrt{x^2 + y^2}, \qquad |z|^2 = zz^* \tag{D-11}$$

and

$$\theta = \arg(z) = \tan^{-1}\frac{y}{x} \tag{D-12}$$

A complex number can thus be written either in the *Cartesian* form of Eq. (D-2) or in the *polar* form,

$$z = |z|(\cos\theta + i\sin\theta) \tag{D-13}$$

where the complex number $\cos\theta + i\sin\theta$, represents a vector of *unit* magnitude at an angle θ to the real axis. In this geometric representation, addition of complex numbers is equivalent to vector addition in the complex plane.

To express Eq. (D-13) in a still more compact form, we digress briefly to discuss how functions of z, now regarded as a complex *variable*, might be defined. Since multiplication is well defined, the integer power z^n is defined, and polynomial functions of z, such as

$$f(z) = (1 + 2i)z^5 - 3iz^2 + 6 + \pi i \tag{D-14}$$

are meaningful. Without raising here the question of convergence, we also know at least intuitively what is meant by an *infinite* power series, and it is natural to *define* such functions as $\sin z$, $\cos z$, and e^z by replacing the real variable in the usual power series (PB-6) by the complex variable z, e.g.,

$$e^z = \sum_{n=0}^{\infty}\frac{z^n}{n!} \tag{D-15}$$

With this definition, let us now examine $e^{i\theta}$; we find that

$$\begin{aligned} e^{i\theta} &= \sum_{n=0}^{\infty}\frac{(i\theta)^n}{n!} = 1 + i\theta - \frac{\theta^2}{2!} - i\frac{\theta^3}{3!} + \cdots \\ &= \left[1 - \frac{\theta^2}{2!} + \cdots\right] + i\left[\theta - \frac{\theta^3}{3!} + \cdots\right] \end{aligned} \tag{D-16}$$

The series emerging in square brackets, however, are the series for $\cos\theta$ and $\sin\theta$, respectively. Thus, we arrive at the *Euler formula*,

$$e^{i\theta} = \cos\theta + i\sin\theta \tag{D-17}$$

which in particular makes the real and imaginary parts of $e^{i\theta}$ explicit when θ is real. The polar form of the complex number z [Eq. (D-13)] can therefore also be written as

$$z = |z|e^{i\theta} \tag{D-18}$$

which is the compact expression sought in this paragraph.

PROBLEMS

PD-1. Working directly with the definition in item (3) above, show that $z_1z_2 = z_2z_1$ and that $z_1(z_2z_3) = (z_1z_2)z_3$.

PD-2. Let $z_1 = 3 + 4i$ and $z_2 = 1 - i$. (a) Reduce z_1z_2 and z_2/z_1 to the form $a + bi$. (b) Write z_2 in polar form.

PD-3. Show that $z = 1 + i$ satisfies $z^2 - 2z + 2 = 0$.

PD-4. Show that $(z_1 + z_2)^* = z_1^* + z_2^*$ and $(z_1z_2)^* = z_1^*z_2^*$.

PD-5. Show that if $z_1z_2 = 0$, either z_1 or z_2 *must* be zero. (Both, of course, *may* be zero.)

PD-6. Prove the *triangle inequality*, $|z_1 + z_2| \leq |z_1| + |z_2|$. Under what conditions does the equality hold?

PD-7. Prove Eq. (D-10).

PD-8. Geometrically, how is z^* positioned in Fig. D-1?

PD-9. Show that $\arg(z_1z_2) = \arg(z_1) + \arg(z_2)$ and $|z_1z_2| = |z_1||z_2|$.

PD-10. Since $e^{in\theta} = (e^{i\theta})^n$, Eq. (D-17) implies that

$$\cos n\theta + i \sin n\theta = (\cos \theta + i \sin \theta)^n$$

for any n. Let $n = 3$ and extract expressions for $\cos 3\theta$ and $\sin 3\theta$ in terms of $\cos \theta$ and $\sin \theta$.

PD-11. By its definition in Eq. (D-15), the exponential function of a complex argument has all the algebraic properties of the exponential function of a real argument. In particular, $e^{z_1}e^{z_2} = e^{z_1+z_2}$. (a) Show that $e^z = e^x(\cos y + i \sin y)$. (b) Find the real and imaginary parts of $\sinh z$, which is defined by $\frac{1}{2}(e^z - e^{-z})$.

PD-12. (a) Given the Euler formula, Eq. (D-17), show that

$$\cos \theta = \frac{e^{i\theta} + e^{-i\theta}}{2} \qquad \sin \theta = \frac{e^{i\theta} - e^{-i\theta}}{2i}$$

(b) Given the definitions

$$\cosh z = \frac{e^z + e^{-z}}{2} \qquad \sinh z = \frac{e^z - e^{-z}}{2}$$

show that

$$\cosh (i\theta) = \cos \theta \qquad \sinh (i\theta) = i \sin \theta$$

$$\cos (iz) = \cosh z \qquad \sin (iz) = i \sinh z$$

and that

$$\cos (x + iy) = \cos x \cosh y - i \sin x \sinh y$$

$$\sin (x + iy) = \sin x \cosh y + i \cos x \sinh y$$

thereby displaying the real and imaginary parts of $\cos (x + iy)$ and $\sin (x + iy)$ when x and y are real.

D-2
Fourier Series

Let $f(x)$ be a periodic function with period $2L$. Since $\cos(n\pi x/L)$ and $\sin(n\pi x/L)$ are also periodic with period $2L$ for any integer n, it is reasonable to expect that $f(x)$ can be expanded in a trigonometric series. We choose to separate the term for $n = 0$ and to write a trial series, called a *Fourier series*, in the form

$$f(x) = \frac{a_0}{2} + \sum_{n=1}^{\infty} \left[a_n \cos \frac{n\pi x}{L} + b_n \sin \frac{n\pi x}{L} \right] \tag{D-19}$$

To determine the constant coefficients in this series, we first note the identities

$$\begin{aligned} &\int_{-L}^{L} \sin \frac{n\pi x}{L} \cos \frac{m\pi x}{L}\, dx = 0 \\ &\int_{-L}^{L} \sin \frac{n\pi x}{L} \sin \frac{m\pi x}{L}\, dx = \int_{-L}^{L} \cos \frac{n\pi x}{L} \cos \frac{m\pi x}{L}\, dx = L\delta_{nm} \end{aligned} \tag{D-20}$$

where $\delta_{nm} = 1$ if $n = m$ and $\delta_{nm} = 0$ if $n \neq m$. These identities apply when $n > 0$, $m > 0$ and can be verified by direct integration (PD-13). Now, assume that term-by-term multiplication and integration of the series in Eq. (D-19) is possible, multiply every term, for example, by $\sin(n\pi x/L)$ and integrate the result term by term over the interval $-L < x < L$. Because of the identities in Eq. (D-20), only one term—that involving b_m—of the entire series survives, and that term gives

$$b_m = \frac{1}{L} \int_{-L}^{L} f(x) \sin \frac{m\pi x}{L}\, dx, \quad m = 1, 2, \ldots \tag{D-21}$$

By similar arguments, we find that

$$a_m = \frac{1}{L} \int_{-L}^{L} f(x) \cos \frac{m\pi x}{L}\, dx, \quad m = 0, 1, 2, \ldots \tag{D-22}$$

Although a_0 must be treated separately from the rest of the a's, Eq. (D-22) nonetheless applies to all of the a's [which explains why the factor of $\frac{1}{2}$ was explicitly inserted with a_0 in Eq. (D-19)]. At least formally, the series in Eq. (D-19) with coefficients given by Eqs. (D-21) and (D-22) ought to converge to $f(x)$. Such is in fact the case, as is stated by the following theorem, whose detailed proof we leave to the references: *In $-L \leq x < L$, let $f(x)$ be defined and bounded. Further let $f(x)$ have no more than a finite number of maxima and minima and only a finite number of discontinuities. Finally let $f(x)$ be defined outside the interval $-L \leq x < L$ by the requirement of periodicity, $f(x + 2L) = f(x)$. Then the Fourier series, Eq. (D-19), with coefficients given by Eqs. (D-21) and (D-22) converges at every x to $\frac{1}{2}[f(x + \epsilon) + f(x - \epsilon)]$, where ϵ is positive and arbitrarily small, and in particular converges to $f(x)$ at* *points where $f(x)$ is continuous.* Essentially every periodic function of interest in physical problems satisfies these conditions.

Equation (D-19) can be written in a complex form if the trigonometric functions in it are reexpressed using the results in PD-12. After combining terms, we find that

$$f(x) = \frac{a_0}{2} + \sum_{n=1}^{\infty}\left[\frac{a_n - ib_n}{2}e^{i(n\pi x/L)} + \frac{a_n + ib_n}{2}e^{-i(n\pi x/L)}\right] \tag{D-23}$$

Thus, if we introduce a new set of coefficients defined by

$$\begin{aligned} c_0 &= \tfrac{1}{2}a_0 \\ c_n &= \tfrac{1}{2}(a_n - ib_n), \qquad n > 0 \\ c_n &= \tfrac{1}{2}(a_{-n} + ib_{-n}), \quad n < 0 \end{aligned} \tag{D-24}$$

Eq. (D-23) can be expressed more compactly as

$$f(x) = \sum_{n=-\infty}^{\infty} c_n e^{i(n\pi x/L)} \tag{D-25}$$

Further, we find by combining Eqs. (D-21) and (D-22) with Eq. (D-24) that

$$c_m = \frac{1}{2L}\int_{-L}^{L} f(x)e^{-i(m\pi x/L)}\,dx \tag{D-26}$$

for all m, positive, negative, and zero. Although our development does not show it, this exponential form of the Fourier series represents $f(x)$ correctly even if $f(x)$ happens to be a complex function of x.

PROBLEMS

PD-13. Verify the identities in Eq. (D-20). *Hint:* Express the sines and cosines in exponential form using the results in PD-12.

PD-14. Find the coefficients in the Fourier series Eq. (D-19) for the function $f(x)$ defined by $f(x) = x$ in $-\pi \leq x < \pi$. Sketch a graph of the function to which the series converges in $-5\pi < x < 5\pi$. To what value does the series converge at $x = \pi$?

PD-15. Suppose that $f(x)$ is an *even* function of x in the interval $-L < x < L$. Show that $b_n = 0$ for all n and hence that

$$f(x) = \frac{a_0}{2} + \sum_{n=1}^{\infty} a_n \cos\frac{n\pi x}{L}; \qquad a_n = \frac{2}{L}\int_0^L f(x)\cos\frac{n\pi x}{L}\,dx \tag{1}$$

Comment: If $f(x)$ is an *odd* function of x in the interval $-L < x < L$, then $a_n = 0$ for all n and

$$f(x) = \sum_{n=1}^{\infty} b_n \sin\frac{n\pi x}{L}; \qquad b_n = \frac{2}{L}\int_0^L f(x)\sin\frac{n\pi x}{L}\,dx \tag{2}$$

More usefully if a series representing $f(x)$ only in the half-interval $0 < x < L$ is needed, $f(x)$ can be extended into the interval $-L < x < 0$ so as to be *either* even *or* odd. Thus, when only the *half*-interval is of interest, any function can be expanded either in a *Fourier cosine series* by Eq. (1) or in a *Fourier sine series* by Eq. (2).

D-3 Fourier Transforms

To represent functions that are not periodic, we might imagine letting the interval $-L < x < L$ of Section D-2 expand to encompass all x; i.e., we let $L \longrightarrow \infty$. To obtain expressions valid in that limit, we combine Eqs. (D-25) and (D-26) into the single expression

$$f(x) = \sum_{n=-\infty}^{\infty} \frac{1}{2L}\left[\int_{-L}^{L} f(x')e^{-i(n\pi x'/L)}\, dx'\right] e^{i(n\pi x/L)} \tag{D-27}$$

where x' has been used for the integration variable to avoid confusion with x. We now introduce the notation, $\kappa_n = \pi n/L$, and $\Delta\kappa = \kappa_{n+1} - \kappa_n = \pi/L$. Then, Eq. (D-27) becomes

$$f(x) = \sum_{n=-\infty}^{\infty} \frac{1}{2\pi}\left[\int_{-L}^{L} f(x')e^{-i\kappa_n x'}\, dx'\right] e^{i\kappa_n x}\, \Delta\kappa \tag{D-28}$$

If we think of the quantity in square brackets temporarily simply as some function of κ, say $g(\kappa)$, evaluated at κ_n, we can interpret the sum on n in Eq. (D-28) as an approximation to the integral of $g(\kappa)$. Thus, as L approaches infinity and $\Delta\kappa$ simultaneously approaches zero, Eq. (D-28) becomes

$$f(x) = \frac{1}{2\pi}\int_{-\infty}^{\infty}\left[\int_{-\infty}^{\infty} f(x')e^{-i\kappa x'}\, dx'\right] e^{i\kappa x}\, d\kappa \tag{D-29}$$

which is one form of the *Fourier integral theorem.* It can be split into two pieces by viewing the integral in square brackets as an integral transform $\tilde{f}(\kappa)$, called the *Fourier* transform, of $f(x)$,

$$\tilde{f}(\kappa) = \int_{-\infty}^{\infty} f(x')e^{-i\kappa x'}\, dx' \tag{D-30}$$

The rest of Eq. (D-29) is then viewed as an *inversion formula,*

$$f(x) = \int_{-\infty}^{\infty} \tilde{f}(\kappa)e^{i\kappa x}\frac{d\kappa}{2\pi} \tag{D-31}$$

In effect, Eq. (D-30) defines a transform of some function and Eq. (D-31) permits recovery of the original function if its transform is known. We can also view Eq. (D-31) to express $f(x)$ as a superposition of elementary functions $e^{i\kappa x}$; in that view, $\tilde{f}(\kappa)$ determines how much of each elementary function is present. Note finally that in separating Eq. (D-29) into two pieces, different authors may elect to distribute the factor of 2π differently between Eqs. (D-30) and (D-31); the symmetry between the transform and its inversion is, for example, enhanced if the factor $1/\sqrt{2\pi}$ multiplies each equation. We nonetheless elect the forms given above.

Two more points: (1) Fourier transforms and inverses in two or more dimensions also occur. By direct analogy with Eqs. (D-30) and (D-31), we infer, for example, in three dimensions that

$$\tilde{f}(\boldsymbol{\kappa}) = \iiint f(\mathbf{r}')e^{-i\boldsymbol{\kappa}\cdot\mathbf{r}'}\,dx'\,dy'\,dz' \tag{D-32}$$

and then that

$$f(\mathbf{r}) = \iiint \tilde{f}(\boldsymbol{\kappa})e^{i\boldsymbol{\kappa}\cdot\mathbf{r}}\frac{d\kappa_x\,d\kappa_y\,d\kappa_z}{(2\pi)^3} \tag{D-33}$$

where $\boldsymbol{\kappa} = \kappa_x\hat{\mathbf{i}} + \kappa_y\hat{\mathbf{j}} + \kappa_z\hat{\mathbf{k}}$ and $\boldsymbol{\kappa}\cdot\mathbf{r} = \kappa_x x + \kappa_y y + \kappa_z z$. (2) *Parseval's theorem*, which relates the function and its transform by

$$\int_{-\infty}^{\infty} |f(x)|^2\,dx = \frac{1}{2\pi}\int_{-\infty}^{\infty} |\tilde{f}(\kappa)|^2\,d\kappa \tag{D-34}$$

is proved in PD-16.

PROBLEMS

PD-16. Prove Parseval's theorem, Eq. (D-34). *Hint:* Write $\int |f(x)|^2\,dx$ as $\int f(x)f(x)^*\,dx$, substitute the complex conjugate of Eq. (D-31) for $f(x)^*$, interchange orders of integration, and interpret the resulting inner integral by Eq. (D-30).

PD-17. Let $f(x) = 1$, $-a \le x \le a$, and $f(x) = 0$ outside this range. (a) Find the Fourier transform $\tilde{f}(\kappa)$ of this function and sketch a graph of $\tilde{f}(\kappa)$ versus κ. (b) Use Parseval's theorem to show that $\int_0^\infty [(\sin^2 \kappa a)/\kappa^2]\,d\kappa = \pi a/2$.

REFERENCES

R. V. Churchill, *Introduction to Complex Variables and Applications* (McGraw-Hill Book Company, New York, 1948).

R. V. Churchill, *Fourier Series and Boundary Value Problems* (McGraw-Hill Book Company, New York, 1963), Second Edition.

I. S. Sokolnikoff and R. M. Redheffer, *Mathematics of Physics and Modern Engineering* (McGraw-Hill Book Company, New York, 1966), Second Edition, Chapter 8.

E

Reference Tables

E-1

Selected Physical Constants

Elementary charge	1.602×10^{-19} C
Electron rest mass	9.109×10^{-31} kg
Proton rest mass	1.672×10^{-27} kg
Neutron rest mass	1.675×10^{-27} kg
Bohr radius	5.292×10^{-11} m
Planck's constant, h	6.626×10^{-34} J·sec
$\hbar = h/2\pi$	1.054×10^{-34} J·sec
Gravitational constant	6.670×10^{-11} N·m²/kg²
Avogadro's number	6.023×10^{-23}
Boltzmann's constant, k	1.380×10^{-23} J/°K
Speed of light, c	2.998×10^{8} m/sec
Permittivity of free space, ϵ_0	8.854×10^{-12} $C^2/N \cdot m^2$
$1/4\pi\epsilon_0$	8.998×10^{9} $N \cdot m^2/C^2$
Permeability of free space, μ_0	$4\pi \times 10^{-7}$ N/A^2
	$(1.257 \times 10^{-6}$ $N/A^2)$

E-2
Abbreviations of Unit Names

A	ampere	J	joule
Å	angstrom (10^{-10} m)	°K	degrees Kelvin
abA	abampere	kg	kilogram
abC	abcoulomb	m	meter
C	coulomb	MeV	10^6 eV
°C	degrees centigrade	mm	millimeter
cm	centimeter	N	newton
dyn	dyne	sec	second
eV	electron volt	statA	statampere
F	farad	statC	statcoulomb
g	gram	T	tesla
G	gauss	V	volt
GeV	10^9 eV	W	watt
H	henry	Wb	weber
Hz	hertz	Ω	ohm

E-3
Dot Products between Unit Vectors

$\downarrow \cdot \rightarrow$	$\hat{\mathbf{e}}_1$	$\hat{\mathbf{e}}_2$	$\hat{\mathbf{e}}_3$
$\hat{\mathbf{e}}_1$	1	0	0
$\hat{\mathbf{e}}_2$	0	1	0
$\hat{\mathbf{e}}_3$	0	0	1

E-4
Cross Products between Unit Vectors

$\downarrow \times \rightarrow$	$\hat{\mathbf{e}}_1$	$\hat{\mathbf{e}}_2$	$\hat{\mathbf{e}}_3$
$\hat{\mathbf{e}}_1$	0	$\hat{\mathbf{e}}_3$	$-\hat{\mathbf{e}}_2$
$\hat{\mathbf{e}}_2$	$-\hat{\mathbf{e}}_3$	0	$\hat{\mathbf{e}}_1$
$\hat{\mathbf{e}}_3$	$\hat{\mathbf{e}}_2$	$-\hat{\mathbf{e}}_1$	0

Answers to Selected Problems

Chapter 0

P0-2. $\mathbf{A}\cdot(\mathbf{B}\times\mathbf{C})$ is the volume of the parallelopiped bounded by **A**, **B**, and **C**.

P0-6. See Tables E-3 and E-4.

P0-8. $\hat{\mathbf{A}} = \frac{1}{7}(3\hat{\mathbf{e}}_1 + 2\hat{\mathbf{e}}_2 - 6\hat{\mathbf{e}}_3)$; $\mathbf{A}\cdot\mathbf{B} = -3$; $\mathbf{A}\times\mathbf{B} = 2\hat{\mathbf{e}}_1 - 9\hat{\mathbf{e}}_2 - 2\hat{\mathbf{e}}_3$; 107.6°.

P0-11. $\mathbf{r}_2 - \mathbf{r}_1$

P0-12. $S(\mathbf{r}) = x^2 + y^2 + z^2 = \imath^2 + z^2 = r^2$

P0-13. $S(\mathbf{r}, \mathbf{r}') = \dfrac{1}{[(x-x')^2 + (y-y')^2 + (z-z')^2]^{1/2}}$

$$\mathbf{Q}(\mathbf{r}, \mathbf{r}') = \frac{(x-x')\hat{\mathbf{i}} + (y-y')\hat{\mathbf{j}} + (z-z')\hat{\mathbf{k}}}{[(x-x')^2 + (y-y')^2 + (z-z')^2]^{3/2}}$$

P0-15. spheres centered at $\mathbf{r}_0$.

P0-18. (a) 1; (b) $\frac{1}{2}$; (c) $\frac{13}{12}$; (d) 1

P0-21. $U(\mathbf{r}) = \dfrac{a}{[x^2 + y^2 + z^2]^{1/2}} = \dfrac{a}{r}$

P0-27. $U(\mathbf{r}) = U(\mathbf{r}_0) - \displaystyle\int_{r_0}^{r} f(r')\,dr'$

Chapter 1

P1-6. (c) 1.60210×10^{-19} C $= 4.80298 \times 10^{-10}$ statC $= 1.60210 \times 10^{-20}$ abC

P1-7. k_3 (modified Gaussian) $= \dfrac{1}{c}$

P1-9. 1.24×10^{36}

P1-10. (c) 55.69°; (d) 2.95×10^{-7} C

P1-11. Maximum fraction of electrons transferred $\approx 7 \times 10^{-14}$

P1-12. $s = \dfrac{\mu_0 I^2 \ell}{2\pi mg \sin\theta}$

Chapter 2

P2-1. (a) $\approx 10^{14}$ years; (b) ≈ 10 light years

P2-3. $\rho = \dfrac{Q}{\frac{4}{3}\pi R^3}$; $\sigma = \dfrac{Q}{4\pi R^2}$; $\lambda = \dfrac{Q}{2\pi R}$

P2-5. 6.25×10^{17} electrons/sec

P2-6. (a) $\omega Q/2\pi$

P2-8. $\mathbf{J}(\mathbf{r}) = \dfrac{qN}{4\pi r^3}\mathbf{r}$

P2-9. $\mathbf{J}(\mathbf{r}) = \dfrac{3\omega Q \imath}{4\pi a^3}\hat{\boldsymbol{\phi}}$

P2-10. $\mathbf{J}\cdot\mathbf{S}$

P2-11. (a) $4\alpha a^2 b$; (b) 0

P2-19. $Q(t) = \frac{1}{2}\pi\alpha a^4 t(1 - e^{-\beta b})$

P2-23. Cookian units: $\nabla\cdot\mathbf{J} + \dfrac{1}{c}\sqrt{\dfrac{K_1}{K_2}}\dfrac{\partial\rho}{\partial t} = 0$

modified Gaussian units: $\nabla\cdot\mathbf{J} + \dfrac{1}{c}\dfrac{\partial\rho}{\partial t} = 0$

P2-24. $\mathbf{J} = \dfrac{\hbar}{2im}(\psi^*\nabla\psi - \psi\nabla\psi^*)$

P2-25. (a) A/m; (c) $\dfrac{\partial}{\partial t}\displaystyle\int \sigma\, dS = -\oint \mathbf{j}\cdot d\boldsymbol{\ell} \times \hat{\mathbf{n}}$

Chapter 3

P3-3. $y = \dfrac{qEa}{mv^2}\left(b - \dfrac{1}{2}a\right)$

P3-6. $v_0 = E/B$

P3-10. (c) static equilibrium at $\theta = 0$ (stable) and $\theta = \pi$ (unstable).

P3-12. $\mathbf{p} = 3qa_0\hat{\mathbf{k}}$

P3-13. static equilibrium at $\theta = 0$ (stable) and $\theta = \pi$ (unstable).

P3-14. $\mathbf{m} = \frac{1}{3}QR^2\boldsymbol{\omega}$

P3-15. With $\omega_0 = qB_0/m$ and $\psi = qdB_0/2mv_{z0}$,

$$x(t) = \frac{v_{x0}}{\omega_0}\sin\omega_0 t;\ y(t) = -\frac{v_{x0}}{\omega_0}(1 - \cos\omega_0 t);\ z(t) = v_{z0}t$$

$$x_d = \frac{v_{x0}}{\omega_0}\sin 2\psi;\ y_d = -\frac{v_{x0}}{\omega_0}(1 - \cos 2\psi)$$

P3-21. (b) $\dfrac{\mathbf{m}(t)}{m_0} = \sin\theta(\cos\omega t\hat{\mathbf{i}} - \sin\omega t\hat{\mathbf{j}}) + \cos\theta\hat{\mathbf{k}}$

Chapter 4

P4-1. 1720 N/C directed away from the vacant point along the diagonal through that point.

P4-3. $\mathbf{F} = \dfrac{Qp}{4\pi\epsilon_0 r^3}(-2\cos\alpha\hat{\mathbf{r}} + \sin\alpha\hat{\boldsymbol{\theta}})$

P4-7. $\mathbf{E}(0, 0, b) = \dfrac{Q}{4\pi\epsilon_0}\dfrac{b\hat{\mathbf{k}} - \frac{1}{2}a\hat{\mathbf{j}}}{[b^2 + a^2]^{3/2}}$

P4-8. (a)

$$\mathbf{E}(0, y, z) = \frac{Q/\pi a^2}{4\pi\epsilon_0}\int_0^{2\pi} d\phi' \int_0^a \imath' \, d\imath' \frac{-\imath'\cos\phi'\hat{\mathbf{i}} + (y - \imath'\sin\phi')\hat{\mathbf{j}} + z\hat{\mathbf{k}}}{[\imath'^2 + y^2 + z^2 - 2y\imath'\sin\phi']^{3/2}}$$

P4-11. $\mathbf{E} = \dfrac{\rho b}{2\epsilon_0}\dfrac{\imath}{b}\hat{\boldsymbol{\imath}}, \imath < b; \cdots = \dfrac{\rho b}{2\epsilon_0}\dfrac{b}{\imath}\hat{\boldsymbol{\imath}}, \imath > b.$

P4-20. $Zq^2/\pi\epsilon_0 m v^2$

P4-21. $2^{2/3}V = 1.587$ V

P4-22. 6×10^5 volts

P4-24. $V(r, \theta) = \dfrac{qa^2}{4\pi\epsilon_0 r^3}(3\cos^2\theta - 1)$

$$\mathbf{E}(r, \theta) = \frac{3qa^2}{4\pi\epsilon_0 r^4}\{(3\cos^2\theta - 1)\hat{\mathbf{r}} + 2\cos\theta\sin\theta\hat{\boldsymbol{\theta}}\}$$

P4-25. (a)

$$V(\mathbf{r}) = \frac{\sigma}{4\pi\epsilon_0}\int_0^{2\pi}\int_0^a \frac{\imath' \, d\imath' \, d\phi'}{[(x - \imath'\cos\phi')^2 + (y - \imath'\sin\phi')^2 + z^2]^{1/2}}$$

where σ is the charge density on the disc.

(b) $V(0, 0, z) = \dfrac{\sigma}{2\epsilon_0}(\sqrt{a^2 + z^2} - |z|)$

P4-26. (a) $V(\imath, z) = \dfrac{\lambda}{4\pi\epsilon_0}\displaystyle\int_{-a}^{a}\dfrac{dz'}{[\imath^2 + (z - z')^2]^{1/2}}$

(b) $\qquad = \dfrac{\lambda}{4\pi\epsilon_0}\ell n\left(\dfrac{z + a}{z - a}\right); z > a$

P4-27. $\mathbf{E} = \dfrac{Q}{4\pi\epsilon_0 r^2}(1 + \alpha r)e^{-\alpha r}\hat{\mathbf{r}}$

$$\rho = -\frac{Q\alpha^2}{4\pi r}e^{-\alpha r} + \left(\begin{matrix}\text{point charge of} \\ \text{strength } Q \text{ at origin}\end{matrix}\right)$$

P4-28. $V(z) = V_0 z/d$; $\mathbf{E} = -(V_0/d)\hat{\mathbf{k}}$

$$\sigma(z = 0) = -\sigma(z = d) = -V_0\epsilon_0/d$$

P4-29. Spheres centered on the charge.

P4-31. 6.7 V/m

P4-36. $W = -\sqrt{2}\,q^2/4\pi\epsilon_0 a$

P4-39. (a) $W = Q^2/8\pi\epsilon_0 a$; (b) $a = Q^2/8\pi\epsilon_0 mc^2 = 1.41 \times 10^{-15}$ m

P4-41. $(Q_{ij}) = \begin{pmatrix} 2qa^2 & 0 & 0 \\ 0 & 2qa^2 & 0 \\ 0 & 0 & -4qa^2 \end{pmatrix}$

To find $V(r, \theta)$ and $\mathbf{E}(r, \theta)$, replace q by $-q$ in the expressions given as the answers to P4-24.

P4-42. (a) $Q = 2\pi a\lambda_0$; $\mathbf{p} = \frac{1}{2}Qa\hat{\mathbf{j}}$

$$(Q_{ij}) = \begin{pmatrix} \frac{1}{2}Qa^2 & 0 & 0 \\ 0 & \frac{1}{2}Qa^2 & 0 \\ 0 & 0 & -Qa^2 \end{pmatrix}$$

(b) $$V(r, \theta) = \frac{Q}{4\pi\epsilon_0 r} + \frac{Qa}{8\pi\epsilon_0 r^2}\sin\theta\sin\phi + \frac{Qa^2}{16\pi\epsilon_0 r^3}(1 - 3\cos^2\theta) + \cdots$$

P4-44. (a) $Q = 2\pi \int_0^\pi \int_0^\infty \rho(r, \theta) r^2 \sin\theta \, dr\, d\theta$

(b) $p_z = 2\pi \int_0^\pi \int_0^\infty \rho(r, \theta) r^3 \cos\theta \sin\theta \, dr\, d\theta$

(c) $Q_{33} = 2\pi \int_0^\pi \int_0^\infty \rho(r, \theta) r^4 (3\cos^2\theta - 1)\sin\theta \, dr\, d\theta$

(d) $V(r, \theta) = \dfrac{Q}{4\pi\epsilon_0 r} + \dfrac{p_z \cos\theta}{4\pi\epsilon_0 r^2} + \dfrac{Q_{33}}{8\pi\epsilon_0 r^3}\left\{\dfrac{1}{2}(3\cos^2\theta - 1)\right\} + \cdots$

(e) $Q = -q$; $\mathbf{p} = 0$; $Q_{33} = 12qa_0^2$

P4-47. (a) $\mathscr{E}(\mathbf{r}, \mathbf{r}') = \dfrac{1}{4\pi\epsilon_0}\dfrac{\mathbf{r} - \mathbf{r}'}{|\mathbf{r} - \mathbf{r}'|^3}$

(c) $\mathbf{E}_{\text{dipole}}(\mathbf{r}) = \dfrac{1}{4\pi\epsilon_0}\left[3\dfrac{\mathbf{p}\cdot(\mathbf{r} - \mathbf{r}_0)}{|\mathbf{r} - \mathbf{r}_0|^5}(\mathbf{r} - \mathbf{r}_0) - \dfrac{\mathbf{p}}{|\mathbf{r} - \mathbf{r}_0|^3}\right]$

P4-48. $\dfrac{q}{4\pi\epsilon_0 r^2}\left[1 + 2\dfrac{r}{a_0} + 2\dfrac{r^2}{a_0^2}\right]e^{-2r/a_0}\hat{\mathbf{r}}$

P4-49. (a) $\mathbf{E}(\mathbf{r}) = \dfrac{Q}{4\pi\epsilon_0 a^2}\dfrac{r}{a}\hat{\mathbf{r}}, \quad r < a$

$= \dfrac{Q}{4\pi\epsilon_0 r^2}\hat{\mathbf{r}}, \quad r > a$

(c) $V(\mathbf{r}) = \dfrac{Q}{4\pi\epsilon_0 a}\left[\dfrac{3}{2} - \dfrac{1}{2}\dfrac{r^2}{a^2}\right], \quad r < a$

$= \dfrac{Q}{4\pi\epsilon_0 r}, \quad r > a$

(d) $\omega = \sqrt{Q^2/4\pi\epsilon_0 a^3 m} \approx 4.5 \times 10^{16}\ \text{sec}^{-1}$

P4-54. Force $= \dfrac{1}{2}\epsilon_0 A\left(\dfrac{\Delta V}{d}\right)^2$

Chapter 5

P5-3. attractive force of magnitude $\frac{\mu_0 I I' h}{2\pi}\left(\frac{1}{a} - \frac{1}{b}\right)$ when I and I' are both positive.

P5-4. (b) $\frac{B_z}{B_0} = \frac{5^{3/2}}{16}\left[\frac{1}{\left\{1 + \left(\frac{z}{a} - \frac{1}{2}\right)^2\right\}^{3/2}} + \frac{1}{\left\{1 + \left(\frac{z}{a} + \frac{1}{2}\right)^2\right\}^{3/2}}\right]$

P5-5. current $= q_e v/2\pi a$; magnetic dipole moment $= \frac{1}{2} q_e v a$; **B**-field at center $= \mu_0 q_e v/4\pi a^2 \approx 1.2 \times 10^5$ G at nucleus of hydrogen atom.

P5-7. $\Phi_m = \frac{\mu_0 I h}{2\pi} \ln \frac{b}{a}$ into page if $I > 0$.

P5-10. $\mathbf{B}(\mathbf{r}) = \frac{\mu_0 I}{2\pi a}\frac{r}{a}\hat{\boldsymbol{\phi}},\ r < a$

$= \frac{\mu_0 I}{2\pi r}\hat{\boldsymbol{\phi}},\ a < r < b$

$= 0, \quad b < r$

P5-13. $\mathbf{A}(\mathbf{r}) = \frac{\mu_0 I \hat{\mathbf{k}}}{4\pi} \ln \left|\frac{b - z + \sqrt{r^2 + (b - z)^2}}{a - z + \sqrt{r^2 + (a - z)^2}}\right|$

P5-16. (b) $\Lambda = \frac{1}{2} B_0 x y$

P5-17. $\mathbf{A} = -\frac{\mu_0 I' \hat{\mathbf{k}}}{2\pi} \ln r$

P5-18. $\mathbf{A} = \frac{\mu_0 n I r}{2}\hat{\boldsymbol{\phi}},\ r < a$; $\mathbf{A} = \frac{\mu_0 n I a^2}{2r}\hat{\boldsymbol{\phi}},\ r > a.$

P5-22. (b) $\mathbf{Q}_{ij}^{(m)} = \begin{pmatrix} 0 & 0 & Q\hat{\mathbf{j}} \\ 0 & 0 & -Q\hat{\mathbf{i}} \\ Q\hat{\mathbf{j}} & -Q\hat{\mathbf{i}} & 0 \end{pmatrix}$ where $Q = 3\pi a^2 I b$

(c) $\mathbf{A} = \frac{\mu_0 Q}{4\pi r^3} \cos\theta \sin\theta \hat{\boldsymbol{\phi}}$

$\mathbf{B} = \frac{\mu_0 Q}{4\pi r^4}[(3\cos^2\theta - 1)\hat{\mathbf{r}} + 2\cos\theta \sin\theta \hat{\boldsymbol{\theta}}]$

P5-26. (b) $V^{(m)} = \mathbf{m}\cdot\mathbf{r}/4\pi r^3$

(c) $V^{(m)} = -I\phi/2\pi$

(d) $V^{(m)} = -\frac{m}{2\pi a^2}\frac{z}{(a^2 + z^2)^{1/2}}$

P5-27. (a) $\mathbf{F} = -\frac{3\mu_0 m m' b}{2\pi(a^2 + b^2)^{5/2}}\hat{\mathbf{k}}$

Chapter 6

P6-3. $v = mgR/w^2B^2$, where g = acceleration of gravity.

P6-4. $\mathscr{E}_{\text{em}}^{\text{mot}} = \frac{1}{2}\omega B R^2$

P6-5. (a) $\mathscr{E} = \omega B A \sin \omega t$, where $t = 0$ when the axis of the loop coincides with the field.

P6-6. (a) clockwise viewed from $z = \infty$.
(b) counterclockwise viewed from $z = \infty$.
(c) no induced emf.

P6-7. (a) $\mathscr{E} = -\dfrac{\mu_0 h}{2\pi}\dfrac{dI}{dt}\,\ell n\,\dfrac{b}{a}$, where the positive direction for flux is into the page.

P6-11. $\dfrac{L}{\mu_0 a} = \displaystyle\int_0^1 \xi\, d\xi \int_0^\pi d\phi' \frac{1 - \xi\cos\phi'}{[1 - 2\xi\cos\phi' + \xi^2]^{3/2}}$

P6-12. $M = \dfrac{\mu_0 h}{2\pi}\,\ell n\,\dfrac{b}{a}$

P6-26. $\mathbf{S} = -\dfrac{EI}{2\pi b}\hat{\boldsymbol{\imath}}$

P6-29. (a) $V(\mathbf{r}) = \begin{cases} \dfrac{Q}{4\pi\epsilon_0 a}\left(\dfrac{3}{2} - \dfrac{1}{2}\dfrac{r^2}{a^2}\right) \\ \dfrac{Q}{4\pi\epsilon_0 r} \end{cases}$; (b) $\mathbf{E}(\mathbf{r}) = \begin{cases} \dfrac{Q\mathbf{r}}{4\pi\epsilon_0 a^3};\ r < a \\ \dfrac{Q\mathbf{r}}{4\pi\epsilon_0 r^3};\ r > a \end{cases}$

where $Q = \frac{4}{3}\pi a^3 \rho_0$.

P6-30. (c) $\mathbf{E} = -(V_0/d)\hat{\mathbf{k}}$
(d) $\sigma(z = 0) = -\epsilon_0 V_0/d;\ \sigma(z = d) = \epsilon_0 V_0/d$

P6-31. Let $\alpha = (V_b - V_a)/\ell n\,(b/a)$; then

$$V = V_a + \alpha\,\ell n\,\frac{\imath}{a};\ \mathbf{E} = -\frac{\alpha}{\imath}\hat{\boldsymbol{\imath}};$$

$$\sigma(\imath = a) = -\epsilon_0\alpha/a;\ \sigma(\imath = b) = \epsilon_0\alpha/b$$

P6-32. Let $\alpha = \dfrac{ab(V_b - V_a)}{a - b}$ and $\beta = \dfrac{aV_a - bV_b}{a - b}$; then

$$V = \beta + \frac{\alpha}{r};\ \mathbf{E} = \frac{\alpha}{r^3}\mathbf{r};$$

$$\sigma(r = a) = \epsilon_0\alpha/a^2;\ \sigma(r = b) = -\epsilon_0\alpha/b^2$$

P6-37. (b) $T_{ij} = \epsilon_0 E_i E_j + \dfrac{1}{\mu_0}B_i B_j - \left(\dfrac{1}{2}\epsilon_0 E^2 + \dfrac{1}{2\mu_0}B^2\right)\delta_{ij}$

P6-39. (b) $\dfrac{d^2 V}{dz^2} = -\dfrac{J}{\epsilon_0}\sqrt{\dfrac{m}{2e}}\,V^{-1/2}$

Chapter 7

P7-4. $\nu_{\text{visible}} \approx 6 \times 10^{14}$ Hz; $\nu_{\text{microwave}} \approx 3 \times 10^9$ Hz.

P7-5. $\mathbf{B} = \dfrac{\kappa E_0}{\omega}[\hat{\mathbf{i}}\sin(\kappa y - \omega t) - \hat{\mathbf{k}}\cos(\kappa y - \omega t)] = \dfrac{\kappa}{\omega}\hat{\mathbf{j}} \times \mathbf{E}$

P7-6. $\mathbf{B} = \dfrac{1}{c}f(z - ct)\hat{\mathbf{j}}$

P7-8. $\langle p_r \rangle = \langle u_{\text{EM}} \rangle$

P7-9. (a) amplitude ≈ 1000 V/m
pressure $\approx 4 \times 10^{-6}$ N/m^2 (4×10^{-11} atmospheres)
(b) $\approx 3.6 \times 10^{21}$ photons/sec$\cdot$m^2

P7-10. $\langle p_r \rangle = \epsilon_0 E_{x0}^2$

P7-11. (a) $\langle p_r \rangle = 2\langle u_{EM} \rangle \cos^2 \theta$

(b) $\langle p_r \rangle = \frac{1}{3}\begin{pmatrix}\text{total energy density}\\ \text{in the fields present}\end{pmatrix}$

P7-12. $\tan \alpha = E_{y0}/E_{x0}$, where α is measured *clockwise* from the positive x axis.

P7-14. With θ measured counterclockwise from the positive x axis, one axis of the ellipse is at an angle satisfying

$$\tan 2\theta = \frac{2E_{x0}E_{y0} \cos \Phi}{(E_{x0})^2 - (E_{y0})^2}$$

and the other axis is at 90° to the first axis.

P7-17. (a) $\Delta = 2\pi/N$; (b) secondary maximum $\approx$ .045 as high as primary maximum.

P7-22. (a) $A(\kappa) = c \int_{-\infty}^{\infty} E_x(0, t)e^{i\kappa ct}\, dt$

(b) $A(\kappa) = \frac{2E_0 \sin \kappa cT}{\kappa}$; $u(\kappa) = \frac{4E_0^2 \sin^2 \kappa cT}{\pi \mu_0 c^2 \kappa^2}$

P7-23. (a) $\frac{|A(\omega/c)|^2}{\pi \mu_0 c^3}$; (b) $\frac{2|A(2\pi/\lambda)|^2}{\mu_0 c^2 \lambda^2}$

P7-24. (a) $\mathcal{E} = E_{y0}\hat{\mathbf{j}}e^{i(\kappa x - \omega t)}$; $\mathcal{B} = \frac{\kappa E_{y0}}{\omega}\hat{\mathbf{k}}e^{i(\kappa x - \omega t)}$

(b) $\mathcal{E} = E_{z0}\hat{\mathbf{k}}e^{i[\kappa(x \cos \theta + y \sin \theta) - \omega t]}$

$\mathcal{B} = \frac{\kappa E_{z0}}{\omega}(\hat{\mathbf{i}} \sin \theta - \hat{\mathbf{j}} \cos \theta)e^{i[\kappa(x \cos \theta + y \sin \theta) - \omega t]}$

(c) $\mathcal{E} = E_0(\hat{\mathbf{j}} - i\hat{\mathbf{k}})e^{i(\kappa x - \omega t)}$

$\mathcal{B} = \frac{\kappa E_0}{\omega} i(\hat{\mathbf{j}} - i\hat{\mathbf{k}})e^{i(\kappa x - \omega t)}$

P7-27. $u(r, t) = \frac{q(r - at)}{r} + \frac{Q(r + at)}{r}$, where q and Q are arbitrary functions of the indicated arguments.

P7-29. $\mathbf{A} = \frac{E_{x0}\hat{\mathbf{i}}}{\omega} \sin (\kappa z - \omega t + \phi)$; $V = 0$

P7-31. $\pi a^2 \langle u_{EM} \rangle$, where $\langle u_{EM} \rangle$ is the energy density in the incident wave.

P7-32. (b) $\begin{pmatrix}\cos \alpha\\ \sin \alpha\end{pmatrix}$; (c) rotates angle of polarization toward the x axis by amount ϕ.

P7-36. $x(t) = \frac{qE_{x0}/m}{(\omega^2 - \omega_0^2) + i(b\omega/m)}e^{-i\omega t}$, where $\omega_0^2 = \frac{k}{m}$

Chapter 8

P8-5. $A_n = \frac{2V_0[1 - (-1)^n]}{n\pi \sinh (n\pi a/b)}$

P8-8. $V(\imath, \phi) = -E_0 \imath\left(1 - \frac{a^2}{\imath^2}\right)\cos\phi;\ \sigma(\phi) = 2\epsilon_0 E_0 \cos\phi$

P8-9. $V(\imath, \phi) = \frac{4V_0}{\pi}\sum_{m=0}^{\infty}\left(\frac{\imath}{a}\right)^{4m+2}\frac{\sin(4m+2)\phi}{2m+1}$

where $0 \leq \phi \leq \frac{1}{2}\pi$.

P8-11. $J_0(x) = \sum_{n=0}^{\infty}\frac{(-1)^n}{(n!)^2}\left(\frac{x}{2}\right)^{2n}$

P8-12. $\sigma(\imath) = \frac{2\epsilon_0 V_0}{\pi}\frac{1}{\sqrt{a^2 - \imath^2}};\ Q = 8\epsilon_0 V_0 a$

P8-15. $\mathbf{p} = 4\pi\epsilon_0 E_0 a^3\hat{\mathbf{k}};\ Q = 3\pi\epsilon_0 E_0 a^2;\ s = \frac{4}{3}a$

$$\mathbf{E} = E_0\left(1 + \frac{2a^3}{r^3}\right)\cos\theta\,\hat{\mathbf{r}} - E_0\left(1 - \frac{a^3}{r^3}\right)\sin\theta\,\hat{\boldsymbol{\theta}}$$

P8-16. Set $b_0 = Q/4\pi\epsilon_0$ in Eqs. (8-46) and (8-47).

P8-17. $V(r, \theta) = \frac{Q}{4\pi\epsilon_0 r}\left[1 - \frac{a^2}{2r^2}P_2(\cos\theta) + \frac{3a^4}{8r^4}P_4(\cos\theta) - \frac{5a^6}{16r^6}P_6(\cos\theta) + \cdots\right]$

P8-19. $\text{Re}\, z^n = \imath^n \cos n\phi;\ \text{Im}\, z^n = \imath^n \sin n\phi$

P8-20. $4\epsilon_0$ on x axis; $-4\epsilon_0$ on y axis.

P8-23. $\sigma(y, z) = -\frac{qd/2\pi}{[d^2 + y^2 + z^2]^{3/2}}$

$$Q = -q;\ \mathbf{F} = -\frac{q^2}{4\pi\epsilon_0(2d)^2}\hat{\mathbf{i}}$$

P8-25. Charges are required at $(a, b, 0)$, $(-a, b, 0)$, $(a, -b, 0)$ and $(-a, -b, 0)$.

P8-27. $\sigma(\theta) = -\frac{q}{4\pi a}\frac{d^2 - a^2}{[a^2 + d^2 - 2ad\cos\theta]^{3/2}}$

P8-30. The charge having the smaller magnitude lies inside the sphere. Let $|q'| < |q|$. Then the radius of the sphere is $a = s\Big/\left[\left|\frac{q}{q'}\right| - \left|\frac{q'}{q}\right|\right]$ and its center is located on the line joining the two charges and a distance $b = \left|\frac{q'}{q}\right| a$ from the charge q'.

P8-36. $V(x, y) = \frac{1}{4}\left[V(x + d, y) + V(x - d, y) + V(x, y + d) + V(x, y - d) + \frac{d^2\rho(x, y)}{\epsilon_0}\right] + O(d^4)$

P8-37. $V(x, y, z) = \frac{16V_0}{\pi^2}\sum_{m=1}^{\infty}\sum_{n=1}^{\infty}\frac{\sinh\alpha_{mn}\pi z}{mn\sinh\alpha_{mn}\pi c}\sin\frac{m\pi x}{a}\sin\frac{n\pi y}{b}$

where m and n assume only *odd* values and $\alpha_{mn} = \sqrt{\frac{m^2}{a^2} + \frac{n^2}{b^2}}$

P8-38. $\begin{pmatrix}\text{product}\\ \text{solution}\end{pmatrix} = J_n(\alpha\imath)\{A\cos n\phi + B\sin n\phi\}\{C\,e^{-\alpha z} + D\,e^{\alpha z}\}$

where $n = 0, 1, 2, \ldots$ and α is positive and real.

P8-39. $\begin{pmatrix}\text{product}\\ \text{solution}\end{pmatrix} = \left\{Ar^n + \frac{B}{r^{n+1}}\right\} P_n^m(\cos\theta)\{C\cos m\phi + D\sin m\phi\}$

P8-41. $V(x, y) = \frac{2}{\pi}\int_0^\infty \frac{\sin\kappa}{\kappa} e^{-\kappa y}\cos\kappa x\, d\kappa$

Chapter 9

P9-1. $(\text{time})^{-1}$

P9-2. 4.8 m

P9-5. $\tau_r = \epsilon_0/g \approx 2.5 \times 10^{-19}$ sec for Al

P9-6. (a) 1.74×10^{-7} m/sec; (b) 1.17×10^5 m/sec; (c) $\approx 7 \times 10^{-15}$ sec; (d) 8.2 Å; (e) $\mu \approx -1.23 \times 10^{-3}$ mks units, E for drift velocity $\approx 1.4 \times 10^{-4}$ V/m, E for thermal velocity $\approx 9.5 \times 10^7$ V/m

P9-7. $\Delta V = V_b - V_a = -wvB$

P9-8. $b = q/\mu$; approach to terminal velocity is characterized by a time constant $m/b \approx 6.8 \times 10^{-15}$ sec

Chapter 10

P10-3. (a) $qd/4\pi\epsilon_0 r^3$; (c) $\alpha_0 \approx 10^{-40}$ mks units;
(d) $\approx 10^{-34}$ C·m; (e) $\approx 6 \times 10^{-16}$ m

P10-4. $\approx 6 \times 10^{11}$ V/m

P10-6. 1.2×10^{-4} C/m^2

P10-9. $\sigma_b = ba$ on all surfaces; $\rho_b = -3b$

P10-10. (a) $\sigma_b(z = \pm\frac{1}{2}L) = \pm P$, $\sigma_b(\text{cylindrical side}) = 0$, $\rho_b = 0$

(b) $$V(0, 0, z) = \frac{PR}{2\epsilon_0}\left[\sqrt{1 + \left(\frac{z}{R} - \frac{L}{2R}\right)^2} - \left|\frac{z}{R} - \frac{L}{2R}\right| - \sqrt{1 + \left(\frac{z}{R} + \frac{L}{2R}\right)^2} + \left|\frac{z}{R} + \frac{L}{2R}\right|\right]$$

(c) $V(0, 0, z) \longrightarrow \frac{PLR^2}{4\epsilon_0 z^2}$

(d) dipole moment $= \pi PLR^2$

P10-11. (a) $\sigma_b(\theta) = P\cos\theta$, $\rho_b = 0$; (b) dipole moment $= \frac{4}{3}\pi a^3 P\hat{\mathbf{k}}$

P10-13. $F_{qq'} = qq'/4\pi\epsilon r^2$

P10-16. $C = \frac{A\epsilon_0}{d}\left|\frac{K_2 - K_1}{\ell n(K_2/K_1)}\right|$

P10-17. (a) $E = \dfrac{\Delta V}{d + \left(\frac{1}{K} - 1\right)t}$; $E' = \frac{1}{K}E$; $D = D' = \epsilon_0 E$

(b) $\sigma_{\text{plate}} = D = \epsilon_0 E$; $\sigma_{\text{dielectric}} = \chi_e E' = \left(1 - \frac{1}{K}\right)\sigma_{\text{plate}}$

(c) $C = \dfrac{\epsilon_0 A/d}{1 + \left(\frac{1}{K} - 1\right)\frac{t}{d}}$

P10-18. (a) $\mathbf{D} = \frac{q}{4\pi r^2}\hat{\mathbf{r}}$; $\mathbf{E} = \frac{1}{\epsilon}\mathbf{D}$; $\mathbf{P} = \frac{\chi_e}{\epsilon}\mathbf{D}$

(b) $\rho_b = 0$; $\sigma_b(a) = -\frac{\chi_e q}{4\pi\epsilon a^2}$; $\sigma_b(b) = \frac{\chi_e q}{4\pi\epsilon b^2}$

(c) $\mathbf{D} = \frac{q}{4\pi r^2}\hat{\mathbf{r}}$; $\mathbf{E} = \frac{1}{\epsilon_0}\mathbf{D}$

(d) $\mathbf{E}(b^+) - \mathbf{E}(b^-) = \frac{\sigma_b(b)}{\epsilon_0}\hat{\mathbf{r}}$

(e) $C = 4\pi\epsilon\frac{ab}{b-a}$

P10-20. $\approx 1.3 \times 10^{-10}$ m

P10-21. (b) $\alpha_t = 2.10 \times 10^{-40}$ mks units; (c) $K_e = 1.48$

P10-24. $\mathbf{D} = \mathbf{E} + 4\pi\mathbf{P}$, $\frac{1}{c^2}\mathbf{E} + 4\pi\mathbf{P}$, $\mathbf{E} + 4\pi\mathbf{P}$, and $\mathbf{E} + \mathbf{P}$ in cgs-esu, cgs-emu, Gaussian units, and Heaviside-Lorentz units, respectively.

Chapter 11

P11-1. $B_m \ll 4 \times 10^6$ G

P11-2. $B_m \approx 5 \times 10^7$ G

P11-3. (c) $M_B = 0.922 \times 10^{-23}$ J/T; $\omega_0 \approx 4.1 \times 10^{16}$ sec^{-1}, assuming a radius of 0.528×10^{-10} m; $B_m \ll 5 \times 10^9$ G

P11-5. current $= M/n$

P11-6. (a) $\mathbf{J}_m = 0$, $\mathbf{j}_m = M \sin\theta\hat{\boldsymbol{\phi}}$; (b) $\mathbf{m} = \frac{4}{3}\pi R^3\mathbf{M}$

P11-13. $\sigma_m(z = \pm\frac{1}{2}L) = \pm M$; no other equivalent poles. $V^{(m)}(0, 0, z)$ is obtained by setting $\epsilon_0 = 1$ and replacing P by M in the answer to P10-10(b).

P11-14. $\rho_m = 0$; $\sigma_m = M\cos\theta$; $\mathbf{m} = \frac{4}{3}\pi R^3\mathbf{M}$

P11-23. (a) $\mathbf{H} = NI\hat{\mathbf{k}}$, $\mathbf{B} = \mu NI\hat{\mathbf{k}}$, $\mathbf{M} = \chi_m NI\hat{\mathbf{k}}$; (c) $\mathbf{j}_m = \chi_m NI\hat{\boldsymbol{\phi}}$

P11-25. (b) $T_c = .361$ °K; (c) $\gamma = 963$

P11-26. (b) .17, 1.10, 1.30 T, respectively.

P11-33. $\mathbf{H} = c^2\mathbf{B} - 4\pi\mathbf{M}$, $\mathbf{B} - 4\pi\mathbf{M}$, $\mathbf{B} - 4\pi\mathbf{M}$, and $\mathbf{B} - \mathbf{M}$ in cgs-esu, cgs-emu, Gaussian units, and Heaviside-Lorentz units, respectively.

Chapter 12

P12-4. $\mathbf{F}_{qq'} = \frac{qq'}{4\pi\epsilon}\frac{\mathbf{r} - \mathbf{r}'}{|\mathbf{r} - \mathbf{r}'|^3}$

P12-6. (b) speed $= 1/\sqrt{\mu\epsilon}$; $n = \sqrt{K_m K_e}$

P12-10. force $= (\epsilon - \epsilon_0)\frac{(\Delta V)^2 b}{2d}$; pulls slab in.

P12-18. $\mu_1 \cot\theta_1 = \mu_2 \cot\theta_2$

P12-19. $H_{\text{material}} = \frac{NI}{2\pi r + \chi_m d}$; $H_{\text{gap}} = K_m H_{\text{material}}$;

$B_{\text{gap}} = B_{\text{material}} = \mu H_{\text{material}}$

P12-20. $\mathbf{E}_{\text{in}} = \dfrac{3\epsilon_0 E_0}{\epsilon + 2\epsilon_0}(\cos\theta\hat{\mathbf{r}} - \sin\theta\hat{\boldsymbol{\theta}})$

$$\mathbf{E}_{\text{out}} = \left(1 + \frac{2(\epsilon - \epsilon_0)}{\epsilon + 2\epsilon_0}\frac{a^3}{r^3}\right)E_0\cos\theta\hat{\mathbf{r}} - \left(1 - \frac{\epsilon - \epsilon_0}{\epsilon + 2\epsilon_0}\frac{a^3}{r^3}\right)E_0\sin\theta\hat{\boldsymbol{\theta}}$$

$$\rho_b = 0;\ \sigma_b = \frac{3\epsilon_0\chi_e}{\epsilon + 2\epsilon_0}E_0\cos\theta$$

$$\text{dipole moment} = 4\pi\epsilon_0 a^3\frac{\epsilon - \epsilon_0}{\epsilon + 2\epsilon_0}\mathbf{E}_0$$

P12-21. $V_{\text{in}}(\imath, \phi) = -\dfrac{2}{K_e + 1}E_0\imath\cos\phi$

$$V_{\text{out}}(\imath, \phi) = -\left(1 - \frac{K_e - 1}{K_e + 1}\frac{a^2}{\imath^2}\right)E_0\imath\cos\phi$$

$$\mathbf{E}_{\text{in}} = \frac{2E_0}{K_e + 1}(\cos\phi\hat{\boldsymbol{\imath}} - \sin\phi\hat{\boldsymbol{\phi}}) = \frac{2E_0}{K_e + 1}\hat{\mathbf{i}}$$

$$\mathbf{E}_{\text{out}} = \left(1 + \frac{K_e - 1}{K_e + 1}\frac{a^2}{\imath^2}\right)E_0\cos\phi\hat{\boldsymbol{\imath}} - \left(1 - \frac{K_e - 1}{K_e + 1}\frac{a^2}{\imath^2}\right)E_0\sin\phi\hat{\boldsymbol{\phi}}$$

P12-24. $\mathbf{B}_{\text{in}} = \dfrac{3K_m}{K_m + 2}\mathbf{B}$

$$\mathbf{B}_{\text{out}} = \mathbf{B} + \frac{\mu_0}{4\pi r^3}[3(\mathbf{m}\cdot\hat{\mathbf{r}})\hat{\mathbf{r}} - \mathbf{m}] \text{ where } \mathbf{m} = \frac{4\pi a^3(K_m - 1)}{\mu_0(K_m + 2)}\mathbf{B}$$

P12-25. $\mathbf{H}_{\text{in}} = -\frac{1}{3}\mathbf{M};\ \mathbf{B}_{\text{out}} = \mu_0\mathbf{H}_{\text{out}};\ \mathbf{B}_{\text{in}} = \frac{2}{3}\mu_0\mathbf{M}$

$$\mathbf{H}_{\text{out}} = \frac{a^3}{3r^3}[3(\mathbf{M}\cdot\hat{\mathbf{r}})\hat{\mathbf{r}} - \mathbf{M}]$$

P12-26. $V^{(m)} = \text{constant} + \dfrac{I}{2}\Big[\dfrac{1}{2}\left(\dfrac{a}{r}\right)^2 P_1(\cos\theta)$

$$- \frac{3}{8}\left(\frac{a}{r}\right)^4 P_3(\cos\theta) + \frac{5}{16}\left(\frac{a}{r}\right)^6 P_5(\cos\theta) - \cdots\Big]$$

P12-29. $L = \frac{1}{3}$

P12-30. $d_1 = d_2 = d;\ q_1 = \dfrac{\epsilon_1 - \epsilon_2}{\epsilon_1 + \epsilon_2}q;\ q_2 = \dfrac{2\epsilon_2}{\epsilon_1 + \epsilon_2}q$

Chapter 13

P13-3. (a) $2.07 \times 10^{16}\ \text{sec}^{-1}$; (b) $4.14 \times 10^{16}\ \text{sec}^{-1}$;
(c) $K_e = 1.0000727$ from (a) $= 1.000291$ from (b)

P13-11. (a) $\omega \ll 6.5 \times 10^9\ \text{sec}^{-1}$; (b) 0.77 m

P13-20. $T \longrightarrow \sqrt{\dfrac{4}{\mu_0 g\omega}}$

P13-21. $\sigma = 0;\ \mathbf{j} = 2\sqrt{\dfrac{\epsilon_i}{\mu_i}}\,\text{Re}(\mathcal{E}_{i0}e^{-i\omega t})\hat{\mathbf{i}}$

P13-26. In the coordinate system of Fig. 13-8, $\sigma = 0$ and

$$\mathbf{j} = 2\frac{\kappa_i\cos\theta_i}{\omega\mu_i}\,\text{Re}\,\{\mathcal{E}_{i0}e^{-i(\kappa_i y\sin\theta_i + \omega t)}\}\hat{\mathbf{i}}$$

P13-27. Let $\kappa' = 2\pi n/\lambda$. Then

$$[T; R] = \frac{[16n^2; 4(n^2-1)^2 \sin^2 \kappa' d]}{16n^2 + 4(n^2-1)^2 \sin^2 \kappa' d}$$

P13-29. (a) $R_\perp = \left(\frac{n_i^2 - n_t^2}{n_i^2 + n_t^2}\right)^2$; (b) $\begin{matrix} R_\perp = .148 \\ \theta_B = 56.31° \end{matrix}$

P13-31. $\sigma_{④} = -\sigma_{②} = A\epsilon_0 \sin\left(\frac{n\pi y}{b}\right)\cos(\kappa_z z - \omega t)$

$$\mathbf{j}_{①} = -\frac{n\pi A}{\omega\mu_0 b}\sin(\kappa_z z - \omega t)\hat{\mathbf{i}}$$

$$\mathbf{j}_{④} = \frac{A}{\omega\mu_0}\left[\kappa_z \sin\left(\frac{n\pi y}{b}\right)\cos(\kappa_z z - \omega t)\hat{\mathbf{k}} + \frac{n\pi}{b}\cos\left(\frac{n\pi y}{b}\right)\sin(\kappa_z z - \omega t)\hat{\mathbf{j}}\right]$$

$$\sigma_{①} = \sigma_{③} = 0; \mathbf{j}_{③} = -(-1)^n \mathbf{j}_{①}; \mathbf{j}_{②} = -\mathbf{j}_{④}$$

P13-33. $\boldsymbol{\mathcal{E}}_{m0}^{\text{TE}}(\mathbf{r}, t) = A\sin\left(\frac{m\pi x}{a}\right)\hat{\mathbf{j}}e^{i(\kappa_z z - \omega t)}$

$$\boldsymbol{\mathcal{H}}_{m0}^{\text{TE}}(\mathbf{r}, t) = \frac{A}{i\omega\mu_0}\left[-i\kappa_z \sin\left(\frac{m\pi x}{a}\right)\hat{\mathbf{i}} + \frac{m\pi}{a}\cos\left(\frac{m\pi x}{a}\right)\hat{\mathbf{k}}\right]e^{i(\kappa_z z - \omega t)}$$

P13-34. Let $\kappa_x = m\pi/a$, $\kappa_y = n\pi/b$. Then

$$\boldsymbol{\mathcal{E}}_{mn}^{\text{TE}}(\mathbf{r}, t) = C[\kappa_y \cos\kappa_x x \sin\kappa_y y\hat{\mathbf{i}} - \kappa_x \sin\kappa_x x\cos\kappa_y y\hat{\mathbf{j}}]e^{i(\kappa_z z - \omega t)}$$

$$\boldsymbol{\mathcal{H}}_{mn}^{\text{TE}}(\mathbf{r}, t) = \frac{C}{i\omega\mu_0}[i\kappa_z\kappa_x \sin\kappa_x x\cos\kappa_y y\hat{\mathbf{i}} + i\kappa_z\kappa_y\cos\kappa_x x\sin\kappa_y y\hat{\mathbf{j}} - (\kappa_x^2 + \kappa_y^2)\cos\kappa_x x\cos\kappa_y y\hat{\mathbf{k}}]e^{i(\kappa_z z - \omega t)}$$

where κ_z and ω satisfy $(\omega/c)^2 = \kappa_x^2 + \kappa_y^2 + \kappa_z^2$.

P13-35. $\boldsymbol{\mathcal{E}}(\mathbf{r}, t) = AJ_1(\kappa_r r)\hat{\boldsymbol{\phi}}e^{i(\kappa_z z - \omega t)}$

$$\boldsymbol{\mathcal{H}}(\mathbf{r}, t) = \frac{A}{i\omega\mu_0}[-i\kappa_z J_1(\kappa_r r)\hat{\mathbf{r}} + \kappa_r J_0(\kappa_r r)\hat{\mathbf{k}}]e^{i(\kappa_z z - \omega t)}$$

where $\kappa_z^2 + \kappa_r^2 = (\omega/c)^2$ and $J_1(\kappa_r a) = 0$.

P13-37. $\boldsymbol{\mathcal{H}}(\mathbf{r}, t) = \boldsymbol{\mathcal{H}}_0(\mathbf{r})e^{-i\omega t}$ where

$$\mathcal{H}_{0x}(x, y, z) = \frac{\kappa_y E_z - \kappa_z E_y}{i\omega\mu_0}\sin\kappa_x x\cos\kappa_y y\cos\kappa_z z$$

$$\mathcal{H}_{0y}(x, y, z) = \frac{\kappa_z E_x - \kappa_x E_z}{i\omega\mu_0}\cos\kappa_x x\sin\kappa_y y\cos\kappa_z z$$

$$\mathcal{H}_{0z}(x, y, z) = \frac{\kappa_x E_y - \kappa_y E_x}{i\omega\mu_0}\cos\kappa_x x\cos\kappa_y y\sin\kappa_z z$$

P13-42. $\frac{v_p}{c} = \frac{1}{A + B/\lambda^2}$; $\frac{v_g}{c} = \frac{A - B/\lambda^2}{A + B/\lambda^2}$

P13-46. $\kappa = \sqrt{\frac{\mu g^2}{2\epsilon}Q}\left[\sqrt{1 + \frac{1}{Q^2}} + 1\right]^{1/2}$; $\alpha = \sqrt{\frac{\mu g^2}{2\epsilon}Q}\left[\sqrt{1 + \frac{1}{Q^2}} - 1\right]^{1/2}$

$$|K| = \sqrt{\frac{\mu g^2}{\epsilon}Q}\left[1 + \frac{1}{Q^2}\right]^{1/4}; \tan\phi = \left[\frac{\sqrt{1 + \frac{1}{Q^2}} - 1}{\sqrt{1 + \frac{1}{Q^2}} + 1}\right]^{1/2}$$

$$\frac{u_E}{u_B} = \left[1 + \frac{1}{Q^2}\right]^{-1/2}$$

P13-47. $\zeta = \frac{\gamma}{\sqrt{2}}\left[\left(\sqrt{1+\frac{1}{Q^2}}+1\right)^{1/2} + i\left(\sqrt{1+\frac{1}{Q^2}}-1\right)^{1/2}\right]$

P13-50. $\mathcal{E}_{r0}^{||} = \frac{\tan(\theta_t - \theta_i)}{\tan(\theta_t + \theta_i)}\mathcal{E}_{i0}^{||}$

$$\mathcal{E}_{t0}^{||} = \frac{2\cos\theta_i \sin\theta_t}{\sin(\theta_t + \theta_i)\cos(\theta_t - \theta_i)}\mathcal{E}_{i0}^{||}$$

P13-55. $\omega_{np} = c\sqrt{\frac{\alpha_n^2}{a^2} + \frac{p^2\pi^2}{b^2}}$ where α_n is the nth root of $J_1(x) = 0$.

P13-56. $\mathcal{E}_{0y} = \frac{i}{\omega\epsilon_0}\left(1 - \frac{\kappa_z^2 c^2}{\omega^2}\right)^{-1}\left(-\frac{\partial \mathcal{H}_{0z}}{\partial x} + \frac{\kappa_z}{\omega\mu_0}\frac{\partial \mathcal{E}_{0z}}{\partial y}\right)$

$$\mathcal{H}_{0x} = \frac{i}{\omega\mu_0}\left(1 - \frac{\kappa_z^2 c^2}{\omega^2}\right)^{-1}\left(\frac{\kappa_z}{\omega\epsilon_0}\frac{\partial \mathcal{H}_{0z}}{\partial x} - \frac{\partial \mathcal{E}_{0z}}{\partial y}\right)$$

$$\mathcal{H}_{0y} = \frac{i}{\omega\mu_0}\left(1 - \frac{\kappa_z^2 c^2}{\omega^2}\right)^{-1}\left(\frac{\kappa_z}{\omega\epsilon_0}\frac{\partial \mathcal{H}_{0z}}{\partial y} + \frac{\partial \mathcal{E}_{0z}}{\partial x}\right)$$

P13-57. (a) $\kappa^2 = \mu_0\epsilon_0(\omega^2 - \omega_p^2)$; (e) $n = \sqrt{1 - \frac{\omega_p^2}{\omega^2}}$

Chapter 14

P14-3. $\begin{Bmatrix}\mathbf{E}(\mathbf{r}, t)\\ \mathbf{B}(\mathbf{r}, t)\end{Bmatrix} = -\frac{\mu_0 p_0 \omega\kappa}{4\pi}\frac{\cos(\kappa r - \omega t)}{r}\sin\theta\begin{Bmatrix}c\hat{\boldsymbol{\theta}}\\ \hat{\boldsymbol{\phi}}\end{Bmatrix}$

P14-7. Let $\mathbf{m}_0 = \pi a^2 \mathcal{I}_0 \hat{\mathbf{k}}$. Then

$$\mathcal{A}_0(\mathbf{r}) = \frac{i\kappa\mu_0 e^{i\kappa r}}{4\pi r}\left(1 - \frac{1}{i\kappa r}\right)(\hat{\mathbf{r}} \times \mathbf{m}_0)$$

$$\mathcal{B}_0(\mathbf{r}) = \frac{\kappa^2\mu_0 e^{i\kappa r}}{4\pi r}[(\hat{\mathbf{r}} \times \mathbf{m}_0) \times \hat{\mathbf{r}}] + \frac{\mu_0 e^{i\kappa r}}{4\pi r^3}(1 - i\kappa r)[3(\hat{\mathbf{r}}\cdot\mathbf{m}_0)\hat{\mathbf{r}} - \mathbf{m}_0]$$

$$\mathcal{E}_0(\mathbf{r}) = \frac{\kappa^2\mu_0 c e^{i\kappa r}}{4\pi r}\left(1 - \frac{1}{i\kappa r}\right)(\mathbf{m}_0 \times \hat{\mathbf{r}})$$

P14-10. (b) $V = \frac{q}{4\pi\epsilon_0 a\left(1 - \frac{v}{c}\cos\theta\right)}$

P14-15. (b) $|\mathbf{E}(r, \phi, z, t = 0)| = \frac{q}{4\pi\epsilon_0 r^2}\frac{1 - \frac{v^2}{c^2}}{\left[1 - \frac{v^2}{c^2}\sin^2\theta\right]^{3/2}}$

P14-17. (b) $\frac{dE}{dt} = -\frac{1}{\tau}E$ where $\tau = \frac{3\pi m^3\epsilon_0 c^3}{q^4 B^2}$

P14-18. $\frac{dP(t')}{d\Omega} = \frac{q^2\dot{v}^2}{16\pi^2\epsilon_0 c^3}\frac{\sin^2\theta}{\left(1 - \frac{v}{c}\cos\theta\right)^5}$

P14-19. (c) $P/(dE/dt) \approx 4 \times 10^{-14}$

P14-20. (b) ≈ 5000 MeV

P14-21. $\frac{dW}{d\Omega\, d\omega} = \frac{q^2 v_0^2 \sin^2\theta}{16\pi^3\epsilon_0 c^3}\cos^2\frac{\omega T}{2}\left[\frac{\pi^2/T^2}{\omega^2 - (\pi^2/T^2)}\right]^2$

P14-22. $\dfrac{dW}{d\omega} = \dfrac{2q^2\omega_0^4 R_0^2}{3\pi^2\epsilon_0 c^3}\left[\dfrac{\sin^2[(\omega-\omega_0)T]}{[(\omega-\omega_0)T]^2} + \dfrac{\sin^2[(\omega+\omega_0)T]}{[(\omega+\omega_0)T]^2}\right]$

P14-23. (c) $\approx 1.3 \times 10^{-11}$ sec

P14-24. (a) 6.3×10^{-24} sec

P14-26. (c) $V(\mathbf{r}, t) = \dfrac{1}{(2\pi)^4\epsilon_0}\displaystyle\int \dfrac{\tilde{\rho}(\kappa, \omega)e^{i(\boldsymbol{\kappa}\cdot\mathbf{r}-\omega t)}}{\kappa^2 - (\omega^2/c^2)}\, d\kappa_x\, d\kappa_y\, d\kappa_z\, d\omega$

P14-27. Let $r_0 = q^2/4\pi\epsilon_0 mc^2$. Then
(a) $\langle dP(t')/d\Omega\rangle = \frac{1}{2}\epsilon_0 E_0^2 c r_0^2\{1 - \sin^2\theta\cos^2(\phi-\psi)\}$
(b) $d\sigma/d\Omega = r_0^2\{1 - \sin^2\theta\cos^2(\phi-\psi)\}$
(c) $(d\sigma/d\Omega)_{\text{unpolarized}} = \frac{1}{2}r_0^2(1+\cos^2\theta)$
(d) $\sigma_{\text{Thomson}} = 8\pi r_0^2/3$

Chapter 15

P15-3. $\left[\left(1 - \dfrac{v^2}{c^2}\right)\dfrac{\partial^2}{\partial x'^2} + \dfrac{\partial^2}{\partial y'^2} + \dfrac{\partial^2}{\partial z'^2} + 2\dfrac{v}{c^2}\dfrac{\partial^2}{\partial t'\,\partial x'} - \dfrac{1}{c^2}\dfrac{\partial^2}{\partial t'^2}\right]u = 0$

P15-7. speed$' = c$; $\tan\theta' = \dfrac{\sqrt{1-\beta^2}\sin\theta}{\cos\theta - \beta}$

P15-22. $\mathbf{v} = c^2\mathbf{E}\times\mathbf{B}/E^2$

P15-24. $\begin{Bmatrix} V \\ \mathbf{A}\end{Bmatrix} = \dfrac{q}{4\pi\epsilon_0}\dfrac{1}{[(x-vt)^2 + (1-\beta^2)(y^2+z^2)]^{1/2}}\begin{Bmatrix} 1 \\ v\hat{\mathbf{i}}/c^2\end{Bmatrix}$

P15-30. $F'_x = F_x - \dfrac{v(V_yF_y + V_zF_z)}{c^2 - V_xv}$; $F'_y = \dfrac{F_y}{\gamma\left(1 - \dfrac{V_xv}{c^2}\right)}$; $F'_z = \dfrac{F_z}{\gamma\left(1 - \dfrac{V_xv}{c^2}\right)}$

P15-34. $$\mathfrak{L} = \begin{pmatrix} \gamma\gamma_1(1+\beta\beta_1) & 0 & 0 & i\gamma\gamma_1(\beta+\beta_1) \\ 0 & 1 & 0 & 0 \\ 0 & 0 & 1 & 0 \\ -i\gamma\gamma_1(\beta+\beta_1) & 0 & 0 & \gamma\gamma_1(1+\beta\beta_1)\end{pmatrix}$$

P15-35. $$\mathfrak{L} = \frac{1}{\beta^2}\begin{pmatrix} \gamma\beta_x^2 + \beta_y^2 & (\gamma-1)\beta_x\beta_y & 0 & i\gamma\beta^2\beta_x \\ (\gamma-1)\beta_x\beta_y & \beta_x^2 + \gamma\beta_y^2 & 0 & i\gamma\beta^2\beta_y \\ 0 & 0 & 1 & 0 \\ -i\gamma\beta^2\beta_x & -i\gamma\beta^2\beta_y & 0 & \gamma\beta^2\end{pmatrix}$$

P15-37. $$\bar{F} = \begin{pmatrix} 0 & -\frac{i}{c}E_z & \frac{i}{c}E_y & B_x \\ \frac{i}{c}E_z & 0 & -\frac{i}{c}E_x & B_y \\ -\frac{i}{c}E_y & \frac{i}{c}E_x & 0 & B_z \\ -B_x & -B_y & -B_z & 0\end{pmatrix}$$

P15-38. $\mathbf{E}' = \dfrac{\lambda\gamma}{2\pi\epsilon_0}\dfrac{y'\hat{\mathbf{j}} + z'\hat{\mathbf{k}}}{y'^2 + z'^2}$

$$\mathbf{B}' = \frac{\mu_0\lambda\gamma v}{2\pi}\frac{z'\hat{\mathbf{j}} - y'\hat{\mathbf{k}}}{y'^2 + z'^2}$$

P15-39. (d) $F^0_{\mu\nu} = i(\kappa_\mu\alpha_\nu - \kappa_\nu\alpha_\mu)$

P15-41. $x(t) = \dfrac{mc^2}{qE}\left\{\sqrt{1 + \left(\dfrac{qE}{mc}t\right)^2} - 1\right\}$

Appendix A

PA-1. (a) equations inconsistent, e.g., $2x + 3y = 5$ and $4x + 6y = 8$.
(b) equations redundant, e.g., $2x + 3y = 5$ and $4x + 6y = 10$.

PA-3. $x_1 = -1, x_2 = 2, x_3 = 5$

PA-4. $x_1 = \frac{2}{3}\alpha, x_2 = \alpha, x_3 = \frac{5}{3}\alpha$ for α arbitrary

PA-7. $\mathbf{x}^T\mathbf{y} = 7$ $\qquad (B\mathbf{x})^T = (-6, 10, -2)$

$$\mathbf{y}\mathbf{x}^T = \begin{pmatrix} -1 & 2 & 5 \\ 1 & -2 & -5 \\ -2 & 4 & 10 \end{pmatrix} \qquad A(A + B) = \begin{pmatrix} 0 & 2 & 2 \\ 8 & 2 & 0 \\ 0 & 2 & 2 \end{pmatrix}$$

PA-11. $A^{-1} = \dfrac{1}{a_{11}a_{22} - a_{21}a_{12}}\begin{pmatrix} a_{22} & -a_{12} \\ -a_{21} & a_{11} \end{pmatrix}$

PA-13. $\lambda_1 = -\sqrt{2}, \mathbf{x}_1^T = \alpha(1, -\sqrt{2}, 1)$
$\lambda_2 = 0, \quad \mathbf{x}_2^T = \beta(1, 0, -1)$
$\lambda_3 = \sqrt{2}, \quad \mathbf{x}_3^T = \gamma(1, \sqrt{2}, 1)$

where α, β, and γ are arbitrary

Appendix B

PB-3. $\dfrac{1}{2}mv^2\left[1 + \dfrac{3}{4}\dfrac{v^2}{c^2} + \dfrac{5}{8}\dfrac{v^4}{c^4} + 0\left(\dfrac{v^6}{c^6}\right)\right]$

PB-4. $|f| < 0.1$

PB-5. See Table 8-1.

PB-6. $\sin x = x - \dfrac{1}{3!}x^3 + \dfrac{1}{5!}x^5 - \cdots$

$$e^x = 1 + x + \frac{1}{2!}x^2 + \cdots$$

$$(1 + x)^q = 1 + qx + \frac{q(q-1)}{2!}x^2 + \cdots$$

Appendix D

PD-2. $z_1 z_2 = 7 + i$; $z_2/z_1 = -\frac{1}{25}(1 + 7i)$; $z_2 = \sqrt{2}\,e^{-i\pi/4}$

PD-10. $\cos 3\theta = 4\cos^3\theta - 3\cos\theta$

$\sin 3\theta = 3\sin\theta - 4\sin^3\theta$

PD-11. $\sinh z = \sinh x \cos y + i \cosh x \sin y$

PD-14. $x = 2\sum_{n=1}^{\infty} \frac{(-1)^{n+1}}{n} \sin nx$

PD-17. $\tilde{f}(\kappa) = 2\frac{\sin \kappa a}{\kappa}$

Index

The user of this index should be aware of the following features:

Not only textual discussions but also *some* of the problems are indexed. Problems are identified by the problem number and page, e.g., the notation P10-12(278) refers to problem P10-12 on page 278. The user seeking problems on a given topic should also scan the problems at the end of any section to which a textual reference is given; not all of those problems are indexed.

A brief index to frequently used information may be found on page 510.

BRIEF INDEX TO FREQUENTLY USED INFORMATION

Operator \ Coordinate System	*Cartesian*	*Cylindrical*	*Spherical*	*General Orthogonal*
Gradient	Eq. (0-47), p. 22	Eq. (0-56), p. 25	Eq. (0-57), p. 25	P0-32, p. 31
Curl	Eq. (0-52), p. 23 P0-22, p. 24	Eq. (0-63), p. 27	Eq. (0-64), p. 27	P0-32, p. 32
Divergence	Eq. (2-24), p. 64	Eq. (2-29), p. 65	Eq. (2-30), p. 65	P2-22, p. 72
Laplacian	Eq. (2-51), p. 71	Eq. (2-52), p. 71	Eq. (2-53), p. 71	P2-22, p. 73